Oxidative Stress and Redox Signalling in Parkinson's Disease

Issues in Toxicology

Series editors:
Diana Anderson, *University of Bradford, UK*
Michael D. Waters, *Michael Waters Consulting, USA*
Timothy C. Marrs, *Edentox Associates, UK*

Editorial advisor:
Alok Dhawan, *CSIR-Indian Institute of Toxicology Research, Lucknow, India*

Titles in the Series:

1: Hair in Toxicology: An Important Bio-Monitor
2: Male-mediated Developmental Toxicity
3: Cytochrome P450: Role in the Metabolism and Toxicity of Drugs and other Xenobiotics
4: Bile Acids: Toxicology and Bioactivity
5: The Comet Assay in Toxicology
6: Silver in Healthcare
7: *In Silico* Toxicology: Principles and Applications
8: Environmental Cardiology
9: Biomarkers and Human Biomonitoring, Volume 1: Ongoing Programs and Exposures
10: Biomarkers and Human Biomonitoring, Volume 2: Selected Biomarkers of Current Interest
11: Hormone-Disruptive Chemical Contaminants in Food
12: Mammalian Toxicology of Insecticides
13: The Cellular Response to the Genotoxic Insult: The Question of Threshold for Genotoxic Carcinogens
14: Toxicological Effects of Veterinary Medicinal Products in Humans: Volume 1
15: Toxicological Effects of Veterinary Medicinal Products in Humans: Volume 2
16: Aging and Vulnerability to Environmental Chemicals: Age-related Disorders and their Origins in Environmental Exposures
17: Chemical Toxicity Prediction: Category Formation and Read-Across
18: The Carcinogenicity of Metals: Human Risk Through Occupational and Environmental Exposure
19: Reducing, Refining and Replacing the Use of Animals in Toxicity Testing
20: Advances in Dermatological Sciences
21: Metabolic Profiling: Disease and Xenobiotics
22: Manganese in Health and Disease
23: Toxicology, Survival and Health Hazards of Combustion Products
24: Masked Mycotoxins in Food: Formation, Occurrence and Toxicological Relevance
25: Aerobiology: The Toxicology of Airborne Pathogens and Toxins

26: Chemical Warfare Toxicology, Volume 1: Fundamental Aspects
27: Chemical Warfare Toxicology, Volume 2: Management of Poisoning
28: Toxicogenomics in Predictive Carcinogenicity
29: Human Stem Cell Toxicology
30: The Comet Assay in Toxicology, 2nd edition
31: Computational Systems Pharmacology and Toxicology
32: Ecotoxicology and Genotoxicology: Non-traditional Terrestrial Models
33: Ecotoxicology and Genotoxicology: Non-traditional Aquatic Models
34: Oxidative Stress and Redox Signalling in Parkinson's Disease

How to obtain future titles on publication:
A standing order plan is available for this series. A standing order will bring delivery of each new volume immediately on publication.

For further information please contact:
Book Sales Department, Royal Society of Chemistry, Thomas Graham House, Science Park, Milton Road, Cambridge, CB4 0WF, UK
Telephone: +44 (0)1223 420066, Fax: +44 (0)1223 420247
Email: booksales@rsc.org
Visit our website at www.rsc.org/books

Oxidative Stress and Redox Signalling in Parkinson's Disease

Edited by

Rodrigo Franco
University of Nebraska-Lincoln, Lincoln, NE, USA
Email: rfrancocruz2@unl.edu

Jonathan A. Doorn
University of Iowa, Iowa City, IA, USA
Email: jonathan-doorn@uiowa.edu

and

Jean-Christophe Rochet
Purdue University, West Lafayette, IN, USA
Email: jrochet@purdue.edu

ROYAL SOCIETY OF CHEMISTRY

THE QUEEN'S AWARDS FOR ENTERPRISE: INTERNATIONAL TRADE 2013

Issues in Toxicology No. 34

Print ISBN: 978-1-78262-188-1
PDF eISBN: 978-1-78262-288-8
EPUB eISBN: 978-1-78801-191-4
ISSN: 1757-7179

A catalogue record for this book is available from the British Library

The Royal Society of Chemistry is a charity, registered in England and Wales, Number 207890, and a company incorporated in England by Royal Charter (Registered No. RC000524), registered office: Burlington House, Piccadilly, London W1J 0BA, UK, Telephone: +44 (0) 207 4378 6556.

For further information see our web site at www.rsc.org

Printed in the United Kingdom by CPI Group (UK) Ltd, Croydon, CR0 4YY, UK

Preface

Parkinson's disease (PD) was first described in detail in 1817 by the London surgeon James Parkinson, who referred to the disorder as 'paralysis agitans' or 'shaking palsy'. PD is now defined as a chronic and progressive neurodegenerative disorder that currently affects ~1 million individuals in the US alone, and ~10 million around the world. Importantly, from 2006 to 2016 the population affected with PD doubled, while only a 10% increase in the US population was reported. The exact prevalence of PD is unclear as the disease is only diagnosed after the pathogenic process is far advanced. Thus, the exact number of individuals with PD is expected to be higher than the actual number of cases diagnosed. Economically, the estimated cost of PD in the US is ~$25 million per year in both treatment and lost income from the inability to work.

PD prevalence and incidence increase exponentially for individuals 65 to 85 years of age. While aging is considered the major risk factor for PD, the etiology of the disease is still largely unclear. Around 10% of diagnosed cases have been linked to inherited (familial) mutations in genes encoding proteins such as α-synuclein, Parkin, PINK1, DJ-1, and LRRK2, while other genetic modifications only increase the risk of developing the disease. The remaining ~90% of PD cases are sporadic, of which ~5% are linked to *de novo* mutations primarily in the α-synuclein and LRRK2 genes. For those sporadic cases without a clear delineated genetic background (~85% of total PD cases), it is hypothesized that environmental or occupational exposures are important contributors to neuronal cell loss. Thus, it is now considered that aging, genetics and environmental/occupational risk factors contribute to PD.

The primary clinical phenotype of PD is what is called parkinsonism, a movement disorder that is characterized by tremor at rest, bradykinesia, rigidity and postural instability, which is directly associated with the depletion of dopamine neurotransmission from degenerated dopaminergic A9

Issues in Toxicology No. 34
Oxidative Stress and Redox Signalling in Parkinson's Disease
Edited by Rodrigo Franco, Jonathan A. Doorn and Jean-Christophe Rochet

Published by the Royal Society of Chemistry, www.rsc.org

neurons in the *substantia nigra pars compacta* (SNpc) innervating the striatum. Neuronal degeneration in PD affects other areas of the brain as the disease progresses, but SNpc dopaminergic neurons have been considered one of the primary and likely more sensitive targeted neuronal populations in PD. Importantly, pathological features in the olfactory bulb and brain stem that correlate with early nonmotor dysfunction in the sense of smell and sleep regulation, as well as the degeneration of dopaminergic neurons in the intestine leading to gastrointestinal dysfunction, are considered important components of the disease. To date, the bulk of research in the PD field has been focused on understanding the mechanisms behind the loss of SNpc dopaminergic neurons, but what makes these cells a primary pathological target in PD, and what makes them particularly sensitive to PD risk factors, remains unclear.

Since the late 1970s and early 1980s, oxidative stress has been recognized as a biomarker of the imbalance in redox homeostasis that occurs during the progressive loss of dopaminergic neurons in the brains of PD patients. Byproducts of protein, lipid and nucleic acid oxidation are found in post-mortem PD brain samples. It has been established that mitochondrial dysfunction, inflammatory processes, the pro-oxidant metabolism of dopamine, and metal-catalyzed free-radical formation combine to generate oxidative damage in dopaminergic neurons. Importantly, recent evidence has demonstrated that aside from non-specific oxidative damage, alterations in redox homeostasis associated with PD result in perturbations of enzymatic and signalling processes that are essential for neuronal physiology. Examples of these perturbations include (i) alterations in protein quality control mechanisms that lead to the accumulation of protein inclusions of α-synuclein (Lewy bodies); and (ii) energy failure that renders nigral dopaminergic neurons (which have high energy demands because of their extensive arborization and active pacemaking activity) very sensitive to ATP depletion.

This book aims to highlight recent advances regarding the role of alterations in redox homeostasis in PD pathogenesis. A complete overview of themes including mitochondrial dysfunction, iron and dopamine metabolism, antioxidants, protein aggregation/oxidation and others is reviewed. From this work emerges a strong sense of the rapid pace of discovery in the PD field, but also of substantial knowledge gaps that must be addressed if we are to meet the over-arching goals of identifying biomarkers for earlier diagnosis and developing disease-altering therapies.

Rodrigo Franco
University of Nebraska-Lincoln, NE, USA

Jonathan Doorn
The University of Iowa, IA, USA

Jean-Christophe Rochet
Purdue University, IN, USA

Contents

Issues in Toxicology No. 34
Oxidative Stress and Redox Signalling in Parkinson's Disease
Edited by Rodrigo Franco, Jonathan A. Doorn and Jean-Christophe Rochet

Published by the Royal Society of Chemistry, www.rsc.org

CHAPTER 1

Etiology and Pathogenesis of Parkinson's Disease

BRIANA R. DE MIRANDA AND J. TIMOTHY GREENAMYRE*

Pittsburgh Institute for Neurodegenerative Diseases and Department of Neurology, University of Pittsburgh, Pittsburgh, PA 15260, USA
*E-mail: jgreena@pitt.edu

1.1 Introduction

Parkinson's disease (PD) is the second most common neurodegenerative disorder (after Alzheimer's disease), and it is estimated that PD affects approximately 10 million individuals worldwide, though many cases may go undiagnosed. With the growth of aging populations, it is estimated that PD will nearly double in incidence over the next 25 years, representing a major social and economic burden to provide long-term treatment and care for those affected.[1] This pressing concern has focused increased attention on the field of neurodegeneration, and while laboratory discoveries have begun translation into the clinic, one critical issue persists; the underlying causes of this progressive disorder remain, for the most part, unidentified. Inherited forms of PD strongly correspond to known genetic mutations in proteins involved with mitochondrial function, oxidative stress, and protein degradation pathways. However, inherited forms of PD only account for about 10% of PD cases, and sporadic PD has a much lower association with single gene mutations that are readily identified by specific protein dysfunction.[2,3] The pathogenesis of both familial and idiopathic PD involves several components;

Issues in Toxicology No. 34
Oxidative Stress and Redox Signalling in Parkinson's Disease
Edited by Rodrigo Franco, Jonathan A. Doorn and Jean-Christophe Rochet

Published by the Royal Society of Chemistry, www.rsc.org

the gross manifestations of the disorder, the underlying neuronal death and cellular pathology, the molecular mechanisms behind progressive degeneration, and the genetic or environmental dysregulation of proteins responsible for cellular dysfunction.

Currently, no curative or 'disease-modifying' therapy is available to slow or stop the inevitable and inexorable progression of PD. While symptomatic treatment options are available, as the disease progresses and medication doses rise, the tolerability of PD drugs may decrease and side effects often become problematic. In order to develop therapeutic strategies that prevent the progressive loss of dopamine neurons, a clear understanding of the mechanisms behind cell death in PD must be elucidated. Many putative pathological mechanisms in PD can be linked to common pathways that converge on the mitochondrial production of oxidative stress. Here, the pathogenesis of oxidative stress in PD is examined in the context of the cellular pathology observed in the disease.

1.2 Clinical Manifestations of Parkinson's Disease

The motor signs and symptoms of PD – bradykinesia, resting tremor, rigidity, and postural instability – together with the patient's history, are the primary means for identifying the disorder, and current guidelines require two of the four main signs of the disease to be present, typically presenting with asymmetrical onset.[4] Most individuals are diagnosed over the age of 45, with only a small percentage (10%) of cases considered early-onset (under the age of 45).[4] Accompanying these movement deficits are non-motor symptoms of PD, such as decreased GI motility, loss of olfactory function, sleep disorders, and cognitive or behavioral changes.[5] Non-motor symptoms often occur prior to the onset of motor symptoms, though their presence alone has proven unreliable for detecting PD.[6] Motor symptoms of PD occur when there is approximately 80% loss of striatal dopamine levels, indicating that significant cell death and damage has occurred prior to emergence of visible symptoms of the disease.[7,8] This 'silent' period of pathogenesis is extremely problematic as it narrows the window for neuroprotective therapeutic intervention to the period after clinical diagnosis.

There is no definitive test for PD outside of the diagnostic criteria in the clinic and the response of a patient to levodopa (L-DOPA), which is the precursor of dopamine and the most efficacious symptomatic treatment for the disease.[9] Unfortunately, maintenance with L-DOPA and dopamine receptor agonists has limitations. On the one hand, dopamine mimetics are useful at abating many of the symptoms in PD such as bradykinesia, rigidity, and tremor, while on the other hand they contribute to a host of iatrogenic symptoms, including dyskinesias, 'wearing off' effects, and hallucinations.[10] In addition, treatment with dopamine agonists may exacerbate impulse control disorders, such as excessive gambling and reward-seeking behavior.[11,12] There are also indications that dopamine-replacement therapy itself contributes to cellular toxicity, possibly enhancing the progressive loss of neurons associated with PD.[13] Several small molecule therapies aimed at inhibiting the progression of PD have entered the pipeline for novel drug development, many of which are anti-inflammatory therapies targeted at limiting oxidative stress.[14] It is critical to note, however, that no compound has been proven successful in Phase III clinical trials for this purpose, and the pursuit for new therapeutic strategies continues.

1.3 Neuropathology

Underlying many of the motor symptoms of PD is the selective loss of dopaminergic neurons of the *substantia nigra pars compacta* and their principal axon projections to the striatum. Degeneration of the nigrostriatal tract is considered the hallmark lesion of the disease; however, several other extranigral sites exhibit pathology: the locus coeruleus and subcoeruleus complex, reticular formation and raphe nuclei, dorsal nuclei of the Vagus, and nucleus basalis of Meynert may all display cell loss.[15] Aptly named for its dark appearance in the adult human brain, the *substantia nigra pars compacta* consists mainly of dopamine neurons that contain the pigment neuromelanin, a product of catecholamine metabolism, which is visibly reduced in intensity in the postmortem PD brain. Dopamine neuron loss does not occur uniformly throughout the *substantia nigra*, and the pattern of neuron death is distinct to the disease, differing from the typical loss of dopamine neurons associated with aging alone, which occurs predominantly in the dorsal tier of the *substantia nigra*.[16] Neurons of the nigrostriatal tract exhibit dieback, with the degenerative process beginning at the nerve terminal and extending retrogradely to the cell body with medioventral loss of the nigral regions showing more early lesions of PD and extending laterally.[17] The distinctive pattern of loss suggests that dopamine neuron death in PD is not merely due to mechanisms of accelerated aging.

The other key pathological feature of PD is the cellular accumulation of Lewy bodies (LBs) and Lewy neurites (LNs), which are protein accumulations within the somal cytoplasm of neurons (LBs) and their processes (LNs), in the *substantia nigra* and other regions.[18] Lewy bodies and neurites consist predominantly of aggregated α-synuclein protein with a variety of post-translational modifications, including phosphorylation, ubiquitination, nitration, and oxidation of several residues.[19]

The role of α-synuclein in the healthy brain remains somewhat unclear, though there is evidence that it participates in synaptic vesicle function.[15,20,21] α-Synuclein point mutations, (wildtype) gene duplications and triplications, and polymorphisms that increase expression of wildtype α-synuclein all cause PD.[22–25] This genetic information, together with the fact that α-synuclein accumulates in Lewy pathology even in idiopathic PD, suggests the central importance of this protein in almost all cases of the disease.[26] Neuropathological staging of PD progression is assessed by immunostaining of accumulated α-synuclein, beginning in the caudal brainstem (or olfactory bulb) and spreading rostrally to the neocortex over the course of the disease.[18] According to this scheme, classical involvement of the nigrostriatal neurons occurs in the middle stages of the disease. It is currently unclear whether the spread of α-synuclein pathology represents prion-like cell-to-cell transfer of the protein (discussed below) or simply reflects the relative vulnerabilities of various neuronal populations over the long course of the disease.

α-Synuclein accumulation and aggregation and neuronal death are accompanied by a glial-driven neuroinflammatory response.[27] The involvement of glial cells in PD pathogenesis is proposed as both a beneficial response in an attempt to preserve damaged neurons, as well as a source of unregulated inflammation that can drive further neuronal damage.[14,28,29] What remains less obvious is whether neuroinflammation might sometimes be an inciting factor in PD, or merely a response to mitigate the damage that has already occurred in neurons.

The resident CNS macrophage cells, microglia, are the key enforcers of the immune response in the brain, regulating inflammatory protein expression and recruiting additional immunological participants, including t-lymphocytes, from the periphery.[30,31] It is also widely recognized that astrocytes, the most abundant cell type in the brain, are essential to the inflammatory response associated with progressive dopamine neuron loss in the *substantia nigra*.[32,33] Postmortem examination of microglia and astrocytes reveals an activated, hypertrophic cell phenotype within brains of individuals with PD, appearing most intensely in the ventral midbrain, but evident throughout the brain.[34,35] Reactive glial cells abundantly express pro-inflammatory proteins, which upregulate the production of many neurotoxic factors, including reactive oxygen species that further contribute to dopamine neuron damage. Several lines of experimental evidence have suggested that the glial-driven inflammatory response within the CNS may contribute to the progression of dopamine neuron death, indicating a possible role for anti-inflammatories as PD therapeutics.[30,31] Epidemiological data describing a link between long-term use of NSAIDs and a decreased risk for developing PD has supported this theory; however, no anti-inflammatory drug has been proven successful in clinical trials to halt the progression of the disease.[36] To this end, it is necessary to investigate the multifactorial nature of idiopathic PD, and the molecular factors that contribute to its etiology.

1.3.1 Selective Vulnerability of the Nigrostriatal Dopamine Neuron

In PD, dopamine neurons within the *substantia nigra* are considered a selectively susceptible population of cells, whereas the adjacent dopamine neurons of the ventral tegmental area (VTA) are much more resistant to degeneration. The vulnerability of the nigrostriatal neurons is due to several factors, including their unique anatomy, physiology, bioenergetic profile, and neurochemistry (reviewed in ref. 37). First, in rat brain, the length of the axon arbor of a single nigrostriatal neuron is up to 80 cm; in humans, this is estimated to be 4 m! In the rat, a single nigrostriatal neuron makes 100 000–240 000 synapses in the striatum. In humans, it is estimated that a nigrostriatal neuron makes 1 000 000–2 400 000 synapses. By contrast, the VTA neuron makes about 10-fold fewer synapses and therefore has a much lower bioenergetic demand. Further compounding the bioenergetic demand of the nigrostriatal neuron is the fact that their axons are unmyelinated. Thus, propagation of each action potential and subsequent repolarization requires much more energy than if the fibers were myelinated.

As discussed later, there is abundant evidence that mitochondria are dysfunctional in PD. If one conservatively estimates that there are 10 mitochondria per nigrostriatal synapse, then it is apparent that each neuron must maintain at least 10 000 000 mitochondria – and must do so over exceedingly long distances. As Paul Bolam has suggested, nigrostriatal neurons may have 'too many mouths to feed'. It seems reasonable to assume that those mitochondria most distant from the soma would be most vulnerable to 'wear and tear' and, as a consequence, they may produce less ATP and more ROS.[37] If so, this might explain why nigrostriatal degeneration begins at the terminals.

Dopamine itself may contribute to the vulnerability of the dopaminergic nigrostriatal neuron, particularly in the setting of mitochondrial impairment. Dopamine, a catecholamine, is produced from tyrosine by the sequential enzymatic actions of tyrosine hydroxylase and amino acid decarboxylase. A redox reactive molecule, dopamine is normally sequestered in synaptic vesicles at concentrations estimated to be in the high millimolar range.[38] Thus, even a small degree of dopamine leakage from vesicles could easily produce local cytosolic concentrations in the micromolar range. It is now clear that (i) mitochondrial impairment (complex I defects) and (ii) increased levels of alpha-synuclein, both of which have been implicated in PD, lead to redistribution of dopamine from vesicles to cytosol.[39] Cytoplasmic dopamine can undergo enzymatic oxidation or non-enzymatic auto-oxidation to produce dopamine quinone species (DAQ).[40] The electron-deficient DAQ readily reacts with cellular nucleophiles, predominantly reduced sulfhydryl groups on free cysteine residues, glutathione, and proteins, forming covalent cysteinyl-dopamine adducts.[41] Since the active sites of many proteins contain cysteine residues, DAQ modification can cause protein inactivation and loss of function.[42]

Dopamine oxidation has been associated mechanistically with (i) mitochondrial impairment, (ii) alpha-synuclein oligomerization, (iii) enhanced NMDA receptor function, and (iv) reduced proteasome function, each of which has been posited to play a central pathogenic role in PD.[39,43–45] In summary, the anatomy and physiology of nigrostriatal neurons predisposes them toward bioenergetic crisis, in contrast to VTA neurons. In the setting of bioenergetic impairment, dopamine may leak from vesicles to the cytosol where it can oxidize to DAQ and exacerbate the neurodegenerative process.

1.3.2 Mitochondrial Dysfunction in PD

It is well established that mitochondrial dysfunction is central to the etiology of dopamine neuron death and dysfunction in PD (reviewed in ref. 46). In addition to energy production *via* aerobic oxidative phosphorylation, mitochondria also regulate intracellular calcium levels, participate in lipid metabolism, as well as steroid, carbohydrate, and amino acid breakdown.[47] Mitochondria are responsible for signalling apoptosis pathways, and are the gatekeepers for apoptotic machinery to complete an organized cell death. The process of mitochondrial respiration within neurons requires aerobic oxidative phosphorylation, a process that naturally produces a high amount of oxidative byproducts, such as hydrogen peroxide (H_2O_2) and superoxide ($O_2^{\bullet-}$).[47] In a healthy neuron, the mitochondria will detoxify these ROS using normal compensatory mechanisms such as antioxidant and reactive oxygen scavenger proteins (ex. glutathione peroxidase).[48] However, even under basal conditions, nigrostriatal dopamine neurons exist in a more oxidized state than other neurons.[49] Under pathological conditions, these compensatory mechanisms can be overwhelmed in the dopamine neuron, and ROS production from the mitochondria becomes a source of oxidative stress for the cell.[50] In addition to oxidative stress, mitochondrial dynamics are also an important factor in dopamine neuron pathology. Mitochondrial fusion and fission, the processes by which the outer membranes of two mitochondria join to form one membrane (fusion) or by which a single mitochondrial membrane becomes two (fission), are also potential sources of dysfunction in the neuron.[51] Mutations in genes regulating mitochondrial dynamics are a known cause of inherited forms of PD (ex. PINK1, Parkin, SNCA).[52] It is not surprising then, that genes controlling mitochondrial dynamics are central to dopamine neuron survival, given that synaptic maintenance, mitochondrial biogenesis, and neurotransmission are all regulated by this process.

1.3.2.1 Mitochondrial DNA Damage

The association of mutations in genes regulating mitochondrial function and PD is further supported by the regulation of mitochondrial DNA (mtDNA). Mitochondrial DNA encodes 37 genes, including 13 protein subunits for complex I-V of the electron transport chain (ETC). mtDNA is located in association with the inner mitochondrial membrane, neighboring where oxidative phosphorylation occurs. Therefore, it is constantly exposed to ROS.[53] The lack of histones and limited repair mechanisms causes mtDNA to be especially vulnerable to damage, such as strand breaks and base modifications; these in turn can lead to mutations.[54] Accordingly, mutations have been found at a higher rate in dopamine neurons of the *substantia nigra* in PD cases than age-matched controls, suggesting a role for mtDNA mutations in the pathogenesis of PD.[55–58]

Aside from frank mutations, mtDNA damage *per se* may be pathogenic, leading for example to blockage of mtDNA replication and cytotoxicity.

An excess of abasic sites (lacking a purine or pyrimidine base) has been detected in dopamine neurons of the *substantia nigra* in postmortem PD tissue; however, such mtDNA damage was not found in cortical neurons, suggesting that mtDNA damage may be specific to the dopaminergic cell pathology.[59] Furthermore, mtDNA abasic sites within dopamine neurons have also been reported in rats following rotenone treatment, a phenomenon that occurs prior to dopamine cell loss.[59] Given these data, it has been proposed that the detection of mtDNA damage, such as abasic sites, may provide a novel biomarker for early pathological changes in PD.[60]

1.3.2.2 Complex I Inhibition

There is a strong connection between mitochondrial complex I dysfunction and death of dopamine neurons in PD. In addition to the observation of diminished complex I activity in postmortem PD tissue, complex I inhibition induced by exogenous chemical agents is the basis for causing selective dopamine neuron death in neurotoxic models of PD.[61,62] First demonstrated with 1-methyl-2-phenyl-1,2,3,6-tetrahydropyradine (MPTP) toxicity, complex I inhibition results in a potent and selective loss of neurons from the *substantia nigra* and a parkinsonian phenotype.[63] MPTP is bioactivated in the astrocyte *via* monamine oxidase B (MAO-B) to its reactive metabolite MPP^+, which has structural specificity for the dopamine transporter (DAT) expressed on the dopamine neuron. MPP^+ accumulates in the mitochondria where it impedes the flow of electrons in the ETC at complex I, resulting in ATP depletion, reactive oxygen species accumulation, and mitochondrial dysfunction.[64]

Another model of PD is based on the classical complex I inhibitor, rotenone. Aside from being a biochemical tool, rotenone is an organic pesticide still employed in the United States for killing invasive aquatic species.[65] It is distinct from MPTP toxicity in that it does not require bioactivation in the astrocyte, nor does it have a specific affinity or requirement for the DAT. The use of rotenone in animal models of PD results in a lesion of the *substantia nigra* and striatum, with accumulation of α-synuclein protein in the cytoplasm of dopamine neurons, and neurobehavioral deficits.[66] This model is additionally useful in demonstrating the selective toxicity of complex I inhibition to dopamine neurons, given that systemic injection of rotenone affects all cells, but results in a nigrostriatal lesion. The rotenone model of PD correctly predicted the transferrin receptor 2-dependent accumulation of iron in the *substantia nigra*, which was later confirmed in human PD cases.[67] It also predicted the accumulation of mtDNA abasic lesions in nigrostriatal neurons in PD.[59] Translational discoveries such as this exemplify the importance of neurotoxin models in the discovery of pathological mechanisms behind PD. It is of course impossible to fully recapitulate the complex disease state of PD using any animal model, and therefore understanding our limitations using these models is equally important in their employment.

1.3.3 Oxidative Stress

Superoxide ($O_2^{\bullet-}$) is the principal reactive species released from the mitochondria, emanating from complex I, when ATP is not being produced and therefore protonmotive force (Δp) is high, or when NADH/NAD$^+$ ratio is high within the mitochondrial matrix.[47] Superoxide can be converted by superoxide dismutase (SOD) to the less toxic hydrogen peroxide (H_2O_2). Mitochondria express their own form of SOD (MnSOD), suggesting that there is a biologically relevant purpose for superoxide production from the mitochondria.[47,53] It is proposed that the production of H_2O_2 from mitochondria serves as a redox signal in cells, by transiently altering protein-thiol redox status.[68] Under normal conditions, H_2O_2 is detoxified by catalase or glutathione peroxidase (GPX), maintaining homeostasis between normal mitochondrial signalling and scavenging enzymes.

Pathological increases in reactive oxygen species may result from aberrant enhanced production or impaired detoxification, or some combination thereof. Although H_2O_2 has low relative toxicity, it is possible to form highly reactive radicals from H_2O_2.[50] Within the brain, nitric oxide (NO) is abundant, and when combined with H_2O_2, leads to production of peroxynitrite ($ONOO^-$) and the hydroxyl radical ($OH^\bullet$). The highly toxic hydroxyl radical is capable of binding to cellular macromolecules causing widespread protein dysfunction, lipid oxidation, and creating strand breaks in DNA. Peroxynitrite attacks tyrosine residues and causes protein dysfunction, which is readily observed in both animal models of PD as well as in the human disease.[69,70] The function of tyrosine hydroxylase within dopamine neurons can be compromised by peroxynitrite, a process that may contribute to the dysfunction of dopamine production even in preserved neurons in the *substantia nigra*.[71]

It is clear that oxidative stress plays a key role in the pathogenesis of this disease. Tissue from PD *substantia nigra* contains evidence of lipid peroxidation, decreased glutathione levels, and increased iron content.[27,72] Iron and other redox sensitive transition metals (copper, manganese) present within the cell contribute to oxidative damage through the Fenton reaction ($Fe^{2+} + H_2O_2 \rightarrow Fe^{3+} + OH^\bullet + OH^-$) resulting in the production of the hydroxyl radical. Several lines of evidence support the theory that iron metabolism is disrupted in PD, leading to both increased accumulation of iron and oxygen radicals in the *substantia nigra*.[73,74] Iron deposits are also observed to interact with α-synuclein, contributing to Lewy body formation.[75] Proteins that store iron in an unreactive state (ferritin, transferrin receptor) are an important component of iron regulation within the brain, and disruption of these proteins may contribute to neurodegenerative processes.[76]

The relative paucity of antioxidants in the CNS is also likely to play a role in the oxidative damage observed in PD. The major detoxifying peptide antioxidant of the CNS, glutathione (GSH), can act to remove ROS either through nonenzymatic reduction or catalysis with glutathione peroxidase, leading to oxidized glutathione (GSSG). The depletion of GSH in the brains of PD patients is one of the earliest biochemical changes that occurs, which suggests that it may contribute to progression of the disease.[48] This may conceivably result from diminished GSH synthesis, due to decreased levels of the GSH precursor, cysteine, circulating levels of which have been reported to decline with age.[77] Additionally, the amino acid carrier 1 (EAAC1), which transports cysteine into neurons has been shown in mouse models to be modified following MPTP treatment.[48,78] EAAC1 may be inactivated through protein nitration (caused by oxidative stress), thereby reducing the amount of GSH production by up to 30% in neurons exposed to MPTP[78] and nearly 50% reduction after rotenone treatment.[79] Because ROS produced by complex I are predominantly detoxified in the neuron by GSH, neuronal damage is widespread after GSH depletion. More recently, Swanson and colleagues reported that EEAC1(-/-) mice show a progressive loss of nigrostriatal dopamine neurons associated with excessive protein nitration. In this model, the cell-permeable GSH precursor, *N*-acetylcysteine, which bypasses EEAC1, was protective.[80]

GSH depletion may also affect the proteasome system, as described in one *in vitro* study that observed inhibition of 26S proteasome activity in cultured SH-SY5Y cells when GSH dropped below 50% of control.[81] Evidence for GSH depletion and inflammatory activation also exists in the c-Jun-N-terminal kinase (JNK) pathway, which becomes activated when GSH levels are decreased.[82] Abnormal activation of cellular inflammatory pathways through this mechanism may contribute to pathologic neuroinflammation that enhances dopamine neuron toxicity. It was shown that gene deletion of JNK2 and JNK3 in mice significantly protected dopamine neurons following MPTP treatment, indicating that JNK pathways stimulated by GSH depletion likely contribute to exacerbated inflammatory responses.[83]

1.3.4 Dopamine Metabolism

Dopamine, an essential neurotransmitter, also has the capacity to produce reactive intermediates, which may cause collateral oxidative damage within the cell.[84] Cellular toxicity resulting from dopamine metabolism has been shown to occur *via* two main pathways; the production of reactive intermediates from enzymatic metabolism of the catecholamine, and the autoxidation of the molecule to a highly electrophilic dopamine quinone.[41] The metabolism of dopamine by the monoamine oxidase-B enzyme (MAO-B) into its primary metabolite 3,4-dihydroxyphenylacetic acid (DOPAC) produces H_2O_2 as a byproduct.[85,86] If cellular compensatory mechanisms (catalase, GSH) cannot readily detoxify this intermediate, Fenton cycling of electrons may occur upon interaction with transition metals present in the brain (Fe^{2+}, Mn^{2+}, Cu^{2+}) resulting in the production of hydroxyl radicals.[87] In addition, autoxidation of the neurotransmitter produces a highly reactive dopamine quinone species, which has been shown to be capable of binding and modifying cellular proteins.[28] Nucleophilic protein residues, such as sulfhydryl groups, are especially vulnerable to covalent modification by the dopamine quinone, a process that likely results in protein dysfunction.

Indeed, intrastriatal dopamine injections in a rat model produced a significant lesion within the striatum, as well as a dose-dependent increase of cysteinyl protein binding by dopamine and DOPAC.[88] These lesions were attenuated by the coadministration of GSH, suggesting that quenching of such reactive species by antioxidants is a key component in maintaining dopamine homeostasis and limiting cellular toxicity.[88,89]

In addition to the production of reactive intermediates, there is evidence that oxidized dopamine may directly contribute to mitochondrial dysfunction, resulting in the swelling of brain mitochondria and opening of the mitochondrial permeability transition pore.[27] Isolated rat mitochondria exposed to dopamine quinone showed characteristics of ETC uncoupling, as measured by an increase in resting state respiration.[27] These data also indicated that the addition of GSH could attenuate dopamine quinone-mediated damage to mitochondria, while reactive oxygen scavengers such as catalase and superoxide dismutase did not, suggesting that formation of the dopamine quinone may directly alter thiol-containing mitochondrial coupling proteins.[41] Another line of evidence suggests that the aldehyde metabolite of dopamine, dihydroxyphenylacetaldehyde (DOPAL), elicits mitochondrial dysfunction and inhibits tyrosine hydroxylase in dopaminergic PC6-3 cells.[90] Together, these data indicate a strong association between the dopamine neurotransmitter and oxidative damage within the cells responsible for producing the molecule.

Given these findings, it is unsurprising that packaging and storage of dopamine is a highly regulated cellular process; sequestration of dopamine into synaptic vesicles provides a storage condition that limits autooxidation of dopamine, by maintaining a low pH and limiting exposure to MAO-B.[15] The vesicular monoamine transporter (VMAT2) has been shown to be critical for the proper handling of dopamine molecules, and knockdown of VMAT2 in mice results in a PD-like progressive loss of dopamine neurons from the *substantia nigra*.[91] There is also evidence that α-synuclein may play a role in forming these synaptic vesicles from early endosomes by interacting with phospholipase D2 (PLD2), and mutations in the α-synuclein gene would likely disrupt the formation of these vesicles.[92] Increases in free dopamine within the presynaptic cytoplasm is a plausible contributor to the amplified amount of oxidative damage observed in PD, as well as the observed reduction of functional dopamine neurotransmitter levels. In addition, it was shown that spontaneously forming dopamine-quinone molecules can modify α-synuclein, creating adducts that inhibit the transformation of synuclein protofibrils to fibrils,[43] which implies that dopamine oxidation may play a role regulating Lewy body formation and accumulation of soluble α-synuclein oligomers.[20] It is clear that the oxidative sequelae related to dopamine dyshomeostasis are critical in the pathogenesis of PD. This also presents a unique challenge in the treatment of the disorder, where on the one hand dopamine replacement is extremely valuable to remediate motor symptoms, but on the other hand may contribute to cellular stress.

1.3.5 Neuroinflammation

Inflammatory processes are associated with a broad spectrum of neurodegenerative diseases including Alzheimer's disease, amyotrophic lateral sclerosis, multiple system atrophy, multiple sclerosis, Huntington's disease, and PD. There is mounting evidence that inflammation-associated oxidative stress plays a central role in the pathogenesis of PD. An increase of pro-inflammatory cytokines in conjunction with decreased glutathione-related genes have been reported in a microarray study of the SN in PD brain tissue.[93] In addition, positron emission tomography (PET) scans assessing *in vivo* microglial activation in patients with PD revealed a significant elevation in inflammatory response within brain regions most affected by the disease.[94] Activated microglia were seen in the basal ganglia, striatum, and neocortical regions of the brain regardless of how advanced the disease, suggesting that microglial activation in PD is an early and continuous process.[94] It is postulated that activated microglia within the midbrain release pathological levels of pro-inflammatory cytokines, leading to an increase in oxidative stress in the already sensitive dopamine neuron population. Microglia contribute to oxidative damage through their antimicrobial defenses, including respiratory bursts, which release superoxide generated from NADPH oxidase enzymatic reactions.[31,50,95] Additionally, pro-inflammatory gene expression within the microglia, such as nuclear factor (NF)-κB, mitogen-activated protein kinases (MAPK), and activator protein (AP)-1, results in secondary production of ROS and nitric oxide (NO).[96]

Inflammatory gene expression within the microglia is dynamically regulated; pro-inflammatory gene transcription must be suppressed under basal conditions, but capable of rapid induction upon occupation of surface receptors for cell damage or pathogens.[97] Redox-sensitive transcription factors, such as NF-κB, are capable of upregulating hundreds of downstream pro-inflammatory mediators, including inducible nitric oxide synthase (iNOS), which converts L-arginine and NADPH to citrulline and NO.[98] In addition, NF-κB activation increases the expression of proteins that promote neurotoxicity including tumor necrosis factor (TNF)-α, interleukin (IL)-1β, and interferon (IFN)-γ.[99] These proteins also activate surrounding astrocytes, which enhance the expression of many pro-inflammatory proteins, including iNOS, resulting in high levels of NO.[100] Several lines of evidence have substantiated that astrocyte-mediated iNOS production is involved in the loss of dopamine neurons from the SN, including *iNOS* gene deletion studies in mouse that result in protection from MPTP-induced toxicity.[101]

Impairment of negative regulation of inflammatory pathways within glia may also be important in PD. Normal inflammatory resolution in the brain involves the expression of anti-inflammatory molecules including TGFβ, IL-10, and glial-derived neurotrophic factors. Suppression of inflammatory transcription is also a target for negative feedback mechanisms involving nuclear corepressors such as histone deacetylaces (HDAC1-3) and nuclear corepressor proteins (NCoR1/2). Saijo *et al.* (2009) suggested a novel role for the TH-regulating protein, Nurr1, in nuclear suppression of inflammatory gene transcription within astrocytes. Their findings indicated that the orphan nuclear receptor Nurr1 is recruited to the p65-NF-κB transcription factors bound to inflammatory gene promoters, where it recruits the CoREST corepressor complex, which removes chromatin bound-NF-κB transcription factors.[102] Conversely, the knockdown of astrocytic Nurr1 in adult mice exacerbated dopamine neuron loss following the peripheral injection of bacterial lipopolysaccharide (LPS).[102] This represents a novel role for Nurr1, which had previously only been characterized as a protein required for the development of tyrosine hydroxylase in dopamine neurons, the genetic mutation of which is also linked to a rare form of late-onset inherited PD.

1.4 Genetics of Parkinson's Disease

A detailed review of the genetics of PD is beyond the scope of this chapter, and readers are referred to recent reviews.[103,104] In brief, a refined knowledge of the genetic factors involved in familial PD has led to better understanding of pathological cellular processes that may be common to all forms of the disease, including mitochondrial dysfunction, oxidative stress, and abnormal protein processing.

The most commonly inherited form of PD is caused by an autosomal dominant mutation in the leucine-rich repeat kinase 2 (LRRK2) protein, a kinase whose substrates have been elusive.[105–107] Mutant LRRK2 has been associated with dysregulation of macroautophagy as well as mitochondrial abnormalities.[108,109] LRRK2 mutations display variable, age-dependent penetrance and have also been found in idiopathic PD, suggesting that other risk factors for the disease affect the risk associated with LRRK2 mutations.[110] The degree to which LRRK2 contributes to idiopathic disease is uncertain, but there is hope that LRRK2 kinase inhibitors may be beneficial for at least some forms of PD.

The first genetic association with PD was a point mutation in α-synuclein; several other point mutations have also been described, all of which are rare.[23] It was subsequently found that PD could be caused by gene duplications or triplications of the *wildtype* gene, indicating that simply producing too much *normal* α-synuclein protein causes neurodegeneration.[24] The presence of α-synuclein in the Lewy pathology of typical idiopathic PD cases further emphasizes its central importance in the disease.[111,112] How α-synuclein causes toxicity is uncertain, but there is evidence that it impairs autophagy[113] or specifically disrupts mitochondrial protein import.[114] Therapeutic strategies aimed at reducing levels of α-synuclein, preventing its aggregation, or inhibiting certain post-translational modifications are under investigation.

Parkin is an E3 ubiquitin ligase, mutations of which cause an autosomal recessive form of PD. The protein is important in proteasome-mediated proteostasis and for lysosomal degradation of proteins. Under certain conditions, Parkin is translocated to mitochondria where it helps to recruit autophagy machinery for mitochondrial quality control.[115] In the same cellular pathway, mutations in PINK1, the protein responsible for recruiting Parkin to the mitochondrial membrane, also result in an early-onset phenotype of PD.[116,117] There is growing interest in Parkin as a therapeutic target, and gene therapy and small molecule Parkin 'activator' approaches are being studied.

Aside from Parkin, several genes associated with inherited PD are related to protein degradation pathways within the cell. As one example, mutations in the glucocerebrosidase (GBA) gene impair autophagic function and are a major risk factor for development of PD.[118,119] Interestingly, Rocha and colleagues have demonstrated that glucocerebrosidase activity is also reduced in idiopathic PD.[120] Small-molecule glucocerebrosidase chaperones to enhance activity are currently under development.

Given the role of oxidative stress in PD, it is not surprising that mutations that affect proteins involved in antioxidant defenses might be associated with PD.[121] 15 is an oncogene that acts as a redox-sensitive protein chaperone and regulates several antioxidant pathways within the cell.[122] Mutations in DJ-1 cause an autosomal recessive, early-onset form of PD.[123] Interestingly, mice with gene deletion of DJ-1 show neuronal impaired mitochondrial complex I function.[124,125] It is noteworthy that the expression of DJ-1 in human brain is higher in astrocytes than neurons, and is elevated in this cell type in postmortem tissue of PD patients.[72,126,127] This suggests that a primary defect in a glial protein – DJ-1 – may result in a neuronal phenotype. Efforts to selectively enhance DJ-1 expression in astrocytes are underway.[153]

In summary, a large number of genetic mutations have been associated with PD. In general, these genes fall into categories affecting mitochondria, proteostasis/autophagy, and oxidative stress. The role of these proteins in idiopathic PD is an area of active investigation and appears to be leading to new therapeutic strategies.

1.5 Environmental Exposures and the Risk of Parkinson's Disease

1.5.1 Pesticides

Until the discovery of α-synuclein mutations in 1998, PD was thought to result primarily from environmental exposures, and genetics was thought to play little, if any, role in pathogenesis. In contrast, it had been reported that early-age exposure to rural environment and well water were associated with enhanced risk of PD.[128] It was hypothesized that well water was a vehicle for a 'causal agent' that might be toxic. In this context, there has been intense interest in the potential role of pesticides in PD. Many correlative studies have been limited in their ability to control for outside variables, leading to only weak associations with pesticides and PD. Recent data from rigorously controlled studies, however, have linked paraquat and rotenone exposure to a measured increase in PD risk.[129,130] Consistent with mechanisms believed to be important in PD, rotenone is a complex I inhibitor[131] and paraquat is a redox cycling pro-oxidant compound;[131] both compounds have been used in rodents to model PD.[132,133] Thus, experimental studies have provided 'biological plausibility' for epidemiological studies.

1.5.2 Metals

A strong correlation exists between occupational metal exposure and neurotoxicity, best exemplified by manganism, a parkinsonian disorder that results from inhalational overexposure to manganese.[134] It has traditionally been believed that manganism differs clinically from idiopathic PD; however, some recent studies suggest that the clinical phenotype of manganism overlaps substantially with that of PD.[135] Many studies investigating the association of transition-metal exposure and PD have suggested a role for mitochondrial dysfunction as a result of metal accumulation in the basal ganglia.[136] In a case-control study of occupational exposure to iron, copper, manganese, zinc, mercury, and lead, there were significant correlations to PD incidence in those exposed to manganese and copper for more than 20 years.[137] Additionally, correlational studies investigating lifelong exposure to airborne manganese in the Valcamonica region of Italy, where ferroalloy plants operated from 1902–2001, indicated that individuals in this population were more likely to develop motor, cognitive, and sensory dysfunction than in surrounding areas.[138] At this point, however, it remains controversial whether manganese exposure causes typical idiopathic PD.

Iron accumulates abnormally in the brains of individuals with PD, although there is little evidence that exposure to iron increases the risk of PD. On the other hand, there is evidence that genetically determined dysregulation of iron may contribute to PD risk.[139] There is also experimental evidence that altered iron homeostasis may contribute to nigrostriatal neurodegeneration.[67,140]

1.5.3 Pathogens

The historical significance of a pathogenic infection and neurodegeneration is most clearly defined by the 1918 Spanish influenza postencephalitic cases of parkinsonism. A large spike in parkinsonism prevalence occurred in the years following the widespread H1N1 influenza infection,[141] although the clinical symptoms were clearly distinct from typical PD. Interestingly, persons *in utero* during the 1918 influenza pandemic had an increased risk for developing PD (2–3 fold), suggesting a role for immune-based neuroinflammation, possibly without direct exposure to viral particles.[142–144]

Intriguingly, it has been reported recently that contemporary strains of influenza virus, such as H5N1, can travel from the peripheral nervous system to the central nervous system and cause neuroinflammation and degeneration of nigrostriatal dopamine neurons.[143,144] Moreover, there is evidence that systemic inflammation may be important. For example, intraperitoneal injection of bacterial lipopolysaccharide (LPS) in pregnant mice can also lead to a decreased expression of dopamine neurons in offspring.[145] It has been suggested that the medical community should be prepared to monitor for the emergence of parkinsonism related to current and future influenza pandemics.[143]

1.6 Gene–Environment Interaction

With only ~10% of PD cases strongly linked to inherited mutations, a widely held hypothesis suggests that low-penetrance susceptibility genes interact with environmental exposures and contribute to the majority of idiopathic PD incidence. One such example may be in populations exposed to large amounts of pesticides, with relative risk determined in part by polymorphisms in enzymes required to detoxify these chemicals (SOD, NADPH, quinone reductase, NQO1, and MAO).[146] The decreased ability to eliminate pesticides linked to PD, such as rotenone or paraquat, could significantly impact dopamine neuron survival, given the selectivity for complex I inhibition that many pesticides exhibit. Indeed in a study examining the odds ratio of polymorphisms in detoxifying enzymes and pesticide exposure, individuals with anomalies in SOD or NQO1 were nearly 2.5 times as likely to develop PD.[146]

Genes involved in the metabolism of xenobiotics may also contribute to polymorphic expression of PD susceptibility. Numerous polymorphisms of the P450 heme-oxygenase enzyme system are well characterized, leading to the altered distribution of chemical compounds normally metabolized by these proteins. Among P450 proteins, certain CYP2D6 polymorphisms have been correlated with a 2–3 fold increased risk of PD.[147,148] As a primary xenobiotic metabolizing enzyme, CYP2D6 is involved in the elimination and detoxification of many chemicals. Therefore, compromised CYP2D6 function may be linked to poor metabolism of these chemicals, leading to increased neurotoxicity from compounds, such as pesticides, known to be toxic to the dopamine system.[149]

Another potential source for heightened susceptibility to neurodegeneration is the differential accumulation of toxic compounds in the brain due to functional polymorphisms in multi-drug resistance transporters (MDR1) in the blood–brain barrier. It has been reported that a polymorphism associated with decreased MDR1 protein expression and function is increased in relation to the severity of PD, with early-onset cases showing a higher frequency of the polymorphism than late onset cases, which in turn, had a higher frequency than controls.[150] Such inter-individual differences in transporter-mediated xenobiotic access to the brain might have significant impact on the long-term consequences of environmental exposures. In this context, polymorphisms affecting blood–brain barrier permeability may influence susceptibility to PD.[151]

Gene–environment interactions involving downstream mechanisms after xenobiotic exposures may also be important. For example, as noted previously, selective mitochondrial DNA (mtDNA) damage in the form of apurinic/apyrimidinic (abasic) sites has been found in the vulnerable nigral neurons in Parkinson's disease (PD). The persistence of abasic sites suggests an ineffective base excision repair (BER) response in PD. In addition, Sanders *et al.* showed that pesticide exposure, which has been linked to PD risk, can cause mtDNA damage.[60] A recent study of 619 PD patients early in disease and 854 population controls found that polymorphisms in BER enzymes, APEX1 and OGG1, were not by themselves associated with PD; however, when combined with pesticide exposures, the polymorphisms markedly increased the risk of PD – and the highest risk was associated with polymorphisms in *both* genes together with pesticide exposure.[152]

In summary, it appears that gene–environment interactions influence both the extent of xenobiotic exposure, as well as the relative efficiency of potential compensatory mechanisms.

1.7 Conclusions

While it is still common practice to discuss 'Parkinson's disease', it is now apparent that PD is actually multiple diseases, with a common phenotype, which may be caused by a variety of distinct genetic mutations, or the cumulative effects of low-penetrance mutations or polymorphisms, or environmental exposures, or some combination of these. We are beginning to understand the key players and cellular pathways leading to degeneration and, as such, are beginning to devise therapeutic strategies to slow or halt the otherwise inevitable progression of the disease. The crucial task ahead is to translate these findings into clinically useful treatments that will impact the lives of those affected by PD.

Acknowledgements

Some of the work described here was supported by NIH grants NS095387, ES020327, ES020718 and training grant T32 NS086749 and the American Parkinson Disease Association, the Blechman Family Foundation, the DSF Charitable Foundation and the Consolidated Anti-Aging Foundation.

References

1. E. R. Dorsey, R. Constantinescu, J. P. Thompson, K. M. Biglan, R. G. Holloway and K. Kieburtz, *et al.*, Projected number of people with Parkinson disease in the most populous nations, 2005 through 2030, *Neurology*, 2007, **68**(5), 384–386.
2. A. Samii, J. G. Nutt and B. R. Ransom, Parkinson's disease, *Lancet*, 2004, **363**(9423), 1783–1793.
3. C. M. Tanner, Is the cause of Parkinson's disease environmental or hereditary? Evidence from twin studies, *Adv. Neurol.*, 2003, **91**, 133–142.

4. A. J. Lees, J. Hardy and T. Revesz, Parkinson's disease, *Lancet*, 2009, **373**(9680), 2055–2066.
5. D. Verbaan, J. Marinus, M. Visser, S. M. van Rooden, A. M. Stiggelbout and J. J. van Hilten, Patient-reported autonomic symptoms in Parkinson disease, *Neurology*, 2007, **69**(4), 333–341.
6. K. R. Chaudhuri, D. G. Healy and A. H. Schapira, Non-motor symptoms of Parkinson's disease: diagnosis and management, *Lancet Neurol.*, 2006, **5**(3), 235–245.
7. H. Bernheimer, W. Birkmayer, O. Hornykiewicz, K. Jellinger and F. Seitelberger, Brain dopamine and the syndromes of Parkinson and Huntington Clinical, morphological and neurochemical correlations, *J. Neurol. Sci.*, 1973, **20**(4), 415–455.
8. P. Riederer and S. Wuketich, Time course of nigrostriatal degeneration in parkinson's disease. A detailed study of influential factors in human brain amine analysis, *J. Neural Transm.*, 1976, **38**(3–4), 277–301.
9. J. Jankovic, Parkinson's disease: clinical features and diagnosis, *J. Neurol., Neurosurg. Psychiatry*, 2008, **79**(4), 368–376.
10. K. Kieburtz, Designing neuroprotection trials in Parkinson's disease, *Ann. Neurol.*, 2003, **53**(S3), S100–S109.
11. M. L. Dodd, K. J. Klos, J. H. Bower, Y. E. Geda, K. A. Josephs and J. E. Ahlskog, Pathological gambling caused by drugs used to treat Parkinson disease, *Arch. Neurol.*, 2005, **62**(9), 1377–1381.
12. E. C. Wolters, Y. D. van der Werf and O. A. van den Heuvel, Parkinson's disease-related disorders in the impulsive-compulsive spectrum, *J. Neurol.*, 2008, **255**(suppl. 5), 48–56.
13. J. Lipski, R. Nistico, N. Berretta, E. Guatteo, G. Bernardi and N. B. Mercuri, l-DOPA: A scapegoat for accelerated neurodegeneration in Parkinson's disease? *Prog. Neurobiol.*, 2011, **94**(4), 389–407.
14. C. K. Glass, K. Saijo, B. Winner, M. C. Marchetto and F. H. Gage, Mechanisms underlying inflammation in neurodegeneration, *Cell*, 2010, **140**(6), 918–934.
15. J. Lotharius and P. Brundin, Pathogenesis of parkinson's disease: dopamine, vesicles and α-synuclein, *Nat. Rev. Neurosci.*, 2002, **3**(12), 932–942.
16. J. M. Fearnley and A. J. Lees, Ageing and Parkinson's disease: Substantia Nigra Regional Selectivity, *Brain*, 1991, **114**(5), 2283–2301.
17. P. Damier, E. C. Hirsch, Y. Agid and A. M. Graybiel, The substantia nigra of the human brain. II. Patterns of loss of dopamine-containing neurons in Parkinson's disease, *Brain*, 1999, **122**(pt 8), 1437–1448.
18. H. Braak, K. Del Tredici, U. Rüb, R. A. I. de Vos, E. N. H. Jansen Steur and E. Braak, Staging of brain pathology related to sporadic Parkinson's disease, *Neurobiol. Aging*, 2003, **24**(2), 197–211.
19. K. Beyer and A. Ariza, Alpha-Synuclein Posttranslational Modification and Alternative Splicing as a Trigger for Neurodegeneration, *Mol. Neurobiol.*, 2012, **47**(2), 509–524.
20. J. Lotharius, Effect of Mutant alpha -Synuclein on Dopamine Homeostasis in a New Human Mesencephalic Cell Line, *J. Biol. Chem.*, 2002, **277**(41), 38884–38894.

21. T. C. Sudhof and J. Rizo, Synaptic Vesicle Exocytosis, *Cold Spring Harbor Perspect. Biol.*, 2011, **3**(12), a005637.
22. R. Kruger, W. Kuhn, T. Muller, D. Woitalla, M. Graeber and S. Kösel, *et al.*, Ala30Pro mutation in the gene encoding alpha-synuclein in Parkinson's disease, *Nat. Genet.*, 1998, **18**(2), 106–108.
23. M. H. Polymeropoulos, Mutation in the -Synuclein Gene Identified in Families with Parkinson's Disease, *Science*, 1997, **276**(5321), 2045–2047.
24. A. B. Singleton, M. Farrer, J. Johnson, A. Singleton, S. Hague and J. Kachergus, *et al.*, alpha-Synuclein locus triplication causes Parkinson's disease, *Science*, 2003, **302**(5646), 841.
25. L. J. Martin, Y. Pan, A. C. Price, W. Sterling, N. G. Copeland and N. A. Jenkins, *et al.*, Parkinson's disease alpha-synuclein transgenic mice develop neuronal mitochondrial degeneration and cell death, *J. Neurosci.*, 2006, **26**(1), 41–50.
26. H. Deng and L. Yuan, Genetic variants and animal models in SNCA and Parkinson disease, *Ageing Res. Rev.*, 2014, **15**, 161–176.
27. M. G. Tansey and M. S. Goldberg, Neuroinflammation in Parkinson's disease: its role in neuronal death and implications for therapeutic intervention, *Neurobiol. Dis.*, 2010, **37**(3), 510–518.
28. H. Yokoyama, H. Uchida, H. Kuroiwa, J. Kasahara and T. Araki, Role of glial cells in neurotoxin-induced animal models of Parkinson's disease, *Neurol. Sci.*, 2010, **32**(1), 1063–1080.
29. A. V. Daniela Rossi, Astrocyte dysfunction: Insights on the role in neurodegeneration, *Brain Res. Bull.*, 2009, **80**(4–5), 1–9.
30. E. Polazzi and B. Monti, Microglia and neuroprotection: From *in vitro* studies to therapeutic applications, *Prog. Neurobiol.*, 2010, **92**(3), 293–315.
31. G. C. Brown and J. J. Neher, Inflammatory Neurodegeneration and Mechanisms of Microglial Killing of Neurons, *Mol. Neurobiol.*, 2010, **41**(2–3), 242–247.
32. E. C. Hirsch and S. Hunot, Neuroinflammation in Parkinson's disease: a target for neuroprotection? *Lancet Neurol.*, 2009, **8**(4), 382–397.
33. H.-M. Gao and J.-S. Hong, Why neurodegenerative diseases are progressive: uncontrolled inflammation drives disease progression, *Trends Immunol.*, 2008, **29**(8), 357–365.
34. A. Hartmann, S. Hunot and E. C. Hirsch, Inflammation and dopaminergic neuronal loss in Parkinson's disease: a complex matter, *Exp. Neurol.*, 2003, **184**(2), 561–564.
35. M. Deleidi and T. Gasser, The role of inflammation in sporadic and familial Parkinson's disease, *Cell. Mol. Life Sci.*, 2013, **70**(22), 4259–4273.
36. H. Chen, E. Jacobs, M. A. Schwarzschild, M. L. McCullough, E. E. Calle and M. J. Thun, *et al.*, Nonsteroidal antiinflammatory drug use and the risk for Parkinson's disease, *Ann. Neurol.*, 2005, **58**(6), 963–967.
37. J. P. Bolam and E. K. Pissadaki, Living on the edge with too many mouths to feed: why dopamine neurons die, *Mov. Disord.*, 2012, **27**(12), 1478–1483.

38. E. N. Pothos, V. Davila and D. Sulzer, Presynaptic Recording of Quanta from Midbrain Dopamine Neurons and Modulation of the Quantal Size, *J. Neurosci.*, 1998, **18**(11), 4106–4118.
39. S. B. Berman and T. G. Hastings, Dopamine Oxidation Alters Mitochondrial Respiration and Induces Permeability Transition in Brain Mitochondria, *J. Neurochem.*, 1999, **73**(3), 1127–1137.
40. M. J. LaVoie and T. G. Hastings, Dopamine quinone formation and protein modification associated with the striatal neurotoxicity of methamphetamine: evidence against a role for extracellular dopamine, *J. Neurosci.*, 1999, **19**(4), 1484–1491.
41. T. G. Hastings, The role of dopamine oxidation in mitochondrial dysfunction: implications for Parkinson's disease, *J. Bioenerg. Biomembr.*, 2009, **41**(6), 469–472.
42. D. N. Hauser, A. A. Dukes, A. D. Mortimer and T. G. Hastings, Dopamine quinone modifies and decreases the abundance of the mitochondrial selenoprotein glutathione peroxidase 4, *Free Radical Biol. Med.*, 2013, **65**, 419–427.
43. K. A. Conway, J.-C. Rochet, R. M. Bieganski and P. T. Lansbury, Kinetic Stabilization of the α-Synuclein Protofibril by a Dopamine-α-Synuclein Adduct, *Science*, 2001, **294**(5545), 1346–1349.
44. A. A. Dukes, K. M. Korwek and T. G. Hastings, The Effect of Endogenous Dopamine in Rotenone-Induced Toxicity in PC12 Cells, *Antioxid. Redox Signaling,* 2005, 7(5–6), 630–638.
45. M. Martinez-Vicente, Z. Talloczy, S. Kaushik, A. C. Massey, J. Mazzulli and E. V. Mosharov, *et al.*, Dopamine-modified α-synuclein blocks chaperone-mediated autophagy, *J. Clin. Invest.*, 2008, **118**(2), 777–788.
46. C. Perier and M. Vila, Mitochondrial Biology and Parkinson's Disease, *Cold Spring Harbor Perspect. Med.*, 2012, **2**(2), a009332.
47. M. P. Murphy, How mitochondria produce reactive oxygen species, *Biochem. J.*, 2009, **417**(1), 1–13.
48. H. L. Martin and P. Teismann, Glutathione–a review on its role and significance in Parkinson's disease, *FASEB J.*, 2009, **23**(10), 3263–3272.
49. M. P. Horowitz, C. Milanese, R. Di Maio, X. Hu, L. M. Montero and L. H. Sanders, *et al.*, Single-cell redox imaging demonstrates a distinctive response of dopaminergic neurons to oxidative insults, *Antioxid. Redox Signaling*, 2011, **15**(4), 855–871.
50. V. Dias, E. Junn and M. M. Mouradian, The role of oxidative stress in Parkinson's disease, *J. Parkinson's Dis.*, 2013, **3**(4), 461–491.
51. B. Su, X. Wang, L. Zheng, G. Perry, M. A. Smith and X. Zhu, Abnormal mitochondrial dynamics and neurodegenerative diseases, *BBA, Mol. Basis Dis.*, 2010, **1802**(1), 135–142.
52. A. M. Pickrell, M. Pinto and C. T. Moraes, Mouse models of Parkinson's disease associated with mitochondrial dysfunction, *Mol. Cell. Neurosci.*, 2013, **55**(C), 87–94.
53. M. T. Lin and M. F. Beal, Mitochondrial dysfunction and oxidative stress in neurodegenerative diseases, *Nature*, 2006, **443**(7113), 787–795.

54. M. F. Alexeyev, Is there more to aging than mitochondrial DNA and reactive oxygen species? *FEBS J.*, 2009, **276**(20), 5768–5787.
55. M. Elstner, S. K. Müller, L. Leidolt, C. Laub, L. Krieg and F. Schlaudraff, *et al.*, Neuromelanin, neurotransmitter status and brainstem location determine the differential vulnerability of catecholaminergic neurons to mitochondrial DNA deletions, *Mol. Brain*, 2011, **4**(1), 43.
56. Y. Kraytsberg, A. Nicholas, P. Caro and K. Khrapko, Single molecule PCR in mtDNA mutational analysis: Genuine mutations *vs.* damage bypass-derived artifacts, *Methods*, 2008, **46**(4), 269–273.
57. D. A. Bender, R. M. Schwarzkopf, A. McMillan, K. J. Krishnan, G. Rieder and M. Neumann, *et al.*, Dopaminergic midbrain neurons are the prime target for mitochondrial DNA deletions, *J. Neurol.*, 2008, **255**(8), 1231–1235.
58. A. Bender, K. J. Krishnan, C. M. Morris, G. A. Taylor, A. K. Reeve and R. H. Perry, *et al.*, High levels of mitochondrial DNA deletions in substantia nigra neurons in aging and Parkinson disease, *Nat. Genet.*, 2006, **38**(5), 515–517.
59. L. H. Sanders, J. McCoy, X. Hu, P. G. Mastroberardino, B. C. Dickinson and C. J. Chang, *et al.*, Mitochondrial DNA damage: Molecular marker of vulnerable nigral neurons in Parkinson's disease, *Neurobiol. Dis.*, 2014, **70**(C), 214–223.
60. L. H. Sanders, E. H. Howlett, J. McCoy and J. T. Greenamyre, Mitochondrial DNA Damage as a Peripheral Biomarker for Mitochondrial Toxin Exposure in Rats, *Toxicol. Sci.*, 2014, **142**(2), 395–402.
61. J. T. Greenamyre, T. B. Sherer, R. Betarbet and A. V. Panov, Complex I and Parkinson's Disease, *IUBMB Life*, 2001, **52**(3–5), 135–141.
62. A. H. Schapira, J. M. Cooper, D. Dexter, J. B. Clark, P. Jenner and C. D. Marsden, Mitochondrial complex I deficiency in Parkinson's disease, *J. Neurochem.*, 1990, **54**(3), 823–827.
63. J. W. Langston, P. Ballard, J. W. Tetrud and I. Irwin, Chronic Parkinsonism in humans due to a product of meperidine-analog synthesis, *Science*, 1983, **219**(4587), 979–980.
64. S. Przedborski, V. Jackson-Lewis, R. Djaldetti, G. Liberatore, M. Vila and S. Vukosavic, *et al.*, The parkinsonian toxin MPTP: action and mechanism, *Restor. Neurol. Neurosci.*, 2000, **16**(2), 135–142.
65. J. T. Greenamyre, R. Betarbet and T. B. Sherer, The rotenone model of Parkinson's disease: genes, environment and mitochondria, *Parkinsonism Relat. Disord.*, 2003, **9**, 59–64.
66. J. R. Cannon, V. Tapias, H. M. Na, A. S. Honick, R. E. Drolet and J. T. Greenamyre, A highly reproducible rotenone model of Parkinson's disease, *Neurobiol. Dis.*, 2009, **34**(2), 279–290.
67. P. G. Mastroberardino, E. K. Hoffman, M. P. Horowitz, R. Betarbet, G. Taylor and D. Cheng, *et al.*, A novel transferrin/TfR2-mediated mitochondrial iron transport system is disrupted in Parkinson's disease, *Neurobiol. Dis.*, 2009, **34**(3), 417–431.

68. S. Collins, J. Pi and E. Yehuda-Shnaidman, Uncoupling and reactive oxygen species (ROS)–a double-edged sword for β-cell function? Moderation in all things, *Best Pract. Res., Clin. Endocrinol. Metab.*, 2012, **26**(6), 753–758.
69. P. F. Good, A. Hsu, P. Werner, D. P. Perl and C. W. Olanow, Protein nitration in Parkinson's disease, *J. Neuropathol. Exp. Neurol.*, 1998, **57**(4), 338–342.
70. R. A. Roberts, R. A. Smith, S. Safe, C. Szabo, R. B. Tjalkens and F. M. Robertson, Toxicological and pathophysiological roles of reactive oxygen and nitrogen species, *Toxicology*, 2010, **276**(2), 85–94.
71. B. Blanchard-Fillion, J. M. Souza, T. Friel, G. C. T. Jiang, K. Vrana and V. Sharov, *et al.*, Nitration and Inactivation of Tyrosine Hydroxylase by Peroxynitrite, *J. Biol. Chem.*, 2001, **276**(49), 46017–46023.
72. A. Hartmann, Postmortem studies in Parkinson's disease, *Dialogues Clin. Neurosci.*, 2004, **6**(3), 281–293.
73. M. P. Horowitz and J. T. Greenamyre, Mitochondrial Iron Metabolism and Its Role in Neurodegeneration, *J. Alzheimer's Dis.*, 2010, **3**(4), 461–491.
74. L. M. Sayre, P. I. Moreira, M. A. Smith and G. Perry, Metal ions and oxidative protein modification in neurological disease, *Ann. Ist. Super Sanita*, 2005, **41**(2), 143–164.
75. K. A. Jellinger, Neuropathological spectrum of synucleinopathies, *Mov. Disord.*, 2003, **18**(S6), 2–12.
76. T. Yoshida, M. Tanaka, A. Sotomatsu, S. Hirai and K. Okamoto, Activated microglia cause iron-dependent lipid peroxidation in the presence of ferritin, *NeuroReport*, 1998, **9**(9), 1929–1933.
77. W. Dröge and H. M. Schipper, Oxidative stress and aberrant signaling in aging and cognitive decline, *Aging Cell*, 2007, **6**(3), 361–370.
78. K. Aoyama, N. Matsumura, M. Watabe and T. Nakaki, Oxidative stress on EAAC1 is involved in MPTP-induced glutathione depletion and motor dysfunction, *Eur. J. Neurosci.*, 2007, **27**(1), 20–30.
79. K. S. Saravanan, K. M. Sindhu, K. S. Senthilkumar and K. P. Mohanakumar, l-deprenyl protects against rotenone-induced, oxidative stress-mediated dopaminergic neurodegeneration in rats, *Neurochem. Int.*, 2006, **49**(1), 28–40.
80. A. E. Berman, W. Y. Chan, A. M. Brennan, R. C. Reyes, B. L. Adler and S. W. Suh, *et al.*, *N*-acetylcysteine prevents loss of dopaminergic neurons in the EAAC1-/- mouse, *Ann. Neurol.*, 2011, **69**(3), 509–520.
81. B. Caneda-Ferrón, L. A. De Girolamo, T. Costa, K. E. Beck, R. Layfield and E. E. Billett, Assessment of the direct and indirect effects of MPP+ and dopamine on the human proteasome: implications for Parkinson's disease aetiology, *J. Neurochem.*, 2008, **105**(1), 225–238.
82. G. D. Zeevalk, R. Razmpour and L. P. Bernard, Glutathione and Parkinson's disease: Is this the elephant in the room? *Biomed. Pharmacother.*, 2008, **62**(4), 236–249.
83. S. Hunot, M. Vila, P. Teismann, R. J. Davis, E. C. Hirsch and S. Przedborski, *et al.*, JNK-mediated induction of cyclooxygenase 2 is required for

neurodegeneration in a mouse model of Parkinson's disease, *Proc. Natl. Acad. Sci.*, 2004, **101**(2), 665–670.
84. A. H. Stokes, T. G. Hastings and K. E. Vrana, Cytotoxic and genotoxic potential of dopamine, *J. Neurosci. Res.*, 1999, **55**(6), 659–665.
85. N. Hattoria, M. Wanga, H. Taka, T. Fujimura, A. Yoritaka and S.-I. Kubo, *et al.*, Toxic effects of dopamine metabolism in Parkinson's disease, *Parkinsonism Relat. Disord.*, 2009, **15**(suppl. 1), S35–S38.
86. J. Meiser, D. Weindl and K. Hiller, Complexity of dopamine metabolism, *Cell Commun. Signaling*, 2013, **11**(1), 34.
87. M. Valko, D. Leibfritz, J. Moncol, M. T. D. Cronin, M. Mazur and J. Telser, Free radicals and antioxidants in normal physiological functions and human disease, *Int. J. Biochem. Cell Biol.*, 2007, **39**(1), 44–84.
88. T. G. Hastings, D. A. Lewis and M. J. Zigmond, Role of oxidation in the neurotoxic effects of intrastriatal dopamine injections, *Proc. Natl. Acad. Sci.*, 1996, **93**(5), 1956–1961.
89. K. R. Hoyt, I. J. Reynolds and T. G. Hastings, Mechanisms of dopamine-induced cell death in cultured rat forebrain neurons: interactions with and differences from glutamate-induced cell death, *Exp. Neurol.*, 1997, **143**(2), 269–281.
90. L. M. M. Vermeer, V. R. Florang and J. A. Doorn, Catechol and aldehyde moieties of 3,4-dihydroxyphenylacetaldehyde contribute to tyrosine hydroxylase inhibition and neurotoxicity, *Brain Res.*, 2012, **1474**(C), 100–109.
91. W. M. Caudle, J. R. Richardson, M. Z. Wang, T. N. Taylor, T. S. Guillot and A. L. McCormack, *et al.*, Reduced Vesicular Storage of Dopamine Causes Progressive Nigrostriatal Neurodegeneration, *J. Neurosci.*, 2007, **27**(30), 8138–8148.
92. J. M. Jenco, A. Rawlingson, B. Daniels and A. J. Morris, Regulation of phospholipase D2: selective inhibition of mammalian phospholipase D isoenzymes by alpha- and beta-synucleins, *Biochemistry*, 1998, **37**(14), 4901–4909.
93. D. C. Duke, L. B. Moran, R. K. B. Pearce and M. B. Graeber, The medial and lateral substantia nigra in Parkinson's disease: mRNA profiles associated with higher brain tissue vulnerability, *Neurogenetics*, 2007, **8**(2), 83–94.
94. A. Gerhard, N. Pavese, G. Hotton, F. Turkheimer, M. Es and A. Hammers, *et al., In vivo* imaging of microglial activation with [11C](R)-PK11195 PET in idiopathic Parkinson's disease, *Neurobiol. Dis.*, 2006, **21**(2), 404–412.
95. B. L. Wilkinson and G. E. Landreth, The microglial NADPH oxidase complex as a source of oxidative stress in Alzheimer's disease, *J. Neuroinflammation*, 2006, **3**(1), 12–30.
96. L. J. Peterson and P. M. Flood, Oxidative Stress and Microglial Cells in Parkinson's Disease, *Mediators Inflammation*, 2012, **2012**(11), 1–12.
97. B. P. Cho, S. Sugama, D. H. Shin, L. A. DeGiorgio, S. S. Kim and Y. S. Kim, *et al.*, Microglial phagocytosis of dopamine neurons at early phases of apoptosis, *Cell. Mol. Neurobiol.*, 2003, **23**(4–5), 551–560.
98. U. Förstermann and W. C. Sessa, Nitric oxide synthases: regulation and function, *Eur. Heart J.*, 2012, **33**(7), 829–837.

99. A. Ghosh, J. Y. Park, C. Fenno and Y. L. Kapila, Porphyromonas gingivalis, gamma interferon, and a proapoptotic fibronectin matrix form a synergistic trio that induces c-Jun N-terminal kinase 1-mediated nitric oxide generation and cell death, *Infect. Immun.*, 2008, **76**(12), 5514–5523.
100. S. J. Hewett, J. A. Corbett, M. L. McDaniel and D. W. Choi, Interferon-γ and interleukin-1β induce nitric oxide formation from primary mouse astrocytes, *Neurosci. Lett.*, 1993, **164**(1–2), 229–232.
101. G. T. Liberatore, V. Jackson-Lewis, S. Vukosavic, A. S. Mandir, M. Vila and W. G. McAuliffe, *et al.*, Inducible nitric oxide synthase stimulates dopaminergic neurodegeneration in the MPTP model of Parkinson disease, *Nat. Med.*, 1999, **5**(12), 1403–1409.
102. K. Saijo, B. Winner, C. T. Carson, J. G. Collier, L. Boyer and M. G. Rosenfeld, *et al.*, A Nurr1/CoREST pathway in microglia and astrocytes protects dopaminergic neurons from inflammation-induced death, *Cell*, 2009, **137**(1), 47–59.
103. M. P. van der Brug, A. Singleton, T. Gasser and P. A. Lewis, Parkinson's disease: From human genetics to clinical trials, *Sci Transl Med.*, 2015, 7(305), 205ps20.
104. A. B. Singleton, M. J. Farrer and V. Bonifati, The genetics of Parkinson's disease: progress and therapeutic implications, *Mov. Disord.*, 2013, **28**(1), 14–23.
105. I. F. Mata, W. J. Wedemeyer, M. J. Farrer, J. P. Taylor and K. A. Gallo, LRRK2 in Parkinson's disease: protein domains and functional insights, *Trends Neurosci.*, 2006, **29**(5), 286–293.
106. L. J. Ozelius, G. Senthil, R. Saunders-Pullman, E. Ohmann, A. Deligtisch and M. Tagliati, *et al.*, LRRK2G2019S as a Cause of Parkinson's Disease in Ashkenazi Jews, *N. Engl. J. Med.*, 2006, **354**(4), 424–425.
107. M. Steger, F. Tonelli, G. Ito, P. Davies, M. Trost and M. Vetter, *et al.*, Phosphoproteomics reveals that Parkinson's disease kinase LRRK2 regulates a subset of Rab GTPases, *eLife*, 2016, **5**, e12813.
108. P.-Y. Pan and Z. Yue, Genetic causes of Parkinson's disease and their links to autophagy regulation, *Parkinsonism Relat. Disord.*, 2015, **20**(S1), S154–S157.
109. L. H. Sanders, J. Laganière, O. Cooper, S. K. Mak, B. J. Vu and Y. A. Huang, *et al.*, LRRK2 mutations cause mitochondrial DNA damage in iPSC-derived neural cells from Parkinson's disease patients: Reversal by gene correction, *Neurobiol. Dis.*, 2014, **62**(C), 381–386.
110. D. G. Healy, M. Falchi, S. S. O'Sullivan, V. Bonifati, A. Durr and S. Bressman, *et al.*, Phenotype, genotype, and worldwide genetic penetrance of LRRK2-associated Parkinson's disease: a case-control study, *Lancet Neurol.*, 2008, **7**(7), 583–590.
111. M. Baba, S. Nakajo, P. H. Tu, T. Tomita, K. Nakaya and V. M. Lee, *et al.*, Aggregation of alpha-synuclein in Lewy bodies of sporadic Parkinson's disease and dementia with Lewy bodies, *Am. J. Pathol.*, 1998, **152**(4), 879–884.
112. M. G. Spillantini, R. A. Crowther, R. Jakes, M. Hasegawa and M. Goedert, alpha-Synuclein in filamentous inclusions of Lewy bodies

from Parkinson's disease and dementia with lewy bodies, *Proc. Natl. Acad. Sci. U. S. A.*, 1998, **95**(11), 6469–6473.

113. A. M. Cuervo, L. Stefanis, R. Fredenburg, P. T. Lansbury and D. Sulzer, Impaired Degradation of Mutant α-Synuclein by Chaperone-Mediated Autophagy, *Science*, 2004, **305**(5688), 1292–1295.
114. R. Di Maio, P. J. Barrett, E. K. Hoffman, C. W. Barrett, A. Zharikov and A. Borah, *et al.*, α-Synuclein binds to TOM20 and inhibits mitochondrial protein import in Parkinson's disease, *Sci. Transl. Med.*, 2016, **8**(342), 342ra78.
115. V. S. Van Laar and S. B. Berman, The interplay of neuronal mitochondrial dynamics and bioenergetics: implications for Parkinson's disease, *Neurobiol. Dis.*, 2013, **51**, 43–55.
116. E. M. Valente, P. M. Abou-Sleiman, V. Caputo, M. M. K. Muqit, K. Harvey and S. Gispert, *et al.*, Hereditary early-onset Parkinson's disease caused by mutations in PINK1, *Science*, 2004, **304**(5674), 1158–1160.
117. V. A. Morais, P. Verstreken, A. Roethig, J. Smet, A. Snellinx and M. Vanbrabant, *et al.*, Parkinson's disease mutations in PINK1 result in decreased Complex I activity and deficient synaptic function, *EMBO Mol. Med.*, 2009, **1**(2), 99–111.
118. E. Sidransky, M. A. Nalls, J. O. Aasly, J. Aharon-Peretz, G. Annesi and E. R. Barbosa, *et al.*, Multicenter Analysis of Glucocerebrosidase Mutations in Parkinson's Disease, *N. Engl. J. Med.*, 2009, **361**(17), 1651–1661.
119. E. Sidransky and G. Lopez, The link between the GBA gene and parkinsonism, *Lancet Neurol.*, 2012, **11**(11), 986–998.
120. E. M. Rocha, G. A. Smith, E. Park, H. Cao, E. Brown and M. A. Hayes, *et al.*, Glucocerebrosidase gene therapy prevents α-synucleinopathy of midbrain dopamine neurons, *Neurobiol Dis*, 2015, **82**, 495–503.
121. D. J. Moore, L. Zhang, T. M. Dawson and V. L. Dawson, A missense mutation (L166P) in DJ-1, linked to familial Parkinson's disease, confers reduced protein stability and impairs homo-oligomerization, *J. Neurochem.*, 2003, **87**(6), 1558–1567.
122. R. M. Canet-Aviles, M. A. Wilson, D. W. Miller, R. Ahmad, C. McLendon and S. Bandyopadhyay, *et al.*, The Parkinson's disease protein DJ-1 is neuroprotective due to cysteine-sulfinic acid-driven mitochondrial localization, *Proc. Natl. Acad. Sci. U. S. A.*, 2004, **101**(24), 9103–9108.
123. V. Bonifati, P. Rizzu, M. J. van Baren, O. Schaap, G. J. Breedveld and E. Krieger, *et al.*, Mutations in the DJ-1 Gene Associated with Autosomal Recessive Early-Onset Parkinsonism, *Science*, 2003, **299**(5604), 256–259.
124. T. Hayashi, C. Ishimori, K. Takahashi-Niki, T. Taira, Y.-C. Kim and H. Maita, *et al.*, DJ-1 binds to mitochondrial complex I and maintains its activity, *Biochem. Biophys. Res. Commun.*, 2009, **390**(3), 667–672.
125. R. H. Kim, P. D. Smith, H. Aleyasin, S. Hayley, M. P. Mount and S. Pownall, *et al.*, Hypersensitivity of DJ-1-deficient mice to 1-methyl-4-phenyl-1,2,3,6-tetrahydropyrindine (MPTP) and oxidative stress, *Proc. Natl. Acad. Sci. U. S. A.*, 2005, **102**(14), 5215–5220.
126. D. N. Hauser and M. R. Cookson, Astrocytes in Parkinson's disease and DJ-1, *J. Neurochem.*, 2011, **117**(3), 357–358.

127. M. R. Cookson, DJ-1, PINK1, and their effects on mitochondrial pathways, *Mov. Disord.*, 2010, **25**(suppl. 1), S44–S48.
128. A. H. Rajput, R. J. Uitti, W. Stern, W. Laverty, K. O'Donnell and D. O'Donnell, *et al.*, Geography, drinking water chemistry, pesticides and herbicides and the etiology of Parkinson's disease, *Can. J. Neurol. Sci.*, 1987, **14**(suppl. 3), 414–418.
129. C. M. Tanner, F. Kamel, G. W. Ross, J. A. Hoppin, S. M. Goldman and M. Korell, *et al.*, Rotenone, Paraquat, and Parkinson's Disease, *Environ. Health Perspect.*, 2011, **119**(6), 866–872.
130. S. Costello, M. Cockburn, J. Bronstein, X. Zhang and B. Ritz, Parkinson's disease and residential exposure to maneb and paraquat from agricultural applications in the central valley of California, *Am. J. Epidemiol.*, 2009, **169**(8), 919–926.
131. D. Di Monte, M. S. Sandy, G. Ekström and M. T. Smith, Comparative studies on the mechanisms of paraquat and 1-methyl-4-phenylpyridine (MPP+) cytotoxicity, *Biochem. Biophys. Res. Commun.*, 1986, **137**(1), 303–309.
132. R. Betarbet, T. B. Sherer, G. MacKenzie, M. Garcia-Osuna, A. V. Panov and J. T. Greenamyre, Chronic systemic pesticide exposure reproduces features of Parkinson's disease, *Nat. Neurosci.*, 2000, **3**(12), 1301–1306.
133. A. L. McCormack, M. Thiruchelvam, A. B. Manning-Bog, C. Thiffault, J. W. Langston and D. A. Cory-Slechta, *et al.*, Environmental risk factors and Parkinson's disease: selective degeneration of nigral dopaminergic neurons caused by the herbicide paraquat, *Neurobiol. Dis.*, 2002, **10**(2), 119–127.
134. D. Mergler, G. Huel, R. Bowler, A. Iregren, S. Bélanger and M. Baldwin, *et al.*, Nervous system dysfunction among workers with long-term exposure to manganese, *Environ. Res.*, 1994, **64**(2), 151–180.
135. B. A. Racette, J. A. Antenor, L. McGee-Minnich, S. M. Moerlein, T. O. Videen and V. Kotagal, *et al.*, [18F]FDOPA PET and clinical features in parkinsonism due to manganism, *Mov. Disord.*, 2005, **20**(4), 492–496.
136. J. A. Moreno, E. C. Yeomans, K. M. Streifel, B. L. Brattin, R. J. Taylor and R. B. Tjalkens, Age-dependent susceptibility to manganese-induced neurological dysfunction, *Toxicol. Sci.*, 2009, **112**(2), 394–404.
137. J. M. Gorell, C. C. Johnson, B. A. Rybicki, E. L. Peterson, G. X. Kortsha and G. G. Brown, *et al.*, Occupational exposures to metals as risk factors for Parkinson's disease, *Neurology*, 1997, **48**(3), 650–658.
138. R. G. Lucchini, S. Guazzetti, S. Zoni, C. Benedetti, C. Fedrighi and M. Peli, *et al.*, NeuroToxicologyNeurofunctional dopaminergic impairment in elderly after lifetime exposure to manganese, *Neurotoxicology*, 2014, **45**, 309–317.
139. S. L. Rhodes, D. D. Buchanan, I. Ahmed, K. D. Taylor, M.-A. Loriot and J. S. Sinsheimer, *et al.*, Pooled analysis of iron-related genes in Parkinson's disease: association with transferrin, *Neurobiol. Dis.*, 2014, **62**, 172–178.
140. P. Matak, A. Matak, S. Moustafa, D. K. Aryal, E. J. Benner and W. Wetsel, *et al.*, Disrupted iron homeostasis causes dopaminergic neurodegeneration in mice, *Proc. Natl. Acad. Sci. U. S. A.*, 2016, **113**, 3428–3435.

141. M. Takahashi, T. Yamada, S. Nakajima, K. Nakajima, T. Yamamoto and H. Okada, The substantia nigra is a major target for neurovirulent influenza A virus, *J. Exp. Med.*, 1995, **181**(6), 2161–2169.
142. C. N. Martyn and C. Osmond, Parkinson's disease and the environment in early life, *J. Neurol. Sci.*, 1995, **132**(2), 201–206.
143. H. Jang, D. Boltz, K. Sturm-Ramirez, K. R. Shepherd, Y. Jiao and R. Webster, *et al.*, Highly pathogenic H5N1 influenza virus can enter the central nervous system and induce neuroinflammation and neurodegeneration, *Proc. Natl. Acad. Sci. U. S. A.*, 2009, **106**(33), 14063–14068.
144. H. Jang, D. Boltz, J. McClaren, A. K. Pani, M. Smeyne and A. Korff, *et al.*, Inflammatory effects of highly pathogenic H5N1 influenza virus infection in the CNS of mice, *J. Neurosci.*, 2012, **32**(5), 1545–1559.
145. Z. Ling, D. A. Gayle, S. Y. Ma, J. W. Lipton, C. W. Tong and J.-S. Hong, *et al.*, *In utero* bacterial endotoxin exposure causes loss of tyrosine hydroxylase neurons in the postnatal rat midbrain, *Mov. Disord.*, 2002, **17**(1), 116–124.
146. C.-S. Fong, R.-M. Wu, J.-C. Shieh, Y.-T. Chao, Y.-P. Fu and C.-L. Kuao, *et al.*, Pesticide exposure on southwestern Taiwanese with MnSOD and NQO1 polymorphisms is associated with increased risk of Parkinson's disease, *Clin. Chim. Acta*, 2007, **378**(1–2), 136–141.
147. J. A. G. Agúndez, F. J. Jiménez Jiménez, A. Luengo, M. L. Bernal, J. A. Molina and L. Ayuso, *et al.*, Association between the oxidative polymorphism and early onset of Parkinson's disease, *Clin. Pharmacol. Ther.*, 1995, **57**(3), 291–298.
148. C. A. Smith, A. C. Gough, P. N. Leigh, B. A. Summers, A. E. Harding and D. M. Maraganore, *et al.*, Debrisoquine hydroxylase gene polymorphism and susceptibility to Parkinson's disease, *Lancet*, 1992, **339**(8806), 1375–1377.
149. A. Elbaz, C. Levecque, J. Clavel, J.-S. Vidal, F. Richard and P. Amouyel, *et al.*, CYP2D6 polymorphism, pesticide exposure, and Parkinson's disease, *Ann. Neurol.*, 2004, **55**(3), 430–434.
150. T. Furuno, M.-T. Landi, M. Ceroni, N. Caporaso, I. Bernucci and G. Nappi, *et al.*, Expression polymorphism of the blood-brain barrier component P-glycoprotein (MDR1) in relation to Parkinson's disease, *Pharmacogenetics*, 2002, **12**(7), 529–534.
151. J. R. Cannon and J. T. Greenamyre, Gene–environment interactions in Parkinson's disease: Specific evidence in humans and mammalian models, *Neurobiol. Dis.*, 2013, **57**(C), 38–46.
152. L. H. Sanders, K. C. Paul, E. H. Howlett, H. Lawal, S. Boppana, J. M. Bronstein, B. Ritz and J. T. Greenamyre, Base excision repair variants and pesticide exposure increase Parkinson's disease risk, *Toxicol. Sci.*, 2017, kfx086, DOI: 10.1093/toxsci/kfx086.
153. B. R. De Miranda, E. A. Burton and J. T. Greenamyre, Astrocyte Specific Expression of DJ-1 Attenuates Rotenone-Induced Neurotoxicity: A Gene Therapy Approach, *Toxicol., Suppl. Toxicol. Sci.*, 2016, **150**(1), 325.

CHAPTER 2

Oxidative Stress and Redox Signalling in the Parkinson's Disease Brain

PABLO HERNANDEZ-FRANCO, ANNANDURAI ANANDHAN, RACHEL M. FOGUTH AND RODRIGO FRANCO*

Redox Biology Center & School of Veterinary Medicine and Biomedical Sciences, University of Nebraska-Lincoln, Lincoln, NE 68583, USA
*E-mail: rfrancocruz2@unl.edu

2.1 Introduction

Dopaminergic neuronal cell loss in the *substantia nigra* (SN) *pars compacta* (SNpc) is considered the pathological hallmark of Parkinson's disease (PD). Since the early 1990s, oxidative stress has been suggested to exert a causative role in the loss of dopaminergic cells.[1] Accordingly, the three major risk factors linked to PD (aging, environmental exposures and genetic alterations) have been linked to an increased generation of reactive oxygen species (ROS) and the accumulation of byproducts of oxidative damage. Indeed, postmortem brain-sample analyses have reported an increased accumulation of oxidized proteins, nucleic acids and lipids in PD brains. In this chapter, we will provide an introductory overview of the pathogenic mechanisms involved in the alteration of redox homeostasis that occurs in PD. We will also discuss the intrinsic properties of SNpc dopaminergic neurons that make them vulnerable to neurodegeneration. The mechanisms by which these alterations

Issues in Toxicology No. 34
Oxidative Stress and Redox Signalling in Parkinson's Disease
Edited by Rodrigo Franco, Jonathan A. Doorn and Jean-Christophe Rochet

Published by the Royal Society of Chemistry, www.rsc.org

in the cellular redox homeostasis contribute to dopaminergic neurodegeneration will be discussed in greater detail in the following chapters.

2.2 Oxidative Stress and Antioxidant Systems

Molecular oxygen (O_2) is vital for the survival of aerobic organisms. In normal cells, 30–32 molecules of ATP are generated per O_2 during aerobic metabolism. During cellular respiration O_2 is consumed by mitochondria, and at the terminal step of the oxidative phosphorylation (OXPHOS), it is reduced to H_2O. Electrons travel along the respiratory electron transport chain (ETC) composed of four complexes: complex I (nicotinamide adenine dinucleotide [NADH] ubiquinone oxidoreductase), complex II (succinate ubiquinone oxidoreductase), complex III (ubiquinone-cytochrome c oxidase), complex IV (cytochrome c oxidase), plus ubiquinone and cytochrome c. The ETC is leaky, and leaked electrons reduce O_2 resulting in the generation of superoxide anion radicals, $O_2^{\bullet-}$ (Figure 2.1).[2]

Oxidative stress is defined then as an increase in the steady-state levels of ROS and its reactive byproducts (including reactive nitrogen species [RNS]) that surpasses cellular antioxidant defenses.[3] Oxidative stress leads to the oxidation of biomolecules (proteins, nucleic acids and lipids). Cells have evolved not only antioxidant defenses, but also a wide range of mechanisms for the removal or repair of oxidative damage that involve reduction (when oxidation is reversible) and degradation of oxidized biomolecules or organelles. When oxidative damage persists it not only affects a single cell, but also the tissue in which it resides, thus having important implications for overall organism physiology.[3,4]

2.2.1 Reactive Oxygen and Nitrogen Species: Sources

ROS include radical (an atom or molecule with a single unpaired electron, •) and non-radical species such as $O_2^{\bullet-}$, singlet $^1O_2$, hydrogen peroxide H_2O_2 and the hydroxyl radical ($^\bullet OH$). Mitochondria metabolism is the primary source for ROS. In the mitochondria, $O_2^{\bullet-}$ is generally considered to be generated in the matrix by electron leakage from complex I,[5] and in both the matrix and the inner membrane space (IMS) by electron leakage from complex III.[6] $O_2^{\bullet-}$ generated in the matrix and the IMS has been proposed to be released to the cytosol *via* the mitochondrial permeability transition pore (mPTP)[7] or the voltage-dependent anion channels (VDAC),[8] respectively (Figure 2.1). However, at least 10 different sites for $O_2^{\bullet-}/H_2O_2$ production in the mitochondria have been described including 2-oxoglutarate dehydrogenase (OGDH), pyruvate dehydrogenase (PDH), complex II (flavin site), glycerol 3-phosphate dehydrogenase (mGPDH), as well as the less well-characterized electron transferring flavoprotein/ETF:Q oxidoreductase (ETF/ETF:QOR) system of fatty acid β-oxidation, proline dehydrogenase, aconitase, dihydroorotate dehydrogenase and monoamine oxidase (MAO). MAO is an enzyme located

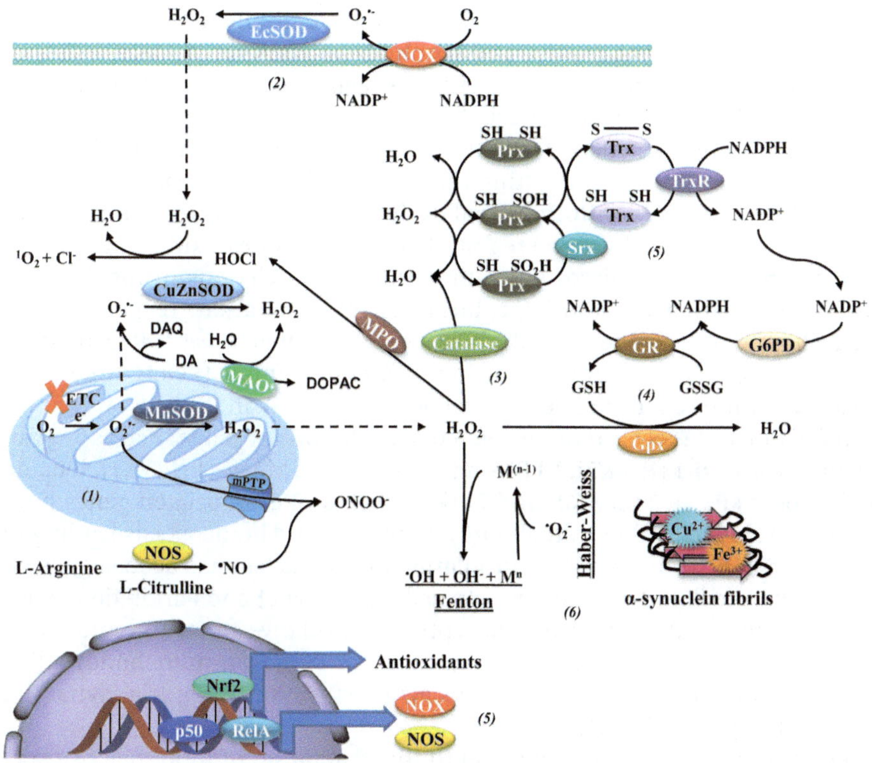

Figure 2.1 Oxidative stress and catalytic antioxidant systems. Cells have intrinsic antioxidant mechanisms to maintain a tight homeostatic control of ROS/RNS generated under physiological conditions, and to detoxify their excessive accumulation in pathological situations. (1) Mitochondria are the primary source of ROS. In PD, a dysfunction in mitochondrial ETC leads to increased formation of $O_2^{\bullet-}$. SODs dismutate $O_2^{\bullet-}$ to O_2 and H_2O_2. MnSOD is localized in the mitochondrial matrix, CuZnSOD in the IMS, peroxysomes, nucleous or cytosol, and EcSOD in the extracellular space (2). (3) Catalase catalyzes the decomposition of H_2O_2 to water and O_2 and is primarily localized in the peroxisomes. (4) Gpx reduce peroxides using GSH as substrate. Gpx isozymes encoded by different genes vary in their cellular location and substrate specificity. GR, which reduces GSSG back to GSH, requires NADPH as electron-donor reductant, and G6PD is indispensable for the regeneration of NADPH from $NADP^+$ in the cytosol. (5) Prxs are ubiquitous thiol peroxidases that scavenge peroxides. Catalysis of H_2O_2 is mediated by the reaction of their active Cys residue with H_2O_2. The catalytic activity of Prxs requires reducing equivalents provided by the Trx/TrxR system. (6) Misfolded aggregates of α-synuclein and metals also contribute to oxidative damage by Fenton reaction and $^{\bullet}OH$ formation. (7) In glial cells transcriptional regulation of NOX and NOS is mediated by inflammatory signalling *via* the activation of the nuclear factor kappa-light-chain-enhancer of activated B cells (NF-κB). Both NF-κB (in neurons) and Nrf2 (glial cells primarily) activation regulates the expression of antioxidant defenses.

on the outer face of the mitochondrial outer membrane that can generate H_2O_2 at high rates. The relative and absolute contribution of specific sites in the mitochondria to the production of ROS seems to depend on the availability of the substrate being oxidized.[9,10]

A decrease in the activity of the mitochondrial ETC is found in the SNpc of patients with PD[11–13] suggesting that increased electron leakage and ROS formation is linked to mitochondrial dysfunction in PD. Mitochondrial DNA (mtDNA) mutations have been found in rare families exhibiting parkinsonism. Somatic (acquired) point mutations and deletions in mtDNA also accumulate with age in the SNpc, but no specific mutations responsible for mitochondrial dysfunction in PD have been identified as yet.[14] Recent studies have suggested that mutations in polymerase gamma (POLG) are linked to acquired mitochondrial defects in complex I and IV in PD.[15]

A fraction of PD occurrence is related to mutations in genes such as α-synuclein (*SNCA*), DJ-1 (*PARK7*), PTEN-induced putative kinase 1 (*PINK1*), leucine rich repeat kinase 2 (*LRRK2*) and Parkin (*PARK2*). PD-associated genes have also been shown to alter mitochondrial function and homeostasis leading to increased ROS formation.[16] Accumulation of α-synuclein has been suggested to trigger mitochondrial oxidative stress.[17] DJ-1, PINK1 and Parkin deficiency or mutations render cells more susceptible to oxidative stress and mitochondrial dysfunction, and impair the turnover of damaged mitochondria.[18–21] Gain-of-function of LRRK2 mutations also disturb mitochondrial dynamics leading to increased ROS formation.[22]

Of importance, one ROS can lead to the generation of more reactive species. $O_2^{\bullet-}$ can be dismutated in an enzymatic or non-enzymatic manner to H_2O_2. The enzymatic generation of H_2O_2 is catalyzed by $O_2^{\bullet-}$ dismutases (SODs). While H_2O_2 is thought to diffuse across membranes, it has also been reported that H_2O_2 transport can be mediated by aquaporins.[23,24] Metals (M) can also promote ROS formation. The Fenton reaction catalyzes the formation of $^{\bullet}OH$ and hydroxide (^{-}OH) by the reaction of H_2O_2 with M^{n-1}. The Haber–Weiss reaction then reduces the metal ($M^n \rightarrow M^{n-1}$) to further promote free-radical formation. Metals such as iron (Fe) and copper (Cu) can generate $^{\bullet}OH$ by Fenton reactions (Figure 2.1).[3,4] RNS also contribute to oxidative (nitrative) damage. Nitric oxide ($^{\bullet}NO$) is a hydrophobic molecule that diffuses freely across membranes and is generated from L-arginine by nitric oxide synthases (NOS).[25] $O_2^{\bullet-}$ reacts with $^{\bullet}NO$, leading to the production of peroxynitrite ($ONOO^-$), which is also able to cross membranes through anion channels in the anionic form, and by passive diffusion in its protonated form, peroxynitrous acid (ONOOH) (Figure 2.1).[26] When ONOOH undergoes decomposition, it generates the one-electron oxidants hydroxyl ($^{\bullet}OH$) and nitrogen dioxide ($^{\bullet}NO_2$). $ONOO^-$ can react with carbon dioxide (CO_2) to produce the carbonate radical anion ($CO_3^{\bullet-}$) and $^{\bullet}NO_2$, which can undergo further reaction with $^{\bullet}NO$ to produce dinitrogen trioxide (N_2O_3). NOS uncoupling has also been recognized as an important source for $O_2^{\bullet-}$.[27]

Other enzymatic mechanisms also mediate reactive species formation. In glial cells, myeloperoxidases (MPO) produce hypochlorous acid (HOCl) from

H_2O_2 and chloride anion (Cl^-) using heme as a cofactor (Figure 2.1).[28] Cyclo-oxygenases (COX) generate ROS as byproducts of arachidonic acid metabolism to prostaglandin G2 (PGG_2), utilizing O_2 and producing peroxyl radicals ($ROO^•$). COXs also possess a heme-containing active site that provides peroxidase activity, converting PGG_2 to prostaglandin H2 (PGH_2) by removing O_2, which is a source of $O_2^{•-}$. In the presence of H_2O_2, the peroxide activity of COXs oxidizes various co-substrates such as NADH and glutathione (GSH), which reduces O_2 to $O_2^{•-}$.[29]

In addition to mitochondria, several organelles within the cell have also been recognized as important sources for ROS, and under certain circumstances, might actually play a bigger role than mitochondria in ROS generation.[30] During protein synthesis, ROS are produced in the endoplasmic reticulum (ER) as a consequence of the activity of oxidoreductases that catalyze protein-folding reactions. Disulfide bond formation in nascent proteins is catalyzed by protein disulfide isomerase (PDI) that accepts electrons from cysteine (Cys) thiol residues in polypeptide chain substrates leading to the formation of disulfide bonds. ERO1 (ER oxidoreductin) transfers electrons from PDI to O_2, resulting in the production of H_2O_2.[31] ER stress and the activation of the unfolded protein response (UPR) has been reported in PD, however, its causative role in oxidative damage has not been established.[32] By virtue of their high Fe content, lysosomes are also another important potential source of ROS by catalysis of Fenton-type reactions.[33] As such mutations in PD-related genes and lysossomal breakdown has been linked to PD pathogenesis.[34] Another important source of ROS production is the nicotinamide adenine dinucleotide phosphate (NAD[P]H) oxidase (NOX) family of enzymes that catalyze the production of $O_2^{•-}$ from O_2 and NADPH (Figure 2.1).[35] Importantly, distinct NOX enzymes have also been reported in distinct brain regions, and their activation in glial cells (as an inflammatory response) is known to contribute to oxidative damage and dopaminergic (DAergic) neurodegeneration.[35,36]

2.2.2 Antioxidant Systems

Oxidative stress is counteracted by cellular antioxidant mechanisms. The most important cellular defenses against $O_2^{•-}$ are SODs. In mammals, there are three isoforms: manganese superoxide dismutase (MnSOD or SOD2) localized in the mitochondrial matrix;[37] copper-zinc SOD (CuZnSOD or SOD1) in the IMS, peroxysomes, nucleus, or cytosol;[38] and extracellular SOD (EcSOD or SOD3) anchored to the cell's surface (Figure 2.1). A number of scavenging mechanisms exist against H_2O_2. Catalase mediates the decomposition of H_2O_2 and is primarily localized in the peroxisomes. Catalase activity was reported reduced in the SN and putamen of PD brains.[39] In contrast, SOD activity was found increased in postmortem PD brains.[40,41]

Peroxiredoxins (Prxs) are ubiquitous thiol peroxidases. Mammals have six Prxs, with Prx1, 2, and 6 found in the cytoplasm; Prx4 in the ER; Prx3 in the mitochondria; and Prx5 found in various compartments within the cell, including peroxisomes and mitochondria.[42] While Prxs can be considered as important antioxidant defenses, they are primarily seen as regulators of peroxide signalling and dynamics, because of their high propensity for inactivation.[43] Except for Prx6, thioredoxins (Trx) provide the reducing equivalents required for the peroxide scavenging activity of peroxiredoxins (Prxs). The Trx redox system itself depends on thiol–disulfide exchange reactions at the active site, and Trx reductase (TrxR) transfers reducing equivalents from NADPH to Trxs reducing them (Figure 2.1). While the Trx1/TrxR1 system is localized in the cytoplasm, the Trx2/TrxR2 system is mitochondrial. In addition, Trxs have also been shown to reduce and regulate the activity of a number of proteins. In the SN, DAergic neurons express Prx2–5 at very low levels, with Prx5 being the isoform with the lowest expression.[44,45] An increase in Prx2 and a decrease in Prx3 and Prx6 levels is observed in postmortem PD brains.[46] Prx2 oxidation has also been reported in PD brains.[47] In the SN both Trx1 and Trx2 mRNA have been detected.[48,49] At the protein level, recent immunohistochemical analyses have demonstrated the strong expression of both Trx1 and Trx2 in the SN of mouse and rat brain, while weaker but consistent detection of TrxR1 and TrxR2 was also found.[44,50]

Glutathione peroxidases (GPx), selenoproteins that reduce peroxides, are present in distinct cellular locations with different substrate specificity. For example, GPx1 is found primarily in the cytoplasm and preferably scavenges H_2O_2, while GPx4 (PhGPx) hydrolyzes lipid hydroperoxides in both cytosolic and mitochondrial compartments. GPxs and Prx6 use the reducing power of GSH (L-γ-glutamyl-L-cysteinyl-glycine) to hydrolyze peroxides (Figure 2.1).[51,52] GSH synthesis is initiated by the generation of γ-glutamylcysteine from glutamate (Glu) and Cys *via* glutamate-cysteine ligase (composed of a catalytic subunit [GCLC] and a modifier subunit [GCLM]), with the subsequent addition of glycine (Gly) by the activity of GSH synthetase (GS) (see Chapter 6 for further details).[51,53–57] A decrease in the expression of GPx has been found in DAergic cells in the SN.[58]

Glutathione S-transferases (GSTs) catalyze the conjugation of GSH to electrophilic carbons and are divided into three main families: cytosolic, mitochondrial, and the membrane-associated microsomal family. Microsomal GSTs are membrane-associated proteins involved in eicosanoid and GSH metabolism (MAPEG) and are structurally distinct from cytosolic GSTs. Cytosolic GSTs are divided into alpha (A), mu (M), omega (O), pi (P), sigma (S), theta (T), and zeta (Z) classes. GSTs participate in the detoxification of a chemically diverse group of compounds. GSTs have also been reported to detoxify ROS *via* a GSH-peroxidase activity and mediate isomerization of ketosteroids and prostaglandins, as well as tyrosine degradation, glutathionylation, and protein–protein interactions with stress-activated kinases (SAPK) to regulate their activation.[59,60] Polymorphisms in GSTs have been associated with PD, particularly when related to pesticide exposure.[61–66] However, the exact role that these enzymes play in DAergic degeneration is far from understood. GST-pi is found highly expressed in PD brains compared to controls.[67] In fact, GST-mu has been reported to delay the onset of PD.[68,69]

The transcription of a variety of antioxidant genes (SOD1, SOD2, GCLM, GCLC, and GSTs) through *cis*-acting sequences known as antioxidant response elements (ARE) is an important mechanism to counteract oxidative stress. Members of the Cap-N-Collar family of transcription factors, including the nuclear factor (erythroid-derived 2)-like 1 (Nrf1) and 2 (Nrf2), have been identified to bind ARE. Under resting conditions, Nrf2 is kept in a cytosolic complex with the kelch-like ECH-associated protein 1 (Keap1), where it is subjected to proteasomal degradation. Oxidants or electrophiles induce a conformational change in Keap1 that allows the translocation of Nrf2 to the nucleus for transduction (Figure 2.1). Keap1 contains Zn-coordinated Cys273 and Cys288 that are sensitive to alkylation. Initial thiylation of Cys is thought to be followed by thiol/disulfide exchange to form internal disulfides within Keap1.[70] Accordingly, a protective Nrf2 haplotype associated with increased transcriptional activity was linked to decreased risk of PD.[71]

2.2.3 Oxidative Damage to Biomolecules

When antioxidants systems are compromised, an increase in the steady-state levels of ROS results in the oxidation of nucleotides, proteins, and lipid biomolecules. In this section, we will revise the evidence linking oxidative damage of biomolecules and repair systems to PD.

2.2.3.1 Oxidative DNA Damage

It has been shown that the levels of oxidized DNA byproducts are significantly increased in mtDNA as well as nuclear DNA in the brains of patients with PD.[72,73] DNA damage associated with oxidative stress can be induced directly by the interaction of DNA with various ROS or indirectly as the result of alterations in DNA repair. Oxidation attack to DNA is induced primarily by ROS but other electrophiles generated as byproducts of oxidative damage can also mediate DNA alterations. For example, DNA-adducts generated by lipid peroxidation products are slowly repaired compared to direct byproducts of oxidative damage to DNA.[74] Oxidative damage to mtDNA is higher than nuclear DNA as mtDNA do not contain introns and do not have the proofing mechanisms that nuclear DNA has, so oxidation is more likely to cause point mutations or deletions in the mitochondria.[75]

DNA lesions arising directly from ROS are apurinic/apyrimidinic (abasic; AP) DNA sites, oxidized purines and pyrimidines, single-stranded DNA breaks (SSBs) and double-stranded DNA breaks (DSBs). However, not all ROS or RNS react with the deoxyribose moieties or nucleobases at the same rate. $^{\bullet}OH$ can cause DNA modification by inducing sugar and base modifications. The sugar DNA backbone can interact with the $^{\bullet}OH$ ending in the formation of carbonyls in the backbone or the addition of protein radicals to the DNA that can lead to strand breaks. The most common lesions due to oxidative base modifications are 8-hydroxy-2-deoxyguanosine (8-OHdG), 8-oxodeoxyadenosine, deoxythymidine glycol and 2,6-diamino-4-hydroxy-5-formamidopyrimidine (FapyG) which are selectively excised from DNA by DNA glycosylases creating an AP site.[76] Expression levels of DNA-glycosylases OGG1, MYH, and MTH1 (MUTYH, a nucleoside triphosphatase that hydrolyzes 8-OHdG preventing its misincorporation into DNA) have been found increased in PD brains.[77,78] Furthermore, MTH1 polymorphisms have been found linked to PD.[79] The base excision repair pathway (BER) repairs AP sites by the activity of AP-endonucleases (Ape1/Ref-1), which incises the DNA strand 5′ to the baseless sugar generating a SSB. Then, short-patch (single nucleotide filling) or long-patch BER repairs/seals the nick(s) *via* the activity of DNA polymerases (POL) and DNA ligases.[80,81] When a SSB is directly induced by ROS, the poly (ADP-ribose) polymerase 1 enzyme (PARP1) is implicated in its recognition and induction of BER.[82] Interestingly, an increase in phosphorylated Ape1 has been found in PD brains, and this modification impairs its endonuclease activity.[83] Furthermore, as mentioned above mutations in POLG1 have been linked to increase risk to develop PD.[84,85]

2.2.3.2 Oxidative Damage to Lipids

Polyunsaturated fatty acids (PUFAs) and esters, constituents of glycerophospholipids and other lipids in plasma and organelle membranes, are the major targets of intracellular oxidizing agents. Highly oxidizable lipid substrates include linoleate, arachidonate, pleate, cholesterol, eicosapentaenoate and docosahexaenoate. Neuronal membrane lipids are particularly rich in PUFAs. Lipid peroxidation occurs through a radical-mediated abstraction of a bis-allylic hydrogen (H) from the polyunsaturated fatty acids. The delocalized lipid radical (L•) reacts then with O_2 leading to the formation of lipid peroxyl radicals (LOO•). LOO• starts a series of radical chain reactions abstracting a H atom from a polyunsaturated lipid to generate another L• and lipid hydroperoxide (LOOH). This chain reaction is terminated by antioxidants (vitamin E) that donate a H to the LOO• species and form a corresponding vitamin E radical that reacts with another LOO• forming non-radical products. LOOH generates a number of byproducts such as malondialdehyde (MDA), acrolein, propanal, hexanal, 4-hydroperoxy-2-nonenal (HPNE), 4-oxo-2-nonenal (ONE) and 4-hydroxy-2-nonenal (4-HNE). These aldehyde products can react with individual nucleotides and nucleophilic amino acids. Most highly reactive aldehydes are inactivated by conjugation to GSH (spontaneously or catalyzed by GST-alpha), oxidation by aldehyde dehydrogenases (ALDH) and reduction by aldo-ketoreductases. Lipids can also be oxidized by lipoxygenases, COX, and cytochrome P450.[86–88]

Increased levels of byproducts of lipid oxidation have been found in PD brains and plasma of PD patients.[89–91] ALDH1A1 is highly expressed in the SN and a decrease in its expression levels is found in PD postmortem brain samples.[92] Genetic variation in ALDH2 increases the risk for PD in individuals exposed to pesticides,[93,94] while an increased activity of ALDH2 was found in the putamen of PD brains.[95] However, whether the role of ALDH enzymes in PD is related to dopamine (DA) metabolism or detoxification of aldehydes byproducts of lipid peroxidation is unclear.

2.2.3.3 *Oxidative Protein Modifications: Redox Sensors and Transducers*

Reversible or irreversible oxidative post-translational modifications result either from direct oxidation of protein amino acid residues or through the formation of reactive intermediates (*i.e.*, reactive carbonyl species) generated during lipid peroxidation, resulting in a change in the structure and/or function of the targeted protein.[96] Specific amino acids residues such as Cys, methionine (Met), and tyrosine (Tyr) are prone to oxidative modification. An important cellular redox target or sensor is the thiol group of Cys. Redox-sensitive Cys (P[protein]-SH[Cys thiol]) undergo reversible and irreversible modifications in response to ROS/RNS thereby modulating protein function, activity or localization. Oxidation and reduction of P-SHs are thought to be one of the major mechanisms by which oxidative stress is integrated into cellular signal transduction pathways. Almost all physiological oxidants react with P-SHs.[97,98] Free radicals such as $O_2^{\bullet-}$, and peroxides (H_2O_2, and $ONOO^-$) mediate one- and two-electron oxidation of P-SHs, respectively, leading to the formation of reactive intermediates including protein sulfenic acids (P-SOH) and thiyl radicals (P-S•). The reaction of P-SOHs with either a neighboring P-SH or GSH will generate a disulfide bond (P-SS-P) or a glutathionylated residue (P-SSG). Both disulfide products can be reduced back by the action of the Trx/TrxR and the glutaredoxins (Grx)/GSH systems. P-SOH conversion to P-SS-P or P-SSG prevents their overoxidation to irreversible protein sulfinic (P-SO_2H), and sulfonic (P-SO_3H) acids.[97–99] Sulfiredoxin (Srx) has been reported to reduce Prx-SO_2H, but no other P-SO_2H target has been reported as a substrate for Srx.[100]

Protein nitros(yl)ation (P-SNO) refers to the reversible covalent adduction of a nitroso group (NO) to a P-SH. P-SNO formation is regulated by $NO^\bullet$, nitros(yl)ating agents such as N_2O_3 or by transition metal-catalyzed addition of NO. Transfer of NO groups between P-SNO and nitrosoglutathione (GSNO) has also been reported as one of the major mechanisms mediating P-SNO (transnitros[yl]ation). GSNO is formed by the reaction of $^\bullet NO$ with GSH, and as a minor byproduct from the oxidation of GSH by $ONOO^-$.[101] Protein denitrosylation can be regulated by Trxs and also through the metabolism of GSNO *via* GSH-dependent formaldehyde dehydrogenase or class III alcohol dehydrogenase, also known as GSNO reductase (GSNOR).[102]

A variety of protein Cys oxidative modifications have been reported in PD and experimental PD models. Recent studies have demonstrated the importance of protein Cys oxidation in PD. DJ-1 mutations are associated with early-onset PD. DJ-1-SO_2H formation at Cys106 has been detected in erythrocytes of PD patients,[103–105] which has been proposed to regulate its protective effects against PD-related insults.[106,107] Sulfonylation and nitros(yl)ation inhibit the E3 ligase activity of the autosomal recessive PD gene Parkin, which regulates ubiquitin-mediated degradation of misfolded proteins, and is found increased in idiopathic PD brains.[108–110] Sulfonylation of the deubiquitinating enzyme ubiquitin carboxyl-terminal hydrolase L1 (UCH-L1), whose mutation is linked to early-onset PD, was also found in PD brains.[111] PDI is nitros(yl)ated in PD brains, which also inhibits its enzymatic activity, leading to the increased accumulation of misfolded proteins and the concomitant activation of the UPR.[112] In addition, increased nitros(yl)ation of Prx2 and the X-linked inhibitor of apoptosis protein (XIAP) is also found in PD brains,[47,113] and is associated with loss of their corresponding activity/function. A decrease in the expression levels of Grx1 has been reported in PD brains.[114]

Met is one of the most vulnerable amino acid residues to oxidation. Addition of O_2 oxidizes Met to methionine sulfoxide (MetSO), and a strong oxidant oxidizes further MetSO to methionine sulfone ($MetSO_2$). Two different stereoisomers of MetSO, L-Met-S-SO and L-Met-R-SO, can be found at physiological conditions and they are reverted by MetSO reductases (Msrs). In humans, there are two different Msrs, MsrA and MsrB. MsrA is specific to the S-stereoisomer of MetSO, whereas MsrB reduces the R-stereoisomer of MetSO.[115] Met oxidation of α-synuclein inhibits its fibrillation.[116,117]

Nitration of the 3-position of the phenol group in Tyr conveys the irreversible generation of nitrotyrosine (3-NT) that is thought to be primarily mediated by $ONOO^-$. However, it has been debated whether $ONOO^-$ is the major source of 3-NT at physiological conditions. In fact, under physiologically relevant $ONOO^-$ levels, the principal change that Tyr suffers is the dimerization of tyrosyl radicals into dityrosine.[118] Different mechanisms for 3-NT formation have been demonstrated. Most mechanisms for 3-NT formation involve the double hit of the phenol group in Tyr by radicals. The first hit generates the tyrosyl radical, and the second one is mediated between the tyrosyl radical and $^{\bullet}NO_2$ to form 3-NT.[119,120] Oxidants such as $CO_3^{\bullet-}$, oxo–metal complexes and $^{\bullet}OH$, promote tyrosyl radical formation. $^{\bullet}NO_2$ formation by heme peroxidases and nitrite (NO_2^-) also generates 3-NT.[121] Interestingly, 3-NT can be enzymatically incorporated post-translationally into α-tubulin by tubulin-Tyr ligase, and cotranslationally by a selected aminoacyl-transfer RNA (tRNA) synthetase/tRNA pair that specifically incorporates 3-NT into purposely encoded sites in proteins.[122] Nitrated α-synuclein and MnSOD have been found increased in aging, PD brains and the cerebrospinal fluid (CSF) of PD patients.[123–126] 3-NT is considered to be an irreversible modification. However, there is biological evidence of the existence of a denitrating system whose identity is still unclear.[127]

Protein carbonylation is an oxidation process that refers to the formation of reactive ketones or aldehydes. While initially considered to be an irreversible process, recent reports suggest that protein carbonylation can be reversed by thiols.[128] Oxidative carbonylation occurs preferentially at prolines (Pro), threonines (Thr), lysines (Lys), and arginines (Arg) through metal-catalyzed activation of H_2O_2 to $^{\bullet}OH$. Protein carbonyl derivatives are also generated through oxidative cleavage of proteins *via* the α-amidation pathway or by oxidation of glutamyl side chains, leading to the formation of a peptide in which the N-terminal amino acid is blocked by a α-ketoacyl derivative. Secondary carbonylation is mediated by the covalent adduction of reactive electrophilic lipid aldehydes (such as 4-HNE or MDA) to the side chains of Lys, histidine

(His) and Cys residues.[129] Increased protein carbonylation has been observed in PD CSF.[130] More recently, a proteomic analysis of human plasma from PD patients identified 4-HNE adduct formation in Lys62 and 32 of DJ-1.[131]

2.3 What Makes the Dopaminergic Neurons in the SNpc Vulnerable?

The pathological hallmark in PD is the accumulation of protein inclusions or Lewy bodies (LB) in different brain regions. LB pathology is seen in peripheral norepinephrin neurons innervating the heart and skin, and DAergic of the enteric nervous system, which is associated with orthostatic hypotension, sweating (hyperhydosis) and constipation. In the central nervous system (CNS), loss of neuromelanin positive (NM) catecholamine neurons of the SNpc and locus ceruleus (LC) is found in nearly all postmortem brains from PD patients, and loss of neurons from the dorsal moor nucleus of the vagus (DMV) is seen in most patients as well. Loss of neurons in the ventral tegmental area (VTA), the retrorubral field (RRF), the pedunculopontine nuclei (PPN), the raphe nuclei RN and the basal nucleus of Meynert (BNM) are common but variable. In later stages, neuronal loss occurs in the neocortex as well. While the loss of DAergic A9 neurons in the SNpc is considered the pathological hallmark of PD and the cause of motor dysfunction in patients, there are diseases with parkinsonian symptoms that do not exhibit SNpc DAergic cell death.[132,133] Thus, one of the major scientific questions in PD research is what makes the SNpc vulnerable. In the next section, we will address potential mechanisms involved in the increased susceptibility of the SNpc DAergic neurons from the redox/oxidative stress and bioenergetics perspective.

2.3.1 Cellular Organization of the SNpc

The brain is the most complex cellular organ of the CNS. It is divided into the forebrain, brainstem, cortex and cerebellum. The telencephalon (cerebral hemispheres and deep structures) and diencephalon (thalamus, hypothalamus and subthalamus) are part of the forebrain, while the midbrain, pons, and medulla are the brainstem. The SN is located in the midbrain and is involved in reward, addiction, learned response to stimuli (reinforcement), spatial learning, temporal processing, and motor control. Within this area, the SNpc acts as an input path to the basal ganglia circuit supplying DA to the striatum. Brain tissue is made up about 100 billion neurons and trillions of glia cells. Neurons are highly polarized cells with subcellular domains that serve different functions: the cell body (or soma), branching dendrites (signal receivers), and the axon that conducts the nerve signal.[134,135] DAergic cells in the SNpc (A9) have dendrites that receive inhibiting signals from the collateral neurons of the pars reticulate and the striatum. Striatum input to SNpc is GABA (γ-aminobutyric acid)ergic (inhibitory).

SNpc DAergic neurons connect their axons to the striatum (both sensorimotor and associative) along the nigrostriatal pathway. SNpc neurons also innervate the basal ganglia system (lateral and medial pallidum), the SN pars reticulata, and the subthalamic nucleus. SNpc neurons that project to the thalamus use GABA as neurotransmitter.

Glial cells make up 90 percent of brain cells. Glial cells regulate a number of physiological processes required for proper neuronal survival and function. Three types of glial cells exist in the CNS: oligodendrocytes, astrocytes, and microglia. Oligodendrocytes are small cells found in rows along myelinated nerve fibers and surround nerve-cell bodies.[136] Oligodendrocytes are responsible for the myelination of nerve fibers, which provides axons with an 'insulating coat' that enhances nerve-impulse conduction. Oligodendrocytes have several processes and form several internodal segments of myelin on axons separated by gaps identified as Ranvier nodes.[137,138] Importantly, in rats, but not in dogs or primates, nigrostriatal fibers composed from the axons from SNpc DAergic neurons are not myelinated, but the implication of axon myelination to the vulnerability of the SNpc in PD is unclear.[139–141]

Astrocytes are small cells with radially arranged processes. Astrocytes expansions also end in blood vessels (perivascular feet), where they cover the external surface of capillaries. Astrocytic processes are also found around the initial segment of most axons and in the bare segments of axons at the nodes of Ranvier.[142] Astrocytes processes cover a defined territory with little overlap between neighboring cells forming highly organized domains interconnected *via* gap junctions. Astrocytes regulate neurotransmitter levels in the synaptic cleft, provide neurons with energetic and antioxidant precursors, play an important role in migration of immature neurons, tissue repair, and regulate blood flow and inflammatory processes by release of molecular mediators.[143] Compared to other brain areas, the SNpc has a low number of astrocytes and it is likely that as a result, SNpc DAergic neurons are more susceptible to oxidative stress and energy failure (see Section 2.3.2).[144] In addition, neurotrophic factors released by astrocytes such as the mesencephalic astrocyte-derived neurotrophic factor (MANF), the glial cell line-derived neurotrophic factor (GDNF), and the brain-derived neurotrophic factor (BDNF) are also thought to contribute to the survival of DAergic neurons. While astrocytes normally serve a neuroprotective role, reactive astrocytes are found in PD brains and are through to contribute to the neuroinflammatory process.[145,146]

Microglial cells are embryologically unrelated to the other glial cells and are resident macrophage-like cells distributed throughout the CNS.[147] They are the smallest of the glial cells. From their small cell bodies arise wavy branching processes that give off numerous spine-like projections. As innate immune cells, they are activated by infection, tissue injury (ischemia or trauma), or neurodegenerative processes. Activated microglia retract their processes and migrate to the site of the lesion, where they proliferate and become antigen presenting cells. Microglia phagocyte degenerating cells that undergo programmed cell death and also act as antigen-presenting cells and sources of immunoregulatory and neuromodulatory factors such as cytokines, chemokines, neurotrophic factors (BDNF or GDNF) and ROS/RNS.[148–150] Microglia can be activated by a diverse selection cell-surface receptors for endotoxin, cytokines, chemokines, misfolded proteins, serum factors and ATP. While mild activation of microglia is a vital immune adaptive response to control the microenvironment, continuous activation or overactivation of microglia is thought to contribute to neurodegeneration. The density of microglial cells is remarkably higher in the SN compared to other brain regions. Increased levels of activated microglia and pro-inflammatory cytokines (interleukin-1β [IL-1β], tumor necrosis-factor α [TNFα], IL-2 and IL-6) have been found in the SNpc of elderly individuals. Furthermore, activation of microglia in PD brains has been proposed to be triggered by released aggregates/fibrils of α-synuclein, ATP and/or signalling proteins released from neurons, NM, infectious events (encephalitis), and environmental exposures (Figure 2.2).[151]

The blood–brain barrier (BBB) refers to a barrier of capillaries that restrict the exchange of solutes between the blood and the brain.[135,152] Endothelial cells in capillaries are connected by gap junctions and regulate the exchange of metabolites and signalling molecules while preventing the entrance of deleterious molecules or agents to the brain. Disruption of the BBB can be induced by oxidative stress, immune cells and pathogens. Breakdown of the BBB leads to a dysregulated ionic transport/clearance, and the initiation of a central inflammatory response proposed to contribute to neuronal dysfunction and neurodegeneration.[153] A disruption in the BBB has been linked to aging and PD, which might involve dysfunction in both endothelial and astrocytic cells.[154,155]

2.3.2 Redox Basis of the Vulnerability of SNpc DAergic Neurons

In the next section, we will highlight the redox basis of the vulnerability of SNpc DAergic neurons. Furthermore, because bioenergetics and redox homeostasis are highly interlinked (though very rarely studied together) we will also revise how alterations in energy metabolism can contribute to oxidative stress in PD.

2.3.2.1 Bioenergetics and Central Carbon Metabolism

The regulation and maintenance of brain function requires high amounts of energy. The brain accounts for 25% of total body energy consumption, and brain energy levels are more resistant to starvation compared to other organs. Glucose is considered the obligatory energy substrate in the adult brain; however, alternate energy sources can fuel energy function under certain energetic stress conditions. Neurons observe high rates of oxidative metabolism compared to glial cells. Interestingly, neurons primarily metabolize

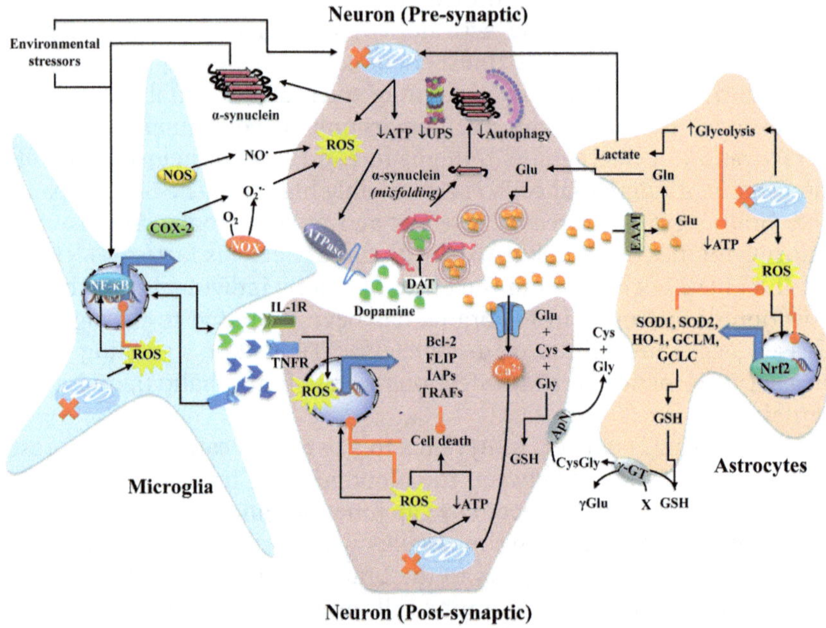

Figure 2.2 Bioenergetics and redox crosstalk in the brain and its relevance to PD. *Pre-synaptic neuron*. Mitochondrial dysfunction linked to genetic, environmental and aging risk factors in PD leads to energy deficiency (↓ATP) and increased ROS steady-state levels. Energy failure impairs a number of neuronal processes including: (1) maintenance of ionic gradients, vesicular neurotransmitter transport, plasma membrane potential and thus, neuronal excitability; (2) protein quality control mechanisms (autophagy and the UPS) leading to the accumulation of misfolded α-synuclein aggregates; (3) mitochondrial biogenesis and dynamics; and (4) antioxidant systems (these last two not depicted). *Astrocytes*. Astrocytes (and other glial cell types) counteract mitochondrial dysfunction and ROS by upregulating glycolysis and activating Nrf2. Furthermore, they supply lactate, GSH precursors (Cys and Gly) and neurotrophic factors, and regulate the extracellular levels of excitatory amino acids (Glu) for the proper survival/function of SNpc dopaminergic neurons. *Microglia*. Environmental toxicants, misfolded protein aggregates, and oxidative stress associated with neurodegeneration induce a NF-κB-dependent pro-inflammatory response in microglia that mediates the increased expression of cytokines and ROS/RNS generating enzymes (NOX, NOS and COX-2). While 'mild' microglia activation can be beneficial, excessive reactive species and released cytokines can promote neurodegeneration. *Postsynaptic neuron*. In neurons, neurotrophic factor- and cytokine-mediated signalling modulates the transcriptional regulation of anti-apoptotic and pro-apoptotic genes. Abbreviations not found in the main text: ATPase, Na^+/K^+-ATPase; IL-1R, interleukin-1 receptor; TNFR, tumor necrosis factor receptor; TRAF, TNF receptor associated factor.

glucose *via* the pentose phosphate pathway (PPP) to provide the reducing equivalents required to maintain antioxidant defenses *via* the production of NADPH.[156] Because neurons do not have the capacity to upregulate glycolysis in response to stress, they depend on the shuttle of lactate from astrocytes and oligodendrocytes as energy fuel for OXPHOS (Figure 2.2).[157,158] As such, neuronal function *ex vivo* can be maintained by lactate in the absence of glucose.[159] Accordingly, extracellular lactate levels are similar to those of glucose (0.5–1.5 mM). To date, the primary neuronal energy substrate (glucose *vs.* lactate) and the relative contributions of glycolysis and OXPHOS to neuronal bioenergetics (ATP) are still unclear. Neurons utilize most of their energy to maintain and restore ionic gradients, and for the uptake and recycling of neurotransmitters.[160]

Glial cells are highly glycolytic and a large portion of glucose metabolism is directed to lactate production and its release to the extracellular space. Glycogen storages in astrocytes can also be broken down to lactate. In addition, Glu can be used as a carbon source for the TCA cycle in both neurons and astrocytes, while acetate is preferentially used by astrocytes.[158,161,162] The creatine/phosphocreatine/creatine kinase system has also been proposed to contribute to the regeneration of ATP in the brain.[163] In contrast to other tissues, the brain metabolizes very poorly hydrogen-rich fatty acids to obtain energy, which has been related to the high O_2 demand and resultant generation of $O_2^{\bullet -}$ from β-oxidation. Ketone bodies and fatty acids (and lactate) can be transported across the BBB during starvation. Astrocytes can oxidize fatty acids and ketone bodies, while neurons and oligodendrocytes can only use ketone bodies. β-oxidation of fatty acids by astrocytes provides ketone bodies for neurons.[164] Interestingly, a ketogenic diet has been reported to exert protective effects in PD.[165]

Energy failure associated with impaired mitochondrial ETC is the hallmark of PD.[166] In addition, alterations in central carbon metabolism have also been reported to occur in PD. For example, a dysfunction in the tricarboxylic acid (TCA or Krebs) cycle has been reported in PD brains.[167,168] Decreased glucose metabolism and abnormally elevated levels of lactate/pyruvate have also been reported in PD patients.[168–171] PD-related genes Parkin, PINK1 and DJ-1 have been reported to regulate glycolysis.[172–174] Interestingly, lactate promotes α-synuclein accumulation.[175] However, the contribution of alterations in central carbon metabolism to PD is unclear.

The primary consequence of mitochondrial dysfunction is energy (ATP) depletion. If all neurons require ATP for their proper function, what would make SNpc DAergic neurons particularly susceptible to energy failure? DAergic neurons in the SNpc consume a significant amount of energy during their pacemaking activity, which leads to increased levels of basal oxidative stress.[176] Their extended (and probably unmyelinated) axon, and the number of striatal synapses established are characteristics expected to exert a high energy stress demand for the maintenance of the membranc potential (ionic gradients), protein/organelle (mitochondria) traffic and homeostasis.[140,177] Importantly, the intrinsic bioenergetic capacity of DAergic and non-DAergic pre-synaptic synaptosomes do not differ from each other.[178] Thus, it is suggested that SNpc DAergic neurons might be particularly sensitive to alterations in bioenergetics. This hypothesis is clearly supported by the reported susceptibility of these cells to mitochondrial toxins.[179] In addition it would be expected that in an attempt to maintain their excitability (including active uptake of neurotransmitters into vesicles) and function, SNpc DAergic neurons might siphon energy from other important ATP-dependent homeostatic processes such as: (1) protein quality control mechanisms (autophagy and the ubiquitin-proteasome system [UPS]) leading to the accumulation of misfolded/oxidized protein aggregates;[180] (2) transport of mitochondria to pre-synaptic and post-synaptic sites (kinesin motors require ATP) exacerbating energy failure;[181] and (3) mitochondrial biogenesis and dynamics (fusion, fission and mitophagy).[182]

While mitochondrial dysfunction is considered to deplete antioxidant defenses primarily by an increase in the steady-state levels of ROS, less recognized is the impact that alterations in bioenergetics and central carbon metabolism have on antioxidant systems. ATP is required for *de novo* GSH synthesis, the reduction of Prxs by Srx, and the transcriptional regulation of antioxidant defenses *via* Nrf2.[55,183] ATP depletion is also expected to impair oxidative DNA-damage repair systems (BER) and the ATP-dependent efflux of GSH-adducts (including those of DA). Down-regulation of PPP enzymes and a failure to increase the antioxidant reserve capacity are early events in the pathogenesis of sporadic PD.[184] As such, impairment in the PPP leads to the depletion of NADPH and dysfunction in the GSH/Grx, Prx/Trx/Trx2 and Catalase systems,[185] as well as the decrease in purine synthesis *via* ribose 5-phoshate.

2.3.2.2 Oxidative Stress

The brain is highly susceptive to oxidative damage due to: (1) its high rate of oxidative metabolism resulting in excessive ROS formation; (2) the high PUFA content of neuronal membranes that makes them vulnerable to oxidative damage; 3) the low intrinsic antioxidant capacity of neurons.[158,186] Neurons in particular display limited antioxidant defense mechanisms compared to astrocytes, which have significantly higher levels of GSH, HO-1, GPx, GST, catalase and TrxR.[158,186–188] Both neurons and astrocytes can synthesize GSH; however, neurons depend on astrocytes for the supply of GSH precursors (Figure 2.2). Extracellular Cys is unstable and has neurotoxic pro-oxidant effects.[189] A major extracellular source for Cys is the dipeptide cystine (the oxidized form of Cys). Cystine uptake occurs by a Na^+ -independent heteroexchange mechanism denominated X_c^-, which is found in immature neurons, astrocytes, microglia, retinal Muller and Bergmann glial cells.[190] Thus, matured neurons have a poor capacity to uptake cystine as a precursor. An inter-relationship exists between astrocytes and neurons where neurons rely on extracellular Cys provided by GSH efflux from astrocytes *via* GAP junction hemichannels, multidrug resistance protein transporters (MRP/ABCC) or EAAT1 (GLAST).[159,191–194] Effluxed GSH is cleaved by γ-glutamyl transpeptidase (γGT) to produce CysGly, which in turn is cleaved by neuronal aminopeptidase N (ApN), producing Gly and Cys, which are subsequently taken up by neurons for GSH synthesis (Figure 2.2). In neurons, Cys uptake is mediated primarily by the Na^+-dependent excitatory amino acid transporter (EAAT) isoform 3, also known as system X_{AG}^-.[195,196] EAAT3 (or EAAC1) can transport Cys at a rate comparable to that of Glu. Interestingly, while mice deficient in X_c^- system have no alterations in brain GSH content, EAAT3-deficient mice have reduced GSH levels.[196,197] The differential antioxidant response of neurons and astrocytes is also a result of the preferential expression/activation of Nrf2 in astrocytes.[198]

While the neuronal properties described above make neurons particularly sensitive to oxidative damage, by themselves they do not explain the selective vulnerability of SNpc DAergic neurons to oxidative damage. Fe, NM and DA content are considered to increase the sensitivity of SNpc DAergic neurons to oxidative stress. We will briefly introduce their role in oxidative stress in PD, but the reader is referred to Chapters 4, 5 and 9 of this book for more detailed information in this regard.

2.3.2.2.1 Iron (Fe) and Neuromelanin. In the brain, Fe is most abundant in areas rich in DAergic neurons. Increased Fe deposition and increased free Fe concentrations have been found in the SNpc of PD brains, which may lead to an increased generation of ˙OH *via* Fenton and Haber–Weiss reactions.[199] Autopsy studies have found an elevated total Fe content in the SNpc of PD patients compared to age-matched controls but other studies have failed to confirm these findings. Fe content in the SNpc seems to be primarily affected in advanced PD patients suggesting that it is not involved in early disease development. Alterations in Fe distribution are also found in PD. Contrary to normal subjects, where more Fe is deposited in the SN *pars reticulata*, in PD, Fe is deposited abundantly in the SNpc containing NM-pigmented neurons.[199] Low levels of ferritin in the SNpc of PD patients and incidental Lewy body disease cases have also been reported.[200] NM pigmented neurons in SNpc are preferentially lost in PD, and Fe accumulation in the SNpc is deposited within NM granules. A specific haplotype of the divalent metal transporter 1 (DMT1/*SLC11A2*) gene was found to occur at greater frequencies in PD patients suggesting that alterations in DMT1 function may contribute to PD.[201] The levels of Fe 'handling' proteins such as ceruloplasmin have also been reported altered in PD.[202,203] In the presence of Fe, α-synuclein induces ˙OH formation,[204] whereas in other studies, oxidative damage induced by Fe is exacerbated primarily by mutant, but not wild-type (WT) α-synuclein (Figure 2.1).[205–207]

2.3.2.2.2 Dopamine (DA). Oxidative stress in PD is also associated with the pro-oxidant properties of DA. Mutant α-synuclein downregulates the vesicular monoamine transporter (VMAT2), and energy depletion is also expected to augment DA levels in thc cytosol by impairing the ATP-dependent proton pump that drives vesicular accumulation of DA.[208–210] In the cytosol DA is either metabolized by MAO to generate H_2O_2, or auto-oxidized in the presence of Fe generating $O_2^{\bullet -}$, H_2O_2 and DA-quinone species (DAQ) (Figure 2.1).[211] α-synuclein also interacts with the DA transporter (DAT), although its effects in DA uptake and toxicity are unclear.[212,213] Furthermore, DAergic cell death induced by mitochondrial toxins has been proposed to depend on DA oxidation.[214,215] DAQ byproducts from DA oxidation have been reported to bind to P-SH.[216] A recent proteomic analysis demonstrates the susceptibility of a variety of proteins to be covalently modified by dopamine including the PD-associated proteins UCH-L1 and DJ-1[217] but the exact amino acid residues involved were not identified.

2.4 Conclusions and Perspectives

While a more detailed analysis of specific redox-related processes and their role in PD will be found in the upcoming chapters of this book, in this chapter we have aimed to summarize the intrinsic bioenergetics and redox properties of the SNpc DAergic neurons that make them susceptible to oxidative damage. Energy failure and oxidative stress in PD are linked primarily to impaired mitochondria function (ETC), and both phenomena are expected to synergistically act to promote neuronal dysfunction and neurodegeneration. The high energy demands that SNpc DAergic neurons have to maintain neuronal homeostasis and excitability, and the pro-oxidant environment (Fe/NM and DA content) are characteristics that make them primary targets for mitochondrial dysfunction.

Acknowledgements

Research in Dr Franco's lab is supported by the National Institutes of Health Grant P20RR17675, Centers of Biomedical Research Excellence (COBRE), the Scientist Development Grant of the American Heart Association (12SDG12090015), and the Office of Research of the University of Nebraska-Lincoln. R.M.F. was supported by grant DBI-1461240 from NSF. P.H.-F. was supported by a Postdoctoral Fellowship (238703) from the National Council of Science and Technology (CONACYT, Mexico).

References

1. M. E. Gotz, A. Freyberger and P. Riederer, Oxidative stress: a role in the pathogenesis of Parkinson's disease, *J. Neural Transm., Suppl.*, 1990, **29**, 241–249.
2. B. Kalyanaraman, Teaching the basics of redox biology to medical and graduate students: Oxidants, antioxidants and disease mechanisms, *Redox Biol.*, 2013, **1**, 244–257.
3. F. A. Villamena, *Molecular Basis of Oxidative Stress Chemistry, Mechanisms, and Disease Pathogenesis*, Wiley, Hoboken, 2013, p. 1, (448 p.).

4. B. Halliwell and J. M. C. Gutteridge, *Free Radicals in Biology and Medicine*, Oxford University Press, Oxford, New York, 4th edn, 2007.
5. V. G. Grivennikova and A. D. Vinogradov, Generation of superoxide by the mitochondrial Complex I, *Biochim. Biophys. Acta*, 2006, **1757**, 553–561.
6. Q. Chen, E. J. Vazquez, S. Moghaddas, C. L. Hoppel and E. J. Lesnefsky, Production of reactive oxygen species by mitochondria: central role of complex III, *J. Biol. Chem.*, 2003, **278**, 36027–36031.
7. W. Wang, H. Fang, L. Groom, A. Cheng, W. Zhang, J. Liu, X. Wang, K. Li, P. Han, M. Zheng, J. Yin, M. P. Mattson, J. P. Kao, E. G. Lakatta, S. S. Sheu, K. Ouyang, J. Chen, R. T. Dirksen and H. Cheng, Superoxide flashes in single mitochondria, *Cell*, 2008, **134**, 279–290.
8. D. Han, F. Antunes, R. Canali, D. Rettori and E. Cadenas, Voltage-dependent anion channels control the release of the superoxide anion from mitochondria to cytosol, *J. Biol. Chem.*, 2003, **278**, 5557–5563.
9. C. L. Quinlan, I. V. Perevoshchikova, M. Hey-Mogensen, A. L. Orr and M. D. Brand, Sites of reactive oxygen species generation by mitochondria oxidizing different substrates, *Redox Biol.*, 2013, **1**, 304–312.
10. A. R. Cardoso, B. Chausse, F. M. da Cunha, L. A. Luevano-Martinez, T. B. Marazzi, P. S. Pessoa, B. B. Queliconi and A. J. Kowaltowski, Mitochondrial compartmentalization of redox processes, *Free Radical Biol. Med.*, 2012, **52**, 2201–2208.
11. A. H. Schapira, Mitochondria in the aetiology and pathogenesis of Parkinson's disease, *Lancet Neurol.*, 2008, **7**, 97–109.
12. C. Henchcliffe and M. F. Beal, Mitochondrial biology and oxidative stress in Parkinson disease pathogenesis, *Nat. Clin. Pract. Neurol.*, 2008, **4**, 600–609.
13. A. H. Schapira, J. M. Cooper, D. Dexter, P. Jenner, J. B. Clark and C. D. Marsden, Mitochondrial complex I deficiency in Parkinson's disease, *Lancet*, 1989, **1**, 1269.
14. J. Clark, Y. Dai and D. K. Simon, Do somatic mitochondrial DNA mutations contribute to Parkinson's disease? *Parkinson's Dis.*, 2011, **2011**, 659694.
15. A. Reeve, M. Meagher, N. Lax, E. Simcox, P. Hepplewhite, E. Jaros and D. Turnbull, The impact of pathogenic mitochondrial DNA mutations on substantia nigra neurons, *J. Neurosci.*, 2013, **33**, 10790–10801.
16. A. Trancikova, E. Tsika and D. J. Moore, Mitochondrial dysfunction in genetic animal models of Parkinson's disease, *Antioxid. Redox Signaling*, 2012, **16**, 896–919.
17. L. Devi, V. Raghavendran, B. M. Prabhu, N. G. Avadhani and H. K. Anandatheerthavarada, Mitochondrial import and accumulation of alpha-synuclein impair complex I in human dopaminergic neuronal cultures and Parkinson disease brain, *J. Biol. Chem.*, 2008, **283**, 9089–9100.
18. T. Taira, Y. Saito, T. Niki, S. M. Iguchi-Ariga, K. Takahashi and H. Ariga, DJ-1 has a role in antioxidative stress to prevent cell death, *EMBO Rep.*, 2004, **5**, 213–218.

19. V. A. Morais, P. Verstreken, A. Roethig, J. Smet, A. Snellinx, M. Vanbrabant, D. Haddad, C. Frezza, W. Mandemakers, D. Vogt-Weisenhorn, R. Van Coster, W. Wurst, L. Scorrano and B. De Strooper, Parkinson's disease mutations in PINK1 result in decreased Complex I activity and deficient synaptic function, *EMBO Mol. Med.*, 2009, **1**, 99–111.
20. J. J. Palacino, D. Sagi, M. S. Goldberg, S. Krauss, C. Motz, M. Wacker, J. Klose and J. Shen, Mitochondrial dysfunction and oxidative damage in parkin-deficient mice, *J. Biol. Chem.*, 2004, **279**, 18614–18622.
21. H. Jiang, Y. Ren, J. Zhao and J. Feng, Parkin protects human dopaminergic neuroblastoma cells against dopamine-induced apoptosis, *Hum. Mol. Genet.*, 2004, **13**, 1745–1754.
22. J. Niu, M. Yu, C. Wang and Z. Xu, Leucine-rich repeat kinase 2 disturbs mitochondrial dynamics *via* Dynamin-like protein, *J. Neurochem.*, 2012, **122**, 650–658.
23. G. P. Bienert, A. L. Moller, K. A. Kristiansen, A. Schulz, I. M. Moller, J. K. Schjoerring and T. P. Jahn, Specific aquaporins facilitate the diffusion of hydrogen peroxide across membranes, *J. Biol. Chem.*, 2007, **282**, 1183–1192.
24. G. P. Bienert, J. K. Schjoerring and T. P. Jahn, Membrane transport of hydrogen peroxide, *Biochim. Biophys. Acta, Biomembr.*, 2006, **1758**, 994–1003.
25. S. Miersch, M. G. Espey, R. Chaube, A. Akarca, R. Tweten, S. Ananvoranich and B. Mutus, Plasma membrane cholesterol content affects nitric oxide diffusion dynamics and signaling, *J. Biol. Chem.*, 2008, **283**, 18513–18521.
26. A. Denicola, J. M. Souza and R. Radi, Diffusion of peroxynitrite across erythrocyte membranes, *Proc. Natl. Acad. Sci. U. S. A.*, 1998, **95**, 3566–3571.
27. S. Luo, H. Lei, H. Qin and Y. Xia, Molecular mechanisms of endothelial NO synthase uncoupling, *Curr. Pharm. Des.*, 2014, **20**, 3548–3553.
28. B. S. van der Veen, M. P. de Winther and P. Heeringa, Myeloperoxidase: molecular mechanisms of action and their relevance to human health and disease, *Antioxid. Redox Signaling*, 2009, **11**, 2899–2937.
29. J. Y. Im, D. Kim, S. G. Paik and P. L. Han, Cyclooxygenase-2-dependent neuronal death proceeds *via* superoxide anion generation, *Free Radical Biol. Med.*, 2006, **41**, 960–972.
30. G. C. Brown and V. Borutaite, There is no evidence that mitochondria are the main source of reactive oxygen species in mammalian cells, *Mitochondrion*, 2012, **12**, 1–4.
31. T. J. Tavender and N. J. Bulleid, Molecular mechanisms regulating oxidative activity of the Ero1 family in the endoplasmic reticulum, *Antioxid. Redox Signaling*, 2010, **13**, 1177–1187.
32. G. Mercado, P. Valdes and C. Hetz, An ERcentric view of Parkinson's disease, *Trends Mol. Med.*, 2013, **19**, 165–175.
33. T. Kurz, J. W. Eaton and U. T. Brunk, Redox activity within the lysosomal compartment: implications for aging and apoptosis, *Antioxid. Redox Signaling*, 2010, **13**, 511–523.

34. B. Dehay, M. Martinez-Vicente, G. A. Caldwell, K. A. Caldwell, Z. Yue, M. R. Cookson, C. Klein, M. Vila and E. Bezard, Lysosomal impairment in Parkinson's disease, *Mov. Disord.*, 2013, **28**, 725–732.
35. D. W. Infanger, R. V. Sharma and R. L. Davisson, NADPH oxidases of the brain: distribution, regulation, and function, *Antioxid. Redox Signaling*, 2006, **8**, 1583–1596.
36. T. Taetzsch and M. L. Block, Pesticides, microglial NOX2, and Parkinson's disease, *J. Biochem. Mol. Toxicol.*, 2013, **27**, 137–149.
37. A. Sutton, H. Khoury, C. Prip-Buus, C. Cepanec, D. Pessayre and F. Degoul, The Ala16Val genetic dimorphism modulates the import of human manganese superoxide dismutase into rat liver mitochondria, *Pharmacogenetics*, 2003, **13**, 145–157.
38. A. Okado-Matsumoto and I. Fridovich, Subcellular distribution of superoxide dismutases (SOD) in rat liver: Cu,Zn-SOD in mitochondria, *J. Biol. Chem.*, 2001, **276**, 38388–38393.
39. L. M. Ambani, M. H. Van Woert and S. Murphy, Brain peroxidase and catalase in Parkinson disease, *Arch. Neurol.*, 1975, **32**, 114–118.
40. I. Ceballos, M. Lafon, F. Javoy-Agid, E. Hirsch, A. Nicole, P. M. Sinet and Y. Agid, Superoxide dismutase and Parkinson's disease, *Lancet*, 1990, **335**, 1035–1036.
41. H. Saggu, J. Cooksey, D. Dexter, F. R. Wells, A. Lees, P. Jenner and C. D. Marsden, A selective increase in particulate superoxide dismutase activity in parkinsonian substantia nigra, *J. Neurochem.*, 1989, **53**, 692–697.
42. A. Hall, P. A. Karplus and L. B. Poole, Typical 2-Cys peroxiredoxins–structures, mechanisms and functions, *FEBS J.*, 2009, **276**, 2469–2477.
43. A. G. Cox, C. C. Winterbourn and M. B. Hampton, Mitochondrial peroxiredoxin involvement in antioxidant defence and redox signalling, *Biochem. J.*, 2010, **425**, 313–325.
44. M. L. Aon-Bertolino, J. I. Romero, P. Galeano, M. Holubiec, M. S. Badorrey, G. E. Saraceno, E. M. Hanschmann, C. H. Lillig and F. Capani, Thioredoxin and glutaredoxin system proteins-immunolocalization in the rat central nervous system, *Biochim. Biophys. Acta*, 2011, **1810**, 93–110.
45. J. Goemaere and B. Knoops, Peroxiredoxin distribution in the mouse brain with emphasis on neuronal populations affected in neurodegenerative disorders, *J. Comp. Neurol.*, 2012, **520**, 258–280.
46. K. Krapfenbauer, E. Engidawork, N. Cairns, M. Fountoulakis and G. Lubec, Aberrant expression of peroxiredoxin subtypes in neurodegenerative disorders, *Brain Res.*, 2003, **967**, 152–160.
47. J. Fang, T. Nakamura, D. H. Cho, Z. Gu and S. A. Lipton, S-nitrosylation of peroxiredoxin 2 promotes oxidative stress-induced neuronal cell death in Parkinson's disease, *Proc. Natl. Acad. Sci. U. S. A.*, 2007, **104**, 18742–18747.
48. A. Lippoldt, C. A. Padilla, H. Gerst, B. Andbjer, E. Richter, A. Holmgren and K. Fuxe, Localization of thioredoxin in the rat brain and functional implications, *J. Neurosci.*, 1995, **15**, 6747–6756.

49. E. Rybnikova, A. E. Damdimopoulos, J. A. Gustafsson, G. Spyrou and M. Pelto-Huikko, Expression of novel antioxidant thioredoxin-2 in the rat brain, *Eur. J. Neurosci.*, 2000, **12**, 1669–1678.
50. J. R. Godoy, M. Funke, W. Ackermann, P. Haunhorst, S. Oesteritz, F. Capani, H. P. Elsasser and C. H. Lillig, Redox atlas of the mouse. Immunohistochemical detection of glutaredoxin-, peroxiredoxin-, and thioredoxin-family proteins in various tissues of the laboratory mouse, *Biochim. Biophys. Acta*, 2011, **1810**, 2–92.
51. R. Franco and J. A. Cidlowski, Apoptosis and glutathione: beyond an antioxidant, *Cell Death Differ.*, 2009, **16**, 1303–1314.
52. R. Franco, O. J. Schoneveld, A. Pappa and M. I. Panayiotidis, The central role of glutathione in the pathophysiology of human diseases, *Arch. Physiol. Biochem.*, 2007, **113**, 234–258.
53. A. Meister, Glutathione biosynthesis and its inhibition, *Methods Enzymol.*, 1995, **252**, 26–30.
54. H. Sies, Glutathione and its role in cellular functions, *Free Radical Biol. Med.*, 1999, **27**, 916–921.
55. R. Franco and J. A. Cidlowski, Glutathione Efflux and Cell Death, *Antioxid. Redox Signaling*, 2012, **17**, 1694–1713.
56. F. Q. Schafer and G. R. Buettner, Redox environment of the cell as viewed through the redox state of the glutathione disulfide/glutathione couple, *Free Radical Biol. Med.*, 2001, **30**, 1191–1212.
57. D. P. Jones, Redefining oxidative stress, *Antioxid. Redox Signaling*, 2006, **8**, 1865–1879.
58. D. C. Duke, L. B. Moran, R. K. Pearce and M. B. Graeber, The medial and lateral substantia nigra in Parkinson's disease: mRNA profiles associated with higher brain tissue vulnerability, *Neurogenetics*, 2007, **8**, 83–94.
59. A. P. Mazzetti, M. C. Fiorile, A. Primavera and M. Lo Bello, Glutathione transferases and neurodegenerative diseases, *Neurochem. Int.*, 2015, **82**, 10–18.
60. P. G. Board and D. Menon, Glutathione transferases, regulators of cellular metabolism and physiology, *Biochim. Biophys. Acta*, 2013, **1830**, 3267–3288.
61. J. B. Wilk, J. E. Tobin, O. Suchowersky, H. A. Shill, C. Klein, G. F. Wooten, M. F. Lew, M. H. Mark, M. Guttman, R. L. Watts, C. Singer, J. H. Growdon, J. C. Latourelle, M. H. Saint-Hilaire, A. L. DeStefano, R. Prakash, S. Williamson, C. J. Berg, M. Sun, S. Goldwurm, G. Pezzoli, B. A. Racette, J. S. Perlmutter, A. Parsian, K. B. Baker, M. L. Giroux, I. Litvan, P. P. Pramstaller, G. Nicholson, D. J. Burn, P. F. Chinnery, P. Vieregge, J. T. Slevin, F. Cambi, M. E. MacDonald, J. F. Gusella, R. H. Myers and L. I. Golbe, Herbicide exposure modifies GSTP1 haplotype association to Parkinson onset age: the GenePD Study, *Neurology*, 2006, **67**, 2206–2210.
62. A. Menegon, P. G. Board, A. C. Blackburn, G. D. Mellick and D. G. Le Couteur, Parkinson's disease, pesticides, and glutathione transferase polymorphisms, *Lancet*, 1998, **352**, 1344–1346.

63. G. S. Longo, M. S. Pinhel, C. L. Sado, M. L. Gregorio, G. S. Amorim, G. S. Florim, C. M. Mazeti, D. P. Martins, F. N. Oliveira, W. A. Tognola, M. A. Nakazone and D. R. Souza, Exposure to pesticides and heterozygote genotype of GSTP1-Alw26I are associated to Parkinson's disease, *Arq. Neuro-Psiquiatr.*, 2013, **71**, 446–452.
64. M. A. Pinhel, C. L. Sado, S. Longo Gdos, M. L. Gregorio, G. S. Amorim, G. M. Florim, C. M. Mazeti, D. P. Martins, N. Oliveira Fde, M. A. Nakazone, W. A. Tognola and D. R. Souza, Nullity of GSTT1/GSTM1 related to pesticides is associated with Parkinson's disease, *Arq. Neuro-Psiquiatr.*, 2013, **71**, 527–532.
65. S. M. Goldman, F. Kamel, G. W. Ross, G. S. Bhudhikanok, J. A. Hoppin, M. Korell, C. Marras, C. Meng, D. M. Umbach, M. Kasten, A. R. Chade, K. Comyns, M. B. Richards, D. P. Sandler, A. Blair, J. W. Langston and C. M. Tanner, Genetic modification of the association of paraquat and Parkinson's disease, *Mov. Disord.*, 2012, **27**, 1652–1658.
66. R. Vilar, H. Coelho, E. Rodrigues, M. J. Gama, I. Rivera, E. Taioli and M. C. Lechner, Association of A313 G polymorphism (GSTP1*B) in the glutathione-S-transferase P1 gene with sporadic Parkinson's disease, *Eur. J. Neurol.*, 2007, **14**, 156–161.
67. M. Shi, J. Bradner, T. K. Bammler, D. L. Eaton, J. Zhang, Z. Ye, A. M. Wilson, T. J. Montine and C. Pan, Identification of glutathione S-transferase pi as a protein involved in Parkinson disease progression, *Am. J. Pathol.*, 2009, **175**, 54–65.
68. C. Perez-Pastene, R. Graumann, F. Diaz-Grez, M. Miranda, P. Venegas, O. T. Godoy, L. Layson, R. Villagra, J. M. Matamala, L. Herrera and J. Segura-Aguilar, Association of GST M1 null polymorphism with Parkinson's disease in a Chilean population with a strong Amerindian genetic component, *Neurosci. Lett.*, 2007, **418**, 181–185.
69. A. Ahmadi, M. Fredrikson, H. Jerregard, A. Akerback, P. A. Fall, A. Rannug, O. Axelson and P. Soderkvist, GSTM1 and mEPHX polymorphisms in Parkinson's disease and age of onset, *Biochem. Biophys. Res. Commun.*, 2000, **269**, 676–680.
70. R. Brigelius-Flohe and L. Flohe, Basic principles and emerging concepts in the redox control of transcription factors, *Antioxid. Redox Signaling*, 2011, **15**, 2335–2381.
71. M. von Otter, S. Landgren, S. Nilsson, D. Celojevic, P. Bergstrom, A. Hakansson, H. Nissbrandt, M. Drozdzik, M. Bialecka, M. Kurzawski, K. Blennow, M. Nilsson, O. Hammarsten and H. Zetterberg, Association of Nrf2-encoding NFE2L2 haplotypes with Parkinson's disease, *BMC Med. Genet.*, 2010, **11**, 36.
72. H. Shimura-Miura, N. Hattori, D. Kang, K. Miyako, Y. Nakabeppu and Y. Mizuno, Increased 8-oxo-dGTPase in the mitochondria of substantia nigral neurons in Parkinson's disease, *Ann. Neurol.*, 1999, **46**, 920–924.
73. Z. I. Alam, A. Jenner, S. E. Daniel, A. J. Lees, N. Cairns, C. D. Marsden, P. Jenner and B. Halliwell, Oxidative DNA damage in the parkinsonian brain: an apparent selective increase in 8-hydroxyguanine levels in substantia nigra, *J. Neurochem.*, 1997, **69**, 1196–1203.

74. J. Woodrick, S. Gupta, P. Khatkar, S. Sarangi, G. Narasimhan, A. Trehan, S. Adhikari and R. Roy, Slow repair of lipid peroxidation-induced DNA damage at p53 mutation hotspots in human cells caused by low turnover of a DNA glycosylase, *Nucleic Acids Res.*, 2014, **42**, 9033–9046.
75. K. B. Beckman and B. N. Ames, Endogenous oxidative damage of mtDNA, *Mutat. Res.*, 1999, **424**, 51–58.
76. M. L. Hegde, A. K. Mantha, T. K. Hazra, K. K. Bhakat, S. Mitra and B. Szczesny, Oxidative genome damage and its repair: implications in aging and neurodegenerative diseases, *Mech. Ageing Dev.*, 2012, **133**, 157–168.
77. J. Fukae, M. Takanashi, S. Kubo, K. Nishioka, Y. Nakabeppu, H. Mori, Y. Mizuno and N. Hattori, Expression of 8-oxoguanine DNA glycosylase (OGG1) in Parkinson's disease and related neurodegenerative disorders, *Acta Neuropathol.*, 2005, **109**, 256–262.
78. Y. Nakabeppu, D. Tsuchimoto, H. Yamaguchi and K. Sakumi, Oxidative damage in nucleic acids and Parkinson's disease, *J. Neurosci. Res.*, 2007, **85**, 919–934.
79. G. Jiang, L. Xu, L. Wang, S. Song and C. Zhu, Association study of human MTH1 Ile45Thr polymorphism with sporadic Parkinson's disease, *Eur. Neurol.*, 2008, **59**, 15–17.
80. S. S. Wallace, Base excision repair: a critical player in many games, *DNA Repair (Amst)*, 2014, **19**, 14–26.
81. H. E. Krokan and M. Bjoras, Base excision repair, *Cold Spring Harbor Perspect. Biol.*, 2013, **5**, a012583.
82. B. A. Gibson and W. L. Kraus, New insights into the molecular and cellular functions of poly(ADP-ribose) and PARPs, *Nat. Rev. Mol. Cell Biol.*, 2012, **13**, 411–424.
83. E. Huang, D. Qu, Y. Zhang, K. Venderova, M. E. Haque, M. W. Rousseaux, R. S. Slack, J. M. Woulfe and D. S. Park, The role of Cdk5-mediated apurinic/apyrimidinic endonuclease 1 phosphorylation in neuronal death, *Nat. Cell Biol.*, 2010, **12**, 563–571.
84. J. W. Taanman and A. H. Schapira, Analysis of the trinucleotide CAG repeat from the DNA polymerase gamma gene (POLG) in patients with Parkinson's disease, *Neurosci. Lett.*, 2005, **376**, 56–59.
85. S. Ylonen, P. Ylikotila, A. Siitonen, S. Finnila, J. Autere and K. Majamaa, Variations of mitochondrial DNA polymerase gamma in patients with Parkinson's disease, *J. Neurol.*, 2013, **260**, 3144–3149.
86. H. Yin, L. Xu and N. A. Porter, Free radical lipid peroxidation: mechanisms and analysis, *Chem. Rev.*, 2011, **111**, 5944–5972.
87. A. Ayala, M. F. Munoz and S. Arguelles, Lipid peroxidation: production, metabolism, and signaling mechanisms of malondialdehyde and 4-hydroxy-2-nonenal, *Oxid. Med. Cell. Longevity*, 2014, **2014**, 360438.
88. F. Gueraud, M. Atalay, N. Bresgen, A. Cipak, P. M. Eckl, L. Huc, I. Jouanin, W. Siems and K. Uchida, Chemistry and biochemistry of lipid peroxidation products, *Free Radical Res.*, 2010, **44**, 1098–1124.
89. T. T. Reed, Lipid peroxidation and neurodegenerative disease, *Free Radical Biol. Med.*, 2011, **51**, 1302–1319.

90. D. Dexter, C. Carter, F. Agid, Y. Agid, A. J. Lees, P. Jenner and C. D. Marsden, Lipid peroxidation as cause of nigral cell death in Parkinson's disease, *Lancet*, 1986, **2**, 639–640.
91. A. Agil, R. Duran, F. Barrero, B. Morales, M. Arauzo, F. Alba, M. T. Miranda, I. Prieto, M. Ramirez and F. Vives, Plasma lipid peroxidation in sporadic Parkinson's disease. Role of the L-dopa, *J. Neurol. Sci.*, 2006, **240**, 31–36.
92. G. Liu, J. Yu, J. Ding, C. Xie, L. Sun, I. Rudenko, W. Zheng, N. Sastry, J. Luo, G. Rudow, J. C. Troncoso and H. Cai, Aldehyde dehydrogenase 1 defines and protects a nigrostriatal dopaminergic neuron subpopulation, *J. Clin. Invest.*, 2014, **124**, 3032–3046.
93. A. G. Fitzmaurice, S. L. Rhodes, M. Cockburn, B. Ritz and J. M. Bronstein, Aldehyde dehydrogenase variation enhances effect of pesticides associated with Parkinson disease, *Neurology*, 2014, **82**, 419–426.
94. A. G. Fitzmaurice, S. L. Rhodes, A. Lulla, N. P. Murphy, H. A. Lam, K. C. O'Donnell, L. Barnhill, J. E. Casida, M. Cockburn, A. Sagasti, M. C. Stahl, N. T. Maidment, B. Ritz and J. M. Bronstein, Aldehyde dehydrogenase inhibition as a pathogenic mechanism in Parkinson disease, *Proc. Natl. Acad. Sci. U. S. A.*, 2013, **110**, 636–641.
95. T. M. Michel, L. Kasbauer, W. Gsell, J. Jecel, A. J. Sheldrick, M. Cortese, T. Nickl-Jockschat, E. Grunblatt and P. Riederer, Aldehyde dehydrogenase 2 in sporadic Parkinson's disease, *Parkinsonism Relat. Disord.*, 2014, **20**(suppl. 1), S68–S72.
96. D. A. Butterfield, M. Perluigi, T. Reed, T. Muharib, C. P. Hughes, R. A. Robinson and R. Sultana, Redox proteomics in selected neurodegenerative disorders: from its infancy to future applications, *Antioxid. Redox Signaling*, 2012, **17**, 1610–1655.
97. L. B. Poole and K. J. Nelson, Discovering mechanisms of signaling-mediated cysteine oxidation, *Curr. Opin. Chem. Biol.*, 2008, **12**, 18–24.
98. C. C. Winterbourn and M. B. Hampton, Thiol chemistry and specificity in redox signaling, *Free Radical Biol. Med.*, 2008, **45**, 549–561.
99. C. E. Paulsen and K. S. Carroll, Orchestrating redox signaling networks through regulatory cysteine switches, *ACS Chem. Biol.*, 2010, **5**, 47–62.
100. W. Jeong, S. H. Bae, M. B. Toledano and S. G. Rhee, Role of sulfiredoxin as a regulator of peroxiredoxin function and regulation of its expression, *Free Radical Biol. Med.*, 2012, **53**, 447–456.
101. M. W. Foster, D. T. Hess and J. S. Stamler, Protein S-nitrosylation in health and disease: a current perspective, *Trends Mol. Med.*, 2009, **15**, 391–404.
102. M. Benhar, M. T. Forrester and J. S. Stamler, Protein denitrosylation: enzymatic mechanisms and cellular functions, *Nat. Rev. Mol. Cell Biol.*, 2009, **10**, 721–732.
103. Y. O. Akazawa, Y. Saito, T. Hamakubo, Y. Masuo, Y. Yoshida, K. Nishio, M. Shichiri, T. Miyasaka, H. Iwanari, Y. Mochizuki, T. Kodama, N. Noguchi and E. Niki, Elevation of oxidized DJ-1 in the brain and erythrocytes of Parkinson disease model animals, *Neurosci. Lett.*, 2010, **483**, 201–205.

104. R. M. Canet-Aviles, M. A. Wilson, D. W. Miller, R. Ahmad, C. McLendon, S. Bandyopadhyay, M. J. Baptista, D. Ringe, G. A. Petsko and M. R. Cookson, The Parkinson's disease protein DJ-1 is neuroprotective due to cysteine-sulfinic acid-driven mitochondrial localization, *Proc. Natl. Acad. Sci. U. S. A.*, 2004, **101**, 9103–9108.
105. Y. Saito, T. Hamakubo, Y. Yoshida, Y. Ogawa, Y. Hara, H. Fujimura, Y. Imai, H. Iwanari, Y. Mochizuki, M. Shichiri, K. Nishio, T. Kinumi, N. Noguchi, T. Kodama and E. Niki, Preparation and application of monoclonal antibodies against oxidized DJ-1. Significant elevation of oxidized DJ-1 in erythrocytes of early-stage Parkinson disease patients, *Neurosci. Lett.*, 2009, **465**, 1–5.
106. M. A. Wilson, The role of cysteine oxidation in DJ-1 function and dysfunction, *Antioxid. Redox Signaling*, 2011, **15**, 111–122.
107. W. Zhou, M. Zhu, M. A. Wilson, G. A. Petsko and A. L. Fink, The oxidation state of DJ-1 regulates its chaperone activity toward alpha-synuclein, *J. Mol. Biol.*, 2006, **356**, 1036–1048.
108. F. Meng, D. Yao, Y. Shi, J. Kabakoff, W. Wu, J. Reicher, Y. Ma, B. Moosmann, E. Masliah, S. A. Lipton and Z. Gu, Oxidation of the cysteine-rich regions of parkin perturbs its E3 ligase activity and contributes to protein aggregation, *Mol. Neurodegener.*, 2011, **6**, 34.
109. K. K. Chung, B. Thomas, X. Li, O. Pletnikova, J. C. Troncoso, L. Marsh, V. L. Dawson and T. M. Dawson, S-nitrosylation of parkin regulates ubiquitination and compromises parkin's protective function, *Science*, 2004, **304**, 1328–1331.
110. D. Yao, Z. Gu, T. Nakamura, Z. Q. Shi, Y. Ma, B. Gaston, L. A. Palmer, E. M. Rockenstein, Z. Zhang, E. Masliah, T. Uehara and S. A. Lipton, Nitrosative stress linked to sporadic Parkinson's disease: S-nitrosylation of parkin regulates its E3 ubiquitin ligase activity, *Proc. Natl. Acad. Sci. U. S. A.*, 2004, **101**, 10810–10814.
111. J. Choi, A. I. Levey, S. T. Weintraub, H. D. Rees, M. Gearing, L. S. Chin and L. Li, Oxidative modifications and down-regulation of ubiquitin carboxyl-terminal hydrolase L1 associated with idiopathic Parkinson's and Alzheimer's diseases, *J. Biol. Chem.*, 2004, **279**, 13256–13264.
112. T. Uehara, T. Nakamura, D. Yao, Z. Q. Shi, Z. Gu, Y. Ma, E. Masliah, Y. Nomura and S. A. Lipton, S-nitrosylated protein-disulphide isomerase links protein misfolding to neurodegeneration, *Nature*, 2006, **441**, 513–517.
113. A. H. Tsang, Y. I. Lee, H. S. Ko, J. M. Savitt, O. Pletnikova, J. C. Troncoso, V. L. Dawson, T. M. Dawson and K. K. Chung, S-nitrosylation of XIAP compromises neuronal survival in Parkinson's disease, *Proc. Natl. Acad. Sci. U. S. A.*, 2009, **106**, 4900–4905.
114. W. M. Johnson, C. Yao, S. L. Siedlak, W. Wang, X. Zhu, G. A. Caldwell, A. L. Wilson-Delfosse, J. J. Mieyal and S. G. Chen, Glutaredoxin deficiency exacerbates neurodegeneration in C. elegans models of Parkinson's disease, *Hum. Mol. Genet.*, 2015, **24**, 1322–1335.

115. B. C. Lee and V. N. Gladyshev, The biological significance of methionine sulfoxide stereochemistry, *Free Radical Biol. Med.*, 2011, **50**, 221–227.
116. W. Zhou, C. Long, S. H. Reaney, D. A. Di Monte, A. L. Fink and V. N. Uversky, Methionine oxidation stabilizes non-toxic oligomers of alpha-synuclein through strengthening the auto-inhibitory intra-molecular long-range interactions, *Biochim. Biophys. Acta*, 2010, **1802**, 322–330.
117. C. B. Glaser, G. Yamin, V. N. Uversky and A. L. Fink, Methionine oxidation, alpha-synuclein and Parkinson's disease, *Biochim. Biophys. Acta*, 2005, **1703**, 157–169.
118. S. Pfeiffer, K. Schmidt and B. Mayer, Dityrosine formation outcompetes tyrosine nitration at low steady-state concentrations of peroxynitrite. Implications for tyrosine modification by nitric oxide/superoxide *in vivo*, *J. Biol. Chem.*, 2000, **275**, 6346–6352.
119. R. Radi, Nitric oxide, oxidants, and protein tyrosine nitration, *Proc. Natl. Acad. Sci. U. S. A.*, 2004, **101**, 4003–4008.
120. M. C. Franco and A. G. Estevez, Tyrosine nitration as mediator of cell death, *Cell. Mol. Life Sci.*, 2014, **71**, 3939–3950.
121. L. H. Jones, Chemistry and biology of biomolecule nitration, *Chem. Biol.*, 2012, **19**, 1086–1092.
122. J. M. Souza, G. Peluffo and R. Radi, Protein tyrosine nitration–functional alteration or just a biomarker? *Free Radical Biol. Med.*, 2008, **45**, 357–366.
123. K. Aoyama, K. Matsubara, Y. Fujikawa, Y. Nagahiro, K. Shimizu, N. Umegae, N. Hayase, H. Shiono and S. Kobayashi, Nitration of manganese superoxide dismutase in cerebrospinal fluids is a marker for peroxynitrite-mediated oxidative stress in neurodegenerative diseases, *Ann. Neurol.*, 2000, **47**, 524–527.
124. J. E. Duda, B. I. Giasson, Q. Chen, T. L. Gur, H. I. Hurtig, M. B. Stern, S. M. Gollomp, H. Ischiropoulos, V. M. Lee and J. Q. Trojanowski, Widespread nitration of pathological inclusions in neurodegenerative synucleinopathies, *Am. J. Pathol.*, 2000, **157**, 1439–1445.
125. B. I. Giasson, J. E. Duda, I. V. Murray, Q. Chen, J. M. Souza, H. I. Hurtig, H. Ischiropoulos, J. Q. Trojanowski and V. M. Lee, Oxidative damage linked to neurodegeneration by selective alpha-synuclein nitration in synucleinopathy lesions, *Science*, 2000, **290**, 985–989.
126. A. L. McCormack, S. K. Mak and D. A. Di Monte, Increased alpha-synuclein phosphorylation and nitration in the aging primate substantia nigra, *Cell Death Dis.*, 2012, **3**, e315.
127. N. Abello, H. A. Kerstjens, D. S. Postma and R. Bischoff, Protein tyrosine nitration: selectivity, physicochemical and biological consequences, denitration, and proteomics methods for the identification of tyrosine-nitrated proteins, *J. Proteome Res.*, 2009, **8**, 3222–3238.
128. C. M. Wong, L. Marcocci, D. Das, X. Wang, H. Luo, M. Zungu-Edmondson and Y. J. Suzuki, Mechanism of protein decarbonylation, *Free Radical Biol. Med.*, 2013, **65**, 1126–1133.

129. I. Dalle-Donne, R. Rossi, D. Giustarini, A. Milzani and R. Colombo, Protein carbonyl groups as biomarkers of oxidative stress, *Clin. Chim. Acta*, 2003, **329**, 23–38.
130. S. Olivieri, A. Conti, S. Iannaccone, C. V. Cannistraci, A. Campanella, M. Barbariga, F. Codazzi, I. Pelizzoni, G. Magnani, M. Pesca, D. Franciotta, S. F. Cappa and M. Alessio, Ceruloplasmin oxidation, a feature of Parkinson's disease CSF, inhibits ferroxidase activity and promotes cellular iron retention, *J. Neurosci.*, 2011, **31**, 18568–18577.
131. X. Lin, T. J. Cook, C. P. Zabetian, J. B. Leverenz, E. R. Peskind, S. C. Hu, K. C. Cain, C. Pan, J. S. Edgar, D. R. Goodlett, B. A. Racette, H. Checkoway, T. J. Montine, M. Shi and J. Zhang, DJ-1 isoforms in whole blood as potential biomarkers of Parkinson disease, *Sci. Rep.*, 2012, **2**, 954.
132. D. W. Dickson, Parkinson's disease and parkinsonism: neuropathology, *Cold Spring Harbor Perspect. Biol.*, 2012, **2**.
133. D. Sulzer and D. J. Surmeier, Neuronal vulnerability, pathogenesis, and Parkinson's disease, *Mov. Disord.*, 2013, **28**, 41–50.
134. W. M. Becker, L. J. Kleinsmith and J. Hardin, *The World of the Cell*, Benjamin/Cummings, San Francisco, 4th edn, 2000.
135. L. R. Squire, *Fundamental Neuroscience*, Elsevier/Academic Press, Amsterdam, Boston, 3rd edn, 2008.
136. W. D. Richardson, N. Kessaris and N. Pringle, Oligodendrocyte wars, *Nat. Rev. Neurosci.*, 2006, **7**, 11–18.
137. N. Baumann and D. Pham-Dinh, Biology of oligodendrocyte and myelin in the mammalian central nervous system, *Physiol. Rev.*, 2001, **81**, 871–927.
138. R. S. Snell, *Clinical Neuroanatomy*, Wolters Kluwer Health/Lippincott Williams & Wilkins, Philadelphia, 7th edn, 2010.
139. H. Braak, E. Ghebremedhin, U. Rub, H. Bratzke and K. Del Tredici, Stages in the development of Parkinson's disease-related pathology, *Cell Tissue Res.*, 2004, **318**, 121–134.
140. E. K. Pissadaki and J. P. Bolam, The energy cost of action potential propagation in dopamine neurons: clues to susceptibility in Parkinson's disease, *Front. Comput. Neurosci.*, 2013, **7**, 13.
141. G. Fiskum, *Neurodegenerative Diseases : Molecular and Cellular Mechanisms and Therapeutic Advances*, Plenum Press, New York, 1996.
142. M. V. Sofroniew and H. V. Vinters, Astrocytes: biology and pathology, *Acta Neuropathol.*, 2010, **119**, 7–35.
143. M. Nedergaard, B. Ransom and S. A. Goldman, New roles for astrocytes: redefining the functional architecture of the brain, *Trends Neurosci.*, 2003, **26**, 523–530.
144. M. A. Mena and J. Garcia de Yebenes, Glial cells as players in parkinsonism: the "good," the "bad," and the "mysterious" glia, *Neuroscientist*, 2008, **14**, 544–560.
145. M. H. Voutilainen, S. Back, E. Porsti, L. Toppinen, L. Lindgren, P. Lindholm, J. Peranen, M. Saarma and R. K. Tuominen, Mesencephalic

astrocyte-derived neurotrophic factor is neurorestorative in rat model of Parkinson's disease, *J. Neurosci.*, 2009, **29**, 9651–9659.
146. P. S. Chen, G. S. Peng, G. Li, S. Yang, X. Wu, C. C. Wang, B. Wilson, R. B. Lu, P. W. Gean, D. M. Chuang and J. S. Hong, Valproate protects dopaminergic neurons in midbrain neuron/glia cultures by stimulating the release of neurotrophic factors from astrocytes, *Mol. Psychiatry*, 2006, **11**, 1116–1125.
147. J. H. Byrne and J. L. Roberts, *From Molecules to Networks : An Introduction to Cellular and Molecular Neuroscience*, Academic Press/Elsevier, Amsterdam, Boston, 2nd edn, 2009.
148. U. K. Hanisch and H. Kettenmann, Microglia: active sensor and versatile effector cells in the normal and pathologic brain, *Nat. Neurosci.*, 2007, **10**, 1387–1394.
149. B. Finsen and T. Owens, Innate immune responses in central nervous system inflammation, *FEBS Lett.*, 2011, **585**, 3806–3812.
150. U. K. Hanisch, Proteins in microglial activation–inputs and outputs by subsets, *Curr. Protein Pept. Sci.*, 2013, **14**, 3–15.
151. C. M. Long-Smith, A. M. Sullivan and Y. M. Nolan, The influence of microglia on the pathogenesis of Parkinson's disease, *Prog. Neurobiol.*, 2009, **89**, 277–287.
152. A. Peters, S. L. Palay and H. D. Webster, *The Fine Structure of the Nervous System : Neurons and Their Supporting Cells*, Oxford University Press, New York, 3rd edn, 1991.
153. B. Obermeier, R. Daneman and R. M. Ransohoff, Development, maintenance and disruption of the blood-brain barrier, *Nat. Med.*, 2013, **19**, 1584–1596.
154. R. Cabezas, M. Avila, J. Gonzalez, R. S. El-Bacha, E. Baez, L. M. Garcia-Segura, J. C. Jurado Coronel, F. Capani, G. P. Cardona-Gomez and G. E. Barreto, Astrocytic modulation of blood brain barrier: perspectives on Parkinson's disease, *Front. Cell Neurosci.*, 2014, **8**, 211.
155. M. T. Gray and J. M. Woulfe, Striatal blood-brain barrier permeability in Parkinson's disease, *J. Cereb. Blood Flow Metab.*, 2015, **35**, 747–750.
156. E. M. Brekke, A. B. Walls, A. Schousboe, H. S. Waagepetersen and U. Sonnewald, Quantitative importance of the pentose phosphate pathway determined by incorporation of 13C from [2-13C]- and [3-13C]glucose into TCA cycle intermediates and neurotransmitter amino acids in functionally intact neurons, *J. Cereb. Blood Flow Metab.*, 2012, **32**, 1788–1799.
157. Y. Lee, B. M. Morrison, Y. Li, S. Lengacher, M. H. Farah, P. N. Hoffman, Y. Liu, A. Tsingalia, L. Jin, P. W. Zhang, L. Pellerin, P. J. Magistretti and J. D. Rothstein, Oligodendroglia metabolically support axons and contribute to neurodegeneration, *Nature*, 2012, **487**, 443–448.
158. M. Belanger, I. Allaman and P. J. Magistretti, Brain energy metabolism: focus on astrocyte-neuron metabolic cooperation, *Cell Metab.*, 2011, **14**, 724–738.

159. S. Fernandez-Fernandez, A. Almeida and J. P. Bolanos, Antioxidant and bioenergetic coupling between neurons and astrocytes, *Biochem. J.*, 2012, **443**, 3–11.
160. A. I. Ivanov, A. E. Malkov, T. Waseem, M. Mukhtarov, S. Buldakova, O. Gubkina, M. Zilberter and Y. Zilberter, Glycolysis and oxidative phosphorylation in neurons and astrocytes during network activity in hippocampal slices, *J. Cereb. Blood Flow Metab.*, 2014, **34**, 397–407.
161. J. L. Stobart and C. M. Anderson, Multifunctional role of astrocytes as gatekeepers of neuronal energy supply, *Front. Cell. Neurosci.*, 2013, **7**, 38.
162. L. Jiang, B. I. Gulanski, H. M. De Feyter, S. A. Weinzimer, B. Pittman, E. Guidone, J. Koretski, S. Harman, I. L. Petrakis, J. H. Krystal and G. F. Mason, Increased brain uptake and oxidation of acetate in heavy drinkers, *J. Clin. Invest.*, 2013, **123**, 1605–1614.
163. E. Beard and O. Braissant, Synthesis and transport of creatine in the CNS: importance for cerebral functions, *J. Neurochem.*, 2010, **115**, 297–313.
164. P. Schonfeld and G. Reiser, Why does brain metabolism not favor burning of fatty acids to provide energy? Reflections on disadvantages of the use of free fatty acids as fuel for brain, *J. Cereb. Blood Flow Metab.*, 2013, **33**, 1493–1499.
165. T. B. Vanitallie, C. Nonas, A. Di Rocco, K. Boyar, K. Hyams and S. B. Heymsfield, Treatment of Parkinson disease with diet-induced hyperketonemia: a feasibility study, *Neurology*, 2005, **64**, 728–730.
166. C. Perier and M. Vila, Mitochondrial biology and Parkinson's disease, *Cold Spring Harbor Perspect. Biol.*, 2012, **2**, a009332.
167. G. E. Gibson, A. E. Kingsbury, H. Xu, J. G. Lindsay, S. Daniel, O. J. Foster, A. J. Lees and J. P. Blass, Deficits in a tricarboxylic acid cycle enzyme in brains from patients with Parkinson's disease, *Neurochem. Int.*, 2003, **43**, 129–135.
168. S. S. Ahmed, W. Santosh, S. Kumar and H. T. Christlet, Metabolic profiling of Parkinson's disease: evidence of biomarker from gene expression analysis and rapid neural network detection, *J. Biomed. Sci.*, 2009, **16**, 63.
169. E. Palombo, L. J. Porrino, K. S. Bankiewicz, A. M. Crane, L. Sokoloff and I. J. Kopin, Local cerebral glucose utilization in monkeys with hemiparkinsonism induced by intracarotid infusion of the neurotoxin MPTP, *J. Neurosci.*, 1990, **10**, 860–869.
170. J. L. Eberling, B. C. Richardson, B. R. Reed, N. Wolfe and W. J. Jagust, Cortical glucose metabolism in Parkinson's disease without dementia, *Neurobiol. Aging.*, 1994, **15**, 329–335.
171. C. Henchcliffe, D. C. Shungu, X. Mao, C. Huang, M. J. Nirenberg, B. G. Jenkins and M. F. Beal, Multinuclear magnetic resonance spectroscopy for *in vivo* assessment of mitochondrial dysfunction in Parkinson's disease, *Ann. N. Y. Acad. Sci.*, 2008, **1147**, 206–220.

172. C. Zhang, M. Lin, R. Wu, X. Wang, B. Yang, A. J. Levine, W. Hu and Z. Feng, Parkin, a p53 target gene, mediates the role of p53 in glucose metabolism and the Warburg effect, *Proc. Natl. Acad. Sci. U. S. A.*, 2011, **108**, 16259–16264.
173. R. Requejo-Aguilar, I. Lopez-Fabuel, E. Fernandez, L. M. Martins, A. Almeida and J. P. Bolanos, PINK1 deficiency sustains cell proliferation by reprogramming glucose metabolism through HIF1, *Nat. Commun.*, 2014, **5**, 4514.
174. R. Requejo-Aguilar, I. Lopez-Fabuel, D. Jimenez-Blasco, E. Fernandez, A. Almeida and J. P. Bolanos, DJ1 represses glycolysis and cell proliferation by transcriptionally up-regulating Pink1, *Biochem. J.*, 2015, **467**, 303–310.
175. P. Jiang, M. Gan, A. S. Ebrahim, M. Castanedes-Casey, D. W. Dickson and S. H. Yen, Adenosine monophosphate-activated protein kinase overactivation leads to accumulation of alpha-synuclein oligomers and decrease of neurites, *Neurobiol. Aging*, 2013, **34**, 1504–1515.
176. J. N. Guzman, J. Sanchez-Padilla, D. Wokosin, J. Kondapalli, E. Ilijic, P. T. Schumacker and D. J. Surmeier, Oxidant stress evoked by pacemaking in dopaminergic neurons is attenuated by DJ-1, *Nature*, 2010, **468**, 696–700.
177. W. Matsuda, T. Furuta, K. C. Nakamura, H. Hioki, F. Fujiyama, R. Arai and T. Kaneko, Single nigrostriatal dopaminergic neurons form widely spread and highly dense axonal arborizations in the neostriatum, *J. Neurosci.*, 2009, **29**, 444–453.
178. S. W. Choi, A. A. Gerencser, D. W. Lee, S. Rajagopalan, D. G. Nicholls, J. K. Andersen and M. D. Brand, Intrinsic bioenergetic properties and stress sensitivity of dopaminergic synaptosomes, *J. Neurosci.*, 2011, **31**, 4524–4534.
179. T. N. Martinez and J. T. Greenamyre, Toxin models of mitochondrial dysfunction in Parkinson's disease, *Antioxid. Redox Signaling*, 2012, **16**, 920–934.
180. C. Cook, C. Stetler and L. Petrucelli, Disruption of protein quality control in Parkinson's disease, *Cold Spring Harbor Perspect. Biol.*, 2012, **2**, a009423.
181. Y. Chu, G. A. Morfini, L. B. Langhamer, Y. He, S. T. Brady and J. H. Kordower, Alterations in axonal transport motor proteins in sporadic and experimental Parkinson's disease, *Brain*, 2012, **135**, 2058–2073.
182. M. K. McCoy and M. R. Cookson, Mitochondrial quality control and dynamics in Parkinson's disease, *Antioxid. Redox Signaling*, 2012, **16**, 869–882.
183. B. Biteau, J. Labarre and M. B. Toledano, ATP-dependent reduction of cysteine-sulphinic acid by S. cerevisiae sulphiredoxin, *Nature*, 2003, **425**, 980–984.
184. L. Dunn, G. F. Allen, A. Mamais, H. Ling, A. Li, K. E. Duberley, I. P. Hargreaves, S. Pope, J. L. Holton, A. Lees, S. J. Heales and R. Bandopadhyay, Dysregulation of glucose metabolism is an early event in sporadic Parkinson's disease, *Neurobiol. Aging*, 2014, **35**, 1111–1115.

185. W. Ying, NAD+/NADH and NADP+/NADPH in cellular functions and cell death: regulation and biological consequences, *Antioxid. Redox Signaling*, 2008, **10**, 179–206.
186. R. Dringen, Metabolism and functions of glutathione in brain, *Prog. Neurobiol.*, 2000, **62**, 649–671.
187. P. J. Magistretti, Role of glutamate in neuron-glia metabolic coupling, *Am. J. Clin. Nutr.*, 2009, **90**, 875S–880S.
188. A. Y. Shih, D. A. Johnson, G. Wong, A. D. Kraft, L. Jiang, H. Erb, J. A. Johnson and T. H. Murphy, Coordinate regulation of glutathione biosynthesis and release by Nrf2-expressing glia potently protects neurons from oxidative stress, *J. Neurosci.*, 2003, **23**, 3394–3406.
189. R. Janaky, V. Varga, A. Hermann, P. Saransaari and S. S. Oja, Mechanisms of L-cysteine neurotoxicity, *Neurochem. Res.*, 2000, **25**, 1397–1405.
190. J. Lewerenz, P. Maher and A. Methner, Regulation of xCT expression and system x (c) (-) function in neuronal cells, *Amino Acids*, 2012, **42**, 171–179.
191. M. H. Stridh, M. Tranberg, S. G. Weber, F. Blomstrand and M. Sandberg, Stimulated efflux of amino acids and glutathione from cultured hippocampal slices by omission of extracellular calcium: likely involvement of connexin hemichannels, *J. Biol. Chem.*, 2008, **283**, 10347–10356.
192. S. Rana and R. Dringen, Gap junction hemichannel-mediated release of glutathione from cultured rat astrocytes, *Neurosci. Lett.*, 2007, **415**, 45–48.
193. M. H. Stridh, F. Correa, C. Nodin, S. G. Weber, F. Blomstrand, M. Nilsson and M. Sandberg, Enhanced glutathione efflux from astrocytes in culture by low extracellular Ca2+ and curcumin, *Neurochem. Res.*, 2010, **35**, 1231–1238.
194. T. Minich, J. Riemer, J. B. Schulz, P. Wielinga, J. Wijnholds and R. Dringen, The multidrug resistance protein 1 (Mrp1), but not Mrp5, mediates export of glutathione and glutathione disulfide from brain astrocytes, *J. Neurochem.*, 2006, **97**, 373–384.
195. R. Dringen and J. Hirrlinger, Glutathione pathways in the brain, *Biol. Chem.*, 2003, **384**, 505–516.
196. K. Aoyama, M. Watabe and T. Nakaki, Regulation of neuronal glutathione synthesis, *J. Pharmacol. Sci.*, 2008, **108**, 227–238.
197. K. Aoyama, S. W. Suh, A. M. Hamby, J. Liu, W. Y. Chan, Y. Chen and R. A. Swanson, Neuronal glutathione deficiency and age-dependent neurodegeneration in the EAAC1 deficient mouse, *Nat. Neurosci.*, 2006, **9**, 119–126.
198. M. R. Vargas and J. A. Johnson, The Nrf2-ARE cytoprotective pathway in astrocytes, *Expert Rev. Mol. Med.*, 2009, **11**, e17.
199. J. Sian-Hulsmann, S. Mandel, M. B. Youdim and P. Riederer, The relevance of iron in the pathogenesis of Parkinson's disease, *J. Neurochem.*, 2011, **118**, 939–957.
200. D. W. Lee and J. K. Andersen, Iron elevations in the aging Parkinsonian brain: a consequence of impaired iron homeostasis? *J. Neurochem.*, 2010, **112**, 332–339.

201. P. Dusek, J. Jankovic and W. Le, Iron dysregulation in movement disorders, *Neurobiol. Dis.*, 2012, **46**, 1–18.
202. L. Jin, J. Wang, L. Zhao, H. Jin, G. Fei, Y. Zhang, M. Zeng and C. Zhong, Decreased serum ceruloplasmin levels characteristically aggravate nigral iron deposition in Parkinson's disease, *Brain*, 2011, **134**, 50–58.
203. H. Hochstrasser, J. Tomiuk, U. Walter, S. Behnke, J. Spiegel, R. Kruger, G. Becker, O. Riess and D. Berg, Functional relevance of ceruloplasmin mutations in Parkinson's disease, *FASEB J.*, 2005, **19**, 1851–1853.
204. S. Turnbull, B. J. Tabner, O. M. El-Agnaf, S. Moore, Y. Davies and D. Allsop, Alpha-Synuclein implicated in Parkinson's disease catalyses the formation of hydrogen peroxide *in vitro*, *Free Radical Biol. Med.*, 2001, **30**, 1163–1170.
205. K. C. Chew, E. T. Ang, Y. K. Tai, F. Tsang, S. Q. Lo, E. Ong, W. Y. Ong, H. M. Shen, K. L. Lim, V. L. Dawson, T. M. Dawson and T. W. Soong, Enhanced autophagy from chronic toxicity of iron and mutant A53T alpha-synuclein: implications for neuronal cell death in Parkinson disease, *J. Biol. Chem.*, 2011, **286**, 33380–33389.
206. F. L. Martin, S. J. Williamson, K. E. Paleologou, R. Hewitt, O. M. El-Agnaf and D. Allsop, Fe(II)-induced DNA damage in alpha-synuclein-transfected human dopaminergic BE(2)-M17 neuroblastoma cells: detection by the Comet assay, *J. Neurochem.*, 2003, **87**, 620–630.
207. N. Ostrerova-Golts, L. Petrucelli, J. Hardy, J. M. Lee, M. Farer and B. Wolozin, The A53T alpha-synuclein mutation increases iron-dependent aggregation and toxicity, *J. Neurosci.*, 2000, **20**, 6048–6054.
208. J. Lotharius, S. Barg, P. Wiekop, C. Lundberg, H. K. Raymon and P. Brundin, Effect of mutant alpha-synuclein on dopamine homeostasis in a new human mesencephalic cell line, *J. Biol. Chem.*, 2002, **277**, 38884–38894.
209. J. Lotharius and P. Brundin, Impaired dopamine storage resulting from alpha-synuclein mutations may contribute to the pathogenesis of Parkinson's disease, *Hum. Mol. Genet.*, 2002, **11**, 2395–2407.
210. C. Anne and B. Gasnier, Vesicular neurotransmitter transporters: mechanistic aspects, *Curr. Top Membr.*, 2014, **73**, 149–174.
211. P. M. Abou-Sleiman, M. M. Muqit and N. W. Wood, Expanding insights of mitochondrial dysfunction in Parkinson's disease, *Nat. Rev. Neurosci.*, 2006, **7**, 207–219.
212. C. Wersinger, D. Prou, P. Vernier and A. Sidhu, Modulation of dopamine transporter function by alpha-synuclein is altered by impairment of cell adhesion and by induction of oxidative stress, *FASEB J.*, 2003, **17**, 2151–2153.
213. F. J. Lee, F. Liu, Z. B. Pristupa and H. B. Niznik, Direct binding and functional coupling of alpha-synuclein to the dopamine transporters accelerate dopamine-induced apoptosis, *FASEB J.*, 2001, **15**, 916–926.
214. M. J. Kang, S. J. Gil and H. C. Koh, Paraquat induces alternation of the dopamine catabolic pathways and glutathione levels in the substantia nigra of mice, *Toxicol. Lett.*, 2009, **188**, 148–152.

215. H. Q. Liu, X. Z. Zhu and E. Q. Weng, Intracellular dopamine oxidation mediates rotenone-induced apoptosis in PC12 cells, *Acta Pharmacol. Sin.*, 2005, **26**, 17–26.
216. T. G. Hastings, D. A. Lewis and M. J. Zigmond, Role of oxidation in the neurotoxic effects of intrastriatal dopamine injections, *Proc. Natl. Acad. Sci. U. S. A.*, 1996, **93**, 1956–1961.
217. V. S. Van Laar, A. J. Mishizen, M. Cascio and T. G. Hastings, Proteomic identification of dopamine-conjugated proteins from isolated rat brain mitochondria and SH-SY5Y cells, *Neurobiol. Dis.*, 2009, **34**, 487–500.

CHAPTER 3

Mitochondrial Dysfunction in Parkinson's Disease

MANISHA PATEL* AND PALLAVI BHUYAN MCELROY

Department of Pharmaceutical Sciences, University of Colorado Anschutz Medical Campus, Aurora, CO 80045, USA
*E-mail: manisha.patel@ucdenver.edu

3.1 Reactive Oxygen Species (ROS)

Reactive oxygen species (ROS) are defined as any oxygen species that is capable of independent existence and contains one or more unpaired electron (e^-). ROS include oxygen (O_2), superoxide ($O_2^{\bullet -}$), hydrogen peroxide (H_2O_2), nitric oxide ($^{\bullet}NO$), hydroxyl radical ($^{\bullet}OH$) and peroxynitrite ($ONOO^-$). They are formed from the gain or loss of an e^- by multiple mechanisms throughout the cell including nitric oxide synthases (NOS), xanthine oxidase, NAD(P)H oxidase (NOX), iron (Fe^{2+})-containing proteins such as aconitase and hemoglobin, autoxidation with molecules like adrenalin, dopamine (DA) and flavin adenine dinucleotide (FAD), as well as, electron (e^-) leak from the electron-transport chain (ETC) within the mitochondria.[1,2] ROS can react with metals, other ROS, proteins, lipids, DNA, or antioxidant enzymes. These interactions can result in activation of cell signalling pathways or oxidative damage, or detoxification to a non-reactive species such as water (H_2O), depending on the type of reaction. ROS are important for many signalling pathways, immune responses and regulation of transcription factors.[1,3]

Issues in Toxicology No. 34
Oxidative Stress and Redox Signalling in Parkinson's Disease
Edited by Rodrigo Franco, Jonathan A. Doorn and Jean-Christophe Rochet

Published by the Royal Society of Chemistry, www.rsc.org

3.1.1 Mitochondria and ROS Production

Mitochondria are a major source of ROS in most mammalian cells.[4] ROS are the byproduct of ATP production by the mitochondria, its primary function in cells, using substrates generated by the tricarboxylic acid (TCA) cycle to feed e^-s into the ETC. Reducing equivalents NADH and $FADH_2$ enter the ETC at complex I and complex III, respectively, and the e^-s are then passed on between complexes until they reduce O_2 to H_2O at complex IV. Simultaneously, protons are pumped across the mitochondrial membrane into the intermembrane space creating an electrochemical gradient or a proton-motive force. This force consists of a pH gradient (ΔpH) and an electrical potential ($\Delta\psi$) and drives ATP synthesis from ADP as protons re-enter the matrix through the ATP synthase or complex V. During the transfer of electrons between the ETC complexes, e^- can leak from NADH dehydrogenase (complex I) or cytochrome C oxidoreductase (complex III), react with O_2, the terminal e^- acceptor, and produce $O_2^{\bullet-}$ through one e^- reduction. Apart from ATP synthesis, mitochondria have other important cellular functions such as fatty acid synthesis, the biogenesis of heme and Fe^{2+}-sulfur proteins, O_2 sensing, calcium (Ca^{2+}) homeostasis and regulation of programmed cell death or apoptosis.[5,6]

Complex I (CI) is one of the two major entry points for reducing equivalents from the TCA into the ETC. NADH is oxidized using coenzyme Q as an e^- acceptor coupled with a transmembrane potential generated by proton transport that drives ATP synthesis.[4] Early studies have shown that isolated CI in the presence of NADH produces $O_2^{\bullet-}$ that can be further increased by rotenone, an inhibitor of this complex. However, despite several mechanistic studies, there is no consensus about the site of $O_2^{\bullet-}$ production, albeit one of the Fe^{2+}-sulfur clusters is a primary choice.[7] Nevertheless, CI is widely accepted as a foremost site of mitochondrial ROS (mtROS) production. There are a couple of critical conditions that are required for ROS production at CI: (i) low ATP production leading to high proton-motive force and reduced ubiquinone pool and (ii) high $NADH/NAD^+$ ratio in the mitochondrial matrix either when the ETC is damaged or ATP demand is low. When mitochondria are actively producing ATP and therefore have a lower proton-motive force and $NADH/NAD^+$ ratio, ROS production is low. However, during reverse e^- transport which occurs during exercise and in hypoxic tissue, where the e^-s are supplied by succinate, glycerol-3-phosphate or fatty acid oxidation, ROS formation at CI is significantly increased under conditions of high proton-motive force. NAD^+ is reduced to NADH and ubiquinol is oxidized to ubiquinone at CI.[8] CI is also known to be a target of ROS due to its sensitivity to oxidative inactivation compared to other components of the ETC.[9] The contribution of complex III towards ROS production is greatly debated although previous research has suggested the formation of an unstable semiquinone in a ubiquinone-binding site, to which the $O_2^{\bullet-}$ production has been attributed. Even though complex III can produce $O_2^{\bullet-}$ when induced with its inhibitor antimycin, under physiological conditions the production is much lower compared to the rate of production from CI.[4]

Approximately, ~0.1–2% of total mitochondrial O_2 consumed leads to the production of $O_2^{\bullet -}$. Superoxide can spontaneously dismutate non-enzymatically or be enzymatically converted to H_2O_2 by the action of superoxide dismutase 2 (SOD2 or MnSOD).[4,10] Other sources of H_2O_2 in the mitochondria include enzymes such as monoamine oxidase (MAO), α-ketoglutarate dehydrogenase, mitochondrial sn-glycerol 3-phosphate dehydrogenase (mGPDH) and aconitase.[11,12] The primary systems for H_2O_2 detoxification in the brain mitochondria are the glutathione (GSH) and the thioredoxin/peroxiredoxin (Trx/Prx) systems. However, liver and heart mitochondria can remove H_2O_2 by the action of the enzyme catalase unlike brain mitochondria.[11,13] When the rate of H_2O_2 production exceeds the capacity of these antioxidant systems, H_2O_2 can diffuse out of the mitochondria into the cytosolic compartment and therefore it is often considered the principal species mediating the effects of mtROS on extramitochondrial components.[14] Additionally, H_2O_2 can participate in the Fe^{2+}-catalyzed Fenton reaction to form the highly reactive $^{\bullet}OH$ and therefore can cause its harmful effects in this way as well.[2] The *in vivo* concentrations of $O_2^{\bullet -}$ and H_2O_2 in the mitochondria are estimated to be in the 1–200 pM range and physiological or pathological conditions can cause their steady-state levels to rise. Additionally, in isolated mitochondria, H_2O_2 production has been determined to be in the 0.5–2 nM min^{-1} per mg protein range, which indicates that H_2O_2 formation not only occurs at levels necessary for redox signalling but must also get detoxified to avoid damage to cellular macromolecules.[15]

3.2 Parkinson's Disease

Parkinson's disease (PD) is the most common movement disorder and the second most common neurodegenerative disorder after Alzheimer's disease, characterized by the loss of dopaminergic (DAergic) neurons in the *substantia nigra* (SN) *pars compacta* (SNpc) leading to a DA deficit in the striatum. It is also characterized by the appearance of Lewy bodies in the rest of the neurons.[8,16] PD is a complex multifunctional disorder and little is known about its etiology. Only an estimated 5–10% of all parkinsonian cases are familial. Epidemiological studies have linked environmental factors such as pesticidal exposure with the incidence of PD. The environmental hypothesis of PD speculates that environmental chemical agents can selectively damage the DAergic neurons contributing to the development of the disease. Although precise mechanisms that contribute to DAergic cell death are unknown, the role of ROS and mitochondria is strongly suggested in its etiology.[17] One widely accepted hypothesis is a 'multi-hit' model that suggests the combination between environmental exposures with the increased susceptibility of DAergic cells to oxidative stress and/or unknown genetic mutations render some people more susceptible to development of PD.[18]

3.3 Mitochondrial Dysfunction in PD

Oxidative stress and mitochondrial dysfunction have been implicated in the pathogenesis of several neurodegenerative disorders including PD. The role of oxidative stress in PD was affirmed by studies demonstrating that DA metabolism generates H_2O_2 and other reactive species. This led to the notion that the DAergic neurons of the SNpc, are more sensitive to ROS than other brain regions. High levels of free Fe^{2+}, high basal levels of O_2, the presence of MAOs that break down DA into highly reactive quinones and semiquinones, as well as relatively low antioxidant capacity, increase the susceptibility of this population of neurons to ROS-induced damage.[19–21] Another factor that makes the DAergic neurons of the SNpc more susceptible to oxidative damage than the neurons in the ventral tegmental area (VTA), is the high number of Cav1.3 channels, a type of L-type voltage-gated Ca^{2+} channel that might result in a higher metabolic activity due their pacemaking function.[22] Cav1.3 channels are open more often than Ca^{2+} channels in other neurons leading to an increased ATP usage in the cells. This increased ATP use is needed in order to pump Ca^{2+} across the membrane up steep concentration gradients, which would lead to increased ETC function, thereby increasing ROS production.[22,23]

As mentioned earlier, mitochondria are one of the major producers of ROS and defects in mitochondria have been identified in PD patient samples.[24,25] Mitochondrial dysfunction occurs when mtROS production happens at a rate faster than it is removed and damage to mitochondrial DNA (mtDNA) and proteins occur. This can ensue due to decreased ATP production and further ROS production from damaged macromolecules. MtDNA codes for 13 polypeptide components of the respiratory chain and mutations can lead to alterations in ETC complexes resulting in decreased respiratory enzyme activity and ATP production.[26,27] Another important target of mtROS is mitochondrial aconitase (*m*-aconitase), an enzyme of the TCA cycle that catalyzes the isomerization of citrate and isocitrate. It is uniquely sensitive to inactivation by $O_2^{\bullet-}$ due to an Fe^{2+} cluster in its active site. Inactivation of *m*-aconitase results in decreased mitochondrial respiration, decreased ATP production and increased H_2O_2 production indicative of mitochondrial dysfunction.[28,29] Nitric oxide ($^{\bullet}NO$) is also a crucial target of mitochondrial $O_2^{\bullet-}$. $^{\bullet}NO$ is a highly diffusible signalling molecule generated by NOS that can be found in neuronal and glial cells. The reaction between $O_2^{\bullet-}$ and $^{\bullet}NO$ occurs at diffusion-limited rates and produces $ONOO^-$, which can exert several deleterious effects such as depletion of thiol antioxidants, DNA damage, lipid peroxidation and protein nitration.[2,17] Other targets for ROS are unsaturated membrane lipids that can result in damage to membranes through the multiple mechanisms such as the generation of more reactive conjugates or lipid peroxidation products.[30] Furthermore, increased mitochondrial ROS is also known to cause apoptosis by cytochrome c release from the mitochondria into the cytosol.[31] Therefore, there is plenty of evidence that increased levels of mitochondrial ROS can mutate mtDNA, damage mitochondrial lipids, proteins,

ETC complexes and alter the conductance of the mitochondrial permeability transition pore, all of which can lead to mitochondrial dysfunction. This can lead to decreased or impaired ATP production, disrupting several key cellular processes such as ion transport, DNA repair and protein synthesis, ultimately leading to cell death and disease such as PD.[32,33]

3.3.1 ETC Complex Deficiency in PD

Perhaps the most compelling evidence for the role of mitochondrial dysfunction in PD is the description of CI deficiency in platelets and frontal cortex in the postmortem SNpc of PD patients, which has been suggested to be one of the fundamental causes of PD.[34] Schapira and colleagues first reported a CI defect in the SN of PD patients in 1989.[35] A 30% reduction was also observed in frontal cortex of PD postmortem brains[24] as well as in peripheral tissues such as platelets and skeletal muscle.[25,36] It has also been shown that the catalytic subunits of CI derived from PD frontal cortex mitochondria are oxidatively damaged, correlating with CI misassembly and dysfunction.[37] Collectively, this evidence suggests a global decrease in mitochondrial CI activity in PD. Such deficiencies can decrease mitochondrial ATP production and increase steady-state $O_2^{\bullet-}$ levels by enhanced electron leak and production of molecular O_2. The ETC itself can be targeted by mtROS[9] or oxidative damage to mtDNA encoding components of the respiratory chain can occur, leading to a vicious cycle of further damage to the mitochondria, ROS production and bioenergetic failure. In addition, this energy failure can disrupt DA storage and lead to an elevation of the concentrations of free DA that is capable of undergoing auto-oxidization.[38] Importantly, the DAergic neurons of the SNpc are uniquely susceptible to CI inhibition compared to other neuronal populations.[39]

One of the earliest indicators of CI dysfunction in the pathogenesis of PD was the link between 1-methyl-4-phenyl-1,2,3,6-tetrahydropyridine (MPTP) toxicity in parkinsonism. Young drug addicts in the late 1970s/early 1980s in the US injected themselves intravenously with what was later identified as trace amounts of MPTP. MPTP was the unintended side product of the illicit production of meperidine, a heroine analog.[40] It was later discovered that MPTP is a 'protoxin' that gets metabolized to the neurotoxin 1-methyl-4-phenylpyridinium ion (MPP^+) by the action of MAO-B. MPP^+ is then actively and selectively transported into DA neurons by dopamine transporters (DATs). Once inside the neuron, it accumulates in the mitochondria in millimolar levels and inhibits respiration at CI. Thus, a selective CI toxin was able to produce a parkinsonian syndrome in humans that was remarkably similar to typical 'idiopathic' PD.[41] This noteworthy insight led other researchers to investigate whether defects in CI and toxicant exposure might play a role in the pathogenesis of PD.

The precise origin and outcome of CI deficiency in PD is however, not very well understood. One of the most plausible mechanisms by which CI becomes dysfunctional is through mtROS-induced mutations in mitochondrial or nuclear DNA.[42] However, studies with rotenone, a CI inhibitor, have shed some light on the cause of this deficiency. Rotenone blocks proton pumping and increases $O_2^{\bullet-}$ generation when pyruvate or glutamate plus malate are the substrates present. However, when succinate is used as a substrate (reverse electron transport), rotenone, can block $O_2^{\bullet-}$ formation. The components of the proton-motive force (ΔpH and $\Delta\psi_m$) across the mitochondrial inner membrane also have a key role to play in $O_2^{\bullet-}$ production with a more alkaline pH or a high membrane potential favoring ROS production.[42,43] In addition to decreased ATP production, mitochondrial depolarization (depending on local substrate availability) might also be a consequence of CI dysfunction. There may also be a secondary increase in lactic acid, which can often be detected in blood or cerebrospinal fluid. If the $\Delta\psi_m$ is impaired, the mitochondrial Ca^{2+} handling is impaired, and neurons may incur Ca^{2+}-mediated damage in the setting of CI dysfunction. Therefore, decreased ATP generation coupled with depolarization, Ca^{2+} dysregulation and oxidative stress can lead to several pathological outcomes including apoptotic or necrotic cell death.[35]

Another probable cause of CI deficiency and mitochondrial dysfunction in PD is the involvement of alpha-synuclein (α-syn), a major component of Lewy bodies, the pathological hallmark of the degenerating PD brain. Recent studies have demonstrated that α-syn can localize to the mitochondria, coinciding with mitochondrial dysfunction.[44] It was found to associate predominantly with the inner mitochondrial membrane and interact with CI leading to its decreased activity.[44] A study illustrated that depletion of mtDNA in yeast by growing BY4741 cells in ethidium bromide for 4 days (rho^0), prevented *α-syn*-induced ROS production and apoptosis.[45] *In vivo* studies have demonstrated that wild-type (WT) α-syn expression exacerbates abnormalities in mitochondrial morphology.[46] More recently, Chinta *et al.* have utilized transgenic animals which overexpress A53T, the familial mutant form of α-syn, selectively within DAergic neurons that revealed that A53T localizes to the mitochondrial membranes as monomers and oligomers.[47] They also reported that this localization coincides with a selective age-related CI inhibition and decreased substrate-specific respiration along with increased mitochondrial autophagy (mitophagy).[47]

Glutathione (GSH) depletion has also been linked to CI deficiency in PD. *In vitro* studies by Hsu *et al.* have shown that inducible GSH depletion in PC12 cells can lead to selective inhibition of CI function, which could be reversed by dithiothreitol (DTT), a thiol reducing agent, implicating reversible oxidation of cysteine (Cys) residues as the probable cause of the dysfunction.[48] This finding corroborates the knowledge that oxidation of GSH to GSSG (glutathione disulfide), can result in increased mtROS production that correlates with glutathionylation of two subunits of complex I (51 and 75 kDa) leading to CI inhibition.[49] GSH depletion-induced decreased mitochondrial glutaredoxin (Grx2) activity has also been implicated in CI inhibition.[50] Another study illustrated that inhibition of CI function as a consequence of GSH depletion by a $^{\bullet}$NO-mediated mechanism, was also accompanied by an increase in mitochondrial S-nitrosothiol levels.[48] $ONOO^-$ can also lead to decreased CI function by forming 3-nitrotyrosine (3-NT) adducts on several subunits of the complex including, 49 kDa subunit (NDUFS2), TYKY (NDUFS8), B17.2 (17.2 kDa differentiation associated protein), B15 (NDUFB4) and B14 (NDUFA6).[48,51] Together, this suggests that modifications induced by oxidative and nitrative stress such as glutathionylation, nitration and thiol-oxidation, can inhibit mitochondrial CI activity and that maintaining the thiol homeostasis in SNpc, is crucial for normal CI function in PD.

A decrease in complex III activity has also been demonstrated in leukocytes and platelets of PD patients. There is also evidence of a link between complex III assembly, ROS production and PD. As mentioned previously, the leaking of e^- from this complex can increase ROS production. Complex III misassembly can also cause a decrease in CI function that can in turn generate ROS.[36] Therefore, there is convincing evidence that deficiency in ETC complexes has a crucial role to play in the pathophysiolgy of PD, which is also a testament to the involvement of mitochondrial dysfunction.

3.3.2 Altered Mitochondrial Morphology

There are a few papers that have described changes in mitochondrial morphology in PD patients. One study found a reduction in mitochondrial number and disruption of the mitochondrial membrane in muscle tissue of PD patients.[25] In a juvenile PD case, there was irregular distribution of mitochondria leading to the formation of small clusters in the SNpc as well as evidence of intramitochondrial inclusions in the stellate ganglion of patients with PD.[52] Other examples include mitochondrial swelling, rarefaction of cristae, discontinuous outer membranes, and, occasionally, intramitochondrial inclusions in human PD cybrids[53] and the presence of mitochondrial abnormalities in dog, monkey, and mouse SNpc after MPTP exposure.[52]

3.3.3 Mitochondrial Ca^{2+} Buffering in PD

As mentioned previously, the DAergic neurons of the SNpc have a higher number of L-type voltage-gated Ca^{2+} channels, which renders them more susceptibility to environmental agent-induced oxidative stress.[22] Administration of carbachol, an inositol 1,4,5-trisphosphate (InsP3)-inducing agonist that causes a transient elevation in cytosolic Ca^{2+}, has been shown to cause altered Ca^{2+} homeostasis in cybrid cells containing mtDNA from PD patients, decreased mitochondrial Ca^{2+} sequestration and slower Ca^{2+} recovery.[54] MPTP can stimulate the release of Ca^{2+} from mitochondria in the presence of DA or 6-hydroxydopamine (6-OHDA), leading to cytosolic Ca^{2+} overload. Rotenone also causes a rise in intracellular free Ca^{2+} due to diminished mitochondrial Ca^{2+} uptake.[55] Moreover, regulation of Ca^{2+} flux could be impaired by decreased mitochondrial function caused by CI inhibitors. Alternately, dysregulation of mitochondrial Ca^{2+} has been linked to altered glucose transporter activity and substrate limitation.[52,56] Calcium dysregulation has been well associated with excitotoxic cell death. In addition, Ca^{2+} influx in the mitochondria leads to alterations on mitochondrial membrane potential and ATP synthesis as well as ROS production contributing to the oxidative damage.[57] MtROS can also disrupt the uptake of Ca^{2+} by mitochondria, elevating intracellular Ca^{2+} levels and exacerbating excitotoxicity. This suggests that mitochondrial-driven or -derived excitotoxicity could be a contributing factor to PD.[36] Mitochondrial dysfunction-induced increase in cytosolic Ca^{2+} can also activate calpains and thereby raise levels of toxic α-syn, providing an etiological tie to PD.

3.3.4 PD-Related Genes and Mitochondrial Dysfunction

Approximately 10% of all PD cases are accounted for by familial variants of the disease. It was first shown in 1996 that PD might be a hereditary disease, when the gene mutations responsible for PD were mapped and identified. Of the 28 chromosomal regions related to PD, only six have genes with mutations that cause monogenic PD (Klein and Westenberger). Our knowledge about PD-related genes confirms the relevance of mitochondrial dysfunction in sporadic PD. Indeed, some PD genes have been associated with mitochondrial integrity and cellular stress response systems. For example, mutant α-syn can be targeted to the inner mitochondrial membrane where it accumulates and can cause CI impairment. Other important genes linked to PD like Parkin, PTEN-induced putative kinase 1 (*PINK1*) and *DJ-1* encode proteins that are important in maintaining mitochondrial function.[36]

The first evidence of the involvement of Parkin in mitochondrial function was provided by the observation that parkin can prevent mitochondrial swelling and cytochrome c release in ceramide-treated cells as well as physically interact with the outer mitochondrial membrane.[8] It has also been shown that Parkin can protect against mitochondrial ROS. PD patients with Parkin mutations have reduced CI activity. Additionally, Parkin null mice have decreased CI and IV activities along with decreased ETC function and increased oxidative stress in the brain.[36] *PINK1* mutations have also been associated with mitochondrial dysfunction in PD and its deficiency can lead to a variety of mitochondrial defects including decreases in membrane potential, activities of CI and IV and ATP production, in addition to increases in ROS production and abnormal ultrastructural mitochondrial morphology.[58,59] It was also demonstrated that ETC complexes are intact in *PINK1* deficient neurons as respiration in these neurons were restored by providing substrates for CI and II.[8] DJ-1 is another key protein associated with familial PD that is thought to protect neurons against oxidative stress and its deficiency makes them more susceptible to oxidative damage.[60–62] In the mitochondria, DJ-1 is localized within the matrix and the intermembrane space.[63] Moreover, isolated mitochondria from *DJ-1* knockout mice demonstrated a twofold increase in H_2O_2 accompanied by a reduction in mitochondrial aconitase activity, which indicates a deficit in scavenging mitochondrial ROS.[64]

As mentioned earlier, the N-terminal of α-syn has a sequence of amino acids that encode a mitochondrial targeting signal that directs it to the inner mitochondrial membrane and associates with CI. This leads to decreased CI activity and increased ROS production in human fetal DAergic primary neuronal cultures overexpressing WT α-syn.[44] Cellular and mouse models overexpressing WT or mutant α-syn provide further evidence of mitochondrial abnormalities including but not limited to decreased complex IV activity, decrease in membrane potential, oxidation of mitochondrial proteins and increased sensitivity to mitochondrial toxins.[65] Also, α-syn knockout mice are resistant to such toxins.[66,67] Therefore, mutations in PD-related genes are associated with several mitochondrial abnormalities and strengthen the case for the role of mitochondrial dysfunction in the etiology of PD.

3.4 Mitochondrial Dysfunction in Toxicant-Induced PD

The strongest support for mitochondrial oxidative stress and dysfunction comes from studies in toxicant based models such as the 6-OHDA, MPTP and paraquat (PQ), all of which have the ability to generate ROS in neurons. Other models of pesticide-induced PD utilize rotenone, dieldrin and maneb, which also support the role of oxidative stress in PD. Therefore, it is critical to understand the mechanisms by which environmental toxins exert oxidative stress and contribute to the etiology of PD. The structures of commonly used pesticides to model parkinsonism are illustrated in Figure 3.1 and the mechanisms by which these toxins exert their toxic effects are listed in Table 3.1.

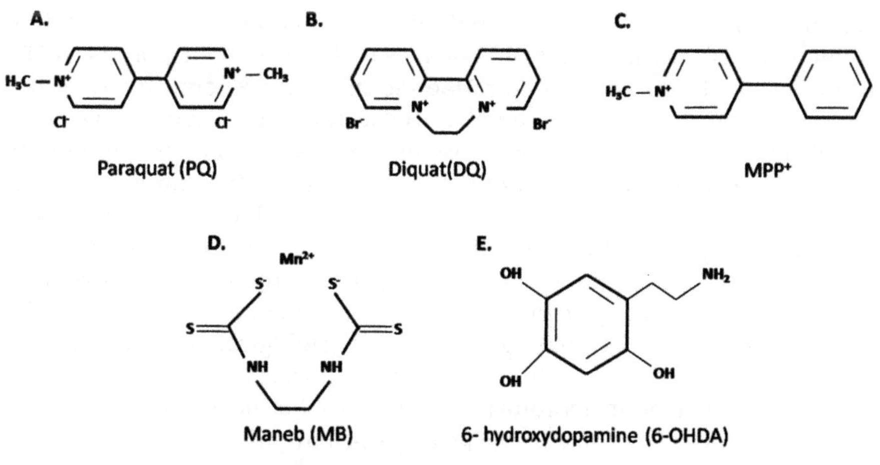

Figure 3.1 Chemical structures of toxins implicated in the development of PD. The structures of common environmental toxins PQ (A), DQ (B), MPP$^+$ (C), MB (D) and 6-OHDA (E) that have been associated with the development of parkinsonism are illustrated here.

Table 3.1 List of pesticides commonly implicated in the development of PD and the mechanism by which they exert toxicity and mitochondrial dysfunction.

Parkinsonian pesticide	Mechanism of action	Mitochondrial mechanism
Paraquat	Redox cycling and $O_2^{\bullet-}$ production	CI and III inhibition
Diquat	Redox cycling and $O_2^{\bullet-}$ production	CI and III inhibition
Rotenone	Binding to CI and inhibition of oxidative phosphorylation	CI inhibition
MPTP	Conversion to toxic metabolite MPP^+	CI inhibition
6-OHDA	ROS production and para- and semiquinone products	CI inhibition
Maneb	ROS production and oxidative stress	Complex III inhibition
Dieldrin	Inhibition of GABA receptors	Complex III inhibition

3.4.1 6-OHDA and MPTP: Classic Toxicant Models

3.4.1.1 6-OHDA: An Experimental Catecholaminergic Neurotoxicant

6-OHDA, a hydroxylated analog of DA, was the first chemical agent discovered that had specific neurotoxic effects on nigrostriatal pathways, including the DA system. It shares some structural similarities with DA and norepinephrine (NE) and has a high affinity to be taken up by catecholaminergic membrane transporters such as DAT and NE transporters (NET). Therefore, 6-OHDA can enter both DAergic and noradrenergic neurons and can cause damage to the peripheral as well as the central nervous system.[68] 6-OHDA causes degeneration of nerve terminals and cell bodies and exerts its toxic effects by inhibition of mitochondrial CI and IV.[69] Since this molecule cannot cross the blood–brain barrier (BBB), production of central neuronal lesions can only be achieved after direct stereotactic administration into the brain. In experimental models of PD, it is injected in the striatum, the SN or the ascending medial forebrain bundle, depleting DA from the striatum, thereby recapitulating the pathophysiological characteristics of PD. Interestingly, some studies have indicated that 6-OHDA can be considered an endogenous neurotoxin. Its presence has been detected in both rat and human brain. Since the SNpc neurons have high levels of DA, H_2O_2 and Fe^{2+}, these components can potentially react to form 6-OHDA non-enzymatically. Manganese can also stimulate the formation of 6-OHDA by auto-oxidation of DA. The fact that this toxin can cause ROS production and mitochondrial dysfunction makes it a useful tool to mimic parkinsonian symptoms in animal models.[70]

6-OHDA exerts neurotoxicity by production of ROS and para- and semiquinone products.[71] Evidence for a role of oxidative stress in mediating the effects of this compound, comes from studies that have demonstrated the ability of antioxidants to protect against 6-OHDA-induced toxicity.[72,73] Moreover, there was decreased cell death in transgenic mice overexpressing SOD[74] and glutathione peroxidase (GPx).[75] ROS generation by 6-OHDA occurs by two mechanisms: (a) deamination by MAO or (b) auto-oxidation. 6-OHDA is a substrate for MAO just like DA and the deamination results in the production of H_2O_2. The involvement of MAO in mediating its toxicity was suggested by the observation that selegiline, an inhibitor of MAO (MAO-I), prevents 6-OHDA toxicity.[70] The auto-oxidation process generates H_2O_2, $^{\bullet}OH$ radical, $O_2^{\bullet-}$, and quinones in a non-enzymatic manner. Additionally, treatment with antioxidants like GSH, Cys and *N*-acetylcysteine (NAC) caused significant reductions in the auto-oxidation of 6-OHDA.[71,76] ROS production by auto-oxidation can be amplified by Fe^{2+} *via* Fenton chemistry. Notably, Fe^{2+} levels are increased in the SNpc and striatum after 6-OHDA injection. Fe^{2+} chelating agents can prevent the deleterious effects of 6-OHDA indicating that Fe^{2+} plays a role in 6-OHDA-induced toxicity.[77] Therefore, MAO activity, auto-oxidation and elevated Fe^{2+} levels, synergistically contribute to 6-OHDA-induced ROS production.

6-OHDA impairs mitochondrial function by several mechanisms targeting the respiratory chain. Specifically, 6-OHDA is able to interact with and inhibit CI activity.[78] This is a completely reversible process independent of ROS as neither antioxidants nor 6-OHDA chelators can rescue mitochondrial dysfunction.[70] However, in whole cells, inhibition of mitochondrial respiration is not the primary mechanism by which 6-OHDA exerts toxicity as it was found that in SH-SY5Y neuroblastoma cells, there was no reduction of ATP production, ATP/ADP ratio or NAD^+ cellular content.[79] On the other hand, 6-OHDA is known to induce a ROS-dependent collapse of the mitochondrial membrane potential and is also an uncoupler of oxidative phosphorylation.[80] Nevertheless, this data implicates mitochondrial dysfunction in 6-OHDA-induced neurotoxicity that not only leads to impaired membrane potential but also defects in DA vesiculation that may increase cytoplasmic levels of DA, and both these events can lead to cell death. Since 6-OHDA needs to be directly injected into the brain, it is considered a poor model of environmental exposure of PD. Despite this drawback, studies involving 6-OHDA have provided the basis for understanding the mechanisms by which environmental agents cause PD and is a widely utilized compound to generate animal models of PD.

3.4.1.2 *MPTP: Evidence for CI and the Toxic Exposure Hypothesis of PD*

As mentioned previously in this chapter, the first evidence of environmental agents causing PD was the case of young drug users in California who exposed themselves to MPTP and developed parkinsonian symptoms. Since then, MPTP has been instrumental in deciphering the mechanisms of

DAergic neuronal death in PD. The MPTP model is the best characterized one yet and has provided the most reliable information on the role of oxidative stress and mitochondrial dysfunction in PD pathogenesis.

Due to its highly lipophilic nature, MPTP can readily cross into the BBB, where it is readily converted to MPP^+, its active toxic form through an unstable intermediate, by the action of MAO-B. This reaction is thought to occur predominantly in glial cells. MPP^+ may be released into the extracellular space where it could be taken up by DATs into DAergic neurons. Besides its uptake by transporters, there are two other methods by which it is thought to accumulate in the cytosol: (i) the neuromelanin that forms a complex with MPP^+ and delays its cytoplasmic release and (ii) the vesicular monoamine transporters (VMAT) that confine it to synaptic vesicles.[70] Subsequently, MPP^+ could enter the mitochondria by an energy-dependent mechanism where it targets CI resides. MPP^+ binds to the NADH binding domain of CI, increasing electron leakage leading to increased ROS production, particularly $O_2^{\bullet-}$.[81–83] The role of mitochondrial $O_2^{\bullet-}$ has been confirmed by studies that have demonstrated that overexpressing SOD2 attenuated MPTP toxicity and partial deficiency in SOD2 exacerbated it.[84,85] CI inhibition by MPP^+ leads to decreased ATP production and eventual cell death.

Complex I inhibition by MPTP not only leads to inhibition of oxidation of NAD^+-linked substrates but also inhibition of mitochondrial aconitase[86] and α-ketoglutarate dehydrogenase.[81] In addition to lower ATP levels, this also leads to loss of mitochondrial membrane potential, alterations of Ca^{2+} homeostasis and radical formation. However, it was also shown that rho^0 cells, which are devoid of a functional ETC, were equally sensitive to MPP^+ as parental cells, suggesting a mitochondria-independent ROS-induced damage by MPP^+.[70] It is noteworthy to mention here that MPP^+ has a reduction potential of approximately −1.180 V at physiological pH, which makes it a poor redox cycling agent.[87] Despite its poor redox potential, there is some evidence in the literature that indicates the ability of MPP^+ to be redox active in the presence of certain e^- donors such as cytochrome P450 reductase, xanthine oxidase and aldehyde dehydrogenase to produce $^{\bullet}OH$. This putative redox cycling mechanism of MPP^+ is thought to involve its reduction to $MPP^{\bullet}$ radical that proceeds to cleave H_2O_2 into $^{\bullet}OH$ and parental MPP^+.[88,89] Regardless of whether MPP^+ can produce ROS *via* redox cycling, it is widely accepted that it can exert neurotoxicity through ROS production.

Early studies have demonstrated ROS production by MPTP using DCDHF-DA, the ROS-dependent fluorescent probe, *in vitro*,[90] and by electron spin resonance method *in vivo*.[91] MPTP was also shown to decrease GSH and metallothionein content[92] and increase 3-NT formation.[93] Drechsel *et al.* have shown MPP^+ to cause a decrease in GSH redox status prior to cell death in rat DAergic N27 cells, which was associated with decreases in the activities of glutamate-cysteine ligase (GCL) and GSH-reductase (GR) and an increase in GPx activity,[94] elucidating the mechanisms by which a parkinsonian toxin like MPP^+ causes GSH depletion and subsequent cell death. Furthermore, the decrease of GSH content potentiated MPTP toxicity.[70] Confirmation of ROS involvement in MPTP was also obtained by studies illustrating the potentiation of MPTP effects by either deficiency in SOD2 or GPx genes.[95,96]

Amongst the several factors that can contribute to MPTP-induced ROS production, Fe^{2+} is of great importance as it can participate in Fenton chemistry in DAergic neurons. MPTP can increase free Fe^{2+} levels in SNpc[97] and its effects are blocked by desferrioxamine, an Fe^{2+} chelator.[98] Liang *et al.* have shown that MPTP administration in mice leads to Fe^{2+} accumulation and DA depletion in addition to *m*-aconitase inactivation and that these effects are attenuated in mice overexpressing *Sod2*, while being exacerbated in mice partially deficient in *Sod2*.[86] Fe^{2+} release can in turn lead to ${}^{\bullet}OH$ production by the Fenton reaction and cause further damage. These results suggest that $O_2^{\bullet -}$ toxicity may play an important role in PD pathogenesis and that *m*-aconitase might be a major source of mitochondrial Fe^{2+} accumulation in experimental PD.

MPTP can also generate ROS by extrinsic mechanisms, in addition to mtROS production. MPP^+-induced release of DA from storage vesicles into the cytosol can lead to ROS production by auto-oxidation of the neurotransmitter. MPTP injection can additionally induce cyclo-oxygenase-2 that can generate DA-quinones (DAQ) *via* its peroxidase activity.[80,99] MPTP-induced ROS production may also be a result of an increase in cytoplasmic Ca^{2+} from neuronal impairment of energy metabolism and indirect excitotoxicity. In that respect, NMDA receptor antagonists[100] and Ca^{2+} channel blockers[101] were able to protect SNpc efficiently against MPTP. A role of Ca^{2+}-dependent NOS activity has been illustrated.[102,103] Reactive micoglia-mediated inflammatory processes are also responsible for MPTP-induced ROS production.[104] These data can together explain the increase in lipid peroxidation products by MPTP.[105]

From the time MPTP was discovered, it has been widely used to produce specific damage to DAergic cells resulting in parkinsonism symptoms in several animal models including mice and non-human primates. Studies utilizing the MPTP model, have laid the groundwork for further investigation into the environmental hypothesis of PD etiology and the mechanisms by which environmental agents exert neurotoxicity and mimic the pathology associated with PD.

3.4.2 Rotenone: A Case for Complex I Inhibition

Rotenone is a naturally occurring complex ketone and a member of the retinoids, which are a family of cytotoxic compounds, derived from the Lonchocarpus species.[106] It is used as a pesticide and to kill fish in lakes and reservoirs that are considered to be pests. The biggest advantage for using rotenone is that it biodegrades in several days even when spread over hundreds of acres of land.[107] It is a high-affinity inhibitor of CI and it specifically interacts with the NADH dehydrogenase, subunit 1 (ND1) and PSST subunits. Rotenone is extremely lipophilic and unlike MPP^+, it does not require any transporters. Rotenone easily crosses the BBB, rapidly enters the brain and accumulates into the mitochondria. Once within the mitochondria, it binds to CI at the ubiquinone-binding site and inhibits oxidative phosphorylation and therefore, it is used as a classical mitochondrial toxin in cell-culture experiments.[17,107] It has also been used for the systemic inhibition of CI *in vivo*. The specificity of rotenone for CI was demonstrated by Sherer *et al.*, where they replaced endogenous CI with the rotenone-insensitive single-subunit NADH dehydrogenase of *Saccharomyces cerevisiae* (NDI1). These cells did not show mitochondrial impairment, oxidative damage, or death in response to rotenone.[106]

Rotenone-induced inhibition of CI leads to ATP deficiency and ROS production, similar to MPP^+. It slows down electron flow at sites upstream that are prone to electron leakage and with the e^-s being available in the site longer than normal, O_2 can undergo one e^- reduction to form $O_2^{\bullet-}$. Interestingly, since the doses of rotenone used to induce parkinsonism in rats do not alter mitochondrial O_2 consumption in isolated brain mitochondria, bioenergetic defects alone cannot account for rotenone-induced neurodegeneration. However, rotenone toxicity has been attributed to oxidative stress. In fact, it is believed that the mechanism by which CI inhibition damages neurons is cumulative oxidative stress.[106] Indeed, chronic exposure to 5 nM rotenone *in vitro* produced progressive depletion of GSH, oxidative damage to proteins and DNA, and release of cytochrome c from mitochondria ultimately leading to apoptosis.[35] Additionally, oxidative damage and DAergic cell death caused by rotenone, was prevented by the antioxidant α-tocopherol (vitamin E), strengthening the case for oxidative stress in rotenone toxicity.[106] Rotenone has also been shown to oxidatively modify DJ-1 and cause α-syn aggregation, both of which events are associated with PD pathogenesis and located to DAergic neurons.

The toxic effects of rotenone are similar to those of MPTP as they share the same primary mechanism of action. Similar to MPTP, rotenone toxicity is also linked to glial cell activation. There is evidence of robust microglial activation in the rat model of rotenone induced parkinsonism, which was found to occur prior to DAergic neuronal loss.[108] Microglial activation by rotenone can also lead to production of ROS, from several sources such as NOX, cyclooxygenase-2 (COX-2), which can generate free radicals as byproducts of prostaglandin synthesis; and iNOS.[109] Superoxide produced by the action of NOX can form $ONOO^-$ that can form 3-NT adducts on proteins in Lewy-body positive neurons.[107] Interestingly, NOX inhibitors prevent the generation of H_2O_2, which suggests that rotenone induces H_2O_2 generation derived from $O_2^{\bullet-}$, either through the inhibition of the NADH dehydrogenase complex, the activation of NOX, or possibly through both mechanisms. This H_2O_2 produced also caused significant disruption of the mitochondrial membrane potential.[110] The role of microglial NOX-derived $O_2^{\bullet-}$ a key contributor to the neurotoxicity of rotenone can be confirmed by three observations from a previous paper: (1) NOX inhibition or scavenging of free radicals afforded significant neuroprotection, (2) rotenone and LPS synergistically stimulated the NOX-mediated release of $O_2^{\bullet-}$ and, (3) in cultures from NADPH oxidase-deficient mice, rotenone and LPS failed to induce the synergistic neurotoxicity as well as the production of $O_2^{\bullet-}$.[111]

In primary mesencephalic cultures, rotenone is selectively toxic to DAergic neurons, compared to GABAergic neurons[112] and this effect is potentiated by the presence of miroglia.[113] It is hypothesized that the DAergic neurons in the nigrostriatal pathway are selectively susceptible to CI inhibition, since rotenone does not appear to utilize a specific transport system. Chronic exposure to low doses of rotenone leads to a 30% decrease in DAergic neuronal loss in the SNpc.[39] However, there are a few complications with using rotenone for animal models. First, rotenone-induced CI inhibition causes significant reductions in non-DAergic neuronal populations in the striatum.[114] In addition, there is also wide variability in the nigrostriatal effects even with the same treatment paradigm as well as technical difficulties with using rotenone in animals.[39] Despite providing robust evidence for the role of CI inhibition in mitochondrial dysfunction in the etiology of PD, the possibility of human exposure is restricted as the half-life of rotenone is between 1 and 3 days in water and soil and it breaks down readily. Nevertheless, the rotenone model has provided valuable mechanistic insights. Many commercially available pesticides and environmental agents, with greater risk of exposure to humans, exert neurotoxicity by inhibiting CI activity. Therefore, knowledge obtained from studies utilizing rotenone is critical in understanding the mechanisms by which such compounds wield their toxic effects.

3.4.3 Paraquat: Redox Cycling Agent

The discovery of the role of ROS and mitochondrial dysfunction in mediating DAergic toxicity by 6-OHDA and MPTP, has led researchers to investigate whether other environmental agents are linked to PD development. During the past three decades, over 40 epidemiological studies have been published that provide data on the relationship between pesticide exposure and the risk of PD. A significant proportion of these studies were instigated after the suggestions made by Barbaeu, who reviewed the pathogenesis of parkinsonism induced by manganese and MPTP intoxication. It was suggested that it was most likely possible that PD incidence was related to the exposure to products with a similar conformation of MPTP occurring naturally or in industrial settings. For example, many herbicides might have a similar mechanism of action. In individuals who are genetically susceptible to developing PD, exposure to such chemicals can trigger DAergic degeneration that is the classic hallmark of PD. Based on the epidemiological studies that identified agricultural chemicals as potential risk factors, pesticides and herbicides have become an increasingly attractive candidate for toxicant-induced models of PD. PQ is one such environmental agent that was initially identified for its potential as specific DAergic neurotoxicant based on exposure data and its structural similarity with MPTP.

3.4.3.1 Characteristics and Risk of Exposure

PQ (1,1′-dimethyl-4-4′-bipyridinium dichloride) is one of the most commonly used herbicides in the world. It belongs to a class of bipyridal herbicides with a structural backbone containing two covalently linked pyridine rings. It was introduced in 1962 for agricultural uses and although its use is banned or restricted in most European countries and the United States, PQ is still extensively used in less developed countries. In the United States, it is available primarily as a solution in various strengths and has 'restricted use', used only by licensed applicators. It is very fast acting, non-selective and non-systemic, and kills green plant tissue on contact. Its main use is for weed control in orchards and plantations. Additionally, PQ is also used as a defoliant and desiccant to facilitate harvesting. PQ loses its biological activity shortly after its application as it is strongly adsorbed to soil particles or decomposed into a non-toxic product by soil bacteria and sunlight.[115]

When ingested, PQ is highly toxic to humans and can lead to acute respiratory distress syndrome (ARDS) including acute alveolitis, widespread fibrosis and fatal hypoxia.[116] The population most at risk from PQ exposure is farmers who spray their crops with PQ and its applicators. Suicide attempts are also common in developing countries, which accounts for approximately 93% of fatalities from PQ poisoning. The diamine transporters in type II epithelial cells of the lung, selectively and actively take up PQ, which contributes to the lung toxicity observed with PQ.[17] Epidemiological exposure data has been obtained from Canada and Germany in the 1990s that links PQ exposure to PD development.[117] A study from Taiwan based on 120 cases, reported increased risk of developing PD after chronic PQ exposure.[118] Bipyridinium herbicides including PQ were first recognized as potential DAergic neurotoxicants because of the structural similarity with MPP^+. However, the mechanisms by which these two compounds exert their harmful effects are unique, although ROS production and mitochondrial dysfunction are shared traits.

3.4.3.2 Mechanism of Toxicity: Redox Cycling and ROS Generation

Despite the structural similarities between PQ and MPP^+, they exert their neurotoxic effects *via* very different mechanisms. PQ crosses the BBB very slowly and inefficiently, although levels have been detected in the central nervous system after systemic injection into rodents. The toxicity of PQ is thought to be mediated by the $O_2^{\bullet-}$ radical that is produced as byproduct of redox cycling by PQ. PQ accepts an e^- from an appropriate donor, forming a PQ radical ($PQ^{\bullet+}$). This radical then reduces O_2 to form $O_2^{\bullet-}$ along with regeneration of the parent compound PQ. $O_2^{\bullet-}$ dismutates to produce H_2O_2 and $^{\bullet}OH$ radical that can cause damage to cellular components like proteins, lipids and DNA. Indeed, $^{\bullet}OH$ radicals have been implicated in the initiation of lipid peroxidation and thereby causing membrane damage during the exposure to PQ *in vitro*[119] and *in vivo*.[120] The redox potential of PQ is −0.450 V that has biological significance as the redox potentials of the NAD^+/NADH and the $NADP^+$/NADPH couples are −0.320 and 0.324 V, respectively. This suggests that under physiological conditions, PQ can accept an e^- from one such redox couple. Many enzymes that can transfer an e^- from NAD(P)H, called diaphorases, have the capacity to reduce PQ to its radical. On the other hand, MPP^+ has a reduction potential of approximately −1.180 V and hence, has a much lower capacity to accepting an e^- from donors.[17,121] Redox cycling denotes the continuous reduction and oxidation of PQ during the process. Therefore, in the presence of O_2, PQ can continuously cycle between the parent compound and the radical form to generate high amounts of $O_2^{\bullet-}$, which serves as the mechanistic basis for the herbicide's toxicity. The role of $O_2^{\bullet-}$ in mediating PQ-induced toxicity is further confirmed by studies where SOD mimetics and SOD overexpression were able to inhibit the effects of PQ.[122,123] The redox cycling mechanism of PQ can deplete the intracellular stores of NAD(P)H since it is oxidized, which can also contribute to its toxicity. Additionally, $PQ^{\bullet+}$ can reduce Fe^{3+} triggering Fenton reaction to produce $^{\bullet}OH$ radicals. However, the brain has high O_2 content, which favors the reaction of PQ with O_2 over Fe^{3+}. The reduction of PQ to the radical is the rate-limiting step for the redox cycling process. The enzymes that catalyze this reaction such as NOX, NADPH-cytochrome c reductase and NOS, can be found in the microsomes, plasma membrane, and cytosol of various organs.[123] Of the various diaphorases, isoforms of NOS have been identified as being specific to PQ and are associated with a loss of NO-producing function of the enzyme.[121]

PQ toxicity has also been shown to be dependent on the activation of NOX in microglial cells. Moreover, there was a decrease in DA uptake and loss of tyrosine hydroxylase positive (TH+) cells in mixed neuron–glia primary cultures treated with PQ. However, cultures that were depleted of microglia, did not exhibit this response. Additionally, the neurotoxic effects of PQ were absent in mutant mice lacking functional NOX. This indicates that microglial activation and consequent induction of NOX might be one of the risk factors for PD by increasing the vulnerability of DAergic cells to toxic injury.[124]

3.4.3.3 *Role of Mitochondria in PQ Toxicity*

With respect to the brain, the notion that mitochondria are involved in ROS production by PQ has emerged in recent years. Specifically, the ETC has emerged as a target for PQ-induced ROS generation and mitochondrial dysfunction. This idea is based on two main observations: (1) the respiratory chain complexes contain e^--carrier sites that can pass on e^-s to PQ that can initiate the redox cycling reaction to generate $O_2^{\bullet-}$ and (2) similarity in structure with MPP^+ suggests that PQ might be able to inhibit complex I activity in a manner comparable to MPP^+. In fact, one of the first studies demonstrating the relationship between PQ and CI inhibition by Fukushima *et al.* showed that PQ might cause ROS production by interacting with purified CI. This was followed by CI inhibition at millimolar levels of PQ, although it was not clear whether this effect was due to PQ directly or was a deleterious consequence of ROS production. This CI inhibition then leads to mitochondrial dysfunction.[125] A role of *m*-aconitase in mediating PQ-induced neurotoxicity has been demonstrated by Cantu *et al.*[28] Treatment of N27 cells with PQ resulted in decreased aconitase activity, increased Fe^{2+} and H_2O_2 production preceding cell death. Furthermore, knockdown of *m*-aconitase led to decreases in H_2O_2 production, Fe^{2+} accumulation and cell death in response to PQ.[28] Together, these findings strengthen the case for mitochondrial mechanisms in PQ-induced neurotoxicity.

More recently, it has been shown that intact, respiring brain mitochondria are a major subcellular source of PQ-induced H_2O_2 production and can take up PQ as well. Furthermore, this ROS production required the presence of respiratory substrates and was dependent on the mitochondrial membrane potential. H_2O_2 production by PQ was also significantly attenuated by a complex III inhibitor, antimycin A, suggesting involvement of this complex in PQ-induced mitochondrial ROS production, compared to CI, as rotenone had a limited effect. These results were also confirmed in primary neuronal culture obtained from the midbrain. Therefore, mitochondria appear to be the main subcellular compartment contributing to PQ-induced ROS generation *via* CI and III of the ETC.[123] Furthermore, it appeared that PQ produces $O_2^{\bullet-}$ that is then converted to H_2O_2 by SOD2 or by spontaneous dismutation. This H_2O_2 can then diffuse through the mitochondrial membrane into the extracellular space which was measured in this study utilizing polarographic and fluorescent methods.[123] Moreover, PQ has also been shown to selectively and reversibly inhibit aconitase activity, which can be used as a marker for mitochondrial $O_2^{\bullet-}$ production and this inactivation preceded cell death in rat primary cortical culture. In the same study, the SOD mimetic MnTBAP significantly protected against PQ-induced aconitase inactivation and neurotoxicity.[126] A study by Drechsel *et al.*, additionally, demonstrated that brain mitochondria can not only produce but also remove/consume H_2O_2 produced by PQ and that this removal was dependent on the thioredoxin/peroxiredoxin (Trx/Prx) system.[127] More recently, it was shown by Lopert *et al.* that inhibiting the Trx/Prx system in the presence of subtoxic concentrations of PQ, sensitized DAergic cells to H_2O_2 production, mitochondrial dysfunction and neuronal death.[128] These data collectively highlight the importance of mitochondrial ROS and dysfunction in mediating PQ-induced DAergic toxicity.

Since DAergic neurons are especially vulnerable to oxidative damage, and mtROS production can exert toxicity through inducing oxidative stress, ROS production and consequent oxidative stress can account in part for PQ-induced selective degeneration of DAergic neurons. PQ can wield oxidative damage to lipids and other macromolecules that has been documented in several organ systems and tissues other than the brain.[129,130] In addition, in cerebellar granule cells, oxidative stress has also been proposed to initiate PQ-induced apoptotic cell death.[131] It has also been shown that PQ caused an increase in lipid peroxidation and protein nitration in the SN of mice.[132]

A direct role of oxidative stress was demonstrated by the fact that overexpressing SOD1 or GPx in mice, prevented PQ-induced decrease in DA metabolite levels and DAergic neurons.[133] Moreover, PQ-mediated neuronal death was attenuated by synthetic SOD or catalase mimetics *in vivo*.[134,135] A recent study by Liang *et al.* administered PQ to *Gclm−/−* mice, where the gene for the modifier subunit of glutamate cysteine ligase (GCLM), the rate-limiting enzyme for GSH synthesis was deleted. This led to decreases in aconitase activity, DA levels, as well as CI inhibition in the striatum, suggesting that chronic deficiency in an antioxidant like GSH, along with PQ exposure, can lead to the development of PD pathology.[136] Observations from all these studies have provided substantiation for the vital role of mtROS and oxidative stress in PQ-induced development of PD.

3.4.3.4 PQ and Excitotoxicity

There is some evidence that excitotoxic injury plays an important role in mediating neurodegeneration in PD. Therefore, it is not surprising that an indirect excitotoxic mechanism has also been proposed, to explain the toxic effects of PQ. *In vivo* studies have demonstrated that PQ stimulates glutamate efflux and induces depolarization of NMDA receptor channels and Ca^{2+} influx into the cells by activation of non-NMDA receptor channels.[137] This influx activates neuronal NOS that in turn can lead to the production of $ONOO^-$ upon reaction with $O_2^{\bullet-}$ produced from redox cycling, leading to excitotoxicity. Although the exact mechanism by which PQ increases extracellular glutamate levels is unknown, this effect is concentration dependent. $ONOO^-$ can also attack lipids, proteins and DNA, as well as render SOD2 nonfunctional, causing further elevation of $O_2^{\bullet-}$ and $ONOO^-$ levels. It can additionally, inhibit enzymes of the ETC, decreasing ATP synthesis. Treatment with glutamate receptor antagonists, a NOS inhibitor and by the MAO inhibitor L-deprenyl, PQ-mediated DAergic neuronal death was prevented, suggesting that excitotoxic mechanisms play a part in increased vulnerability of DAergic neurons in the nigrostriatal system, as a consequence of chronic exposure to low doses of PQ.[137] Moreover, a study in PC12 cells illustrated inactivation of TH, the rate-limiting enzyme in DA synthesis, by $ONOO^-$-mediated nitration, and this loss of activity parallels the decline in DA levels.[138] Therefore, RNS-mediated toxicity induced by glutamate has been suggested in DAergic neuronal death by PQ, leading to PD development.[139]

3.4.3.5 Using PQ for Developing Animal Models of PD

The first evidence of the poor ability of PQ to cross the BBB came from early studies in animal models where systemic exposure of PQ failed to show any changes to the nigrostriatal pathway including no changes to striatal DA levels.[140] A successful model was eventually developed that recapitulated the chronic and progressive nature of PD. This consisted of a successive sub-lethal dosing regimen, administered over a three-week period. Stereological counting, uncovered decreases in DAergic cells in the SNpc in a dose-dependent manner, along with a decrease in striatal DA nerve terminal density. This dosing regimen also caused an increase in α-synuclein levels in the brain and accumulation of α-syn containing lesions in the SNpc, in addition to inducing neurobehavioral syndrome characterized by reduced ambulatory activity.[141] Additionally, the DAergic neurons in the SN and striatum were particularly susceptible to PQ, while other subpopulations were unaffected.[132] There is evidence of enhanced DA turnover, which is indicated by increased TH activity[132] and altered levels of DA metabolites,[133] although no significant depletion of DA itself has been observed with PQ treatment *in vivo*.[132] Even though PQ exhibits poor BBB penetrability, there is evidence of paraquat entering the brain *via* the assistance of neutral amino acid transporters, as pretreatment of animals with L-valine or L-phenylalanine completely prevented PQ-mediated neurodegeneration in mice, and is transported into cells in a sodium (Na^+)-dependent manner.[68] It is therefore, expected that PQ is taken up into the mitochondria with a similar transport system. Even more evidence of PQ being an appropriate parkinsonian toxin comes from the observation that mice treated with PQ have an upregulation and aggregation of α-syn in the SNpc of these mice.[142] The advantages of using PQ as an environmental toxin-induced PD model are its widespread use as a herbicide and presence in the environment. Moreover, the chronic treatment paradigm provides a more accurate representation of the progressive nature of PD as well as disease pathology, even though the relevance of this model to human exposure and PD risk is yet to be assessed completely.

3.4.3.6 Diquat Use with PQ: Example of Another Redox Cycling Agent

PQ is often mixed with diquat (DQ), another bipyridyl herbicide for agricultural purposes. DQ is a non-selective contact herbicide that has been available commercially since the 1950s. About a third of DQ is sold as a contact herbicide for control of aquatic weeds and broad-leaved weeds among various crops. The other two-thirds of DQ are used as a preharvest desiccant for seed and fodder crops. Its toxic effects on the environment are minimized by swift photochemical degradation and rapid adsorption to aquatic weeds and clay particles in soil.[143] Exposure to DQ can cause extensive damage to the CNS and has been associated with increased risk of PD.

The mechanism by DQ exerts neurotoxicity is similar to PQ in that it is also capable of redox cycling to form high levels of $O_2^{\bullet -}$. DQ can undergo a one e^- reduction to form the DQ radical ($DQ^{\bullet +}$) that then reacts with molecular O_2 to form $O_2^{\bullet -}$ and the parent compound. It is important to note that PQ produces $O_2^{\bullet -}$ in the mitochondria at a rate higher and is more toxic than DQ in the lung, liver and kidneys.[144] Interestingly, in isolated rat brain mitochondria and primary midbrain cultures, H_2O_2 production by DQ was found to be higher than PQ, which was attenuated by inhibition of CI and III, emphasizing the role of mitochondrial mechanisms in DQ toxicity.[145] DQ does not induce DA depletion in mouse striatum, although it exacerbated motor impairment, which might due to the decreased DA turnover and mild nigrostriatal neurodegeneration.[146] Therefore, DQ has very limited toxicity on its own, but when combined with PQ can have a synergistic effect on ROS production and DAergic cell death.

3.4.4 Maneb: A Role of Complex III in PD Pathogenesis

The majority of the work involving DAergic toxicants and their role in the etiology of PD has been achieved by examining mechanisms and risks associated with one chemical agent. The multiple-hit hypothesis of PD suggests that the brain might be able to withstand one chemical insult targeting the DAergic system. However, a combination of environmental risk factors can increase the susceptibility of DAergic neurons to injury due to compromised defense systems and this notion is supported by several experimental observations. These findings are crucial since humans are most likely to be exposed to a complex mixture of environmental agents in their surroundings, and these interactions are much more dynamic. Consequently, there might be a lot of variability in terms of potency, effects and time of onset when toxic effects are the result of exposure to multiple chemicals *versus* a single agent. Most studies supporting the multiple-hit hypothesis have focused on complex mixtures of pesticides that are often applied in overlapping geographical areas. An example of this is the simultaneous use of PQ and diethyldithiocarbamates like maneb.[139]

Dithiocarbamate (DTC) pesticides have been used for over 50 years with steadily increasing use, on a broad range of crops worldwide. They can have adverse effects on both the glutamate and DA systems and have been linked to several neurobehavioral abnormalities such as hyperactivity, ataxia, convulsions, and sedation.[147] Their mechanism of toxicity is thought to involve catalysis of DA oxidation and chelation of metals which can lead to oxidative stress *via* altered redox status.[148,149] Diethyl-DTC is a SOD inhibitor and can enhance MPTP toxicity. Maneb or MB belongs to the DTC class and is an organomanganese fungicide containing the active fungicidal component, manganese ethylene-bis-dithiocarbamate (Mn-EBDC). It is largely used in agricultural areas to control field crop pathologies. Chronic exposure to MB was shown to cause permanent parkinsonism in a 37-year old man, 2 years after exposure had ceased, supporting the role of MB in the etiology of PD.[150] Mice treated with MB and PQ are characterized with reduced motor activity and increased damage in the nigrostriatal system in a synergistic manner. The levels of DA and its metabolites and the efficiency of DA turnover were shown to increase immediately post-injection and return to control levels within 48 h. Furthermore, only animals with the combined PQ+MB exposure had the reductions in TH immunoreactivity measured 3 days after the last injection.[133,151] These results point to the fact that a combination of agents could be potential environmental risk factors for PD.

MB is thought to cross the BBB based on the observation that it can disrupt the nigrostriatal system and lead to neurochemical and behavioral changes. However, the exact transporter system has not yet been identified.[152] Although, not much is known about the mechanism of toxicity of MB, it has been demonstrated that the parent compound is responsible for exerting neurotoxicity, and not its metabolites.[153] Mn-EBDC is relatively stable but it may break down into Mn and EBDC, both of which are potential neurotoxicants. Chronic exposure to Mn can lead to brain damage, including PD-like symptoms.[154] However, when the Mn was replaced with zinc (Zn-EBDC), there was no reduction in toxicity in primary neuronal cultures, which suggests that EBDC is the primary component responsible for MB's toxicity. Additionally, EBDC was also shown to enhance MPTP neurotoxicity.[17,107] Administration of Mn-EBDC directly into rat lateral ventricles led to selective DAergic neurodegeneration and DA efflux in the striatum, comparable to that caused by MPP^+. Furthermore, in isolated rat brain mitochondria, this compound directly inhibited complex III, leading to ROS production and mitochondrial dysfunction.[155]

The case for MB's role in inducing toxicity by mitochondrial dysfunction is strengthened by the observations in primary mesencephalic cultures, where non-toxic doses of MB decreased ATP levels, inhibited NADH-linked state 3 respiration, in addition to causing mild to moderate uncoupling.[153] Non-toxic concentrations of MB also led to increases in GSH and GSSG levels in PC12 cells and primary mesencephalic cultures. Additionally, when PC12 cells were treated with L-buthionine-(S,R)-sulfoximine (BSO) to deplete cellular GSH, treatment with MB acted as an additional insult to the system and prevented the recovery of antioxidant defenses. Levels of heme-oxygenase-1 (HO-1) were also altered indicating that MB induces an oxidative stress response in DAergic neurons.[147] MB can also cause the stoichiometric loss of thiols in Trx1 and reversibly dimerizes the protein leading to inhibition of enzyme activity. Decrease in Trx activity has been shown to cause ROS production, mitochondrial dysfunction and eventual cell death,[128] providing further evidence of the role of MB in causing neurotoxicity by disrupting redox circuits. Additional proof of the involvement of oxidative stress in mediating MB-induced neurotoxicity comes from studies overexpressing SOD1 or GPx in mice. The transgenic mice showed no significant changes in behavioral and pathological markers such as lipid peroxidation and DAergic cell number in the SNpc, when treated with PQ+MB, compared to the non-transgenic controls. These data taken together, support the notion that neurodegeneration caused by MB or PQ+MB exposure involves ROS induction and oxidative stress.

3.4.5 Other Environmental Toxins

There are environmental toxins other than pesticides that have also been associated with the development of parkinsonism. Occupational exposures to metals have been suggested as a potential risk factor. Risk of developing

PD was found to be significantly elevated in individuals exposed to metals for longer than 20 years which suggest a long latency and progressive mechanism of accumulation and toxicity. These exposures were a combination of lead (Pb^{2+})-copper (Cu^{2+}), Pb^{2+}–Fe^{2+} and Fe^{2+}–Cu^{2+}.[156] One of the foremost examples of metals in relation to PD is manganese toxicity. The pathophysiology of Mn^{2+}-induced PD is different from that of idiopathic PD and therefore unlikely to contribute to the typical PD degeneration process. However, Mn^{2+} exposure could accelerate the progression of PD symptoms and therefore long-term studies are necessary with low-level Mn exposure and the development of PD.[157] Another class of environmental toxins that has been suggested as parkinsonian toxins and most studied is MPTP analogs such as tetrahydroisoquinoline (TIQ) and β-carboline (β-C). They are naturally occurring alkaloids in a variety of foods.[158] They have structures and mechanisms similar to MPTP in that they can undergo metabolic activation to form corresponding toxic metabolites. These products have the ability to interfere with mitochondrial respiration.[157]

3.5 Concluding Remarks

Parkinson's disease represents the best example among several age-related neurodegenerative disorders where mitochondrial dysfunction and oxidative stress have been strongly implicated. Even though the exact etiology is unknown, a combination of factors such as age-related lowering in antioxidant capacity, lifetime exposure to low levels of environmental agents and increased susceptibility of DAergic neurons to oxidative damage has been hypothesized to be involved the pathogenesis of PD. Decreased CI activity, increased ROS production, decreased GSH levels and increased mitochondrial dysfunction has been observed in postmortem SNpc tissue from PD patients. Plasma from PD patients have also displayed decreases in CI activity. This chapter examined the different mechanisms by which mitochondrial dysfunction can contribute to PD pathogenesis and focused on environmental agents used to model PD *in vitro* and *in vivo* in that context. A summary of the various mitochondrial mechanisms involved in the pathogenesis of toxicant-induced PD is illustrated in Figure 3.2.

Mitochondrial ROS production and consequent dysfunction have gained significant attention with respect to pesticide exposure and PD. Both epidemiological and toxicological studies have provided substantial evidence that pesticides can cause DAergic toxicity and lead to the development of PD. Each of the toxicants reviewed in this chapter gives us critical information regarding the mitochondrial mechanisms underlying neurodegeneration in PD.

However, there are certain limitations of studying these chemicals, including, but not limited to the fact that one chemical cannot produce all the characteristics of idiopathic PD. The information obtained from these models should be carefully considered along with genetics and epidemiology in order to achieve a comprehensive understanding of the disease process. Additionally, the effect on brain regions other than the nigrostriatal system

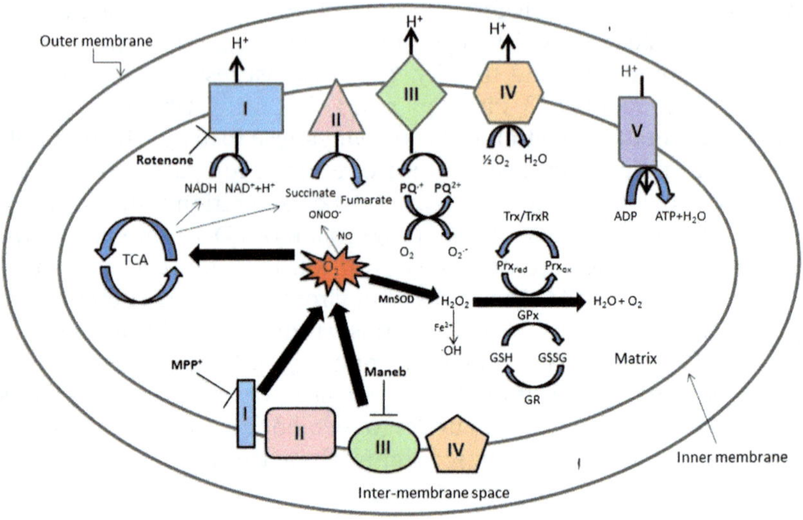

Figure 3.2 Summary of various mitochondrial mechanisms contributing to toxicant-induced PD pathogenesis. The primary role of the mitochondria is to produce ATP involving the process of electron (e^-) transfer through the different complexes of the electron-transport chain (ETC). The e^-s are provided by reducing equivalents such as NADH from the TCA cycle. During the transfer some e^-s might leak out of CI and III and react with molecular oxygen (O_2) to form superoxide radical ($O_2^{\cdot -}$). $O_2^{\cdot -}$ can then react with nitric oxide ($^{\cdot}NO$) to form peroxynitrite ($ONOO^-$) or be dismutated by the action of manganese superoxide dismutase (MnSOD) to form hydrogen peroxide (H_2O_2), which can be further detoxified to water by the action of the Trx/Prx system or the GSH system. H_2O_2 can also undergo Fenton chemistry leading to the formation of the highly reactive hydroxyl ($^{\cdot}OH$) radical that can have damaging effects on mitochondrial lipids, proteins and DNA. The various environmental toxins like paraquat (PQ), diquat (DQ), MPTP, rotenone and maneb (MB) have inhibitory effects on different ETC complexes by radical production, leading to mitochondrial dysfunction and dopaminergic cell death. PQ and DQ can also redox cycle and form $O_2^{\cdot -}$, causing mitochondrial damage and DAergic neuronal death.

should also be examined. Lastly, extrapolation of data from *in vitro* and *in vivo* models to human cases must be exercised with caution. In contrast to experimental models, humans are exposed to low, chronic doses of toxicants, over the course of several years. Therefore, the concentrations of the agents should be correlated with scrutiny in terms of their involvement in DAergic cell death.

In conclusion, despite the drawbacks of the environmental agent-induced models of PD, they have provided invaluable information that helps understand the mechanistic underpinnings of PD pathogenesis. Specifically, the role of mitochondrial ROS generation and dysfunction have been identified as crucial factors contributing to the development and progression of PD.

Additional studies are needed to further elucidate the precise mechanisms underlying the etiology of PD and develop targeted therapeutics to combat this neurodegenerative disease.

References

1. N. S. Chandel and G. R. S. Budinger, The cellular basis for diverse responses to oxygen, *Free Radical Biol. Med.*, 2007, **42**, 165–174.
2. B. Halliwell and J. Gutteridge, *Free Radicals in Biology and Medicine*, Oxford University Press, 2007.
3. E. A. Veal, A. M. Day and B. A. Morgan, Hydrogen peroxide sensing and signaling, *Mol. Cell*, 2007, **26**, 1–14.
4. M. P. Murphy, How mitochondria produce reactive oxygen species, *Biochem. J.*, 2009, **417**, 1–13.
5. Y. Collins, E. T. Chouchani, A. M. James, K. E. Menger, H. M. Cocheme and M. P. Murphy, Mitochondrial redox signalling at a glance, *J. Cell Sci.*, 2012, **125**, 801–806.
6. S. W. G. Tait and D. R. Green, Mitochondria and cell signalling, *J. Cell Sci.*, 2012, **125**, 807–815.
7. L. Kussmaul and J. Hirst, The mechanism of superoxide production by NADH: ubiquinone oxidoreductase (complex I) from bovine heart mitochondria, *Proc. Natl. Acad. Sci.*, 2006, **103**, 7607–7612.
8. K. F. Winklhofer and C. Haass, Mitochondrial dysfunction in Parkinson's disease, *Biochim. Biophys. Acta*, 2010, **1802**, 29–44.
9. Y. Zhang, O. Marcillat, C. Giulivi, L. Ernster and K. J. Davies, The Oxidative Inactivation of Mitochondrial Electron Transport Chain Components and ATPase, *J. Biol. Chem.*, 1990, **265**, 16330–16336.
10. B. A. Boveris and B. Chance, The Mitochondrial Generation of Hydrogen Peroxide, *Biochem. J.*, 1973, **134**, 707–716.
11. J. F. Turrens, Mitochondrial formation of reactive oxygen species, *J. Physiol.*, 2003, **552**, 335–344.
12. A. L. Orr, C. L. Quinlan, I. V. Perevoshchikova and M. D. Brand, A refined analysis of superoxide production by mitochondrial sn-glycerol 3-phosphate dehydrogenase, *J. Biol. Chem.*, 2012, **287**, 42921–42935.
13. I. Mavelli, A. Rigo, R. Federico, M. R. Ciriolo and G. Rotilio, Superoxide dismutase, glutathione peroxidase and catalase in developing rat brain, *Biochem. J.*, 1982, **204**, 535–540.
14. B. A. Stanley, V. Sivakumaran, S. Shi, I. McDonald, D. Lloyd, W. H. Watson, M. A. Aon and N. Paolocci, Thioredoxin reductase-2 is essential for keeping low levels of H_2O_2 emission from isolated heart mitochondria, *J. Biol. Chem.*, 2011, **286**, 33669–33677.
15. E. Cadenas, K. Davies and L. Ernster, Mitochondrial free radical generation, oxidative stress and aging, *Free Radical Biol. Med.*, 2000, **29**, 222–230.
16. J. Fukae, Y. Mizuno and N. Hattori, Mitochondrial dysfunction in Parkinson's disease, *Mitochondrion*, 2007, **7**, 58–62.

17. D. A. Drechsel and M. Patel, Role of reactive oxygen species in the neurotoxicity of environmental agents implicated in Parkinson's disease, *Free Radical Biol. Med.*, 2008, **44**, 1873–1886.
18. D. Sulzer, Multiple hit hypotheses for dopamine neuron loss in Parkinson's disease, *Trends Neurosci.*, 2007, **30**, 244–250.
19. E. Rybnikova, A. E. Damdimopoulos, J. A. Gustafsson, G. Spyrou and M. Pelto-Huikko, Expression of novel antioxidant thioredoxin-2 in the rat brain, *Eur. J. Neurosci.*, 2000, **12**, 1669–1678.
20. P. Jenner, Oxidative stress in Parkinson's disease, *Ann. Neurol.*, 2003, **53**, 26–38.
21. J. Andersen, Iron dysregulation and Parkinson's disease, *J. Alzheimers Dis.*, 2004, **6**, S47–S52.
22. J. N. Guzman, J. Sánchez-Padilla, C. S. Chan and D. J. Surmeier, Robust pacemaking in substantia nigra dopaminergic neurons, *J. Neurosci.*, 2009, **29**, 11011–11019.
23. D. J. Surmeier and P. T. Schumacker, Calcium, bioenergetics, and neuronal vulnerability in Parkinson's disease, *J. Biol. Chem.*, 2013, **288**, 10736–10741.
24. W. D. Parker Jr, J. Parks and R. Swerdlow, Complex I Deficiency in Parkinson's Disease Frontal Cortex, *Brain Res.*, 2009, **1189**, 215–218.
25. L. Bindoff, M. Birch-Machin, N. E. Cartlidge, W. D. Parker Jr and D. Turnbull, Respiratory chain abnormalities in skeletal muscle from patients with Parkinson's disease, *J. Neurol. Sci.*, 1991, **104**, 203–208.
26. M. T. Lin and M. F. Beal, Mitochondrial dysfunction and oxidative stress in neurodegenerative diseases, *Nature*, 2006, **443**, 787–795.
27. A. Trifunovic, Mitochondrial DNA and ageing, *Biochim. Biophys. Acta*, 2006, **1757**, 611–617.
28. D. Cantu, R. E. Fulton, D. A. Drechsel and M. Patel, Mitochondrial aconitase knockdown attenuates paraquat-induced dopaminergic cell death *via* decreased cellular metabolism and release of iron and H_2O_2, *J. Neurochem.*, 2011, **118**, 79–92.
29. L. P. Liang and M. Patel, Seizure-induced changes in mitochondrial redox status, *Free Radical Biol. Med.*, 2006, **40**, 316–322.
30. R. Radi, J. S. Beckman, K. M. Bush and B. A. Freeman, Peroxynitrite-induced membrane lipid peroxidation: The cytotoxic potential of superoxide and nitric oxide, *Arch. Biochem. Biophys.*, 1991, **288**, 481–487.
31. G. Petrosillo, F. M. Ruggiero and G. Paradies, Role of reactive oxygen species and cardiolipin in the release of cytochrome c from mitochondria, *FASEB J.*, 2003, **17**, 2202–2208.
32. S. Marchi, C. Giorgi, J. M. Suski, C. Agnoletto, A. Bononi, M. Bonora, E. D. Marchi, S. Missiroli, S. Patergnani, F. Poletti, A. Rimessi, J. Duszynski, M. R. Wieckowski and P. Pinton, Mitochondria-ros crosstalk in the control of cell death and aging, *J. Signal Transduction*, 2012, **2012**, 329635.
33. D. J. Pagliarini and J. Rutter, Hallmarks of a new era in mitochondrial biochemistry, *Genes Dev.*, 2013, **27**, 2615–2627.

34. Y. Mizuno, S. Ohta, M. Tanaka, S. Takamiya, K. Suzuki, T. Sato, H. Oya, T. Ozawa and Y. Kagawa, Deficiencies in complex I subunits of the respiratory chain in Parkinson's disease, *Biochem. Biophys. Res. Commun.*, 1989, **163**, 1450–1455.
35. J. T. Greenamyre, T. B. Sherer, R. Betarbet and A. V. Panov, Complex I and Parkinson's disease, *IUBMB Life*, 2001, **52**, 135–141.
36. P. C. Keane, M. Kurzawa, P. G. Blain and C. M. Morris, Mitochondrial dysfunction in Parkinson's disease, *Parkinsons Dis.*, 2011, **2011**, 716871.
37. P. M. Keeney, J. Xie, R. A. Capaldi and J. P. Bennett, Parkinson's disease brain mitochondrial complex I has oxidatively damaged subunits and is functionally impaired and misassembled, *J. Neurosci.*, 2006, **26**, 5256–5264.
38. W. Dauer and S. Przedborski, Parkinson's Disease: Mechanisms and Models, *Neuron*, 2003, **39**, 889–909.
39. R. Betarbet, T. B. Sherer, G. MacKenzie, M. Garcia-Osuna, A. V. Panov and J. T. Greenamyre, Chronic systemic pesticide exposure reproduces features of Parkinson's disease, *Nature*, 2000, **26**, 1301–1306.
40. J. W. L. Langston, P. Ballard, J. W. Tetrud and I. Irwin, Chronic Parkinsonism in humans due to a product of meperidine-analog synthesis, *Science*, 1983, **219**, 979–980.
41. J. W. Langston, The etiology of Parkinson's disease with emphasis on the MPTP story, *Neurology*, 1996, **47**, 153S–160S.
42. A. J. Lambert and M. D. Brand, Inhibitors of the quinone-binding site allow rapid superoxide production from mitochondrial NADH:ubiquinone oxidoreductase (complex I), *J. Biol. Chem.*, 2004, **279**, 39414–39420.
43. A. Y. Andreyev, Y. E. Kushnareva and A. A. Starkov, Mitochondrial metabolism of reactive oxygen species, *Biochemistry*, 2005, **70**, 200–214.
44. L. Devi, V. Raghavendran, B. M. Prabhu, N. G. Avadhani and H. Anandatheerthavarada, Mitochondrial import and accumulation of alpha-synuclein impair complex I in human dopaminergic neuronal cultures and Parkinson disease brain, *J. Biol. Chem.*, 2008, **283**, 9089–9100.
45. S. Büttner, A. Bitto, J. Ring, M. Augsten, P. Zabrocki, T. Eisenberg, H. Jungwirth, S. Hutter, D. Carmona-Gutierrez, G. Kroemer, J. Winderickx and F. Madeo, Functional mitochondria are required for alpha-synuclein toxicity in aging yeast, *J. Biol. Chem.*, 2008, **283**, 7554–7560.
46. D. D. Song, C. W. Shults, A. Sisk, E. Rockenstein and E. Masliah, Enhanced substantia nigra mitochondrial pathology in human alpha-synuclein transgenic mice after treatment with MPTP, *Exp. Neurol.*, 2004, **186**, 158–172.
47. S. J. Chinta, J. K. Mallajosyula, A. Rane and J. K. Andersen, Mitochondrial α-synuclein accumulation impairs complex I function in dopaminergic neurons and results in increased mitophagy *in vivo*, *Neurosci. Lett.*, 2010, **486**, 235–239.
48. M. Hsu, B. Srinivas, J. Kumar, R. Subramanian and J. Andersen, Glutathione depletion resulting in selective mitochondrial complex I inhibition

in dopaminergic cells is via an NO-mediated pathway not involving peroxynitrite: implications for Parkinson's disease, *J. Neurochem.*, 2005, **92**, 1091–1103.
49. E. R. Taylor, F. Hurrell, R. J. Shannon, T. Lin, J. Hirst and M. P. Murphy, Reversible glutathionylation of complex I increases mitochondrial superoxide formation, *J. Biol. Chem.*, 2003, **278**, 19603–19610.
50. D. W. Lee, D. Kaur, S. J. Chinta, S. Rajagopalan and J. K. Andersen, A disruption in iron-sulfur biogenesis *via* inhibiton of mitochondrial dithiol glutaredoxin 2 may contribute to mitochondrial and cellular iron dysregulation in mammalian glutathione-depleted dopaminergic cells: implications for Parkinson's disease, *Antioxid. Redox Signaling*, 2009, **11**, 2083–2094.
51. S. J. Chinta and J. K. Andersen, Nitrosylation and nitration of mitochondrial complex I in Parkinson's disease, *Free Radic. Res.*, 2011, **45**, 53–58.
52. J. Zhu and C. T. Chu, Mitochondrial dysfunction in Parkinson's disease, *J. Alzheimers Dis.*, 2010, **20**(suppl. 2), S325–S334.
53. P. A. Trimmer, R. H. Swerdlow, J. K. Parks, P. Keeney, J. P. Bennett Jr, S. W. Miller, R. E. Davis and W. D. Parker Jr, Abnormal mitochondrial morphology in sporadic Parkinson's and Alzheimer's disease cybrid cell lines, *Exp. Neurol.*, 2000, **162**, 37–50.
54. J. P. Sheehan, R. H. Swerdlow, W. D. Parker, S. W. Miller, R. E. Davis and J. B. Tuttle, Altered Calcium Homeostasis in Cells Transformed by Mitochondria from Individuals with Parkinson's Disease, *J. Neurochem.*, 2002, **68**, 1221–1233.
55. X. J. Wang and J. X. Xu, Possible involvement of Ca2+ signaling in rotenone-induced apoptosis in human neuroblastoma SH-SY5Y cells, *Neurosci. Lett.*, 2005, **376**, 127–132.
56. H. J. Visch, G. A. Rutter, W. J. Koopman, J. B. Koenderink, S. Verkaart, T. de Groot, A. Varadi, K. J. Mitchell, L. P. van den Heuvel, J. A. Smeitink and P. H. Willems, Inhibition of mitochondrial Na+–Ca2+ exchange restores agonist-induced ATP production and Ca2+ handling in human complex I deficiency, *J. Biol. Chem.*, 2004, **279**, 40328–40336.
57. D. G. Nicholls and S. L. Budd, Neuronal excitotoxicity: the role of mitochondria, *BioFactors*, 1998, **8**, 287–299.
58. M. E. Haque, K. J. Thomas, C. D'Souza, S. Callaghan, T. Kitada, R. S. Slack, P. Fraser, M. R. Cookson, A. Tandon and D. S. Park, Cytoplasmic Pink1 activity protects neurons from dopaminergic neurotoxin MPTP, *Proc. Natl. Acad. Sci.*, 2008, **105**, 1716–1721.
59. A. Wood-Kaczmar, S. Gandhi, Z. Yao, A. S. Y. Abramov, E. A. Miljan, G. Keen, L. Stanyer, I. Hargreaves, K. Klupsch, E. Deeas, J. Downward, L. Mansfield, P. Jat, J. Taylor, S. Heales, M. R. Duchen, D. Latchman, S. J. Tabrizi and N. W. Wood, PINK1 is necessary for long term survival and mitochondrial function in human dopaminergic neurons, *PLoS One*, 2008, **3**, e2455.
60. J. C. Paterna, A. Leng, E. Weber, J. Feldon and H. Büeler, DJ-1 and Parkin modulate dopamine-dependent behavior and inhibit MPTP-induced nigral dopamine neuron loss in mice, *Mol. Ther.*, 2007, **15**, 698–704.

61. R. H. Kim, P. D. Smith, H. Aleyasin, S. Hayley, M. P. Mount, S. Pownall, A. Wakeham, A. J. You-Ten, S. K. Kalia, P. Horne, D. Westaway, A. M. Lozano, H. Anisman, D. S. Park and T. W. Mak, Hypersensitivity of DJ-1-deficient mice to 1-methyl-4-phenyl-1,2,3,6-tetrahydropyrindine (MPTP) and oxidative stress, *Proc. Natl. Acad. Sci.*, 2005, **102**, 5215–5220.
62. T. Taira, Y. Saito, T. Niki, S. M. Iguchi-Ariga, K. Takahashi and H. Ariga, DJ-1 has a role in antioxidative stress to prevent cell death, *EMBO Rep.*, 2004, **5**, 213–218.
63. L. Zhang, M. Shimoji, B. Thomas, D. J. Moore, S. W. Yu, N. I. Marupudi, R. Torp, I. A. Torgner, O. P. Ottersen, T. M. Dawson and V. L. Dawson, Mitochondrial localization of the Parkinson's disease related protein DJ-1: Implications for pathogenesis, *Hum. Mol. Genet.*, 2005, **14**, 2063–2073.
64. E. Andres-Mateos, C. Perier, L. Zhang, B. Blanchard-Fillion, T. M. Greco, B. Thomas, H. S. Ko, M. Sasaki, H. Ischiropoulos, S. Przedborski, T. M. Dawson and V. L. Dawson, DJ-1 gene deletion reveals that DJ-1 is an atypical peroxiredoxin-like peroxidase, *Proc. Natl. Acad. Sci.*, 2007, **104**, 14807–14812.
65. L. J. Hsu, Y. Sagara, A. Arroyo, E. Rockenstein, A. Sisk, M. Mallory, J. Wong, T. Takenouchi, M. Hashimoto and E. Masliah, Alpha-Synuclein promotes mitochondrial deficit and oxidative stress, *Am. J. Pathol.*, 2000, **157**, 401–410.
66. P. Klivenyi, D. Siwek, G. Gardian, L. Yang, A. Starkov, C. Cleren, R. J. Ferrante, N. W. Kowall, A. Abeliovich and M. F. Beal, Mice lacking alpha-synuclein are resistant to mitochondrial toxins, *Neurobiol. Dis.*, 2006, **21**, 541–548.
67. W. Dauer, N. Kholodilov, M. Vila, A. C. Trillat, R. Goodchild, K. E. Larsen, R. Staal, K. Tieu, Y. Schmitz, C. A. Yuan, M. Rocha, V. Jackson-Lewis, S. Hersch, D. Sulzer, S. Przedborski, R. Burke and R. Hen, Resistance of alpha-synuclein null mice to the parkinsonian neurotoxin MPTP, *Proc. Natl. Acad. Sci.*, 2002, **99**, 14524–14529.
68. J. Bové, D. Prou, C. Perier and S. Przedborski, Toxin-induced models of Parkinson's disease, *NeuroRx*, 2005, **2**, 484–494.
69. R. Deumens, A. Blokland and J. Prickaerts, Modeling Parkinson's disease in rats: an evaluation of 6-OHDA lesions of the nigrostriatal pathway, *Exp. Neurol.*, 2002, **175**, 303–317.
70. D. Blum, S. Torch, N. Lambeng, M. F. Nissou, A. L. Benabid, R. Sadoul and J. M. Verna, Molecular pathways involved in the neurotoxicity of 6-OHDA, dopamine and MPTP: contribution to the apoptotic theory in Parkinson's disease, *Prog. Neurobiol.*, 2001, **65**, 135–172.
71. R. Soto-Otero, E. Mendez-Alvarez, A. Hermida-Ameijeiras, A. M. Munoz-Patino and J. L. Labandeira-Garcia, Autoxidation and neurotoxicity of 6-hydroxydopamine in the presence of some antioxidants: potential implication in relation to the pathogenesis of Parkinson's disease, *J. Neurochem.*, 2000, **74**, 1605–1612.
72. K. Yamada, H. Umegaki, I. Maezawa, A. Iguchi, T. Kameyama and T. Nabeshima, Possible Involvement of Catalase in the Protective Effect of

Interleukin-6 Against 6-Hydroxydopamine Toxicity in PC12 Cells, *Brain Res. Bull.*, 1997, **43**, 573–577.

73. D. Blum, S. Torch, M. F. Nissou, A. L. Benabid and J. M. Verna, Extracellular toxicity of 6-hydroxydopamine on PC12 cells, *Neurosci. Lett.*, 2000, **283**, 193–196.
74. M. Asanuma, H. Hirata and J. Cadet, Attenuation of 6-hydroxydopamine-induced dopaminergic nigrostriatal lesions in superoxide dismutase transgenic mice, *Neuroscience*, 1998, **85**, 907–917.
75. J. C. Bensadoun, O. Mirochnitchenko, M. Inouye, P. Aebischer and A. D. Zurn, Attenuation of 6-OHDA-induced neurotoxicity in glutathione peroxidase transgenic mice, *Eur. J. Neurosci.*, 1998, **10**, 3231–3236.
76. G. Seitz and H. Stegmann, Neuroblastoma cells expressing the noradrenaline transporter are destroyed more selectively by 6-fluorodopamine than by 6-hydroxydopamine, *J. Neurochem.*, 2000, **75**, 511–520.
77. Y. He, P. S. Thong, T. Lee, S. K. Leong, C. Y. Shi, P. T. Wong, S. Y. Yuan and F. Watt, Increased iron in the substantia nigra of 6-OHDA induced parkinsonian rats: a nuclear microscopy study, *Brain Res.*, 1996, **735**, 149–153.
78. Y. Y. Glinka and M. B. H. Youdim, Inhibition of mitochondrial complexes I and IV by 6-hydroxydopamine, *Eur. J. Pharmacol.*, 1995, **292**, 329–332.
79. A. Storch, A. Kaftan, K. Burkhardt and J. Schwarz, 6-Hydroxydopamine toxicity towards human SH-SY5Y dopaminergic neuroblastoma cells: independent of mitochondrial energy metabolism, *J. Neural Transm.*, 2000, **107**, 0281–0293.
80. J. Lotharius, L. L. Dugan and K. L. O'Malley, Distinct mechanisms underlie neurotoxin-mediated cell death in cultured dopaminergic neurons, *J. Neurosci.*, 1999, **19**, 1284–1293.
81. W. J. Nicklas, I. Vyas and R. E. Heikkila, Inhibition of NADH-linked oxidation in brain mitochondria by 1-methyl-4-phenyl-pyridine, a metabolite of the neurotoxin, 1-methyl-4-phenyl-1,2,5,6-tetrahydropyridine, *Life Sci.*, 1985, **36**, 2503–2508.
82. R. E. Heikkila, A. Hess and R. C. Duvoisin, Dopaminergic neurotoxicity of 1-methyl-4-phenyl-1,2,5,6-tetrahydropyridine in mice, *Science*, 1984, **224**, 1451–1453.
83. E. Hasegama, K. Takeshige, T. Oishi, Y. Murai and S. Minakami, 1-Methyl-4-phenylpyridinium (MPP+) induces NADH-dependent superoxide formation and enhances NADH-dependent lipid peroxidation in bovine heart submitochondrial particles, *Biochem. Biophys. Res. Commun.*, 1990, **170**, 1049–1055.
84. P. Klivenyi, D. St Clair, M. Wermer and H. C. Yen, Manganese Superoxide Dismutase Overexpression Attenuates MPTP Toxicity, *Neurobiol. Dis.*, 1998, **5**, 253–258.
85. O. A. Andreassen, R. J. Ferrante, A. Dedeoglu, D. W. Albers, P. Klivenyi, E. J. Carlson, C. J. Epstein and M. F. Beal, Mice with a partial deficiency of manganese superoxide dismutase show increased vulnerability to the

mitochondrial toxins malonate, 3-nitropropionic acid, and MPTP, *Exp. Neurol.*, 2001, **167**, 189–195.

86. L. Liang and M. Patel, Iron-sulfur enzyme mediated mitochondrial superoxide toxicity in experimental Parkinson's disease, *J. Neurochem.*, 2004, **90**, 1076–1084.
87. J. N. Chacon, M. R. Chedekel, E. J. Land and T. G. Truscott, Chemically induced Parkinson's disease: Intermediates in the oxidation of 1-methyl-4-phenyl-1,2,3,6-tetrahydropyridine to the 1-methyl-4-phenyl-pyridinium ion, *Biochem. Biophys. Res. Commun.*, 1987, **144**, 957–964.
88. J. D. Adams, L. K. Klaidman and E. Cadenas, MPP+ redox cycling: A new mechanism involving hydride transfer, *Ann. N. Y. Acad. Sci.*, 1992, **648**, 239–240.
89. L. K. Klaidman, J. D. Adams Jr, A. C. Leung, S. S. Kim and E. Cadenas, Redox cycling of MPP+: Evidence for a new mechanism involving hydride transfer with xanthine oxidase, aldehyde dehydrogenase and lipoamide dehydrogenase, *Free Radical Biol. Med.*, 1993, **15**, 169–179.
90. Y. Kitamura, T. Kosaka, J. I. Kakimura, Y. Matsuoka, Y. Kohno, Y. Nomura and T. Taniguchi, Protective effects of the antiparkinsonian drugs talipexole and pramipexole against 1-methyl-4-phenylpyridinium-induced apoptotic death in human neuroblastoma SH-SY5Y cells, *Mol. Pharmacol.*, 1998, **1054**, 1046–1054.
91. Z. Rossetti, A. Sotqiu, D. Sharp, M. Hadjiconstantinou and N. Neff, 1-Methyl-4-phenyl-1,2,3,6-tetrahydropyridine (MPTP) and free radicals *in vitro*, *Biochem. Pharmacol.*, 1988, **37**, 4573–4574.
92. M. S. Desole, G. Esposito, L. Fresu, R. Migheli, P. Enrico, M. Miele, G. De Natale and E. Miele, Correlation between 1-methyl-4-phenylpyridinium ion (MPP+) levels, ascorbic acid oxidation and glutathione levels in the striatal synaptosomes of the 1-methyl-4-phenyl-l,2,3-6-tetrahydropyridine (MPTP)-treated rat, *Neurosci. Lett.*, 1993, **161**, 121–123.
93. S. Pennathur, V. Jackson-Lewis, S. Przedborski and J. W. Heinecke, Mass spectrometric quantification of 3-nitrotyrosine, ortho-tyrosine, and o,o'-dityrosine in brain tissue of a model of oxidative stress in Parkinson's Disease, *J. Biol. Chem.*, 1999, **274**, 34621–34628.
94. D. A. Drechsel, L. P. Liang and M. Patel, 1-Methyl-4-Phenylpyridinium-induced alterations of glutathione status in immortalized rat dopaminergic neurons, *Toxicol. Appl. Pharmacol.*, 2007, **220**, 341–348.
95. P. Klivenyi and O. Andreassen, Mice deficient in cellular glutathione peroxidase show increased vulnerability to malonate, 3-nitropropionic acid, and 1-methyl-4-phenyl-1,2,5,6-tetrahydropyridine, *J. Neurosci.*, 2000, **20**, 1–7.
96. J. Zhang, D. Graham, T. Montine and Y. S. Ho, Enhanced N-methyl-4-phenyl-1,2,3,6-tetrahydropyridine toxicity in mice deficient in CuZn-superoxide dismutase or glutathione peroxidase, *J. Neuropathol. Exp. Neurol.*, 2000, **59**, 53–61.
97. H. Mochizuki, H. Imai, K. Endo, K. Yokomizo, Y. Murata, N. Hattori and Y. Mizuno, Iron accumulation in the substantia nigra of

1-methyl-4-phenyl-1,2,3,6-tetrahydropyridine (MPTP)-induced hemi-parkinsonian monkeys, *Neurosci. Lett.*, 1994, **168**, 251–253.
98. J. Lan and D. H. Jiang, Desferrioxamine and vitamin E protect against iron and MPTP-induced neurodegeneration in mice, *J. Neural Transm.*, 1997, **104**, 469–481.
99. P. Teismann, K. Tieu, D. K. Choi, D. C. Wu, A. Naini, S. Hunot, M. Vila, V. Jackson-Lewis and S. Przedborski, Cyclooxygenase-2 is instrumental in Parkinson's disease neurodegeneration, *Proc. Natl. Acad. Sci.*, 2003, **100**, 5473–5478.
100. L. Turski, K. Bressler, L. J. Rettig, P. A. Loschmann and H. Wachtel, Protection of substantia nigra from MPP+ neurotoxicity by *N*-methyl-D-aspartate antagonists, *Nature*, 1991, **349**, 414–418.
101. A. Kupsch, J. Sautter, J. Schwarz, P. Riederer, M. Gerlach and W. H. Oertel, 1-Methyl-4-phenyl-1,2,3,6-tetrahydropyridine-induced neurotoxicity in non-human primates is antagonized by pretreatment with nimodipine at the nigral, but not at the striatal level, *Brain Res.*, 1996, **741**, 185–196.
102. S. Przedborski, V. Jackson-Lewis, R. Yokoyama, T. Shibata, V. L. Dawson and T. M. Dawson, Role of neuronal nitric oxide in 1-methyl-4-phenyl-1,2,3,6-tetrahydropyridine (MPTP)-induced dopaminergic neurotoxicity, *Proc. Natl. Acad. Sci.*, 1996, **93**, 4565–4571.
103. T. Dehmer, J. Lindenau, S. Haid, J. Dichgans and J. B. Schulz, Deficiency of inducible nitric oxide synthase protects against MPTP toxicity *in vivo*, *J. Neurochem.*, 2008, **74**, 2213–2216.
104. I. Kurkowska-Jastrze, A. Wronska, M. Kohutnicka, A. Czlonkowski and A. Czlonkowska, The inflammatory reaction following 1-methyl-4-phenyl-1,2,3,6-tetrahydropyridine intoxication in mouse, *Exp. Neurol.*, 1999, **61**, 50–61.
105. C. Rios and R. Tapia, Changes in lipid peroxidation induced by 1-methyl-4-phenyl-1,2,3,6-tetrahydropyridine and 1-methyl-4-phenylpyridinium in mouse brain homogenates, *Neurosci. Lett.*, 1987, **77**, 321–326.
106. T. B. Sherer, R. Betarbet, C. M. Testa, B. B. Seo, J. R. Richardson, J. H. Kim, G. W. Miller, T. Yagi, A. Matsuno-Yagi and J. T. Greenamyre, Mechanism of toxicity in rotenone models of Parkinson's disease, *J. Neurosci.*, 2003, **23**, 10756–10764.
107. V. N. Uversky, Neurotoxicant-induced animal models of Parkinson's disease: understanding the role of rotenone, maneb and paraquat in neurodegeneration, *Cell Tissue Res.*, 2004, **318**, 225–241.
108. T. B. Sherer, R. Betarbet, J. H. Kim and J. T. Greenamyre, Selective microglial activation in the rat rotenone model of Parkinson's disease, *Neurosci. Lett.*, 2003, **341**, 87–90.
109. M. F. Beal, Mitochondria, oxidative damage and inflammation in Parkinson's disease, *Ann. N. Y. Acad. Sci.*, 2003, **991**, 120–131.
110. S. Tada-Oikawa, Y. Hiraku, M. Kawanishi and S. Kawanishi, Mechanism for generation of hydrogen peroxide and change of mitochondrial membrane potential during rotenone-induced apoptosis, *Life Sci.*, 2003, **73**, 3277–3288.

111. H. Gao, J. Hong, W. Zhang and B. Liu, Synergistic dopaminergic neurotoxicity of the pesticide rotenone and inflammogen lipopolysaccharide: relevance to the etiology of Parkinson's disease, *J. Neurosci.*, 2003, **23**, 1228–1236.
112. I. Marey-Semper, M. Gelam and M. Levi-Strauss, A selective toxicity toward cultured mesencephalic dopaminergic neurons is induced by the synergistic effects of energetic metabolism impairment and NMDA receptor, *J. Neurosci.*, 1995, **15**, 5912–5918.
113. H. Gao, J. Hong, W. Zhang and B. Liu, Distinct role for microglia in rotenone-induced degeneration of dopaminergic neurons, *J. Neurosci.*, 2002, **22**, 782–790.
114. G. U. Höglinger, J. Feger, A. Prigent, P. P. Michel, K. Parain, P. Champy, M. Ruberg, W. H. Oertel and E. C. Hirsch, Chronic systemic complex I inhibition induces a hypokinetic multisystem degeneration in rats, *J. Neurochem.*, 2003, **84**, 491–502.
115. T. Roberts, J. Dyson and M. Lane, Deactivation of the biological activity of paraquat in the soil environment: a review of long-term environmental fate, *J. Agric. Food Chem.*, 2002, **50**, 3623–3631.
116. M. Corasaniti and M. Strongoli, Paraquat: a useful tool for the *in vivo* study of mechanisms of neuronal cell death, *Pharmacol. Toxicol.*, 1998, **83**, 1–7.
117. C. Berry, C. La Vecchia and P. Nicotera, Paraquat and Parkinson's disease, *Cell Death Differ.*, 2010, **17**, 1115–1125.
118. H. H. Liou, M. C. Tsai, C. J. Chen, J. S. Jeng, Y. C. Chang, S. Y. Chen and R. C. Chen, Environmental risk factors and Parkinson's disease: A case-control study in Taiwan, *Neurology*, 1997, **48**, 1583–1588.
119. A. Busch, U. Karbach, D. Miska, V. Gorboulev, A. Akhoundova, C. Volk, P. Arndt, J. C. Ulzheimer, M. S. Sonders, C. Baumann, S. Waldegger, F. Lang and H. Loepsell, Human neurons express the polyspecific cation transporter hOCT2, which translocates monoamine neurotransmitters, amantadine, and memantine, *Mol. Pharmacol.*, 1998, **352**, 342–352.
120. R. F. Burk, R. A. Lawrence and J. M. Lane, Liver necrosis and lipid peroxidation in the rat as the result of paraquat and diquat administration, *J. Clin. Invest.*, 1980, **65**, 1–8.
121. B. J. Day, M. Patel, L. Calavetta, L. Y. Chang and J. S. Stamler, A mechanism of paraquat toxicity involving nitric oxide synthase, *Proc. Natl. Acad. Sci.*, 1999, **96**, 12760–12765.
122. B. J. Day and J. D. Crapo, A metalloporphyrin superoxide dismutase mimetic protects against paraquat-induced lung injury *in vivo*, *Toxicol. Appl. Pharmacol.*, 1996, **100**, 94–100.
123. P. R. Castello, D. A. Drechsel and M. Patel, Mitochondria are a major source of paraquat-induced reactive oxygen species production in the brain, *J. Biol. Chem.*, 2007, **282**, 14186–14193.
124. M. G. Purisai, A. L. McCormack, S. Cuminc, J. Li, M. Z. Isla and D. A. Di Monte, Microglial activation as a priming event leading to paraquat-induced dopaminergic cell degeneration, *Neurobiol. Dis.*, 2008, **25**, 392–400.

125. T. Fukushima, K. Tanaka, H. Lim and M. Moriyama, Mechanism of cytotoxicity of paraquat, *Environ. Health Prev. Med.*, 2002, **81**, 89–94.
126. M. Patel, B. J. Day, J. D. Crapo, I. Fridovich and J. O. McNamara, Requirement for superoxide in excitotoxic cell death, *Neuron*, 1996, **16**, 345–355.
127. D. A. Drechsel and M. Patel, Respiration-dependent H2O2 removal in brain mitochondria *via* the thioredoxin/peroxiredoxin system, *J. Biol. Chem.*, 2010, **285**, 27850–27858.
128. P. Lopert, B. J. Day and M. Patel, Thioredoxin reductase deficiency potentiates oxidative stress, mitochondrial dysfunction and cell death in dopaminergic cells, *PLoS One*, 2012, **7**, e50683.
129. J. Tamarit, E. Cabiscol and J. Ros, Identification of the major oxidatively damaged proteins in Escherichia coli cells exposed to oxidative stress, *J. Biol. Chem.*, 1998, **273**, 3027–3032.
130. W. Cheng, Y. Fu, J. Porres, D. Ross and X. Lei, Selenium-dependent cellular glutathione peroxidase protects mice against a pro-oxidant-induced oxidation of NADPH, NADH, lipids, and protein, *FASEB J.*, 1999, **13**, 1467–1475.
131. R. A. González-Polo, A. Rodriguez-Martin, J. M. Moran, M. Niso, G. Soler and J. M. Fuentes, Paraquat-induced apoptotic cell death in cerebellar granule cells, *Brain Res.*, 2004, **1011**, 170–176.
132. A. L. McCormack, J. G. Atienza, L. C. Johnston, J. K. Andersen, S. Vu and D. A. Di Monte, Role of oxidative stress in paraquat-induced dopaminergic cell degeneration, *J. Neurochem.*, 2005, **93**, 1030–1037.
133. M. Thiruchelvam, E. Richfield, R. B. Baggs, A. W. Tank and D. Cory-Slechta, The nigrostriatal dopaminergic system as a preferential target of repeated exposures to combined paraquat and maneb: implications for Parkinson's disease, *J. Neurosci.*, 2000, **20**, 9207–9214.
134. V. Mollace, C. Muscoli, E. Palma and T. Granato, The role of oxidative stress in paraquat-induced neurotoxicity in rats: protection by non peptidyl superoxide dismutase mimetic, *Neurosci. Lett.*, 2003, **335**, 163–166.
135. J. Peng, F. F. Stevenson, S. R. Doctrow and J. K. Andersen, Superoxide dismutase/catalase mimetics are neuroprotective against selective paraquat-mediated dopaminergic neuron death in the substantial nigra: implications for Parkinson disease, *J. Biol. Chem.*, 2005, **280**, 29194–29198.
136. L. P. Liang, T. J. Kavanagh and M. Patel, Glutathione deficiency in Gclm null mice results in complex I inhibition and dopamine depletion following paraquat administration, *Toxicol. Sci.*, 2013, **134**, 366–373.
137. K. Shimizu, K. Matsubara, K. Ohtaki and H. Shiono, Paraquat leads to dopaminergic neural vulnerability in organotypic midbrain culture, *Neurosci. Res.*, 2003, **46**, 523–532.
138. J. Ara, S. Przedborski, A. B. Naini, V. Jackson-Lewis, R. R. Trifiletti, J. Horwitz and H. Ischiropoulos, Inactivation of tyrosine hydroxylase by nitration following exposure to peroxynitrite and 1-methyl-4-phenyl-1,2,3, 6-tetrahydropyridine (MPTP), *Proc. Natl. Acad. Sci.*, 1998, **95**, 7659–7663.

139. R. J. Dinis-Oliveira, F. Remiao, H. Carmo, J. A. Duarte, A. S. Navarro, M. L. Bastos and F. Carvalho, Paraquat exposure as an etiological factor of Parkinson's disease, *Neurotoxicology*, 2006, **27**, 1110–1122.
140. T. L. Perry, V. W. Yong, R. A. Wall and K. Jones, Paraquat and two endogenous analogues of the neurotoxic substance *N*-methyl-4-phenyl-1,2,3,6-tetrahydropyridine do not dmaage dopaminergic nigrostriatal neurons in the mouse, *Neurosci. Lett.*, 1986, **69**, 285–289.
141. A. Brooks, C. Chadwick, H. Gelbard, D. Cory-Slechta and H. Federoff, Paraquat elicited neurobehavioral syndrome caused by dopaminergic neuron loss, *Brain Res.*, 1999, **823**, 1–10.
142. A. B. Manning-Bog, A. L. McCormack, J. Li, V. N. Uversky, A. L. Fink and D. A. Di Monte, The herbicide paraquat causes up-regulation and aggregation of alpha-synuclein in mice: paraquat and alpha-synuclein, *J. Biol. Chem.*, 2002, **277**, 1641–1644.
143. G. M. Jones and J. A. Vale, Mechanisms of toxicity, clinical features, and management of diquat poisoning: a review, *Clin. Toxicol.*, 2000, **38**, 123–128.
144. M. Tomita, Comparison of one-electron reduction activity against the bipyridylium herbicides, paraquat and diquat, in microsomal and mitochondrial fractions of liver, lung and kidney (*in vitro*), *Biochem. Pharmacol.*, 1991, **42**, 303–309.
145. D. A. Drechsel and M. Patel, Differential contribution of the mitochondrial respiratory chain complexes to reactive oxygen species production by redox cycling agents implicated in parkinsonism, *Toxicol. Sci.*, 2009, **112**, 427–434.
146. S. S. Karuppagounder, M. Ahuja, M. Buabeid, K. Parameshwaran, E. Abdel-Rehman, V. Suppiramaniam and M. Dhanasekaran, Investigate the chronic neurotoxic effects of diquat, *Neurochem. Res.*, 2012, **37**, 1102–1111.
147. B. K. Barlow, D. W. Lee, D. A. Cory-Slechta and L. A. Opanashuk, Modulation of antioxidant defense systems by the environmental pesticide maneb in dopaminergic cells, *Neurotoxicology*, 2005, **26**, 63–75.
148. V. A. Fitsanakis, V. Amarnath, J. T. Moore, K. S. Montine, J. Zhang and T. J. Montine, Catalysis of catechol oxidation by metal-dithiocarbamate complexes in pesticides, *Free Radical Biol. Med.*, 2002, **33**, 1714–1723.
149. S. Orrenius, C. Nobel, D. Van den Dobbelsteen, M. Burkitt and A. Slater, Dithiocarbamates and the redox regulation of cell death, *Biochem. Soc. Trans.*, 1995, **24**, 197–202.
150. G. Meco, V. Bonifati, N. Vanacore and E. Fabrizio, Parkinsonism after chronic exposure to the fungicide maneb (manganese ethylene-bis-dithiocarbamate), *Scand. J. Work, Environ. Health*, 1994, **20**, 301–305.
151. M. Thiruchelvam, B. Brockel, E. Richfield, R. Baggs and D. Cory-Slechta, Potentiated and preferential effects of combined paraquat and maneb on nigrostriatal dopamine systems: environmental risk factors for Parkinson's disease? *Brain Res.*, 2000, **873**, 225–234.

152. M. Thiruchelvam, E. K. Richfield, B. M. Goodman, R. B. Baggs and D. A. Cory-Slechta, Developmental Exposure to the Pesticides Paraquat and Maneb and the Parkinson's Disease Phenotype, *Neurotoxicology*, 2002, **23**, 621–633.
153. L. M. Domico, G. D. Zeevalk, L. P. Bernard and K. R. Cooper, Acute neurotoxic effects of mancozeb and maneb in mesencephalic neuronal cultures are associated with mitochondrial dysfunction, *Neurotoxicology*, 2006, **27**, 816–825.
154. G. Gerber, A. Leonard and P. Hantson, Carcinogenicity, mutagenicity and teratogenicity of manganese compounds, *Crit. Rev. Oncol. Hematol.*, 2002, **42**, 25–34.
155. J. Zhang, V. A. Fitsanakis, G. Gu, D. Jing, M. Ao, V. Amarnath and T. J. Montine, Manganese ethylene-bis-dithiocarbamate and selective dopaminergic neurodegeneration in rat: a link through mitochondrial dysfunction, *J. Neurochem.*, 2003, **84**, 336–346.
156. D. A. Di Monte, The role of environmental agents in Parkinson's disease, *Clin. Neurosci. Res.*, 2001, **1**, 419–426.
157. D. A. Di Monte, M. Lavasani and A. B. Manning-Bog, Environmental factors in Parkinson's disease, *Neurotoxicology*, 2002, **23**, 487–502.
158. Y. Makino, S. Ohta, O. Tachikawa and M. Hirobe, Presence of tetrahydroisoquinoline and 1-methyl-tetrahydro-isoquioline in foods: Compounds related to Parkinson's disease, *Life Sci.*, 1988, **43**, 373–378.

CHAPTER 4

Dopamine Metabolism and the Generation of a Reactive Aldehyde

JOSEPHINE H. SCHAMP AND JONATHAN A. DOORN*

Department of Pharmaceutical Sciences and Experimental Therapeutics, College of Pharmacy, The University of Iowa, Iowa City, IA 52242-1112, USA
*E-mail: jonathan-doorn@uiowa.edu

4.1 The Life of Dopamine: Synthesis, Storage and Metabolism

Dopamine (DA) was once considered to be simply a precursor in the biosynthetic scheme for norepinephrine (NE) and epinephrine (E) but is now known to be a vital neurotransmitter important for numerous physiological functions and behavior, such as motor control, reward and addiction and hormone (prolactin) release.[1]

DA biosynthesis primarily occurs at the nerve terminal[2,3] and begins with the hydroxylation of L-tyrosine to L-3,4-dihydroxyphenylalanine (L-DOPA), catalyzed by the enzyme tyrosine hydroxylase (TH). L-Aromatic amino acid decarboxylase catalyzes decarboxylation of L-DOPA to DA. The newly synthesized DA is rapidly taken up into vesicles *via* the action of the vesicular monoamine transporter (VMAT2), for the purpose of synaptic signalling but also, importantly, cellular protection against reactive intermediates that can be generated *via* free, cytosolic DA. The lower pH of the vesicle prevents the

Issues in Toxicology No. 34
Oxidative Stress and Redox Signalling in Parkinson's Disease
Edited by Rodrigo Franco, Jonathan A. Doorn and Jean-Christophe Rochet

Published by the Royal Society of Chemistry, www.rsc.org

oxidation of DA to a toxic quinone and the production of reactive oxygen species (ROS) such as superoxide anion. Catabolism of DA generally takes place in the cells in which the neurotransmitter is produced and occurs for free, nonvesicular DA (*i.e.* cytosolic).[2–4] It is thought that DA catabolism occurs primarily for the DA molecules that leak from vesicles, passively or *via* the action of toxicants; in addition, reuptake of DA from the synapse may also contribute to this pathway of degradation.[2,4]

Shown in Figure 4.1 is the metabolism of DA, which involves: (1) oxidative deamination, (2) carbonyl oxidation or reduction; and (3) catechol methylation.[2,5] In the first step, cytosolic DA undergoes oxidative deamination, a reaction catalyzed by the flavin-containing enzyme monoamine oxidase (MAO), which is attached to the outer mitochondrial membrane on the cytosolic side and requires molecular oxygen for catalytic activity.[2,4] There are two forms of MAO (*i.e.* A and B) that are known, and the contribution of each toward neurotransmitter metabolism has been extensively studied in mammals and previously reported.[6–8] Both MAO A and B exist in the mammalian brain in distinct locations, and it is the A isoform that is found in the dopaminergic terminals.[9–11] While MAO B is mainly located in glial cells, the contribution

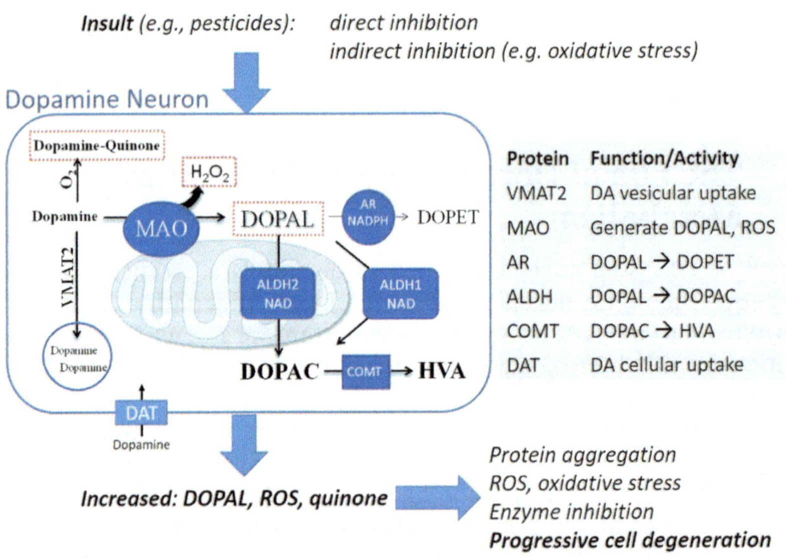

Figure 4.1 Metabolism and trafficking of DA. DA undergoes metabolism *via* MAO, yielding DOPAL and hydrogen peroxide, followed by oxidation to DOPAC *via* ALDH enzymes using the cofactor NAD. As a minor pathway, DOPAL is reduced to DOPET *via* AR and NADPH. DOPAC is methylated *via* COMT, yielding HVA. Cellular and vesicular uptake of DA is accomplished by DAT and VMAT2, respectively. Oxidation of cytosolic DA yields the quinone. The disruption of DA metabolism and/or trafficking, directly or indirectly, *via* insult (*e.g.*, pesticide) is predicted to yield elevated DOPAL, ROS and quinones that could damage the cell by various mechanisms.

of MAO B toward DA deamination should not be underestimated or minimized as the activity of this isoform has been demonstrated to be important for catecholamine metabolism, especially after L-DOPA treatment.[7,8] In addition, the effectiveness of MAO B inhibitors for the treatment of Parkinson's disease (PD) has been demonstrated *via* several clinical studies.[12,13]

As noted above, MAO is an oxidase, requiring oxygen as the electron acceptor in amine oxidation. The MAO-mediated oxidation of the primary amine produces a terminal imine, which is hydrolyzed to an aldehyde *via* water. Three products are produced *via* the MAO catalyzed reaction: (1) ammonia, (2) the ROS hydrogen peroxide, and (3) 3,4-dihydroxyphenylacetaldehyde (DOPAL).[14]

The ammonia produced is taken up for urea synthesis in the mitochondria.[14] Hydrogen peroxide undergoes decomposition *via* the activity of catalase, which catalyzes conversion of the hydrogen peroxide to water and molecular oxygen. As noted above, hydrogen peroxide is a ROS, and its production (*e.g.*, over-production) can be concerning, especially in the presence of transition metals such as iron (Fe^{2+}) that can initiate Fenton chemistry and the generation of the very toxic hydroxyl radical.

The other product, DOPAL, contains an aldehyde that undergoes carbonyl oxidation to 3,4-dihydroxyphenylacetic acid (DOPAC) or reduction to an alcohol, 3,4-dihydroxyphenylethanol (DOPET). DOPAL metabolism to DOPAC and DOPET occurs *via* various aldehyde dehydrogenase (ALDH) enzymes or aldehyde/aldose reductase (AR), respectively.[2,4,15] ALDH1 and ALDH2 represent cytosolic and mitochondrial, respectively, ALDH enzymes that are thought to be major contributors to DOPAL oxidation.[15] ALDH metabolism requires the cofactor NAD for oxidation, which could be limiting in cases of complex 1 inhibition. Interestingly, a knockout mouse lacking ALDH1 and ALDH2 still retains significant ALDH activity (~40%) towards DOPAL, also demonstrating the presence of other ALDH enzymes important for DOPAL metabolism. DOPAC subsequently undergoes methylation on the 3-hydroxyl position, which is catalyzed by catechol-*O*-methyl transferase (COMT), yielding homovanillic acid (HVA). The enzyme COMT requires the cofactor S-adenosyl methionine as the methyl donor and is located postsynaptically.[16]

As noted above, DOPAL is also metabolized *via* AR, which catalyzes reduction of aldehydes using the cofactor NADPH.[17,18] The product is DOPET (also known as hydroxytyrosol), known to be an antioxidant.[19,20] AR-mediated reduction of DOPAL is cytosolic and thought to constitute a secondary pathway for metabolism, with the ALDH pathway being primary.

DA in the presence of molecular oxygen will auto-oxidize to a highly reactive and toxic quinone that can rapidly modify protein thiols (*e.g.*, Cys).[21,22] The protein adducts of DA quinone can redox cycle and produce ROS.[23]

Several transporters are involved in the DA catabolic pathway in addition to VMAT2 (Figure 4.1).[2,4] The DA transporter (DAT) is responsible for the uptake of synaptic DA following release of vesicular DA. DOPAL was found to be an inhibitor for uptake of DA, but not the neurotransmitter γ-aminobutyric acid into striatal synaptosomes, suggesting an interaction for DOPAL and DAT.[24]

Subsequent work showed that mazindol, a selective DAT inhibitor, at μM levels inhibited uptake of DOPAL into rat striatal synpatosomes, implicating uptake of the biogenic aldehyde *via* DAT, at least to some extent.[24]

Previous studies demonstrated the presence of a transporter (efflux) in dopaminergic cells for acidic DA metabolites (*i.e.*, DOPAC and HVA) that was inhibited by the sulfonylurea drug glipizide.[25,26] Both DOPAC and HVA were found to be actively transported to the extracellular compartment for PC12 cells; however, the interaction of DOPAL with this transporter is unknown.

4.2 3,4-Dihydroxyphenylacetaldehyde (DOPAL) and Biogenic Aldehydes Derived from Neurotransmitters

As described above, the MAO-catalyzed oxidative deamination of monoamine neurotransmitters such as DA yields biogenic aldehydes. MAO-mediated activity was first described approximately 100 years ago, and it was found that the product is an aldehyde.[27–30] The review of Blaschko (1952) proposed aldehyde metabolites of the amine oxidases (*e.g.*, DOPAL), to be toxic to cells in which they are produced.[14]

DOPAL was of interest to the field of alcohol/ethanol addiction and hypothesized to cause central nervous system effects similar to those observed for ethanol.[31,32] It was proposed that acetaldehyde, a metabolite of ethanol, competitively inhibited the ALDH step of DA metabolism, yielding elevated levels of DOPAL followed by condensation to morphine-like alkaloids.[31] DA was found to condense with aldehydes such as acetaldehyde and DOPAL *via* a Pictet–Spengler-type reaction to pharmacologically and toxicologically active alkaloids, such as tetrahydropapaveroline in the case of DA.[33,34] Both acetaldehyde and DOPAL are substrates for mitochondrial ALDH, and therefore, it was reasoned that high enough levels of acetaldehyde could impair mitochondrial ALDH (ALDH2) activity towards DOPAL. Interestingly, follow up studies found that physiological concentrations of acetaldehyde did not inhibit ALDH activity towards DOPAL and therefore did not impact DOPAL oxidation.[35,36] Subsequent work demonstrated the presence of multiple isoforms of ALDH that may compensate for DOPAL metabolism should one isoform of the enzyme be inhibited, and in addition, AR enzymes could compensate for ALDH inhibition by catalyzing reduction of DOPAL as a second line of metabolism.[15,37,38]

As will be discussed later in this chapter, much evidence has been acquired over the last 15 years demonstrating DOPAL to be toxic to dopaminergic cells, both *in vitro* and *in vivo*.[39,40] Findings from numerous studies implicate the aldehyde metabolite DOPAL as a pathological factor or chemical trigger in disease, such as PD, to explain the selective degeneration of DA neurons. DOPAL is more toxic to dopaminergic cells than DA, which is considered to be an endogenous neurotoxin, and several DA metabolites (*e.g.* DOPAC, DOPET).

Of note, monoamine neurotransmitters other than DA, *e.g.*, serotonin (5HT) and norepinephrine (NE), undergo MAO-mediated metabolism, producing biogenic aldehydes.[2] Biotransformation of 5HT and NE *via* MAO yields 5-hydroxyindoleacetaldehyde (5HIAL) and 3,4-dihydroxyphenylglycolaldehyde (DOPEGAL). Compared with DOPAL, much less information is known regarding protein reactivity and toxicity of DOPEGAL and 5HIAL. While there is much less information in the scientific literature for these two biogenic aldehydes compared to DOPAL, what is known is that both 5HIAL and DOPEGAL are significantly more reactive and toxic than DOPAL.[39,41]

4.3 Generation of DOPAL and Biogenic Aldehydes at Aberrant Levels

As shown in Figure 4.1, DA metabolism generates DOPAL with subsequent carbonyl oxidation/reduction occurring as part of the life cycle for neurotransmitters, so it is a question of how/why DOPAL could be generated at levels toxic to DA neurons. The level for DOPAL at which cytotoxicity has been observed in DA cells (*i.e.*, ~7 μM) is not that much higher than measured *in vivo* concentrations of the DA metabolite (*i.e.*, ~2 μM).[39,42,43] Given this apparently small difference between the two, one might predict that impairing DOPAL metabolism to even a minor extent could elevate levels of the DA metabolite to toxic levels.

4.3.1 Mechanisms for Elevation of DOPAL

Previously published work has shown that carbonyl metabolism in dopaminergic cells is sensitive to products of oxidative stress. Of note, oxidative stress and lipid peroxidation are known endpoints following exposure *via* several environmental agents/pesticides, including paraquat and dieldrin.[44,45] Oxidative stress initiates lipid peroxidation and the production of several lipid-derived aldehydes, including 4-hydroxy-2-nonenal (4HNE) and malondialdehyde (MDA). Both 4HNE and MDA are potent inhibitors of ALDH enzymes at low μM, physiologic levels, yielding decreased DOPAC and elevated DOPAL concentrations.[46–48] Interestingly, 4HNE and MDA appear to inhibit the complete battery of ALDH isoforms, which was not achieved by acetaldehyde, as noted earlier in this chapter. Should ALDH be inhibited, the AR enzymes may compensate;[49] however, it was found that MDA but not 4HNE is a potent inactivator of the compensatory and cytosolic AR enzymes.[46] Such findings link oxidative stress and lipid peroxidation, known to occur in PD, with generation of elevated DOPAL and present a potential mechanism *via* which environmental insult produces oxidative stress and an endogenous neurotoxin (*i.e.*, DOPAL) in DA neurons.

Several recent papers have identified agents (*e.g.*, environmental) that may directly inhibit carbonyl metabolism of DOPAL without the need for subsequent oxidative stress and lipid peroxidation. The fungicide benomyl is metabolized to a potent and irreversible ALDH inhibitor known to impair DOPAL biotransformation.[50,51] Related to this, human exposure to benomyl was demonstrated to be a risk factor for PD incidence, and based on such data, Fitzmaurice *et al.* proposed ALDH inhibition as a pathogenic mechanism for PD.[51]

One other example is trichloroethylene (TCE), a solvent used as a metal degreaser. *In vitro* and *in vivo*, a major metabolite of TCE is trichloroacetaldehyde, a reversible but potent inhibitor of ALDH.[52] Interestingly, several studies have shown an increased risk of PD following exposure to the metal degreaser TCE.[53,54]

Also, inhibition of complex 1 *via* rotenone will increase the level of DOPAL.[36,55,56] ALDH2, the ALDH found in the mitochondrial matrix, plays a prominent role in DOPAL metabolism and requires the cofactor NAD.[36] Therefore, decreasing NAD production would limit the ability of ALDH to metabolize DOPAL.

One other potential mechanism for the accumulation of DOPAL involves interference with DA trafficking, such as the inhibition of VMAT2-mediated sequestration of cytosolic DA. Goldstein *et al.* showed that inhibition of vesicular DA uptake rapidly increased the intracellular content of DOPAL.[57] The pesticide dieldrin may impair vesicular uptake of DA[58] along with the release of DA into the cytosol,[45,59] yielding an increase in the DOPAL concentration.[60]

4.3.2 Relevance of Altered Dopamine Metabolism/Trafficking to PD

Several animal models have established that interference with DA trafficking and/or metabolism yields injury to DA neurons and signs/symptoms in animals relevant to PD. Mice with low vesicular monoamine transporter 2 (VMAT2; 5% control/normal activity) experience loss of DA neurons[61,62] with observed motor deficits and interestingly, nonmotor symptoms.[63,64] The decrease in VMAT2 will yield elevated cytosolic DA and may increase DOPAL concentrations.[65] Conversely, overexpression of VMAT2 is neuroprotective.[66]

Overexpression of MAO B in astrocytes yields selective and progressive loss of dopaminergic neurons in the *substantia nigra*, correlating with decreased locomotor activity.[67] An increase in MAO is predicted to yield elevated DOPAL and hydrogen peroxide levels.[68]

Mice with increased cytosolic DA (*i.e.*, unregulated and not sequestered in vesicles) due to expression of the DA transporter (DAT) on the postsynaptic neuron experience neurodegeneration.[69] Note that postsynaptic neurons lack VMAT2 and a means to properly sequester and store DA, and therefore, the cytosolic DA may undergo auto-oxidation to DA quinone, produce ROS, and generate elevated DOPAL following MAO metabolism. In addition, a recent study showed that overexpression of DAT in DA neurons of mice yields increased markers of oxidative stress and loss of midbrain dopaminergic cells.[70]

Mice lacking ALDH1 (cytosolic) and ALDH2 (mitochondrial) (ALDH1/2 knockout), the major ALDH enzymes responsible for DOPAL oxidation to DOPAC, demonstrate behavior and pathology relevant to parkinsonism.[62,71]

DA metabolism, specifically, ALDH activity, may be impaired in PD brains based on evidence from PD postmortem tissue. Goldstein *et al.* found that ratios of DA metabolites were significantly different in PD *versus* control brain tissue, demonstrating a decrease in ALDH activity and an increase in DA metabolism/turnover to DOPAL (*i.e.*, decrease in DOPAC : DOPAL ratio for PD demonstrating lower ALDH activity; increase in DOPAL : DA ratio for PD showing greater turnover of DA to DOPAL).[62,72] Other studies demonstrated a decrease in gene expression or transcription of ALDH 1A1 for PD patients.[73,74]

α-Synuclein can form toxic aggregates and is thought to be a contributor to PD pathogenesis; however, ALDH1A1 was found to be important for protecting DA neurons against α-synuclein-mediated loss.[75] In addition, ALDH activation (*i.e.*, pharmacological increase in ALDH activity) protects against rotenone-mediated damage to DA neurons.[76] Of note, rotenone is a complex 1 inhibitor and used as a model neurotoxicant causing parkinsonism in rodents.[77]

4.4 Toxicity and Protein Reactivity of DOPAL and Biogenic Aldehydes

Collectively, these data (above) demonstrate the sensitivity of DA neurons to changes in DA trafficking and/or metabolism and indicate a role for ALDH activity in neuron health and survival. A question arises as to what is/are the harmful intermediate(s) resulting from alteration of DA metabolism and/or trafficking?

ROS such as hydrogen peroxide or superoxide anion are produced from MAO-mediated oxidative deamination of DA and other monoamines or auto-oxidation of catechols, respectively. These species could oxidatively damage proteins or DNA and eventually initiate lipid peroxidation if activated to the hydroxyl radical (*i.e.*, hydrogen peroxide activation to hydroxyl radical *via* Fenton chemistry). Therefore, an increase in MAO expression/activity or elevation in cytosolic DA is predicted to elevate ROS.

As noted previously, cytosolic, unsequestered DA (in the presence of molecular oxygen) auto-oxidizes to a reactive and toxic quinone that can modify protein thiols (*e.g.*, Cys).[21,22] The auto-oxidation process generates superoxide anion as molecular oxygen catalyzes two rounds of one-electron oxidation. Protein adducts of DA quinone can redox cycle and produce ROS.[23]

Inhibition of ALDH in dopaminergic cells will generate elevated levels of biogenic aldehydes, such as DOPAL, the MAO metabolite of DA. The biogenic aldehyde DOPAL is more toxic than DA or other DA metabolites (DOPAC, DOPET) both *in vitro* and *in vivo*.[24,39,40,78] This toxicity is believed, in part, to be due to the production of ROS through protein reactivity and auto-oxidation.

4.4.1 Mechanisms of Toxicity

There are several mechanisms possible for the toxicity of DOPAL, including: generation of ROS or toxic radicals, induction of mitochondrial pore transition, and protein modification.[39,40] DOPAL can auto-oxidize to an ortho-quinone and produce ROS.[79] It was previously reported that DOPAL, but not DA or DOPAC, reacts with hydrogen peroxide to form the highly reactive and toxic hydroxyl radical.[80] Kristal *et al.* found that DOPAL at physiological levels targets the mitochondria, inducing mitochondrial permeability transition, and that DOPAL toxicity could be attenuated using the mitochondrial permeability transition pore inhibitor cyclosporine A.[43] In addition, numerous publications have demonstrated protein reactivity for DOPAL and the ability of the DA metabolite to cause protein aggregation.[81–83]

4.4.2 Protein Reactivity and Targets

Several published papers have reported and described the protein reactivity of DOPAL, which contains an amine-reactive aldehyde plus a catechol group, which could auto-oxidize to an ortho-quinone. According to previous reports, DOPAL reacts with proteins primarily *via* amines with the aldehyde.[81] Protein modification by DOPAL appears to involve a Schiff base reaction with the product being an imine based on the observed mass of the adduct; however, some findings were made suggesting unique chemistry for the DOPAL reactivity: (1) DOPAL readily forms stable adducts with primary amines that do not require reduction for stability; (2) the imine reducing agent, sodium cyanoborohydride, decreased the rate of reactivity for DOPAL with *N*-acetyl Lys; (3) the rate of DOPAL reactivity with amines increases with increasing pH; and (4) the reaction of DOPAL with protein amines generates ROS.[81,84]

The reaction of DOPAL with protein amines produces ROS and could be blocked *via* antioxidants, as was previously demonstrated.[84,85] The antioxidants glutathione, *N*-acetyl Cys and ascorbate were found to block the reaction of DOPAL with protein amines. Similarly, it was found that sodium cyanoborohydride, a reducing agent, also decreased the protein reactivity of DOPAL.[81] Given that protein reactivity by DOPAL is dependent on antioxidant status, it is conceivable that under conditions such as low antioxidant (*e.g.*, glutathione) levels, protein modification by DOPAL might be enhanced and DOPAL-toxicity augmented compared to a control (*i.e.*, normal/basal level of cellular antioxidants). Anderson *et al.*, also reported that the reaction of DOPAL with *N*-acetyl Lys produced a semiquinone radical and an *N*-oxyradical of the tricine buffer used.[84] Of note, tricine is an *N*-oxyradical scavenger.[86]

If the reaction of DOPAL with proteins involves Schiff base chemistry, then why are ROS produced and what is the adduct structure? Anderson *et al.* proposed that following amine modification by DOPAL, there is an oxidative step for the catechol and rearrangement to an indole-type structure.[84] Such a proposed mechanism would explain the stability of the DOPAL adduct, why reduction is not required for stability, the pH dependence (*i.e.*, an increase in pH increases the rate of reactivity) and the origin of the ROS.[84]

Similar to DA, DOPAL can undergo auto-oxidation or enzyme-catalyzed oxidation to a quinone.[79] As with DA, the auto-oxidation reaction produced a superoxide anion and the rate of auto-oxidation was increased in the presence of superoxide dismutase. Comproportionation of DOPAL with the quinone to a semiquinone radical was observed. The resulting DOPAL quinone caused protein aggregation; however, the sites of modification and cross-links were not identified. In addition, the DOPAL quinone was found to be very unstable at physiologic conditions and prone to rapid rearrangement to other products. Such a finding, as well as the report that DOPAL generates ROS when reacting with proteins, suggests that DOPAL production (or over-production) may contribute to oxidative stress.

Numerous protein targets for DOPAL have been reported, including ALDH and TH.[83,87,88] Modification and inhibition of ALDH by DOPAL could be self-propagating for levels of DOPAL. With regards to TH as a target, Vermeer *et al.* found that TH was reversibly and potently inhibited by low, physiologic concentrations of DOPAL (<2 μM), perhaps as a feedback mechanism/sensor preventing DA synthesis in the event of elevated DOPAL levels.[87,89]

α-Synuclein was demonstrated to be covalently modified by DOPAL, yielding protein aggregation.[82,90] DOPAL was found to react with α-synuclein (*i.e.*, Lys) *via* Michael addition and Schiff base chemistry and yield Met oxidation. In addition, α-synuclein oligomer formation was found to be dependent on DOPAL oxidation and could be prevented with the antioxidants glutathione and ascorbate.[90]

Proteins with affinity for DA could be selectively modified or inhibited by DOPAL, such as DAT or VMAT2. In addition, proteins targeted by DA could also be selectively reactive towards DOPAL, such as parkin or DJ-1.[91,92]

4.5 The Role of Biogenic Aldehydes in Disease

Biogenic aldehydes generated *via* catecholamine metabolism, such as DOPAL, are hypothesized to be endogenous neurotoxins and 'autotoxic' to DA neurons.[93] This concept has been termed as the 'catecholaldehyde hypothesis' in recent reports.[94,95] While harmful biogenic aldehydes (*e.g.*, DOPAL) derived from catecholamines are generated in DA neurons on a regular basis (Figure 4.1), their levels are not high enough to cause toxicity given proper neurotransmitter storage (*i.e.*, vesicular trafficking) and metabolism (*e.g.*, aldehyde biotransformation). Pathways for trafficking and metabolism of neurotransmitters, such as DA, are highly vulnerable to disruption by environmental agents both directly (*e.g.*, ALDH inhibitors) and indirectly (*e.g.*, oxidative stress). The disruption of neurotransmitter trafficking and metabolism is predicted to be a mechanistic link or 'trigger' between exposure to environmental neurotoxicants and harmful conditions relevant to PD. Elevated levels of a biogenic aldehyde, such as DOPAL, could lead to inhibition of enzymes, induction of ROS production and/or cause aggregation of α-synuclein.[93,96] In addition, it should be kept in mind that DOPAL's toxic effects may include both neurons and glia.[97] Further study and knowledge of this mechanistic link is predicted to yield therapeutic targets and strategies for health intervention and prevention of disease.

With regards to health intervention and prevention, there are several targets and strategies based on the above information:

- Minimize exposure to environmental agents, especially with those containing structural motifs (thiocarbamates) predicted to be ALDH inhibitors.[50,51,98,99] In addition, metabolism of pesticides should be considered for bioactivation to an ALDH inactivator.[100] Benzimidizole and thiocarbamate structures are present in many environmental agents (*e.g.*, molinate) and represent ALDH inhibitors.[50,51]
- ALDH activation (*e.g.*, alda-1) to increase ALDH activity and protect against elevated levels of biogenic aldehydes, such as DOPAL, resulting from environmental insult and/or pathologic conditions.[76,101]
- Carbonyl scavengers to reduce levels of biogenic aldehydes and aid in increased excretion of harmful aldehydes.[102] Such an approach has shown some utility in protecting against acrolein-mediated toxicity.
- MAO-B inhibition is predicted to decrease cellular concentrations of DOPAL, other biogenic aldehydes and ROS.[103] Previous work has demonstrated some degree of neuroprotection against PD and the dopaminergic neurotoxicant 1-methyl-4-phenyl-tetrahydropyridine (MPTP) for MAO inhibitors.[104,105]
- Therapeutic strategies to enhance vesicular uptake and storage of cytosolic DA. Lohr *et al.* have demonstrated that overexpression of VMAT2 yields protection for DA neurons against the neurotoxicant MPTP.[66]

In addition, loss of ALDH activity and/or elevated biogenic aldehydes could serve as biomarkers for earlier diagnosis of diseases such as PD.[73,74] These biomarkers could be especially important for individuals with significant exposure to environmental agents known to predispose to PD. Altered ratios of DA or metabolites in bodily fluid (*e.g.*, serum, cerebral spinal fluid) might indicate impairment of DA trafficking and/or metabolism, including ALDH inhibition. It is conceivable that elevated levels of protein adducts of DOPAL or biogenic aldehydes could be detected in bodily fluids (*e.g.*, serum, cerebral spinal fluid) *via* an antibody or ELISA approach and indicate ALDH inhibition.

While this chapter focuses on the role of DOPAL in disease, it should be noted that PD involves loss of not just dopaminergic cells of the *substantia nigra pars compact* but also neurons in the *locus ceruleus*, which is a major source of norepinephrine. Therefore, other monoamine-derived biogenic aldehydes (*i.e.*, 5HIAL, DOPEGAL) could play a role in neurodegeneration like DOPAL.[39,106,107]

4.6 Summary

In summary, biogenic aldehydes such as DOPAL are produced *via* the activity of MAO and undergo further metabolism. The generation of DOPAL and other monoamine-derived aldehyde metabolites is concerning given protein reactivity and toxicity. Much evidence has accumulated demonstrating the importance of proper DA trafficking and metabolism, especially the role of ALDH enzymes, in health and survival of DA neurons. Disruption of DA metabolism and trafficking could be accomplished *via* the direct effects of agents on critical proteins (*e.g.*, enzyme inhibition) or indirectly (*e.g.*, generation of oxidative stress with inhibition of carbonyl metabolism). Therefore, generation of DOPAL at aberrant levels may be a 'chemical trigger' and represent a mechanistic link between exposure (*e.g.*, pesticides) and conditions relevant to neurodegenerative disease. Therapeutic and preventive strategies are proposed to prevent ALDH inhibition and increased concentration of toxic biogenic aldehydes.

References

1. C. A. Marsden, Dopamine: the rewarding years, *Br. J. Pharmacol.*, 2006, **147**(suppl. 1), S136–S144.
2. G. Eisenhofer, I. J. Kopin and D. S. Goldstein, Catecholamine metabolism: a contemporary view with implications for physiology and medicine, *Pharmacol. Rev.*, 2004, **56**, 331–349.
3. I. J. Kopin, Storage and Metabolism of Catecholamines: the Role of Monoamine Oxidase, *Pharmacol. Rev.*, 1964, **16**, 179–191.
4. J. D. Elsworth and R. H. Roth, Dopamine synthesis, uptake, metabolism, and receptors: relevance to gene therapy of Parkinson's disease, *Exp. Neurol.*, 1997, **144**, 4–9.
5. A. J. Rivett, B. J. Eddy and J. A. Roth, Contribution of sulfate conjugation, deamination, and O-methylation to metabolism of dopamine and norepinephrine in human brain, *J. Neurochem.*, 1982, **39**, 1009–1016.
6. I. A. Paterson, A. V. Juorio, M. D. Berry and M. Y. Zhu, Inhibition of monoamine oxidase-B by (-)-deprenyl potentiates neuronal responses to dopamine agonists but does not inhibit dopamine catabolism in the rat striatum, *J. Pharmacol. Exp. Ther.*, 1991, **258**, 1019–1026.
7. D. A. Di Monte, L. E. DeLanney, I. Irwin, J. E. Royland, P. Chan, M. W. Jakowec and J. W. Langston, Monoamine oxidase-dependent metabolism of dopamine in the striatum and substantia nigra of L-DOPA-treated monkeys, *Brain Res.*, 1996, **738**, 53–59.
8. V. Glover, M. Sandler, F. Owen and G. J. Riley, Dopamine is a monoamine oxidase B substrate in man, *Nature*, 1977, **265**, 80–81.
9. C. J. Fowler, A. Wiberg, L. Oreland, J. Marcusson and B. Winblad, The effect of age on the activity and molecular properties of human brain monoamine oxidase, *J. Neural Transm.*, 1980, **49**, 1–20.
10. K. N. Westlund, R. M. Denney, L. M. Kochersperger, R. M. Rose and C. W. Abell, Distinct monoamine oxidase A and B populations in primate brain, *Science*, 1985, **230**, 181–183.
11. L. W. Thorpe, K. N. Westlund, L. M. Kochersperger, C. W. Abell and R. M. Denney, Immunocytochemical localization of monoamine oxidases A and B in human peripheral tissues and brain, *J. Histochem. Cytochem.*, 1987, **35**, 23–32.

12. J. W. Tetrud and J. W. Langston, The effect of deprenyl (selegiline) on the natural history of Parkinson's disease, *Science*, 1989, **245**, 519–522.
13. O. Weinreb, T. Amit, O. Bar-Am and M. B. Youdim, Rasagiline: a novel anti-Parkinsonian monoamine oxidase-B inhibitor with neuroprotective activity, *Prog. Neurobiol.*, 2010, **92**, 330–344.
14. H. Blaschko, Amine oxidase and amine metabolism, *Pharmacol. Rev.*, 1952, **4**, 415–458.
15. S. A. Marchitti, R. A. Deitrich and V. Vasiliou, Neurotoxicity and metabolism of the catecholamine-derived 3,4-dihydroxyphenylacetaldehyde and 3,4-dihydroxyphenylglycolaldehyde: the role of aldehyde dehydrogenase, *Pharmacol. Rev.*, 2007, **59**, 125–150.
16. P. T. Mannisto and S. Kaakkola, Catechol-O-methyltransferase (COMT): biochemistry, molecular biology, pharmacology, and clinical efficacy of the new selective COMT inhibitors, *Pharmacol. Rev.*, 1999, **51**, 593–628.
17. B. Tabakoff and V. G. Erwin, Purification and characterization of a reduced nicotinamide adenine dinucleotide phosphate-linked aldehyde reductase from brain, *J. Biol. Chem.*, 1970, **245**, 3263–3268.
18. B. Tabakoff, R. Anderson and S. G. Alivisatos, Enzymatic reduction of "biogenic" aldehydes in brain, *Mol. Pharmacol.*, 1973, **9**, 428–437.
19. I. Fernandez-Pastor, A. Fernandez-Hernandez, F. Rivas, A. Martinez, A. Garcia-Granados and A. Parra, Synthesis and Antioxidant Activity of Hydroxytyrosol Alkyl-Carbonate Derivatives, *J. Nat. Prod.*, 2016, **79**, 1737–1745.
20. D. S. Goldstein, Y. Jinsmaa, P. Sullivan, C. Holmes, I. J. Kopin and Y. Sharabi, 3,4-Dihydroxyphenylethanol (Hydroxytyrosol) Mitigates the Increase in Spontaneous Oxidation of Dopamine During Monoamine Oxidase Inhibition in PC12 Cells, *Neurochem. Res.*, 2016, **41**, 2173–2178.
21. D. N. Hauser, A. A. Dukes, A. D. Mortimer and T. G. Hastings, Dopamine quinone modifies and decreases the abundance of the mitochondrial selenoprotein glutathione peroxidase 4, *Free Radical Biol. Med.*, 2013, **65**, 419–427.
22. V. S. Van Laar, A. J. Mishizen, M. Cascio and T. G. Hastings, Proteomic identification of dopamine-conjugated proteins from isolated rat brain mitochondria and SH-SY5Y cells, *Neurobiol. Dis.*, 2009, **34**, 487–500.
23. M. Akagawa, Y. Ishii, T. Ishii, T. Shibata, M. Yotsu-Yamashita, K. Suyama and K. Uchida, Metal-catalyzed oxidation of protein-bound dopamine, *Biochemistry*, 2006, **45**, 15120–15128.
24. M. B. Mattammal, J. H. Haring, H. D. Chung, G. Raghu and R. Strong, An endogenous dopaminergic neurotoxin: implication for Parkinson's disease, *Neurodegeneration*, 1995, **4**, 271–281.
25. I. Lamensdorf, L. He, A. Nechushtan, J. Harvey-White, G. Eisenhofer, R. Milan, E. Rojas and I. J. Kopin, Effect of glipizide on dopamine synthesis, release and metabolism in PC12 cells, *Eur. J. Pharmacol.*, 2000, **388**, 147–154.
26. I. Lamensdorf, C. Hrycyna, L. P. He, A. Nechushtan, O. Tjurmina, J. Harvey-White, G. Eisenhofer, E. Rojas and I. J. Kopin, Acidic dopamine metabolites are actively extruded from PC12 cells by a novel

sulfonylurea-sensitive transporter, *Naunyn-Schmiedeberg's Arch. Pharmacol.*, 2000, **361**, 654–664.

27. M. L. Hare, Tyramine oxidase: A new enzyme system in liver, *Biochem. J.*, 1928, **22**, 968–979.
28. M. L. Bernheim, Tyramine Oxidase II: The Course of the Oxidation, *J. Biol. Chem.*, 1931, **93**, 299–309.
29. H. I. Kohn, Tyramine oxidase, *Biochem. J.*, 1937, **31**, 1693–1704.
30. D. Richter, Adrenaline and amine oxidase, *Biochem. J.*, 1937, **31**, 2022–2028.
31. R. A. Deitrich and V. G. Erwin, Involvement of biogenic amine metabolism in ethanol addiction, *Fed. Proc.*, 1975, **34**, 1962–1968.
32. V. E. Davis and M. J. Walsh, Alcohol, amines, and alkaloids: a possible biochemical basis for alcohol addiction, *Science*, 1970, **167**, 1005–1007.
33. H. Weiner, Relationship between 3,4-dihydroxyphenylacetaldehyde levels and tetrahydropapaveroline formation, *Alcohol.: Clin. Exp. Res.*, 1978, **2**, 127–131.
34. M. Sandler, S. B. Carter, K. R. Hunter and G. M. Stern, Tetrahydroisoquinoline alkaloids: *in vivo* metabolites of L-dopa in man, *Nature*, 1973, **241**, 439–443.
35. A. W. Tank and H. Weiner, Ethanol-induced alteration of dopamine metabolism in rat liver, *Biochem. Pharmacol.*, 1979, **28**, 3139–3147.
36. A. W. Tank, H. Weiner and J. A. Thurman, Enzymology and subcellular localization of aldehyde oxidation in rat liver. Oxidation of 3,4-dihydroxyphenylacetaldehyde derived from dopamine to 3,4-dihydroxyphenylacetic acid, *Biochem. Pharmacol.*, 1981, **30**, 3265–3275.
37. M. T. Ryzlak and R. Pietruszko, Purification and characterization of aldehyde dehydrogenase from human brain, *Arch. Biochem. Biophys.*, 1987, **255**, 409–418.
38. M. T. Ryzlak and R. Pietruszko, Human brain glyceraldehyde-3-phosphate dehydrogenase, succinic semialdehyde dehydrogenase and aldehyde dehydrogenase isozymes: substrate specificity and sensitivity to disulfiram, *Alcohol.: Clin. Exp. Res.*, 1989, **13**, 755–761.
39. W. J. Burke, S. W. Li, H. D. Chung, D. A. Ruggiero, B. S. Kristal, E. M. Johnson, P. Lampe, V. B. Kumar, M. Franko, E. A. Williams and D. S. Zahm, Neurotoxicity of MAO metabolites of catecholamine neurotransmitters: role in neurodegenerative diseases, *Neurotoxicology*, 2004, **25**, 101–115.
40. W. J. Burke, 3,4-dihydroxyphenylacetaldehyde: a potential target for neuroprotective therapy in Parkinson's disease, *Curr. Drug Targets: CNS Neurol. Disord.*, 2003, **2**, 143–148.
41. G. E. Nilsson and O. Tottmar, Biogenic aldehydes in brain: on their preparation and reactions with rat brain tissue, *J. Neurochem.*, 1987, **48**, 1566–1572.
42. W. J. Burke, H. D. Chung and S. W. Li, Quantitation of 3,4-dihydroxyphenylacetaldehyde and 3,4-dihydroxyphenylglycolaldehyde, the monoamine oxidase metabolites of dopamine and noradrenaline, in human tissues by microcolumn high-performance liquid chromatography, *Anal. Biochem.*, 1999, **273**, 111–116.

43. B. S. Kristal, A. D. Conway, A. M. Brown, J. C. Jain, P. A. Ulluci, S. W. Li and W. J. Burke, Selective dopaminergic vulnerability: 3,4-dihydroxyphenylacetaldehyde targets mitochondria, *Free Radical Biol. Med.*, 2001, **30**, 924–931.
44. W. Yang and E. Tiffany-Castiglioni, The bipyridyl herbicide paraquat produces oxidative stress-mediated toxicity in human neuroblastoma SH-SY5Y cells: relevance to the dopaminergic pathogenesis, *J. Toxicol. Environ. Health, Part A*, 2005, **68**, 1939–1961.
45. A. G. Kanthasamy, M. Kitazawa, A. Kanthasamy and V. Anantharam, Dieldrin-induced neurotoxicity: relevance to Parkinson's disease pathogenesis, *Neurotoxicology*, 2005, **26**, 701–719.
46. Y. Jinsmaa, V. R. Florang, J. N. Rees, D. G. Anderson, S. Strack and J. A. Doorn, Products of oxidative stress inhibit aldehyde oxidation and reduction pathways in dopamine catabolism yielding elevated levels of a reactive intermediate, *Chem. Res. Toxicol.*, 2009, **22**, 835–841.
47. J. N. Rees, V. R. Florang, D. G. Anderson and J. A. Doorn, Lipid peroxidation products inhibit dopamine catabolism yielding aberrant levels of a reactive intermediate, *Chem. Res. Toxicol.*, 2007, **20**, 1536–1542.
48. V. R. Florang, J. N. Rees, N. K. Brogden, D. G. Anderson, T. D. Hurley and J. A. Doorn, Inhibition of the oxidative metabolism of 3,4-dihydroxyphenylacetaldehyde, a reactive intermediate of dopamine metabolism, by 4-hydroxy-2-nonenal, *Neurotoxicology*, 2007, **28**, 76–82.
49. A. J. Turner, J. A. Illingworth and K. F. Tipton, Simulation of biogenic amine metabolism in the brain, *Biochem. J.*, 1974, **144**, 353–360.
50. J. E. Casida, B. Ford, Y. Jinsmaa, P. Sullivan, A. Cooney and D. S. Goldstein, Benomyl, aldehyde dehydrogenase, DOPAL, and the catecholaldehyde hypothesis for the pathogenesis of Parkinson's disease, *Chem. Res. Toxicol.*, 2014, **27**, 1359–1361.
51. A. G. Fitzmaurice, S. L. Rhodes, A. Lulla, N. P. Murphy, H. A. Lam, K. C. O'Donnell, L. Barnhill, J. E. Casida, M. Cockburn, A. Sagasti, M. C. Stahl, N. T. Maidment, B. Ritz and J. M. Bronstein, Aldehyde dehydrogenase inhibition as a pathogenic mechanism in Parkinson disease, *Proc. Natl. Acad. Sci. U. S. A.*, 2013, **110**, 636–641.
52. L. H. Lash, J. W. Fisher, J. C. Lipscomb and J. C. Parker, Metabolism of trichloroethylene, *Environ. Health Perspect.*, 2000, **108**(suppl. 2), 177–200.
53. S. M. Goldman, P. J. Quinlan, G. W. Ross, C. Marras, C. Meng, G. S. Bhudhikanok, K. Comyns, M. Korell, A. R. Chade, M. Kasten, B. Priestley, K. L. Chou, H. H. Fernandez, F. Cambi, J. W. Langston and C. M. Tanner, Solvent exposures and Parkinson disease risk in twins, *Ann. Neurol.*, 2012, **71**, 776–784.
54. D. M. Gash, K. Rutland, N. L. Hudson, P. G. Sullivan, G. Bing, W. A. Cass, J. D. Pandya, M. Liu, D. Y. Choi, R. L. Hunter, G. A. Gerhardt, C. D. Smith, J. T. Slevin and T. S. Prince, Trichloroethylene: Parkinsonism and complex 1 mitochondrial neurotoxicity, *Ann. Neurol.*, 2008, **63**, 184–192.
55. I. Lamensdorf, G. Eisenhofer, J. Harvey-White, Y. Hayakawa, K. Kirk and I. J. Kopin, Metabolic stress in PC12 cells induces the formation of the

endogenous dopaminergic neurotoxin, 3,4-dihydroxyphenylacetaldehyde, *J. Neurosci. Res.*, 2000, **60**, 552–558.

56. I. Lamensdorf, G. Eisenhofer, J. Harvey-White, A. Nechustan, K. Kirk and I. J. Kopin, 3,4-Dihydroxyphenylacetaldehyde potentiates the toxic effects of metabolic stress in PC12 cells, *Brain Res.*, 2000, **868**, 191–201.
57. D. S. Goldstein, P. Sullivan, A. Cooney, Y. Jinsmaa, R. Sullivan, D. J. Gross, C. Holmes, I. J. Kopin and Y. Sharabi, Vesicular uptake blockade generates the toxic dopamine metabolite 3,4-dihydroxyphenylacetaldehyde in PC12 cells: relevance to the pathogenesis of Parkinson's disease, *J. Neurochem.*, 2012, **123**, 932–943.
58. G. W. Miller, M. L. Kirby, A. I. Levey and J. R. Bloomquist, Heptachlor alters expression and function of dopamine transporters, *Neurotoxicology*, 1999, **20**, 631–637.
59. M. Kitazawa, V. Anantharam and A. G. Kanthasamy, Dieldrin-induced oxidative stress and neurochemical changes contribute to apoptopic cell death in dopaminergic cells, *Free Radical Biol. Med.*, 2001, **31**, 1473–1485.
60. E. M. Allen, V. R. Florang, L. L. Davenport, Y. Jinsmaa and J. A. Doorn, Cellular localization of dieldrin and structure-activity relationship of dieldrin analogues in dopaminergic cells, *Chem. Res. Toxicol.*, 2013, **26**, 1043–1054.
61. W. M. Caudle, J. R. Richardson, M. Z. Wang, T. N. Taylor, T. S. Guillot, A. L. McCormack, R. E. Colebrooke, D. A. Di Monte, P. C. Emson and G. W. Miller, Reduced vesicular storage of dopamine causes progressive nigrostriatal neurodegeneration, *J. Neurosci.*, 2007, **27**, 8138–8148.
62. D. S. Goldstein, P. Sullivan, C. Holmes, G. W. Miller, S. Alter, R. Strong, D. C. Mash, I. J. Kopin and Y. Sharabi, Determinants of buildup of the toxic dopamine metabolite DOPAL in Parkinson's disease, *J. Neurochem.*, 2013, **126**, 591–603.
63. T. N. Taylor, W. M. Caudle and G. W. Miller, VMAT2-Deficient Mice Display Nigral and Extranigral Pathology and Motor and Nonmotor Symptoms of Parkinson's Disease, *Parkinson's Dis.*, 2011, **2011**, 124165.
64. T. N. Taylor, W. M. Caudle, K. R. Shepherd, A. Noorian, C. R. Jackson, P. M. Iuvone, D. Weinshenker, J. G. Greene and G. W. Miller, Nonmotor symptoms of Parkinson's disease revealed in an animal model with reduced monoamine storage capacity, *J. Neurosci.*, 2009, **29**, 8103–8113.
65. D. S. Goldstein, P. Sullivan, A. Cooney, Y. Jinsmaa, R. Sullivan, D. J. Gross, C. Holmes, I. J. Kopin and Y. Sharabi, Vesicular Uptake Blockade Generates the Toxic Dopamine Metabolite 3,4-Dihydroxyphenylacetaldehyde in PC12 Cells: Relevance to the Pathogenesis of Parkinson Disease, *J. Neurochem.*, 2012, **123**, 932–943.
66. K. M. Lohr, A. I. Bernstein, K. A. Stout, A. R. Dunn, C. R. Lazo, S. P. Alter, M. Wang, Y. Li, X. Fan, E. J. Hess, H. Yi, L. M. Vecchio, D. S. Goldstein, T. S. Guillot, A. Salahpour and G. W. Miller, Increased vesicular monoamine transporter enhances dopamine release and opposes Parkinson disease-related neurodegeneration *in vivo*, *Proc. Natl. Acad. Sci. U. S. A.*, 2014, **111**, 9977–9982.

67. J. K. Mallajosyula, D. Kaur, S. J. Chinta, S. Rajagopalan, A. Rane, D. G. Nicholls, D. A. Di Monte, H. Macarthur and J. K. Andersen, MAO-B elevation in mouse brain astrocytes results in Parkinson's pathology, *PLoS One*, 2008, **3**, e1616.
68. A. Siddiqui, J. K. Mallajosyula, A. Rane and J. K. Andersen, Ability to delay neuropathological events associated with astrocytic MAO-B increase in a Parkinsonian mouse model: implications for early intervention on disease progression, *Neurobiol. Dis.*, 2010, **40**, 444–448.
69. L. Chen, Y. Ding, B. Cagniard, A. D. Van Laar, A. Mortimer, W. Chi, T. G. Hastings, U. J. Kang and X. Zhuang, Unregulated cytosolic dopamine causes neurodegeneration associated with oxidative stress in mice, *J. Neurosci.*, 2008, **28**, 425–433.
70. S. T. Masoud, L. M. Vecchio, Y. Bergeron, M. M. Hossain, L. T. Nguyen, M. K. Bermejo, B. Kile, T. D. Sotnikova, W. B. Siesser, R. R. Gainetdinov, R. M. Wightman, M. G. Caron, J. R. Richardson, G. W. Miller, A. J. Ramsey, M. Cyr and A. Salahpour, Increased expression of the dopamine transporter leads to loss of dopamine neurons, oxidative stress and l-DOPA reversible motor deficits, *Neurobiol. Dis.*, 2015, **74**, 66–75.
71. M. C. Wey, E. Fernandez, P. A. Martinez, P. Sullivan, D. S. Goldstein and R. Strong, Neurodegeneration and motor dysfunction in mice lacking cytosolic and mitochondrial aldehyde dehydrogenases: implications for Parkinson's disease, *PLoS One*, 2012, **7**, e31522.
72. D. S. Goldstein, P. Sullivan, C. Holmes, I. J. Kopin, M. J. Basile and D. C. Mash, Catechols in post-mortem brain of patients with Parkinson disease, *Eur. J. Neurol.*, 2011, **18**, 703–710.
73. S. Mandel, E. Grunblatt, P. Riederer, N. Amariglio, J. Jacob-Hirsch, G. Rechavi and M. B. Youdim, Gene expression profiling of sporadic Parkinson's disease substantia nigra pars compacta reveals impairment of ubiquitin-proteasome subunits, SKP1A, aldehyde dehydrogenase, and chaperone HSC-70, *Ann. N. Y. Acad. Sci.*, 2005, **1053**, 356–375.
74. L. Molochnikov, J. M. Rabey, E. Dobronevsky, U. Bonucelli, R. Ceravolo, D. Frosini, E. Grunblatt, P. Riederer, C. Jacob, J. Aharon-Peretz, Y. Bashenko, M. B. Youdim and S. A. Mandel, A molecular signature in blood identifies early Parkinson's disease, *Mol. Neurodegener.*, 2012, **7**, 26.
75. G. Liu, J. Yu, J. Ding, C. Xie, L. Sun, I. Rudenko, W. Zheng, N. Sastry, J. Luo, G. Rudow, J. C. Troncoso and H. Cai, Aldehyde dehydrogenase 1 defines and protects a nigrostriatal dopaminergic neuron subpopulation, *J. Clin. Invest.*, 2014, **124**, 3032–3046.
76. C. C. Chiu, T. H. Yeh, S. C. Lai, Y. H. Wu-Chou, C. H. Chen, D. Mochly-Rosen, Y. C. Huang, Y. J. Chen, C. L. Chen, Y. M. Chang, H. L. Wang and C. S. Lu, Neuroprotective effects of aldehyde dehydrogenase 2 activation in rotenone-induced cellular and animal models of parkinsonism, *Exp. Neurol.*, 2015, **263**, 244–253.
77. J. R. Cannon, V. Tapias, H. M. Na, A. S. Honick, R. E. Drolet and J. T. Greenamyre, A highly reproducible rotenone model of Parkinson's disease, *Neurobiol. Dis.*, 2009, **34**, 279–290.

78. S. A. Marchitti, R. A. Deitrich and V. Vasiliou, Neurotoxicity and metabolism of the catecholamine-derived 3,4-dihydroxyphenylacetaldehyde and 3,4-dihydroxyphenylglycolaldehyde: the role of aldehyde dehydrogenase, *Pharmacol. Rev.*, 2007, **59**, 125–150.
79. D. G. Anderson, S. V. Mariappan, G. R. Buettner and J. A. Doorn, Oxidation of 3,4-dihydroxyphenylacetaldehyde, a toxic dopaminergic metabolite, to a semiquinone radical and an ortho-quinone, *J. Biol. Chem.*, 2011, **286**, 26978–26986.
80. S. W. Li, T. S. Lin, S. Minteer and W. J. Burke, 3,4-Dihydroxyphenylacetaldehyde and hydrogen peroxide generate a hydroxyl radical: possible role in Parkinson's disease pathogenesis, *Brain Res. Mol. Brain Res.*, 2001, **93**, 1–7.
81. J. N. Rees, V. R. Florang, L. L. Eckert and J. A. Doorn, Protein reactivity of 3,4-dihydroxyphenylacetaldehyde, a toxic dopamine metabolite, is dependent on both the aldehyde and the catechol, *Chem. Res. Toxicol.*, 2009, **22**, 1256–1263.
82. W. J. Burke, V. B. Kumar, N. Pandey, W. M. Panneton, Q. Gan, M. W. Franko, M. O'Dell, S. W. Li, Y. Pan, H. D. Chung and J. E. Galvin, Aggregation of alpha-synuclein by DOPAL, the monoamine oxidase metabolite of dopamine, *Acta Neuropathol.*, 2008, **115**, 193–203.
83. Y. Jinsmaa, V. R. Florang, J. N. Rees, L. M. Mexas, L. L. Eckert, E. M. Allen, D. G. Anderson and J. A. Doorn, Dopamine-derived biological reactive intermediates and protein modifications: Implications for Parkinson's disease, *Chem.-Biol. Interact.*, 2011, **192**, 118–121.
84. D. G. Anderson, V. R. Florang, J. S. Schamp, G. R. Buettner and J. A. Doorn, Antioxidant-mediated modulation of protein reactivity for 3,4-dihydroxyphenylacetaldehyde, a toxic dopamine metabolite, *Chem. Res. Toxicol.*, 2016, **29**, 1098–1107.
85. F. Ungar, B. Tabakoff and S. G. Alivisatos, Inhibition of binding of aldehydes of biogenic amines in tissues, *Biochem. Pharmacol.*, 1973, **22**, 1905–1913.
86. H. J. Grande and K. R. van der Ploeg, Tricine radicals as formed in the presence of peroxide producing enzymes, *FEBS Lett.*, 1978, **95**, 352–356.
87. L. M. Mexas, V. R. Florang and J. A. Doorn, Inhibition and covalent modification of tyrosine hydroxylase by 3,4-dihydroxyphenylacetaldehyde, a toxic dopamine metabolite, *Neurotoxicology*, 2011, **32**, 471–477.
88. A. D. MacKerell Jr and R. Pietruszko, Chemical modification of human aldehyde dehydrogenase by physiological substrate, *Biochim. Biophys. Acta*, 1987, **911**, 306–317.
89. L. M. Vermeer, V. R. Florang and J. A. Doorn, Catechol and aldehyde moieties of 3,4-dihydroxyphenylacetaldehyde contribute to tyrosine hydroxylase inhibition and neurotoxicity, *Brain Res.*, 2012, **1474**, 100–109.
90. C. Follmer, E. Coelho-Cerqueira, D. Y. Yatabe-Franco, G. D. Araujo, A. S. Pinheiro, G. B. Domont and D. Eliezer, Oligomerization and Membrane-binding Properties of Covalent Adducts Formed by the

Interaction of alpha-Synuclein with the Toxic Dopamine Metabolite 3,4-Dihydroxyphenylacetaldehyde (DOPAL), *J. Biol. Chem.*, 2015, **290**, 27660–27679.

91. S. Girotto, M. Sturlese, M. Bellanda, I. Tessari, R. Cappellini, M. Bisaglia, L. Bubacco and S. Mammi, Dopamine-derived quinones affect the structure of the redox sensor DJ-1 through modifications at Cys-106 and Cys-53, *J. Biol. Chem.*, 2012, **287**, 18738–18749.
92. M. J. LaVoie, B. L. Ostaszewski, A. Weihofen, M. G. Schlossmacher and D. J. Selkoe, Dopamine covalently modifies and functionally inactivates parkin, *Nat. Med.*, 2005, **11**, 1214–1221.
93. D. S. Goldstein, I. J. Kopin and Y. Sharabi, Catecholamine autotoxicity. Implications for pharmacology and therapeutics of Parkinson disease and related disorders, *Pharmacol. Ther.*, 2014, **144**, 268–282.
94. D. S. Goldstein, Biomarkers, mechanisms, and potential prevention of catecholamine neuron loss in Parkinson disease, *Adv. Pharmacol.*, 2013, **68**, 235–272.
95. D. S. Goldstein, Biomarkers, mechanisms, and potential prevention of catecholamine neuron loss in Parkinson disease, *Adv. Pharmacol.*, 2013, **68**, 235–272.
96. W. J. Burke, V. B. Kumar, N. Pandey, W. M. Panneton, Q. Gan, M. W. Franko, M. O'Dell, S. W. Li, Y. Pan, H. D. Chung and J. E. Galvin, Aggregation of alpha-synuclein by DOPAL, the monoamine oxidase metabolite of dopamine, *Acta Neuropathol.*, 2008, **115**, 193–203.
97. Y. Jinsmaa, P. Sullivan, Y. Sharabi and D. S. Goldstein, DOPAL is transmissible to and oligomerizes alpha-synuclein in human glial cells, *Auton. Neurosci.*, 2016, **194**, 46–51.
98. A. G. Fitzmaurice, S. L. Rhodes, M. Cockburn, B. Ritz and J. M. Bronstein, Aldehyde dehydrogenase variation enhances effect of pesticides associated with Parkinson disease, *Neurology*, 2014, **82**, 419–426.
99. D. S. Goldstein, P. Sullivan, A. Cooney, Y. Jinsmaa, I. J. Kopin and Y. Sharabi, Rotenone decreases intracellular aldehyde dehydrogenase activity: implications for the pathogenesis of Parkinson's disease, *J. Neurochem.*, 2015, **133**, 14–25.
100. E. M. Allen, D. G. Anderson, V. R. Florang, M. Khanna, T. D. Hurley and J. A. Doorn, Relative inhibitory potency of molinate and metabolites with aldehyde dehydrogenase 2: implications for the mechanism of enzyme inhibition, *Chem. Res. Toxicol.*, 2010, **23**, 1843–1850.
101. C. H. Chen, G. R. Budas, E. N. Churchill, M. H. Disatnik, T. D. Hurley and D. Mochly-Rosen, Activation of aldehyde dehydrogenase-2 reduces ischemic damage to the heart, *Science*, 2008, **321**, 1493–1495.
102. P. C. Burcham, F. R. Fontaine, L. M. Kaminskas, D. R. Petersen and S. M. Pyke, Protein adduct-trapping by hydrazinophthalazine drugs: mechanisms of cytoprotection against acrolein-mediated toxicity, *Mol. Pharmacol.*, 2004, **65**, 655–664.

103. D. S. Goldstein, Y. Jinsmaa, P. Sullivan, C. Holmes, I. J. Kopin and Y. Sharabi, Comparison of Monoamine Oxidase Inhibitors in Decreasing Production of the Autotoxic Dopamine Metabolite 3,4-Dihydroxyphenylacetaldehyde in PC12 Cells, *J. Pharmacol. Exp. Ther.*, 2016, **356**, 483–492.
104. A. Kupsch, J. Sautter, M. E. Gotz, W. Breithaupt, J. Schwarz, M. B. Youdim, P. Riederer, M. Gerlach and W. H. Oertel, Monoamine oxidase-inhibition and MPTP-induced neurotoxicity in the non-human primate: comparison of rasagiline (TVP 1012) with selegiline, *J. Neural Transm.*, 2001, **108**, 985–1009.
105. O. Weinreb, T. Amit, P. Riederer, M. B. Youdim and S. A. Mandel, Neuroprotective profile of the multitarget drug rasagiline in Parkinson's disease, *Int. Rev. Neurobiol.*, 2011, **100**, 127–149.
106. T. N. Taylor, S. P. Alter, M. Wang, D. S. Goldstein and G. W. Miller, Reduced vesicular storage of catecholamines causes progressive degeneration in the locus ceruleus, *Neuropharmacology*, 2014, **76**(pt A), 97–105.
107. Y. Jinsmaa, A. Cooney, P. Sullivan, Y. Sharabi and D. S. Goldstein, The serotonin aldehyde, 5-HIAL, oligomerizes alpha-synuclein, *Neurosci. Lett.*, 2015, **590**, 134–137.

CHAPTER 5

Dopamine Oxidation and Parkinson's Disease

CAITLYN W. BARRETT[a,b], MEGHAN L. BUCHER[b,c] AND TERESA G. HASTINGS*[a,b,c]

[a]Department of Neurology, University of Pittsburgh School of Medicine, Pittsburgh, PA 15260, USA; [b]Pittsburgh Institute for Neurodegenerative Diseases, University of Pittsburgh, Pittsburgh, PA 15260, USA; [c]Department of Neuroscience, University of Pittsburgh, Pittsburgh, PA 15260, USA
*E-mail: hasttx@upmc.edu

5.1 Oxidative Stress and Susceptibility of Dopaminergic Neurons in Parkinson's Disease

PD is the second most common neurodegenerative disease and leading movement disorder, affecting 1–2% of the population over age 65. PD is pathologically characterized by the presence of Lewy bodies, large ubiquitinated eosinophilic inclusions containing α-synuclein. Various groups of neurons, including neurons in the lower brainstem, olfactory bulb, and *locus coeruleus*[1] display Lewy body pathology, but the most severely impacted neurons are the dopaminergic neurons of the SN. At least 50%, and up to 80%, of the neuromelanin-pigmented DA neurons in the SN are lost in PD.[2] Loss of these neurons results in depletion of DA in the caudate and putamen and contributes to the motor symptoms commonly seen in PD including tremor, postural instability, bradykinesia (slowing of movements), and rigidity. In a subset of PD cases, genetic mutations in the proteins α-synuclein, DJ-1,

Issues in Toxicology No. 34
Oxidative Stress and Redox Signalling in Parkinson's Disease
Edited by Rodrigo Franco, Jonathan A. Doorn and Jean-Christophe Rochet

Published by the Royal Society of Chemistry, www.rsc.org

Parkin, PINK1, and LRRK2 have been implicated in familial PD,[3–6] but a large proportion of PD patients present with idiopathic Parkinsonism for which the cause is not well understood, but increasing evidence points to several contributing factors including oxidative stress, mitochondrial dysfunction, and improper DA handling. These factors will be discussed in subsequent sections of this chapter.

Even under non-pathological conditions, the SN is especially susceptible to oxidative stress. Basal DA catabolism *via* monoamine oxidase (MAO) produces H_2O_2, a reactive oxygen species. In postmortem SN of PD patients, there is a decrease in levels of glutathione peroxidase (GPx),[7] a glutathione-dependent antioxidant enzyme that limits ROS detoxification in this region. There are several indications of increased oxidative stress in brains of PD patients including increased lipid peroxidation,[8] decreased antioxidant capacity, as demonstrated by decreases in antioxidant defenses such as glutathione (GSH),[9,10] GPx,[11] and catalase,[12] as well as increases in pro-oxidant iron[13–15] that catalyzes formation of the hydroxyl radical ($^{\bullet}OH$) from hydrogen peroxide (H_2O_2).[16] Of interest, the ratio between reduced and oxidized glutathione (GSH/GSSG) is further decreased during degeneration, enhancing the formation of toxic hydroxyl radicals.[17,18] Additionally, oxidative DNA damage is apparent in PD brains as assessed by the presence of 8-hydroxy-2′deoxy-guanosine,[8] demonstrating that there is a functional consequence to the high oxidative load in the SN of PD brains.

As mentioned previously, although other brain regions are known to be involved in PD, none are quite so adversely affected as the dopaminergic neurons of the SN. This enhanced degeneration, as well as the increase in oxidative stress that is observed in PD, suggests that the neurotransmitter DA may play a role in PD progression.[19–21] Treatment of rodents with pesticides such as rotenone and paraquat, which are chemicals that cause systemic oxidative stress, also lead to selective dopaminergic degeneration in the SN. Rotenone inhibits Complex I of the electron transport chain (ETC), and paraquat catalyzes the formation of ROS by initiating redox cycling. Despite the fact that both are whole-body accessible, these pesticides trigger selective degeneration of nigral dopaminergic neurons.[22,23] These examples of experimental modulation of ROS will be further discussed in subsequent sections and a link to DA oxidation will be considered.

5.2 Dopamine Regulation and Metabolism

The motor symptoms of PD are a direct result of the loss of DA's influence in the striatum. It is for this reason that treatments for the motor symptoms of PD are targeted to enhance activity in the DA system. The necessity of DA synthesis in proper motor functioning has been explored through studies that decreased the production of DA resulting in hypoactivity.[24,25] In addition, dysregulation in the packaging or release of DA can cause motor dysfunction, indicating the necessity of DA for properly executed movement.

Under normal conditions, DA is synthesized in the cytosol through the initial conversion of L-tyrosine to L-3,4-dihydroxyphenylalanine (L-DOPA) by the enzyme tyrosine hydroxylase. L-DOPA is the immediate precursor to DA, and it is converted to DA by the aromatic L-amino acid decarboxylase. Once synthesized, DA is immediately loaded into synaptic vesicles by the vesicular monoamine transporter 2 (VMAT2),[26] a process that depends upon the acidic vesicular interior and electrochemical gradient established and maintained by vacuolar-type proton ATPase activity. Under conditions of oxidative stress, as are seen in PD, the unique chemical structure of DA is capable of forming endogenous toxins that can contribute to further oxidative stress and mitochondrial dysfunction. Of note, DA is relatively stable in the acidic environment of the vesicles, however, the reactivity of DA is increased at physiological pH and cytoplasmic DA is readily oxidized, generating ROS and reactive DA quinone. For this reason, DA that has been newly synthesized, or transported back into the cell from the synapse, is rapidly sequestered into low pH synaptic vesicles by VMAT2. These vesicles provide a stable environment for DA as they do not contain any of the enzymes or cofactors responsible for the creation or breakdown of DA.

Electrochemical cytometry—a form of cytometry where individual vesicles are separated by capillary electrophoresis, lysed chemically, and detected using a carbon-fiber detecting electrode—in brain tissue was used to count DA molecules in individual synaptic vesicles. Individual mouse striatal vesicles contain an average of 33 000 DA molecules, an amount considerably greater than that estimated using the measure of DA quantal release from cultured neurons.[27] Moreover, treatment of mice with L-DOPA, the precursor to DA and used for PD treatment, within the therapeutic dose, causes a large increase in the average number of DA molecules per vesicle, indicating that vesicles are capable of considerable plasticity.[27] These data indicate that DA vesicles within the cell are capable of holding—and releasing—large quantities of DA. As has been alluded to previously, mishandled DA is capable of producing reactive metabolites, indicating that proper DA metabolism and storage are likely important for the maintenance of cellular health.

5.3 Pathological Dopamine Oxidation

If DA levels increase in the cytoplasm, due to improper uptake or storage in synaptic vesicles as a result of increased oxidative stress, exposure to weak base compounds such as methamphetamine, or the presence of α-synuclein protofibrils, DA is capable of forming reactive metabolites by two distinct pathways:[28,29] (1) metabolism by monoamine oxidase (MAO) and (2) autooxidation or enzymatic oxidation of the catechol ring. The formation of reactive metabolites of DA may contribute to the vulnerability of DA neurons in the SN.

5.3.1 Metabolism of Dopamine by Monoamine Oxidase

Cytoplasmic DA can be metabolized by MAO to 3,4-dihydroxyphenylacetaldehyde (DOPAL), another potentially reactive dopamine metabolite that is quickly oxidized by aldehyde dehydrogenase to 3,4-dihydroxyphenylacetic acid (DOPAC). DOPAC is then converted to homovallinic acid (HVA) by the catechol-*O*-methyl-transferase (COMT) in a reaction that likely occurs outside of DA neurons. In an alternate but similar pathway, DA is first catalyzed to 3-methoxytyramine (3-MT) by COMT and subsequently converted to HVA by MAO, both reactions ultimately resulting in the formation of HVA and the byproduct hydrogen peroxide (H_2O_2).[30–33] H_2O_2, if not immediately reduced by cellular antioxidant mechanisms including GSH and GPx, can react with transition metals such as iron to form the hydroxyl radical ($^{\bullet}OH$).[34] $^{\bullet}OH$ is a highly reactive free radical that, though short-lived, will immediately react with any DNA, lipids, and susceptible amino acids in proteins that are within its vicinity, ultimately resulting in cellular damage.[34]

Key determinants in the destructiveness of this process include MAO, cellular antioxidants, and iron. It has already been noted that decreased antioxidant capacity[9–11] and increased iron[13–15] are inherent features of the Parkinsonian SN. Polymorphisms in MAO have been linked to Parkinson's disease[35] and selective inhibitors for MAO-B have been under consideration for treatment of PD.[36] The associations between the MAO-driven metabolism of DA and drivers of PD strongly suggest that the aberrant breakdown of DA contributes to toxicity in dopaminergic cells.

5.3.2 Oxidation of the Catechol Ring of Dopamine

At neutral pH, the catechol ring of DA can be oxidized to form the electron-deficient DA-quinone as well as superoxide radicals and hydrogen peroxide.[32] This oxidation is thought to be facilitated by the presence of transition-metal ions, such as iron, or when DA acts as a co-substrate in an enzyme-catalyzed reaction.[37] In fact, there is evidence that non-selective enzymatic oxidation of DA by prostaglandin H synthase (COX), lipoxygenase, tyrosinase, and xanthine oxidase may contribute significantly to the production of DA-quinone.[32,37–41] DA-quinone itself is highly reactive, reacting with cellular nucleophiles such as free cysteine or cysteinyl residues on proteins,[42] but a secondary consequence of the oxidation of DA is the generation of ROS *via* redox cycling consuming antioxidants in the process.[43] One of the byproducts of DA oxidation, superoxide, can be converted to H_2O_2 by superoxide dismutase (SOD) or into labile peroxynitrite radicals by nitric oxide.[44] As mentioned earlier, H_2O_2, in the presence of iron, can be broken down into cytotoxic hydroxyl radicals,[45] further contributing to the oxidative load. Finally, free cysteinyl-DA conjugates are capable of further oxidation to form 7-(2-aminoethyl)-3,4-dihydro-2H-1,4-benzothiazine-3-carboxylic acid (DHBT-1), which is a known mitochondrial toxin.[46]

5.3.3 Modification of Protein by Oxidized Dopamine

Electron-deficient DA-quinones are capable of reacting with cellular nucleophiles, such as the reduced sulfhydryl group on cysteinyl residues of small peptides (*e.g.* the tripeptide GSH[38]) and proteins.[47–49] This type of reaction results in covalent binding of thiols by DA-quinone to form 5-cysteinyl-DA.[47,50,51]

In response to these modifications, proteins can be functionally or structurally modified and/or targeted for degradation. As cysteinyl residues on proteins are often within the active sites of proteins, covalent modification of proteins by DA-quinone can result in inactivation of protein function.

Though we have long understood the capability of DA to modify proteins, we have generally taken a protein-by-protein approach to the discovery of DA-conjugated proteins. By this method, it was discovered that the proteins α-synuclein,[52] tyrosine hydroxylase (TH),[14,15] and parkin[53] are covalently modified by DA. In fact, the DA-quinone-α-synuclein adduct inhibits fibril formation by stabilizing the protofibrils that are proposed to be the cytotoxic forms of the protein.[52] Furthermore, the rate-limiting enzyme in catecholamine biosynthesis, TH, is covalently modified and inactivated by DA-quinone.[54,55]

Recently, proteomic techniques were utilized to identify ^{14}C-dopamine conjugated proteins in an unbiased manner.[56] This study was performed in both isolated brain mitochondria and in SH-SY5Y cells, a differentiated dopaminergic cell line. Of interest, proteins associated with the mitochondrial matrix and intermembrane space of the mitochondria were identified as being modified by DA-quinone, indicating that the electrophilic DA-quinone is able to access both compartments of the intact mitochondria at physiological pH.[56] Amongst the mitochondrial proteins identified as being directly conjugated with ^{14}C-dopamine were chaperonin, ubiquinol-cytochrome c reductase core protein 1, glucose-regulated protein 75/mitochondrial HSP70/mortalin, mitofilin, Rieske Fe–S protein subunit of Complex III, the 75 kDa and 30 kDa subunits of Complex I, and mitochondrial creatine kinase. PD-associated proteins ubiquitin carboxy-terminal hydrolase L1 (UCH-L1) and DJ-1 were also identified in both isolated mitochondria and SH-SY5Y cells.[56]

Several processes pivotal to cell health may be disrupted by DA modification of protein, including but not limited to mitochondrial protein import, mitochondrial respiration, cytoskeletal organization, and oxidative defenses. As mitochondrial dysfunction is proposed to play a significant role in PD etiology,[57] it is interesting to note that a large percentage of the proteins covalently modified by DA-quinone at the level of the whole cell were mitochondrial proteins.

Thirteen of the proteins necessary for mitochondrial function are encoded by the mitochondrial genome leaving the other, over 99%, to be synthesized in the cytosol and imported into the mitochondria. Thus, mitochondrial protein import is essential for functioning and survival of mitochondria. Protein import and processing may be impacted by the DA-quinone modification of two mitochondrial protein processing chaperones, heat shock protein (HSP) 60 and GRP75/mtHSP70/mortalin,[56] both of which are required for functional import of mitochondrial proteins. HSP60 is responsible for matrix protein folding,[58] while GRP75/mtHSP70/mortalin participates directly in mitochondrial protein import.[59] Of interest, GRP75/mtHSP70/mortalin protein is substantially decreased in PD brains and in a cellular model of PD. Moreover, modulation of GRP75/mtHSP70/mortalin expression in DA neurons resulted in alterations in proteasomal and oxidative stress pathways that contributed to variations in toxicity,[60] supporting the idea that alterations in protein import machinery can contribute to PD.

The mitochondria are the powerhouses of the cell as they are responsible for the production of ATP *via* the activity of the electron-transport chain (ETC). The ETC is composed of a series of complexes that transfer electrons and couple that transfer with the formation of an electrochemical proton gradient that drives the synthesis of ATP. Subunits of both Complex I (75 kDa and 30 kDa subunits) and Complex III (ubiquinol-cytochrome c reductase core protein 1 and Rieske Fe–S protein) of the ETC are targets of covalent DA modification.[56] This is telling data as studies have illustrated that treatment of isolated brain mitochondria with DA or DA-quinone inhibits mitochondrial respiration.[61–63] Additionally, mitochondrial creatine kinase (mtCK), a critical protein in the maintenance ATP levels,[64] and nucleoside diphosphate kinase A, an enzyme that maintains balance between ADP, GDP, ATP, and GTP levels, are both modified by DA-quinone.[65,66] The loss of mtCK and nucleoside diphosphate kinase A activity may carry further implications for energy maintenance and mitochondrial function. DA-quinone-induced modification of proteins critical to the energy-maintenance function of the mitochondria likely either inhibits their activity and/or targets them for degradation, potentially leading to a decrease in energy production and an increase in mitochondrial dysfunction.

Interestingly, despite the mitochondrially targeted nature of the proteomics study aiming to determine proteins covalently modified by DA-quinone, cytoskeletal proteins were identified as targets of DA modification in isolated brain mitochondria treated with ^{14}C-DA-quinone. Furthermore, the actin-interacting protein tropomyosin was also covalently modified by DA in differentiated SH-SY5Y cells.[56] Though little is understood about the cytoskeletal involvement in PD pathology, there is some evidence that it may be implicated. The cytoskeletal GTP-binding protein Sept4/H5 is associated with Lewy bodies in PD[67] and mutation of the neurofilament M gene has been identified in PD connecting cytoskeletal alterations to neuronal degeneration.[68] Oxidative-induced damage to cytoskeletal proteins could negatively impact not only cellular structure, but also motility and localization of cellular organelles such as the mitochondria. Aberrant localization of the mitochondria could greatly diminish the natural function of the dopaminergic neuron, potentially leading to its demise.

Oxidative stress is a major theme of this chapter and PD in general. As can be expected based on the presence of highly reactive thiols in many antioxidant proteins, this class of proteins has several targets of DA modification. Mitochondrial SOD2 and peroxiredoxin are two such proteins.[56] Both of these enzymes are integral to the maintenance of ROS levels. Loss of function of either of these proteins could contribute to the oxidative stress that, in turn, catalyzes DA oxidation and ultimately results in further oxidative stress and DA modification of proteins, initiating a cycle of oxidative stress that is difficult to break. More recently, a family of proteins with antioxidant capacity has been the focus of DA oxidation studies due to the highly nucleophilic selenocysteine residues that exist in their active sites. It is hypothesized that the high reactive potential of the selenocysteine residue in selenoproteins would make them targets for covalent DA-quinone modification and, as many of them are central to antioxidant systems, this modification would be detrimental to the health of the cell. One selenoprotein of interest is the critically important phospholipid hydroperoxide glutathione peroxidase (GPx4) which is localized primarily to the inner mitochondrial membrane and catalyzes the reduction of H_2O_2, organic hydroperoxides and, uniquely, phospholipid hydroperoxides.[69] GPx4 is, in fact, modified by DA-quinone and, as a result, is decreased in abundance and activity.[70] The discovery of the physical disruption of active sites in antioxidant enzymes that results in functional changes as well as the rapid degradation of those modified proteins is an indication of the detrimental impact of DA oxidation. Not only does it create oxidative stress directly, but it functionally inactivates the defense system developed to combat oxidative stress, potentially leading to damage that is beyond the repair capabilities of the cell.

5.3.4 Mitochondrial Dysfunction and Dopamine Oxidation

Mitochondrial processes can be disrupted by oxidants such as ROS and quinones. Mitochondrial respiration, which is responsible for the generation of ATP through oxidative phosphorylation, is one such process. Several enzymes pivotal to the proper functioning of the ETC have been identified as targets of ROS and sulfhydryl-modifying agents.[71–74] Because both ROS and quinones are capable of modifying critical sulfhydryl groups of proteins, these enzymes may be susceptible to damage by DA oxidation products. In fact, that has been demonstrated directly for several members of Complexes I and III.[56] DA and L-DOPA, the DA precursor, display prominent effects on the mitochondrial respiratory chain, inhibiting Complex I of the ETC in a rat model[75] and inhibiting complex IV in the neuroblastoma NB69 cell line.[76]

Mitochondrial respiration is conventionally classified into several states. Each state is measured based on oxygen consumption.[77] The two most important states being state 3 and state 4. State 3 represents the rate of respiration (O_2 consumption) coupled to ATP synthesis in the presence of ADP and substrate. When all of the ADP is converted to ATP, respiration decreases to a slower resting rate, state 4. Finally, when oxygen has been completely depleted, its uptake ceases and the anoxic state resulting is called state 5.[78] The ratio of state 3 to state 4 can be used to evaluate the functional health of the mitochondria by indicating the degree to which respiration is coupled to ATP synthesis, which is energetically favorable. This ratio is altered in response to DA oxidation in isolated mitochondria. MAO-catalyzed oxidation of DA, leading to the production of H_2O_2, results in a reduction of 24% in state 3 respiration.[61] Interestingly, the addition of an MAO inhibitor (pargyline) or catalase, which is an H_2O_2-scavenger, during the incubation with DA is able to completely prevent the state 3 inhibition, suggesting that it is the production of H_2O_2 and not DA-quinone that is responsible for the state 3 effect.[61] The Complex of the ETC most likely responsible for this phenotype could be determined by utilizing compounds that donate electrons directly to specific points along the ETC, bypassing one or more complexes. The ubiquinone-Complex III component was thus determined to be the most likely target of the H_2O_2-induced inhibition.[61] Though telling, this inhibition of Complex III could not be explained by direct modification of components of the ETC by DA-quinone. To examine the effects of the products resulting from the oxidation of the catechol ring on mitochondrial respiration, the enzyme tyrosinase was used to directly oxidize DA. In response to this direct formation of DA-quinone, state 3 respiration was unchanged, but a dramatic increase was seen in state 4 respiration. This increase of state 4 respiration indicates uncoupling of the mitochondria due to an increase in proton leak across the mitochondrial membrane. Ultimately, after exposure to DA-quinone, respiration in the mitochondria proceeds without coupling to ATP synthesis, which is a wasteful use of cellular energy. To determine the mechanism by which this increase in state 4 respiration was occurring, GSH, capable of either reducing the DA-quinone or acting as a nucleophile to bind DA,[38] was included in the incubation with DA and tyrosinase and prevented the DA oxidation-induced increase in state 4 respiration.[61] This suggests that DA-quinone, and likely its modification of ETC proteins, is involved in the increase in resting respiration. SOD and catalase were used, as before, to determine if detoxification of ROS might protect from the increase in state 4 respiration, but they were unable to do so, further supporting the proposal that the DA-quinone is responsible for the uncoupling of respiration from ATP production.[61]

A decrease in ATP availability is likely to negatively impact many cellular processes. One downstream effect of impaired mitochondrial respiration is the ablation of the proton gradient across the membrane of synaptic vesicles. DA loading into synaptic vesicles relies on the activity of V-ATPase to establish a proton gradient across the vesicular membrane. Without proper mitochondrial functioning, the lack of ATP can disrupt the ability of the V-ATPase to establish a proton gradient, therefore leading to an increase in cytosolic DA oxidation. In addition, the V-ATPase is a target of DA-quinone.[79] From this data it is hypothesized that DA dysregulation can occur through direct inactivation of the V-ATPase or through decreased V-ATPase activity due to ATP deficiency, further propagating the generation of oxidative stress and mitochondrial dysfunction.

The permeability transition pore (PTP) is a calcium-dependent proteinaceous pore that allows the motion of solutes of <1500 daltons (Da) across the normally impermeable mitochondrial inner membrane. Importantly, sulfhydryl modification has been shown to induce PTP opening.[80] As the result of PTP opening, the transmembrane potential is depolarized, resulting in release of small solutes and proteins, osmotic swelling, decreased mitochondrial import, and disruption of oxidative phosphorylation.[81] As mtCK is a component of the PTP and has been identified as a target of DA-quinone modification,[56] and sulfhydryl modification has been shown to induce PTP opening, it is possible that the PTP is a target of DA oxidation products.[80,82–84] Importantly, aberrant opening of the PTP seems to be a pathological event as it has been implicated in apoptosis, excitotoxicity, and ischemia. PTP opening has also been associated with toxicity caused by the PD-associated neurotoxin MPTP.[85–90] DA oxidation to DA-quinone causes an increase in swelling of isolated brain and liver mitochondria that could be inhibited by either the pore inhibitor cyclosporine A or the antioxidant GSH, signifying that the swelling is the consequence of pore opening *via* DA-quinone formation.[61] These data implicate DA oxidation in a pathological mechanism that has been associated with neuronal toxicity and a model of PD, indicating that this is another mechanism by which DA oxidation can cause mitochondrial dysfunction in PD.

5.4 The Role of Dopamine in Toxin-Induced Toxicity

The understanding of the etiology of non-familial PD has been enhanced by the identification of chemical inhibitors of Complex I. The first evidence of the ability of Complex I inhibition to cause selective neurodegeneration came from the discovery of 1-methyl-4-phenyl-1,2,3,6-tetrahydropyridine (MPTP). In the brain, MPTP is converted by MAO-B in glial cells to MPP^+[91,92] that is taken into DA neurons through the DA transporter (DAT). MPP^+ accumulates in mitochondria, where it is a potent complex I inhibitor, and selectively kills dopaminergic neurons in the SN.[93] After the discovery of MPTP's mode of action, it was discovered that there is a systemic Complex I inhibition in patients with PD.[94–98] With that knowledge, systemic administration of a known Complex I inhibitor, the pesticide rotenone, was tested in rats as a potential model of PD.[22] It was found that rotenone, which inhibits Complex I throughout the body, was able to cause selective degeneration of the dopaminergic neurons of the SN.[22] While the Complex I inhibition caused by these toxins has been extensively studied, we are only beginning to understand their impact on DA redistribution and, consequently, DA oxidation.

MPP^+ is a substrate for both DAT and VMAT2, and it can induce a redistribution of DA from vesicles to the cytoplasm, resulting in an increase in DA oxidation.[99] In mice that overexpress VMAT2, MPP^+-induced toxicity was decreased, likely due to enhanced sequestration of MPP^+ removing it from the cytosol, and thus protecting the mitochondria.[100] In mesencephalic cultures, MPP^+-induced cell death could be blocked by depleting DA and administering succinate, a complex II substrate. Conversely, addition of exogenous DA enhanced cellular death in response to MPP^+.[99] Of interest, mice lacking DA in dopaminergic neurons either due to genetic (knockout of TH) or pharmacologic (treatment with alpha-methyl-p-tyrosine) DA depletion were not protected against loss of dopaminergic neurons in response to acute (7 day) administration of MPTP.[101] These data indicate that MPP^+ neurotoxicity is complex and may be dependent on several different mechanisms. Though decreases in DA were not protective against MPP^+ over a short time course, it may be that chronic MPP^+ toxicity is dependent on DA or that the MPP^+ mechanism of action is not dependent on DA.

Rotenone toxicity may also be partially dependent on DA oxidation. Rotenone downregulates expression of the vesicular monoamine transporter (VMAT) and this has been shown to result in the accumulation of DA in PC12 cells.[102] DA depletion with reserpine significantly reduced ROS accumulation and caspase-driven apoptosis in response to rotenone.[103] Interestingly, a direct impact of rotenone treatment on DA-induced toxicity has not been demonstrated, as DA depletion prior to administration of rotenone in differentiated PC12 cells did not protect against rotenone-induced toxicity.[104] However, rotenone toxicity could be potentiated by methamphetamine-induced increases in cytoplasmic DA that was reduced significantly by DA depletion prior to cotreatment of methamphetamine and rotenone,[104] further suggesting that cytoplasmic DA plays a role in rotenone-induced toxicity. Both the MPP^+ and rotenone models, originally thought to be driven primarily by mitochondrial dysfunction, are also influenced by DA oxidation, indicating that DA oxidation may contribute to the selective loss of dopaminergic neurons in response to systemic Complex I inhibition. Likewise, the redox cycler and pesticide paraquat can inhibit Complex I of the ETC[105] and is structurally similar to MPP+, suggesting that it may also be able to increase cystoplasmic DA *via* alteration of vesicular storage,[106] though that has not been demonstrated conclusively.

Epidemiological studies have linked exposure to the fungicide benomyl to increased risk of PD.[107] Benomyl, upon bioactivation, is converted to *S*-methyl *N*-butylthiocarbamate sulfoxide, an aldehyde dehydrogenase inhibitor. Aldehyde dehydrogenase detoxifies MAO-produced DOPAL, converting it to DOPAC within the cytoplasm (refer to Section 5.3.1). When aldehyde dehydrogenase function is reduced, DOPAL levels increase. Increases in DOPAL are postulated to lead to significant neurotoxicity, as suggested by the 'catecholaldehyde hypothesis'.[108–111] In fact, intraperitoneal administration of benomyl in mice resulted in increased DOPAL levels[30] and stereotaxic injection of DOPAL into the rat SN resulted in loss of striatal DA terminals and triggered rotational asymmetry (a behavioral phenotype that suggests disruption of the dopaminergic system).[112] These data suggest that toxins can also target processes designed to detoxify reactive metabolites of cytoplasmic DA.

5.5 Dopamine Toxicity

A common thread runs through both *in vitro* (cell culture) and *in vivo* (animal) models of PD; many induce a shift from vesicular to cytoplasmic DA. As a result of this shift, DA oxidation is significantly increased and, in many of these models, has been shown to contribute significantly to toxicity. The following experiments explore both the impact of cytoplasmic DA and the potential to ameliorate its effects by either depleting DA or administering antioxidants.

5.5.1 *In vitro* Exogenous Dopamine and L-DOPA Treatment

The most direct method of determining the impact of DA oxidation on cells in culture is to treat them with exogenous DA or the DA precursor L-DOPA. Cells capable of DA uptake, such as differentiated rat (PC12) and human

(SH-SY5Y) catecholaminergic cells lines, have been frequently utilized in DA toxicity studies. Administration of DA and L-DOPA in both of these cell types results in significant increases in toxicity; therefore, labs have aimed to determine whether this toxicity is caused by quinoprotein adduct formation or the ROS produced during the course of DA oxidation. SH-SY5Y cells treated with varying concentrations of DA were analyzed, and as early as 24 h after treatment, displayed a decrease in viability. Importantly, quinoprotein adducts increased in a dose-dependent manner in response to DA and correlated well with the decrease in viability, while lipid peroxidation remained unchanged, indicating that quinoprotein adduct formation has a stronger correlation with cell death than oxidative stress as measured by lipid peroxidation.[113] Similarly, in PC12 cells, DA-induced mitochondrial dysfunction and cell death is primarily due to the formation of quinoprotein adducts as opposed to ROS formation.[114]

Prostaglandin H synthase exerts cyclooxygenase (COX) activity as well as peroxidase activity that is thought to couple with oxidation of DA to DA-quinone and may also participate in neuromelanin formation.[37] For this reason, COX inhibitors have been proposed to suppress DA-quinone formation. Using the L-DOPA model of *in vitro* toxicity, the effects of various COX inhibitors on L-DOPA-induced quinoprotein formation and neurotoxicity in dopaminergic CATH.a cells were assessed.[115] Amongst aspirin, acetaminophen, indomethacin, and meloxicam, only aspirin was able to inhibit excess methyl-L-DOPA-induced quinoprotein formation and toxicity, while indomethacin and meloxicam both amplified these alterations, suggesting that COX shows little contribution to DA oxidation and the protective effects of aspirin are not dependent on its COX inhibition, but rather a different activity.[115]

5.5.2 Exogenous Dopamine Administration *in vivo*

If cytoplasmic DA is a major contributor to SN degeneration, administration of excess exogenous DA should be neurotoxic and the toxicity should be specific to dopaminergic neurons as they are the only ones capable of DA uptake. In fact, intrastriatal injections of exogenous DA do cause selective neurotoxicity to dopaminergic terminals.[49,116] Importantly, the extent of DA-induced damage can be directly linked to intracellular DA oxidation and, ultimately, the amount of quinoprotein production.[49,116] Moreover, the impact of DA on terminals could be reduced with the administration of antioxidants, indicating that DA oxidation to ROS and DA-quinone contributes to the loss of DA terminals.[49,116] Moreover, genetic and pharmacological mechanisms of cytoplasmic DA reduction provide neuroprotection.[117] These data suggest that high levels of DA and its toxic metabolites are sufficient to initiate neurodegeneration in rodent models.

The neurotoxin 6-hydroxydopamine (6-OHDA) is taken up by dopaminergic neurons where it is oxidized. As a result of this oxidation, hydrogen peroxide, superoxide, and hydroxyl radicals are formed.[118] Additionally, 6-OHDA, likely through ROS production, has been shown to inhibit mitochondrial ETC components Complex I and Complex IV *in vitro*.[119] 6-OHDA injection into the striatum of rats results in symptoms reminiscent of PD including impairments in movement initiation and in skilled paw use. Concomitant with behavioral deficits, rats demonstrate decreased TH fiber density in the striatum and a reduction in dopaminergic neurons in the SN.[120] An early event in 6-OHDA toxicity seems to be the generation of oxidative stress as oxidative stress markers 4-hydroxynonenal and protein carbonyls are augmented prior to loss of DA in the striatum or SN.[121] 6-OHDA is also oxidized to its *p*-quinone at physiological pH, becoming highly reactive. Although this electron-deficient quinone has been shown to covalently bind to nucleophilic groups of biological macromolecules in a mechanism similar to DA-quinone, 6-OHDA is thought to induce toxicity primarily *via* ROS production and continuous redox cycling.[122] In combination, intrastriatal DA and 6-OHDA injections demonstrate the toxic contributions of DA.

5.5.3 Dysregulation of Dopamine Handling *in vivo*

The psychostimulant methamphetamine is selectively taken into dopaminergic terminals by DAT.[123] Methamphetamine, once taken into the cell, abolishes the proton-motive force necessary for uptake of DA into vesicles and results in cytoplasmic accumulation of DA.[124,125] As such, it serves as a model for establishing the impact of cytoplasmic DA oxidation on neurotoxicity. Treatment with methamphetamine and its derivatives induce the degeneration of DA terminals *in vivo*[126,127] and this effect is dependent on DA as the toxicity can be blocked by DA depletion or treatment with antioxidants.[128–131] Additionally, the magnitude of DA release in response to methamphetamine is considered to be proportional to its long-term toxicity.[132] In an attempt to understand the exact mechanism of methamphetamine-induced toxicity, whether ROS-dependent or DA-quinone-dependent, measurement of the binding of DA-quinones to protein cysteinyl residues has been assessed. A three-fold increase in protein cysteinyl-DA is associated with toxicity resulting from methamphetamine injections in the dopaminergic and serotonergic terminals of the rat striatum.[133] In support of a role for cytoplasmic DA in this phenotype, administration of methamphetamine at 5 °C, when it is not cytotoxic, prevented the indices of DA oxidation and toxicity,[133] linking the two processes.

As previously mentioned, DA is synthesized through the conversion of tyrosine to L-DOPA by tyrosine hydroxylase, and L-DOPA to dopamine by the aromatic L-amino acid decarboxylase. Newly produced DA is packaged into vesicles by VMAT2 for synaptic release but also to prevent pathological DA oxidation. After release into the synapse, DA can be recycled by reuptake *via* the DA transporter (DAT) and then restored in synaptic vesicles *via* VMAT2. If any of these processes is subverted, the disruption in DA sequestration and subsequent DA oxidation can contribute to the dopaminergic neuronal death that we have discussed previously in this chapter.

In rodents, VMAT2 hypomorphs or overexpression of DAT on neurons that are not normally dopaminergic contribute to DA-induced toxicity.[134,135] In addition, the acidic interior of the vesicle is critical for sequestration of DA,[136] and disruptions in vesicular loading result in neurodegeneration.[134,137] Mice expressing only 5% of the normal level of VMAT2 show evidence of DA oxidation and the loss of DA neurons in the SN with age. Additionally, α-synuclein deposition is noted in these mice. This phenotype is considered to be the consequence of increased cytoplasmic DA levels resulting from a loss of sequestration of DA into vesicles.[134,138] These mice also replicate the motor and non-motor symptoms of Parkinson's disease[139] as well as increased susceptibility to methamphetamine-induced toxicity.[140] Conversely, increased DA uptake facilitated by overexpression of DAT on striatal neurons that lacked the ability to sequester DA also resulted in an increase in cytoplasmic DA and ultimately neurotoxicity,[135] providing evidence that alterations to DA sequestration can contribute to PD-like symptoms, specifically loss of dopaminergic neurons of the SN.

5.6 Dopamine and α-Synuclein

Thus far, we have been able to show that environmental toxins linked to PD are capable of increasing DA oxidation and that DA, at least in part, contributes to the toxicity of dopaminergic neurons *in vitro*. One must then consider the familial forms of PD and how DA oxidation might contribute to PD in patients with genetic mutations in proteins such as α-synuclein. The normal function of α-synuclein is not well understood as α-synuclein knockout mice display no overt phenotype. They do have an alteration in DA release in response to stimulus that is hypothesized to be the result of either modulation of presynaptic inhibition of neurotransmitter release or replenishment of synaptic vesicles at DA synapses.[141] Under pathological conditions, such as seen with α-synuclein overexpression or mutation, α-synuclein can cause neurodegeneration of nigral DA neurons as well as hallmark pathological changes including α-synuclein-positive cytoplasmic inclusions and dystrophic neurites.[142] Importantly, assessment of α-synuclein overexpression in non-dopaminergic neurons revealed that α-synuclein inclusion bodies and dystrophic neurites were absent.[142] In an effort to understand the link between PD-associated mutations and DA oxidation, the impact of α-synuclein mutation and overexpression on DA vesicular release and oxidation was assessed. Human embryonic mesencephalic cultures, transfected with wild-type, A53T, and A30P forms of α-synuclein demonstrated increased cell death. Interestingly, this increase in cytotoxicity could be blocked by depletion of intracellular DA with the tyrosine hydroxylase inhibitor α-methyl-*p*-tyrosine or by antioxidant treatment.[143] These data implicate DA oxidation in pathogenic α-synuclein toxicity. The mechanism by which this might occur is through dysregulation of synaptic vesicle recycling which α-synuclein seems to have a role in regulating[141,144] or permeabilization of synaptic vesicles by α-synuclein protofibrils.[145,146] Thus, α-synuclein seems to contribute to release of DA into the cytoplasm, contributing directly to aberrant DA oxidation.

Of note, DA-quinone modifies α-synuclein[147,148] and seems to stabilize the small α-synuclein oligomers that are thought to be the more toxic form of α-synuclein. These α-synuclein protofibrils have been shown to cause membrane damage, which is augmented in the setting of pathogenic α-synuclein mutation.[145] This membrane damage could lead to the permeabilization of DA vesicles that was mentioned previously, initiating a cycle of protein modification and DA release, quickly amplifying the deleterious effects of α-synuclein. Additionally, DA-modified α-synuclein is postulated to inhibit chaperone-mediated autophagy[149] and overexpression of α-synuclein contributes to mitochondrial dysfunction,[150] further exacerbating the vicious cycle of DA-α-synuclein accumulation and resulting damage.

Why is it that the dopaminergic neurons of the SN degenerate while the dopaminergic neurons in other regions, such as the ventral tegmental area (VTA) are relatively spared? If DA is a significant contributing factor to neuronal loss in PD, degeneration of all dopaminergic neurons would be expected. It is hypothesized that there is a fundamental difference between DA handling in dopaminergic neurons of the SN and other dopaminergic neurons that make SN neurons more susceptible.[117] Assessment of cytosolic DA *via* electrochemistry revealed a 2–3-fold higher induction of cytosolic DA in SN neurons relative to VTA neurons in response to L-DOPA. Importantly, this increase in cytosolic DA resulted in greater susceptibility of SN neurons to L-DOPA-induced toxicity.[117] Notably, reduction of intracellular calcium by treatment with the Cav1.2/Cav1.3 blockers nimodipine and nitrendipine specifically reduced L-DOPA-stimulated cytoplasmic DA in SN neurons and decreased L-DOPA-induced neurodegeneration,[117] suggesting that differential regulation of intracellular Ca^{2+} between SN and VTA neurons[151–153] may contribute to differential susceptibility. In order to determine the role that α-synuclein might play in susceptibility, ventral midbrain neurons from α-synuclein knockout mice were treated with L-DOPA and assessed for cell survival. Absence of α-synuclein protected dopaminergic neurons from L-DOPA-induced toxicity but did not alter cytosolic DA concentration.[117] Though cytosolic DA levels were assessed in this study, DA oxidation was not directly measured so care must be taken in extending the results to proof of increased DA oxidation. Taken in combination, the results of these studies from the Sulzer lab support a 'multiple-hit' hypothesis in which high cytoplasmic Ca^{2+}, augmentation of cytosolic DA, and α-synuclein expression all contribute to selective degeneration of dopaminergic neurons of the SN.[117,154]

5.7 Dopamine Storage Disruption in PD

In vitro and *in vivo* models of PD suggest an important role of cytoplasmic DA in pathogenesis of PD. Evidence of reactive DA metabolites has also been gathered from human PD samples. An increase in DA turnover has been identified in PD,[155] and DA oxidation products have been detected at increased levels in the SN of postmortem brain tissue from PD patients[156,157] both suggesting increased cytosolic DA. Increased DA oxidation in postmortem PD SN has been validated by increases in cysteinyl-catechol derivatives.[157]

This increase in DA oxidation may be due to a selective reduction in DA storage vesicles. *In vivo* positron emission tomography (PET) imaging of PD patients reveals more significant reductions in VMAT2 than the signal arising from ^{18}F-DOPA (6-[^{18}F]fluoro-L-DOPA) labeled DA precursor, indicating that VMAT2 is reduced beyond what could be explained by a loss of dopaminergic neurons within the SN.[158] Analysis of the striatum in postmortem tissue from PD patients shows a decrease in VMAT2 immunoreactivity[159] within surviving terminals of the striatum, as well as decreased VMAT2 activity in isolated synaptic vesicles from postmortem PD brains.[160] These data suggest that DA oxidation does occur in the brains of parkinsonian patients and that the mechanism may be due, at least in part, to selective reductions in DA-storage sites that allow DA levels in the cytoplasm to escalate.

Further evidence of DA mishandling in PD is provided by analysis of patients harboring a mutation in the gene coding for VMAT2 that results in infantile-onset motor symptoms including Parkinsonism, in addition to a variety of other developmental dysfunctions.[161] Conversely, gain of function haplotypes confer protection against developing PD in females,[162] and SNPs in the promoter region for VMAT2 have been implicated in decreased likelihood of developing PD.[163] The increase in transcriptional activity resulting from these promoter SNPs is hypothesized to result in additional VMAT2 protein, therefore promoting enhanced DA sequestration, allowing DA to be loaded into vesicles and stabilized. These studies implicate the importance of proper regulation of DA for maintenance of neuronal health.

5.8 Summary and Conclusions

As evidenced by the genetic and environmental conditions that combine to initiate Lewy body pathology and the degeneration of dopaminergic neurons of the *substantia nigra*, PD is a complex disease that is dependent on several causes. This chapter presented evidence that abnormal DA metabolism and/or storage results in pathological DA oxidation and may play a significant role in the demise of the dopaminergic neurons of the SN. In addition to cytosolic DA, SN neurons express high levels of α-synuclein and use L-type Ca^{2+} channels to maintain autonomous pacemaking, resulting in increased cytoplasmic Ca^{2+} concentrations in SN neurons compared to other dopaminergic neurons such as those of the VTA, characteristics that can further enhance cytoplasmic DA release. Additionally, dopaminergic neurons of the SN are susceptible to oxidative stress and antioxidant capacity is reduced in PD, indicating that reactive DA metabolites and ROS resulting from DA oxidation contribute to a high oxidative stress load in PD. Each of these factors contributes to a self-propagating vicious cycle of α-synuclein modification, release of vesicular DA to the cytoplasm, oxidative stress, and mitochondrial dysfunction that likely contribute significantly to the dopaminergic cell death and α-synuclein aggregates that characterize PD. While it is difficult to decipher all of the factors that contribute to PD, having this information enables us to seek novel targets for the treatment and prevention of PD.

References

1. H. Braak, J. R. Bohl, C. M. Muller, U. Rub, R. A. de Vos and K. Del Tredici, Stanley Fahn Lecture 2005: The staging procedure for the inclusion body pathology associated with sporadic Parkinson's disease reconsidered, *Mov. Disord.*, 2006, **21**, 2042–2051.
2. D. C. German, K. Manaye, W. K. Smith, D. J. Woodward and C. B. Saper, Midbrain dopaminergic cell loss in Parkinson's disease: computer visualization, *Ann. Neurol.*, 1989, **26**, 507–514.
3. S. Lesage and A. Brice, Parkinson's disease: from monogenic forms to genetic susceptibility factors, *Hum. Mol. Genet.*, 2009, **18**, R48–R59.
4. T. Gasser, Mendelian forms of Parkinson's disease, *Biochim. Biophys. Acta*, 2009, **1792**, 587–596.
5. C. Klein and A. Westenberger, Genetics of Parkinson's disease, *Cold Spring Harbor Perspect. Med.*, 2012, **2**, a008888.
6. D. N. Hauser and T. G. Hastings, Mitochondrial dysfunction and oxidative stress in Parkinson's disease and monogenic parkinsonism, *Neurobiol. Dis.*, 2013, **51**, 35–42.
7. P. Damier, E. C. Hirsch, P. Zhang, Y. Agid and F. Javoy-Agid, Glutathione peroxidase, glial cells and Parkinson's disease, *Neuroscience*, 1993, **52**, 1–6.
8. D. T. Dexter, C. J. Carter, F. R. Wells, F. Javoy-Agid, Y. Agid, A. Lees, P. Jenner and C. D. Marsden, Basal lipid peroxidation in substantia nigra is increased in Parkinson's disease, *J. Neurochem.*, 1989, **52**, 381–389.
9. H. L. Martin and P. Teismann, Glutathione–a review on its role and significance in Parkinson's disease, *FASEB J.*, 2009, **23**, 3263–3272.
10. P. Riederer, E. Sofic, W. D. Rausch, B. Schmidt, G. P. Reynolds, K. Jellinger and M. B. Youdim, Transition metals, ferritin, glutathione, and ascorbic acid in parkinsonian brains, *J. Neurochem.*, 1989, **52**, 515–520.
11. S. J. Kish, C. Morito and O. Hornykiewicz, Glutathione peroxidase activity in Parkinson's disease brain, *Neurosci. Lett.*, 1985, **58**, 343–346.
12. L. M. Ambani, M. H. Van Woert and S. Murphy, Brain peroxidase and catalase in Parkinson disease, *Arch. Neurol.*, 1975, **32**, 114–118.
13. D. T. Dexter, F. R. Wells, A. J. Lees, F. Agid, Y. Agid, P. Jenner and C. D. Marsden, Increased nigral iron content and alterations in other metal ions occurring in brain in Parkinson's disease, *J. Neurochem.*, 1989, **52**, 1830–1836.
14. E. C. Hirsch, J. P. Brandel, P. Galle, F. Javoy-Agid and Y. Agid, Iron and aluminum increase in the substantia nigra of patients with Parkinson's disease: an X-ray microanalysis, *J. Neurochem.*, 1991, **56**, 446–451.
15. P. F. Good, D. P. Perl, L. M. Bierer and J. Schmeidler, Selective accumulation of aluminum and iron in the neurofibrillary tangles of Alzheimer's disease: a laser microprobe (LAMMA) study, *Ann. Neurol.*, 1992, **31**, 286–292.
16. C. W. Olanow, An introduction to the free radical hypothesis in Parkinson's disease, *Ann. Neurol.*, 1992, **32**, S2–S9.

17. E. Sofic, K. W. Lange, K. Jellinger and P. Riederer, Reduced and oxidized glutathione in the substantia nigra of patients with Parkinson's disease, *Neurosci. Lett.*, 1992, **142**, 128–130.
18. J. Sian, D. T. Dexter, A. J. Lees, S. Daniel, Y. Agid, F. Javoy-Agid, P. Jenner and C. D. Marsden, Alterations in glutathione levels in Parkinson's disease and other neurodegenerative disorders affecting basal ganglia, *Ann. Neurol.*, 1994, **36**, 348–355.
19. J. T. Greenamyre and T. G. Hastings, Biomedicine. Parkinson's–divergent causes, convergent mechanisms, *Science*, 2004, **304**, 1120–1122.
20. N. Ogawa, M. Asanuma, I. Miyazaki, F. J. Diaz-Corrales and K. Miyoshi, L-DOPA treatment from the viewpoint of neuroprotection. Possible mechanism of specific and progressive dopaminergic neuronal death in Parkinson's disease, *J. Neurol.*, 2005, **252**(suppl. 4), IV23–IV31.
21. A. H. Stokes, T. G. Hastings and K. E. Vrana, Cytotoxic and genotoxic potential of dopamine, *J. Neurosci. Res.*, 1999, **55**, 659–665.
22. R. Betarbet, T. B. Sherer, G. MacKenzie, M. Garcia-Osuna, A. V. Panov and J. T. Greenamyre, Chronic systemic pesticide exposure reproduces features of Parkinson's disease, *Nat. Neurosci.*, 2000, **3**, 1301–1306.
23. A. L. McCormack, M. Thiruchelvam, A. B. Manning-Bog, C. Thiffault, J. W. Langston, D. A. Cory-Slechta and D. A. Di Monte, Environmental risk factors and Parkinson's disease: selective degeneration of nigral dopaminergic neurons caused by the herbicide paraquat, *Neurobiol. Dis.*, 2002, **10**, 119–127.
24. M. S. Szczypka, M. A. Rainey, D. S. Kim, W. A. Alaynick, B. T. Marck, A. M. Matsumoto and R. D. Palmiter, Feeding behavior in dopamine-deficient mice, *Proc. Natl. Acad. Sci. U. S. A.*, 1999, **96**, 12138–12143.
25. Q. Y. Zhou and R. D. Palmiter, Dopamine-deficient mice are severely hypoactive, adipsic, and aphagic, *Cell*, 1995, **83**, 1197–1209.
26. R. G. Staal, E. V. Mosharov and D. Sulzer, Dopamine neurons release transmitter *via* a flickering fusion pore, *Nat. Neurosci.*, 2004, **7**, 341–346.
27. D. M. Omiatek, A. J. Bressler, A. S. Cans, A. M. Andrews, M. L. Heien and A. G. Ewing, The real catecholamine content of secretory vesicles in the CNS revealed by electrochemical cytometry, *Sci. Rep.*, 2013, **3**, 1447.
28. W. M. Caudle, R. E. Colebrooke, P. C. Emson and G. W. Miller, Altered vesicular dopamine storage in Parkinson's disease: a premature demise, *Trends Neurosci.*, 2008, **31**, 303–308.
29. T. G. Hastings, The role of dopamine oxidation in mitochondrial dysfunction: implications for Parkinson's disease, *J. Bioenerg. Biomembr.*, 2009, **41**, 469–472.
30. J. E. Casida, B. Ford, Y. Jinsmaa, P. Sullivan, A. Cooney and D. S. Goldstein, Benomyl, aldehyde dehydrogenase, DOPAL, and the catecholaldehyde hypothesis for the pathogenesis of Parkinson's disease, *Chem. Res. Toxicol.*, 2014, **27**, 1359–1361.
31. G. Eisenhofer, I. J. Kopin and D. S. Goldstein, Catecholamine metabolism: a contemporary view with implications for physiology and medicine, *Pharmacol. Rev.*, 2004, **56**, 331–349.

32. D. G. Graham, Oxidative pathways for catecholamines in the genesis of neuromelanin and cytotoxic quinones, *Mol. Pharmacol.*, 1978, **14**, 633–643.
33. H. S. Maker, C. Weiss, D. J. Silides and G. Cohen, Coupling of dopamine oxidation (monoamine oxidase activity) to glutathione oxidation *via* the generation of hydrogen peroxide in rat brain homogenates, *J. Neurochem.*, 1981, **36**, 589–593.
34. B. Halliwell, Reactive oxygen species and the central nervous system, *J. Neurochem.*, 1992, **59**, 1609–1623.
35. H. Checkoway, G. M. Franklin, P. Costa-Mallen, T. Smith-Weller, J. Dilley, P. D. Swanson and L. G. Costa, A genetic polymorphism of MAO-B modifies the association of cigarette smoking and Parkinson's disease, *Neurology*, 1998, **50**, 1458–1461.
36. J. J. Chen, D. M. Swope and K. Dashtipour, Comprehensive review of rasagiline, a second-generation monoamine oxidase inhibitor, for the treatment of Parkinson's disease, *Clin. Ther.*, 2007, **29**, 1825–1849.
37. T. G. Hastings, Enzymatic oxidation of dopamine: the role of prostaglandin H synthase, *J. Neurochem.*, 1995, **64**, 919–924.
38. D. C. Tse, R. L. McCreery and R. N. Adams, Potential oxidative pathways of brain catecholamines, *J. Med. Chem.*, 1976, **19**, 37–40.
39. C. Foppoli, R. Coccia, C. Cini and M. A. Rosei, Catecholamines oxidation by xanthine oxidase, *Biochim. Biophys. Acta*, 1997, **1334**, 200–206.
40. W. Korytowski, T. Sarna, B. Kalyanaraman and R. C. Sealy, Tyrosinase-catalyzed oxidation of dopa and related catechol(amine)s: a kinetic electron spin resonance investigation using spin-stabilization and spin label oximetry, *Biochim. Biophys. Acta*, 1987, **924**, 383–392.
41. M. A. Rosei, C. Blarzino, C. Foppoli, L. Mosca and R. Coccia, Lipoxygenase-catalyzed oxidation of catecholamines, *Biochem. Biophys. Res. Commun.*, 1994, **200**, 344–350.
42. D. Sulzer and L. Zecca, Intraneuronal dopamine-quinone synthesis: a review, *Neurotoxic. Res.*, 2000, **1**, 181–195.
43. J. L. Bolton, M. A. Trush, T. M. Penning, G. Dryhurst and T. J. Monks, Role of quinones in toxicology, *Chem. Res. Toxicol.*, 2000, **13**, 135–160.
44. G. R. Buettner, Superoxide dismutase in redox biology: the roles of superoxide and hydrogen peroxide, *Anticancer Agents Med. Chem.*, 2011, **11**, 341–346.
45. S. Croft, B. C. Gilbert, J. R. Smith and A. C. Whitwood, An E.S.R. investigation of the reactive intermediate generated in the reaction between FeII and H2O2 in aqueous solution. Direct evidence for the formation of the hydroxyl radical, *Free Radical Res. Commun.*, 1992, **17**, 21–39.
46. H. Li and G. Dryhurst, Irreversible inhibition of mitochondrial complex I by 7-(2-aminoethyl)-3,4-dihydro-5-hydroxy-2H-1,4-benzothiazine-3-carboxylic acid (DHBT-1): a putative nigral endotoxin of relevance to Parkinson's disease, *J. Neurochem.*, 1997, **69**, 1530–1541.

47. D. G. Graham, S. M. Tiffany, W. R. Bell Jr and W. F. Gutknecht, Autoxidation *versus* covalent binding of quinones as the mechanism of toxicity of dopamine, 6-hydroxydopamine, and related compounds toward C1300 neuroblastoma cells *in vitro*, *Mol. Pharmacol.*, 1978, **14**, 644–653.
48. T. G. Hastings and M. J. Zigmond, Identification of catechol-protein conjugates in neostriatal slices incubated with [3H]dopamine: impact of ascorbic acid and glutathione, *J. Neurochem.*, 1994, **63**, 1126–1132.
49. T. G. Hastings, D. A. Lewis and M. J. Zigmond, Role of oxidation in the neurotoxic effects of intrastriatal dopamine injections, *Proc. Natl. Acad. Sci. U. S. A.*, 1996, **93**, 1956–1961.
50. T. G. Hastings, D. A. Lewis and M. J. Zigmond, Reactive dopamine metabolites and neurotoxicity: implications for Parkinson's disease, *Adv. Exp. Med. Biol.*, 1996, **387**, 97–106.
51. S. Ito, T. Kato and K. Fujita, Covalent binding of catechols to proteins through the sulphydryl group, *Biochem. Pharmacol.*, 1988, **37**, 1707–1710.
52. K. A. Conway, J. C. Rochet, R. M. Bieganski and P. T. Lansbury Jr, Kinetic stabilization of the alpha-synuclein protofibril by a dopamine-alpha-synuclein adduct, *Science*, 2001, **294**, 1346–1349.
53. M. J. LaVoie, B. L. Ostaszewski, A. Weihofen, M. G. Schlossmacher and D. J. Selkoe, Dopamine covalently modifies and functionally inactivates parkin, *Nat. Med.*, 2005, **11**, 1214–1221.
54. Y. Xu, A. H. Stokes, R. Roskoski Jr and K. E. Vrana, Dopamine, in the presence of tyrosinase, covalently modifies and inactivates tyrosine hydroxylase, *J. Neurosci. Res.*, 1998, **54**, 691–697.
55. D. M. Kuhn, R. E. Arthur Jr, D. M. Thomas and L. A. Elferink, Tyrosine hydroxylase is inactivated by catechol-quinones and converted to a redox-cycling quinoprotein: possible relevance to Parkinson's disease, *J. Neurochem.*, 1999, **73**, 1309–1317.
56. V. S. Van Laar, A. J. Mishizen, M. Cascio and T. G. Hastings, Proteomic identification of dopamine-conjugated proteins from isolated rat brain mitochondria and SH-SY5Y cells, *Neurobiol. Dis.*, 2009, **34**, 487–500.
57. J. T. Greenamyre, G. MacKenzie, T. I. Peng and S. E. Stephans, Mitochondrial dysfunction in Parkinson's disease, *Biochem. Soc. Symp.*, 1999, **66**, 85–97.
58. R. Wadhwa, S. Takano, K. Kaur, S. Aida, T. Yaguchi, Z. Kaul, T. Hirano, K. Taira and S. C. Kaul, Identification and characterization of molecular interactions between mortalin/mtHsp70 and HSP60, *Biochem. J.*, 2005, **391**, 185–190.
59. T. Yaguchi, S. Aida, S. C. Kaul and R. Wadhwa, Involvement of mortalin in cellular senescence from the perspective of its mitochondrial import, chaperone, and oxidative stress management functions, *Ann. N. Y. Acad. Sci.*, 2007, **1100**, 306–311.
60. J. Jin, C. Hulette, Y. Wang, T. Zhang, C. Pan, R. Wadhwa and J. Zhang, Proteomic identification of a stress protein, mortalin/mthsp70/GRP75: relevance to Parkinson disease, *Mol. Cell. Proteomics*, 2006, **5**, 1193–1204.

61. S. B. Berman and T. G. Hastings, Dopamine oxidation alters mitochondrial respiration and induces permeability transition in brain mitochondria: implications for Parkinson's disease, *J. Neurochem.*, 1999, **73**, 1127–1137.
62. G. Cohen, R. Farooqui and N. Kesler, Parkinson disease: a new link between monoamine oxidase and mitochondrial electron flow, *Proc. Natl. Acad. Sci. U. S. A.*, 1997, **94**, 4890–4894.
63. M. Gluck, J. Ehrhart, E. Jayatilleke and G. D. Zeevalk, Inhibition of brain mitochondrial respiration by dopamine: involvement of H(2)O(2) and hydroxyl radicals but not glutathione-protein-mixed disulfides, *J. Neurochem.*, 2002, **82**, 66–74.
64. M. Eder, M. Stolz, T. Wallimann and U. Schlattner, A conserved negatively charged cluster in the active site of creatine kinase is critical for enzymatic activity, *J. Biol. Chem.*, 2000, **275**, 27094–27099.
65. R. C. Cumming, N. L. Andon, P. A. Haynes, M. Park, W. H. Fischer and D. Schubert, Protein disulfide bond formation in the cytoplasm during oxidative stress, *J. Biol. Chem.*, 2004, **279**, 21749–21758.
66. E. J. Song, Y. S. Kim, J. Y. Chung, E. Kim, S. K. Chae and K. J. Lee, Oxidative modification of nucleoside diphosphate kinase and its identification by matrix-assisted laser desorption/ionization time-of-flight mass spectrometry, *Biochemistry*, 2000, **39**, 10090–10097.
67. M. Ihara, H. Tomimoto, H. Kitayama, Y. Morioka, I. Akiguchi, H. Shibasaki, M. Noda and M. Kinoshita, Association of the cytoskeletal GTP-binding protein Sept4/H5 with cytoplasmic inclusions found in Parkinson's disease and other synucleinopathies, *J. Biol. Chem.*, 2003, **278**, 24095–24102.
68. C. Lavedan, S. Buchholtz, R. L. Nussbaum, R. L. Albin and M. H. Polymeropoulos, A mutation in the human neurofilament M gene in Parkinson's disease that suggests a role for the cytoskeleton in neuronal degeneration, *Neurosci. Lett.*, 2002, **322**, 57–61.
69. N. E. Savaskan, C. Ufer, H. Kuhn and A. Borchert, Molecular biology of glutathione peroxidase 4: from genomic structure to developmental expression and neural function, *Biol. Chem.*, 2007, **388**, 1007–1017.
70. D. N. Hauser, A. A. Dukes, A. D. Mortimer and T. G. Hastings, Dopamine quinone modifies and decreases the abundance of the mitochondrial selenoprotein glutathione peroxidase 4, *Free Radical Biol. Med.*, 2013, **65**, 419–427.
71. W. C. Kenney, The reaction of N-ethylmaleimide at the active site of succinate dehydrogenase, *J. Biol. Chem.*, 1975, **250**, 3089–3094.
72. T. Yagi and Y. Hatefi, Thiols in oxidative phosphorylation: thiols in the F0 of ATP synthase essential for ATPase activity, *Arch. Biochem. Biophys.*, 1987, **254**, 102–109.
73. Y. Zhang, O. Marcillat, C. Giulivi, L. Ernster and K. J. Davies, The oxidative inactivation of mitochondrial electron transport chain components and ATPase, *J. Biol. Chem.*, 1990, **265**, 16330–16336.

74. O. Benard and K. A. Balasubramanian, Effect of oxidized glutathione on intestinal mitochondria and brush border membrane, *Int. J. Biochem. Cell Biol.*, 1995, **27**, 589–595.
75. S. Przedborski, V. Jackson-Lewis, U. Muthane, H. Jiang, M. Ferreira, A. B. Naini and S. Fahn, Chronic levodopa administration alters cerebral mitochondrial respiratory chain activity, *Ann. Neurol.*, 1993, **34**, 715–723.
76. B. Pardo, M. A. Mena and J. G. de Yebenes, L-dopa inhibits complex IV of the electron transport chain in catecholamine-rich human neuroblastoma NB69 cells, *J. Neurochem.*, 1995, **64**, 576–582.
77. B. Chance and G. R. Williams, The respiratory chain and oxidative phosphorylation, *Adv. Enzymol. Relat. Subj. Biochem.*, 1956, **17**, 65–134.
78. B. Chance and G. R. Williams, Respiratory enzymes in oxidative phosphorylation. III. The steady state, *J. Biol. Chem.*, 1955, **217**, 409–427.
79. O. Terland, B. Almas, T. Flatmark, K. K. Andersson and M. Sorlie, One-electron oxidation of catecholamines generates free radicals with an *in vitro* toxicity correlating with their lifetime, *Free Radical Biol. Med.*, 2006, **41**, 1266–1271.
80. V. G. Valle, M. M. Fagian, L. S. Parentoni, A. R. Meinicke and A. E. Vercesi, The participation of reactive oxygen species and protein thiols in the mechanism of mitochondrial inner membrane permeabilization by calcium plus prooxidants, *Arch. Biochem. Biophys.*, 1993, **307**, 1–7.
81. T. E. Gunter and D. R. Pfeiffer, Mechanisms by which mitochondria transport calcium, *Am. J. Physiol.*, 1990, **258**, C755–C786.
82. P. Bernardi, S. Vassanelli, P. Veronese, R. Colonna, I. Szabo and M. Zoratti, Modulation of the mitochondrial permeability transition pore. Effect of protons and divalent cations, *J. Biol. Chem.*, 1992, **267**, 2934–2939.
83. V. Petronilli, P. Costantini, L. Scorrano, R. Colonna, S. Passamonti and P. Bernardi, The voltage sensor of the mitochondrial permeability transition pore is tuned by the oxidation-reduction state of vicinal thiols. Increase of the gating potential by oxidants and its reversal by reducing agents, *J. Biol. Chem.*, 1994, **269**, 16638–16642.
84. S. Golshani-Hebroni, Mg(++) requirement for MtHK binding, and Mg(++) stabilization of mitochondrial membranes *via* activation of MtHK & MtCK and promotion of mitochondrial permeability transition pore closure: A hypothesis on mechanisms underlying Mg(++)'s antioxidant and cytoprotective effects, *Gene*, 2016, **581**, 1–13.
85. H. Uchino, E. Elmer, K. Uchino, O. Lindvall and B. K. Siesjo, Cyclosporin A dramatically ameliorates CA1 hippocampal damage following transient forebrain ischaemia in the rat, *Acta Physiol. Scand.*, 1995, **155**, 469–471.
86. A. L. Nieminen, T. G. Petrie, J. J. Lemasters and W. R. Selman, Cyclosporin A delays mitochondrial depolarization induced by N-methyl-D-aspartate in cortical neurons: evidence of the mitochondrial permeability transition, *Neuroscience*, 1996, **75**, 993–997.

87. M. A. Packer, R. Miesel and M. P. Murphy, Exposure to the parkinsonian neurotoxin 1-methyl-4-phenylpyridinium (MPP+) and nitric oxide simultaneously causes cyclosporin A-sensitive mitochondrial calcium efflux and depolarisation, *Biochem. Pharmacol.*, 1996, **51**, 267–273.
88. A. F. Schinder, E. C. Olson, N. C. Spitzer and M. Montal, Mitochondrial dysfunction is a primary event in glutamate neurotoxicity, *J. Neurosci.*, 1996, **16**, 6125–6133.
89. N. Zamzami, S. A. Susin, P. Marchetti, T. Hirsch, I. Gomez-Monterrey, M. Castedo and G. Kroemer, Mitochondrial control of nuclear apoptosis, *J. Exp. Med.*, 1996, **183**, 1533–1544.
90. D. S. Cassarino, C. P. Fall, T. S. Smith and J. P. Bennett Jr, Pramipexole reduces reactive oxygen species production *in vivo* and *in vitro* and inhibits the mitochondrial permeability transition produced by the parkinsonian neurotoxin methylpyridinium ion, *J. Neurochem.*, 1998, **71**, 295–301.
91. J. W. Langston, I. Irwin, E. B. Langston and L. S. Forno, Pargyline prevents MPTP-induced parkinsonism in primates, *Science*, 1984, **225**, 1480–1482.
92. J. A. Javitch and S. H. Snyder, Uptake of MPP(+) by dopamine neurons explains selectivity of parkinsonism-inducing neurotoxin, MPTP, *Eur. J. Pharmacol.*, 1984, **106**, 455–456.
93. J. T. Greenamyre, R. Betarbet and T. B. Sherer, The rotenone model of Parkinson's disease: genes, environment and mitochondria, *Parkinsonism Relat. Disord.*, 2003, **9**(suppl. 2), S59–S64.
94. Y. Mizuno, S. Ohta, M. Tanaka, S. Takamiya, K. Suzuki, T. Sato, H. Oya, T. Ozawa and Y. Kagawa, Deficiencies in complex I subunits of the respiratory chain in Parkinson's disease, *Biochem. Biophys. Res. Commun.*, 1989, **163**, 1450–1455.
95. A. H. Schapira, J. M. Cooper, D. Dexter, P. Jenner, J. B. Clark and C. D. Marsden, Mitochondrial complex I deficiency in Parkinson's disease, *Lancet*, 1989, **1**, 1269.
96. W. D. Parker Jr, S. J. Boyson and J. K. Parks, Abnormalities of the electron transport chain in idiopathic Parkinson's disease, *Ann. Neurol.*, 1989, **26**, 719–723.
97. R. Benecke, P. Strumper and H. Weiss, Electron transfer complex I defect in idiopathic dystonia, *Ann. Neurol.*, 1992, **32**, 683–686.
98. R. H. Haas, F. Nasirian, K. Nakano, D. Ward, M. Pay, R. Hill and C. W. Shults, Low platelet mitochondrial complex I and complex II/III activity in early untreated Parkinson's disease, *Ann. Neurol.*, 1995, **37**, 714–722.
99. J. Lotharius and K. L. O'Malley, The parkinsonism-inducing drug 1-methyl-4-phenylpyridinium triggers intracellular dopamine oxidation. A novel mechanism of toxicity, *J. Biol. Chem.*, 2000, **275**, 38581–38588.
100. K. M. Lohr, A. I. Bernstein, K. A. Stout, A. R. Dunn, C. R. Lazo, S. P. Alter, M. Wang, Y. Li, X. Fan, E. J. Hess, H. Yi, L. M. Vecchio, D. S. Goldstein, T. S. Guillot, A. Salahpour and G. W. Miller, Increased vesicular

monoamine transporter enhances dopamine release and opposes Parkinson disease-related neurodegeneration *in vivo*, *Proc. Natl. Acad. Sci. U. S. A.*, 2014, **111**, 9977–9982.

101. D. M. Hasbani, F. A. Perez, R. D. Palmiter and K. L. O'Malley, Dopamine depletion does not protect against acute 1-methyl-4-phenyl-1,2,3,6-tetrahydropyridine toxicity *in vivo*, *J. Neurosci.*, 2005, **25**, 9428–9433.
102. Y. Sai, Q. Wu, W. Le, F. Ye, Y. Li and Z. Dong, Rotenone-induced PC12 cell toxicity is caused by oxidative stress resulting from altered dopamine metabolism, *Toxicol. In Vitro*, 2008, **22**, 1461–1468.
103. H. Q. Liu, X. Z. Zhu and E. Q. Weng, Intracellular dopamine oxidation mediates rotenone-induced apoptosis in PC12 cells, *Acta Pharmacol. Sin.*, 2005, **26**, 17–26.
104. A. A. Dukes, K. M. Korwek and T. G. Hastings, The effect of endogenous dopamine in rotenone-induced toxicity in PC12 cells, *Antioxid. Redox Signaling*, 2005, **7**, 630–638.
105. T. Fukushima, K. Yamada, A. Isobe, K. Shiwaku and Y. Yamane, Mechanism of cytotoxicity of paraquat. I. NADH oxidation and paraquat radical formation *via* complex I, *Exp. Toxicol. Pathol.*, 1993, **45**, 345–349.
106. J. Lotharius and P. Brundin, Pathogenesis of Parkinson's disease: dopamine, vesicles and alpha-synuclein, *Nat. Rev. Neurosci.*, 2002, **3**, 932–942.
107. A. G. Fitzmaurice, S. L. Rhodes, A. Lulla, N. P. Murphy, H. A. Lam, K. C. O'Donnell, L. Barnhill, J. E. Casida, M. Cockburn, A. Sagasti, M. C. Stahl, N. T. Maidment, B. Ritz and J. M. Bronstein, Aldehyde dehydrogenase inhibition as a pathogenic mechanism in Parkinson disease, *Proc. Natl. Acad. Sci. U. S. A.*, 2013, **110**, 636–641.
108. W. J. Burke, S. W. Li, E. A. Williams, R. Nonneman and D. S. Zahm, 3,4-Dihydroxyphenylacetaldehyde is the toxic dopamine metabolite *in vivo*: implications for Parkinson's disease pathogenesis, *Brain Res.*, 2003, **989**, 205–213.
109. J. N. Rees, V. R. Florang, L. L. Eckert and J. A. Doorn, Protein reactivity of 3,4-dihydroxyphenylacetaldehyde, a toxic dopamine metabolite, is dependent on both the aldehyde and the catechol, *Chem. Res. Toxicol.*, 2009, **22**, 1256–1263.
110. D. G. Anderson, S. V. Mariappan, G. R. Buettner and J. A. Doorn, Oxidation of 3,4-dihydroxyphenylacetaldehyde, a toxic dopaminergic metabolite, to a semiquinone radical and an ortho-quinone, *J. Biol. Chem.*, 2011, **286**, 26978–26986.
111. D. S. Goldstein, P. Sullivan, C. Holmes, I. J. Kopin, M. J. Basile and D. C. Mash, Catechols in post-mortem brain of patients with Parkinson disease, *Eur. J. Neurol.*, 2011, **18**, 703–710.
112. W. M. Panneton, V. B. Kumar, Q. Gan, W. J. Burke and J. E. Galvin, The neurotoxicity of DOPAL: behavioral and stereological evidence for its role in Parkinson disease pathogenesis, *PLoS One*, 2010, **5**, e15251.
113. N. Wang, Y. Wang, G. Yu, C. Yuan and J. Ma, Quinoprotein adducts accumulate in the substantia nigra of aged rats and correlate with dopamine-induced toxicity in SH-SY5Y cells, *Neurochem. Res.*, 2011, **36**, 2169–2175.

114. S. Jana, M. Sinha, D. Chanda, T. Roy, K. Banerjee, S. Munshi, B. S. Patro and S. Chakrabarti, Mitochondrial dysfunction mediated by quinone oxidation products of dopamine: Implications in dopamine cytotoxicity and pathogenesis of Parkinson's disease, *Biochim. Biophys. Acta*, 2011, **1812**, 663–673.
115. M. Asanuma, I. Miyazaki, Y. Kikkawa, N. Kimoto, M. Takeshima, S. Murakami and K. Miyoshi, Cyclooxygenase-independent neuroprotective effects of aspirin against dopamine quinone-induced neurotoxicity, *Neurochem. Res.*, 2012, **37**, 1944–1951.
116. A. D. Rabinovic, D. A. Lewis and T. G. Hastings, Role of oxidative changes in the degeneration of dopamine terminals after injection of neurotoxic levels of dopamine, *Neuroscience*, 2000, **101**, 67–76.
117. E. V. Mosharov, K. E. Larsen, E. Kanter, K. A. Phillips, K. Wilson, Y. Schmitz, D. E. Krantz, K. Kobayashi, R. H. Edwards and D. Sulzer, Interplay between cytosolic dopamine, calcium, and alpha-synuclein causes selective death of substantia nigra neurons, *Neuron*, 2009, **62**, 218–229.
118. G. Cohen and R. E. Heikkila, The generation of hydrogen peroxide, superoxide radical, and hydroxyl radical by 6-hydroxydopamine, dialuric acid, and related cytotoxic agents, *J. Biol. Chem.*, 1974, **249**, 2447–2452.
119. Y. Y. Glinka and M. B. Youdim, Inhibition of mitochondrial complexes I and IV by 6-hydroxydopamine, *Eur. J. Pharmacol.*, 1995, **292**, 329–332.
120. D. Kirik, C. Rosenblad and A. Bjorklund, Characterization of behavioral and neurodegenerative changes following partial lesions of the nigrostriatal dopamine system induced by intrastriatal 6-hydroxydopamine in the rat, *Exp. Neurol.*, 1998, **152**, 259–277.
121. M. P. Smith and W. A. Cass, Oxidative stress and dopamine depletion in an intrastriatal 6-hydroxydopamine model of Parkinson's disease, *Neuroscience*, 2007, **144**, 1057–1066.
122. A. Saner and H. Thoenen, Model experiments on the molecular mechanism of action of 6-hydroxydopamine, *Mol. Pharmacol.*, 1971, **7**, 147–154.
123. S. R. Jones, R. R. Gainetdinov, R. M. Wightman and M. G. Caron, Mechanisms of amphetamine action revealed in mice lacking the dopamine transporter, *J. Neurosci.*, 1998, **18**, 1979–1986.
124. D. Sulzer and S. Rayport, Amphetamine and other psychostimulants reduce pH gradients in midbrain dopaminergic neurons and chromaffin granules: a mechanism of action, *Neuron*, 1990, **5**, 797–808.
125. D. Sulzer, T. K. Chen, Y. Y. Lau, H. Kristensen, S. Rayport and A. Ewing, Amphetamine redistributes dopamine from synaptic vesicles to the cytosol and promotes reverse transport, *J. Neurosci.*, 1995, **15**, 4102–4108.
126. L. S. Seiden, D. L. Commins, G. Vosmer, K. Axt and G. Marek, Neurotoxicity in dopamine and 5-hydroxytryptamine terminal fields: a regional analysis in nigrostriatal and mesolimbic projections, *Ann. N. Y. Acad. Sci.*, 1988, **537**, 161–172.

127. P. K. Sonsalla, N. D. Jochnowitz, G. D. Zeevalk, J. A. Oostveen and E. D. Hall, Treatment of mice with methamphetamine produces cell loss in the substantia nigra, *Brain Res.*, 1996, **738**, 172–175.
128. C. J. Schmidt, J. K. Ritter, P. K. Sonsalla, G. R. Hanson and J. W. Gibb, Role of dopamine in the neurotoxic effects of methamphetamine, *J. Pharmacol. Exp. Ther.*, 1985, **233**, 539–544.
129. M. J. De Vito and G. C. Wagner, Methamphetamine-induced neuronal damage: a possible role for free radicals, *Neuropharmacology*, 1989, **28**, 1145–1150.
130. J. L. Cadet, B. Ladenheim, I. Baum, E. Carlson and C. Epstein, CuZn-superoxide dismutase (CuZnSOD) transgenic mice show resistance to the lethal effects of methylenedioxyamphetamine (MDA) and of methylenedioxymethamphetamine (MDMA), *Brain Res.*, 1994, **655**, 259–262.
131. J. W. Gibb and F. J. Kogan, Influence of dopamine synthesis on methamphetamine-induced changes in striatal and adrenal tyrosine hydroxylase activity, *Naunyn-Schmiedeberg's Arch. Pharmacol.*, 1979, **310**, 185–187.
132. S. J. O'Dell, F. B. Weihmuller and J. F. Marshall, Multiple methamphetamine injections induce marked increases in extracellular striatal dopamine which correlate with subsequent neurotoxicity, *Brain Res.*, 1991, **564**, 256–260.
133. M. J. LaVoie and T. G. Hastings, Dopamine quinone formation and protein modification associated with the striatal neurotoxicity of methamphetamine: evidence against a role for extracellular dopamine, *J. Neurosci.*, 1999, **19**, 1484–1491.
134. W. M. Caudle, J. R. Richardson, M. Z. Wang, T. N. Taylor, T. S. Guillot, A. L. McCormack, R. E. Colebrooke, D. A. Di Monte, P. C. Emson and G. W. Miller, Reduced vesicular storage of dopamine causes progressive nigrostriatal neurodegeneration, *J. Neurosci.*, 2007, **27**, 8138–8148.
135. L. Chen, Y. Ding, B. Cagniard, A. D. Van Laar, A. Mortimer, W. Chi, T. G. Hastings, U. J. Kang and X. Zhuang, Unregulated cytosolic dopamine causes neurodegeneration associated with oxidative stress in mice, *J. Neurosci.*, 2008, **28**, 425–433.
136. T. S. Hnasko, N. Chuhma, H. Zhang, G. Y. Goh, D. Sulzer, R. D. Palmiter, S. Rayport and R. H. Edwards, Vesicular glutamate transport promotes dopamine storage and glutamate corelease *in vivo*, *Neuron*, 2010, **65**, 643–656.
137. M. Larhammar, K. Patra, M. Blunder, L. Emilsson, C. Peuckert, E. Arvidsson, D. Ronnlund, J. Preobraschenski, C. Birgner, C. Limbach, J. Widengren, H. Blom, R. Jahn, A. Wallen-Mackenzie and K. Kullander, SLC10A4 is a vesicular amine-associated transporter modulating dopamine homeostasis, *Biol. Psychiatry*, 2015, **77**, 526–536.
138. R. E. Colebrooke, T. Humby, P. J. Lynch, D. P. McGowan, J. Xia and P. C. Emson, Age-related decline in striatal dopamine content and motor performance occurs in the absence of nigral cell loss in a genetic mouse model of Parkinson's disease, *Eur. J. Neurosci.*, 2006, **24**, 2622–2630.

139. T. N. Taylor, W. M. Caudle and G. W. Miller, VMAT2-Deficient Mice Display Nigral and Extranigral Pathology and Motor and Nonmotor Symptoms of Parkinson's Disease, *Parkinson's Dis.*, 2011, **2011**, 124165.
140. T. S. Guillot, K. R. Shepherd, J. R. Richardson, M. Z. Wang, Y. Li, P. C. Emson and G. W. Miller, Reduced vesicular storage of dopamine exacerbates methamphetamine-induced neurodegeneration and astrogliosis, *J. Neurochem.*, 2008, **106**, 2205–2217.
141. A. Abeliovich, Y. Schmitz, I. Farinas, D. Choi-Lundberg, W. H. Ho, P. E. Castillo, N. Shinsky, J. M. Verdugo, M. Armanini, A. Ryan, M. Hynes, H. Phillips, D. Sulzer and A. Rosenthal, Mice lacking alpha-synuclein display functional deficits in the nigrostriatal dopamine system, *Neuron*, 2000, **25**, 239–252.
142. D. Kirik, C. Rosenblad, C. Burger, C. Lundberg, T. E. Johansen, N. Muzyczka, R. J. Mandel and A. Bjorklund, Parkinson-like neurodegeneration induced by targeted overexpression of alpha-synuclein in the nigrostriatal system, *J. Neurosci.*, 2002, **22**, 2780–2791.
143. J. Xu, S. Y. Kao, F. J. Lee, W. Song, L. W. Jin and B. A. Yankner, Dopamine-dependent neurotoxicity of alpha-synuclein: a mechanism for selective neurodegeneration in Parkinson disease, *Nat. Med.*, 2002, **8**, 600–606.
144. D. E. Cabin, K. Shimazu, D. Murphy, N. B. Cole, W. Gottschalk, K. L. McIlwain, B. Orrison, A. Chen, C. E. Ellis, R. Paylor, B. Lu and R. L. Nussbaum, Synaptic vesicle depletion correlates with attenuated synaptic responses to prolonged repetitive stimulation in mice lacking alpha-synuclein, *J. Neurosci.*, 2002, **22**, 8797–8807.
145. M. J. Volles, S. J. Lee, J. C. Rochet, M. D. Shtilerman, T. T. Ding, J. C. Kessler and P. T. Lansbury Jr, Vesicle permeabilization by protofibrillar alpha-synuclein: implications for the pathogenesis and treatment of Parkinson's disease, *Biochemistry*, 2001, **40**, 7812–7819.
146. H. A. Lashuel, B. M. Petre, J. Wall, M. Simon, R. J. Nowak, T. Walz and P. T. Lansbury Jr, Alpha-synuclein, especially the Parkinson's disease-associated mutants, forms pore-like annular and tubular protofibrils, *J. Mol. Biol.*, 2002, **322**, 1089–1102.
147. J. C. Rochet, T. F. Outeiro, K. A. Conway, T. T. Ding, M. J. Volles, H. A. Lashuel, R. M. Bieganski, S. L. Lindquist and P. T. Lansbury, Interactions among alpha-synuclein, dopamine, and biomembranes: some clues for understanding neurodegeneration in Parkinson's disease, *J. Mol. Neurosci.*, 2004, **23**, 23–34.
148. E. H. Norris, B. I. Giasson, R. Hodara, S. Xu, J. Q. Trojanowski, H. Ischiropoulos and V. M. Lee, Reversible inhibition of alpha-synuclein fibrillization by dopaminochrome-mediated conformational alterations, *J. Biol. Chem.*, 2005, **280**, 21212–21219.
149. M. Martinez-Vicente, Z. Talloczy, S. Kaushik, A. C. Massey, J. Mazzulli, E. V. Mosharov, R. Hodara, R. Fredenburg, D. C. Wu, A. Follenzi, W. Dauer, S. Przedborski, H. Ischiropoulos, P. T. Lansbury, D. Sulzer and A. M. Cuervo, Dopamine-modified alpha-synuclein blocks chaperone-mediated autophagy, *J. Clin. Invest.*, 2008, **118**, 777–788.

150. R. Ved, S. Saha, B. Westlund, C. Perier, L. Burnam, A. Sluder, M. Hoener, C. M. Rodrigues, A. Alfonso, C. Steer, L. Liu, S. Przedborski and B. Wolozin, Similar patterns of mitochondrial vulnerability and rescue induced by genetic modification of alpha-synuclein, parkin, and DJ-1 in Caenorhabditis elegans, *J. Biol. Chem.*, 2005, **280**, 42655–42668.
151. C. R. Gerfen, K. G. Baimbridge and J. J. Miller, The neostriatal mosaic: compartmental distribution of calcium-binding protein and parvalbumin in the basal ganglia of the rat and monkey, *Proc. Natl. Acad. Sci. U. S. A.*, 1985, **82**, 8780–8784.
152. D. J. Surmeier, Calcium, ageing, and neuronal vulnerability in Parkinson's disease, *Lancet Neurol.*, 2007, **6**, 933–938.
153. L. Thompson, P. Barraud, E. Andersson, D. Kirik and A. Bjorklund, Identification of dopaminergic neurons of nigral and ventral tegmental area subtypes in grafts of fetal ventral mesencephalon based on cell morphology, protein expression, and efferent projections, *J. Neurosci.*, 2005, **25**, 6467–6477.
154. D. Sulzer, Multiple hit hypotheses for dopamine neuron loss in Parkinson's disease, *Trends Neurosci.*, 2007, **30**, 244–250.
155. H. Bernheimer, W. Birkmayer, O. Hornykiewicz, K. Jellinger and F. Seitelberger, Brain dopamine and the syndromes of Parkinson and Huntington. Clinical, morphological and neurochemical correlations, *J. Neurol. Sci.*, 1973, **20**, 415–455.
156. B. Fornstedt, A. Brun, E. Rosengren and A. Carlsson, The apparent autoxidation rate of catechols in dopamine-rich regions of human brains increases with the degree of depigmentation of substantia nigra, *J. Neural Transm.: Parkinson's Dis. Dementia Sect.*, 1989, **1**, 279–295.
157. J. P. Spencer, P. Jenner, S. E. Daniel, A. J. Lees, D. C. Marsden and B. Halliwell, Conjugates of catecholamines with cysteine and GSH in Parkinson's disease: possible mechanisms of formation involving reactive oxygen species, *J. Neurochem.*, 1998, **71**, 2112–2122.
158. C. S. Lee, A. Samii, V. Sossi, T. J. Ruth, M. Schulzer, J. E. Holden, J. Wudel, P. K. Pal, R. de la Fuente-Fernandez, D. B. Calne and A. J. Stoessl, *In vivo* positron emission tomographic evidence for compensatory changes in presynaptic dopaminergic nerve terminals in Parkinson's disease, *Ann. Neurol.*, 2000, **47**, 493–503.
159. G. W. Miller, J. D. Erickson, J. T. Perez, S. N. Penland, D. C. Mash, D. B. Rye and A. I. Levey, Immunochemical analysis of vesicular monoamine transporter (VMAT2) protein in Parkinson's disease, *Exp. Neurol.*, 1999, **156**, 138–148.
160. C. Pifl, A. Rajput, H. Reither, J. Blesa, C. Cavada, J. A. Obeso, A. H. Rajput and O. Hornykiewicz, Is Parkinson's disease a vesicular dopamine storage disorder? Evidence from a study in isolated synaptic vesicles of human and nonhuman primate striatum, *J. Neurosci.*, 2014, **34**, 8210–8218.

161. J. J. Rilstone, R. A. Alkhater and B. A. Minassian, Brain dopamine-serotonin vesicular transport disease and its treatment, *N. Engl. J. Med.*, 2013, **368**, 543–550.
162. C. E. Glatt, A. D. Wahner, D. J. White, A. Ruiz-Linares and B. Ritz, Gain-of-function haplotypes in the vesicular monoamine transporter promoter are protective for Parkinson disease in women, *Hum. Mol. Genet.*, 2006, **15**, 299–305.
163. L. Brighina, C. Riva, F. Bertola, E. Saracchi, S. Fermi, S. Goldwurm and C. Ferrarese, Analysis of vesicular monoamine transporter 2 polymorphisms in Parkinson's disease, *Neurobiol. Aging*, 2013, **34**, 1712.e9–1712.e13.

CHAPTER 6

Glutathione and Thiol Redox Signalling in Parkinson's Disease

MICHELLE SMEYNE AND RICHARD JAY SMEYNE*

Department of Neuroscience, Vickie & Jack Farber Institute for Neuroscience, Thomas Jefferson University, 900 Walnut Street, Philadelphia, PA 19107, USA
*E-mail: richard.smeyne@jefferson.edu

6.1 Introduction

Oxidative stress plays a major role in the pathogenesis and progression of Parkinson's disease (PD). Evidence of oxidative stress in the *substantia nigra* (SN) of PD patients includes enhanced basal lipid peroxidation,[1] decreased glutathione (GSH) level (40%)[2,3] and elevated levels of gamma γ-glutamyl transpeptidase (γGT),[4] a degradative enzyme for the recycling of GSH. GSH is a ubiquitous thiol tripeptide, which functions as an antioxidant and is present in abundance in most mammalian cells (0.5–10 mM),[5] but is lower in concentration in the brain (1–3 mM).[6] GSH is composed of the amino acids L-cysteine (L-Cys), L-glutamate, and glycine.[5] Free radicals, generated by oxidative stress are damaging to neurons, and the *substantia nigra pars compacta* (SNpc) is a region of the brain that is especially sensitive to oxidative stress.

Reactive oxygen species (ROS) are generated by a number of cellular processes, including the consumption of molecular oxygen during oxidative

Issues in Toxicology No. 34
Oxidative Stress and Redox Signalling in Parkinson's Disease
Edited by Rodrigo Franco, Jonathan A. Doorn and Jean-Christophe Rochet

Published by the Royal Society of Chemistry, www.rsc.org

phosphorylation in the production of ATP by the mitochondrial electron transport chain. Neurons, and in particular, the dopaminergic (DAergic) neurons of the SNpc are sensitive to ROS, while the brain requires greater energy to function, compared to other organs.[7] The loss of DAergic neurons located in the SNpc (A9) is the lesion most characteristic of PD, although other regions of the central, peripheral and enteric nervous systems also show considerable cell loss.[8–11] Within the central nervous system (CNS), it is not entirely clear why the SNpc is so significantly affected, although this region does have a number of characteristics that make it particularly vulnerable to oxidative stress. These factors include (but are not limited to) the presence of endogenous dopamine (DA), iron, and neuromelanin.[12–14] Additionally, the intrinsic antioxidant defenses in this structure are more vulnerable than in other brain regions due to lower levels of GSH[2,15] and γ-glutamylcysteine ligase (GCL) activity,[16] higher microglial:astrocyte ratios,[17,18] and less abundant oxidoreductases such as thioredoxin (Trx) and peroxiredoxins (Prx) that help to maintain redox homeostasis.[19,20]

6.2 Glutathione

GSH provides protection from oxidative stress-induced damage through the reduction of ROS and reactive nitrogen species (RNS).[3] GSH participates in reactions that mediate antioxidant defense, xenobiotic and eicosanoid metabolism, cell-cycle regulation, and gene expression. GSH acts alone or in concert with other enzymes to reduce superoxide radicals ($O_2^{\bullet-}$), hydroxyl radicals ($^{\bullet}OH$) and peroxynitrites.[21] GSH provides a storage and transfer form for Cys, maintains cellular redox potential by keeping sulfhydryl proteins in a reduced state, contributes reducing equivalents for enzyme catalysis of redox reactions, and detoxifies xenobiotics.[5,22–25]

The redox state of GSH is labile, as it is readily transformed from reduced GSH to oxidized glutathione disulfide (GSSG), and back again.[26] Glutathione reductase (GR) uses hydrogen from reduced nicotinamide adenine dinucleotide phosphate (NADPH) to reduce GSSG to GSH. NADPH is maintained predominantly by the pentose phosphate shunt. GSSG is usually <1% of the GSH pool, since it is rapidly reduced by GR.[27] GSSG may also exchange sulfur groups with protein sulfhydryls to produce mixed protein glutathione disulfides (P-SSG) that have a longer half-life than GSSG alone. GSSG content is also reduced by an ATP-dependent transport system through export. GSH conjugates reactive electrophiles alone or by catalysis with the enzymes glutathione-S-transferases (GSTs). Glutathione peroxidases (Gpx) use GSH in the reduction of hydroperoxides. GPxs are a family of selenoproteins that vary in the hydroperoxide substrate specificity. GPx and GSTs are discussed below. GSH also conjugates small molecules that are excreted from cells and acts in concert with $O_2^{\bullet-}$ dismutase (SOD) to prevent oxidative damage.[28]

6.3 Thiol Redox Signalling and Thiol–Disulfide Exchange

The most basic level of thiol signalling begins with the hydrosulfide ion, two of which form a thiol group that readily reacts with other compounds. Two thiol groups (–SH) are easily oxidized to form disulfide links (–S–S–). Mild oxidizing agents (such as hydrogen peroxide [H_2O_2]) help to drive this reaction. The contribution of reducing equivalents, by NADPH, GSH, or other Cys residues may reverse this reaction.[29]

Peptide chains and protein residues contain disulfide links, so this reversible oxidation–reduction reaction functions to manipulate their (tertiary) structure.[28,29] The thiol-disulfide pool is important for antioxidant defenses and the redox regulation of cell signalling.[30] Some aspects of thiol signalling are conserved throughout evolution. In this chapter, discussion of GSH, and thiol redox signalling will be focused mainly on the information pertaining to mammalian cells and DAergic neurons and cell lines.

The sulfur-containing amino acid Cys, occurs in 10 different sulfur oxidation states in living organisms, leading to a wide range of Cys modifications in proteins. Cysteine acts as a potent nucleophile in the active site of Cys proteases. The valence electrons of sulfur are highly flexible, and in the presence of a neighboring histine (which provides an imidazole ring) a 'charge relayed dyad' is formed. This is a delocalized electron cloud whose effect is to lower the pKa of Cys, allowing enzyme activity to occur over a wide pH range (pH 3–11). The active site Cys also provides sulfur as a ligand for catalytic metal in metalloproteases.[31] Cysteine's provision of sulfur in a catalytic function, as well as a ligand for catalytic metal ions, also functions in another enzyme group, oxidoreductases.

Cysteine catalysis for redox reactions occurs through a number of exchanges. Thiol-disulfide exchange may occur through thiolation and glutathionylation reactions that do not include electron transfer, such as the exchange between glutathione and glutathione disulfide. Glutathionylation involves the formation of a mixed disulfide by the interaction of a Cys residue of GSH and a protein Cys moiety.[32] Electron transfer to disulfides by GR or thioredoxin reductase (TR) uses electrons donated by flavin adenine dinucleotide ($FADH_2$) and NADPH, respectively, to reduce GSSG.[5,33,34] Radical formation and transfer such as the redox coupling of thiol-thiyl radical, involving the formation of a radical from thiol by the one-electron transfer and loss of H^+ or the hydrogen atom (radical) abstraction from the thiol, is distinct from electron transfer. Also, the transfer of oxygen atoms from oxidizing species to Cys leads to the formation of sulfur acids (RSOH).[31]

Thiol-disulfide exchange plays a major role in modulating redox potential and in preventing oxidative damage.[35–37] In cellular systems, thiol-disulfide interchange and thiol oxidation–reduction reactions are kinetically controlled and enzymes play essential roles in these processes. The modification of protein thiols and the location within the cell is important in the cell's response to oxidative stress. These modifications have been visualized *in vivo*, and their functions have been highlighted.[35] Endoplasmic reticulum and periplasm depend on disulfide bonds to support correct folding and the structural stability of proteins. Cytosolic proteins present in the reducing environment use Cys residues that are reduced and are located in binding pockets of substrates, coenzymes, or metal cofactors; or are present in the catalytic sites of enzymes where they participate directly in the catalytic reaction. The activity of the cytosolic enzymes is dependent on the reduced state of the Cys residues involved.[37,38]

The properties that make Cys an excellent amino acid for redox reactions also make it vulnerable to oxidation by ROS.[36] RNS are the products of reactions involving nitric oxide (˙NO) compounds and are also targets of redox reactions. ROS and RNS interact in both chemical reactions and physical interactions.[39] An overabundance of either ROS or RNS is deleterious to the cell. ROS are natural byproducts of cellular respiration and they accumulate during oxidative stress and inflammation. Some thiol-containing proteins use ROS as a mediator to quickly regulate their protein activity. Some examples of these proteins include Hsp33,[40] glyceraldehyde-3-phosphate dehydrogenase,[41] protein kinase C (PKC),[42] and phosphatase and tensin homolog (PTEN).[35,43,44]

Nitric oxide is involved in a number of physiological processes. There are several related forms of ˙NO: the nitrosonium cation (NO^+), nitric oxide with a single unpaired electron (˙NO), and the nitroxyl anion (NO^-). ˙NO reacts rapidly with $O_2^{\bullet-}$ to form $ONOO^-$, a highly reactive free radical.[45] The production of ˙NO is catalyzed by nitric oxide synthase (NOS). Several forms of NOS have been described; endothelial NOS (eNOS),[46] inducible NOS (iNOS) that is produced mainly in microglia and astrocytes,[47,48] and neuronal NOS (nNOS) that is calcium dependent and produced in neurons.[49,50] *S*-nitros(yl)ation is a reversible posttranslational modification of a protein where a NO group is covalently attached to a Cys thiol group to form a *S*-nitroso derivative (R-SNO).[50] Modification by *S*-nitros(yl)ation often affects the activity of the modified protein, and *S*-nitro(sy)lated residues (*S*-nitrosocysteine, *S*-nitrosoglutathione) provide storage forms and a longer half-life for ˙NO.[50,51] Proteins may also be denitro(sy)lated by redox sensitive enzymes in the Trx system and *S*-nitroso-glutathione reductase system.[52]

Cysteine residues surrounded by particular amino acids, referred to as an 'SNO motif' are nitros(yl)ated by groups of amino acids with nucleophilic residues (acid or base), due to their tertiary or quaternary structure.[50,53] One example of a conformation-dependent action of ˙NO is seen with hemoglobin, where a basic histidine located near its Cysβ93 promotes its *S*-nitros(yl)ation and oxygen affinity is favored. When changes in hemoglobin conformation occur, they bring CysB93 in proximity to an acidic asparate side chain, where denitros(yl)ation is favored; promoting low oxygen affinity.[54] *S*-nitros(yl)ation affects several proteins necessary for normal function of neurons: Parkin,[55,56] Prx2,[50,57] x-linked inhibitor of apoptosis (XIAP),[50,58] glyceraldehyde-3-phosphate dehydrogenase (GAPDH),[50,59] protein disulfide isomerase (PDI),[50,60] and DJ-1[50,58] are neuroprotective proteins with functions that are impaired by SNO-mediated stress.

Thiol to disulfide conversion may occur by direct substitution or by a series of redox reactions. Direct substitution is the rate-limiting factor in the folding of proteins that form by disulfide bonds. Enzyme catalysis accelerates these reactions *in vivo* that occur in the endoplasmic reticulum[61,62] and mitochondrial inter membrane space[63] in eukaryotes. Oxidoreductases mediate the interchange of thiols and disulfides of the folding polypeptides using molecular oxygen.[37]

6.4 Cellular Reductases

There are several categories of cellular reductases that use thiol groups to reduce ROS, RNS and oxidized protein residues. Trxs, glutaredoxins (Grx), and Prxs have been investigated for their role in maintaining redox homeostasis and in particular, their role in the etiology of PD. The general reactions can be seen in Figure 6.1.

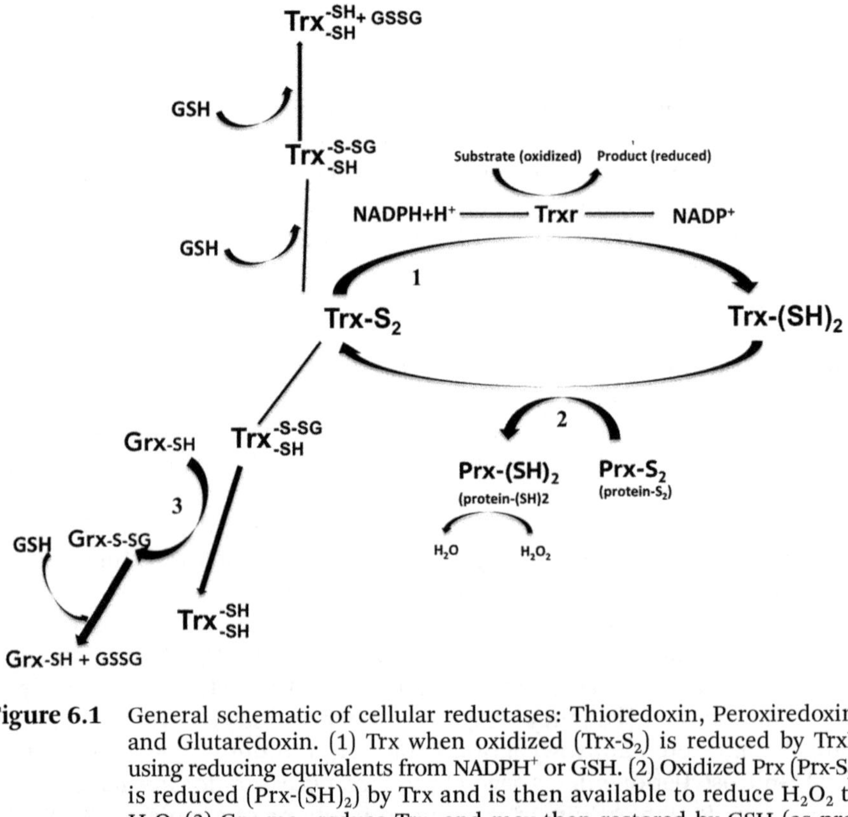

Figure 6.1 General schematic of cellular reductases: Thioredoxin, Peroxiredoxin, and Glutaredoxin. (1) Trx when oxidized (Trx-S_2) is reduced by TrxR using reducing equivalents from NADPH$^+$ or GSH. (2) Oxidized Prx (Prx-S_2) is reduced (Prx-$(SH)_2$) by Trx and is then available to reduce H_2O_2 to H_2O. (3) Grx may reduce Trx, and may then restored by GSH (as proposed by Du *et al.*, 2013).[33,79]

6.4.1 Thioredoxins

Trx is a small protein with 2 Cys in its active site. It is the smallest of a protein family containing a Trx fold, and functions in antioxidant defense, protein folding, and signal transduction.[20,28,33] Trx participates in oxidation–reduction reactions through the reversible oxidation of its thiol active center to disulfide and it catalyzes dithiol–disulfide reactions. Trx also contributes to the response to intracellular •NO, by reversibly nitros(yl)ating Cys residues of target proteins.[64] Trx works with thioredoxin reductases (TrxR) that are homodimeric, selenium-containing, flavoproteins. TrxR functions by transfer of reducing equivalents from NADPH to Trx to reduce the substrate. Trx are the major protein disulfide reductases present in organisms from bacteria to mammals. Their active site has the amino acid sequence CGPC and they serve as electron donors for enzymes such as ribonucleotide reductases, Prxs (thioredoxin peroxidases), and methionine sulfoxide reductases (Msrs).[33] There are specific Trx systems present in the cytoplasm and mitochondria of cells. The Trx1/TrxR1 system is found in the cytoplasm of cells, while mitochondria have a distinct Trx system, Trx2/TrxR2.[20] Trxs are essential for redox regulation of protein function and signalling. This system is necessary for defense against oxidative stress,[65–67] control of growth and apoptosis,[68,69] and in reduction reactions necessary for the binding of some transcription factors to DNA.[33,70] Trx1 reduces and activates certain transcription factors and acts as a chaperone protein, independent of its redox function.[20] Lack of TrxR is embryonically lethal.[33,71,72] In the SNpc, strong immunocytochemical staining of Trx1 and Trx2 is found, while lighter staining of TrxR1/TrxR2 is observed.[20]

The TrxR inhibitor auranofin (Aur) has been used to study the role of the Trx system in protecting DAergic cells from oxidative stress. Treatment of primary mesencephalic cultures with Aur reduces TrxR activity significantly, promotes minor increases in hydrogen peroxide (H_2O_2) production, and significantly increases cell death, as measured by lactate dehydrogenase (LDH) production.[73] Inhibition of Trx in mitochondria isolated from rat brain attenuates the removal of H_2O_2 (by 80%), and increases the production of H_2O_2 when combined with paraquat (PQ) treatment.[67] LDH assay of mesencephalic cells reveals an increase in cell death when treated with either Aur (50–90%) or PQ alone (80–150%), and an even greater increase when the treatments are combined (287–463%).[73] In addition to a decrease in removal of H_2O_2 by Trx inhibition in isolated mitochondria, H_2O_2 removal is attenuated in N27 DAergic cells.[67] Similar results are reported for the specific inhibition of mitochondrial TrxR2 by shRNA in N27 cells, where H_2O_2 production and LDH release are increased with Aur or PQ treatment, compared to mock infected control cells. Exposure of N27 cells to 6-OHDA when Trx is inhibited or TrxR2 is reduced, leads to reduced respiratory capacity and increased cell death.[73] The overexpression of Trx1 in transgenic mice has also been shown to be protective against apoptosis induced by 1-methyl-4-phenyl-1,2,3,6-tetrahydropyridine (MPTP).[65]

Trx1 associates with apoptosis signal-regulating kinase 1 (ASK1), a mitogen-activated protein (MAP) kinase kinase kinase (MAPKKK) that activates c-Jun kinase (JNK)/p38 MAP kinase (MAPK) mediated apoptosis. In the presence of reduced but not oxidized conditions, Trx1 interacts with the N-terminal portion of ASK1 and inhibits ASK1 activity.[74] Inhibition of Trx results in activation of ASK1, suggesting it is a negative regulator of ASK1.[74] Reduction of ASK1 by Trx1 or mutation of the oxidation sensitive Cys of ASK1 prevents JNK mediated apoptosis, while ASK1 is still activated.[75]

6.4.2 Glutaredoxins

The Grx system has GSSG oxidoreductase activity using NADPH for reducing GSH in the presence of GR. Grxs in turn reduce low molecular weight disulfides and protein residues[34] (Figure 6.2). Three categories of Grxs have been classified; classical Grx, the GST-like Grx with a structure similar to GSTs, and those with a monothiol active site. The classical Grxs are 10 kD proteins with a CXXC motif as the active site and a Grx/Trx fold.[34] Examples of these are Grx1 and Grx3, and as mentioned above, they are electron donors for enzymes such as ribonucleotide reductase.[76] The GST-like Grxs contain a 2-domain structure with a Trx/Grx fold that includes the active-site residues and a second domain that has an alpha helical structure. These Grxs differ from GSTs in the active-site sequence that is CPYC and confers the Grx activity. Grxs in this category include Grx2. The third category has a monothiol active site. An example of a monothiol Grxs is the Protein kinase C-interacting cousin of thioredoxin (PICOT) that is expressed in a number of tissues in humans.[34] In T-cells overexpression of PICOT inhibits the activation of JNK, the transcription factors Activator Protein-1 (AP-1) and nuclear factor kB (NF-κB).[77] Grx2 also functions to protect cells by reducing Trx in the presence of oxidizing conditions.[78,79]

Decreases in GSH inhibit the activity of Grx2,[80] and increase the sensitivity of Grx1 to inhibition by L-Dopa,[66] demonstrated in experimental models using

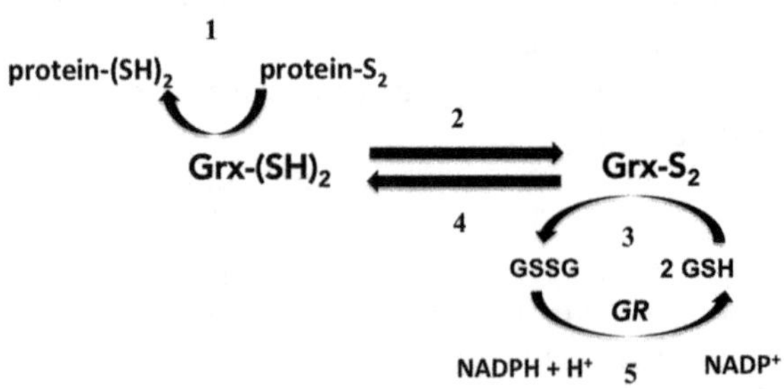

Figure 6.2 Glutaredoxin general mechanisms. (1) Oxidized Cys residues (protein-S_2) are reduced (protein-$(SH)_2$) by Grx (Grx-$(SH)_2$). (2) Cysteine residues of Grx donate H^+ (Grx-S_2). (3) GSH provides reducing equivalents to (4) restore Grx-S_2 to Grx-$(SH)_2$. (5) GSSG is reduced to GSH by GR using $NADPH^+$.

DAergic cell lines. N27 DAergic cells treated with L-buthionine-(S,R)-sulfoximine (BSO) showed a concentration-dependent inhibition of Grx2. siRNA knockdown of GSH modulating enzymes provides further support for the critical role of GSH as a free radical reducing agent. Lee *et al.*,[80] used siRNA directed against Grx2, a protein that is critical for controlling redox signalling in the mitochondria by modulating the interactions between the GSH pools and protein thiols[81] to reduce Complex I activity.[82] Reduction of Grx2 levels results in a loss of Complex I (30%) and *m*-aconitase (60%) activity, decreased ferritin levels and a subsequent increase in mitochondrial iron;[80] each leading to increased oxidative stress. Using SH-SY5Y cells, Saben *et al.* (2010) demonstrated Grx sensitivity to L-Dopa treatment. Grx activity, but not Grx content is reduced in an L-Dopa dose dependent manner, while GR activity is insensitive to L-Dopa, up to 100 μM concentrations. Decreases in GSH level with BSO treatment increases Grx sensitivity to L-Dopa. Grx is also inactivated by oxidized L-Dopa, but is protected with the addition of GSH, while GR, Trx, and TrxR activities are not inactivated by oxidized L-Dopa. Trx and TrxR activity is also reduced with L-Dopa treatment in a dose-dependent manner, and TrxR is more sensitive to inactivation by L-Dopa when GSH is depleted. Reduction of Grx1 activity in SH-SY5Y cells by siRNA (50%) leads to 30% apoptosis, compared to control. In addition, the Cys22 residue of Grx1, required for its catalytic activity, is modified by oxidized L-Dopa.[66] L-Dopa treatment has also been shown to result in phosphorylation of the MAPKs p38 and JNK, and inhibition of p38 or JNK alone reduces apoptosis with L-Dopa treatment, while inhibition of both p38 and JNK activity provides better protection. Reduction of ASK1 activity with siRNA also reduces L-Dopa-mediated apoptosis, proportional to the level of ASK1 reduction.[69]

6.4.3 Peroxiredoxins

Prxs were first known as thioredoxin peroxidases, and are enzymes that reduce peroxides, including H_2O_2, organic hydroperoxides and $ONOO^-$[19] using Trx.[83] Prxs are grouped in 3 subfamilies: typical 2-Cys, atypical 2-Cys, and 1-Cys.[84,85] These three subfamily classifications are based on the number of cysteinyl residues involved in catalysis or peroxidase reactions.[84,86,87] There are six isoforms of Prxs (1–6) found in mammalian cells, in several cellular compartments (mitochondria, peroxisome, nuclei, cytosol, and extracellularly).[88,89] Prx1-4 are typical 2-Cys, Prx5 is in the subfamily atypical 2-Cys, and Prx6 is classified as 1-Cys.[90]

Mapping of the Prxs by immunocytochemistry shows Prxs 1,2,5, and 6 located in the cytosol, Prxs3 and 5 staining localized to mitochondria, and Prx5 in peroxisomes. Prx4 is found in the endoplasmic reticulum and is secreted.[19,90] Prx isoforms are also found in specific cell types. C57BL/6 and 129.SvJKist mouse strains reveal similar cell localization of Prx isoforms in the brain.[19,86] Prx1 is reported in oligodendrocytes,[19,86] astrocytes, and microglia[86] and is not colocalized with NeuN.[19] Prx2–5 are found in neurons[19,86,91] and although immunostaining is low, it is colocalized with tyrosine hydroxylase (TH) staining in the SN.[19] Weak staining of ependymal cells by Prx2 in the basal ganglia has also been shown.[19,86] Prx4 is reported in oligodendrocytes and astrocytes as well as neurons.[19] Prx6 does not stain NeuN cell populations, and is colocalized with GFAP in astrocytes.[19,86]

When H_2O_2 or an organic peroxide are reduced by Prx, the peroxidatic Cys may be further oxidized to sulfinic acid and subsequently to sulfonic acid, which is irreversible and Prx loses its peroxidase activity. Typical 2-Cys Prxs as sulfonic derivatives may be reduced to sulfinic acid by sulfiredoxin (Srx) in the presence of MgATP. These sulfinic derivatives in eukaryotes have the potential to continue to react with H_2O_2, and when they are hyperoxidized, form high-MW oligomeric structures that gain chaperone activity. This chaperone activity is protective against thermally induced aggregation and denaturation of proteins.[83,92]

The importance of Prx function to balance redox homeostasis is illustrated in studies in which Prx knockout or knockdown in mice leads to increased levels of ROS/RNS.[93–95] In addition, modified Prxs have been found in PD brains and in experimental PD models.[19,57,96,97] Further support for the role of Prxs in the protection of DAergic neurons is supplied by the description of one of the signalling pathways modulated by Prx. Lentivirus expression of Prx2 is protective to DAergic neurons that are treated with 6-OHDA.[91] Prx2 provides the most protection against 6-OHDA, with Prx1 and 4 also increasing MN9D cell viability.[97] Overexpression of Prx2 provides protection of DAergic neurons *in vivo* as well as *in vitro*. Prx2 also suppresses apoptosis that occurs through the ASK1 induction of JNK/p38 apoptotic paths by reduced phosphorylation of c-Jun. Prx2 reduces oxidation of Trx, preventing its dissociation from ASK1 and thereby inhibiting apoptosis.[91] Another report describes the PD-related protein DJ-1, which has been shown to assist in maintaining redox homeostasis,[98] to be an atypical Prx-like peroxidase that scavenges H_2O_2. Knockout of DJ-1 in these mice increases the oxidation of Cys106 to sulfinic acid when MPTP is administered.[99]

An example of the physiological interaction of Prx, Trx and Grx is shown in HeLa cells. The oxidation of Trx1 by Prx1 in the presence of H_2O_2 stimulates the reduction of Trx1 by Grx1 in the presence of GSH, NADPH, and GR.[79,100] In this instance, Trx1 functions to signal oxidative stress and activate redox reactions by Grx. Grx signal transduction is also demonstrated as a regulator of neuronal and vascular network formation in zebrafish.[101,102]

6.5 Glutathione Synthesis in the Brain

The GSH tripeptide is synthesized as the product of two successive reactions (Figure 6.3). The GSH precursor molecules may be toxic at high concentrations,[103–106] making the lower GSH levels in the brain a protective mechanism.

The first step in GSH synthesis is the formation of γ-glutamylcysteine (γGC) from the substrates L-glutamate and L-Cys by GCL (also referred to as γ-glutamylcysteine synthetase, γGCS).[107] GCL is the enzyme in the rate-limiting step of GSH synthesis. GCL is a heterodimer containing both disulfide and

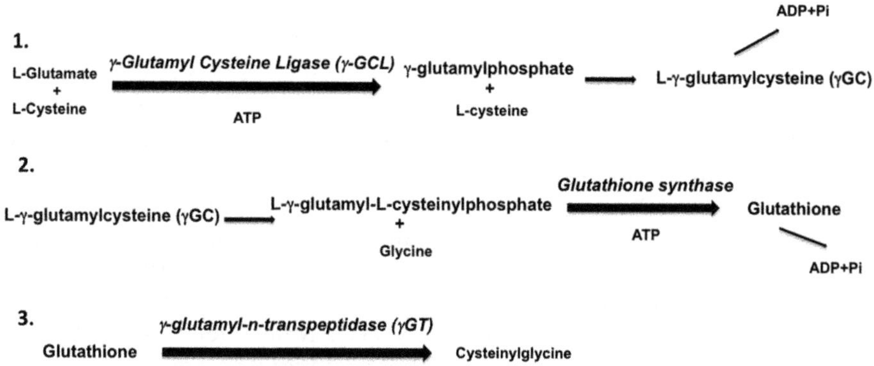

Figure 6.3 Glutathione synthesis. (1) γ-GCL catalyzes the formation of the intermediate γ-glutamylphosphate and L-Cys then L-γ-glutamylcysteine, using ATP. (2) Glycine is added to the intermediate L-γ-glutamyl-L-cysteinylphosphate by GS using ATP. (3) Glutathione is recycled by γ-GT for the reuse of cysteinylglycine.

noncovalent bonds between its heavy and light chain subunits.[26] The GCL heterodimer dissociates into a catalytic heavy subunit (GCLC) and a modulatory light subunit (GCLM). GCLC is 73 kDa, has all of the catalytic activity of GCL and is the site of GSH feedback inhibition. GCLM is 31 kD, is a monomer with no catalytic activity, and has a modulatory effect on GCLC when associated. Intracellular GSH levels are regulated by negative feedback of GSH[108] as well as the availability of Cys.[5] One function of GCLM is to elevate the K_i for GSH so that negative feedback inhibition is decreased. GCL holoenzyme activity may also be regulated positively or negatively by *S*-nitros(yl)ation, phosphorylation, and oxidation.[26] Increased activity usually involves transcription leading to increased production. The Gclc promoter contains many potential *cis* acting elements including consensus sequence recognition sites for binding of activator protein 1 (TRE),[109] TRE-like,[110] NF-κB[111] and the electrophile response unit (EpRE aka antioxidant response element ARE).[28,112,113] The Gclm promoter contains many of the same elements present in the Gclc promoter. Gclc induction is initiated by ARE and TRE sites, and the activation of transcription factors may differ, even by the same compound, according to the signalling components and metabolism of the cell.[28,114,115] An intermediate product of this synthesis, γ-glutamylphosphate, is produced by the reaction of L-glutamate and ATP, which reacts with L-Cys to form L-γ-glutamylcysteine. In the second step of GSH synthesis, L-γ-glutamylcysteine and ATP combine to form L-γ-glutamyl-L-cysteinylphosphate that is combined with glycine in a reaction catalyzed by GSH synthetase (GS) to form GSH.[108,116] Clinical phenotypes exist with missense mutations in GS.[117]

The ectoenzyme γGT catalyzes the transfer of the γ-glutamyl moiety from GSH or a GSH conjugate to an acceptor molecule (CysGly or CysGly conjugate) to regenerate GSH from its metabolized component peptides. GSSG, 2-oxothiadolazine-4-carboxylate (OTC), and *N*-acetylcysteine (NAC) can also serve as precursors for the synthesis of GSH.[118] Alternatively, GR uses NADPH as an electron donor to reduce GSSG and regenerate GSH.[26]

Synthesis of GSH occurs in neurons and glial cells, although astrocytes synthesize GSH more effectively than neurons based on their ability to utilize a wider variety of precursor substrates.[119] Neurons use glutamine, glutamate, Cys and glycine or conjugates of these amino acids provided by the breakdown of GSH by γGT to synthesize GSH. Astrocytes, however, have the ability to utilize a far greater number of substrates in the synthesis of GSH including the amino acids: glutamine, glutamate, Cys, glycine, aspartate, asparagine, ornithine, and proline, conjugates of these amino acids as well as GSSG, 2-oxothiadolazine-4-carboxylate (OTC), *N*-acetylcysteine (NAC) and GSH itself,[21,118,120,121] although each of these components are ultimately converted to glutamate, Cys or glycine prior to GSH synthesis.[122] Imaging of GSH content in individual cells with monochlorobimane fluorescence demonstrates higher GSH content in astrocytes than in neurons.[123,124]

6.6 Glutathione and Models of Oxidative Stress in Dopaminergic Neurons

The antioxidant characteristics of GSH have been demonstrated in a number of models of oxidative stress including depletion of GSH with BSO,[125–129] ethacrynic acid (EA),[130] or reduction of GSH synthesis using antisense directed at GCL[131–134] or Grx2.[80] In these studies, diminished levels of GSH increase oxidative stress in whole cells as well as in mitochondrial fractions, and increase lipid peroxidation, intracellular calcium, and γGT activity.

Several studies discussed below illustrate these points by utilizing DAergic systems. Depletion of GSH by BSO, an irreversible inhibitor of GCL, does not by itself induce nigrostriatal damage *in vivo*,[135] but potentiates the amount of 1-methyl-4-phenyl-1,2,3,6-tetrahydropyridine (MPTP)–induced TH+ neuron death in the SNpc (48.6% cell death compared to 30.1% cell death).[125,135] Additionally, in conditions of increased oxidative stress such as when mesencephalic cells are placed in culture or during normal ageing *in vivo*, decreasing GSH levels causes neuron loss.[132]

The reduction of GSH activity by EA has also been shown to increase cell sensitivity to free radicals. Astrocytes exposed to EA and 3-morpholinosydnonimine (SIN-1, a compound that generates $ONOO^-$) show significant increases in LDH, decreased ATP levels and decreased mitochondrial membrane potential. The critical role of GSH in this system is shown by experiments where excess GSH monoester is included with EA and SIN-1, and LDH activity is inhibited.[130] Treatment with a thiol ester compound (gamma-glutamylcysteinyl ethyl ester) or exogenous GSH while examining cellular models of diminished GSH levels in concert with Complex I inhibitors 1-methyl-4-phenylpyridinium (MPP^+), MPTP, or 7-(2-aminoethyl)-3,4-dihydro-5-hydroxy-2H-1,4-benzothiazine-3-carboxylic acid (DHBT-1), restores mitochondrial Complex I activity, inhibits cell loss and protects against cell and striatal DA loss.[129,131,136]

Similarly, a 50% reduction of GSH by antisense knockdown of GCLC[133] increases cellular oxidative stress in PC12 cells. Using this paradigm, Jha *et al.*[134] noted a reduction in mitochondrial performance using three different measures; a 70% reduction of ATP levels, a 65% reduction in mitochondrial activity, and a 60% reduction in O_2 consumption. It was also demonstrated that this effect was specific to Complex I, and not Complex II, III, or IV activity. Addition of dithiothreitol (DTT, a thiol reducing agent) restored Complex I activity to control levels in GCL-diminished cells, suggesting that it is the oxidation of sulfhydryl groups that results in the inhibition of Complex I. In PC12 cells, GSH depletion decreases TH activity and DA uptake.[133] Additionally, mitochondrial Complex I activity, which is reduced in PD and is critical to DAergic neuron survival[137,138] is compromised when GSH levels are reduced in combination with exposure to NO and generation of peroxynitrite radicals.[129,130]

In the SN, the reduced form of GSH is an important mediator of oxidative stress. Studies of postmortem CNS of PD patients reveal lower levels of GSH in the SN (40%), but not in other regions of the brain, compared to age-matched controls. In contrast, GSH levels in the SN from patients with other neurodegenerative diseases that involve the basal ganglia, such as Multiple System Atrophy and Supranuclear Palsy, are unchanged.[4]

6.7 Glutathione Conjugating Enzymes

6.7.1 Glutathione Peroxidase

GPxs are a group of 8 (GPx1–8) enzymes that are important for reducing H_2O_2 to water. GPxs are selenocysteine enzymes that use GSH as a reducing agent and require selenium[139,140] for their antioxidant function. In the brain, the selenoproteins GPx-1–3 exist as tetrameric proteins, composed of four identical subunits with each monomer having a molecular weight of 22–23 kDa,[141] while GPx4 has activity as a monomer.[142] GPx1 and 4 are found in the mitochondria, nucleus and the cytosol,[143] and GPx1, also known as GSHPx, is present in both neurons and glial cells.[144,145] Overexpression of GPx decreases the amount of neuron loss, reduces H_2O_2 accumulation and lipid peroxidation in neurotoxic conditions.[146] GPx and catalase reduce H_2O_2 when acting alone, but more effectively decrease the toxicity of exogenous H_2O_2 when acting together.[147]

In an experimental model of PD, mice that lack GPx1 under normal oxidative conditions show no apparent neuropathological lesions compared to wild-type (WT) mice. However, when challenged with MPTP, a toxin that induces oxidative stress, DA, DOPAC, and HVA levels in these mice are decreased.[148] Additionally, microarray analysis of mRNA expression in the lateral *versus* medial tiers of the SN reveals downregulation of Gpx1 and Gpx3 in the lateral tier compared to the medial tier of the *substantia nigra*;[149] this is complementary to the observation of greater cell loss in the lateral SN of PD patients.[150] Savaskan *et al.* have shown that GPx4 protein level is increased following a neurodegenerative lesion of the entorhinal cortex. Subsequent immunocytochemical analysis shows that the location of this increase occurs in astrocytes.[151] In cortical samples taken from PD patients, GPx3 and GPx4 protein are also elevated compared to control subjects.[152] An immunocytochemical study of GPx1 expression shows that DAergic neurons in the SNpc express low levels of this protein, while other regions not affected in PD, such the ventral tegmental area, express higher levels.[144] In addition, GPx1 immunoreactive microglia were reported in samples of SN taken from patients with PD as well as those with dementia with Lewy bodies disease (DLB).[145]

Treatment with dopamine quinone (DAQ), which binds Cys residues, decreases GPx4 activity and protein level in rat brain mitochondria, an effect that is blocked with the addition of GSH before DAQ exposure.[153] In DAergic neurons of the SN, GPx4 is colocalized with neuromelanin. Examination of SN in brains from PD patients shows that volume-density immunolabeling of GPx4 is reduced; however, relative to the remaining cell density, GPx4 levels are increased compared to control subjects.[149] Further study of Selenoprotein 1 (Sepp1, a transport protein and source of selenium for selenoproteins) and GPx4 in the SN and putamen shows an association between Sepp1 and GPx4 localization in the putamen of control subjects, while the correlation is lost in PD patients. No correlation of Sepp1 and GPx4 immunoreactivity is seen in the SN of PD or control samples.[154]

6.7.2 Glutathione S-Transferases

GSTs are a class of abundant proteins[155] that function as xenobiotic metabolizing enzymes[156–158] in eukaryotes. This class of enzymes may be viewed as a cellular defense against numerous artificial and naturally occurring environmental agents. GSTs function by catalyzing the conjugation of GSH to various electrophiles and xenobiotics. Additionally, certain GSTs have been shown to have other functions, including modulation of cell-survival pathways.[159,160]

In humans, there are three distinct classes of GSTs: cytosolic, mitochondrial and microsomal. Within the brain, the cytosolic forms of GST are predominant, and will be the focus here. Each of the cytosolic GSTs are configured with different combinations of monomers that are 199–244 amino acids in length.[161,162] The ultimate composition of the holoenzyme is critical to the function, since each of the monomers contributes half of the overall GSH:electrophile active site;[163,164] this dimerization occurs in the presence of increased levels of oxidative stress.[165]

At this time, seven classes of cytosolic GSTs have been identified, and are named alpha, mu, pi, sigma, theta, omega, and zeta.[166] Within each GST molecule, there are two characteristic domains: I and II. Domain I comprises the N-terminal residues (1–80) of the protein whose structure consists of a series of beta pleated sheets and alpha helices. Domain II comprises the remaining residues (81–209 $\pm$ 11) and is also referred to as the hydrophobic site (H site).[163] The GSH binding domain is found in domain I and is structurally conserved in each of the isoforms.[167] Structural differences among the isoforms are found within domain II. The variable residues in this domain contribute to the array of substrate specificity found among the GSTs.[167,168]

For instance, mu and pi classes contain more polar domain II regions that enable them to react with charged substrates,[169] whereas the alpha isoform contains a more hydrophobic domain II.[167]

In the brain, the active GSTs are composed of dimers containing alpha, mu, or pi class GST monomers.[170–174] GST-mu is the most highly expressed isoform in the brain, followed by GST-pi, then GST-alpha.[172] Cellular localization studies show that GST-pi and GST-mu are expressed in both neurons and astrocytes.[172,175] Interestingly, in the SN, the CNS structure most affected in PD patients, only GST-pi, but not GST-mu, is found in the A9 DAergic neurons.[172] The distribution of GSTs in the brain appears to also be age dependent.[173] GST-pi is the only GST isoform that appears to be expressed in human fetal brain,[173,176] while the alpha, mu and pi classes are found in adult.[173]

ASK1, which is activated upon oxidative challenge,[74,177,178] facilitates the downstream activation of JNK and the stress-activated protein kinase p38.[179,180] GST-mu inhibits ASK1 signalling through a protein–protein interaction.[181] Furthermore, following stress, downstream activation of p38 requires the dissociation of GST-mu from ASK1, suggesting that GST-mu regulates the signal cascade mediated by ASK1.[179,181]

GST-pi has two distinct functions related to oxidative stress.[160] Under non-stressed conditions, GST-pi exists predominantly in a monomeric form.[165] In this conformation, sequences at the C-terminus of GST-pi have been shown to bind to JNK; which inhibits the interaction of JNK with c-jun and reduces downstream apoptotic signalling.[182–184] As oxidative stress increases within the cell, GST-pi subunits form dimers. This interferes with the C-terminus interaction with JNK, allowing progression of cell-death signalling.[165] The GSH and GST systems collaborate to maintain oxidative homeostasis. When the cellular environment is stress free (*i.e.* there are low levels of free radicals), GSH levels alone are sufficient to maintain redox balance[185] and monomeric GST-pi bind JNK.[165] As the number of free radicals within the cell increases, GST-pi dimerizes and in this conformation interacts with GSH to more efficiently reduce these free radicals,[186] bringing the cell back to homeostasis.

A number of studies have examined the correlation of GST sequence polymorphisms with PD and with one exception, no associations were found (although see De Palma *et al.*, 1998[187]). Although polymorphisms in GSTs alone do not appear to correlate with PD etiology, they do appear to have some effect on other aspects of PD, including age of onset and interactions with environmental agents. For example, Golbe *et al.* reported that age of onset in individuals carrying the A53T alpha-synuclein (*SNCA/PARK1*) mutation is positively associated with a GST-pi G31A nucleotide substitution,[188] while McCormick *et al.* identified polymorphisms in GST-pi that correlate with increased risk of PD after exposure to pesticides.[189] Additionally, DePalma *et al.* reported a marked increase in association of PD with a GST-theta1*0 polymorphism when correlated with other PD risk factors, such as rural living and well-water consumption.[187] The lack of a consistent correlation between GST polymorphisms and PD does not rule out a significant role for this enzyme in the etiology of PD, as there could be other regulatory influences on these proteins, including those controlling synthesis and breakdown.

6.8 GSH and Transport in the Brain: Multidrug Resistance Proteins (MDRP) and the Blood–Brain Barrier (BBB)

Free radicals that have been reduced by thiol conjugation are transported out of the brain through multidrug resistance proteins (MDRP) at the gliovascular interface.[190–193] MDRPs are a family of ATP-binding cassette (ABC) transporters that actively transport amphipathic, electrophilic, and hydrophobic substances across cell membranes.[194] This large family of proteins (MDR1 (also known as P-glycoprotein), Mrp1-Mrp6, Mrp8, BSEP, and BCRP), is encoded by 10 genes (*ABCB1*, *ABCC1-ABCC6*, *ABCC11*, *ABCB11*, and *ABCG2*).[195,196] Each member of the ABC transporter family has a unique pattern of expression in tissue, which may vary between and within species.[197–199] In particular, the localization of transporters is specific to the luminal or basolateral membrane of endothelial cells and at astrocytic endfeet of the BBB.[196,198,200,201] In addition to the BBB, these ABC transporters are also found at the blood/cerebrospinal fluid interface, which forms both a physical and enzymatic barrier between the tight ependymal cell junctions of the choroid plexus and cerebrospinal fluid,[202,203] and also expresses proteins critical to modulating ROS detoxification. The transport of free-radical conjugates through MDRPs across the BBB is both ATP and sodium (Na^+) dependent.[190,191]

P-glycoprotein (P-gp, Mdr1), Mrp1, Mrp2 and Mrp5 (Mrps) are the most widely studied of the transporters in the brain. These transmembrane proteins alter their conformation by utilizing ATP to translocate substances, including unconjugated xenobiotics (P-gp) as well as products conjugated with GSH (Mrp1), across membranes.[204–209] P-gp transports hydrophobic and amphipathic compounds, and many P-gp substrates are also substrates for cytochrome P450 drug-metabolizing enzymes.[197,200] Mrps preferentially transport GSH-conjugated anions[200] with greater efficiency in the presence of GSH,[210–212] while P-gp is less dependent upon GSH levels.[206] However, P-gp expression is sensitive to GSH levels and oxidative stress.[191,213,214] Traditionally, it was believed that the expression of these transporters was found solely on capillary endothelial cells;[215–217] however, confocal studies suggest that some of these transporters may be localized to astrocytes.[218]

A number of experiments demonstrate that the expression of these efflux transport proteins are correlated to intracellular oxidative stress levels.[191,213,219–222] For example, Hong *et al.*[191] demonstrated that BSO depletion of GSH increases P-gp levels, and conversely, treatment with the antioxidant NAC decreases its expression. In models of the BBB, oxidative stress generated by H_2O_2[213] and GSH depletion[191,222] result in increased expression of P-gp. These studies demonstrate the cellular coordination of ROS detoxification between GSH or GST regulation and cellular efflux. In conditions of high ROS, where free GSH levels are lower due to binding of electrophiles, MDRP expression is increased. Conversely, when cells are at homeostasis, free GSH levels are elevated and MDRP levels decrease.

Numerous studies demonstrate alterations in MDRPs in PD. Expression of P-gp in blood vessels is reduced in the striatum of PD patients compared to control subjects.[223] Additionally, studies that examined the uptake of [$^{(11)}$C]-verapamil, a compound that is usually extruded from the brain by P-gp, demonstrate a significant elevation of [$^{(11)}$C]-verapamil in PD patients compared to controls suggesting that P-gp dysfunction contributes to PD pathogenesis.[224,225] Polymorphisms in P-gp have also been examined as they relate to PD risk. Westerlund[226] showed that a 1236C/T, but not 2677G/T/A or 3435C/T increases the risk of PD in a Swedish population. The lack of a direct effect of the 3435T polymorphism is supported by Funke *et al.*[227] and Furano *et al.*;[228] however, it is implicated as a contributing risk factor in the development of PD when combined with exposure to pesticides.[229,230]

6.9 Parkinson's Disease Genetic Models and Redox Signalling

Several proteins with disparate functions have been studied with regard to their role in oxidative stress and PD. Genetic mutations as well as biochemical alterations of these proteins lead to increased oxidative stress and decreased neuroprotection.

6.9.1 Glutathione-S-Transferase

The role of GSTs in the protection of DAergic neurons has been examined using several models of PD. Several groups have mutated the Parkin gene in Drosophila, and find subtle but significant loss of neurons in the protocerebral posterior lateral (PPL) DAergic neuron cluster[231] as well as an increase in oxidative stress[232] and a downregulation of GSTO1, which has high thiol transferase activity.[233] Examination of these mutants reveals that when GSTS1 is deleted the subtle loss of DAergic neurons in the PPL is enhanced. Conversely, overexpression of GSTS1 ameliorates this neurodegeneration.[231] When GSTO1, which has two distinct alleles, GSTO1A and GSTO1B, is deleted, there is an increased sensitivity to the xenobiotic paraquat.[232,234] The re-expression of the GSTO1A in the null mutant eliminates the sensitivity of the PPL DAergic neurons to paraquat and suppresses phospho-JNK activity, which is implicated in apoptosis.[233]

The effects of loss of GST-pi on MPP^+ or MPTP-induced DAergic neuron death have been examined. Using primary cultured DAergic cells harvested from the SN of MPTP-resistant Swiss-Webster mice, inhibition of GST by EA or siRNA directed against GST-pi increases the amount of MPP^+-induced DAergic neuron death to levels that mimic neurons from an MPTP-sensitive C57BL/6 strain. Similarly, when MPTP is administered to Swiss–Webster mice carrying a null mutation of GST-pi SNpc DAergic neuron loss increases to levels seen in the MPTP-sensitive C57Bl/6 strain.[172] In addition, 7 days after MPTP administration, GST-pi-null mice on a C57BL/6 background have a significantly greater TH+ neuron loss than WT C57BL/6 mice.[183] Administration of MPTP to mice lacking GSTpi also shows cellular ubiquitination and increased susceptibility to ubiquitin proteosome system (UPS) damage and inactivation.[235] Shi *et al.*[236] demonstrates that overexpression of GST-pi1 in Neuro2A cells as well as in primary cortical neurons confers protection from oxidative stress induced by addition of rotenone.

6.9.2 DJ-1

Mutations in the *PARK7* gene lead to loss of function of a ubiquitous, highly conserved protein, DJ-1. The mutation of this gene was first localized in families with autosomal recessive early-onset parkinsonism.[237] DJ-1 is a 189 amino acid protein containing 11 beta strands and 8 alpha helices[238] and is active in dimer formation.[238–240] DJ-1 is an evolutionarily conserved protein that shares sequence homology with bacterial intracellular proteases and a chaperone protein that has protease activity.[241] Studies have shown that reduction or total loss of DJ-1 causes increased sensitivity to oxidative stress and loss of neurons,[98,242–245] while overexpression of DJ-1 provides protection from oxidative insult.[98,242,245]

Functions of DJ-1 include transcriptional regulation, antioxidant activity and mitochondrial function,[99,246–248] as well as protease and chaperone activity.[249,250] It is localized in the cytoplasm, mitochondria,[251] and nucleus of the cell.[237] DJ-1 may be altered by phosphorylation, *S*-nitro(sy)lation, and sumoylation.[238] Cys46 and Cys54 may be *S*-nitros(yl)ated or oxidized and Cys106 may be oxidized but is not regulated by *S*-nitros(yl)ation.[239,246,252,253] The sequential oxidation of Cys106 by SOH to SO_2H and finally SO_3H inactivates DJ-1.[254] However, oxidation of Cys106 is necessary for nuclear translocation of DJ-1. DJ-1 provides protection against oxidative stress in part by functioning as a Cys protease.[31,255] DJ-1 is a zymogen with protease activation occurring upon the cleavage of a 15 amino acid peptide from the C-terminal. This Cys protease uses a Cys106, His126 dyad for catalysis, confirmed by site-directed mutagenesis.[255] Mild oxidative stress by H_2O_2, MPP^+, or rotenone activates DJ-1 protease activity in SH-SY5Y DAergic cells and is protective against apoptosis in these cells. DJ-1 is also protective when expressed in midbrain cultures from DJ-1−/− mice, and the C-terminally cleaved form provides greater protection than the WT form.[255] Mitochondria from DJ-1−/− mice have increased rates of H_2O_2 removal, increased Trx2/TrxR2 and Grx activity, and decreased GR activity, while GSSG levels are also increased, compared to DJ-1+/+ mice.[256] Trx1 is also necessary for the protective function of DJ-1 during oxidative stress, through the activation of NF-E2-related factor (Nrf2) mediated transcription.[257] In addition, DJ-1 has also been shown to bind to GPx4 mRNA, and its binding dissociates with oxidative stress.[258]

6.9.3 PTEN-Induced Putative Kinase 1 (PINK1)

A rare form of familial PD involves *PARK6* and the associated mutation in PTEN-induced kinase 1 (PINK1).[259] Decreases in PINK1 in human midbrain neurons results in higher ROS and lower GSH levels compared to control midbrain neurons,[260] and GSH turnover is greater in fibroblasts from PD patients with a *PARK6* mutation.[261]

6.9.4 Parkin

Parkin, the product of the *PARK2* gene, is responsible for a familial recessive form of PD where it functions as an E3 ubiquitin ligase.[262] Ubiquitination serves to tag misfolded, damaged, or dysfunctional proteins for proteosomal degradation. Failure to degrade and clear dysfunctional proteins may lead to toxic aggregates, as seen in the Lewy bodies that are found in the brains of patients with PD or Lewy body dementia.[56]

Nitros(yl)ation of Parkin by environmental factors results in a syndrome that phenocopies familial recessive PD caused by a Parkin mutation. Pesticides, herbicides, and fungicides that generate ROS/RNS affect the enzymatic activity of Parkin by altering the solubility and aggregation of the protein. Parkin has Cys residues in its RING (really interesting new gene) domain and elsewhere that react with NO and form SNO-Parkin.[55] Nitros(yl)ative stress leads to an initial increase in its E3 ubiquitin ligase activity, autoubiquitination of Parkin, and subsequent inhibition of its activity.[55,56] Treatment with rotenone[55] or MPTP[55,56] results in SNO-Parkin formation.

6.10 Free Radicals as Messengers to Modulate Transcription Factors: Effects on Thiol Redox Regulation

Control of gene expression in response to oxidative stress is due to oxidation of the proteins that control activation and localization of transcription factors, rather than oxidation of the transcription factors.[30] Examples of proteins regulated by oxidation that have an impact on DAergic neuron survival are kelch-like ECH-associated protein 1 (Keap1), which interacts with Nrf2[263] and Iκβ (inhibitor of kappa light polypeptide gene enhancer in B-cells, kinase beta) that regulates NF-κB.[264–266] Nrf2 is important to neuroprotection through its actions on enzymes involved in thiol redox regulation. The transcription factor Nrf2 regulates genes that contain ARE, a *cis*-acting regulatory element that modulates the expression of phase II detoxifying enzymes.[267,268] In normal, unstressed conditions, Nrf2 is bound to the protein Keap1 that is anchored to the actin cytoskeleton. Keap1 is a cytoplasmic protein that functions as an adaptor for Nrf2 and the Cullen 3-based E3 ligase.[269,270] When a cell is not stressed, Nrf2 is ubiquinated and degraded rapidly by the ubiquitin-proteosome system.[271] Cellular oxidative stress disrupts the interaction of Keap1 and Nrf2, and Nrf2 translocates to the nucleus and binds to the ARE region of phase II detoxifying genes.[269,270] Functions of antioxidant enzyme systems regulated by Nrf2 include redox regulation (SOD, catalase, Srx, Trx, and Prx), GSH synthesis and metabolism (GPx, GR, GCL, and GS), quinone recycling (NAD(P)H quinone oxidoreducase) and iron homeostasis (heme oxygenase 1).[267]

Overexpression of Nrf2 in neurons as well as in astrocytes increases antioxidant expression,[272] although normal expression of ARE-mediated genes in neurons is limited compared to this (antioxidant) gene expression in astrocytes, likely due to lower levels of Nrf2 protein in neurons.[273] In addition, total intracellular

GSH is increased and neurons are protected from peroxide induced oxidative stress by Nrf2 overexpression.[272] Overexpression of Nrf2 in astrocytes alone *in vivo*, also confers protection from MPTP.[274] Repression of Nrf2 expression by microRNAs in SH-SY5Y DAergic cells results in reduced Nrf2 concentration, efficacy of translocation, and ultimately the decreased expression of redox-sensitive enzymes GCLC and GR and GSH/GSSG ratio.[275] Inhibition of Keap1 in astrocytes *in vitro* increases the level of Nrf2-driven ARE modulated genes (GLC subunits GCLC and GCLM) and upregulates the Nrf2-ARE pathway; and protects against the loss of TH protein with MPTP treatment *in vivo*.[263] Activation of the Nrf2-ARE pathway in DJ-1-null as well as WT mice, and overexpression of Nrf2 in astrocytes from DJ-1 knockout mice protects neurons from oxidative stress.[276]

6.11 Conclusions

It is well established that oxidative stress is detrimental to neuronal survival. GSH is an important source of reducing equivalents to counteract the effects of ROS/RNS in the brain. For reasons unique to the SNpc, these DAergic neurons are especially susceptible to oxidative insult, which if unaltered, leads to their demise in the course of PD. Thiol redox reactions and the antioxidant GSH play an important role in maintaining redox homeostasis as well as in signalling to initiate pathways for gene induction or enzymatic reactions involved in homeostasis. Functional interactions of oxidoreductases, including Trx, Prx, and Grx through the oxidation and reduction of Cys residues, serve to maintain this balance within the cell. When these and other enzymes (GPx and GST) are overwhelmed, oxidative stress serves as a signal to activate proteins that interact with transcription factors (such as Nrf2) that induce genes to further assist in rescuing the cell from irreversible damage by ROS, or to initiate cell death. The redox system in the brain is complex, complicated by its heterogeneous cell populations, specifically in the SN. The complexity may serve to provide numerous points of intervention for the maintenance of DAergic neuron health and survival in advance of, and during, the aging process. The ubiquity of cellular GSH is an indication of its importance to neuron function and maintenance.

References

1. P. Jenner, D. T. Dexter, J. Sian, A. H. Schapira and C. D. Marsden, Oxidative stress as a cause of nigral cell death in Parkinson's disease and incidental Lewy body disease. The Royal Kings and Queens Parkinson's Disease Research Group, *Ann. Neurol.*, 1992, **32**, S82–S87.
2. J. Sian, D. T. Dexter, A. J. Lees, S. Daniel, Y. Agid, F. Javoy-Agid, P. Jenner and C. D. Marsden, Alterations in glutathione levels in Parkinson's disease and other neurodegenerative disorders affecting basal ganglia [see comments], *Ann. Neurol.*, 1994, **36**, 348–355.

3. H. L. Martin and P. Teismann, Glutathione–a review on its role and significance in Parkinson's disease, *FASEB J.*, 2009, **23**, 3263–3272.
4. J. Sian, D. T. Dexter, A. J. Lees, S. Daniel, P. Jenner and C. D. Marsden, Glutathione-related enzymes in brain in Parkinson's disease, *Ann. Neurol.*, 1994, **36**, 356–361.
5. A. Meister and M. E. Anderson, Glutathione, *Annu. Rev. Biochem.*, 1983, **52**, 711–760.
6. A. J. L. Cooper, *The Molecular and Genetic Basis of Neurological Disease*, Butterworth-Heinemann, Newton, MA, 1997, pp. 1195–1230.
7. M. E. Raichle and D. A. Gusnard, Appraising the brain's energy budget, *Proc. Natl. Acad. Sci. U. S. A.*, 2002, **99**, 10237–10239.
8. K. A. Jellinger, Post mortem studies in Parkinson's disease–is it possible to detect brain areas for specific symptoms? [Review] [225 refs], *J. Neural Transm., Suppl.*, 1999, **56**, 1–29.
9. N. I. Bohnen and R. L. Albin, The cholinergic system and Parkinson disease, *Behav. Brain Res.*, 2011, **221**, 564–573.
10. L. T. Grinberg, U. Rueb, A. T. Alho and H. Heinsen, Brainstem pathology and non-motor symptoms in PD, *J. Neurol. Sci.*, 2010, **289**, 81–88.
11. C. Singaram, W. Ashraf, E. A. Gaumnitz, C. Torbey, A. Sengupta, R. Pfeiffer and E. M. Quigley, Dopaminergic defect of enteric nervous system in Parkinson's disease patients with chronic constipation, *Lancet*, 1995, **346**, 861–864.
12. D. Ben-Shachar and M. B. Youdim, Iron, melanin and dopamine interaction: relevance to Parkinson's disease, *Prog. Neuro-Psychopharmacol. Biol. Psychiatry*, 1993, **17**, 139–150.
13. M. Gerlach, K. L. Double, D. Ben-Shachar, L. Zecca, M. B. Youdim and P. Riederer, Neuromelanin and its interaction with iron as a potential risk factor for dopaminergic neurodegeneration underlying Parkinson's disease, *Neurotoxic. Res.*, 2003, **5**, 35–44.
14. M. E. Gotz, K. Double, M. Gerlach, M. B. Youdim and P. Riederer, The relevance of iron in the pathogenesis of Parkinson's disease, *Ann. N. Y. Acad. Sci.*, 2004, **1012**, 193–208.
15. R. K. Pearce, A. Owen, S. Daniel, P. Jenner and C. D. Marsden, Alterations in the distribution of glutathione in the substantia nigra in Parkinson's disease, *J. Neural Transm.*, 1997, **104**, 661–677.
16. Y. Kang, V. Viswanath, N. Jha, X. Qiao, J. Q. Mo and J. K. Andersen, Brain gamma-glutamyl cysteine synthetase (GCS) mRNA expression patterns correlate with regional-specific enzyme activities and glutathione levels, *J. Neurosci. Res.*, 1999, **58**, 436–441.
17. W. G. Kim, R. P. Mohney, B. Wilson, G. H. Jeohn, B. Liu and J. S. Hong, Regional difference in susceptibility to lipopolysaccharide-induced neurotoxicity in the rat brain: role of microglia, *J. Neurosci.*, 2000, **20**, 6309–6316.
18. M. Smeyne, Y. Jiao, K. R. Shepherd and R. J. Smeyne, Glia cell number modulates sensitivity to MPTP in mice, *Glia*, 2005, **52**, 144–152.

19. J. Goemaere and B. Knoops, Peroxiredoxin distribution in the mouse brain with emphasis on neuronal populations affected in neurodegenerative disorders, *J. Comp. Neurol.*, 2012, **520**, 258–280.
20. A. Garcia-Garcia, L. Zavala-Flores, H. Rodriguez-Rocha and R. Franco, Thiol-redox signaling, dopaminergic cell death, and Parkinson's disease, *Antioxid. Redox Signaling*, 2012, **17**, 1764–1784.
21. R. Dringen, Metabolism and functions of glutathione in brain, *Prog. Neurobiol.*, 2000, **62**, 649–671.
22. G. A. Thompson and A. Meister, Utilization of L-cystine by the gamma-glutamyl transpeptidase-gamma-glutamyl cyclotransferase pathway, *Proc. Natl. Acad. Sci. U. S. A.*, 1975, **72**, 1985–1988.
23. C. D. Puglia and S. R. Powell, Inhibition of cellular antioxidants: a possible mechanism of toxic cell injury, *Environ. Health Perspect.*, 1984, **57**, 307–311.
24. M. E. Anderson, Glutathione: an overview of biosynthesis and modulation, *Chem.-Biol. Interact.*, 1998, **111–112**, 1–14.
25. R. Dringen, Oxidative and antioxidative potential of brain microglial cells, *Antioxid. Redox Signaling*, 2005, **7**, 1223–1233.
26. O. W. Griffith, Biologic and pharmacologic regulation of mammalian glutathione synthesis, *Free Radic. Biol. Med.*, 1999, **27**, 922–935.
27. D. A. Dickinson and H. J. Forman, Glutathione in defense and signaling: lessons from a small thiol, *Ann. N. Y. Acad. Sci.*, 2002, **973**, 488–504.
28. D. A. Dickinson and H. J. Forman, Cellular glutathione and thiols metabolism, *Biochem. Pharmacol.*, 2002, **64**, 1019–1026.
29. H. Hart, *Organic Chemistry: A Short Course*, Houghton Mifflin Co, Chicago, 1983.
30. C. C. Winterbourn and M. B. Hampton, Thiol chemistry and specificity in redox signaling, *Free Radic. Biol. Med.*, 2008, **45**, 549–561.
31. N. M. Giles, G. I. Giles and C. Jacob, Multiple roles of cysteine in biocatalysis, *Biochem. Biophys. Res. Commun.*, 2003, **300**, 1–4.
32. E. Stroher and A. H. Millar, The biological roles of glutaredoxins, *Biochem. J.*, 2012, **446**, 333–348.
33. E. S. Arner and A. Holmgren, Physiological functions of thioredoxin and thioredoxin reductase, *Eur. J. Biochem.*, 2000, **267**, 6102–6109.
34. A. P. Fernandes and A. Holmgren, Glutaredoxins: glutathione-dependent redox enzymes with functions far beyond a simple thioredoxin backup system, *Antioxid. Redox Signaling*, 2004, **6**, 63–74.
35. L. I. Leichert and U. Jakob, Protein thiol modifications visualized *in vivo*, *PLoS Biol.*, 2004, **2**, e333.
36. S. B. Wall, J. Y. Oh, A. R. Diers and A. Landar, Oxidative modification of proteins: an emerging mechanism of cell signaling, *Front. Physiol.*, 2012, **3**, 369.
37. P. Nagy, Kinetics and mechanisms of thiol-disulfide exchange covering direct substitution and thiol oxidation-mediated pathways, *Antioxid. Redox Signaling*, 2013, **18**, 1623–1641.

38. J. M. Held and B. W. Gibson, Regulatory control or oxidative damage? Proteomic approaches to interrogate the role of cysteine oxidation status in biological processes, *Mol. Cell. Proteomics*, 2012, **11**, R111 013037.
39. M. C. Martinez and R. Andriantsitohaina, Reactive nitrogen species: molecular mechanisms and potential significance in health and disease, *Antioxid. Redox Signaling*, 2009, **11**, 669–702.
40. U. Jakob, W. Muse, M. Eser and J. C. Bardwell, Chaperone activity with a redox switch, *Cell*, 1999, **96**, 341–352.
41. I. A. Cotgreave, R. Gerdes, I. Schuppe-Koistinen and C. Lind, *S*-glutathionylation of glyceraldehyde-3-phosphate dehydrogenase: role of thiol oxidation and catalysis by glutaredoxin, *Methods Enzymol.*, 2002, **348**, 175–182.
42. R. Gopalakrishna and S. Jaken, Protein kinase C signaling and oxidative stress, *Free Radic. Biol. Med.*, 2000, **28**, 1349–1361.
43. N. R. Leslie, D. Bennett, Y. E. Lindsay, H. Stewart, A. Gray and C. P. Downes, Redox regulation of PI 3-kinase signalling *via* inactivation of PTEN, *EMBO J.*, 2003, **22**, 5501–5510.
44. Y. J. Suzuki, H. J. Forman and A. Sevanian, Oxidants as stimulators of signal transduction, *Free Radic. Biol. Med.*, 1997, **22**, 269–285.
45. J. S. Stamler, D. J. Singel and J. Loscalzo, Biochemistry of nitric oxide and its redox-activated forms, *Science*, 1992, **258**, 1898–1902.
46. J. S. Pollock, U. Forstermann, J. A. Mitchell, T. D. Warner, H. H. Schmidt, M. Nakane and F. Murad, Purification and characterization of particulate endothelium-derived relaxing factor synthase from cultured and native bovine aortic endothelial cells, *Proc. Natl. Acad. Sci. U. S. A.*, 1991, **88**, 10480–10484.
47. C. J. Lowenstein, C. S. Glatt, D. S. Bredt and S. H. Snyder, Cloned and expressed macrophage nitric oxide synthase contrasts with the brain enzyme, *Proc. Natl. Acad. Sci. U. S. A.*, 1992, **89**, 6711–6715.
48. E. Galea, D. L. Feinstein and D. J. Reis, Induction of calcium-independent nitric oxide synthase activity in primary rat glial cultures, *Proc. Natl. Acad. Sci. U. S. A.*, 1992, **89**, 10945–10949.
49. D. S. Bredt, C. E. Glatt, P. M. Hwang, M. Fotuhi, T. M. Dawson and S. H. Snyder, Nitric oxide synthase protein and mRNA are discretely localized in neuronal populations of the mammalian CNS together with NADPH diaphorase, *Neuron*, 1991, **7**, 615–624.
50. M. W. Akhtar, C. R. Sunico, T. Nakamura and S. A. Lipton, Redox Regulation of Protein Function *via* Cysteine *S*-Nitrosylation and Its Relevance to Neurodegenerative Diseases, *Int. J. Cell Biol.*, 2012, **2012**, 463756.
51. D. Nikitovic and A. Holmgren, *S*-nitrosoglutathione is cleaved by the thioredoxin system with liberation of glutathione and redox regulating nitric oxide, *J. Biol. Chem.*, 1996, **271**, 19180–19185.
52. M. Benhar, M. T. Forrester and J. S. Stamler, Protein denitrosylation: enzymatic mechanisms and cellular functions, *Nat. Rev. Mol. Cell Biol.*, 2009, **10**, 721–732.

53. D. T. Hess, A. Matsumoto, S. O. Kim, H. E. Marshall and J. S. Stamler, Protein *S*-nitrosylation: purview and parameters, *Nat. Rev. Mol. Cell Biol.*, 2005, **6**, 150–166.
54. J. S. Stamler, E. J. Toone, S. A. Lipton and N. J. Sucher, (S)NO signals: translocation, regulation, and a consensus motif, *Neuron*, 1997, **18**, 691–696.
55. D. Yao, Z. Gu, T. Nakamura, Z. Q. Shi, Y. Ma, B. Gaston, L. A. Palmer, E. M. Rockenstein, Z. Zhang, E. Masliah, T. Uehara and S. A. Lipton, Nitrosative stress linked to sporadic Parkinson's disease: *S*-nitrosylation of parkin regulates its E3 ubiquitin ligase activity, *Proc. Natl. Acad. Sci. U. S. A.*, 2004, **101**, 10810–10814.
56. K. K. Chung, B. Thomas, X. Li, O. Pletnikova, J. C. Troncoso, L. Marsh, V. L. Dawson and T. M. Dawson, *S*-nitrosylation of parkin regulates ubiquitination and compromises parkin's protective function, *Science*, 2004, **304**, 1328–1331.
57. J. Fang, T. Nakamura, D. H. Cho, Z. Gu and S. A. Lipton, *S*-nitrosylation of peroxiredoxin 2 promotes oxidative stress-induced neuronal cell death in Parkinson's disease, *Proc. Natl. Acad. Sci. U. S. A.*, 2007, **104**, 18742–18747.
58. A. H. Tsang, Y. I. Lee, H. S. Ko, J. M. Savitt, O. Pletnikova, J. C. Troncoso, V. L. Dawson, T. M. Dawson and K. K. Chung, *S*-nitrosylation of XIAP compromises neuronal survival in Parkinson's disease, *Proc. Natl. Acad. Sci. U. S. A.*, 2009, **106**, 4900–4905.
59. M. R. Hara, N. Agrawal, S. F. Kim, M. B. Cascio, M. Fujimuro, Y. Ozeki, M. Takahashi, J. H. Cheah, S. K. Tankou, L. D. Hester, C. D. Ferris, S. D. Hayward, S. H. Snyder and A. Sawa, *S*-nitrosylated GAPDH initiates apoptotic cell death by nuclear translocation following Siah1 binding, *Nat. Cell Biol.*, 2005, **7**, 665–674.
60. T. Uehara, T. Nakamura, D. Yao, Z. Q. Shi, Z. Gu, Y. Ma, E. Masliah, Y. Nomura and S. A. Lipton, *S*-nitrosylated protein-disulphide isomerase links protein misfolding to neurodegeneration, *Nature*, 2006, **441**, 513–517.
61. N. J. Bulleid and L. Ellgaard, Multiple ways to make disulfides, *Trends Biochem. Sci.*, 2011, **36**, 485–492.
62. N. J. Bulleid, Disulfide bond formation in the mammalian endoplasmic reticulum, *Cold Spring Harbor Perspect. Biol.*, 2012, **4**.
63. J. M. Herrmann and J. Riemer, Mitochondrial disulfide relay: redox-regulated protein import into the intermembrane space, *J. Biol. Chem.*, 2012, **287**, 4426–4433.
64. www.uniprot.org, P10599-THIO_HUMAN, http://www.uniprot.org/uniprot/P10599, accessed 2/27/2015.
65. J. Bai, H. Nakamura, Y. W. Kwon, M. Tanito, S. Ueda, T. Tanaka, I. Hattori, S. Ban, T. Momoi, Y. Kitao, S. Ogawa and J. Yodoi, Does thioredoxin-1 prevent mitochondria- and endoplasmic reticulum-mediated neurotoxicity of 1-methyl-4-phenyl-1,2,3,6-tetrahydropyridine? *Antioxid. Redox Signaling*, 2007, **9**, 603–608.

66. E. A. Sabens, A. M. Distler and J. J. Mieyal, Levodopa deactivates enzymes that regulate thiol-disulfide homeostasis and promotes neuronal cell death: implications for therapy of Parkinson's disease, *Biochemistry*, 2010, **49**, 2715–2724.
67. D. A. Drechsel and M. Patel, Respiration-dependent H2O2 removal in brain mitochondria *via* the thioredoxin/peroxiredoxin system, *J. Biol. Chem.*, 2010, **285**, 27850–27858.
68. J. Soerensen, C. Jakupoglu, H. Beck, H. Forster, J. Schmidt, W. Schmahl, U. Schweizer, M. Conrad and M. Brielmeier, The role of thioredoxin reductases in brain development, *PLoS One*, 2008, **3**, e1813.
69. E. A. Liedhegner, K. M. Steller and J. J. Mieyal, Levodopa activates apoptosis signaling kinase 1 (ASK1) and promotes apoptosis in a neuronal model: implications for the treatment of Parkinson's disease, *Chem. Res. Toxicol.*, 2011, **24**, 1644–1652.
70. H. Schenk, M. Klein, W. Erdbrugger, W. Droge and K. Schulze-Osthoff, Distinct effects of thioredoxin and antioxidants on the activation of transcription factors NF-kappa B and AP-1, *Proc. Natl. Acad. Sci. U. S. A.*, 1994, **91**, 1672–1676.
71. C. Jakupoglu, G. K. Przemeck, M. Schneider, S. G. Moreno, N. Mayr, A. K. Hatzopoulos, M. H. de Angelis, W. Wurst, G. W. Bornkamm, M. Brielmeier and M. Conrad, Cytoplasmic thioredoxin reductase is essential for embryogenesis but dispensable for cardiac development, *Mol. Cell. Biol.*, 2005, **25**, 1980–1988.
72. M. Conrad, Transgenic mouse models for the vital selenoenzymes cytosolic thioredoxin reductase, mitochondrial thioredoxin reductase and glutathione peroxidase 4, *Biochim. Biophys. Acta*, 2009, **1790**, 1575–1585.
73. P. Lopert, B. J. Day and M. Patel, Thioredoxin reductase deficiency potentiates oxidative stress, mitochondrial dysfunction and cell death in dopaminergic cells, *PLoS One*, 2012, **7**, e50683.
74. M. Saitoh, H. Nishitoh, M. Fujii, K. Takeda, K. Tobiume, Y. Sawada, M. Kawabata, K. Miyazono and H. Ichijo, Mammalian thioredoxin is a direct inhibitor of apoptosis signal-regulating kinase (ASK) 1, *EMBO J.*, 1998, **17**, 2596–2606.
75. P. J. Nadeau, S. J. Charette, M. B. Toledano and J. Landry, Disulfide Bond-mediated multimerization of Ask1 and its reduction by thioredoxin-1 regulate H(2)O(2)-induced c-Jun NH(2)-terminal kinase activation and apoptosis, *Mol. Biol. Cell*, 2007, **18**, 3903–3913.
76. M. Luthman, S. Eriksson, A. Holmgren and L. Thelander, Glutathione-dependent hydrogen donor system for calf thymus ribonucleoside-diphosphate reductase, *Proc. Natl. Acad. Sci. U. S. A.*, 1979, **76**, 2158–2162.
77. S. Witte, M. Villalba, K. Bi, Y. Liu, N. Isakov and A. Altman, Inhibition of the c-Jun N-terminal kinase/AP-1 and NF-kappaB pathways by PICOT, a novel protein kinase C-interacting protein with a thioredoxin homology domain, *J. Biol. Chem.*, 2000, **275**, 1902–1909.

78. H. Zhang, Y. Du, X. Zhang, J. Lu and A. Holmgren, Glutaredoxin 2 reduces both thioredoxin 2 and thioredoxin 1 and protects cells from apoptosis induced by auranofin and 4-hydroxynonenal, *Antioxid. Redox Signaling*, 2014, **21**, 669–681.
79. Y. Du, H. Zhang, X. Zhang, J. Lu and A. Holmgren, Thioredoxin 1 is inactivated due to oxidation induced by peroxiredoxin under oxidative stress and reactivated by the glutaredoxin system, *J. Biol. Chem.*, 2013, **288**, 32241–32247.
80. D. W. Lee, D. Kaur, S. J. Chinta, S. Rajagopalan and J. K. Andersen, A disruption in iron-sulfur center biogenesis *via* inhibition of mitochondrial dithiol glutaredoxin 2 may contribute to mitochondrial and cellular iron dysregulation in mammalian glutathione-depleted dopaminergic cells: implications for Parkinson's disease, *Antioxid. Redox Signaling*, 2009, **11**, 2083–2094.
81. S. M. Beer, E. R. Taylor, S. E. Brown, C. C. Dahm, N. J. Costa, M. J. Runswick and M. P. Murphy, Glutaredoxin 2 catalyzes the reversible oxidation and glutathionylation of mitochondrial membrane thiol proteins: implications for mitochondrial redox regulation and antioxidant DEFENSE, *J. Biol. Chem.*, 2004, **279**, 47939–47951.
82. S. Karunakaran, U. Saeed, S. Ramakrishnan, R. C. Koumar and V. Ravindranath, Constitutive expression and functional characterization of mitochondrial glutaredoxin (Grx2) in mouse and human brain, *Brain Res.*, 2007, **1185**, 8–17.
83. J. S. Chae, S. Gil Hwang, D. S. Lim and E. J. Choi, Thioredoxin-1 functions as a molecular switch regulating the oxidative stress-induced activation of MST1, *Free Radic. Biol. Med.*, 2012, **53**, 2335–2343.
84. S. G. Rhee, H. Z. Chae and K. Kim, Peroxiredoxins: a historical overview and speculative preview of novel mechanisms and emerging concepts in cell signaling, *Free Radic. Biol. Med.*, 2005, **38**, 1543–1552.
85. S. G. Rhee, H. A. Woo, I. S. Kil and S. H. Bae, Peroxiredoxin functions as a peroxidase and a regulator and sensor of local peroxides, *J. Biol. Chem.*, 2012, **287**, 4403–4410.
86. M. H. Jin, Y. H. Lee, J. M. Kim, H. N. Sun, E. Y. Moon, M. H. Shong, S. U. Kim, S. H. Lee, T. H. Lee, D. Y. Yu and D. S. Lee, Characterization of neural cell types expressing peroxiredoxins in mouse brain, *Neurosci. Lett.*, 2005, **381**, 252–257.
87. H. Z. Chae, K. Robison, L. B. Poole, G. Church, G. Storz and S. G. Rhee, Cloning and sequencing of thiol-specific antioxidant from mammalian brain: alkyl hydroperoxide reductase and thiol-specific antioxidant define a large family of antioxidant enzymes, *Proc. Natl. Acad. Sci. U. S. A.*, 1994, **91**, 7017–7021.
88. T. J. Tavender, A. M. Sheppard and N. J. Bulleid, Peroxiredoxin IV is an endoplasmic reticulum-localized enzyme forming oligomeric complexes in human cells, *Biochem. J.*, 2008, **411**, 191–199.
89. Z. A. Wood, E. Schroder, J. Robin Harris and L. B. Poole, Structure, mechanism and regulation of peroxiredoxins, *Trends Biochem. Sci.*, 2003, **28**, 32–40.

90. H. Zhu, A. Santo and Y. Li, The antioxidant enzyme peroxiredoxin and its protective role in neurological disorders, *Exp. Biol. Med. (Maywood)*, 2012, **237**, 143–149.
91. X. Hu, Z. Weng, C. T. Chu, L. Zhang, G. Cao, Y. Gao, A. Signore, J. Zhu, T. Hastings, J. T. Greenamyre and J. Chen, Peroxiredoxin-2 protects against 6-hydroxydopamine-induced dopaminergic neurodegeneration *via* attenuation of the apoptosis signal-regulating kinase (ASK1) signaling cascade, *J. Neurosci.*, 2011, **31**, 247–261.
92. H. Z. Chae, H. Oubrahim, J. W. Park, S. G. Rhee and P. B. Chock, Protein glutathionylation in the regulation of peroxiredoxins: a family of thiol-specific peroxidases that function as antioxidants, molecular chaperones, and signal modulators, *Antioxid. Redox Signaling*, 2012, **16**, 506–523.
93. T. H. Lee, S. U. Kim, S. L. Yu, S. H. Kim, D. S. Park, H. B. Moon, S. H. Dho, K. S. Kwon, H. J. Kwon, Y. H. Han, S. Jeong, S. W. Kang, H. S. Shin, K. K. Lee, S. G. Rhee and D. Y. Yu, Peroxiredoxin II is essential for sustaining life span of erythrocytes in mice, *Blood*, 2003, **101**, 5033–5038.
94. L. Li, W. Shoji, H. Takano, N. Nishimura, Y. Aoki, R. Takahashi, S. Goto, T. Kaifu, T. Takai and M. Obinata, Increased susceptibility of MER5 (peroxiredoxin III) knockout mice to LPS-induced oxidative stress, *Biochem. Biophys. Res. Commun.*, 2007, **355**, 715–721.
95. H. N. Sun, S. U. Kim, S. M. Huang, J. M. Kim, Y. H. Park, S. H. Kim, H. Y. Yang, K. J. Chung, T. H. Lee, H. S. Choi, J. S. Min, M. K. Park, S. K. Kim, S. R. Lee, K. T. Chang, S. H. Lee, D. Y. Yu and D. S. Lee, Microglial peroxiredoxin V acts as an inducible anti-inflammatory antioxidant through cooperation with redox signaling cascades, *J. Neurochem.*, 2010, **114**, 39–50.
96. D. Qu, J. Rashidian, M. P. Mount, H. Aleyasin, M. Parsanejad, A. Lira, E. Haque, Y. Zhang, S. Callaghan, M. Daigle, M. W. Rousseaux, R. S. Slack, P. R. Albert, I. Vincent, J. M. Woulfe and D. S. Park, Role of Cdk5-mediated phosphorylation of Prx2 in MPTP toxicity and Parkinson's disease, *Neuron*, 2007, **55**, 37–52.
97. Y. M. Lee, S. H. Park, D. I. Shin, J. Y. Hwang, B. Park, Y. J. Park, T. H. Lee, H. Z. Chae, B. K. Jin, T. H. Oh and Y. J. Oh, Oxidative modification of peroxiredoxin is associated with drug-induced apoptotic signaling in experimental models of Parkinson disease, *J. Biol. Chem.*, 2008, **283**, 9986–9998.
98. R. H. Kim, P. D. Smith, H. Aleyasin, S. Hayley, M. P. Mount, S. Pownall, A. Wakeham, A. J. You-Ten, S. K. Kalia, P. Horne, D. Westaway, A. M. Lozano, H. Anisman, D. S. Park and T. W. Mak, Hypersensitivity of DJ-1-deficient mice to 1-methyl-4-phenyl-1,2,3,6-tetrahydropyrindine (MPTP) and oxidative stress, *Proc. Natl. Acad. Sci. U. S. A.*, 2005, **102**, 5215–5220.
99. E. Andres-Mateos, C. Perier, L. Zhang, B. Blanchard-Fillion, T. M. Greco, B. Thomas, H. S. Ko, M. Sasaki, H. Ischiropoulos, S. Przedborski, T. M. Dawson and V. L. Dawson, DJ-1 gene deletion reveals that DJ-1 is an atypical peroxiredoxin-like peroxidase, *Proc. Natl. Acad. Sci. U. S. A.*, 2007, **104**, 14807–14812.

100. Y. Du, H. Zhang, J. Lu and A. Holmgren, Glutathione and glutaredoxin act as a backup of human thioredoxin reductase 1 to reduce thioredoxin 1 preventing cell death by aurothioglucose, *J. Biol. Chem.*, 2012, **287**, 38210–38219.
101. L. Brautigam, L. D. Jensen, G. Poschmann, S. Nystrom, S. Bannenberg, K. Dreij, K. Lepka, T. Prozorovski, S. J. Montano, O. Aktas, P. Uhlen, K. Stuhler, Y. Cao, A. Holmgren and C. Berndt, Glutaredoxin regulates vascular development by reversible glutathionylation of sirtuin 1, *Proc. Natl. Acad. Sci. U. S. A.*, 2013, **110**, 20057–20062.
102. L. Brautigam, L. D. Schutte, J. R. Godoy, T. Prozorovski, M. Gellert, G. Hauptmann, A. Holmgren, C. H. Lillig and C. Berndt, Vertebrate-specific glutaredoxin is essential for brain development, *Proc. Natl. Acad. Sci. U. S. A.*, 2011, **108**, 20532–20537.
103. M. Yudkoff, Y. Daikhin, I. Nissim, O. Horyn, B. Luhovyy and A. Lazarow, Brain amino acid requirements and toxicity: the example of leucine, *J. Nutr.*, 2005, **135**, 1531S–1538S.
104. V. Calabrese, N. Ragusa, A. Antico, S. Mangiameli and V. Rizza, Cysteine-induced enhancement of lipid peroxidation in substantia nigra: comparative effect with exogenous administration of reduced glutathione, *Drugs Exp. Clin. Res.*, 1997, **23**, 25–31.
105. D. Schubert and D. Piasecki, Oxidative glutamate toxicity can be a component of the excitotoxicity cascade, *J. Neurosci.*, 2001, **21**, 7455–7462.
106. X. F. Wang and M. S. Cynader, Pyruvate released by astrocytes protects neurons from copper-catalyzed cysteine neurotoxicity, *J. Neurosci.*, 2001, **21**, 3322–3331.
107. C. S. Huang, W. R. Moore and A. Meister, On the active site thiol of gamma-glutamylcysteine synthetase: relationships to catalysis, inhibition, and regulation, *Proc. Natl. Acad. Sci. U. S. A.*, 1988, **85**, 2464–2468.
108. O. W. Griffith and R. T. Mulcahy, The enzymes of glutathione synthesis: gamma-glutamylcysteine synthetase, *Adv. Enzymol. Relat. Areas Mol. Biol.*, 1999, **73**, 209–267, xii.
109. R. T. Mulcahy, M. A. Wartman, H. H. Bailey and J. J. Gipp, Constitutive and beta-naphthoflavone-induced expression of the human gamma-glutamylcysteine synthetase heavy subunit gene is regulated by a distal antioxidant response element/TRE sequence, *J. Biol. Chem.*, 1997, **272**, 7445–7454.
110. H. Zhang and H. J. Forman, Reexamination of the electrophile response element sequences and context reveals a lack of consensus in gene function, *Biochim. Biophys. Acta*, 2010, **1799**, 496–501.
111. H. Yang, N. Magilnick, X. Ou and S. C. Lu, Tumour necrosis factor alpha induces co-ordinated activation of rat GSH synthetic enzymes *via* nuclear factor kappaB and activator protein-1, *Biochem. J.*, 2005, **391**, 399–408.
112. R. T. Mulcahy and J. J. Gipp, Identification of a putative antioxidant response element in the 5'-flanking region of the human gamma-glutamylcysteine synthetase heavy subunit gene, *Biochem. Biophys. Res. Commun.*, 1995, **209**, 227–233.

113. H. R. Moinova and R. T. Mulcahy, An electrophile responsive element (EpRE) regulates beta-naphthoflavone induction of the human gamma-glutamylcysteine synthetase regulatory subunit gene. Constitutive expression is mediated by an adjacent AP-1 site, *J. Biol. Chem.*, 1998, **273**, 14683–14689.
114. Y. M. Go, J. J. Gipp, R. T. Mulcahy and D. P. Jones, H2O2-dependent activation of GCLC-ARE4 reporter occurs by mitogen-activated protein kinase pathways without oxidation of cellular glutathione or thioredoxin-1, *J. Biol. Chem.*, 2004, **279**, 5837–5845.
115. A. C. Wild and R. T. Mulcahy, Regulation of gamma-glutamylcysteine synthetase subunit gene expression: insights into transcriptional control of antioxidant defenses, *Free Radic. Res.*, 2000, **32**, 281–301.
116. P. P. Trotta, K. E. Platzer, R. H. Haschemeyer and A. Meister, Glutamine-binding subunit of glutamate synthase and partial reactions catalyzed by this glutamine amidotransferase, *Proc. Natl. Acad. Sci. U. S. A.*, 1974, **71**, 4607–4611.
117. R. Njalsson, E. Ristoff, K. Carlsson, A. Winkler, A. Larsson and S. Norgren, Genotype, enzyme activity, glutathione level, and clinical phenotype in patients with glutathione synthetase deficiency, *Hum. Genet.*, 2005, **116**, 384–389.
118. R. Dringen, B. Pfeiffer and B. Hamprecht, Synthesis of the antioxidant glutathione in neurons: supply by astrocytes of CysGly as precursor for neuronal glutathione, *J. Neurosci.*, 1999, **19**, 562–569.
119. R. Dringen and J. Hirrlinger, Glutathione pathways in the brain, *Biol. Chem.*, 2003, **384**, 505–516.
120. R. Dringen and B. Hamprecht, Glutathione content as an indicator for the presence of metabolic pathways of amino acids in astroglial cultures, *J. Neurochem.*, 1996, **67**, 1375–1382.
121. R. Dringen, O. Kranich, P. A. Loschmann and B. Hamprecht, Use of dipeptides for the synthesis of glutathione by astroglia-rich primary cultures, *J. Neurochem.*, 1997, **69**, 868–874.
122. A. Meister, Glutathione metabolism and its selective modification, *J. Biol. Chem.*, 1988, **263**, 17205–17208.
123. X. Sun, A. Y. Shih, H. C. Johannssen, H. Erb, P. Li and T. H. Murphy, Two-photon imaging of glutathione levels in intact brain indicates enhanced redox buffering in developing neurons and cells at the cerebrospinal fluid and blood-brain interface, *J. Biol. Chem.*, 2006, **281**, 17420–17431.
124. D. E. Bragin, B. Zhou, P. Ramamoorthy, W. S. Muller, J. A. Connor and H. Shi, Differential changes of glutathione levels in astrocytes and neurons in ischemic brains by two-photon imaging, *J. Cereb. Blood Flow Metab.*, 2010, **30**, 734–738.
125. U. Wullner, P. A. Loschmann, J. B. Schulz, A. Schmid, R. Dringen, F. Eblen, L. Turski and T. Klockgether, Glutathione depletion potentiates MPTP and MPP+ toxicity in nigral dopaminergic neurones, *NeuroReport*, 1996, 7, 921–923.

126. S. Canals, M. J. Casarejos, S. de Bernardo, E. Rodriguez-Martin and M. A. Mena, Glutathione depletion switches nitric oxide neurotrophic effects to cell death in midbrain cultures: implications for Parkinson's disease, *J. Neurochem.*, 2001, **79**, 1183–1195.
127. J. K. Andersen, J. Q. Mo, D. G. Hom, F. Y. Lee, P. Harnish, R. W. Hamill and T. H. McNeill, Effect of buthionine sulfoximine, a synthesis inhibitor of the antioxidant glutathione, on the murine nigrostriatal neurons, *J. Neurochem.*, 1996, **67**, 2164–2171.
128. M. Ibi, H. Sawada, T. Kume, H. Katsuki, S. Kaneko, S. Shimohama and A. Akaike, Depletion of intracellular glutathione increases susceptibility to nitric oxide in mesencephalic dopaminergic neurons, *J. Neurochem.*, 1999, **73**, 1696–1703.
129. S. J. Chinta and J. K. Andersen, Reversible inhibition of mitochondrial complex I activity following chronic dopaminergic glutathione depletion *in vitro*: implications for Parkinson's disease, *Free Radic. Biol. Med.*, 2006, **41**, 1442–1448.
130. H. Muyderman, M. Nilsson and N. R. Sims, Highly selective and prolonged depletion of mitochondrial glutathione in astrocytes markedly increases sensitivity to peroxynitrite, *J. Neurosci.*, 2004, **24**, 8019–8028.
131. S. J. Chinta, J. M. Kumar, H. Zhang, H. J. Forman and J. K. Andersen, Up-regulation of gamma-glutamyl transpeptidase activity following glutathione depletion has a compensatory rather than an inhibitory effect on mitochondrial complex I activity: implications for Parkinson's disease, *Free Radic. Biol. Med.*, 2006, **40**, 1557–1563.
132. S. J. Chinta, M. J. Kumar, M. Hsu, S. Rajagopalan, D. Kaur, A. Rane, D. G. Nicholls, J. Choi and J. K. Andersen, Inducible alterations of glutathione levels in adult dopaminergic midbrain neurons result in nigrostriatal degeneration, *J. Neurosci.*, 2007, **27**, 13997–14006.
133. O. P. Jurma, D. G. Hom and J. K. Andersen, Decreased glutathione results in calcium-mediated cell death in PC12, *Free Radic. Biol. Med.*, 1997, **23**, 1055–1066.
134. N. Jha, O. Jurma, G. Lalli, Y. Liu, E. H. Pettus, J. T. Greenamyre, R. M. Liu, H. J. Forman and J. K. Andersen, Glutathione depletion in PC12 results in selective inhibition of mitochondrial complex I activity. Implications for Parkinson's disease, *J. Biol. Chem.*, 2000, **275**, 26096–26101.
135. S. Toffa, G. M. Kunikowska, B. Y. Zeng, P. Jenner and C. D. Marsden, Glutathione depletion in rat brain does not cause nigrostriatal pathway degeneration, *J. Neural Transm.*, 1997, **104**, 67–75.
136. H. Li, X. M. Shen and G. Dryhurst, Brain mitochondria catalyze the oxidation of 7-(2-aminoethyl)-3,4-dihydro-5-hydroxy-2H-1,4-benzothiazine-3-carboxylic acid (DHBT-1) to intermediates that irreversibly inhibit complex I and scavenge glutathione: potential relevance to the pathogenesis of Parkinson's disease, *J. Neurochem.*, 1998, **71**, 2049–2062.
137. P. Jenner, Oxidative stress as a cause of Parkinson's disease, *Acta Neurol. Scand., Suppl.*, 1991, **136**, 6–15.
138. J. T. Greenamyre, T. B. Sherer, R. Betarbet and A. V. Panov, Complex I and Parkinson's disease, *IUBMB Life*, 2001, **52**, 135–141.

139. J. T. Rotruck, A. L. Pope, H. E. Ganther, A. B. Swanson, D. G. Hafeman and W. G. Hoekstra, Selenium: biochemical role as a component of glutathione peroxidase, *Science*, 1973, **179**, 588–590.
140. L. Flohe, W. A. Gunzler and H. H. Schock, Glutathione peroxidase: a selenoenzyme, *FEBS Lett.*, 1973, **32**, 132–134.
141. O. Epp, R. Ladenstein and A. Wendel, The refined structure of the selenoenzyme glutathione peroxidase at 0.2-nm resolution, *Eur. J. Biochem.*, 1983, **133**, 51–69.
142. K. M. Brown, K. Pickard, F. Nicol, G. J. Beckett, G. G. Duthie and J. R. Arthur, Effects of organic and inorganic selenium supplementation on selenoenzyme activity in blood lymphocytes, granulocytes, platelets and erythrocytes, *Clin. Sci. (Lond.)*, 2000, **98**, 593–599.
143. R. Margis, C. Dunand, F. K. Teixeira and M. Margis-Pinheiro, Glutathione peroxidase family - an evolutionary overview, *FEBS J.*, 2008, **275**, 3959–3970.
144. G. Trepanier, D. Furling, J. Puymirat and M. E. Mirault, Immunocytochemical localization of seleno-glutathione peroxidase in the adult mouse brain, *Neuroscience*, 1996, **75**, 231–243.
145. J. H. Power and P. C. Blumbergs, Cellular glutathione peroxidase in human brain: cellular distribution, and its potential role in the degradation of Lewy bodies in Parkinson's disease and dementia with Lewy bodies, *Acta Neuropathol.*, 2009, **117**, 63–73.
146. H. Wang, E. Cheng, S. Brooke, P. Chang and R. Sapolsky, Overexpression of antioxidant enzymes protects cultured hippocampal and cortical neurons from necrotic insults, *J. Neurochem.*, 2003, **87**, 1527–1534.
147. R. Dringen and B. Hamprecht, Involvement of glutathione peroxidase and catalase in the disposal of exogenous hydrogen peroxide by cultured astroglial cells, *Brain Res.*, 1997, **759**, 67–75.
148. P. Klivenyi, O. A. Andreassen, R. J. Ferrante, A. Dedeoglu, G. Mueller, E. Lancelot, M. Bogdanov, J. K. Andersen, D. Jiang and M. F. Beal, Mice deficient in cellular glutathione peroxidase show increased vulnerability to malonate, 3-nitropropionic acid, and 1-methyl-4-phenyl-1,2,5,6-tetrahydropyridine, *J. Neurosci.*, 2000, **20**, 1–7.
149. F. P. Bellinger, M. T. Bellinger, L. A. Seale, A. S. Takemoto, A. V. Raman, T. Miki, A. B. Manning-Bog, M. J. Berry, L. R. White and G. W. Ross, Glutathione Peroxidase 4 is associated with Neuromelanin in Substantia Nigra and Dystrophic Axons in Putamen of Parkinson's brain, *Mol. Neurodegener.*, 2011, **6**, 8.
150. W. R. Gibb and A. J. Lees, Anatomy, pigmentation, ventral and dorsal subpopulations of the substantia nigra, and differential cell death in Parkinson's disease, *J. Neurol., Neurosurg. Psychiatry*, 1991, **54**, 388–396.
151. N. E. Savaskan, A. Borchert, A. U. Brauer and H. Kuhn, Rolc for glutathione peroxidase-4 in brain development and neuronal apoptosis: specific induction of enzyme expression in reactive astrocytes following brain injury, *Free Radic. Biol. Med.*, 2007, **43**, 191–201.

152. J. Blackinton, R. Kumaran, M. P. van der Brug, R. Ahmad, L. Olson, D. Galter, A. Lees, R. Bandopadhyay and M. R. Cookson, Post-transcriptional regulation of mRNA associated with DJ-1 in sporadic Parkinson disease, *Neurosci. Lett.*, 2009, **452**, 8–11.
153. D. N. Hauser, A. A. Dukes, A. D. Mortimer and T. G. Hastings, Dopamine quinone modifies and decreases the abundance of the mitochondrial selenoprotein glutathione peroxidase 4, *Free Radic. Biol. Med.*, 2013, **65**, 419–427.
154. F. P. Bellinger, A. V. Raman, R. H. Rueli, M. T. Bellinger, A. S. Dewing, L. A. Seale, M. A. Andres, J. H. Uyehara-Lock, L. R. White, G. W. Ross and M. J. Berry, Changes in Selenoprotein P in Substantia Nigra and Putamen in Parkinson's Disease, *J. Parkinsons Dis.*, 2012, **2**, 115–126.
155. T. D. Boyer, The glutathione S-transferases: an update, *Hepatology*, 1989, **9**, 486–496.
156. M. A. Philbert, C. M. Beiswanger, M. M. Manson, J. A. Green, R. F. Novak, T. Primiano, K. R. Reuhl and H. E. Lowndes, Glutathione S-transferases and gamma-glutamyl transpeptidase in the rat nervous systems: a basis for differential susceptibility to neurotoxicants, *NeuroToxicology*, 1995, **16**, 349–362.
157. E. L. Abel, T. K. Bammler and D. L. Eaton, Biotransformation of methyl parathion by glutathione S-transferases, *Toxicol. Sci.*, 2004, **79**, 224–232.
158. B. Drukarch and F. L. van Muiswinkel, Drug treatment of Parkinson's disease. Time for phase II, *Biochem. Pharmacol.*, 2000, **59**, 1023–1031.
159. K. D. Tew and D. M. Townsend, Glutathione-S-Transferases As Determinants of Cell Survival and Death, *Antioxid. Redox Signaling*, 2012, **17**, 1728–1737.
160. K. D. Tew and D. M. Townsend, Regulatory functions of glutathione S-transferase P1-1 unrelated to detoxification, *Drug Metab. Rev.*, 2011, **43**, 179–193.
161. J. D. Hayes, J. U. Flanagan and I. R. Jowsey, Glutathione transferases, *Annu. Rev. Pharmacol. Toxicol.*, 2005, **45**, 51–88.
162. M. C. Wilce and M. W. Parker, Structure and function of glutathione S-transferases, *Biochim. Biophys. Acta*, 1994, **1205**, 1–18.
163. H. Dirr, P. Reinemer and R. Huber, X-ray crystal structures of cytosolic glutathione S-transferases. Implications for protein architecture, substrate recognition and catalytic function, *Eur. J. Biochem.*, 1994, **220**, 645–661.
164. B. Ketterer and L. G. Christodoulides, Enzymology of cytosolic glutathione S-transferases, *Adv. Pharmacol.*, 1994, **27**, 37–69.
165. V. Adler, Z. Yin, S. Y. Fuchs, M. Benezra, L. Rosario, K. D. Tew, M. R. Pincus, M. Sardana, C. J. Henderson, C. R. Wolf, R. J. Davis and Z. Ronai, Regulation of JNK signaling by GSTp, *EMBO J.*, 1999, **18**, 1321–1334.
166. L. G. Higgins and J. D. Hayes, Mechanisms of induction of cytosolic and microsomal glutathione transferase (GST) genes by xenobiotics and pro-inflammatory agents, *Drug Metab. Rev.*, 2011, **43**, 92–137.
167. R. N. Armstrong, Structure, catalytic mechanism, and evolution of the glutathione transferases, *Chem. Res. Toxicol.*, 1997, **10**, 2–18.

168. P. C. Babbitt, Reengineering the glutathione S-transferase scaffold: a rational design strategy pays off, *Proc. Natl. Acad. Sci. U. S. A.*, 2000, **97**, 10298–10300.
169. Y. C. Awasthi, S. V. Singh, R. S. Shen, C. W. Abell, W. Gessner and A. Brossi, MPTP metabolites inhibit rat brain glutathione S-transferases, *Neurosci. Lett.*, 1987, **81**, 159–164.
170. A. E. Salinas and M. G. Wong, Glutathione S-transferases–a review, *Curr. Med. Chem.*, 1999, **6**, 279–309.
171. W. Shang, W. H. Liu, X. H. Zhao, Q. J. Sun, J. Z. Bi and Z. F. Chi, Expressions of glutathione S-transferase alpha, mu, and pi in brains of medically intractable epileptic patients, *BMC Neurosci.*, 2008, **9**, 67.
172. M. Smeyne, J. Boyd, K. Raviie Shepherd, Y. Jiao, B. B. Pond, M. Hatler, R. Wolf, C. Henderson and R. J. Smeyne, GSTpi expression mediates dopaminergic neuron sensitivity in experimental parkinsonism, *Proc. Natl. Acad. Sci. U. S. A.*, 2007, **104**, 1977–1982.
173. P. J. Carder, R. Hume, A. A. Fryer, R. C. Strange, J. Lauder and J. E. Bell, Glutathione S-transferase in human brain, *Neuropathol. Appl. Neurobiol.*, 1990, **16**, 293–303.
174. T. Tchaikovskaya, V. Fraifeld, T. Urphanishvili, J. H. Andorfer, P. Davies and I. Listowsky, Glutathione S-transferase hGSTM3 and ageing-associated neurodegeneration: relationship to Alzheimer's disease, *Mech. Ageing Dev.*, 2005, **126**, 309–315.
175. M. Castro-Caldas, A. Neves Carvalho, I. Peixeiro, E. Rodrigues, M. C. Lechner and M. J. Gama, GSTpi expression in MPTP-induced dopaminergic neurodegeneration of C57BL/6 mouse midbrain and striatum, *J. Mol. Neurosci.*, 2009, **38**, 114–127.
176. E. M. van Lieshout, M. F. Knapen, W. P. Lange, E. A. Steegers and W. H. Peters, Localization of glutathione S-transferases alpha and pi in human embryonic tissues at 8 weeks gestational age, *Hum. Reprod.*, 1998, **13**, 1380–1386.
177. O. Vasieva, The many faces of glutathione transferase pi, *Curr. Mol. Med.*, 2011, **11**, 129–139.
178. H. Ichijo, E. Nishida, K. Irie, P. ten Dijke, M. Saitoh, T. Moriguchi, M. Takagi, K. Matsumoto, K. Miyazono and Y. Gotoh, Induction of apoptosis by ASK1, a mammalian MAPKKK that activates SAPK/JNK and p38 signaling pathways, *Science*, 1997, **275**, 90–94.
179. J. Matsukawa, A. Matsuzawa, K. Takeda and H. Ichijo, The ASK1-MAP kinase cascades in mammalian stress response, *J. Biochem.*, 2004, **136**, 261–265.
180. M. Soga, A. Matsuzawa and H. Ichijo, Oxidative Stress-Induced Diseases *via* the ASK1 Signaling Pathway, *Int. J. Cell Biol.*, 2012, **2012**, 439587.
181. S. G. Cho, Y. H. Lee, H. S. Park, K. Ryoo, K. W. Kang, J. Park, S. J. Eom, M. J. Kim, T. S. Chang, S. Y. Choi, J. Shim, Y. Kim, M. S. Dong, M. J. Lee, S. G. Kim, H. Ichijo and E. J. Choi, Glutathione S-transferase mu modulates the stress-activated signals by suppressing apoptosis signal-regulating kinase 1, *J. Biol. Chem.*, 2001, **276**, 12749–12755.

182. V. Adler and M. R. Pincus, Effector peptides from glutathione-S-transferase-pi affect the activation of jun by jun-N-terminal kinase, *Ann. Clin. Lab. Sci.*, 2004, **34**, 35–46.
183. M. Castro-Caldas, A. N. Carvalho, E. Rodrigues, C. Henderson, C. R. Wolf and M. J. Gama, Glutathione S-transferase pi mediates MPTP-induced c-Jun N-terminal kinase activation in the nigrostriatal pathway, *Mol. Neurobiol.*, 2012, **45**, 466–477.
184. T. Asakura, Y. Hashizume, K. Tashiro, Y. Searashi, K. Ohkawa, J. Nishihira, M. Sakai and T. Shibasaki, Suppression of GST-P by treatment with glutathione-doxorubicin conjugate induces potent apoptosis in rat hepatoma cells, *Int. J. Cancer*, 2001, **94**, 171–177.
185. M. L. Circu and T. Y. Aw, Glutathione and apoptosis, *Free Radic. Res.*, 2008, **42**, 689–706.
186. B. Coles, I. Wilson, P. Wardman, J. A. Hinson, S. D. Nelson and B. Ketterer, The spontaneous and enzymatic reaction of N-acetyl-p-benzoquinonimine with glutathione: a stopped-flow kinetic study, *Arch. Biochem. Biophys.*, 1988, **264**, 253–260.
187. G. De Palma, P. Mozzoni, A. Mutti, S. Calzetti and A. Negrotti, Case-control study of interactions between genetic and environmental factors in Parkinson's disease, *Lancet*, 1998, **352**, 1986–1987.
188. L. I. Golbe, G. Di Iorio, K. Markopoulou, A. Athanassiadou, S. Papapetropoulos, R. L. Watts, J. M. Vance, V. Bonifati, T. A. Williams, J. R. Spychala, E. S. Stenroos and W. G. Johnson, Glutathione S-transferase polymorphisms and onset age in alpha-synuclein A53T mutant Parkinson's disease, *Am. J. Med. Genet., Part B*, 2007, **144**, 254–258.
189. A. L. McCormack, M. Thiruchelvam, A. B. Manning-Bog, C. Thiffault, J. W. Langston, D. A. Cory-Slechta and D. A. Di Monte, Environmental risk factors and Parkinson's disease: Selective degeneration of nigral dopaminergic neurons caused by the herbicide paraquat, *Neurobiol. Dis.*, 2002, **10**, 119–127.
190. R. Kannan, A. Mittur, Y. Bao, T. Tsuruo and N. Kaplowitz, GSH transport in immortalized mouse brain endothelial cells: evidence for apical localization of a sodium-dependent GSH transporter, *J. Neurochem.*, 1999, **73**, 390–399.
191. H. Hong, Y. Lu, Z. N. Ji and G. Q. Liu, Up-regulation of P-glycoprotein expression by glutathione depletion-induced oxidative stress in rat brain microvessel endothelial cells, *J. Neurochem.*, 2006, **98**, 1465–1473.
192. K. D. Tew, J. T. Boyd, Z. J. Chen, W. Davis Jr, F. Fazilev, V. Findlay, L. Gate, K. Ile, A. Soulika and D. M. Townsend, Glutathione and ABC transporters as determinants of sensitivity to oxidative and nitrosative stress, *J. Nutr.*, 2004, **134**, 3205S–3206S.
193. J. Hirrlinger, J. Konig, D. Keppler, J. Lindenau, J. B. Schulz and R. Dringen, The multidrug resistance protein MRP1 mediates the release of glutathione disulfide from rat astrocytes during oxidative stress, *J. Neurochem.*, 2001, **76**, 627–636.

194. F. J. Sharom, The P-glycoprotein multidrug transporter, *Essays Biochem.*, 2011, **50**, 161–178.
195. K. W. Scotto, Transcriptional regulation of ABC drug transporters, *Oncogene*, 2003, **22**, 7496–7511.
196. W. Loscher and H. Potschka, Blood-brain barrier active efflux transporters: ATP-binding cassette gene family, *NeuroRx*, 2005, **2**, 86–98.
197. D. M. Hermann, E. Kilic, A. Spudich, S. D. Kramer, H. Wunderli-Allenspach and C. L. Bassetti, Role of drug efflux carriers in the healthy and diseased brain, *Ann. Neurol.*, 2006, **60**, 489–498.
198. A. Soontornmalai, M. L. Vlaming and J. M. Fritschy, Differential, strain-specific cellular and subcellular distribution of multidrug transporters in murine choroid plexus and blood-brain barrier, *Neuroscience*, 2006, **138**, 159–169.
199. G. Lee, S. Dallas, M. Hong and R. Bendayan, Drug transporters in the central nervous system: brain barriers and brain parenchyma considerations, *Pharmacol. Rev.*, 2001, **53**, 569–596.
200. J. M. Scherrmann, Expression and function of multidrug resistance transporters at the blood-brain barriers, *Expert Opin. Drug Metab. Toxicol.*, 2005, **1**, 233–246.
201. M. Smeyne and R. J. Smeyne, Glutathione metabolism and Parkinson's disease, *Free Radic. Biol. Med.*, 2013, **62**, 13–25.
202. N. Strazielle and J. F. Ghersi-Egea, Demonstration of a coupled metabolism-efflux process at the choroid plexus as a mechanism of brain protection toward xenobiotics, *J. Neurosci.*, 1999, **19**, 6275–6289.
203. J. F. Ghersi-Egea, N. Strazielle, A. Murat, A. Jouvet, A. Buenerd and M. F. Belin, Brain protection at the blood-cerebrospinal fluid interface involves a glutathione-dependent metabolic barrier mechanism, *J. Cereb. Blood Flow Metab.*, 2006, **26**, 1165–1175.
204. Z. E. Sauna and S. V. Ambudkar, About a switch: how P-glycoprotein (ABCB1) harnesses the energy of ATP binding and hydrolysis to do mechanical work, *Mol. Cancer Ther.*, 2007, **6**, 13–23.
205. C. Bellarosa, G. Bortolussi and C. Tiribelli, The role of ABC transporters in protecting cells from bilirubin toxicity, *Curr. Pharm. Des.*, 2009, **15**, 2884–2892.
206. G. J. Zaman, J. Lankelma, O. van Tellingen, J. Beijnen, H. Dekker, C. Paulusma, R. P. Oude Elferink, F. Baas and P. Borst, Role of glutathione in the export of compounds from cells by the multidrug-resistance-associated protein, *Proc. Natl. Acad. Sci. U. S. A.*, 1995, **92**, 7690–7694.
207. B. Bauer, A. M. Hartz, J. R. Lucking, X. Yang, G. M. Pollack and D. S. Miller, Coordinated nuclear receptor regulation of the efflux transporter, Mrp2, and the phase-II metabolizing enzyme, GSTpi, at the blood-brain barrier, *J. Cereb. Blood Flow Metab.*, 2008, **28**, 1222–1234.
208. G. Lee and R. Bendayan, Functional expression and localization of P-glycoprotein in the central nervous system: relevance to the pathogenesis and treatment of neurological disorders, *Pharm. Res.*, 2004, **21**, 1313–1330.

209. P. L. Golden and G. M. Pollack, Blood-brain barrier efflux transport, *J. Pharm. Sci.*, 2003, **92**, 1739–1753.
210. M. Muller, C. Meijer, G. J. Zaman, P. Borst, R. J. Scheper, N. H. Mulder, E. G. de Vries and P. L. Jansen, Overexpression of the gene encoding the multidrug resistance-associated protein results in increased ATP-dependent glutathione S-conjugate transport, *Proc. Natl. Acad. Sci. U. S. A.*, 1994, **91**, 13033–13037.
211. D. W. Loe, R. G. Deeley and S. P. Cole, Characterization of vincristine transport by the M(r) 190,000 multidrug resistance protein (MRP): evidence for cotransport with reduced glutathione, *Cancer Res.*, 1998, **58**, 5130–5136.
212. J. Renes, E. G. de Vries, E. F. Nienhuis, P. L. Jansen and M. Muller, ATP- and glutathione-dependent transport of chemotherapeutic drugs by the multidrug resistance protein MRP1, *Br. J. Pharmacol.*, 1999, **126**, 681–688.
213. R. A. Felix and M. A. Barrand, P-glycoprotein expression in rat brain endothelial cells: evidence for regulation by transient oxidative stress, *J. Neurochem.*, 2002, **80**, 64–72.
214. D. S. Miller, B. Bauer and A. M. Hartz, Modulation of P-glycoprotein at the blood-brain barrier: opportunities to improve central nervous system pharmacotherapy, *Pharmacol. Rev.*, 2008, **60**, 196–209.
215. V. V. Rao, J. L. Dahlheimer, M. E. Bardgett, A. Z. Snyder, R. A. Finch, A. C. Sartorelli and D. Piwnica-Worms, Choroid plexus epithelial expression of MDR1 P glycoprotein and multidrug resistance-associated protein contribute to the blood-cerebrospinal-fluid drug-permeability barrier, *Proc. Natl. Acad. Sci. U. S. A.*, 1999, **96**, 3900–3905.
216. C. Cordon-Cardo, J. P. O'Brien, D. Casals, L. Rittman-Grauer, J. L. Biedler, M. R. Melamed and J. R. Bertino, Multidrug-resistance gene (P-glycoprotein) is expressed by endothelial cells at blood-brain barrier sites, *Proc. Natl. Acad. Sci. U. S. A.*, 1989, **86**, 695–698.
217. X. Decleves, A. Regina, J. L. Laplanche, F. Roux, B. Boval, J. M. Launay and J. M. Scherrmann, Functional expression of P-glycoprotein and multidrug resistance-associated protein (Mrp1) in primary cultures of rat astrocytes, *J. Neurosci. Res.*, 2000, **60**, 594–601.
218. P. L. Golden and W. M. Pardridge, P-Glycoprotein on astrocyte foot processes of unfixed isolated human brain capillaries, *Brain Res.*, 1999, **819**, 143–146.
219. C. Ziemann, A. Burkle, G. F. Kahl and K. I. Hirsch-Ernst, Reactive oxygen species participate in mdr1b mRNA and P-glycoprotein overexpression in primary rat hepatocyte cultures, *Carcinogenesis*, 1999, **20**, 407–414.
220. K. I. Hirsch-Ernst, T. Kietzmann, C. Ziemann, K. Jungermann and G. F. Kahl, Physiological oxygen tensions modulate expression of the mdr1b multidrug-resistance gene in primary rat hepatocyte cultures, *Biochem. J.*, 2000, **350**(pt 2), 443–451.

221. F. Thevenod, J. M. Friedmann, A. D. Katsen and I. A. Hauser, Up-regulation of multidrug resistance P-glycoprotein *via* nuclear factor-kappaB activation protects kidney proximal tubule cells from cadmium- and reactive oxygen species-induced apoptosis, *J. Biol. Chem.*, 2000, **275**, 1887–1896.
222. J. Wu, H. Ji, Y. Y. Wang, Y. Wang, Y. Q. Li, W. G. Li, Y. Long, Y. Z. Xia and H. Hong, Glutathione depletion upregulates P-glycoprotein expression at the blood-brain barrier in rats, *J. Pharm. Pharmacol.*, 2009, **61**, 819–824.
223. M. Westerlund, A. C. Belin, L. Olson and D. Galter, Expression of multidrug resistance 1 mRNA in human and rodent tissues: reduced levels in Parkinson patients, *Cell Tissue Res.*, 2008, **334**, 179–185.
224. R. Kortekaas, K. L. Leenders, J. C. van Oostrom, W. Vaalburg, J. Bart, A. T. Willemsen and N. H. Hendrikse, Blood-brain barrier dysfunction in parkinsonian midbrain *in vivo*, *Ann. Neurol.*, 2005, **57**, 176–179.
225. A. L. Bartels, A. T. Willemsen, R. Kortekaas, B. M. de Jong, R. de Vries, O. de Klerk, J. C. van Oostrom, A. Portman and K. L. Leenders, Decreased blood-brain barrier P-glycoprotein function in the progression of Parkinson's disease, PSP and MSA, *J. Neural Transm.*, 2008, **115**, 1001–1009.
226. M. Westerlund, A. C. Belin, A. Anvret, A. Hakansson, H. Nissbrandt, C. Lind, O. Sydow, L. Olson and D. Galter, Association of a polymorphism in the ABCB1 gene with Parkinson's disease, *Parkinsonism Relat. Disord.*, 2009, **15**, 422–424.
227. C. Funke, A. S. Soehn, J. Tomiuk, O. Riess and D. Berg, Genetic analysis of coding SNPs in blood-brain barrier transporter MDR1 in European Parkinson's disease patients, *J. Neural Transm.*, 2009, **116**, 443–450.
228. T. Furuno, M. T. Landi, M. Ceroni, N. Caporaso, I. Bernucci, G. Nappi, E. Martignoni, E. Schaeffeler, M. Eichelbaum, M. Schwab and U. M. Zanger, Expression polymorphism of the blood-brain barrier component P-glycoprotein (MDR1) in relation to Parkinson's disease, *Pharmacogenetics*, 2002, **12**, 529–534.
229. K. Zschiedrich, I. R. Konig, N. Bruggemann, N. Kock, M. Kasten, K. L. Leenders, V. Kostic, P. Vieregge, A. Ziegler, C. Klein and K. Lohmann, MDR1 variants and risk of Parkinson disease. Association with pesticide exposure? *J. Neurol.*, 2009, **256**, 115–120.
230. M. Drozdzik, M. Bialecka, K. Mysliwiec, K. Honczarenko, J. Stankiewicz and Z. Sych, Polymorphism in the P-glycoprotein drug transporter MDR1 gene: a possible link between environmental and genetic factors in Parkinson's disease, *Pharmacogenetics*, 2003, **13**, 259–263.
231. A. J. Whitworth, D. A. Theodore, J. C. Greene, H. Benes, P. D. Wes and L. J. Pallanck, Increased glutathione S-transferase activity rescues dopaminergic neuron loss in a Drosophila model of Parkinson's disease, *Proc. Natl. Acad. Sci. U. S. A.*, 2005, **102**, 8024–8029.
232. Y. Pesah, T. Pham, H. Burgess, B. Middlebrooks, P. Verstreken, Y. Zhou, M. Harding, H. Bellen and G. Mardon, Drosophila parkin mutants have decreased mass and cell size and increased sensitivity to oxygen radical stress, *Development*, 2004, **131**, 2183–2194.

233. K. Kim, S. H. Kim, J. Kim, H. Kim and J. Yim, Glutathione s-transferase omega 1 activity is sufficient to suppress neurodegeneration in a Drosophila model of Parkinson disease, *J. Biol. Chem.*, 2012, **287**, 6628–6641.
234. J. C. Greene, A. J. Whitworth, L. A. Andrews, T. J. Parker and L. J. Pallanck, Genetic and genomic studies of Drosophila parkin mutants implicate oxidative stress and innate immune responses in pathogenesis, *Hum. Mol. Genet.*, 2005, **14**, 799–811.
235. A. N. Carvalho, C. Marques, E. Rodrigues, C. J. Henderson, C. R. Wolf, P. Pereira and M. J. Gama, Ubiquitin-Proteasome System Impairment and MPTP-Induced Oxidative Stress in the Brain of C57BL/6 Wild-type and GSTP Knockout Mice, *Mol. Neurobiol.*, 2012, **47**, 662–672.
236. M. Shi, J. Bradner, T. K. Bammler, D. L. Eaton, J. Zhang, Z. Ye, A. M. Wilson, T. J. Montine, C. Pan and J. Zhang, Identification of glutathione S-transferase pi as a protein involved in Parkinson disease progression, *Am. J. Pathol.*, 2009, **175**, 54–65.
237. V. Bonifati, P. Rizzu, M. J. van Baren, O. Schaap, G. J. Breedveld, E. Krieger, M. C. Dekker, F. Squitieri, P. Ibanez, M. Joosse, J. W. van Dongen, N. Vanacore, J. C. van Swieten, A. Brice, G. Meco, C. M. van Duijn, B. A. Oostra and P. Heutink, Mutations in the DJ-1 gene associated with autosomal recessive early-onset parkinsonism, *Science*, 2003, **299**, 256–259.
238. X. Tao and L. Tong, Crystal structure of human DJ-1, a protein associated with early onset Parkinson's disease, *J. Biol. Chem.*, 2003, **278**, 31372–31379.
239. M. A. Wilson, J. L. Collins, Y. Hod, D. Ringe and G. A. Petsko, The 1.1-A resolution crystal structure of DJ-1, the protein mutated in autosomal recessive early onset Parkinson's disease, *Proc. Natl. Acad. Sci. U. S. A.*, 2003, **100**, 9256–9261.
240. V. Bonifati, P. Rizzu, F. Squitieri, E. Krieger, N. Vanacore, J. C. van Swieten, A. Brice, C. M. van Duijn, B. Oostra, G. Meco and P. Heutink, DJ-1(PARK7), a novel gene for autosomal recessive, early onset parkinsonism, *Neurol. Sci.*, 2003, **24**, 159–160.
241. J. I. Lucas and I. Marin, A new evolutionary paradigm for the Parkinson disease gene DJ-1, *Mol. Biol. Evol.*, 2007, **24**, 551–561.
242. T. Yokota, K. Sugawara, K. Ito, R. Takahashi, H. Ariga and H. Mizusawa, Down regulation of DJ-1 enhances cell death by oxidative stress, ER stress, and proteasome inhibition, *Biochem. Biophys. Res. Commun.*, 2003, **312**, 1342–1348.
243. C. Martinat, S. Shendelman, A. Jonason, T. Leete, M. F. Beal, L. Yang, T. Floss and A. Abeliovich, Sensitivity to oxidative stress in DJ-1-deficient dopamine neurons: an ES- derived cell model of primary Parkinsonism, *PLoS Biol.*, 2004, **2**, e327.
244. H. Aleyasin, M. W. Rousseaux, P. C. Marcogliese, S. J. Hewitt, I. Irrcher, A. P. Joselin, M. Parsanejad, R. H. Kim, P. Rizzu, S. M. Callaghan, R. S. Slack, T. W. Mak and D. S. Park, DJ-1 protects the nigrostriatal axis from the neurotoxin MPTP by modulation of the AKT pathway, *Proc. Natl. Acad. Sci. U. S. A.*, 2010, **107**, 3186–3191.

245. H. Aleyasin, M. W. Rousseaux, M. Phillips, R. H. Kim, R. J. Bland, S. Callaghan, R. S. Slack, M. J. During, T. W. Mak and D. S. Park, The Parkinson's disease gene DJ-1 is also a key regulator of stroke-induced damage, *Proc. Natl. Acad. Sci. U. S. A.*, 2007, **104**, 18748–18753.
246. R. M. Canet-Aviles, M. A. Wilson, D. W. Miller, R. Ahmad, C. McLendon, S. Bandyopadhyay, M. J. Baptista, D. Ringe, G. A. Petsko and M. R. Cookson, The Parkinson's disease protein DJ-1 is neuroprotective due to cysteine-sulfinic acid-driven mitochondrial localization, *Proc. Natl. Acad. Sci. U. S. A.*, 2004, **101**, 9103–9108.
247. E. Giaime, H. Yamaguchi, C. A. Gautier, T. Kitada and J. Shen, Loss of DJ-1 does not affect mitochondrial respiration but increases ROS production and mitochondrial permeability transition pore opening, *PLoS One*, 2012, **7**, e40501.
248. A. P. Joselin, S. J. Hewitt, S. M. Callaghan, R. H. Kim, Y. H. Chung, T. W. Mak, J. Shen, R. S. Slack and D. S. Park, ROS-dependent regulation of Parkin and DJ-1 localization during oxidative stress in neurons, *Hum. Mol. Genet.*, 2012, **21**, 4888–4903.
249. S. Shendelman, A. Jonason, C. Martinat, T. Leete and A. Abeliovich, DJ-1 is a redox-dependent molecular chaperone that inhibits alpha-synuclein aggregate formation, *PLoS Biol.*, 2004, **2**, e362.
250. W. Zhou, M. Zhu, M. A. Wilson, G. A. Petsko and A. L. Fink, The oxidation state of DJ-1 regulates its chaperone activity toward alpha-synuclein, *J. Mol. Biol.*, 2006, **356**, 1036–1048.
251. L. Zhang, M. Shimoji, B. Thomas, D. J. Moore, S. W. Yu, N. I. Marupudi, R. Torp, I. A. Torgner, O. P. Ottersen, T. M. Dawson and V. L. Dawson, Mitochondrial localization of the Parkinson's disease related protein DJ-1: implications for pathogenesis, *Hum. Mol. Genet.*, 2005, **14**, 2063–2073.
252. G. Ito, H. Ariga, Y. Nakagawa and T. Iwatsubo, Roles of distinct cysteine residues in *S*-nitrosylation and dimerization of DJ-1, *Biochem. Biophys. Res. Commun.*, 2006, **339**, 667–672.
253. S. Girotto, M. Sturlese, M. Bellanda, I. Tessari, R. Cappellini, M. Bisaglia, L. Bubacco and S. Mammi, Dopamine-derived quinones affect the structure of the redox sensor DJ-1 through modifications at Cys-106 and Cys-53, *J. Biol. Chem.*, 2012, **287**, 18738–18749.
254. S. Ishikawa, T. Taira, T. Niki, K. Takahashi-Niki, C. Maita, H. Maita, H. Ariga and S. M. Iguchi-Ariga, Oxidative status of DJ-1-dependent activation of dopamine synthesis through interaction of tyrosine hydroxylase and 4-dihydroxy-L-phenylalanine (L-DOPA) decarboxylase with DJ-1, *J. Biol. Chem.*, 2009, **284**, 28832–28844.
255. J. Chen, L. Li and L. S. Chin, Parkinson disease protein DJ-1 converts from a zymogen to a protease by carboxyl-terminal cleavage, *Hum. Mol. Genet.*, 2010, **19**, 2395–2408.
256. P. Lopert and M. Patel, Brain mitochondria from DJ-1 knockout mice show increased respiration-dependent hydrogen peroxide consumption, *Redox Biol.*, 2014, **2**, 667–672.

257. J. Y. Im, K. W. Lee, J. M. Woo, E. Junn and M. M. Mouradian, DJ-1 induces thioredoxin 1 expression through the Nrf2 pathway, *Hum. Mol. Genet.*, 2012, **21**, 3013–3024.
258. M. P. van der Brug, J. Blackinton, J. Chandran, L. Y. Hao, A. Lal, K. Mazan-Mamczarz, J. Martindale, C. Xie, R. Ahmad, K. J. Thomas, A. Beilina, J. R. Gibbs, J. Ding, A. J. Myers, M. Zhan, H. Cai, N. M. Bonini, M. Gorospe and M. R. Cookson, RNA binding activity of the recessive parkinsonism protein DJ-1 supports involvement in multiple cellular pathways, *Proc. Natl. Acad. Sci. U. S. A.*, 2008, **105**, 10244–10249.
259. E. M. Valente, P. M. Abou-Sleiman, V. Caputo, M. M. Muqit, K. Harvey, S. Gispert, Z. Ali, D. Del Turco, A. R. Bentivoglio, D. G. Healy, A. Albanese, R. Nussbaum, R. Gonzalez-Maldonado, T. Deller, S. Salvi, P. Cortelli, W. P. Gilks, D. S. Latchman, R. J. Harvey, B. Dallapiccola, G. Auburger and N. W. Wood, Hereditary early-onset Parkinson's disease caused by mutations in PINK1, *Science*, 2004, **304**, 1158–1160.
260. A. Wood-Kaczmar, S. Gandhi, Z. Yao, A. Y. Abramov, E. A. Miljan, G. Keen, L. Stanyer, I. Hargreaves, K. Klupsch, E. Deas, J. Downward, L. Mansfield, P. Jat, J. Taylor, S. Heales, M. R. Duchen, D. Latchman, S. J. Tabrizi and N. W. Wood, PINK1 is necessary for long term survival and mitochondrial function in human dopaminergic neurons, *PLoS One*, 2008, **3**, e2455.
261. H. H. Hoepken, S. Gispert, B. Morales, O. Wingerter, D. Del Turco, A. Mulsch, R. L. Nussbaum, K. Muller, S. Drose, U. Brandt, T. Deller, B. Wirth, A. P. Kudin, W. S. Kunz and G. Auburger, Mitochondrial dysfunction, peroxidation damage and changes in glutathione metabolism in PARK6, *Neurobiol. Dis.*, 2007, **25**, 401–411.
262. B. Thomas and M. F. Beal, Parkinson's disease, *Hum. Mol. Genet.*, 2007, **16**(2), R183–R194.
263. T. P. Williamson, D. A. Johnson and J. A. Johnson, Activation of the Nrf2-ARE pathway by siRNA knockdown of Keap1 reduces oxidative stress and provides partial protection from MPTP-mediated neurotoxicity, *Neurotoxicology*, 2012, **33**, 272–279.
264. S. Hunot, B. Brugg, D. Ricard, P. P. Michel, M. P. Muriel, M. Ruberg, B. A. Faucheux, Y. Agid and E. C. Hirsch, Nuclear translocation of NF-kappaB is increased in dopaminergic neurons of patients with parkinson disease, *Proc. Natl. Acad. Sci. U. S. A.*, 1997, **94**, 7531–7536.
265. T. Dehmer, M. T. Heneka, M. Sastre, J. Dichgans and J. B. Schulz, Protection by pioglitazone in the MPTP model of Parkinson's disease correlates with I kappa B alpha induction and block of NF kappa B and iNOS activation, *J. Neurochem.*, 2004, **88**, 494–501.
266. T. A. Tran, A. D. Nguyen, J. Chang, M. S. Goldberg, J. K. Lee and M. G. Tansey, Lipopolysaccharide and tumor necrosis factor regulate Parkin expression *via* nuclear factor-kappa B, *PLoS One*, 2011, **6**, e23660.
267. G. Joshi and J. A. Johnson, The Nrf2-ARE pathway: a valuable therapeutic target for the treatment of neurodegenerative diseases, *Recent Pat. CNS Drug Discovery*, 2012, **7**, 218–229.

268. N. Yamamoto, H. Sawada, Y. Izumi, T. Kume, H. Katsuki, S. Shimohama and A. Akaike, Proteasome inhibition induces glutathione synthesis and protects cells from oxidative stress: relevance to Parkinson disease, *J. Biol. Chem.*, 2007, **282**, 4364–4372.
269. A. T. Dinkova-Kostova, W. D. Holtzclaw, R. N. Cole, K. Itoh, N. Wakabayashi, Y. Katoh, M. Yamamoto and P. Talalay, Direct evidence that sulfhydryl groups of Keap1 are the sensors regulating induction of phase 2 enzymes that protect against carcinogens and oxidants, *Proc. Natl. Acad. Sci. U. S. A.*, 2002, **99**, 11908–11913.
270. A. Kobayashi, M. I. Kang, H. Okawa, M. Ohtsuji, Y. Zenke, T. Chiba, K. Igarashi and M. Yamamoto, Oxidative stress sensor Keap1 functions as an adaptor for Cul3-based E3 ligase to regulate proteasomal degradation of Nrf2, *Mol. Cell. Biol.*, 2004, **24**, 7130–7139.
271. M. McMahon, K. Itoh, M. Yamamoto and J. D. Hayes, Keap1-dependent proteasomal degradation of transcription factor Nrf2 contributes to the negative regulation of antioxidant response element-driven gene expression, *J. Biol. Chem.*, 2003, **278**, 21592–21600.
272. A. Y. Shih, D. A. Johnson, G. Wong, A. D. Kraft, L. Jiang, H. Erb, J. A. Johnson and T. H. Murphy, Coordinate regulation of glutathione biosynthesis and release by Nrf2-expressing glia potently protects neurons from oxidative stress, *J. Neurosci.*, 2003, **23**, 3394–3406.
273. T. H. Murphy, J. Yu, R. Ng, D. A. Johnson, H. Shen, C. R. Honey and J. A. Johnson, Preferential expression of antioxidant response element mediated gene expression in astrocytes, *J. Neurochem.*, 2001, **76**, 1670–1678.
274. P. C. Chen, M. R. Vargas, A. K. Pani, R. J. Smeyne, D. A. Johnson, Y. W. Kan and J. A. Johnson, Nrf2-mediated neuroprotection in the MPTP mouse model of Parkinson's disease: Critical role for the astrocyte, *Proc. Natl. Acad. Sci. U. S. A.*, 2009, **106**, 2933–2938.
275. M. Narasimhan, D. Patel, D. Vedpathak, M. Rathinam, G. Henderson and L. Mahimainathan, Identification of novel microRNAs in post-transcriptional control of Nrf2 expression and redox homeostasis in neuronal, SH-SY5Y cells, *PLoS One*, 2012, **7**, e51111.
276. L. Gan, D. A. Johnson and J. A. Johnson, Keap1-Nrf2 activation in the presence and absence of DJ-1, *Eur. J. Neurosci.*, 2010, **31**, 967–977.

CHAPTER 7

Neuroinflammation and Oxidative Stress in Models of Parkinson's Disease and Protein-Misfolding Disorders

RONALD B. TJALKENS*[a], KARIN M. STREIFEL[a] AND JULIE A. MORENO[b]

[a]Department of Environmental and Radiological Health Sciences, 1680 Campus Delivery, Fort Collins, CO 80523-1680, USA; [b]Department of Microbiology, Immunology and Pathology, Colorado State University, 1680 Campus Delivery, Fort Collins, CO 80523-1680, USA
*E-mail: ron.tjalkens@colostate.edu

7.1 Introduction

Several mechanisms are implicated in the etiology of sporadic Parkinson's disease (PD) including glutamatergic toxicity, misfolding of alpha-synuclein and tau, decreased production of neurotrophic factors, exposure to environmental and infectious insults, mitochondrial dysfunction, oxidative stress and neuroinflammation.[1,2] In particular, recent research has focused on the interactions between protein misfolding, oxidative stress and neuroinflammation due to evidence in experimental models linking these pathways in the progression of neuronal loss.[3–5] Activated microglia and astrocytes are fundamental to the pathophysiology of these mechanisms as they are the primary

Issues in Toxicology No. 34
Oxidative Stress and Redox Signalling in Parkinson's Disease
Edited by Rodrigo Franco, Jonathan A. Doorn and Jean-Christophe Rochet

Published by the Royal Society of Chemistry, www.rsc.org

sources of inflammatory and oxidative factors and can both be activated by and induce production of misfolded proteins.[6] This chapter will review current literature regarding the role of oxidative stress and inflammatory mediators in promoting loss of dopaminergic neurons in animal models of PD and protein-misfolding disorders. The relevance to clinical PD and potential sources of inflammatory signalling molecules will also be discussed.

7.2 Molecular Pathways Regulating Neuroinflammation in Glial Cells

Neuroinflammation is a pathological hallmark of PD and parallels the progression and severity of dopamine neuron loss. Neuroinflammation is also closely related to the spread and progression of misfolded α-synuclein protein in PD, which begins in the brainstem and hindbrain and ascends through the midbrain and cortex in a highly conserved manner.[6,7] Astrogliosis and microgliosis are consistently observed during examination of *postmortem* tissues from PD patients as well as elevations in cytokines in CSF.[8,9] Furthermore, infusion of the bacterial product lipopolysaccharide (LPS), a potent inducer of inflammation, into rodent brain is known to lead to specific loss of dopamine neurons in the *substantia nigra pars compacta* (SN) after initiation of glial activation in a manner similar to the progressive loss of neurons in PD.[8] Experimental models of PD, such as the toxin 1-methyl-4-phenyl-1,2,3,6-tetrahydropyridine (MPTP), show extensive glial activation proximal to loss of neurons.[3,9–11] Several glial-derived inflammatory factors, including nitric oxide (NO) and tumor necrosis factor (TNFα), are closely linked to the loss neurons in patients and in animals models of PD, as suppression of NOS2 or deletion of TNFα receptors protect mice from neurodegeneration induced by injection with the dopaminergic neurotoxicant, 1-methyl-4-phenyl-1,2,3,6-tetrahydropyridine (MPTP).[12–15] The cycle of glial activation and neuronal injury may represent a deleterious feedforward process, with the products released from stressed or damaged neurons, such as neuromelanin and alpha-synuclein, can further induce gliosis, creating a cycle of interdependent neuronal loss and sustained glial activation.[16] Neuroinflammation is also closely tied to protein aggregation in PD and related disorders, whereby inflammatory phenotype within the SN is permissive to impaired proteosomal function and increased aggregation of α-synuclein.

7.2.1 Regulation of Inflammatory Genes in Glial Cells by NF-κB

Given the potentially damaging effects of inflammatory mediators on neuronal function and survival, the expression of inflammatory genes is tightly regulated in glia. In this regard, the nuclear factor kappa-light-chain-enhancer of activated B cells (NF-κB) is one of the key transcriptional activators of inflammatory genes in microglia and astrocytes. NF-κB is activated following stimulation of cell-surface receptors that integrate extracellular stress and inflammatory cues to activation of IKK, the IκB kinase complex.[17] Major pathways that induce NF-κB include the tumor necrosis factor (TNF) receptor (TNFR) and Toll-like receptor (TLR)–interleukin-1 (IL-1) receptor (IL-1R) superfamilies, which cause activation of IKK.[18] Following activation of IKK, the classical pathway for NF-κB activation involves phosphorylation of the NF-κB inhibitory subunit, IκBα, by IKK2 (IKKβ) at Ser32 and Ser36, allowing its polyubiquitination at Lys21 and Lys22 and subsequent degradation by the 26S proteasome. The active NF-κB dimer (typically p65/p50) can then translocate to the nucleus, bind DNA and activate gene transcription at *cis*-acting promoter elements.[18]

NF-κB is constitutively suppressed in glia cells under resting conditions by both cytoplasmic proteins, such as IκBα, and by nuclear co-repressors, such as NCoR2, CoREST and HDAC2/3.[19,20] Overproduction of NF-κB-regulated inflammatory mediators such as nitric oxide (NO), cytokines (TNFα, IL-1β) and chemokines (CCL2, CCL5) is acutely damaging to neurons.[21,22] Thus, these signalling pathways are tightly regulated in glial cells by negative feedback mechanisms involving anti-inflammatory proteins such as IL-10 and by nuclear co-repressor proteins that antagonize the transcriptional activity of NF-κB/p65 on inflammatory gene promoters.[23] Activation of inflammatory signalling pathways in glia during neurotoxic injury or neurodegenerative disease can produce pathologic levels of neurotoxic inflammatory mediators contributing to loss of neurons.[24] This is underscored by the capacity of the inflammatory cytokines IL1β and TNFα to stimulate NF-κB-dependent expression of CCL2 and CCL7 in primary astrocytes,[22] which can enhance activation of microglia and thereby exacerbate a damaging neuroinflammatory phenotype. The net effect of NF-κB-regulated inflammatory gene expression in activated glial cells is the overproduction of damaging reactive oxygen and nitrogen species, as well as cytokines and chemokines that enhance cellular stress in neurons, including the unfolded protein response (UPR). Such effects are broadly observed in PD and PD-like disorders. In an immunohistochemical study of postmortem brain tissue from patients diagnosed with Dementia with Lewy Bodies (DLB) and the Lewy body variant of Alzheimer's disease (LBVAD), extensive nitration of α-synuclein-positive Lewy bodies was observed throughout the cortex and midbrain, including the SN.[25] These findings are consistent with extensive NOS2 expression in microglia and astrocytes and subsequent overproductive of NO and reactive oxygen species (ROS), leading to formation of peroxynitrite ($ONOO^-$). The authors concluded that loss of neuronal or glial oxidative protective mechanisms leading to overproduction of $NO^•/ONOO^-$ has deleterious effects on the normal functions or metabolism of α-synuclein that contribute to the onset and/or progression of neurodegenerative synucleinopathies.[25]

7.2.2 Nuclear Regulation of NF-κB Function in Glial Cells

Because activation of NF-κB in glial cells can have such deleterious effects on neurons, there exist both cytoplasmic and nuclear regulators that heavily modulate its activity. A critical regulator of NF-κB in glial cells is the nuclear receptor NR4A2 (Nurr1, HZF-3, RNR1, NOT, DHR38), which also has important functions in regulating energy metabolism, atherosclerosis, T-cell receptor (TCR)-mediated apoptosis, immune responses and regulation of the hypothalamic–pituitary axis (HPA).[26] Nurr1 also plays a significant role in the development and homeostasis of the central nervous system, particularly midbrain dopaminergic neurons, and has been associated with functional working memory as well as neurological disorders such as PD.[1,27] Like NR4A1 and A3, NR4A2/Nurr1 modulates target gene transcription by binding as a monomer, homodimer or heterodimer with RXR to *cis*-acting response elements such as the NGFI-B-responsive element (NBRE) located in gene promoter regions, which exhibits a canonical AAAGGTCA consensus sequence.[28,29] Additionally, NR4A2-RXR heterodimers bind to a related sequence motif, DR5 (5′-AGGTCANNNAAAGGTCA-3′) in the presence of 9-*cis* retinoic acid, whereas Nurr1 homodimers bind to another related sequence with the palindromic structure, 5′-TGACCTTTNNNNNAAAGGTCA-3′.[30]

The transcriptional activity of Nurr1 is not limited to transactivation but also includes transcriptional repression or 'transrepression' by a mechanism involving recruitment of nuclear co-repressor proteins that stabilize histone-DNA binding and suppress gene expression.[10,31] Genome-wide transcriptomic and ChIP-on-ChIP studies have described a large number of target genes that are regulated by Nurr1 but the effects of Nurr1 activation on gene expression are highly dependent on cell type and on the nature of the signalling event studied. Interestingly, Nurr1 appears to have both constitutive and inducible anti-inflammatory activity in monocyte/macrophage lineage immune cells, as well as in brain glial cells including both astrocytes and microglia. This anti-inflammatory activity appears to be directed towards the NF-κB signalling pathway in response to inflammatory stimuli such as tumor necrosis factor alpha (TNFα) and bacterial lipopolysaccharide (LPS). Following exposure to TNFα or LPS, the p65 (RelA) subunit of NF-κB rapidly translocates to the nucleus, where its histone deacetylase activity and subsequent phosphorylation by GSK3β facilitates opening of chromatin and removal of constitutively bound nuclear co-repressor complexes.[10] These studies also report that sumoylation of Nurr1 on K577 of the LBD is essential for this transrepressive activity. Earlier studies support this conclusion, because K577 in the Nurr1 LBD is part of a consensus SUMO-modification sequence and mutation of this lysine to arginine results in decreased transcriptional activity, suggesting that sumoylation of K577 is important for transcriptional modulation by Nurr1.[32] In brain glial cells, the anti-inflammatory effects of Nurr1 are mediated by docking to NF-κB-p65 on target inflammatory gene promoters, followed by recruitment of the CoREST co-repressor complex, resulting in clearance of NF-κB/p65 and transcriptional repression.[10] Pharmacologic activation of Nurr1 in BV-2 microglial cells inhibits expression of numerous inflammatory genes, including *Nos2*, *Tnf*, *Il1b* and *Il6*, by a mechanism involving stabilization of the nuclear co-repressors NCoR2 and HDAC at NF-κB response elements.[33] Because inflammatory activation of glial cells is critical to the progressive loss of dopaminergic neurons in PD, these studies suggest that Nurr1 protects against neuronal loss in part by limiting the production of neurotoxic mediators by microglia and astrocytes. Interestingly, Nurr1 regulates expression of the Forkhead transcription factor Foxp3, which is important for differentiation of regulatory T cells (Treg cells).[34] Induction of Nurr1, along with other soluble and cell-surface mediators with anti-inflammatory activities such as IL-10, TGF-b and resolvins, is important to attenuate inflammation in peripheral immune cells.[1] Based on these and other studies, Nurr1 appears to be broadly important for regulating both inflammation and resolution of inflammatory signalling in activated immune cells and glial cells relevant to neuronal injury in PD and other neurodegenerative disorders.

7.3 Neurotoxic Models of Parkinson's Disease: Reactive Oxygen Species and Neuroinflammatory Mechanisms

Neurodegenerative mechanisms associated with reactive molecules are implicated in the pathogenesis of Parkinson's disease (PD), including mitochondrial dysfunction, production of oxidative stress and inflammation among others. The progressive nature of neuronal injury manifests in the classical motor symptoms of PD, which are typically not present until a loss of 75% of the dopaminergic neurons in the *substantia nigra pars compacta* (SNpc) concomitant with a reduction in neurotransmitter levels. Additionally, present within the neurons are intracytoplasmic protein aggregates of alpha-synuclein, ubiquitin and neurofilaments called Lewy bodies.[35,36] The classical toxin-induced PD models, MPTP (1-methyl-1,2,3,6-tetrahydropiridine) and 6-OHDA (6-hydroxydopamine), have been useful for understanding the mechanistic cascade of cell death that underpins the progression of the disease. Additionally, inflammagens such as the toll-like receptor ligands LPS (TLR4) and Poly(I:C) (TLR3) have been used to study the progressive neuroinflammatory response in PD. However, the translation of the models to clinical PD is not always representative of the behavior and pathophysiological markers of idiopathic PD, specifically the lack of Lewy-body development and non-motor symptoms.[37] Considering the ability to manipulate the genetics of rodents to advance the understanding of the mechanistic pathways, the rodent model can greatly benefit research advancement in the pathogenesis of dopaminergic cell death. This section will discuss classical toxin induced-models and their strengths and weaknesses in understanding the progression of cell death.

7.3.1 1-Methyl-4-Phenyl-1,2,3,6-Tetrahydropyridine (MPTP)

The incidental discovery of the meperidine derivative meperidine (MPTP) has led to many advances in understanding the pathogenesis of PD. MPTP selectively targets dopaminergic neurons in the basal ganglia inducing motor symptoms comparable to PD in non-human primate and mouse models.[38,39]

Non-human primate models are considered the most translatable as they have demonstrated clinical relevance in motor impairments, loss of monoamine neurotransmitters and bilateral loss of tyrosine hydroxylase-positive (TH) neurons in the basal ganglia, as well as responsiveness to Levodopa treatment and development of the same Levodopa-induced dyskinesia observed in humans.[40] Mouse models have a greater presence in research primarily due to the accessibility and cost efficiency, however, depending on the dosing regimen reproducibility can be variable and only modest behavioral evident with acute exposure to MPTP. Interestingly, some rodent strains do not present neuropathological or behavioral deficits with MPTP treatment because of varying levels of metabolic enzymes and transporters across species and strain.[41,42] Additionally, aggregation of alpha-synuclein is generally absent or minimal in these models,[37,43] with the development of neuronal inclusions seemingly dependent on the route and duration of exposure.

MPTP is a lipophilic protoxin that readily crosses the blood–brain barrier after intravenous injection. MPTP is taken up by astrocytes expressing monoamine oxidase B (MAO-B),[44] where it is metabolized into its active form, 1-methyl-4-phenylpyridinium (MPP+). MPP+ is then released through the organic cation transport 3 (OCT-3) into the extracellular space[45] and accumulates proximal to dopaminergic neurons in the nigro-striatal pathway. MPP+ can enter the neuronal terminal by the dopamine transporter (DAT) and can inhibit of complex I in the mitochondrial electron transport chain.[46] This inhibition leads to a reduction in ATP and the formation of reactive oxygen species (ROS), specifically superoxide ($O_2^{-\bullet}$).[47] Superoxide is a critical mediator of MPTP-neurotoxicity, because overexpression of superoxide dismutase 1 (SOD1) in transgenic mice prevents loss of striatal dopamine and neurotoxicity.[48] The metabolism of MPTP and subsequent oxidative of dopamine also produces excess hydrogen peroxide (H_2O_2) leading to a reduction in glutathione (GSH) that propagates oxidative injury in neurons.[49] By contrast, sequestration of MPP+ by the vesicular monoamine transporter (VMAT2) decreases its availability to mitochondria and attenuates MPP+-induced mitochondrial dysfunction and oxidative stress.[50–52] However, VMAT2 can also release dopamine from the vesicle into the cytosol, creating oxidative metabolites such as 3,4-dihydroxyphenylacetaldehyde (DOPAL), reactive quinones and 6-OHDA that have similar oxidative stress pathways and neurotoxic mechanisms within neurons.[53,54] These reactive molecules induced by MPTP are a part of the initial oxidative stress cascade detrimental to vulnerable dopaminergic neurons.

Amongst these oxidative stress pathways is the activation of the high-density population of microglia cells resident in the *substantia nigra pars compacta* (SNpc).[55,56] As shown in Figure 7.1, the activation of microglia and astrocytes are important in mediating the progression of PD and PD-like models due to their abundance and inflammatory phenotype. Microglial activation peaks 24 h after MPTP administration preceding astrogliosis and dopaminergic cell death, suggesting that microglia initiate the neurodegenerative cascade.[55] Microglial activation results in an elevation of many

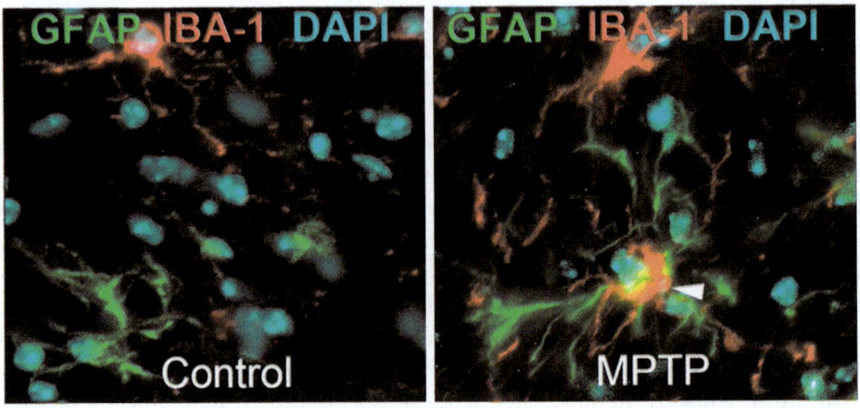

Figure 7.1 Glial activation during MPTP neurotoxicity. Mice were treated with four doses of MPTP (20 mg kg^{-1}) plus probenecid (200 mg kg^{-1}) over the course of one week and then examined one week later for glial activation and neuronal injury in the *substantia nigra*. In control animals, astrocyte and microglial morphology is consistent with a highly branched, resting phenotype, whereas in MPTP-treated mice, both astrocytes (upper arrow) and microglia are hypertrophic, with amoeboid microglia evident surrounding dying neurons (lower arrow). Green (GFAP); red, IBA1; blue nuclear staining (DAPI).

pro-inflammatory molecules, such as tumor necrosis factor alpha (TNFα) and nitric oxide (NO) that are associated with the pathogenesis of clinical PD. TNFα is increased with MPTP treatment prior to the upregulation of GFAP, whereas expression of mutant TNF receptors in mice attenuates both the subsequent GFAP induction and dopaminergic neurotoxicity, implicating TNFα as a critical initiating activator in the inflammatory cascade.[57,58] Furthermore, the microglial inhibitor minocycline mitigates glial activation, cytokine production and dopaminergic cell loss in a dose-dependent manner,[59,60] although its protection has been variable according to the administration of the anti-inflammatory agent.[61]

The production of TNFα activates the nuclear factor kappa B (NF-κB) pathway, resulting in the coordinate expression of numerous inflammatory genes in glial cells and subsequent neuronal apoptosis. Nuclear translocation of NF-κB increases gene transcription for pro-inflammatory cytokines such as COX-2 and inducible nitric oxide synthase (iNOS/NOS2). Inhibition of COX-2 by acetylsalicylic acid and salicylate, as well as COX-2 knock mice, are protected from MPTP neurotoxicity through reduced formation of ROS.[62,63] NOS2 produces large amounts of the free radial, NO, and in the presence of superoxide forms peroxynitrite ($ONOO^-$) leading to the nitrosylation of cellular proteins, an important biomarker of oxidative stress. There are increases in NOS2 expression with MPTP treatment, whereas NOS2-deficient mice are resistant to nitrotyrosine adduct formation and have reduced TH cell loss in the SNpc.[55] Interestingly, nuclear receptors that regulate the transcriptional activity of NF-κB, such as the nerve growth factor-related receptor, NR4A2

(Nurr1) can act as co-repressors of NF-κB in glial cells and thereby decrease the expression of neuroinflammatory genes including TNFα and NOS2.[64] This may be critical to the neurotoxicity of MPTP, given that TNF receptor knockout mice are resistant to MPTP-induced loss of dopaminergic neurons.[14] TNF is one of numerous NF-κB-regulated inflammatory genes and it is therefore not surprising that a reduction in Nurr1 protein corresponds to the loss of TH neurons,[65,66] implying this pro-inflammatory cascade role in MPTP toxicity. These studies with MPTP demonstrate that injury to dopaminergic neurons in this model results from the interaction between oxidative stress and neuroinflammation, both of which are critical mechanisms implicated in the progression of clinical PD.

7.3.2 6-Hydroxydopamine (6-OHDA)

The addition of a hydroxyl group to dopamine results in a selective catecholaminergic neurotoxic molecule, 6-hydroxydopamine (6-OHDA). 6-OHDA gains access to neurons through the dopamine and norepinephrine transporters and administration of norepinephrine transporter inhibitors decreases its neurotoxicity as well as the loss of dopaminergic neurons.[67] Although most species are sensitive to 6-OHDA, rat models are frequently used due to the ease of administration compared to mice.[39] Because 6-OHDA is incapable of crossing the blood–brain barrier, it must be stereotactically injected into the *substantia nigra* or medial forebrain bundle for a significant reduction in the number of TH neurons.[39] Injection into the striatum results in subtle degeneration in a retrograde manner that is relevant to the progression and symptomatic features of clinical PD.[68] Depending on the dose and injection site, administration can lead to full or partial lesions of the TH neurons with a reduction in dopamine, however, there is a general paucity in Lewy-body formation.[41,69] Motor symptoms appear when approximately 70% of the dopaminergic neurons are lost.[41] Unilateral injection of 6-OHDA serves as an internal control due to lack of pathophysiological effects in the contralateral hemisphere. The administration of dopamine agonists with this regimen such as apomorphine or amphetamine results in rotational behavior corresponding to the severity of the lesion.[39] Such rotational behavior makes this model well suited to the testing of putative neuroprotective drugs, which, if effective, should improve motor deficits and reduce ligand-induced rotational behavior resulting from 6-OHDA-induced loss of dopaminergic neurons.

Once in the cytoplasm of the neuron, the postulated neurotoxic mechanism of 6-OHDA is through the formation of ROS, quinones and the inhibition of mitochondrial respiratory complexes,[70] all of which propagates oxidative stress leading to cell death. 6-OHDA inhibits complexes I and IV,[71] uncoupling respiration, which decreases the mitochondrial membrane potential and enhances transfer of electrons to molecular oxygen, resulting in production of ROS. Dopaminergic neurons are innately vulnerable to the endogenous formation of 6-OHDA, due to the oxidation of dopamine by

MAO-B, which generates hydrogen peroxide that can, in the presence of free iron, cause auto-oxidative of dopamine to 6-OHDA.[72–74] The production of ROS by 6-OHDA is critical to its neurotoxicity, evidence for which is provided by studies reporting that treatment with antioxidants or overexpression of SOD1 attenuates the neurotoxicity 6-OHDA in rats and transgenic mice, respectively.[75–77] Inhibitors of MAO-B have demonstrated partial neuroprotective effects with the administration of 6-OHDA,[78] although these results have been inconsistent. The increased production ROS in the 6-OHDA model precedes the unfolded protein response and induction of dopaminergic apoptosis,[35] whereas antioxidant treatment rescues the cells from protein oxidation and toxicity.[79]

The activation of glial cells is also apparent with 6-OHDA exposure. After 6-OHDA administration, microglial cells become activated concomitant to increased levels of TNFα in the striatum and SN ipsilateral to the injection.[80,81] It was reported that blockade of TNFα signalling *in vivo* attenuates the 6-OHDA-induced loss of dopaminergic neurons by 50%, although microglial activation was still present.[82] This suggests that microglial-derived TNFα is critical to loss of dopaminergic neurons in the 6-OHDA model but is not solely responsible for dopaminergic loss, likely due to additional production of the cytokine by activated astrocytes. TNFα binds to and activates the TNF receptor complex, leading to p38 MAPK activation *via* TRAF and stimulation of IKK-dependent activation of NF-κB, with subsequent increases in expression of NOS2 and COX-2. NF-κB activation accompanies 6-OHDA-induced apoptosis in dopaminergic neurons, whereas the inhibition of COX-2 by celecoxib demonstrates a reduction in microglia activation and neuroprotection.[83] The 6-OHDA-mediated inflammatory processes in glial cells are thought to be antagonized by the nuclear factor (erythroid-derived 2)-like 2, also known as NFE2L2 or Nrf2, which the transcription of anti-oxidant genes through the *cis*-acting antioxidant response element (ARE).[84] 6-OHDA administration increases the activation of ARE-responsive genes through Nrf2 *in vivo* and *in vitro*, with Nrf2 knockout mice displaying exacerbated 6-OHDA induced neuronal loss.[85] Furthermore, transgenic mice with astrocyte-specific overexpression of Nrf2 are significantly protected against 6-OHDA-induced loss of TH immunoreactivity in the striatum,[85] supporting the antioxidant role of this factor in preventing neuronal injury. This also suggests that protection against the damaging effects of ROS in the 6-OHDA model is critically regulated by anti-oxidant gene expression in astrocytes and that dysregulation of glial function during neuroinflammation may have profound effects on neuronal survival.

7.3.3 Lipopolysaccharide (LPS)

The bacterial endotoxin lipopolysaccharide (LPS) is known to induce neuronal injury in part through the activation of glial cells and therefore the induction of reactive oxygen species (ROS) and inflammatory cytokine production. Although LPS treatment has been used to model dopaminergic cells loss and inflammation in PD, it is not as widely used as either MPTP or 6-OHDA, in part because it does not cause as robust behavioral and neurochemical dysfunction.[86,87] However, intranigral administration of LPS in mice replicates some the behavioral symptoms and neuropathological loss dopaminergic neurons.[88] Microglia initiate LPS-mediated inflammation and astrocytes propagate the production of neurotoxic factors such as NO, superoxide and TNFα. These inflammatory mediators are largely produced through the activation of Nf-κB pathway, which is activated by LPS through TLR4. LPS treatment leads to a dose-dependent increase in ROS, which is mitigated by SOD1 overexpression that significantly reduces levels of both $O_2^{-\bullet}$ and $NO/ONOO^-$.[64] Furthermore inhibiting the activity of NOS2 reduced LPS-induced neurodegeneration both *in vitro* and *in vivo*,[89,90] indicating that both $O_2^{-\bullet}$ and $NO/ONOO^-$ are important to neuronal injury from LPS. The source of ROS and reactive nitrogen species (RNS) following LPS exposure may also come from glial cells. Direct application of LPS to astrocyte-derived C6 glioma cells caused a decrease in mitochondrial membrane potential and an increase in cellular ROS, suggesting that alterations in mitochondrial function directly stimulate production of reactive oxygen and nitrogen species in glial cells that may be harmful to adjacent neurons.[91]

Collectively, these neurotoxin-based models of PD pathogenesis demonstrate that dopaminergic neurotoxicity is susceptible to the convergent pathways of oxidative stress and neuroinflammation. These models recapitulate certain features of PD and have been useful for studying the pathogenesis and progression of the disease, as well as providing platforms for testing the efficacy of potential therapeutic agents. Factors associated with oxidative stress and inflammation, including TNFα expression and the presence of 3-nitrotyrosine protein adducts in Lewy bodies, have been observed in postmortem PD brain tissue and support the use of neurotoxic models for understanding the events leading to the death of dopaminergic neurons. The cascade of neurotoxic events leading to neuronal apoptosis include oxidative and nitrosative stress, mitochondrial dysfunction and inflammation, all of which are critical to neuronal injury.

7.4 Neuroinflammation and Protein Aggregation in PD and Protein-Misfolding Disorders

A common theme in neurodegenerative diseases including Parkinson's disease (PD), Alzheimer's disease (AD), amyotrophic lateral sclerosis (ALS), frontotemporal dementia (FTD) and prion diseases is the accumulation and aggregation of misfolded disease-specific proteins in the brain followed by irreversible loss of neurons. These disorders have been identified as protein-misfolding disorders each with a specific misfolded 'toxic' protein causing cellular stress. Many of these disorders also share a 'prion-like' propagation of misfolded proteins with a spreading pathology of aggregates as reviewed in the following.[92–94] Although clinically and pathologically these disorders are distinct, biochemical changes such as neuroinflammation, oxidative stress, and activation of specific cellular stress pathways such as the unfolded protein response (UPR) are common amongst them.

7.4.1 Neuroinflammation in Protein-Misfolding Disorders

In protein-misfolding disorders neuroinflammation is an early phenomenon that continues throughout the disease in both patients and mouse models. Neuroinflammation is characterized pathologically by activation of both microglia and astrocytes, the release of pro-inflammatory cytokines and activation of inflammatory pathways including MHC class II, cyclo-oxygenase-2 (COX-2) and inducible nitric oxide synthase (iNOS/NOS2). Pathologically, Alzheimer's disease is characterized by amyloid plaques and neurofibrillary tangles (NFTs) with accumulation of amyloid and misfolded tau, respectively.[95] Both pathologies have a strong glial component with reactive astrocytes and microglia, however, NFTs and glial activation parallel the severity of clinical disease in patients more so than amyloid plaques.[96] Activation of glial cells in AD increases levels of inflammatory pathways in cells and in mouse models of disease.[97] A recent study comparing microarray data of RNA expression from five transgenic AD mouse models of amyloid and tau with human genome wide association studies show immune and inflammatory pathways activated during disease that correlate closely with the extent of neuropathology.[98]

Mutated superoxide dismutase 1 (SOD1) is known to cause ALS pathogenesis including the accumulation of misfolded SOD1 in human and mouse models with this mutation.[99,100] Another mutated protein is RNA binding protein TDP-43 found to aggregate in both ALS and FTD patients.[101,102] Activation of astrocytes and microglia and the subsequent downstream inflammatory pathways are common features of ALS pathology.[103,104] Inhibition of neuroninflammation by Withaferin A, an inhibitor of NF-κB activity, protects neuronal toxicity due to both misfolded TDP-43 [105] and SOD1 [106] in transgenic mouse models.

Similar to other neurodegenerative diseases, in prion diseases, both microgliosis and astrogliosis occur early and coincide with rising levels of the misfolded prion protein, PrP^{Sc}.[107,108] Cellular macrophage treatment with human misfolded prion protein caused upregulation of NOS2 and activation of NF-κB pathway with downstream pro-inflammatory cytokine production.[109] In a recent study a number of inflammatory genes and markers were identified early in prion-diseased mice, prior to clinical signs being observed.[110] Numerous inflammatory genes and pathways are activated due to the accumulation of misfolded proteins such as Aβ, tau, SOD1, TDP-43 and PrP. Targeting the inflammatory phenotype resulting from accumulation of toxic oligomeric forms of these proteins may therefore be part of the strategy to protect neurons and prevent or slow the spread of misfolded proteins.

7.4.2 Oxidative Stress in Protein-Misfolding Disorders

Oxidative stress or damage in protein-misfolding disorders is detrimental to neurons. As neuronal damage increases during disease there is a marked rise in levels of reactive oxygen species (ROS) and/or reactive nitrogen species (RNS), causing oxidation and nitration of proteins, lipids and DNA.[111] Accumulation of misfolded proteins in neurodegenerative diseases and the role of oxidative stress has been intensely studied for decades and yet no viable therapeutic approach has been discovered to date. In AD patient brains, high levels of oxidative damage and nitrated proteins colocalize with NFTs but, interestingly, there is no prerequisite for intact NFTs in the neuronal cytoplasm and nuclei to stimulate oxidative stress, suggesting that smaller oligomeric forms of these proteins may direct stimulate the formation of ROS/RNS.[112,113] Oxidative stress including ROS and nitric oxide (NO) increases with aggregation of Aβ and formation of plaques.[1] Aβ toxicity forms neuronal protein nitration and accelerated aggregation of Aβ plaques in an AD mouse model with mutations in both amyloid precursor protein (APP) and persenlin (PS1). Further, injection of 3Ntyr-A into young APP/PS1 mice accelerated disease and amyloid aggregation.[114] A recent study using a double transgenic mouse model of AD, where mice expressing human APP (a mouse model of AD) crossed with mice expressing a dominant negative mutant of mitochondrial aldehyde dehydrogenase 2 (ALDH2) (a mouse model of increased oxidative stress) were assessed. Acceleration of cognitive decline, accumulation of amyloid and phosphorylation of tau and gliosis were all identified in these double transgenic AD/oxidative stress mice indicating a role for oxidative stress in disease pathogenesis.[115] The repressor element 1-silencing transcription factor (REST) is a transcription factor important in regulating cellular death and stress genes that was found to be lost in AD and FTD patient brains and induction of REST protects neurons from oxidative stress Aβ toxicity.[116] This study emphasizes the role of oxidative stress in neurodegenerative disease and how modulating key genes and/or pathways involved in oxidative damage is key to the protection of neurons.

Oxidative stress and production of nitric oxide (NO) have been shown to cause both glial activation and neuronal death in ALS, similarly to PD and AD.[117] This increased production of ROS and NO causes nitration of neurons, which leads to oxidative stress and neuronal death in mouse models of ALS.[118,119] In mtSOD1 expressing mice and cells, an induction of NOS2 leads to the production of NO and S-nitrosylation protein disulfide isomerase (PDI) in parallel with rising levels of misfolded SOD1.[120] This oxidative stress pathway can then be inhibited by both inhibiting NOS2 and rescue the downstream aggregation of mtSOD1 seen in ALS.[120] In patients with Creutzfeldt–Jakob disease (CJD; human prion disease) and in mouse models of prion disease oxidative damage has been identified as an early event in disease.[121–123] One oxidative stress pathway identified in both CJD patients and experiment mouse prion disease is the NADPH oxidase (NOX2) pathway that localizes in the microglial cells. Knocking out NOX2 in mice then exposing to prion disease modulated progression of prion disease by reducing ROS, glia activation motor deficits and increases survival days.[124]

Oxidative stress and damage can cause improper protein folding in the ER that in turn can further stimulate the production of additional ROS and oxidative damage in a feedforward cycle that increases cellular stress, oxidation of macromolecules and organelle dysfunction.[125] To deal with such cellular stressors as neuroinflammation, oxidative stress and accumulation of misfolded proteins, there are quality-control pathways regulating ER stress through activation of the UPR.[126] Interestingly, numerous PARK locus genes that cause autosomal dominant, early-onset PD are linked to regulation of protein degradation and processing.

7.4.3 Unfolded Protein Response in Protein-Misfolding Disorders

During disease, proteins with abundant alpha-helical secondary structure such as alpha-synuclein, amyloid beta and prion protein (PrP), undergo a conformational change to β-sheet rich structures that accumulate and cause endoplasmic reticulum (ER) stress and activation of quality-control pathways like the unfolded protein response (UPR). Chaperones regulate the proper folding of proteins in normal conditions, however, during cellular stress proteostatic mechanisms are activated to refold, inhibit or degrade misfolded proteins. Proteins are degraded through proteosomal or autophagic pathways[127,128] and inhibited by translational control pathways such as the UPR.[129] The UPR is a normal protective cellular stress response pathway that is activated with accumulation of misfolded proteins in the ER. Bip the master regulator of ER stress, and once it detects a rise in misfolded proteins, it then dissociates from the sensors for activation.[130] There are three branches of the UPR that act as ER sensors protein kinase RNA-like ER kinase (PERK), activating transcription factor-6 (ATF6) and inositol-requiring enzyme 1a (IRE1α). Activation of the UPR has been identified in brains of patients with multiple neurodegenerative diseases including AD, PD, ALS and prion disease.[131–133]

In prion-diseased mice accumulation of misfolded prion protein leads to neurodegeneration through sustained overactivation of the PERK/eIF2α branch of the UPR. High levels of eIF2α phosphorylation reduce protein synthesis of key proteins, ultimately leading to neuronal death. Through focal genetic manipulation of this pathway protein synthesis was restored protecting neuronal loss and behavior in prion-diseased mice.[134] The UPR was then targeted in a more global manner, throughout the brain, and in a non-surgical way using an orally bioavailable inhibitor of PERK phosphorylation, the key kinase for eIF2,[135] that allowed for whole-brain modulation of the UPR. Clinical signs, behavior, whole-brain neuropathology and neuronal loss were protected in prion-infected mice at both an early time point, as misfolding of the prion protein begins, and later when brain neuropathological changes are detected.[136] In a *Drosophila* ALS model and primary neuronal cell culture expressing the misfolded protein TDP-43, the same compound, an inhibitor of PERK phosphorylation, was found to inhibit TDP-43 toxicity in both models.[137] Critically, in both studies the neuroprotection by inhibiting PERK phosphorylation is downstream of the misfolded or toxic protein that causes pathogenesis. Therefore, one would predict that targeting the UPR in a spectrum of neurodegenerative diseases would be neuroprotective, irrespective of the disease-specific misfolded protein.

The UPR pathway has also been shown to be involved in models of AD, PD and manganese neurotoxicity. Activation of the UPR has been described in mouse models expressing multiple genetic mutations of AD (5XAD) and in mice with tau accumulation.[138,139] Genetic deletion of PERK rescued synaptic plasticity and cognitive function in a transgenic mouse model of AD and in cultured hippocampal brain slices treated *ex vivo* with misfolded Aβ protein.[140] The accumulation of misfolded alpha-synuclein is a hallmark pathologic feature of PD.[141] Accumulation of alpha-synuclein in both cell and mouse models causes activation of the UPR[142–144] through induction of ER-stress in dopaminergic neurons in the *substantia nigra* (SN). The overexpression of Bip (the master molecular chaperone for the UPR) inhibits UPR activation and alpha-synuclein neurotoxicity.[144] PERK/eIF2 and another branch of the UPR, ATF6 was activated in a model of PD using chronic exposure to MPTP.[145] In related studies, ATF6 knockout mice have accelerated neuronal loss following MPTP injections but less astrocytic and UPR pathway activation, suggesting that in this MPTP model ATF6 is required to protect neurons and activate astroglial cells.[145] In a C. *elegans* model of manganese toxicity a rise in Bip due to loss of calcium levels and subsequent ER stress was reported.[146] The effect of misfolded proteins on production of ROS and RNS also suggests that the UPR is a critical cellular stress in neurons that likely contributes both to neuronal injury and to the stimulation of reactive glial responses that exacerbate injury. Collectively, these studies identify that UPR activation is critical in neurodegenerative diseases and targeting this generic pathway may be a therapeutic approach used for multiple protein-misfolding disorders.

7.5 Conclusions

Glial cells play a key role in tissue homeostasis in the brain and are critical to the response to infection and injury. Microglia, as the resident tissue macrophages of the CNS, are the primary surveyors of the brain microenvironment and can produce a large number of factors that activate surrounding astrocytes. Astrocytes, by contrast, conduct much of the metabolic and homestatic functions of the CNS and also regulate synaptic activity, vascular function and circadian rhythms.[147] However, both astrocytes and microglia can acquire an activated, inflammatory phenotype when the severity of stress, infection or injury overwhelms their ability to maintain homeostasis in the CNS microenvironment. Under such conditions, the cascade of inflammatory factors produced by activated glial cells can be damaging to neurons and is clearly linked to the progression of PD and associated synucleinopathies, as well as AD and related disorders. In models of PD, as well as in clinical disease, inflammatory mediators are consistently implicated in the deprecation of neuronal function leading to protein aggregation, loss of mitochondrial function and cell death. These include classical cytokines and chemokines associated with an M1 macrophage response such as TNFα and IL1β, as well as reactive oxygen and nitrogen species that directly damage cellular macromolecules and organelles in neurons. Understanding the molecular pathways that regulate the production of inflammatory mediators in glial cells may therefore offer new insights into targets for therapeutic intervention to slow the progression of PD and related neurodegenerative disorders.

References

1. C. K. Glass, K. Saijo, B. Winner, M. C. Marchetto and F. H. Gage, Mechanisms underlying inflammation in neurodegeneration, *Cell*, 2010, **140**, 918–934.
2. W. G. Meissner, M. Frasier, T. Gasser, C. G. Goetz, A. Lozano, P. Piccini, J. A. Obeso, O. Rascol, A. Schapira, V. Voon, D. M. Weiner, F. Tison and E. Bezard, Priorities in Parkinson's disease research, *Nat. Rev. Drug Discovery*, 2011, **10**, 377–393.
3. E. C. Hirsch and S. Hunot, Neuroinflammation in Parkinson's disease: a target for neuroprotection? *Lancet Neurol.*, 2009, **8**, 382–397.
4. C. M. Lema Tome, T. Tyson, N. L. Rey, S. Grathwohl, M. Britschgi and P. Brundin, Inflammation and alpha-synuclein's prion-like behavior in Parkinson's disease–is there a link? *Mol. Neurobiol.*, 2013, **47**, 561–574.
5. R. L. Mosley, E. J. Benner, I. Kadiu, M. Thomas, M. D. Boska, K. Hasan, C. Laurie and H. E. Gendelman, Neuroinflammation, Oxidative Stress and the Pathogenesis of Parkinson's Disease, *Clin. Neurosci. Res.*, 2006, **6**, 261–281.
6. S. J. Lee, Origins and effects of extracellular alpha-synuclein: implications in Parkinson's disease, *J. Mol. Neurosci.*, 2008, **34**, 17–22.
7. H. Braak, E. Ghebremedhin, U. Rub, H. Bratzke and K. Del Tredici, Stages in the development of Parkinson's disease-related pathology, *Cell Tissue Res.*, 2004, **318**, 121–134.
8. M. L. Block and J. S. Hong, Microglia and inflammation-mediated neurodegeneration: multiple triggers with a common mechanism, *Prog. Neurobiol.*, 2005, **76**, 77–98.
9. T. Nagatsu and M. Sawada, Inflammatory process in Parkinson's disease: role for cytokines, *Curr. Pharm. Des.*, 2005, **11**, 999–1016.
10. K. Saijo, B. Winner, C. T. Carson, J. G. Collier, L. Boyer, M. G. Rosenfeld, F. H. Gage and C. K. Glass, A Nurr1/CoREST pathway in microglia and astrocytes protects dopaminergic neurons from inflammation-induced death, *Cell*, 2009, **137**, 47–59.
11. S. Sugama, L. Yang, B. P. Cho, L. A. DeGiorgio, S. Lorenzl, D. S. Albers, M. F. Beal, B. T. Volpe and T. H. Joh, Age-related microglial activation in 1-methyl-4-phenyl-1,2,3,6-tetrahydropyridine (MPTP)-induced dopaminergic neurodegeneration in C57BL/6 mice, *Brain Res.*, 2003, **964**, 288–294.

12. H. M. Gao, H. Zhou, F. Zhang, B. C. Wilson, W. Kam and J. S. Hong, HMGB1 acts on microglia Mac1 to mediate chronic neuroinflammation that drives progressive neurodegeneration, *J. Neurosci.*, 2011, **31**, 1081–1092.
13. M. F. McCarty, Down-regulation of microglial activation may represent a practical strategy for combating neurodegenerative disorders, *Med. Hypotheses*, 2006, **67**, 251–269.
14. K. Sriram, J. M. Matheson, S. A. Benkovic, D. B. Miller, M. I. Luster and J. P. O'Callaghan, Deficiency of TNF receptors suppresses microglial activation and alters the susceptibility of brain regions to MPTP-induced neurotoxicity: role of TNF-alpha, *FASEB J.*, 2006, **20**, 670–682.
15. T. Dehmer, J. Lindenau, S. Haid, J. Dichgans and J. B. Schulz, Deficiency of inducible nitric oxide synthase protects against MPTP toxicity *in vivo*, *J. Neurochem.*, 2000, **74**, 2213–2216.
16. T. Wyss-Coray and L. Mucke, Inflammation in neurodegenerative disease–a double-edged sword, *Neuron*, 2002, **35**, 419–432.
17. J. A. DiDonato, M. Hayakawa, D. M. Rothwarf, E. Zandi and M. Karin, A cytokine-responsive IkappaB kinase that activates the transcription factor NF-kappaB, *Nature*, 1997, **388**, 548–554.
18. G. Bonizzi and M. Karin, The two NF-kappaB activation pathways and their role in innate and adaptive immunity, *Trends Immunol.*, 2004, **25**, 280–288.
19. S. Ghisletti, W. Huang, K. Jepsen, C. Benner, G. Hardiman, M. G. Rosenfeld and C. K. Glass, Cooperative NCoR/SMRT interactions establish a corepressor-based strategy for integration of inflammatory and anti-inflammatory signaling pathways, *Genes Dev.*, 2009, **23**, 681–693.
20. D. L. Carbone, J. A. Moreno and R. B. Tjalkens, Nuclear factor kappa-B mediates selective induction of neuronal nitric oxide synthase in astrocytes during low-level inflammatory stimulation with MPTP, *Brain Res.*, 2008, **1217**, 1–9.
21. R. Bianchi, I. Giambanco and R. Donato, S100B/RAGE-dependent activation of microglia *via* NF-kappaB and AP-1 Co-regulation of COX-2 expression by S100B, IL-1beta and TNF-alpha, *Neurobiol. Aging*, 2010, **31**, 665–677.
22. W. L. Thompson and L. J. Van Eldik, Inflammatory cytokines stimulate the chemokines CCL2/MCP-1 and CCL7/MCP-3 through NFkB and MAPK dependent pathways in rat astrocytes [corrected], *Brain Res.*, 2009, **1287**, 47–57.
23. T. N. Collingwood, F. D. Urnov and A. P. Wolffe, Nuclear receptors: coactivators, corepressors and chromatin remodeling in the control of transcription, *J. Mol. Endocrinol.*, 1999, **23**, 255–275.
24. H. M. Gao and J. S. Hong, Why neurodegenerative diseases are progressive: uncontrolled inflammation drives disease progression, *Trends Immunol.*, 2008, **29**, 357–365.
25. J. F. Duda, B. I. Giasson, Q. Chen, T. L. Gur, H. I. Hurtig, M. B. Stern, S. M. Gollomp, H. Ischiropoulos, V. M. Lee and J. Q. Trojanowski, Widespread nitration of pathological inclusions in neurodegenerative synucleinopathies, *Am. J. Pathol.*, 2000, **157**, 1439–1445.

26. A. A. Hamers, R. N. Hanna, H. Nowyhed, C. C. Hedrick and C. J. de Vries, NR4A nuclear receptors in immunity and atherosclerosis, *Curr. Opin. Lipidol.*, 2013, **24**, 381–385.
27. J. D. Hawk and T. Abel, The role of NR4A transcription factors in memory formation, *Brain Res. Bull.*, 2011, **85**, 21–29.
28. H. S. Ranhotra, The NR4A orphan nuclear receptors: mediators in metabolism and diseases, *J. Recept. Signal Transduction Res.*, 2015, **35**, 184–188.
29. P. Aarnisalo, C. H. Kim, J. W. Lee and T. Perlmann, Defining requirements for heterodimerization between the retinoid X receptor and the orphan nuclear receptor Nurr1, *J. Biol. Chem.*, 2002, **277**, 35118–35123.
30. K. S. Kim, C. H. Kim, D. Y. Hwang, H. Seo, S. Chung, S. J. Hong, J. K. Lim, T. Anderson and O. Isacson, Orphan nuclear receptor Nurr1 directly transactivates the promoter activity of the tyrosine hydroxylase gene in a cell-specific manner, *J. Neurochem.*, 2003, **85**, 622–634.
31. G. Pascual, A. L. Fong, S. Ogawa, A. Gamliel, A. C. Li, V. Perissi, D. W. Rose, T. M. Willson, M. G. Rosenfeld and C. K. Glass, A SUMOylation-dependent pathway mediates transrepression of inflammatory response genes by PPAR-gamma, *Nature*, 2005, **437**, 759–763.
32. N. Volakakis, M. Malewicz, B. Kadkhodai, T. Perlmann and G. Benoit, Characterization of the Nurr1 ligand-binding domain co-activator interaction surface, *J. Mol. Endocrinol.*, 2006, **37**, 317–326.
33. B. R. De Miranda, K. A. Popichak, S. L. Hammond, B. A. Jorgensen, A. T. Phillips, S. Safe and R. B. Tjalkens, The Nurr1 Activator 1,1-Bis(3'-Indolyl)-1-(p-Chlorophenyl)Methane Blocks Inflammatory Gene Expression in BV-2 Microglial Cells by Inhibiting Nuclear Factor kappaB, *Mol. Pharmacol.*, 2015, **87**, 1021–1034.
34. M. A. Pearen and G. E. O. Muscat, Minireview: Nuclear hormone receptor 4A signaling: implications for metabolic disease, *Mol. Endocrinol.*, 2010, **24**, 1891–1903.
35. W. Dauer and S. Przedborski, Parkinson's disease: mechanisms and models, *Neuron*, 2003, **39**, 889–909.
36. K. G. Lloyd, CNS compensation to dopamine neuron loss in Parkinson's disease, *Adv. Exp. Med. Biol.*, 1977, **90**, 255–266.
37. K. A. Jellinger, Neuropathology of sporadic Parkinson's disease: evaluation and changes of concepts, *Mov. Disord.*, 2012, **27**, 8–30.
38. J. W. Langston, P. Ballard, J. W. Tetrud and I. Irwin, Chronic Parkinsonism in humans due to a product of meperidine-analog synthesis, *Science*, 1983, **219**, 979–980.
39. K. Tieu, A guide to neurotoxic animal models of Parkinson's disease, *Cold Spring Harbor Perspect. Med.*, 2011, **1**, a009316.
40. S. H. Fox and J. M. Brotchie, The MPTP-lesioned non-human primate models of Parkinson's disease. Past, present, and future, *Prog. Brain Res.*, 2010, **184**, 133–157.
41. S. Duty and P. Jenner, Animal models of Parkinson's disease: a source of novel treatments and clues to the cause of the disease, *Br. J. Pharmacol.*, 2011, **164**, 1357–1391.

42. R. G. Staal, K. A. Hogan, C. L. Liang, D. C. German and P. K. Sonsalla, *In vitro* studies of striatal vesicles containing the vesicular monoamine transporter (VMAT2): rat *versus* mouse differences in sequestration of 1-methyl-4-phenylpyridinium, *J. Pharmacol. Exp. Ther.*, 2000, **293**, 329–335.
43. L. S. Forno, J. W. Langston, L. E. DeLanney, I. Irwin and G. A. Ricaurte, Locus ceruleus lesions and eosinophilic inclusions in MPTP-treated monkeys, *Ann. Neurol.*, 1986, **20**, 449–455.
44. K. Chiba, A. Trevor and N. Castagnoli Jr, Metabolism of the neurotoxic tertiary amine, MPTP, by brain monoamine oxidase, *Biochem. Biophys. Res. Commun.*, 1984, **120**, 574–578.
45. M. Cui, R. Aras, W. V. Christian, P. M. Rappold, M. Hatwar, J. Panza, V. Jackson-Lewis, J. A. Javitch, N. Ballatori, S. Przedborski and K. Tieu, The organic cation transporter-3 is a pivotal modulator of neurodegeneration in the nigrostriatal dopaminergic pathway, *Proc. Natl. Acad. Sci. U. S. A.*, 2009, **106**, 8043–8048.
46. Y. Mizuno, N. Sone and T. Saitoh, Effects of 1-methyl-4-phenyl-1,2,3,6-tetrahydropyridine and 1-methyl-4-phenylpyridinium ion on activities of the enzymes in the electron transport system in mouse brain, *J. Neurochem.*, 1987, **48**, 1787–1793.
47. D. A. Drechsel and M. Patel, Role of reactive oxygen species in the neurotoxicity of environmental agents implicated in Parkinson's disease, *Free Radical Biol. Med.*, 2008, **44**, 1873–1886.
48. S. Przedborski, V. Kostic, V. Jackson-Lewis, A. B. Naini, S. Simonetti, S. Fahn, E. Carlson, C. J. Epstein and J. L. Cadet, Transgenic mice with increased Cu/Zn-superoxide dismutase activity are resistant to N-methyl-4-phenyl-1,2,3,6-tetrahydropyridine-induced neurotoxicity, *J. Neurosci.*, 1992, **12**, 1658–1667.
49. V. Dias, E. Junn and M. M. Mouradian, The role of oxidative stress in Parkinson's disease, *J. Parkinson's Dis.*, 2013, **3**, 461–491.
50. T. S. Guillot and G. W. Miller, Protective actions of the vesicular monoamine transporter 2 (VMAT2) in monoaminergic neurons, *Mol. Neurobiol.*, 2009, **39**, 149–170.
51. M. K. Chen, H. Kuwabara, Y. Zhou, R. J. Adams, J. R. Brasic, J. L. McGlothan, T. Verina, N. C. Burton, M. Alexander, A. Kumar, D. F. Wong and T. R. Guilarte, VMAT2 and dopamine neuron loss in a primate model of Parkinson's disease, *J. Neurochem.*, 2008, **105**, 78–90.
52. K. M. Lohr, A. I. Bernstein, K. A. Stout, A. R. Dunn, C. R. Lazo, S. P. Alter, M. Wang, Y. Li, X. Fan, E. J. Hess, H. Yi, L. M. Vecchio, D. S. Goldstein, T. S. Guillot, A. Salahpour and G. W. Miller, Increased vesicular monoamine transporter enhances dopamine release and opposes Parkinson disease-related neurodegeneration *in vivo*, *Proc. Natl. Acad. Sci. U. S. A.*, 2014, **111**, 9977–9982.
53. W. J. Burke, S. W. Li, H. D. Chung, D. A. Ruggiero, B. S. Kristal, E. M. Johnson, P. Lampe, V. B. Kumar, M. Franko, E. A. Williams and D. S. Zahm, Neurotoxicity of MAO metabolites of catecholamine neurotransmitters: role in neurodegenerative diseases, *Neurotoxicology*, 2004, **25**, 101–115.

54. Y. Jinsmaa, V. R. Florang, J. N. Rees, L. M. Mexas, L. L. Eckert, E. M. Allen, D. G. Anderson and J. A. Doorn, Dopamine-derived biological reactive intermediates and protein modifications: Implications for Parkinson's disease, *Chem.-Biol. Interact.*, 2011, **192**, 118–121.
55. G. T. Liberatore, V. Jackson-Lewis, S. Vukosavic, A. S. Mandir, M. Vila, W. G. McAuliffe, V. L. Dawson, T. M. Dawson and S. Przedborski, Inducible nitric oxide synthase stimulates dopaminergic neurodegeneration in the MPTP model of Parkinson disease, *Nat. Med.*, 1999, **5**, 1403–1409.
56. W. G. Kim, R. P. Mohney, B. Wilson, G. H. Jeohn, B. Liu and J. S. Hong, Regional difference in susceptibility to lipopolysaccharide-induced neurotoxicity in the rat brain: role of microglia, *J. Neurosci.*, 2000, **20**, 6309–6316.
57. B. Ferger, A. Leng, A. Mura, B. Hengerer and J. Feldon, Genetic ablation of tumor necrosis factor-alpha (TNF-alpha) and pharmacological inhibition of TNF-synthesis attenuates MPTP toxicity in mouse striatum, *J. Neurochem.*, 2004, **89**, 822–833.
58. K. Sriram, J. M. Matheson, S. A. Benkovic, D. B. Miller, M. I. Luster and J. P. O'Callaghan, Mice deficient in TNF receptors are protected against dopaminergic neurotoxicity: implications for Parkinson's disease, *FASEB J.*, 2002, **16**, 1474–1476.
59. Y. Du, Z. Ma, S. Lin, R. C. Dodel, F. Gao, K. R. Bales, L. C. Triarhou, E. Chernet, K. W. Perry, D. L. Nelson, S. Luecke, L. A. Phebus, F. P. Bymaster and S. M. Paul, Minocycline prevents nigrostriatal dopaminergic neurodegeneration in the MPTP model of Parkinson's disease, *Proc. Natl. Acad. Sci. U. S. A.*, 2001, **98**, 14669–14674.
60. C. Zhao, Z. Ling, M. B. Newman, A. Bhatia and P. M. Carvey, TNF-alpha knockout and minocycline treatment attenuates blood-brain barrier leakage in MPTP-treated mice, *Neurobiol. Dis.*, 2007, **26**, 36–46.
61. C. J. Barnum and M. G. Tansey, Modeling neuroinflammatory pathogenesis of Parkinson's disease, *Prog. Brain Res.*, 2010, **184**, 113–132.
62. P. Teismann and B. Ferger, Inhibition of the cyclooxygenase isoenzymes COX-1 and COX-2 provide neuroprotection in the MPTP-mouse model of Parkinson's disease, *Synapse*, 2001, **39**, 167–174.
63. P. Teismann, K. Tieu, D. K. Choi, D. C. Wu, A. Naini, S. Hunot, M. Vila, V. Jackson-Lewis and S. Przedborski, Cyclooxygenase-2 is instrumental in Parkinson's disease neurodegeneration, *Proc. Natl. Acad. Sci. U. S. A.*, 2003, **100**, 5473–5478.
64. K. Saijo, B. Winner, C. T. Carson, J. G. Collier, L. Boyer, M. G. Rosenfeld, F. H. Gage and C. K. Glass, A Nurr1/CoREST pathway in microglia and astrocytes protects dopaminergic neurons from inflammation-induced death, *Cell*, 2009, **137**, 47–59.
65. W. Le, O. M. Conneely, Y. He, J. Jankovic and S. H. Appel, Reduced Nurr1 expression increases the vulnerability of mesencephalic dopamine neurons to MPTP-induced injury, *J. Neurochem.*, 1999, **73**, 2218–2221.

66. C. Gibrat, M. Saint-Pierre, M. Bousquet, D. Levesque, C. Rouillard and F. Cicchetti, Differences between subacute and chronic MPTP mice models: investigation of dopaminergic neuronal degeneration and alpha-synuclein inclusions, *J. Neurochem.*, 2009, **109**, 1469–1482.
67. J. Luthman, A. Fredriksson, E. Sundstrom, G. Jonsson and T. Archer, Selective lesion of central dopamine or noradrenaline neuron systems in the neonatal rat: motor behavior and monoamine alterations at adult stage, *Behav. Brain Res.*, 1989, **33**, 267–277.
68. S. Przedborski, M. Levivier, H. Jiang, M. Ferreira, V. Jackson-Lewis, D. Donaldson and D. M. Togasaki, Dose-dependent lesions of the dopaminergic nigrostriatal pathway induced by intrastriatal injection of 6-hydroxydopamine, *Neuroscience*, 1995, **67**, 631–647.
69. J. Blesa, S. Phani, V. Jackson-Lewis and S. Przedborski, Classic and new animal models of Parkinson's disease, *J. Biomed. Biotechnol.*, 2012, **2012**, 845618.
70. G. Cohen and R. E. Heikkila, The generation of hydrogen peroxide, superoxide radical, and hydroxyl radical by 6-hydroxydopamine, dialuric acid, and related cytotoxic agents, *J. Biol. Chem.*, 1974, **249**, 2447–2452.
71. Y. Glinka, M. Gassen and M. B. Youdim, Mechanism of 6-hydroxydopamine neurotoxicity, *J. Neural Transm., Suppl.*, 1997, **50**, 55–66.
72. D. Blum, S. Torch, N. Lambeng, M. Nissou, A. L. Benabid, R. Sadoul and J. M. Verna, Molecular pathways involved in the neurotoxicity of 6-OHDA, dopamine and MPTP: contribution to the apoptotic theory in Parkinson's disease, *Prog. Neurobiol.*, 2001, **65**, 135–172.
73. M. Gesi, A. Santinami, R. Ruffoli, G. Conti and F. Fornai, Novel aspects of dopamine oxidative metabolism (confounding outcomes take place of certainties), *Pharmacol. Toxicol.*, 2001, **89**, 217–224.
74. A. H. Stokes, T. G. Hastings and K. E. Vrana, Cytotoxic and genotoxic potential of dopamine, *J. Neurosci. Res.*, 1999, **55**, 659–665.
75. M. Asanuma, H. Hirata and J. L. Cadet, Attenuation of 6-hydroxydopamine-induced dopaminergic nigrostriatal lesions in superoxide dismutase transgenic mice, *Neuroscience*, 1998, **85**, 907–917.
76. J. L. Cadet, M. Katz, V. Jackson-Lewis and S. Fahn, Vitamin E attenuates the toxic effects of intrastriatal injection of 6-hydroxydopamine (6-OHDA) in rats: behavioral and biochemical evidence, *Brain Res.*, 1989, **476**, 10–15.
77. R. Soto-Otero, E. Mendez-Alvarez, A. Hermida-Ameijeiras, A. M. Munoz-Patino and J. L. Labandeira-Garcia, Autoxidation and neurotoxicity of 6-hydroxydopamine in the presence of some antioxidants: potential implication in relation to the pathogenesis of Parkinson's disease, *J. Neurochem.*, 2000, **74**, 1605–1612.
78. B. Cutillas, S. Ambrosio and M. Unzeta, Neuroprotective effect of the monoamine oxidase inhibitor PF 9601N [N-(2-propynyl)-2-(5-benzyloxy-indolyl) methylamine] on rat nigral neurons after 6-hydroxydopamine-striatal lesion, *Neurosci. Lett.*, 2002, **329**, 165–168.

79. W. A. Holtz, J. M. Turetzky, Y. J. Jong and K. L. O'Malley, Oxidative stress-triggered unfolded protein response is upstream of intrinsic cell death evoked by parkinsonian mimetics, *J. Neurochem.*, 2006, **99**, 54–69.
80. F. Cicchetti, A. L. Brownell, K. Williams, Y. I. Chen, E. Livni and O. Isacson, Neuroinflammation of the nigrostriatal pathway during progressive 6-OHDA dopamine degeneration in rats monitored by immunohistochemistry and PET imaging, *Eur. J. Neurosci.*, 2002, **15**, 991–998.
81. M. Mogi, A. Togari, K. Tanaka, N. Ogawa, H. Ichinose and T. Nagatsu, Increase in level of tumor necrosis factor-alpha in 6-hydroxydopamine-lesioned striatum in rats is suppressed by immunosuppressant FK506, *Neurosci. Lett.*, 2000, **289**, 165–168.
82. M. K. McCoy, T. N. Martinez, K. A. Ruhn, D. E. Szymkowski, C. G. Smith, B. R. Botterman, K. E. Tansey and M. G. Tansey, Blocking soluble tumor necrosis factor signaling with dominant-negative tumor necrosis factor inhibitor attenuates loss of dopaminergic neurons in models of Parkinson's disease, *J. Neurosci.*, 2006, **26**, 9365–9375.
83. R. Sanchez-Pernaute, A. Ferree, O. Cooper, M. Yu, A. L. Brownell and O. Isacson, Selective COX-2 inhibition prevents progressive dopamine neuron degeneration in a rat model of Parkinson's disease, *J. Neuroinflammation*, 2004, **1**, 6.
84. J. C. Tobon-Velasco, J. H. Limon-Pacheco, M. Orozco-Ibarra, M. Macias-Silva, G. Vazquez-Victorio, E. Cuevas, S. F. Ali, A. Cuadrado, J. Pedraza-Chaverri and A. Santamaria, 6-OHDA-induced apoptosis and mitochondrial dysfunction are mediated by early modulation of intracellular signals and interaction of Nrf2 and NF-kappaB factors, *Toxicology*, 2013, **304**, 109–119.
85. R. J. Jakel, J. T. Kern, D. A. Johnson and J. A. Johnson, Induction of the protective antioxidant response element pathway by 6-hydroxydopamine *in vivo* and *in vitro*, *Toxicol. Sci.*, 2005, **87**, 176–186.
86. R. M. Santiago, J. Barbieiro, M. M. Lima, P. A. Dombrowski, R. Andreatini and M. A. Vital, Depressive-like behaviors alterations induced by intranigral MPTP, 6-OHDA, LPS and rotenone models of Parkinson's disease are predominantly associated with serotonin and dopamine, *Prog. Neuro-Psychopharmacol. Biol. Psychiatry*, 2010, **34**, 1104–1114.
87. A. J. Herrera, A. Castano, J. L. Venero, J. Cano and A. Machado, The single intranigral injection of LPS as a new model for studying the selective effects of inflammatory reactions on dopaminergic system, *Neurobiol. Dis.*, 2000, **7**, 429–447.
88. D. B. Hoban, E. Connaughton, C. Connaughton, G. Hogan, C. Thornton, P. Mulcahy, T. C. Moloney and E. Dowd, Further characterisation of the LPS model of Parkinson's disease: a comparison of intra-nigral and intra-striatal lipopolysaccharide administration on motor function, microgliosis and nigrostriatal neurodegeneration in the rat, *Brain, Behav., Immun.*, 2013, **27**, 91–100.

89. M. M. Iravani, K. Kashefi, P. Mander, S. Rose and P. Jenner, Involvement of inducible nitric oxide synthase in inflammation-induced dopaminergic neurodegeneration, *Neuroscience*, 2002, **110**, 49–58.
90. T. Arimoto and G. Bing, Up-regulation of inducible nitric oxide synthase in the substantia nigra by lipopolysaccharide causes microglial activation and neurodegeneration, *Neurobiol. Dis.*, 2003, **12**, 35–45.
91. R. Barhoumi, J. Faske, X. Liu and R. B. Tjalkens, Manganese potentiates lipopolysaccharide-induced expression of NOS2 in C6 glioma cells through mitochondrial-dependent activation of nuclear factor kappaB, *Brain Res. Mol. Brain Res.*, 2004, **122**, 167–179.
92. B. Frost and M. I. Diamond, Prion-like mechanisms in neurodegenerative diseases, *Nat. Rev. Neurosci.*, 2010, **11**, 155–159.
93. M. Jucker and L. C. Walker, Self-propagation of pathogenic protein aggregates in neurodegenerative diseases, *Nature*, 2013, **501**, 45–51.
94. P. Smethurst, K. C. Sidle and J. Hardy, Review: Prion-like mechanisms of transactive response DNA binding protein of 43 kDa (TDP-43) in amyotrophic lateral sclerosis (ALS), *Neuropathol. Appl. Neurobiol.*, 2015, **41**, 578–597.
95. A. Serrano-Pozo, M. P. Frosch, E. Masliah and B. T. Hyman, Neuropathological alterations in Alzheimer disease, *Cold Spring Harbor Perspect. Med.*, 2011, **1**, a006189.
96. A. Serrano-Pozo, M. L. Mielke, T. Gomez-Isla, R. A. Betensky, J. H. Growdon, M. P. Frosch and B. T. Hyman, Reactive glia not only associates with plaques but also parallels tangles in Alzheimer's disease, *Am. J. Pathol.*, 2011, **179**, 1373–1384.
97. M. Orre, W. Kamphuis, L. M. Osborn, A. H. Jansen, L. Kooijman, K. Bossers and E. M. Hol, Isolation of glia from Alzheimer's mice reveals inflammation and dysfunction, *Neurobiol. Aging*, 2014, **35**, 2746–2760.
98. M. Matarin, D. A. Salih, M. Yasvoina, D. M. Cummings, S. Guelfi, W. Liu, M. A. Nahaboo Solim, T. G. Moens, R. M. Paublete, S. S. Ali, M. Perona, R. Desai, K. J. Smith, J. Latcham, M. Fulleylove, J. C. Richardson, J. Hardy and F. A. Edwards, A Genome-wide Gene-Expression Analysis and Database in Transgenic Mice during Development of Amyloid or Tau Pathology, *Cell Rep.*, 2015, **10**, 633–644.
99. R. Rakhit, J. Robertson, C. Vande Velde, P. Horne, D. M. Ruth, J. Griffin, D. W. Cleveland, N. R. Cashman and A. Chakrabartty, An immunological epitope selective for pathological monomer-misfolded SOD1 in ALS, *Nat. Med.*, 2007, **13**, 754–759.
100. A. Al-Chalabi, A. Jones, C. Troakes, A. King, S. Al-Sarraj and L. H. van den Berg, The genetics and neuropathology of amyotrophic lateral sclerosis, *Acta Neuropathol.*, 2012, **124**, 339–352.
101. M. Neumann, D. M. Sampathu, L. K. Kwong, A. C. Truax, M. C. Micsenyi, T. T. Chou, J. Bruce, T. Schuck, M. Grossman, C. M. Clark, L. F. McCluskey, B. L. Miller, E. Masliah, I. R. Mackenzie, H. Feldman, W. Feiden, H. A. Kretzschmar, J. Q. Trojanowski and V. M. Lee, Ubiquitinated TDP-43 in frontotemporal lobar degeneration and amyotrophic lateral sclerosis, *Science*, 2006, **314**, 130–133.

102. T. Arai, M. Hasegawa, H. Akiyama, K. Ikeda, T. Nonaka, H. Mori, D. Mann, K. Tsuchiya, M. Yoshida, Y. Hashizume and T. Oda, TDP-43 is a component of ubiquitin-positive tau-negative inclusions in frontotemporal lobar degeneration and amyotrophic lateral sclerosis, *Biochem. Biophys. Res. Commun.*, 2006, **351**, 602–611.
103. E. D. Hall, J. A. Oostveen and M. E. Gurney, Relationship of microglial and astrocytic activation to disease onset and progression in a transgenic model of familial ALS, *Glia*, 1998, **23**, 249–256.
104. G. Olmos and J. Llado, Tumor necrosis factor alpha: a link between neuroinflammation and excitotoxicity, *Mediators Inflammation*, 2014, **2014**, 861231.
105. V. Swarup, D. Phaneuf, N. Dupre, S. Petri, M. Strong, J. Kriz and J. P. Julien, Deregulation of TDP-43 in amyotrophic lateral sclerosis triggers nuclear factor kappaB-mediated pathogenic pathways, *J. Exp. Med.*, 2011, **208**, 2429–2447.
106. P. Patel, J. P. Julien and J. Kriz, Early-stage treatment with Withaferin A reduces levels of misfolded superoxide dismutase 1 and extends lifespan in a mouse model of amyotrophic lateral sclerosis, *Neurotherapeutics*, 2015, **12**, 217–233.
107. M. Jeffrey, W. G. Halliday, J. Bell, A. R. Johnston, N. K. MacLeod, C. Ingham, A. R. Sayers, D. A. Brown and J. R. Fraser, Synapse loss associated with abnormal PrP precedes neuronal degeneration in the scrapie-infected murine hippocampus, *Neuropathol. Appl. Neurobiol.*, 2000, **26**, 41–54.
108. M. Jeffrey, S. Martin, J. Barr, A. Chong and J. R. Fraser, Onset of accumulation of PrPres in murine ME7 scrapie in relation to pathological and PrP immunohistochemical changes, *J. Comp. Pathol.*, 2001, **124**, 20–28.
109. Y. Lu, A. Liu, X. Zhou, M. Kouadir, W. Zhao, S. Zhang, X. Yin, L. Yang and D. Zhao, Prion peptide PrP106-126 induces inducible nitric oxide synthase and proinflammatory cytokine gene expression through the activation of NF-kappaB in macrophage cells, *DNA Cell Biol.*, 2012, **31**, 833–838.
110. J. A. Carroll, J. F. Striebel, B. Race, K. Phillips and B. Chesebro, Prion infection of mouse brain reveals multiple new upregulated genes involved in neuroinflammation or signal transduction, *J. Virol.*, 2015, **89**, 2388–2404.
111. R. Thanan, S. Oikawa, Y. Hiraku, S. Ohnishi, N. Ma, S. Pinlaor, P. Yongvanit, S. Kawanishi and M. Murata, Oxidative stress and its significant roles in neurodegenerative diseases and cancer, *Int. J. Mol. Sci.*, 2015, **16**, 193–217.
112. M. A. Smith, G. Perry, P. L. Richey, L. M. Sayre, V. E. Anderson, M. F. Beal and N. Kowall, Oxidative damage in Alzheimer's, *Nature*, 1996, **382**, 120–121.
113. M. A. Smith, P. L. Richey Harris, L. M. Sayre, J. S. Beckman and G. Perry, Widespread peroxynitrite-mediated damage in Alzheimer's disease, *J. Neurosci.*, 1997, **17**, 2653–2657.

114. M. P. Kummer, M. Hermes, A. Delekarte, T. Hammerschmidt, S. Kumar, D. Terwel, J. Walter, H. C. Pape, S. Konig, S. Roeber, F. Jessen, T. Klockgether, M. Korte and M. T. Heneka, Nitration of tyrosine 10 critically enhances amyloid beta aggregation and plaque formation, *Neuron*, 2011, **71**, 833–844.
115. T. Kanamaru, N. Kamimura, T. Yokota, K. Iuchi, K. Nishimaki, S. Takami, H. Akashiba, Y. Shitaka, K. Katsura, K. Kimura and S. Ohta, Oxidative stress accelerates amyloid deposition and memory impairment in a double-transgenic mouse model of Alzheimer's disease, *Neurosci. Lett.*, 2015, **587**, 126–131.
116. T. Lu, L. Aron, J. Zullo, Y. Pan, H. Kim, Y. Chen, T. H. Yang, H. M. Kim, D. Drake, X. S. Liu, D. A. Bennett, M. P. Colaiacovo and B. A. Yankner, REST and stress resistance in ageing and Alzheimer's disease, *Nature*, 2014, **507**, 448–454.
117. D. A. Drechsel, A. G. Estevez, L. Barbeito and J. S. Beckman, Nitric oxide-mediated oxidative damage and the progressive demise of motor neurons in ALS, *Neurotoxic. Res.*, 2012, **22**, 251–264.
118. M. F. Beal, R. J. Ferrante, S. E. Browne, R. T. Matthews, N. W. Kowall and R. H. Brown Jr, Increased 3-nitrotyrosine in both sporadic and familial amyotrophic lateral sclerosis, *Ann. Neurol.*, 1997, **42**, 644–654.
119. R. J. Ferrante, L. A. Shinobu, J. B. Schulz, R. T. Matthews, C. E. Thomas, N. W. Kowall, M. E. Gurney and M. F. Beal, Increased 3-nitrotyrosine and oxidative damage in mice with a human copper/zinc superoxide dismutase mutation, *Ann. Neurol.*, 1997, **42**, 326–334.
120. X. Chen, X. Zhang, C. Li, T. Guan, H. Shang, L. Cui, X. M. Li and J. Kong, S-nitrosylated protein disulfide isomerase contributes to mutant SOD1 aggregates in amyotrophic lateral sclerosis, *J. Neurochem.*, 2013, **124**, 45–58.
121. B. Van Everbroeck, I. Dobbeleir, M. De Waele, E. De Leenheir, U. Lubke, J. J. Martin and P. Cras, Extracellular protein deposition correlates with glial activation and oxidative stress in Creutzfeldt-Jakob and Alzheimer's disease, *Acta Neuropathol.*, 2004, **108**, 194–200.
122. M. Guentchev, S. L. Siedlak, C. Jarius, F. Tagliavini, R. J. Castellani, G. Perry, M. A. Smith and H. Budka, Oxidative damage to nucleic acids in human prion disease, *Neurobiol. Dis.*, 2002, **9**, 275–281.
123. S. W. Yun, M. Gerlach, P. Riederer and M. A. Klein, Oxidative stress in the brain at early preclinical stages of mouse scrapie, *Exp. Neurol.*, 2006, **201**, 90–98.
124. S. Sorce, M. Nuvolone, A. Keller, J. Falsig, A. Varol, P. Schwarz, M. Bieri, H. Budka and A. Aguzzi, The Role of the NADPH Oxidase NOX2 in Prion Pathogenesis, *PLoS Pathog.*, 2014, **10**, e1004531.
125. B. Bhandary, A. Marahatta, H. R. Kim and H. J. Chae, An involvement of oxidative stress in endoplasmic reticulum stress and its associated diseases, *Int. J. Mol. Sci.*, 2012, **14**, 434–456.
126. E. Rath and D. Haller, Inflammation and cellular strcss: a mechanistic link between immune-mediated and metabolically driven pathologies, *Eur. J. Nutr.*, 2011, **50**, 219–233.

127. A. L. Goldberg, Protein degradation and protection against misfolded or damaged proteins, *Nature*, 2003, **426**, 895–899.
128. S. J. Orenstein and A. M. Cuervo, Chaperone-mediated autophagy: molecular mechanisms and physiological relevance, *Semin. Cell Dev. Biol.*, 2010, **21**, 719–726.
129. D. Ron and P. Walter, Signal integration in the endoplasmic reticulum unfolded protein response, *Nat. Rev. Mol. Cell Biol.*, 2007, **8**, 519–529.
130. Y. Ma and L. M. Hendershot, ER chaperone functions during normal and stress conditions, *J. Chem. Neuroanat.*, 2004, **28**, 51–65.
131. J. J. Hoozemans, E. S. van Haastert, P. Eikelenboom, R. A. de Vos, J. M. Rozemuller and W. Scheper, Activation of the unfolded protein response in Parkinson's disease, *Biochem. Biophys. Res. Commun.*, 2007, **354**, 707–711.
132. J. J. Hoozemans, E. S. van Haastert, D. A. Nijholt, A. J. Rozemuller, P. Eikelenboom and W. Scheper, The unfolded protein response is activated in pretangle neurons in Alzheimer's disease hippocampus, *Am. J. Pathol.*, 2009, **174**, 1241–1251.
133. J. D. Atkin, M. A. Farg, A. K. Walker, C. McLean, D. Tomas and M. K. Horne, Endoplasmic reticulum stress and induction of the unfolded protein response in human sporadic amyotrophic lateral sclerosis, *Neurobiol. Dis.*, 2008, **30**, 400–407.
134. J. A. Moreno, H. Radford, D. Peretti, J. R. Steinert, N. Verity, M. G. Martin, M. Halliday, J. Morgan, D. Dinsdale, C. A. Ortori, D. A. Barrett, P. Tsaytler, A. Bertolotti, A. E. Willis, M. Bushell and G. R. Mallucci, Sustained translational repression by eIF2alpha-P mediates prion neurodegeneration, *Nature*, 2012, **485**, 507–511.
135. J. M. Axten, J. R. Medina, Y. Feng, A. Shu, S. P. Romeril, S. W. Grant, W. H. Li, D. A. Heerding, E. Minthorn, T. Mencken, C. Atkins, Q. Liu, S. Rabindran, R. Kumar, X. Hong, A. Goetz, T. Stanley, J. D. Taylor, S. D. Sigethy, G. H. Tomberlin, A. M. Hassell, K. M. Kahler, L. M. Shewchuk and R. T. Gampe, Discovery of 7-methyl-5-(1-{[3-(trifluoromethyl)phenyl] acetyl}-2,3-dihydro-1H-indol-5-yl)-7H-pyrrolo[2,3-d]pyrimidin-4-amine (GSK2606414), a potent and selective first-in-class inhibitor of protein kinase R (PKR)-like endoplasmic reticulum kinase (PERK), *J. Med. Chem.*, 2012, **55**, 7193–7207.
136. J. A. Moreno, M. Halliday, C. Molloy, H. Radford, N. Verity, J. M. Axten, C. A. Ortori, A. E. Willis, P. M. Fischer, D. A. Barrett and G. R. Mallucci, Oral treatment targeting the unfolded protein response prevents neurodegeneration and clinical disease in prion-infected mice, *Sci. Transl. Med.*, 2013, **5**, 206ra138.
137. H. J. Kim, A. R. Raphael, E. S. LaDow, L. McGurk, R. A. Weber, J. Q. Trojanowski, V. M. Lee, S. Finkbeiner, A. D. Gitler and N. M. Bonini, Therapeutic modulation of eIF2alpha phosphorylation rescues TDP-43 toxicity in amyotrophic lateral sclerosis disease models, *Nat. Genet.*, 2014, **46**, 152–160.

138. J. F. Abisambra, U. K. Jinwal, L. J. Blair, J. C. O'Leary 3rd, Q. Li, S. Brady, L. Wang, C. E. Guidi, B. Zhang, B. A. Nordhues, M. Cockman, A. Suntharalingham, P. Li, Y. Jin, C. A. Atkins and C. A. Dickey, Tau accumulation activates the unfolded protein response by impairing endoplasmic reticulum-associated degradation, *J. Neurosci.*, 2013, **33**, 9498–9507.
139. L. Devi and M. Ohno, PERK mediates eIF2alpha phosphorylation responsible for BACE1 elevation, CREB dysfunction and neurodegeneration in a mouse model of Alzheimer's disease, *Neurobiol. Aging*, 2014, **35**, 2272–2281.
140. T. Ma, M. A. Trinh, A. J. Wexler, C. Bourbon, E. Gatti, P. Pierre, D. R. Cavener and E. Klann, Suppression of eIF2alpha kinases alleviates Alzheimer's disease-related plasticity and memory deficits, *Nat. Neurosci.*, 2013, **16**, 1299–1305.
141. H. A. Lashuel, C. R. Overk, A. Oueslati and E. Masliah, The many faces of alpha-synuclein: from structure and toxicity to therapeutic target, *Nat. Rev. Neurosci.*, 2013, **14**, 38–48.
142. A. Bellucci, L. Navarria, M. Zaltieri, E. Falarti, S. Bodei, S. Sigala, L. Battistin, M. Spillantini, C. Missale and P. Spano, Induction of the unfolded protein response by alpha-synuclein in experimental models of Parkinson's disease, *J. Neurochem.*, 2011, **116**, 588–605.
143. E. Colla, P. Coune, Y. Liu, O. Pletnikova, J. C. Troncoso, T. Iwatsubo, B. L. Schneider and M. K. Lee, Endoplasmic reticulum stress is important for the manifestations of alpha-synucleinopathy *in vivo*, *J. Neurosci.*, 2012, **32**, 3306–3320.
144. M. S. Gorbatyuk, A. Shabashvili, W. Chen, C. Meyers, L. F. Sullivan, M. Salganik, J. H. Lin, A. S. Lewin, N. Muzyczka and O. S. Gorbatyuk, Glucose regulated protein 78 diminishes alpha-synuclein neurotoxicity in a rat model of Parkinson disease, *Mol. Ther.*, 2012, **20**, 1327–1337.
145. K. Hashida, Y. Kitao, H. Sudo, Y. Awa, S. Maeda, K. Mori, R. Takahashi, M. Iinuma and O. Hori, ATF6alpha promotes astroglial activation and neuronal survival in a chronic mouse model of Parkinson's disease, *PLoS One*, 2012, **7**, e47950.
146. S. Angeli, T. Barhydt, R. Jacobs, D. W. Killilea, G. J. Lithgow and J. K. Andersen, Manganese disturbs metal and protein homeostasis in Caenorhabditis elegans, *Metallomics*, 2014, **6**, 1816–1823.
147. A. Araque, G. Carmignoto, P. G. Haydon, S. H. Oliet, R. Robitaille and A. Volterra, Gliotransmitters travel in time and space, *Neuron*, 2014, **81**, 728–739.

CHAPTER 8

Redox Signalling in Dopaminergic Cell Death and Survival

AJIT RAY, ADITI VERMA AND VIJAYALAKSHMI RAVINDRANATH*

Centre for Neuroscience, Indian Institute of Science, Bangalore–560012, India
*E-mail: viji@cns.iisc.ernet.in

8.1 Neuronal Degeneration in Parkinson's Disease

Neurodegenerative diseases, such as Parkinson's (PD), Alzheimer's and motor neuron disease affect specific cell populations selectively in certain brain regions. Since there is poor understanding of the molecular pathways that lead to selective degeneration, there is a need to examine survival and death signalling pathways within specific cell populations. Oxidative stress has been implicated in a variety of neurodegenerative disorders, but it is debated whether it is a cause or consequence of the degenerative process. However, in Parkinson's disease, wherein selective loss of DArgic (DAergic) neurons is seen, oxidative stress has been implied to be causative in nature.

Issues in Toxicology No. 34
Oxidative Stress and Redox Signalling in Parkinson's Disease
Edited by Rodrigo Franco, Jonathan A. Doorn and Jean-Christophe Rochet

Published by the Royal Society of Chemistry, www.rsc.org

8.1.1 Relatively Selective Dopaminergic Degeneration

The classical neuropathological hallmark of PD is loss of dopaminergic (DAergic) neurons in the ventral midbrain coupled with formation of cytoplasmic proteinaceous inclusion bodies, referred to as Lewy bodies.[1] While non-DAergic neurons are also reported to be affected in the disease process, such as neurons from the anterior olfactory nucleus, dorsal nucleus of the vagal nerve, nucleus basalis of Meynert, locus coeruleus and raphe nuclei, and may partly be responsible for the extranigral symptoms exhibited in PD patients,[2,3] the extent of cell death in these brain structures appears to be more variable with respect to severity of core symptoms of the disease.[4,5] Indeed, neuronal loss in some of these non-DAergic nuclei has been reported in other neurodegenerative conditions as well such as Alzheimer's disease, dementia with Lewy bodies and incidental Lewy-body disease.[6]

Among the several DAergic populations in the brain, the neurons situated in the *substantia nigra pars compacta* region (SNpc; A9 group) of the ventral midbrain appear to be the most vulnerable[7] while the nearby DAergic nuclei, such as the A8 (midbrain reticular formation) and A10 (ventral tegmental area; VTA) are relatively spared till later in the disease process.[8,9] Even within the SNpc, the ventrolateral tier shows the greatest neuronal loss (70–90%), followed by the medioventral and dorsal tiers (25–70%).[8,10] The loss of cell bodies in the *substantia nigra* (SN) is associated with loss of DAergic fibers in the striatum, along with severe reduction in DA levels. DAergic cell death in PD is accompanied and often preceded by a gamut of pathological events, such as increased oxidative stress, deficits and dysfunction of mitochondrial OXPHOS, deficits in protein turnover, dysfunction of ubiquitin proteasomal system, autophagy dysregulation, endoplasmic reticulum stress, and inflammation.[11–14] While such stressors could indeed lead to cell death, it is difficult to determine why only certain neuronal populations of the brain would die during the disease. Such specificity or selective vulnerability could arise from both cell autonomous and noncell autonomous mechanisms, with DAergic neurons possessing an idiosyncratic molecular phenotype rendering them more susceptible to such stressors and the extracellular milieu of the cells generating higher levels of such stress, respectively. In this chapter, we discuss how oxidative stress and the resultant activation of cellular signalling pathways in DAergic neurons can contribute to selective neuronal cell death seen in PD.

8.1.2 Sources of Oxidative Stress and Selective Vulnerability

In this section, we describe the factors that could play a part in the remarkable specificity of cell death observed in PD pathology. While the list is not exhaustive in nature, we have attempted to discuss factors that directly or indirectly result in increase of reactive oxygen species (ROS) and/or inability to handle oxidative stress in a cell type-specific manner, thus providing insight to understand the causes and effects of differential susceptibility (Figure 8.1).

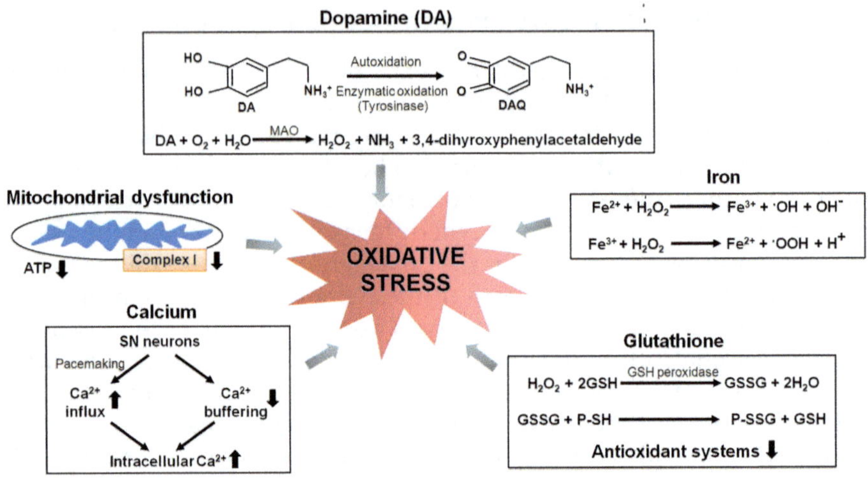

Figure 8.1 Sources of oxidative stress in dopaminergic neurons. DAergic neurons face increased oxidative stress and ROS generation due to several factors. Firstly, auto-oxidation or enzymatic oxidation of DA yields DAQ, which can undergo redox cycling. DA metabolism through MAO also leads to generation of H_2O_2. Secondly, Fe^{2+} accumulation in DAergic neurons increases the generation of hydroperoxyl and hydroxyl radicals *via* Fenton reaction. Thirdly, GSH levels are lower in the SN and they further decrease with age and/or progression of PD, which can impair antioxidant capability and thiol homeostasis in affected neurons. This can lead to the formation of GSSG as well as glutathionylated protein residues (P-SSG). Fourthly, pacemaking in SNpc neurons is accompanied with constant Ca^{2+} entry into these cells. SN neurons also have poor Ca^{2+} buffering systems. These two factors together lead to an increase in Ca^{2+} levels in the cells and lead to ROS generation. Finally, mitochondrial dysfunction caused by complex I inhibition causes ATP loss and also further generation of ROS.

8.1.2.1 Dopamine Metabolism

The preferential loss of DAergic neurons compared to other populations directly indicates that the chemical nature of the neurotransmitter itself could be partly responsible for the pathology. Dopamine (DA) has a catechol group that renders it amenable to redox cycling and generation of ROS. Further, metabolism of DA by monoamine oxidase (MAO) generates one mole of hydrogen peroxide (H_2O_2) per mole of DA turned over. Thus, DAergic neurons can potentially generate ROS through two major pathways – enzymatic oxidation through the DA metabolism pathway involving breakdown of DA to 3,4-dihydroxyphenylacetic acid and homovanillic acid by MAO leading to formation of H_2O_2 as a byproduct,[15–17] and through spontaneous auto-oxidation of DA to DA-*o*-quinone and dopaminochromes that undergo redox cycling and generate superoxide ($O_2^{\bullet-}$) as well.[18,19] DA can also directly target cysteinyl residues in proteins and antioxidants, such as glutathione (GSH) forming DA-cysteinyl target adducts leading to inactivation of proteins including tyrosine hydroxylase (TH), and Parkin, and can even alter α-synuclein fibrillization.[20] Indeed, the load from the oxidative metabolism of DA may increase during PD pathogenesis since DA turnover increases in surviving neurons to maintain the neurotransmitter levels as seen from studies using both human autopsy samples and tissue from animal models.[21,22] From these studies, it appears that DAergic neurons in the brain face higher basal oxidative stress and this could contribute to their vulnerability.

On the other hand, it is less clear as to why certain sub-populations of these neurons are rendered vulnerable. This can potentially stem from differential expression of two DA uptake-related proteins. DA transporter (DAT) is responsible for reuptake of extrasynaptic DA back into DAergic cells. Interestingly, differential DAT expression is seen in the midbrain DAergic populations with the ventral tier in SNpc expressing higher levels of the transporter.[23–25] This increased reuptake could render these cells susceptible to DA-induced ROS generation, as described earlier. Indeed, DAergic neuronal populations from DAT knockout animals are protected from DAergic toxins, such as the complex I inhibitor 1-methyl-4-phenyl-1,2,3,6-tetrahydropyridine (MPTP).[26,27] While DAT leads to DA build-up in the cytoplasm of DAergic cells, vesicular monoamine transporter (VMAT) sequesters DA into synaptic vesicles from the cytoplasm, preventing an increase of cytosolic DA that could be a target of metabolism.[28,29] Remarkably, unlike DAT, the dorsal SNpc neurons express higher levels of VMAT2.[24] This could, in turn, make less DA available for redox cycling and/or limit the ROS production to the vesicles. Supporting this hypothesis, VMAT2 knockout animals have been demonstrated to show enhanced DAergic degeneration when treated with MPTP.[30,31] Further evidence suggests differential expression of DAT and VMAT2 expression in SNpc and VTA neurons in the rat midbrain.[32,33] It will be interesting to examine whether these differences could contribute to differential intracellular DA load in these cells as well. Thus, it can be seen that expression of DA transporters coupled with the oxidative nature of DA metabolism might render differential vulnerability not only of the broad neurotransmitter-type neurons, but its sub-populations as well. DA oxidation is covered in Chapter 5 in greater detail.

8.1.2.2 Mitochondrial Dysfunction

Mitochondria, the powerhouses of the cell, encode and express several electron carriers that participate in the generation of adenosine triphosphate (ATP) by employing oxidative phosphorylation (OXPHOS) through sequential transfer of electrons in the electron-transport chain (ETC).[34] Among the several cellular pathways regulated by mitochondria, they play essential roles in calcium (Ca^{2+}) homeostasis and regulation of cell death through apoptosis. The link between PD and dysfunctional mitochondria was first established when people abusing the drug meperidine accidentally administered themselves with the synthetic byproduct MPTP, leading to the manifestation of Parkinsonian symptoms and degeneration of SN neurons.[35]

1-methyl-4-phenyl pyridinium (MPP^+), the toxic metabolite of MPTP, was later found to inhibit mitochondrial complex I (ETC component).[36,37] Later, Schapira *et al.* also observed a specific decrease in the activity of complex I in the SN from postmortem samples from sporadic PD patients.[38] Complex I inhibition can cause loss of ATP as well as leakage of electrons from complex I directly to O_2, resulting in generation of ROS.[39–41] In turn, complex I itself can be inhibited further through ROS-mediated oxidative modifications,[42,43] setting up a feedback cycle. The initial inhibition of complex I activity may also spring from ROS generated through DA redox cycling of DA-*o*-quinones, as discussed earlier.[44] Many of the familial PD associated gene mutations including those in PTEN-induced putative kinase 1 (PINK1), E3 ubiquitin protein ligase (Parkin), DJ-1 (PARK7), α-synuclein, and leucine-rich repeat kinase 2 (LRRK2) result in altered protein functions that negatively impact on mitochondrial function.[45] This load may also lead to generation of ROS from the affected mitochondria. In histochemical studies from postmortem tissue of PD patients, marked decrease of cytochrome c oxidase (COX) has been reported as compared to age-matched controls.[46] The study further reported an increase in the levels of mitochondrial DNA (mtDNA) deletions in laser microdissected SN neurons from the patients of PD as compared to age-matched controls. Interestingly, this study showed that among the neurons of SN from patients of PD, the COX-deficient neurons show greater levels of somatic mitochondrial DNA (mtDNA) deletions as compared to neurons with normal COX activity. Thus, the dysregulation of mitochondria in PD is probably a mechanism specific to DAergic neurons in SN, and may explain the increased oxidative load in these cells. Mitochondrial dysfunction has been discussed in greater detail in Chapter 3.

8.1.2.3 Iron Accumulation

Increased iron (Fe^{2+}) has been linked to the etiology of PD as it can stimulate ROS generation through Fenton reaction.[47,48] Increased Fe^{2+} levels have been demonstrated in parkinsonian SN,[49,50] and in individual DAergic neurons.[51] Indeed, in the MPTP mouse model of PD, drug treatment leads to an increase in Fe^{2+} levels in SNpc neurons, while VTA neurons remained unaffected.[52] Levels of the Fe^{2+}-binding protein ferritin are suggested to be a cause of increased Fe^{2+} storage and are reported to be lowered during the disease process.[49,53] This change in ferritin expression could release highly reactive divalent iron in cells.[54] High levels of neuromelanin in the SNpc neurons have been implicated in Fe^{2+} accumulation in this region. Neuromelanin granules sequester iron in its ferric form and could potentially prevent the redox cycling and generation of ROS.[55,56] The drastic decrease in neuromelanin levels during PD disease process could release this Fe^{2+} and cause oxidative stress in the neurons.[57] The high levels of iron in the SN and the differences reported between SNpc and VTA neurons could contribute to the differential vulnerability seen amongst neuronal populations in the disease process. Iron metabolism in PD will be covered in more detail in Chapter 9.

8.1.2.4 Calcium Channel Expression and Calcium Homeostasis

Both A9 neurons of SNpc and A10 neurons in VTA exhibit pacemaker-like activity in rodents.[58,59] While VTA neurons use non-voltage and voltage-dependent sodium (Na^+) channels during their pacemaking,[60] adult A9 neurons use L-type Ca^{2+} channels, specifically $Ca_v1.3$ type Ca^{2+} channels.[61] Increased intracellular Ca^{2+} from the pacemaker-induced influx as well as release from intracellular stores in response to other stimuli, such as activation of metabotropic glutamate receptors[62,63] can lead to increased Ca^{2+} load in DAergic neurons. Coupled with this, studies have shown that dying SNpc neurons also have lower levels of Ca^{2+}-binding proteins, such as calbindin, parvalbumin and calretinin suggesting poor Ca^{2+} buffering.[64,65] To handle this excess, the cells extrude Ca^{2+} consuming ATP generated primarily through OXPHOS from the mitochondria.[66,67] This process is inherently linked with generation of ROS, such as $O_2^{\bullet-}$.[68,69] Increased oxidative load can in turn lead to cell death and could potentially explain the differential vulnerability of SNpc DAergic neurons compared to VTA seen in PD. Further evidence for pathological effects of L-type Ca^{2+} channel-dependent pacemaking comes from reliance of two other extranigral nuclei adversely affected in PD.[70,71] While no increase in $Ca_v1.3$ messenger mRNA was seen in SNpc from PD patient brains compared to controls, SN did show the highest levels of this channel expression compared to other brain regions studied.[72] Epidemiological studies show that use of Ca^{2+}-channel antagonists for treating hypertension is associated with lower incidence of PD,[73–75] and a phase III clinical trial is currently underway (STEADY-PD).

8.1.2.5 Glutathione Loss and Associated Thiol Perturbations

While the earlier factors explain how increased oxidative stress may occur in DAergic neurons as a result of ROS generation, the same may also occur through a decreased antioxidant capacity of the cells. GSH is the most abundant non-protein thiol in the cell and one of the most critical components of the antioxidant defense systems of the cell. A dysfunction of this system can render the cells unable to handle increased ROS levels, and the oxidative stress generated could activate adverse cellular pathways, possibly triggering cell death. Neurons contain lower levels of GSH compared to glia, specifically astrocytes, and depend on astrocytes for the supply of substrates for GSH synthesis,[76,77] and this synthesis decreases with age in the brain.[78] The lower levels of GSH can thus make neurons more vulnerable to aging-related increased oxidative stress compared to glia. As described earlier, DA metabolism generates ROS such as H_2O_2. These ROS are detoxified by cells using GSH.[16,79] Further sources of GSH consumption in the neuronal cells can arise from oxidative and nitrosative stress due to mitochondrial dysfunction and impaired complex I activity (described earlier), Fe^{2+} accumulation (described earlier), increase in Nicotinamide adenine dinucleotide phosphate (NADPH) oxidase (NOX)[80] and nitric oxide synthase (NOS)[81] activities in microglia during PD pathogenesis and many other factors.

Interestingly, the human SN has lower levels of GSH compared to other brain regions and this is drastically reduced in PD patients.[82,83] Indeed, not only do the total levels of GSH change but the ratio of reduced:oxidized glutathione (GSH/GSSG) is also decreased in autopsy tissues from PD patients and PD models.[77,84–87] Moreover, the activity of GSH peroxidase, an enzyme that detoxifies peroxides in the cell using GSH as a substrate, is also reduced in PD patient brains.[88] Furthermore, GSH peroxidase mRNA expression is also higher in medial SNpc neurons compared to lateral SNpc neurons that are more susceptible to death.[89] SNpc neurons, which have low levels of GSH, may be more adversely affected by microglial ROS and nitric oxide (˙NO) generation in response to an inflammatory stimulus compared to other neuronal populations, as this brain region has the highest density of microglia.[90] GSH is also important for the normal functioning of other antioxidant systems, such as glutaredoxin (Grx) and thioredoxin (Trx). In later sections, we describe how these systems may be compromised in PD. It is thus apparent that maintenance of GSH levels is critical to neuronal survival; more so in DAergic neurons that have intrinsically enhanced levels of oxidative stress. Lowering of GSH in these neurons through multiple mechanisms renders them more susceptible to both generators and targets of oxidative stress-induced cell death. Chapter 6 discusses GSH and thiol redox signalling in more detail.

8.1.3 Increased Oxidative Stress and Selective Cell Death

While it is increasingly becoming clear that oxidative stress plays a role in both upstream and downstream events during PD pathogenesis, its role in the selective degeneration of DAergic neurons is yet unclear. Thus, oxidative stress could negatively impact a variety of cellular processes, such as ubiquitin-proteasomal dysfunction, endoplasmic reticulum stress, autophagy dysfunction and inflammation, in addition to those mentioned above. Based on their reactivity and charge, ROS and reactive nitrogen species (RNS) can cross membranes resulting in diffuse downstream effects of ROS across subcellular compartments and indeed across cells (paracrine effects).[91–93] If oxidative stress has such diffuse effects, it is difficult to ascertain how it may cause selective cell death. Several lines of evidence have demonstrated the neuroprotective efficacy of antioxidant treatment in animal models of PD, such as coenzyme Q_{10} in the MPTP model[94] and tocopherol in the 6-hydroxydopamine (6-OHDA) model.[95] However, it is interesting to note here that the major antioxidant therapies so far tested in PD patients, such as ascorbate, tocopherol, coenzyme Q_{10} and N-acetylcysteine have met with little success.[96–99] Thus, ROS inhibition alone does not appear to prevent further DAergic cell death once the disease process ensues. The question thus arises as to how ROS and indeed oxidative stress, can trigger signalling pathways in a cell-specific fashion, and if this is related to DAergic degeneration in PD and PD models.

8.2 Redox Signalling and Cell Survival/Death

Amino acid residues on proteins are amongst several targets of ROS and RNS and their oxidation results in altering their structure and function.[100] Oxidative stress is handled by cellular antioxidant defense systems including non-protein thiols, such as GSH and enzymes, such as superoxide dismutases (SOD), catalase, several peroxidases and thiol disulfide oxidoreductases.[91,101] Indeed, many of these proteins act as the first line of defense against such stressors because of their highly reactive cysteine (Cys) residues.[91,100] These proteins not only act as enzymes to reduce ROS and ROS-related oxidative modification of proteins in general, but also play important roles in cellular signalling.[102–105] This is achieved by regulating the oxidation status of critical Cys residues on key signalling proteins (collectively referred to as redox signalling). Oxidative modification of these proteins can either enhance or inhibit their function and thus alter downstream signalling potentially shifting cellular fate from survival to death.[106]

We are increasingly finding evidence for such events in normal cellular function and in disease.[107,108] Earlier in the chapter, we have described how brain regions vulnerable to PD-associated degeneration, especially DAergic neurons of the SNpc, face high levels of oxidative stress. While other cells are also exposed to such stress due to long-range effects of ROS, they may be differentially affected due to altered distinct antioxidant and redox signalling system. Redox signalling proteins can act as gate-keepers to initiate signalling cascades in a cell-type specific manner. Death of DAergic neurons can occur both due to suppression of cell-survival mechanisms and/or activation of cell-death pathways. For the remainder of the chapter, we focus on how redox signalling can potentially modify two candidate pathways, which are involved in selective degeneration of DAergic neurons, namely, inhibition of Akt-mediated cell survival signalling and activation of the death signaling pathway by apoptosis signal-regulating kinase 1 (ASK1) MAP3K.

8.3 Redox Regulation of DAergic Cell-Survival Pathways

The Akt signalling pathway is a major cell-survival mechanism of the cell.[109] Oxidative stress is known to repress this pathway, thereby inducing cell death. Dysfunction of this pathway has been extensively shown in PD brain autopsy samples as well as in PD models.[110–112] Therefore, we describe Akt structure, function and regulation in greater detail in the subsequent sections with special emphasis on its redox regulation and involvement in DAergic cell death.

8.3.1 Akt Structure and Function

Akt (also known as protein kinase B) is a serine/threonine kinase belonging to the AGC family of kinases and is a homolog of the AKT8 viral oncogene v-akt.[113,114] It plays important roles in cell survival, growth, proliferation, and metabolism. The N-terminal of Akt has a pleckstrin homology domain (PHD) followed by a short kinase domain and a C-terminal regulatory domain.[109] There are three known forms of Akt kinases – Akt1, Akt2 and Akt3, also known as PKBα, PKBβ and PKBγ respectively – that are encoded by three genes.[115,116] The proteins Akt2 and Akt3 show 81% and 83% homology with Akt1 respectively.[116] Each of the three isoforms has the N-terminal PHD, the kinase domain and the C-terminal regulatory domain. Of the three isoforms, Akt1 has been shown to be most abundantly expressed in the SN.[117]

Akt1 is phosphorylated upon activation of phosphatidylinositol (PI)-3-kinase (PI3K) on the plasma membrane in response to insulin and various growth factors.[118,119] The enzyme PI3K catalyzes the conversion of PI-3,4-bisphosphate (PIP_2) to the lipid secondary messenger PI-3,4,5-trisphosphate (PIP_3). PIP_3 then recruits Akt1 to the plasma membrane through interaction with its PHD[120] and thus initiates its activation. However, the phosphorylation of Akt1 at threonine (Thr) 308 in the activation loop of the kinase domain and serine (Ser) 473 in the hydrophobic motif of the regulatory domain is crucial for its activation.[116,118] This discovery was made in view of the fact that Akt1 is inactivated upon dephosphorylation by serine/threonine-specific protein phosphatase 2A (PP2A) but not tyrosine-specific protein phosphatases.[121] The phosphorylation of Akt1 at Thr308 is mediated by 3-phosphoinositide-dependent kinase 1 (PDK1) that is also recruited to the plasma membrane and activated through interaction of its PH domain with PIP_3.[116,122,123] This leads to partial activation of Akt.[124] The phosphorylation of the Ser473 in the regulatory domain hydrophobic motif is catalyzed by the mammalian target of rapamycin complex, mTORC2[125] (Figure 8.2).

Inhibition of lipid phosphatase and tensin homolog on chromosome 10 (PTEN) is known to lead to hyperactivation of Akt1 in tumor cells through increase in PIP_3 levels.[126] PTEN inhibition through redox modification can lead to an increase in Akt1 activity in mouse embryonic fibroblasts.[127] Interestingly, a study in PTEN null mouse embryonic fibroblasts has shown that reducing levels of $O_2^{\bullet-}$ increases dephosphorylation of cytosolic Akt at Thr308 and Ser473 through its interaction with PP2A.[128] Zhang *et al.* demonstrated that PTEN deletion enhances DAergic survival in the 'MitoPark' mouse model of PD, which has mitochondrial transcription factor A (TFAM) selectively knocked out in DAergic neurons.[129]

Various pathways downstream of Akt1 help mediate its role in cell survival. The negative regulation of the proapoptotic Bcl–2 homology domain 3 (BH3)-only proteins, such as Bad through direct phosphorylation or indirectly through the transcription factors such as FOXO family and p53, help in cell survival. Other well-known prosurvival pathways include inhibiting glycogen synthase kinase 3 (GSK3) isoforms, activating nuclear factor-κB (NF-κB) signalling or inhibition of c-Jun kinase (JNK)/p38 signalling.[109] Nuclear translocation of NF-κB is increased in postmortem tissues from patients of PD as compared to their age-matched controls.[130] However, activation of NF-κB by Akt directly or indirectly by phosphorylation of IκB, leading to its dissociation and nuclear translocation of NF-κB, has been implicated in cell survival.[101]

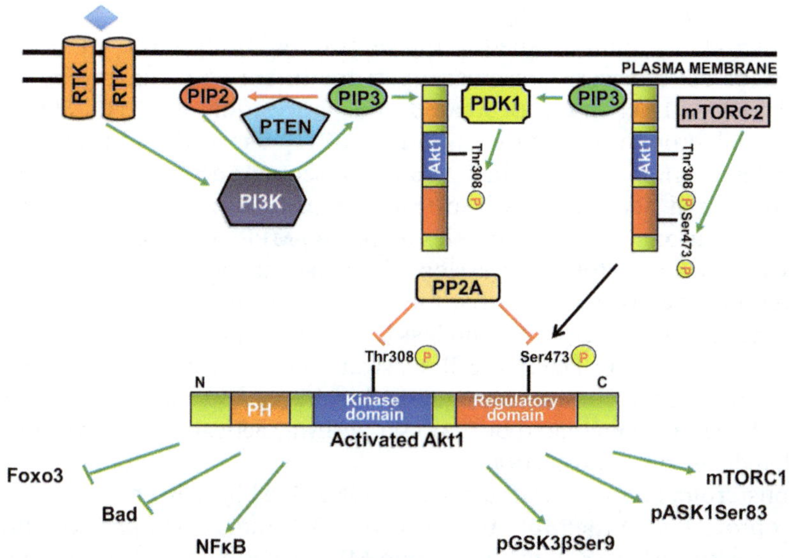

Figure 8.2 Regulation of survival through Akt1 activation. Receptor tyrosine kinases (RTK) are activated in response to insulin and other growth factors. This leads to activation of PI3K that converts phosphatidylinositol-3,4-bisphosphate (PIP_2) to phosphatidylinositol-3,4,5-trisphosphate (PIP_3) and recruits Akt1 and PDK1 to the plasma membrane. PDK1 phosphorylates Akt1 at Thr308. Complete activation of Akt1 is achieved by subsequent phosphorylation of Ser473 residue by mTORC2. Activated Akt1 can regulate cell survival through several downstream molecules, such as Foxo3, GSK3β, NF-κB, ASK1 and mTORC1. Dephosphorylation of PIP_3 by PTEN and dephosphorylation of Akt1 directly by PP2A inactivates Akt1-signalling. N and C at the ends of the Akt1 molecule indicate the N- and C-termini of the protein. PH in the Akt1 molecule refers to the pleckstrin homology domain.

GSK3 inhibition is known to be protective in PD.[131] Further, FOXO3A phosphorylation is abrogated in 6-OHDA model of PD and this induces dysregulation of proapoptotic Bcl-2, Bax and caspase (cysteine-aspartic protease) 3.[132]

8.3.2 Evidence of Akt1 Involvement in PD

Akt is a critical factor for neuronal survival.[133,134] The levels of Akt phosphorylation (pAkt) at Ser473[112] and also at Thr308 have been shown to be diminished in the melanized neurons from autopsy brain tissue of PD.[111] Furthermore, a decrease in pAkt levels was concomitant with an increase in the levels of RTP801/REDD1 (*re*gulated in *d*evelopment and *D*NA *d*amage responses), a negative regulator of mTOR activity, which is involved in Akt phosphorylation. Loss of Akt activity was also confirmed by the decrease in the levels of its downstream effector molecules pSer136 Bad and pSer253

FOXO3A.[111] Further work by the same group has shown that the levels of pThr308 Akt were diminished in the ventral midbrain in the MPTP mouse model of PD but were preserved upon treatment with rapamycin, which blocks RTP801 activity.[135] Decreased phosphorylation of mTOR, S6K1 (p70 ribosomal protein S6 kinase 1) and 4E-BP1 (eukaryotic translation initiation factor 4E-binding protein 1) and subsequent activation of caspase 3 and increased neuronal death in response to treatment of PC12 cells and primary neurons with PD mimetics (6-OHDA, MPP^+ and rotenone) was partially reversed by ectopic expression of a constitutively active form of Akt.[136] Expression of constitutively active form of Akt in 6-OHDA mouse model of PD protected against DAergic neuron loss that further reinforces the importance of the role of Akt in DAergic cell survival.[117] Furthermore, in the 6-OHDA model, neuroprotection by estrogen and insulin growth factor-1 (IGF-1) has been shown to be mediated by PI3K/Akt pathway activation as Akt inhibition abolishes the neuroprotection.[137]

Nonsteroidal anti-inflammatory drugs (NSAIDs) have been reported to be neuroprotective in the MPP^+ model. The NSAID meloxicam protects human neuroblastoma SH-SY5Y cell loss upon MPP^+ treatment by suppressing the loss of pAkt. Meloxicam also ameliorated behavioral symptoms of PD and reversed the loss of DAergic neurons by preventing the decrease in pAkt levels in mouse midbrain after chronic treatment with MPTP.[138] The neuroprotective effects of caffeine are also suggested to be mediated by activation of the PI3K/Akt pathway in 6-OHDA treated SH-SY5Y cells.[139] Thus, Akt1 signalling is downregulated in PD, and we next discuss the mechanism(s) underlying the dysfunction of this important cell-survival pathway.

8.3.3 Redox Regulation of Akt1 Activity

The study of the crystal structure of inactive Akt2 showed the presence of a redox-sensitive disulfide bond in the kinase domain of Akt2.[140] Formation of a disulfide bond between Cys297 towards the N-terminus of the activation loop of the kinase domain and the Cys311 towards the C-terminus in the human Akt2 was proposed, which was reduced when the crystal was soaked in 20M DTT.[140] Different forms of Akt could potentially have different redox sensitivities.[141] Further work in cardiomyocytes, H9c2 cells, has shown the formation of disulfide bond between the aforementioned Cys residues in response to oxidative stress. The disulfide bond formation is accompanied by dephosphorylation due to increased association of Akt with PP2A. Oxidation of Akt and its increased association with PP2A could be reversed by overexpression of the thiol disulfide oxidoreductase Grx1 in the system.[142] Furthermore, exposure to estrogen 17β-estradiol (E_2) protected these cells from oxidative stress-mediated apoptosis by inducing the expression of Grx1 and maintaining the reduced state of Akt, thereby preserving its activity.[143] Similarly, 3,4-dihyroxybenzalacetone was found to be protective against 6-OHDA-mediated neurotoxicity in SH-SY5Y cells by activating Akt and inducing GSH synthesis.[144]

Does a similar redox modification take place in DAergic neurons? Indeed, redox modification of Akt1 in DAergic neurons in the MPTP mouse model of PD leads to inactivation of Akt1. In the ventral midbrain from MPTP-treated mice, we have observed a decrease in phosphorylated levels of Akt1 at both Thr308 and Ser473 positions, as well as its kinase activity, which attenuated downstream signalling, such as phosphorylation of ASK1 at Ser83 and GSK3β at Ser9 positions, resulting in reduced nuclear β-catenin. The loss of pAkt1 was triggered by its increased oxidation, resulting in enhanced association with PP2A. Interestingly, coimmunohistochemistry showed that pAkt levels were decreased in DAergic neurons in the SNpc following a single dose of MPTP to mice. Furthermore, pretreatment of animals with antioxidants, such as GSH-ethyl ester prior to MPTP treatment provided protection against Akt1 oxidation and attenuated its association with PP2A (Figure 8.3), thus maintaining the cell survival signalling of Akt1.[110]

When human primary neurons derived from progenitor cells were exposed to MPP^+, Akt1 phosphorylation was decreased in a manner similar to that seen in mice. However, overexpression of Grx1 prevented the loss of pAkt1.[110] Thus, in PD models, Akt1 signalling is diminished in DAergic neurons through the oxidation of Cys residues in Akt1, leading to its increased association with the phosphatase PP2A. The underlying

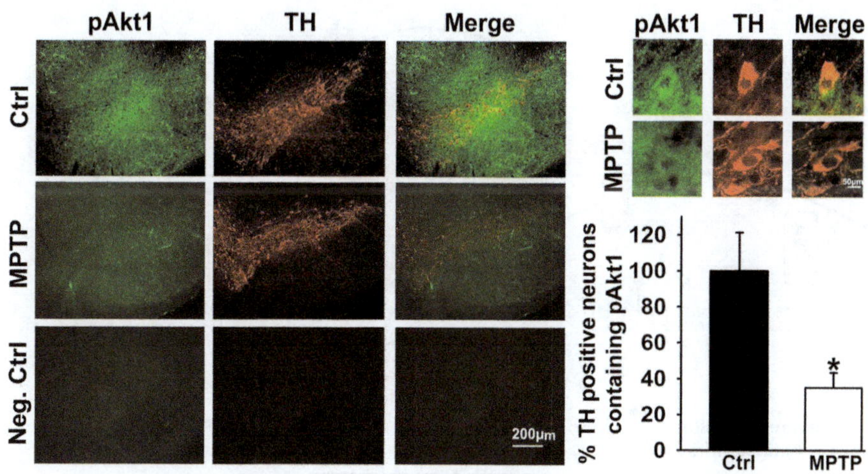

Figure 8.3 Loss of pAkt1 in DAergic neurons of SNpc after MPTP. Animals were sacrificed 12 h after MPTP or vehicle (Ctrl) injection and brains post-fixed in paraformaldehyde were sectioned through the midbrain and processed for TH and phospho-Akt1 (pAkt1) immunohistochemistry. Neg. Ctrl refers to negative control for staining. The staining in SNpc is shown in the figure at both low magnification (scale bar = 200 µm) and high magnification (scale bar = 50 µm). The quantification of staining for pAkt1 in TH-positive neurons shows a smaller number of DAergic neurons having detectable pAkt1 following MPTP administration. Data are represented as mean ± SD ($n = 4$). Asterisk indicates $p < 0.05$. The authors thank Dr Lalitha Durgadoss for providing the unpublished data for this figure.

mechanism of the loss of pAkt was shown to be due to oxidative modification of two critical Cys296 and Cys310 in the activation loop of the kinase domain of Akt1.[145] Using mouse-derived neuroblastoma cells (Neuro-2a), this study also showed that the mutation of these Cys residues to Ser *per se* did not affect the kinase activity of Akt1. In addition, when the cells were treated with cadmium, a potent inhibitor of thiol disulfide oxidoreductases, such as Grx1, the mutation of Akt1 cysteines prevented the oxidation and hence its increased association with PP2A. Inhibition of Grx1 by cadmium also led to increased proteasomal degradation of Akt1, which was rescued when the critical Cys in Akt1 were mutated to Ser. This indicates that Akt1 is also degraded by proteasome in a redox-dependent manner in Neuro-2a cells, although this was not observed in mouse brain SNpc following MPTP treatment. These studies demonstrate the importance of the critical Cys296 and 310 in Akt1 in maintaining its signalling[145] (Figure 8.4).

We can conclude that Akt1 kinase activity is essential for survival of DAergic neurons. Inhibition of this pathway in PD models and human PD brain tissue through oxidation of critical cysteines can lead to attenuation of Akt1 kinase activity thus downregulating cell-survival signalling, potentially culminating in DAergic cell death.

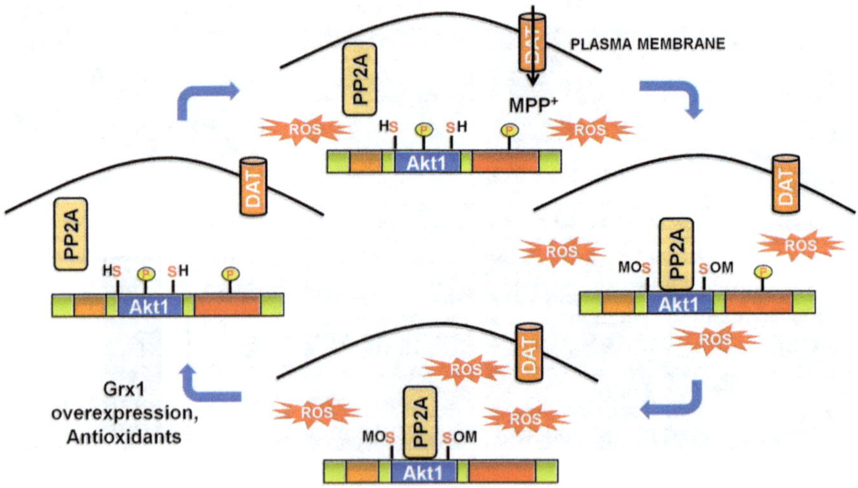

Figure 8.4 Schematic representation of redox modification of Akt1. Under normal physiological conditions, Akt1 remains in the active phosphorylated state and the critical Cys296 and Cys310 residues of Akt1 are in the reduced state (-SH). Increased production of ROS in the cell [such as ROS induced by entry of MPP^+ in dopaminergic neurons through DAT upon systemic administration of MPTP] results in oxidative modification (-S-OM) of Cys residues in Akt1 that leads to its increased association with PP2A resulting in dephosphorylation. Overexpression of Grx1 and administration of antioxidants, such as GSH-ethyl ester or α-lipoic acid have been shown to prevent Akt1 dephosphorylation.

8.4 Redox Regulation of DAergic Cell-Death Pathways

Amongst the several cell-death signalling pathways involved in pathogenesis of PD leading to DAergic cell death, ASK1-initiated mitogen-activated protein kinase (MAPK) signalling has gained prominence as an important pathway linking oxidative stress generated in the disease process to activation of cell death effectors specifically in DAergic cells. In autopsy samples from PD patients and models of PD (*in vitro* and *in vivo*),[146–148] we observe selective activation of this pathway in DAergic neurons. ASK1-MAPK pathway is one of the key stress/death-signalling MAPK pathways in the cell. Like all prototypical cellular MAPK pathways, ASK1 is the apex kinase relaying phosphorylation through its target MAP2Ks and MAPKs. The activation of the MAPKs has several downstream effects, such as phosphorylation of transcription factors, transcription of proapoptotic genes, induction of inflammatory cytokine signalling, and induction of cell death, *etc.* In the following sections, we describe the key components of this pathway (ASK1, JNK and p38) with reference to their redox sensitivity and their role in PD pathology from experiments performed on models and in PD patient tissues.

8.4.1 Redox Regulation of MAP3K-ASK1

ASK1 is encoded by the *MAP3K5* gene located on the human chromosome 6 at locus 6q22.33, and is translated into a 1374 amino acid protein. It is a Ser/Thr kinase that activates downstream kinases in a Raf (rapidly accelerated fibrosarcoma)-independent manner. ASK1 was the first MAPK pathway member demonstrated to be sensitive to the redox environment of the cell. The protein was shown to trigger downstream signalling based on its molecular association with the redox-sensitive thiol disulfide oxidoreductase, Trx1.[149] ASK1 is a member of the MAPKKK or MAP3K family and phosphorylates the downstream MAPKs – p38 and JNK – by relaying the phosphorylation signal through the intermediate MAP2Ks (MAPKK) – MKK3/6 and MKK4/7. The downstream MAPKs have been shown to be involved in a multitude of cellular phenotypes ranging from cell growth and differentiation to cell stress and/or death signalling.[150–152] While the upstream activators and the duration of activation of these MAPKs can differ in varying cellular contexts, ASK1-mediated phosphorylation has been primarily shown to elicit cell stress/death-signalling response in cells following oxidative stress,[149,153] heat shock,[154] osmotic stress,[155] endoplasmic reticulum stress,[156] UV-irradiation stress,[157] cytokine-mediated inflammation,[149] and excessive Ca^{2+} influx.[158]

In basal conditions, the protein remains primarily inactivated in an oligomerized state through interactions mediated by the C-terminal coiled-coil (CCC) domain of ASK1. The activation of ASK1 is further controlled by interactions of its N-terminal coiled-coil domain with other ASK1 molecules as well as signal-activation partners, leading to phosphorylation of one of its Thr residues in the kinase domain at the 838 and 845 positions in humans

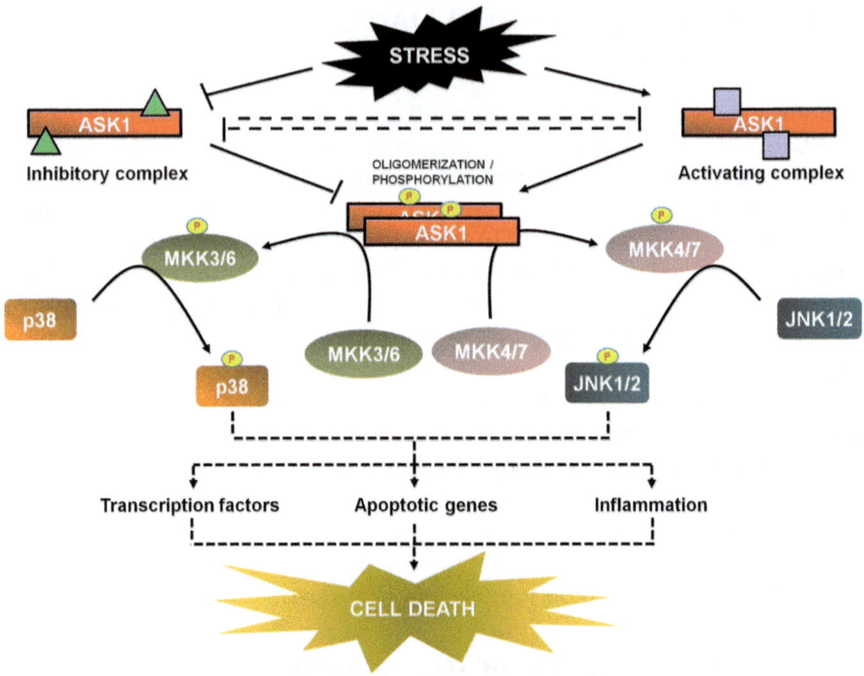

Figure 8.5 ASK1-induced MAPK cascade. Cellular stress can trigger activation of the MAP3K, ASK1, by dissociating inhibitory protein partners and/or associating activation partners to induce ASK1 oligomerization and phosphorylation. This phosphorylation signal is relayed through the MAPK kinases (MAP2K) – MKK3/6 and MKK4/7. These MAP2Ks, in turn, phosphorylate MAPKs, p38 and JNK respectively, which can trigger cell-death signalling through various downstream effectors, such as phosphorylation of transcription factors, activation of pro-apoptotic genes and promoting inflammatory signalling.

and mice respectively[159] (Figure 8.5). As ASK1 responds to a wide range of stress stimuli, research groups have investigated and confirmed a range of proteins that can regulate activation of ASK1 in response to these stimuli, as described below. The regulatory proteins of ASK1 can be divided in to two major classes – inhibitory proteins and activating proteins.

The inhibitory proteins include Trx1,[149] Grx1,[160] GSTM1,[161] the PD-related protein DJ-1,[162] Prx1,[163] Ca^{2+} and integrin-binding protein 1 (CIB1),[158] 14-3-3 proteins,[164] protein phosphatase 5 (PP5),[165] heat-shock protein 72 (Hsp72)[166] and Akt1.[167] Most of these proteins prevent ASK1 phosphorylation by association with the inactive ASK1 oligomer, thus preventing Thr838 (Thr845 in mouse) phosphorylation. PP5 phosphatase dephosphorylates active ASK1 at Thr838,[165] while 14-3-3 proteins and Akt1 interaction with ASK1 lead to inhibitory phosphorylation occurring at Ser967 and Ser83 positions, respectively.[164,167] ASK1-activating molecules include ROS,[168] tumor necrosis factor (TNF-α),[149] TNF receptor-associated factor (TRAF2/6),[169] ASK1-interacting protein 1,[170] and Daxx.[171]

As mentioned earlier, several interactors of ASK1 have been reported in literature and their respective roles in regulating this molecule have been elucidated across models of different forms of cellular stress. Phosphorylation of ASK1 and the resultant activation has been demonstrated to be redox sensitive. Cysteine residues, including the critical Cys250 residue, in ASK1 are oxidized upon exposure of cells, to H_2O_2 and form interchain disulfide bond linked multimers.[168] Specifically, the Cys250 residue was shown to be critical for relaying the downstream phosphorylation signal to JNK.[172] So far, evidence of ASK1 oxidation *in vivo* and indeed in context of PD models is lacking. Nevertheless, due to the increased oxidative load in the DAergic neurons, it will be interesting to see whether such a mechanism could be responsible for activation of the ASK1 molecule, and determine whether such selectivity extends even to different subpopulations of DAergic neurons. In the subsequent sections, we describe the major ASK1 interacting partners that are directly or indirectly regulated by oxidative stress and could potentially regulate the activation of the ASK1-MAPK pathway in limited populations in the brain, specifically DAergic neurons affected in PD (Figure 8.6).

8.4.1.1 Inhibitory Redox-Regulated ASK1 Partners

8.4.1.1.1 Thioredoxin. Thioredoxin (Trx1) was the first ASK1-inhibitory protein to be described.[149] It is a thiol disulfide oxidoreductase catalyzing reversible exchange of Cys residues in proteins between their free thiol and disulfide forms, regulating their structure and function in response to redox-related changes.[173,174] Trx1 inhibits the kinase activity of ASK1 by directly binding to the N-terminal region of ASK1. The redox-active site of Trx1 has two Cys32 and Cys35 residues that are essential for its function as a thiol disulfide oxidoreductase. The sulfhydryl groups of this protein can undergo reversible oxidation, between a disulfide (-S-S-) and a free thiol $(\text{-SH})_2$ forms. While the free thiol form of Trx1 is able to associate with ASK1, ROS or other oxidants can oxidize these Cys residues leading to its dissociation from the complex, thereby leading to phosphorylation of Thr845 in the ASK1 molecule through autophosphorylation.[149,175] Trx1 is also known to regulate the levels of ASK1 protein in the cell by promoting ubiquitination and degradation of ASK1 in a redox-independent manner.[175] Landry's group has demonstrated that Trx1 covalently associates with ASK1 forming disulfide bonds between its own and the ASK1 molecule's Cys residues.[168]

The rat SN shows moderate levels of Trx1 mRNA expression.[176] Using immunohistochemistry, Godoy *et al.* have concluded that though Trx1 protein levels are moderately high in mouse SN neurons, they express very low levels of thioredoxin reductase 1 (TrxR1 reduces Trx1 from disulfide to free thiol form), which are primarily expressed in glia.[177] Low levels of TrxR1 can inactivate the protein during increased oxidative stress in these neurons. Indeed, Trx1 oxidation has been shown in a cell-culture model of PD treating MN9D DAergic cells with 6-OHDA. Further, Trx1 overexpressing mice are partially protected from DAergic loss seen after MPTP treatment.[178] Interestingly,

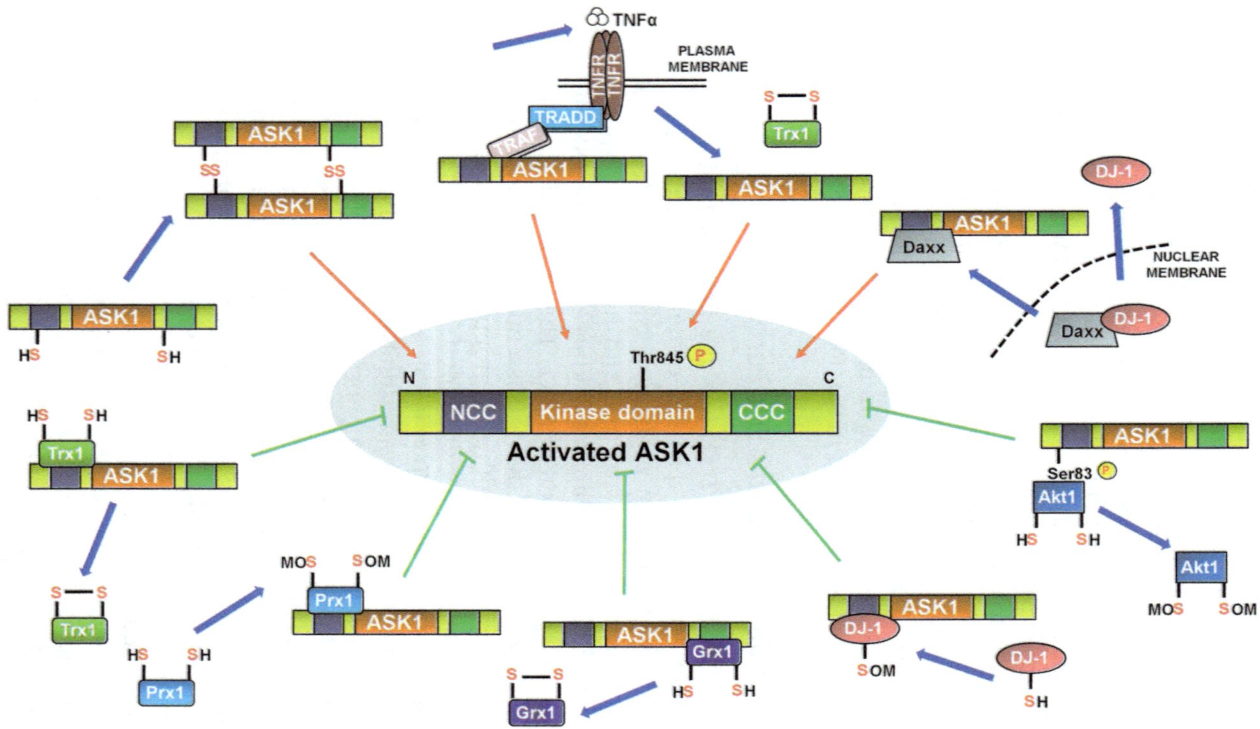

Figure 8.6 Redox regulatory mechanisms of ASK1 protein activation. Different kinds of oxidative stressors can lead to ASK1 activation leading to phosphorylation of Thr845 residue (Thr838 in humans), which lies in the kinase domain housed between the N-terminal coiled coil (NCC) and C-terminal coiled coil (CCC) domains. N and C at the ends of the ASK1 molecule denote the N- and C-termini of the protein. While proteins such as Trx1 and Grx1 bind to ASK1 in their reduced forms, they dissociate from ASK1 when they are oxidized, thus allowing activation of ASK1 through other mechanisms, such as direct oxidation of thiol residues on ASK1 forming covalently linked ASK1 oligomers, activation by TNF signalling and/or Daxx association. DJ-1 and Prx1 inhibit ASK1 by associating with the molecule in their oxidized states. TNFα signalling can lead to ASK1 phosphorylation both through ASK1 signalosome formation and/or indirectly through production of ROS. During oxidative stress, Daxx dissociates from its nuclear complex with DJ-1 and translocates to the cytoplasm associating and activating ASK1. Red and green arrows indicate activation and inhibition of ASK1, respectively. -SH and -SOM denote reduced thiol and oxidatively modified thiol groups respectively. Blue arrows indicate changes induced by oxidative stress/ROS formation.

lowered levels of Trx1 as well as TrxR1 were observed in PD patient *substantia nigra* recently.[179] While it is not known whether Trx1 expression is different between SNpc and VTA, it is nevertheless possible that increased oxidative stress can cause inactivation of this system, rendering activation of ASK1.

8.4.1.1.2 Glutaredoxin. Grx1 was shown to be another thiol disulfide oxidoreductase inhibiting ASK1 activation by associating with it. Unlike Trx1, Grx1 binds to the C-terminal end of ASK1.[160] The reduced Cys22/25 dithiol motif of Grx1 was found to be essential for inhibiting ASK1 activation in cells deprived of glucose. Similar to Trx1, Grx1 catalyzes reversible protein thiol-disulfide exchanges using GSH as a substrate and has been described to have both glutathionylating and deglutathionylating functions.[180,181]

Grx1 expression has been detected in mouse SNpc neurons but in lower amounts than hippocampus and cortex.[177,182] Unfortunately, further resolution of this expression into different neuronal types has not been carried out so far to give us insights into its role in differential vulnerability. While direct involvement of Grx1 in regulating ASK1 activity in PD-relevant model systems is lacking, it has increasingly become apparent that Grx1 is critical in maintaining the health of DAergic neurons in the brain. Earlier work from our laboratory has shown that activity of this enzyme is lowered acutely after MPTP exposure and is responsible for the early mitochondrial complex I activity loss seen in the MPTP mouse model.[42] Female mice were shown to have higher basal expression of Grx1 and are protected from MPTP toxicity as compared to male mice.[183] Overexpression and knockdown studies of Grx1 in human SK-N-SH cells established the protective effects of this protein when cells were exposed to the parkinsonian toxins, 6-OHDA and paraquat.[184] More recently, using different transgenic models of *Caenorhabditis elegans* overexpressing mutants of LRRK2, human α-synuclein or TH that display DAergic degeneration, Johnson *et al.* have shown that Grx1 homolog knockouts carrying such transgenes showed exacerbated neuronal death, while selectively reintroducing the homolog's expression in DAergic neurons attenuated the phenotype.[185] They also confirmed the decrease of Grx1 protein expression in PD postmortem brains with DAergic neurons showing higher loss. These evidences suggest that altered levels of functional Grx1 in certain neurons can make them susceptible to cell death, possibly through dysregulation of ASK1 activity and indeed other cellular-signalling pathways in PD.

8.4.1.1.3 DJ-1. The autosomal recessive PD-related gene DJ-1 has been shown to inhibit ASK1 activity directly by binding at the N-terminal site, forming Cys mixed disulfide bonds.[186] This association was found to be dependent on the DJ-1 Cys residue since Cys106Ala mutant DJ-1 could not associate with ASK1, while Cys106 oxidation mimetic Cys106Asp-Asp DJ-1 could still form a non-covalent association and inhibit ASK1 activation.[162] Cys106 of DJ-1 has been independently shown to be highly redox sensitive in several studies including cell-culture studies and PD postmortem brain tissue.[187–189] Oxidation of this Cys residue alters the subcellular location of

DJ-1 and offers cytoprotection from redox imbalance.[190] Im *et al.* described an indirect mechanism of DJ-1 regulating ASK1 phosphorylation. In HEK-293T cells, they found that overexpressed DJ-1 prevents dissociation of the Trx1-ASK1 complex upon treating cells with H_2O_2. On the other hand, knocking out DJ-1 in fibroblasts led to increased dissociation of the complex during oxidative stress.[191] While DJ-1 expression is ubiquitous in the brain and has been shown to have some degree of antioxidant properties,[189,192,193] it is possible that deficiency or functional inactivation of the protein either through degradation, post-translational modifications or mutations could render neurons that are most exposed to oxidative stress to be more susceptible to cell death.

8.4.1.1.4 Peroxiredoxin. Prx1 is one of the members of a family of thiol-based peroxidase enzymes, and acts as a signalling molecule and controls peroxide-mediated protein oxidation.[194,195] Using two non-neuronal cell lines – HEK293 and Hela – Kim *et al.* demonstrated that Prx1 can bind and inhibit ASK1 activation. Using Cys mutants, they further demonstrated the importance of redox state of Prx1 in regulating ASK1, and suggested that oxidized Prx1 generated during peroxide detoxification binds and inhibits the molecule.[163] On the other hand, another study from Ledgerwood's laboratory concluded that Prx2, but not Prx1, inhibited ROS-induced ASK1 oxidation.[196] They reported that Prx1 disulfide formation upon H_2O_2 stimulation causes ASK1 oxidation and resultant activation, while Trx1 inhibits this reaction by reducing Prx1 disulfide. These studies point to the differential role of Prx1 in regulating ASK1-mediated signalling.

So far, involvement of Prxs in regulating ASK1 activity in PD-relevant scenarios has limited evidence. While Prx1 protein expression is high in neurons of the mouse SN compared to many brain regions, other Prxs have much weaker expression profiles and are primarily expressed in non-neuronal cells.[177,197] If this is indeed the case, it is conceivable that DAergic populations may be more vulnerable to ROS compared to other cells. Alteration in Prx levels and their oxidation has been observed in PD postmortem brains.[198,199] Using the 6-OHDA mouse model, Hu *et al.* have clearly demonstrated the importance of Prx2 in preventing ASK1-cascade activation by modulating the levels of this protein using viral vectors. From their studies, they propose that Prx2 inhibits ASK1 phosphorylation by preventing oxidation of Trx1.[146] Such studies throw light on the putatively important role of Prxs in death signalling and further studies may help us understand their involvement in differential susceptibility of DAergic neurons in PD.

8.4.1.1.5 Akt1. The role of Akt1 in inhibiting ASK1 activation has been discussed earlier in this chapter, wherein Akt1-mediated phosphorylation of ASK at Ser83 attenuates ASK1 signalling. Indeed, the decrease in ASK1 inhibitory phosphorylation at Ser83 due to reduced pAkt1 signalling after MPTP treatment in the mouse ventral midbrain[110] raises the interesting question whether, under conditions of oxidative stress, selective oxidation of Akt1 leading to loss of its kinase activity in certain neurons could potentiate activation of ASK1 by its other partners. Based on research described in this section, it increasingly becomes clear that the thiol status of ASK1-inhibiting proteins is critical in regulating ASK1 activation and selective switching of this redox signal in certain neuronal populations can activate cell-death pathways in a cell-specific manner.

8.4.1.2 Activating Redox-Regulated ASK1 Partners

In this section, we discuss two key molecules that have been linked to ASK1 activation and have been reported to do so through redox-regulated mechanisms. Increased tumor necrosis factor alpha (TNF-α) signalling has been extensively linked to cellular death in PD[200,201] and we discuss how this molecule regulates ASK1 activity. Further, we also describe Daxx, a death-associated protein, which through an indirect oxidative event can potentially activate ASK1 in PD-relevant scenarios.

8.4.1.2.1 Tumor Necrosis Factor-α. TNF-α has been extensively proven to activate ASK1.[149,202] TNF-α can activate ASK1 directly by increasing production of ROS downstream[203,204] or indirectly through TNF-α-receptor dependent pathway.[159,205] TNF-α acts through binding to its two receptors, TNFR1/2. These receptors recruit the adaptor protein TRADD (TNFR1-associated death domain). TRADD can in turn activate ASK1-MAPK pathway by associating with TRAF2/6 (TNF-α receptor associated factors), or binding to FADD (Fas-associated protein with death domain) leading to caspase-dependent cell death. Importantly, TNF-α/TRAF2/6 signalling activates ASK1 by a ROS-dependent release of Trx1 inhibitory association, and subsequent association of TRAF2/6 with ASK1. This association leads to the induction of ASK1 molecular homophilic interactions and autophosphorylation of ASK1 at Thr845.[206] While neurons can also express TNF-α, microglia are the major sources of TNF-α synthesis in the brain.[207] SN has the highest microglial density,[90] which could render it more vulnerable to age- or toxin- related inflammatory signalling compared to other regions of the brain.

While direct involvement of TNF-α in ASK1 activation in PD models has yet to be ascertained, there is evidence for TNF-α playing an important role in neurodegeneration. The DAergic terminals but not the cell bodies are protected from MPTP toxicity in TNFR double knockout mice[208–210] or in TNF knockout mice.[211,212] On the other hand, expressing dominant negative TNF-α virally or exposure to a systemic inhibitor of TNF-α offers significant DAergic neuroprotection in 6-OHDA injected rats.[213,214]

8.4.1.2.2 Daxx. Chang *et al.* first demonstrated Daxx (death-domain associated protein), to be involved in the ASK1 signalosome, further activating ASK1.[171] In cells deprived of glucose, Daxx translocates from its physiological localization in the nucleus to the cytoplasm where it binds ASK1, leading to its subsequent oligomerization and activation.[215] DJ-1 interacts with Daxx in a redox-dependent manner and prevents its translocation from the nucleus.[216]

Translocation of Daxx from the nucleus occurs in response to oxidative stress generated by exposure to H_2O_2[217] or downregulation of Grx1.[216] We have also shown increased cytoplasmic Daxx localization in cultured DAergic cells after MPP^+ treatment, and in mouse SNpc but not VTA neurons after MPTP treatment.[147] Overall, even though Daxx does not directly undergo any redox modifications, it can potentially lead to selective activation of ASK1 in certain cells that have a highly oxidizing intracellular milieu.

8.4.1.3 Evidences of ASK1 Involvement in PD

ASK1 activation and its role in the cell death seen in PD have been shown extensively in various models as well as in autopsy brain tissue from PD patients. The first suggestion of ASK1 in DAergic cell death came from an *in vitro* study carried out by Ouyang and Shen.[148] They found direct evidence of ROS-dependent ASK1-MAPK activation after exposing DAergic SH-SY5Y cells to 6-OHDA. Further, they also elucidated the important role of ASK1 in activating p38 and JNK in this model by downregulating ASK1 or overexpressing the dominant negative kinase-dead ASK1 mutant in these cells and attenuating MAPK phosphorylation.

Work from our laboratory soon established the activation of this pathway in the MPTP model.[147,218] Upon treating C57BL/6 mice with MPTP, we observed increased phosphorylation of the downstream targets of ASK1, p38 and JNK in SNpc housing the DAergic cell bodies. The projection area for these neurons, striatum, showed relatively less perturbation of the pathway members, indicating that only the DAergic neurons show dysregulation of the pathway while the striatum, primarily containing γ-amino butyric acid-expressing neurons (GABAergic) remains unaffected. Protection offered by α-lipoic acid in MPTP-treated mice further confirmed the redox-sensitive activation of this pathway.[147] Our results also showed remarkable specificity of signal transduction even in the ventral midbrain between SNpc and VTA, which we describe in the downstream effectors section under JNK and p38. Confirmatory evidence of the role of ASK1 in MPTP-induced cell death arrived when death of DAergic neurons was attenuated in ASK1 knockout mice. This was attributed to reduced microglial activation in response to the toxin,[219] possibly through suppression of ASK1-p38 cascade mediated inflammatory signalling.

In other models of PD, paraquat exposure leads to activation of ASK1 pathway in DAergic SH-SY5Y cells in a redox-dependent manner, as vitamin E coincubation or Nrf2 (nuclear factor (erythroid-derived 2)-like 2 transcription factor involved in mounting an antioxidant defense response) induction using sulforaphane prevented this response.[220] Hu *et al.* also concluded the critical role of ASK1 in PD-related DAergic cell death using 6-OHDA in a mouse model. They demonstrated the phosphorylation of ASK1 in SNpc DAergic neurons coupled with downstream activation of JNK, p38 and transcriptional factor c-Jun. Further, knock-down of ASK1 in the SNpc using viral transduction of ASK1 shRNA (short-hairpin siRNA) resulted in protection of DAergic neurons,[146] indicating the direct role of ASK1 in cell death.

Apart from studies using neurotoxin models of PD, ASK1 has also been found to be responsible for neuroinflammation and cell death in α-synuclein overexpressing mice. ASK1 knockout did not alter the aggregation characteristics of α-synuclein in either cells or in mice brains but attenuated neuronal damage and improved motor behavioral outcome of these mice.[221] Finally, increased phosphorylation of ASK1 has been found in surviving SNpc neurons and they colocalize with Lewy bodies or abnormal α-synuclein staining in PD SN brain sections.[146] Oxidative stress-induced activation of ASK1 plays a key role in the initiation of DAergic cell death and could possibly occur differentially between SNpc and VTA neurons. In the next section, we describe the downstream effectors of the ASK1 pathway and their roles in PD.

8.4.2 MAPKs – p38 and JNK

p38 and JNK MAPKs form two major sub-groups of the mammalian Ser–Thr MAPKs.[222] They are involved in a multitude of cellular processes from differentiation, proliferation to cell death in response to extracellular signals, and have been extensively studied as cell-stress/death regulators. Amongst the several upstream pathways that are known to lead to phosphorylation and activation of these MAPKs, we have primarily focused on the ASK1 MAP3K pathway. Below, we describe the two MAPKs function and regulation in greater detail.

8.4.2.1 JNK

JNKs were originally identified to phosphorylate c-Jun on Ser63 and Ser73 residues. As seen with ASK1 and p38, JNKs are also responsive to a host of stimuli including oxidative stress, heat shock, inflammation, *etc.*[223] They play important roles in triggering apoptosis in response to extracellular signals/stressors. JNKs have several different isoforms generated through splice variation of mRNA derived from three genes – JNK1, JNK2 and JNK3. Depending on the processing of the 3′ coding region of the mRNA, each gene may be further expressed as either a 46 kDa or 55 kDa protein kinase. All JNK isoforms carry the characteristic Thr-proline (Pro)-tyrosine (Tyr) (TPY) dual phosphorylation motif.[224] While JNK1 and JNK2 expression is observed across all cells and tissue types, JNK3 expression is mainly restricted to the brain, apart from heart and testes.[225,226]

The molecule is activated by two MAP kinase kinases (MAP2Ks or MKKs), MKK4 and MKK7. Similar to MKK3/6 for p38, these MAP2Ks also relay signals differentially based on cellular and stress context.[227,228] ASK1 was demonstrated to phosphorylate JNK under conditions of oxidative stress and TNF-α activation.[149,153] JNK, in turn, activates several transcription factors including c-Jun, activating transcription factor 2 (ATF2), ETS domain-containing protein (ELK1) and p53.[229] Apart from upstream redox-regulated ASK1 regulation, JNK activity can be directly regulated by peroxiredoxin in a redox-sensitive manner.[230]

Evidence for JNK phosphorylation has been seen in both cellular and animal models of PD[147,231–234] and in PD patient brains.[235] This aberrant JNK activation has been associated with downstream effects including phosphorylation of p53[236] and phosphorylation of Bim and Bax translocation downstream.[237] These molecules can then initiate cell-death processes. While several *in vitro* studies using cultured cells demonstrated neuroprotection offered by JNK inhibitors, a clinical trial with the mixed lineage kinase inhibitor CEP1347 (suppressing JNK signalling downstream) in PD has failed.[238] To understand the potential causes of discrepancy between *in vitro* studies and the failure of the clinical trial, we carried out experiments in mice treating them with MPTP, which triggers the ASK1 cascade. Interestingly, we observed that while JNK phosphorylation occurred in the microglial cells in SN, DAergic neurons primarily showed increased p38 phosphorylation.[218] Thus, it appears that the MAPK signalling cascade does not activate similarly between different cell populations in the brain and could contribute to differential cellular fates in response to oxidative stress.

8.4.2.2 p38 MAPKs

p38 MAPKs are also described as stress-activated protein kinases (SAPK). As seen with ASK1, p38 activation occurs in response to extracellular stimuli such as oxidative stress, UV-irradiation, heat shock, osmotic shock and inflammatory cytokines such as TNF-α and interleukin-1 (IL-1).[150,239] Four splice variants of the p38 family have been identified – p38α, p38β,[240] p38γ (ERK6, SAPK3),[241,242] and p38δ (SAPK4).[243] While p38α and p38β are ubiquitously expressed, p38γ and p38δ are expressed differentially in different tissues. All the isoforms carry Thr-glycine (Gly)-Tyr (TGY) dual phosphorylation motif.[244] The plethora of activators and the cell-type and isoform-type specific expression of this family paint a complex picture of their functional role.

p38 kinases are activated by the dual kinases MAP2K. The two major MAP2Ks known to activate p38 are MKK3 and MKK6. These MAP2Ks can differentially regulate p38 isoforms.[245–247] Concurrent with JNK activation, p38 phosphorylation was demonstrated to be downstream of ASK1 phosphorylation in response to TNF-α and H_2O_2.[149,153] Increased p38 phosphorylation has been observed in animal models of PD including MPTP,[218] rotenone[248] and 6-OHDA,[146] as well as in PD postmortem tissue.[218] Unlike JNK activation seen in microglia after MPTP, our studies have shown that p38 is selectively activated in DAergic neurons with the mouse SNpc neurons showing higher levels of p38 phosphorylation than VTA DAergic neurons.[218] Such differential activation between the populations may potentially underlie the differences seen in cell death between these two regions.

The downstream targets of p38 include transcription factors, ATF2,[249] NF-κB,[250] and p53,[251,252] all of which have been shown to be dysregulated in PD-relevant model systems or in PD. Specifically, p53 has been shown to be a very important mediator of DAergic cell death as p53 knockouts are protected in PD models.[236,253] By inhibiting p38 phosphorylation using a p38 inhibitor, we could prevent p53 phosphorylation, its subsequent activation and nuclear translocation,[218] which have been earlier shown to be essential for p53 to exert its cell death-related effects.[254,255] From these results, we can note that p38 phosphorylation is a remarkably specific pathogenic event occurring primarily in SNpc DAergic neurons and could potentially be responsible for cell death through initiation of downstream signalling in the PD context.

Finally, based on the several lines of evidence in favor of activation of ASK1-MAPK cascade through redox signalling, we can conclude that oxidative stress can trigger cell-death signalling cascades in a cell-type specific manner in the brain. Cell-specific activation of this cascade may thereby help us understand the selective vulnerability of DAergic neurons observed in PD.

8.5 Conclusions

In this chapter, we have described how generalized oxidative stress-induced redox signalling can potentially trigger cell-specific activation of signalling pathways in DAergic neurons, which has potential implications in PD pathogenesis. In this context, we have discussed how the cell-survival pathway, namely Akt signalling can be suppressed while cell-death signalling mechanisms, namely the ASK1-MAPK pathway, are activated. The inhibition of the protective survival pathway and upregulation of cell death processes can render DAergic neurons highly vulnerable to other stressors as well as leading to their ultimate demise. While we have limited our discussion to these two pathways, it is needless to state that potentially unidentified pathways may also contribute to the ultimate cell death in a cell-specific manner. Some of the different classes of proteins that are known to be redox regulated are listed in Table 8.1. These include transcription factors, kinases, phosphatases, mitochondrial proteins, *etc.* While redox modification and the consequences of such modification have been elucidated, often using cultured cell lines; it is yet to be determined if these modifications occur in PD models and if so, do they contribute to the selective cell DAergic death in SNpc seen in PD.

Further, it is important to note that such signalling events may help us link pathological mechanism(s) of degeneration between sporadic and familial PD. Oxidative stress generated by various means could ultimately regulate similar signalling mechanisms in a cell-specific manner, thus leading to a final common pathology of DAergic cell death. Many of the mutations and dysfunctions in PD-associated gene products such as PINK1, Parkin, DJ-1, α-synuclein and LRRK2 are directly or indirectly linked to generation of oxidative stress amongst other cellular effects.[264–267] Even if they are not directly responsible, the increased basal oxidative load of the DAergic cells as mentioned in the initial section, coupled with dysregulation of these proteins in familial PD may overwhelm the defense system of the cell and lead to adverse effects.

Table 8.1 Redox modification of cysteine residues of cellular proteins.

Redox modulated molecules[a]
Transcription factors
Activator protein 1 (AP-1)[256]
cAMP response element binding protein (CREB)[256]
Heat shock factor protein 1(HSF-1)[257,259]
Hypoxia inducible factor-1α (HIF-1α)[256]
HIF-1α like factor (HLF)[256]
Nuclear factor κ-light-chain-enhancer of activated B cells (NF-κB)[256]
Nuclear factor erythroid 2 (NF-E2)-related factor 2 (Nrf2)[256,259]
p53[256]
Kinases
Proto-oncogene tyrosine protein kinase Src (Src)[256]
Proto-oncogene tyrosine protein kinase Yes (Yes)[256]
MAP/ERK kinase 1 (MEKK1)[101]
Phosphatases
T-cell protein tyrosine phosphatase (TC-PTP)[258]
Calpain[258,259]
Lipid phosphatase and tensin homolog on chromosome 10 (PTEN)[258]
Apoptotic proteins
Caspases[101]
B-cell lymphoma 2 (Bcl-2)[101]
Mitochondrial proteins
Complex I[260,261]
Voltage dependent anion channel (VDAC)[260,262]
Adenine nucleotide translocator (ANT)[260,263]
Cyclophilin D[260]

[a]This table provides a representative but not exhaustive list of proteins that are redox regulated. Their roles in dopaminergic cell death in PD through redox modification of cysteine residues is yet to be demonstrated in model systems.

Finally, while redox signalling and its downstream effects appear to be major players in determination of selective cell death in PD and indeed in other neurodegenerative disorders, it is necessary to note that there are many other aspects to PD pathological mechanisms including the important features of proteinopathy and inflammation (refer to Chapter 7 for greater detail). It will be interesting and important to study how such pathways may converge. Further, it would be important to identify the signalling pathways involved in cause and effect-related changes in neuronal death and design multifaceted therapeutic strategies to counter the disease.

Acknowledgements

The authors thank Dr Lalitha Durgadoss for providing the unpublished data shown in Figure 8.3. Funding from the Department of Science & Technology, India, is gratefully acknowledged.

References

1. D. W. Dickson, H. Braak, J. E. Duda, C. Duyckaerts, T. Gasser, G. M. Halliday, J. Hardy, J. B. Leverenz, K. Del Tredici, Z. K. Wszolek and I. Litvan, Neuropathological assessment of Parkinson's disease: refining the diagnostic criteria, *Lancet Neurol.*, 2009, **8**, 1150–1157.
2. H. Braak and K. Del Tredici, Nervous system pathology in sporadic Parkinson disease, *Neurology*, 2008, **70**, 1916–1925.
3. H. Braak, K. Del Tredici, U. Rub, R. A. de Vos, E. N. Jansen Steur and E. Braak, Staging of brain pathology related to sporadic Parkinson's disease, *Neurobiol. Aging*, 2003, **24**, 197–211.
4. G. Halliday, M. Hely, W. Reid and J. Morris, The progression of pathology in longitudinally followed patients with Parkinson's disease, *Acta Neuropathol.*, 2008, **115**, 409–415.
5. S. Y. Lim, S. H. Fox and A. E. Lang, Overview of the extranigral aspects of Parkinson disease, *Arch. Neurol.*, 2009, **66**, 167–172.
6. C. Zarow, S. A. Lyness, J. A. Mortimer and H. C. Chui, Neuronal loss is greater in the locus coeruleus than nucleus basalis and substantia nigra in Alzheimer and Parkinson diseases, *Arch. Neurol.*, 2003, **60**, 337–341.
7. J. H. Kordower, C. W. Olanow, H. B. Dodiya, Y. Chu, T. G. Beach, C. H. Adler, G. M. Halliday and R. T. Bartus, Disease duration and the integrity of the nigrostriatal system in Parkinson's disease, *Brain*, 2013, **136**, 2419–2431.
8. J. M. Fearnley and A. J. Lees, Ageing and Parkinson's disease: substantia nigra regional selectivity, *Brain*, 1991, **114**(pt 5), 2283–2301.
9. D. A. McRitchie, H. R. Cartwright and G. M. Halliday, Specific A10 dopaminergic nuclei in the midbrain degenerate in Parkinson's disease, *Exp. Neurol.*, 1997, **144**, 202–213.
10. P. Damier, E. C. Hirsch, Y. Agid and A. M. Graybiel, The substantia nigra of the human brain. II. Patterns of loss of dopamine-containing neurons in Parkinson's disease, *Brain*, 1999, **122**(pt 8), 1437–1448.
11. W. Dauer and S. Przedborski, Parkinson's disease: mechanisms and models, *Neuron*, 2003, **39**, 889–909.
12. R. A. Gonzalez-Polo, J. M. Fuentes, M. Niso-Santano and L. Alvarez-Erviti, Implication of autophagy in Parkinson's disease, *Parkinson's Dis.*, 2013, **2013**, 436481.
13. T. Omura, M. Kaneko, Y. Okuma, K. Matsubara and Y. Nomura, Endoplasmic reticulum stress and Parkinson's disease: the role of HRD1 in averting apoptosis in neurodegenerative disease, *Oxid. Med. Cell. Longevity*, 2013, **2013**, 239854.
14. J. M. Shulman, P. L. De Jager and M. B. Feany, Parkinson's disease: genetics and pathogenesis, *Annu. Rev. Pathol.: Mech. Dis.*, 2011, **6**, 193–222.
15. M. Asanuma, I. Miyazaki, F. J. Diaz-Corrales and N. Ogawa, Quinone formation as dopaminergic neuron-specific oxidative stress in the pathogenesis of sporadic Parkinson's disease and neurotoxin-induced parkinsonism, *Acta Med. Okayama*, 2004, **58**, 221–233.

16. A. Hald and J. Lotharius, Oxidative stress and inflammation in Parkinson's disease: is there a causal link? *Exp. Neurol.*, 2005, **193**, 279–290.
17. J. Segura-Aguilar, I. Paris, P. Munoz, E. Ferrari, L. Zecca and F. A. Zucca, Protective and toxic roles of dopamine in Parkinson's disease, *J. Neurochem.*, 2014, **129**, 898–915.
18. D. G. Graham, S. M. Tiffany, W. R. Bell Jr and W. F. Gutknecht, Autoxidation *versus* covalent binding of quinones as the mechanism of toxicity of dopamine, 6-hydroxydopamine, and related compounds toward C1300 neuroblastoma cells *in vitro*, *Mol. Pharmacol.*, 1978, **14**, 644–653.
19. W. Linert, E. Herlinger, R. F. Jameson, E. Kienzl, K. Jellinger and M. B. Youdim, Dopamine, 6-hydroxydopamine, iron, and dioxygen–their mutual interactions and possible implication in the development of Parkinson's disease, *Biochim. Biophys. Acta*, 1996, **1316**, 160–168.
20. I. Miyazaki and M. Asanuma, Dopamine, 6-hydroxydopamine, iron, and dioxygen–their mutual interactions and possible implication in the development of Parkinson's disease, *Acta Med. Okayama*, 2008, **62**, 141–150.
21. C. A. Altar, M. R. Marien and J. F. Marshall, Time course of adaptations in dopamine biosynthesis, metabolism, and release following nigrostriatal lesions: implications for behavioral recovery from brain injury, *J. Neurochem.*, 1987, **48**, 390–399.
22. O. Hornykiewicz and S. J. Kish, Biochemical pathophysiology of Parkinson's disease, *Adv. Neurol.*, 1987, **45**, 19–34.
23. M. J. Bannon and C. J. Whitty, Age-related and regional differences in dopamine transporter mRNA expression in human midbrain, *Neurology*, 1997, **48**, 969–977.
24. T. Gonzalez-Hernandez, P. Barroso-Chinea, I. De La Cruz Muros, M. Del Mar Perez-Delgado and M. Rodriguez, Expression of dopamine and vesicular monoamine transporters and differential vulnerability of mesostriatal dopaminergic neurons, *J. Comp. Neurol.*, 2004, **479**, 198–215.
25. S. N. Haber, H. Ryoo, C. Cox and W. Lu, Subsets of midbrain dopaminergic neurons in monkeys are distinguished by different levels of mRNA for the dopamine transporter: comparison with the mRNA for the D2 receptor, tyrosine hydroxylase and calbindin immunoreactivity, *J. Comp. Neurol.*, 1995, **362**, 400–410.
26. E. Bezard, C. E. Gross, M. C. Fournier, S. Dovero, B. Bloch and M. Jaber, Absence of MPTP-induced neuronal death in mice lacking the dopamine transporter, *Exp. Neurol.*, 1999, **155**, 268–273.
27. R. R. Gainetdinov, F. Fumagalli, S. R. Jones and M. G. Caron, Dopamine transporter is required for *in vivo* MPTP neurotoxicity: evidence from mice lacking the transporter, *J. Neurochem.*, 1997, **69**, 1322–1325.
28. G. Eisenhofer, I. J. Kopin and D. S. Goldstein, Catecholamine metabolism: a contemporary view with implications for physiology and medicine, *Pharmacol. Rev.*, 2004, **56**, 331–349.

29. M. J. Nirenberg, Y. Liu, D. Peter, R. H. Edwards and V. M. Pickel, The vesicular monoamine transporter 2 is present in small synaptic vesicles and preferentially localizes to large dense core vesicles in rat solitary tract nuclei, *Proc. Natl. Acad. Sci. U. S. A.*, 1995, **92**, 8773–8777.
30. R. R. Gainetdinov, F. Fumagalli, Y. M. Wang, S. R. Jones, A. I. Levey, G. W. Miller and M. G. Caron, Increased MPTP neurotoxicity in vesicular monoamine transporter 2 heterozygote knockout mice, *J. Neurochem.*, 1998, **70**, 1973–1978.
31. T. N. Taylor, W. M. Caudle and G. W. Miller, VMAT2-deficient mice display nigral and extranigral pathology and motor and nonmotor symptoms of Parkinson's disease, *Parkinson's Dis.*, 2011, **2011**, 124165.
32. A. L. Brown, T. A. Day, C. V. Dayas and D. W. Smith, Purity and enrichment of laser-microdissected midbrain dopamine neurons, *BioMed Res. Int.*, 2013, **2013**, 747938.
33. W. Lu and M. E. Wolf, Expression of dopamine transporter and vesicular monoamine transporter 2 mRNAs in rat midbrain after repeated amphetamine administration, *Brain Res. Mol. Brain Res.*, 1997, **49**, 137–148.
34. M. T. Lin and M. F. Beal, Mitochondrial dysfunction and oxidative stress in neurodegenerative diseases, *Nature*, 2006, **443**, 787–795.
35. J. W. Langston, P. Ballard, J. W. Tetrud and I. Irwin, Chronic Parkinsonism in humans due to a product of meperidine-analog synthesis, *Science*, 1983, **219**, 979–980.
36. Y. Mizuno, N. Sone and T. Saitoh, Effects of 1-methyl-4-phenyl-1,2,3,6-tetrahydropyridine and 1-methyl-4-phenylpyridinium ion on activities of the enzymes in the electron transport system in mouse brain, *J. Neurochem.*, 1987, **48**, 1787–1793.
37. W. J. Nicklas, S. K. Youngster, M. V. Kindt and R. E. Heikkila, MPTP, MPP^+ and mitochondrial function, *Life Sci.*, 1987, **40**, 721–729.
38. A. H. Schapira, J. M. Cooper, D. Dexter, P. Jenner, J. B. Clark and C. D. Marsden, Mitochondrial complex I deficiency in Parkinson's disease, *Lancet*, 1989, **1**, 1269.
39. N. Exner, A. K. Lutz, C. Haass and K. F. Winklhofer, Mitochondrial dysfunction in Parkinson's disease: molecular mechanisms and pathophysiological consequences, *EMBO J.*, 2012, **31**, 3038–3062.
40. A. J. Lambert and M. D. Brand, Inhibitors of the quinone-binding site allow rapid superoxide production from mitochondrial NADH:ubiquinone oxidoreductase (complex I), *J. Biol. Chem.*, 2004, **279**, 39414–39420.
41. M. P. Murphy, How mitochondria produce reactive oxygen species, *Biochem. J.*, 2009, **417**, 1–13.
42. R. S. Kenchappa and V. Ravindranath, Glutaredoxin is essential for maintenance of brain mitochondrial complex I: studies with MPTP, *FASEB J.*, 2003, **17**, 717–719.
43. A. I. Padrao, R. M. Ferreira, R. Vitorino, R. M. Alves, M. J. Neuparth, J. A. Duarte and F. Amado, OXPHOS susceptibility to oxidative modifications: the role of heart mitochondrial subcellular location, *Biochim. Biophys. Acta*, 2011, **1807**, 1106–1113.

44. F. H. Khan, T. Sen, A. K. Maiti, S. Jana, U. Chatterjee and S. Chakrabarti, Inhibition of rat brain mitochondrial electron transport chain activity by dopamine oxidation products during extended *in vitro* incubation: implications for Parkinson's disease, *Biochim. Biophys. Acta*, 2005, **1741**, 65–74.
45. P. C. Keane, M. Kurzawa, P. G. Blain and C. M. Morris, Mitochondrial dysfunction in Parkinson's disease, *Parkinson's Dis.*, 2011, **2011**, 716871.
46. A. Bender, K. J. Krishnan, C. M. Morris, G. A. Taylor, A. K. Reeve, R. H. Perry, E. Jaros, J. S. Hersheson, J. Betts, T. Klopstock, R. W. Taylor and D. M. Turnbull, High levels of mitochondrial DNA deletions in substantia nigra neurons in aging and Parkinson disease, *Nat. Genet.*, 2006, **38**, 515–517.
47. D. B. Kell, Towards a unifying, systems biology understanding of large-scale cellular death and destruction caused by poorly liganded iron: Parkinson's, Huntington's, Alzheimer's, prions, bactericides, chemical toxicology and others as examples, *Arch. Toxicol.*, 2010, **84**, 825–889.
48. C. Thomas, M. M. Mackey, A. A. Diaz and D. P. Cox, Hydroxyl radical is produced *via* the Fenton reaction in submitochondrial particles under oxidative stress: implications for diseases associated with iron accumulation, *Redox Rep.*, 2009, **14**, 102–108.
49. V. M. Mann, J. M. Cooper, S. E. Daniel, K. Srai, P. Jenner, C. D. Marsden and A. H. Schapira, Complex I, iron, and ferritin in Parkinson's disease substantia nigra, *Ann. Neurol.*, 1994, **36**, 876–881.
50. J. Sian-Hulsmann, S. Mandel, M. B. Youdim and P. Riederer, The relevance of iron in the pathogenesis of Parkinson's disease, *J. Neurochem.*, 2011, **118**, 939–957.
51. A. E. Oakley, J. F. Collingwood, J. Dobson, G. Love, H. R. Perrott, J. A. Edwardson, M. Elstner and C. M. Morris, Individual dopaminergic neurons show raised iron levels in Parkinson disease, *Neurology*, 2007, **68**, 1820–1825.
52. Z. Lv, H. Jiang, H. Xu, N. Song and J. Xie, Increased iron levels correlate with the selective nigral dopaminergic neuron degeneration in Parkinson's disease, *J. Neural Transm.*, 2011, **118**, 361–369.
53. D. Kaur, F. Yantiri, S. Rajagopalan, J. Kumar, J. Q. Mo, R. Boonplueang, V. Viswanath, R. Jacobs, L. Yang, M. F. Beal, D. DiMonte, I. Volitaskis, L. Ellerby, R. A. Cherny, A. I. Bush and J. K. Andersen, Genetic or pharmacological iron chelation prevents MPTP-induced neurotoxicity *in vivo*: a novel therapy for Parkinson's disease, *Neuron*, 2003, **37**, 899–909.
54. A. Friedman, J. Galazka-Friedman and D. Koziorowski, Iron as a cause of Parkinson disease–a myth or a well established hypothesis? *Parkinsonism Relat. Disord.*, 2009, **15**(suppl. 3), S212–S214.
55. P. F. Good, C. W. Olanow and D. P. Perl, Neuromelanin-containing neurons of the substantia nigra accumulate iron and aluminum in Parkinson's disease: a LAMMA study, *Brain Res.*, 1992, **593**, 343–346.

56. M. Zareba, A. Bober, W. Korytowski, L. Zecca and T. Sarna, The effect of a synthetic neuromelanin on yield of free hydroxyl radicals generated in model systems, *Biochim. Biophys. Acta*, 1995, **1271**, 343–348.
57. L. Zecca, R. Fariello, P. Riederer, D. Sulzer, A. Gatti and D. Tampellini, The absolute concentration of nigral neuromelanin, assayed by a new sensitive method, increases throughout the life and is dramatically decreased in Parkinson's disease, *FEBS Lett.*, 2002, **510**, 216–220.
58. S. Nedergaard, J. A. Flatman and I. Engberg, Nifedipine- and omega-conotoxin-sensitive Ca^{2+} conductances in guinea-pig substantia nigra pars compacta neurons, *J. Physiol.*, 1993, **466**, 727–747.
59. M. Puopolo, E. Raviola and B. P. Bean, Roles of subthreshold calcium current and sodium current in spontaneous firing of mouse midbrain dopamine neurons, *J. Neurosci.*, 2007, **27**, 645–656.
60. Z. M. Khaliq and B. P. Bean, Pacemaking in dopaminergic ventral tegmental area neurons: depolarizing drive from background and voltage-dependent sodium conductances, *J. Neurosci.*, 2010, **30**, 7401–7413.
61. C. S. Chan, J. N. Guzman, E. Ilijic, J. N. Mercer, C. Rick, T. Tkatch, G. E. Meredith and D. J. Surmeier, 'Rejuvenation' protects neurons in mouse models of Parkinson's disease, *Nature*, 2007, **447**, 1081–1086.
62. H. Morikawa, F. Imani, K. Khodakhah and J. T. Williams, Inositol 1,4,5-triphosphate-evoked responses in midbrain dopamine neurons, *J. Neurosci.*, 2000, **20**, RC103.
63. V. Seutin, F. Mkahli, L. Massotte and A. Dresse, Calcium release from internal stores is required for the generation of spontaneous hyperpolarizations in dopaminergic neurons of neonatal rats, *J. Neurophysiol.*, 2000, **83**, 192–197.
64. D. A. McRitchie, C. D. Hardman and G. M. Halliday, Cytoarchitectural distribution of calcium binding proteins in midbrain dopaminergic regions of rats and humans, *J. Comp. Neurol.*, 1996, **364**, 121–150.
65. T. J. Collier, N. M. Kanaan and J. H. Kordower, Ageing as a primary risk factor for Parkinson's disease: evidence from studies of non-human primates, *Nat. Rev. Neurosci.*, 2013, **12**, 359–366.
66. T. M. Dawson and V. L. Dawson, Molecular pathways of neurodegeneration in Parkinson's disease, *Science*, 2003, **302**, 819–822.
67. S. Orrenius, B. Zhivotovsky and P. Nicotera, Regulation of cell death: the calcium-apoptosis link, *Nat. Rev. Mol. Cell Biol.*, 2003, **4**, 552–565.
68. J. N. Guzman, J. Sanchez-Padilla, D. Wokosin, J. Kondapalli, E. Ilijic, P. T. Schumacker and D. J. Surmeier, Oxidant stress evoked by pacemaking in dopaminergic neurons is attenuated by DJ-1, *Nature*, 2010, **468**, 696–700.
69. E. V. Mosharov, K. E. Larsen, E. Kanter, K. A. Phillips, K. Wilson, Y. Schmitz, D. E. Krantz, K. Kobayashi, R. H. Edwards and D. Sulzer, Interplay between cytosolic dopamine, calcium, and alpha-synuclein causes selective death of substantia nigra neurons, *Neuron*, 2009, **62**, 218–229.

70. J. A. Goldberg, J. N. Guzman, C. M. Estep, E. Ilijic, J. Kondapalli, J. Sanchez-Padilla and D. J. Surmeier, Calcium entry induces mitochondrial oxidant stress in vagal neurons at risk in Parkinson's disease, *Nat. Neurosci.*, 2012, **15**, 1414–1421.
71. J. Sanchez-Padilla, J. N. Guzman, E. Ilijic, J. Kondapalli, D. J. Galtieri, B. Yang, S. Schieber, W. Oertel, D. Wokosin, P. T. Schumacker and D. J. Surmeier, Mitochondrial oxidant stress in locus coeruleus is regulated by activity and nitric oxide synthase, *Nat. Neurosci.*, 2014, **17**, 832–840.
72. M. J. Hurley, S. M. Gentleman and D. T. Dexter, Calcium Ca_V1 channel subtype mRNA expression in Parkinson's disease examined by *in situ* hybridization, *J. Mol. Neurosci.*, 2015, **55**, 715–724.
73. C. Becker, S. S. Jick and C. R. Meier, Use of antihypertensives and the risk of Parkinson disease, *Neurology*, 2008, **70**, 1438–1444.
74. B. Pasternak, H. Svanstrom, N. M. Nielsen, L. Fugger, M. Melbye and A. Hviid, Use of calcium channel blockers and Parkinson's disease, *Am. J. Epidemiol.*, 2012, **175**, 627–635.
75. B. Ritz, S. L. Rhodes, L. Qian, E. Schernhammer, J. H. Olsen and S. Friis, L-type calcium channel blockers and Parkinson disease in Denmark, *Ann. Neurol.*, 2010, **67**, 600–606.
76. R. Dringen, J. M. Gutterer and J. Hirrlinger, Glutathione metabolism in brain metabolic interaction between astrocytes and neurons in the defense against reactive oxygen species, *Eur. J. Biochem.*, 2000, **267**, 4912–4916.
77. G. D. Zeevalk, R. Razmpour and L. P. Bernard, Glutathione and Parkinson's disease: is this the elephant in the room? *Biomed. Pharmacother.*, 2008, **62**, 236–249.
78. Y. Zhu, P. M. Carvey and Z. Ling, Age-related changes in glutathione and glutathione-related enzymes in rat brain, *Brain Res.*, 2006, **1090**, 35–44.
79. M. B. Spina and G. Cohen, Dopamine turnover and glutathione oxidation: implications for Parkinson disease, *Proc. Natl. Acad. Sci. U. S. A.*, 1989, **86**, 1398–1400.
80. D. C. Wu, P. Teismann, K. Tieu, M. Vila, V. Jackson-Lewis, H. Ischiropoulos and S. Przedborski, NADPH oxidase mediates oxidative stress in the 1-methyl-4-phenyl-1,2,3,6-tetrahydropyridine model of Parkinson's disease, *Proc. Natl. Acad. Sci. U. S. A.*, 2003, **100**, 6145–6150.
81. S. Przedborski, V. Jackson-Lewis, R. Yokoyama, T. Shibata, V. L. Dawson and T. M. Dawson, Role of neuronal nitric oxide in 1-methyl-4-phenyl-1,2,3,6-tetrahydropyridine (MPTP)-induced dopaminergic neurotoxicity, *Proc. Natl. Acad. Sci. U. S. A.*, 1996, **93**, 4565–4571.
82. T. L. Perry, D. V. Godin and S. Hansen, Parkinson's disease: a disorder due to nigral glutathione deficiency? *Neurosci. Lett.*, 1982, **33**, 305–310.
83. J. Sian, D. T. Dexter, A. J. Lees, S. Daniel, Y. Agid, F. Javoy-Agid, P. Jenner and C. D. Marsden, Alterations in glutathione levels in Parkinson's disease and other neurodegenerative disorders affecting basal ganglia, *Ann. Neurol.*, 1994, **36**, 348–355.

84. S. J. Chinta, S. Rajagopalan, D. A. Butterfield and J. K. Andersen, *In vitro* and *in vivo* neuroprotection by γ-glutamylcysteine ethyl ester against MPTP: relevance to the role of glutathione in Parkinson's disease, *Neurosci. Lett.*, 2006, **402**, 137–141.
85. K. S. Saravanan, K. M. Sindhu, K. S. Senthilkumar and K. P. Mohanakumar, L-deprenyl protects against rotenone-induced, oxidative stress-mediated dopaminergic neurodegeneration in rats, *Neurochem. Int.*, 2006, **49**, 28–40.
86. E. Sofic, K. W. Lange, K. Jellinger and P. Riederer, Reduced and oxidized glutathione in the substantia nigra of patients with Parkinson's disease, *Neurosci. Lett.*, 1992, **142**, 128–130.
87. K. Sriram, K. S. Pai, M. R. Boyd and V. Ravindranath, Evidence for generation of oxidative stress in brain by MPTP: *in vitro* and *in vivo* studies in mice, *Brain Res.*, 1997, **749**, 44–52.
88. S. J. Kish, C. Morito and O. Hornykiewicz, Glutathione peroxidase activity in Parkinson's disease brain, *Neurosci. Lett.*, 1985, **58**, 343–346.
89. D. C. Duke, L. B. Moran, R. K. Pearce and M. B. Graeber, The medial and lateral substantia nigra in Parkinson's disease: mRNA profiles associated with higher brain tissue vulnerability, *Neurogenetics*, 2007, **8**, 83–94.
90. W. G. Kim, R. P. Mohney, B. Wilson, G. H. Jeohn, B. Liu and J. S. Hong, Regional difference in susceptibility to lipopolysaccharide-induced neurotoxicity in the rat brain: role of microglia, *J. Neurosci.*, 2000, **20**, 6309–6316.
91. V. V. Belousov, G. N. Enikolopov and N. M. Mishina, Compartmentalization of ROS-mediated signal transduction, *Bioorg. Khim.*, 2013, **39**, 383–399.
92. G. P. Bienert, A. L. Moller, K. A. Kristiansen, A. Schulz, I. M. Moller, J. K. Schjoerring and T. P. Jahn, Specific aquaporins facilitate the diffusion of hydrogen peroxide across membranes, *J. Biol. Chem.*, 2007, **282**, 1183–1192.
93. J. M. Denu and K. G. Tanner, Specific and reversible inactivation of protein tyrosine phosphatases by hydrogen peroxide: evidence for a sulfenic acid intermediate and implications for redox regulation, *Biochemistry*, 1998, **37**, 5633–5642.
94. M. F. Beal, R. T. Matthews, A. Tieleman and C. W. Shults, Coenzyme Q10 attenuates the 1-methyl-4-phenyl-1,2,3,tetrahydropyridine (MPTP) induced loss of striatal dopamine and dopaminergic axons in aged mice, *Brain Res.*, 1998, **783**, 109–114.
95. M. Roghani and G. Behzadi, Neuroprotective effect of vitamin E on the early model of Parkinson's disease in rat: behavioral and histochemical evidence, *Brain Res.*, 2001, **892**, 211–217.
96. S. Fahn, A pilot trial of high-dose alpha-tocopherol and ascorbate in early Parkinson's discase, *Ann. Neurol.*, 1992, **32**, S128–S132.
97. B. J. Snow, F. L. Rolfe, M. M. Lockhart, C. M. Frampton, J. D. O'Sullivan, V. Fung, R. A. Smith, M. P. Murphy and K. M. Taylor, A double-blind,

placebo-controlled study to assess the mitochondria-targeted antioxidant MitoQ as a disease-modifying therapy in Parkinson's disease, *Mov. Disord.*, 2010, **25**, 1670–1674.
98. C. A. Weber and M. E. Ernst, Antioxidants, supplements, and Parkinson's disease, *Ann. Pharmacother.*, 2006, **40**, 935–938.
99. S. M. Zhang, M. A. Hernan, H. Chen, D. Spiegelman, W. C. Willett and A. Ascherio, Intakes of vitamins E and C, carotenoids, vitamin supplements, and PD risk, *Neurology*, 2002, **59**, 1161–1169.
100. R. Kohen and A. Nyska, Oxidation of biological systems: oxidative stress phenomena, antioxidants, redox reactions, and methods for their quantification, *Toxicol. Pathol.*, 2002, **30**, 620–650.
101. D. Trachootham, W. Lu, M. A. Ogasawara, R. D. Nilsa and P. Huang, Redox regulation of cell survival, *Antioxid. Redox Signaling*, 2008, **10**, 1343–1374.
102. G. Fujino, T. Noguchi, K. Takeda and H. Ichijo, Thioredoxin and protein kinases in redox signaling, *Semin. Cancer Biol.*, 2006, **16**, 427–435.
103. T. Fukai and M. Ushio-Fukai, Superoxide dismutases: role in redox signaling, vascular function, and diseases, *Antioxid. Redox Signaling*, 2011, **15**, 1583–1606.
104. M. M. Gallogly and J. J. Mieyal, Mechanisms of reversible protein glutathionylation in redox signaling and oxidative stress, *Curr. Opin. Pharmacol.*, 2007, **7**, 381–391.
105. J. Park, S. Lee, S. Lee and S. W. Kang, 2-cys peroxiredoxins: emerging hubs determining redox dependency of Mammalian signaling networks, *Int. J. Cell Biol.*, 2014, **2014**, 715867.
106. C. Nathan, Specificity of a third kind: reactive oxygen and nitrogen intermediates in cell signaling, *J. Clin. Invest.*, 2003, **111**, 769–778.
107. T. Finkel, Redox-dependent signal transduction, *FEBS Lett.*, 2000, **476**, 52–54.
108. Y. J. Suzuki, H. J. Forman and A. Sevanian, Oxidants as stimulators of signal transduction, *Free Radical Biol. Med.*, 1997, **22**, 269–285.
109. B. D. Manning and L. C. Cantley, AKT/PKB signaling: navigating downstream, *Cell*, 2007, **129**, 1261–1274.
110. L. Durgadoss, P. Nidadavolu, R. K. Valli, U. Saeed, M. Mishra, P. Seth and V. Ravindranath, Redox modification of Akt mediated by the dopaminergic neurotoxin MPTP, in mouse midbrain, leads to down-regulation of pAkt, *FASEB J.*, 2012, **26**, 1473–1483.
111. C. Malagelada, Z. H. Jin and L. A. Greene, RTP801 is induced in Parkinson's disease and mediates neuron death by inhibiting Akt phosphorylation/activation, *J. Neurosci.*, 2008, **28**, 14363–14371.
112. S. Timmons, M. F. Coakley, A. M. Moloney and C. O'Neill, Akt signal transduction dysfunction in Parkinson's disease, *Neurosci. Lett.*, 2009, **467**, 30–35.
113. D. P. Brazil and B. A. Hemmings, Ten years of protein kinase B signalling: a hard Akt to follow, *Trends Biochem. Sci.*, 2001, **26**, 657–664.

114. S. P. Staal, J. W. Hartley and W. P. Rowe, Isolation of transforming murine leukemia viruses from mice with a high incidence of spontaneous lymphoma, *Proc. Natl. Acad. Sci. U. S. A.*, 1977, **74**, 3065–3067.
115. E. Gonzalez and T. E. McGraw, The Akt kinases: isoform specificity in metabolism and cancer, *Cell Cycle*, 2009, **8**, 2502–2508.
116. B. Vanhaesebroeck and D. R. Alessi, The PI3K-PDK1 connection: more than just a road to PKB, *Biochem. J.*, 2000, **346**(pt 3), 561–576.
117. V. Ries, C. Henchcliffe, T. Kareva, M. Rzhetskaya, R. Bland, M. J. During, N. Kholodilov and R. E. Burke, Oncoprotein Akt/PKB induces trophic effects in murine models of Parkinson's disease, *Proc. Natl. Acad. Sci. U. S. A.*, 2006, **103**, 18757–18762.
118. D. R. Alessi, M. Andjelkovic, B. Caudwell, P. Cron, N. Morrice, P. Cohen and B. A. Hemmings, Mechanism of activation of protein kinase B by insulin and IGF-1, *EMBO J.*, 1996, **15**, 6541–6551.
119. T. F. Franke, S. I. Yang, T. O. Chan, K. Datta, A. Kazlauskas, D. K. Morrison, D. R. Kaplan and P. N. Tsichlis, The protein kinase encoded by the Akt proto-oncogene is a target of the PDGF-activated phosphatidylinositol 3-kinase, *Cell*, 1995, **81**, 727–736.
120. M. Frech, M. Andjelkovic, E. Ingley, K. K. Reddy, J. R. Falck and B. A. Hemmings, High affinity binding of inositol phosphates and phosphoinositides to the pleckstrin homology domain of RAC/protein kinase B and their influence on kinase activity, *J. Biol. Chem.*, 1997, **272**, 8474–8481.
121. M. Andjelkovic, T. Jakubowicz, P. Cron, X. F. Ming, J. W. Han and B. A. Hemmings, Activation and phosphorylation of a pleckstrin homology domain containing protein kinase (RAC-PK/PKB) promoted by serum and protein phosphatase inhibitors, *Proc. Natl. Acad. Sci. U. S. A.*, 1996, **93**, 5699–5704.
122. D. R. Alessi, S. R. James, C. P. Downes, A. B. Holmes, P. R. Gaffney, C. B. Reese and P. Cohen, Characterization of a 3-phosphoinositide-dependent protein kinase which phosphorylates and activates protein kinase Bα, *Curr. Biol.*, 1997, **7**, 261–269.
123. R. A. Currie, K. S. Walker, A. Gray, M. Deak, A. Casamayor, C. P. Downes, P. Cohen, D. R. Alessi and J. Lucocq, Role of phosphatidylinositol 3,4,5-trisphosphate in regulating the activity and localization of 3-phosphoinositide-dependent protein kinase-1, *Biochem. J.*, 1999, **337**(pt 3), 575–583.
124. B. A. Hemmings and D. F. Restuccia, PI3K-PKB/Akt pathway, *Cold Spring Harbor Perspect. Biol.*, 2012, **4**, a011189.
125. D. D. Sarbassov, D. A. Guertin, S. M. Ali and D. M. Sabatini, Phosphorylation and regulation of Akt/PKB by the rictor-mTOR complex, *Science*, 2005, **307**, 1098–1101.
126. V. Stambolic, A. Suzuki, J. L. de la Pompa, G. M. Brothers, C. Mirtsos, T. Sasaki, J. Ruland, J. M. Penninger, D. P. Siderovski and T. W. Mak, Negative regulation of PKB/Akt-dependent cell survival by the tumor suppressor PTEN, *Cell*, 1998, **95**, 29–39.

127. S. Lim and M. V. Clement, Phosphorylation of the survival kinase Akt by superoxide is dependent on an ascorbate-reversible oxidation of PTEN, *Free Radical Biol. Med.*, 2007, **42**, 1178–1192.
128. L. Luo, J. Kaur Kumar and M. V. Clement, Redox control of cytosolic Akt phosphorylation in PTEN null cells, *Free Radical Biol. Med.*, 2012, **53**, 1697–1707.
129. Y. Zhang, A. C. Granholm, K. Huh, L. Shan, O. Diaz-Ruiz, N. Malik, L. Olson, B. J. Hoffer, C. R. Lupica, A. F. Hoffman and C. M. Backman, PTEN deletion enhances survival, neurite outgrowth and function of dopamine neuron grafts to MitoPark mice, *Brain*, 2012, **135**, 2736–2749.
130. S. Hunot, B. Brugg, D. Ricard, P. P. Michel, M. P. Muriel, M. Ruberg, B. A. Faucheux, Y. Agid and E. C. Hirsch, Nuclear translocation of NF-κB is increased in dopaminergic neurons of patients with Parkinson disease, *Proc. Natl. Acad. Sci. U. S. A.*, 1997, **94**, 7531–7536.
131. P. Lei, S. Ayton, A. I. Bush and P. A. Adlard, GSK-3 in neurodegenerative diseases, *Int. J. Alzheimer's Dis.*, 2011, **2011**, 189246.
132. Y. Jia, S. J. Mo, Q. Q. Feng, M. L. Zhan, L. S. OuYang, J. C. Chen, Y. X. Ma, J. J. Wu and W. L. Lei, EPO-dependent activation of PI3K/Akt/FoxO3a signalling mediates neuroprotection in *in vitro* and *in vivo* models of Parkinson's disease, *J. Mol. Neurosci.*, 2014, **53**, 117–124.
133. H. Dudek, S. R. Datta, T. F. Franke, M. J. Birnbaum, R. Yao, G. M. Cooper, R. A. Segal, D. R. Kaplan and M. E. Greenberg, Regulation of neuronal survival by the serine-threonine protein kinase Akt, *Science*, 1997, **275**, 661–665.
134. T. F. Franke, D. R. Kaplan and L. C. Cantley, PI3K: downstream AKTion blocks apoptosis, *Cell*, 1997, **88**, 435–437.
135. C. Malagelada, Z. H. Jin, V. Jackson-Lewis, S. Przedborski and L. A. Greene, Rapamycin protects against neuron death in *in vitro* and *in vivo* models of Parkinson's disease, *J. Neurosci.*, 2010, **30**, 1166–1175.
136. Y. Xu, C. Liu, S. Chen, Y. Ye, M. Guo, Q. Ren, L. Liu, H. Zhang, C. Xu, Q. Zhou, S. Huang and L. Chen, Activation of AMPK and inactivation of Akt result in suppression of mTOR-mediated S6K1 and 4E-BP1 pathways leading to neuronal cell death in *in vitro* models of Parkinson's disease, *Cell. Signalling*, 2014, **26**, 1680–1689.
137. A. Quesada, B. Y. Lee and P. E. Micevych, PI3 kinase/Akt activation mediates estrogen and IGF-1 nigral DA neuronal neuroprotection against a unilateral rat model of Parkinson's disease, *Dev. Neurobiol.*, 2008, **68**, 632–644.
138. Y. Tasaki, J. Yamamoto, T. Omura, T. Sakaguchi, N. Kimura, K. Ohtaki, T. Ono, M. Suno, M. Asari, T. Ohkubo, T. Noda, T. Awaya, K. Shimizu and K. Matsubara, Meloxicam ameliorates motor dysfunction and dopaminergic neurodegeneration by maintaining Akt-signaling in a mouse Parkinson's disease model, *Neurosci. Lett.*, 2012, **521**, 15–19.
139. K. Nakaso, S. Ito and K. Nakashima, Caffeine activates the PI3K/Akt pathway and prevents apoptotic cell death in a Parkinson's disease model of SH-SY5Y cells, *Neurosci. Lett.*, 2008, **432**, 146–150.

140. X. Huang, M. Begley, K. A. Morgenstern, Y. Gu, P. Rose, H. Zhao and X. Zhu, Crystal structure of an inactive Akt2 kinase domain, *Structure*, 2003, **11**, 21–30.
141. R. Wani, J. Qian, L. Yin, E. Bechtold, S. B. King, L. B. Poole, E. Paek, A. W. Tsang and C. M. Furdui, Isoform-specific regulation of Akt by PDGF-induced reactive oxygen species, *Proc. Natl. Acad. Sci. U. S. A.*, 2011, **108**, 10550–10555.
142. H. Murata, Y. Ihara, H. Nakamura, J. Yodoi, K. Sumikawa and T. Kondo, Glutaredoxin exerts an antiapoptotic effect by regulating the redox state of Akt, *J. Biol. Chem.*, 2003, **278**, 50226–50233.
143. Y. Urata, Y. Ihara, H. Murata, S. Goto, T. Koji, J. Yodoi, S. Inoue and T. Kondo, 17β-estradiol protects against oxidative stress-induced cell death through the glutathione/glutaredoxin-dependent redox regulation of Akt in myocardiac H9c2 cells, *J. Biol. Chem.*, 2006, **281**, 13092–13102.
144. K. Gunjima, R. Tomiyama, K. Takakura, T. Yamada, K. Hashida, Y. Nakamura, T. Konishi, S. Matsugo and O. Hori, 3,4-dihydroxybenzalacetone protects against Parkinson's disease-related neurotoxin 6-OHDA through Akt/Nrf2/glutathione pathway, *J. Cell. Biochem.*, 2014, **115**, 151–160.
145. F. Ahmad, P. Nidadavolu, L. Durgadoss and V. Ravindranath, Critical cysteines in Akt1 regulate its activity and proteasomal degradation: implications for neurodegenerative diseases, *Free Radical Biol. Med.*, 2014, **74**, 118–128.
146. X. Hu, Z. Weng, C. T. Chu, L. Zhang, G. Cao, Y. Gao, A. Signore, J. Zhu, T. Hastings, J. T. Greenamyre and J. Chen, Peroxiredoxin-2 protects against 6-hydroxydopamine-induced dopaminergic neurodegeneration *via* attenuation of the apoptosis signal-regulating kinase (ASK1) signaling cascade, *J. Neurosci.*, 2011, **31**, 247–261.
147. S. Karunakaran, L. Diwakar, U. Saeed, V. Agarwal, S. Ramakrishnan, S. Iyengar and V. Ravindranath, Activation of apoptosis signal regulating kinase 1 (ASK1) and translocation of death-associated protein, Daxx, in substantia nigra pars compacta in a mouse model of Parkinson's disease: protection by α-lipoic acid, *FASEB J.*, 2007, **21**, 2226–2236.
148. M. Ouyang and X. Shen, Critical role of ASK1 in the 6-hydroxydopamine-induced apoptosis in human neuroblastoma SH-SY5Y cells, *J. Neurochem.*, 2006, **97**, 234–244.
149. M. Saitoh, H. Nishitoh, M. Fujii, K. Takeda, K. Tobiume, Y. Sawada, M. Kawabata, K. Miyazono and H. Ichijo, Mammalian thioredoxin is a direct inhibitor of apoptosis signal-regulating kinase (ASK) 1, *EMBO J.*, 1998, **17**, 2596–2606.
150. S. A. Correa and K. L. Eales, The role of p38 MAPK and its substrates in neuronal plasticity and neurodegenerative disease, *J. Signal Transduction*, 2012, **2012**, 649079.
151. M. Karin and E. Gallagher, From JNK to pay dirt: jun kinases, their biochemistry, physiology and clinical importance, *IUBMB Life*, 2005, **57**, 283–295.

152. A. R. Nebreda and A. Porras, p38 MAP kinases: beyond the stress response, *Trends Biochem. Sci.*, 2000, **25**, 257–260.
153. H. Ichijo, E. Nishida, K. Irie, P. ten Dijke, M. Saitoh, T. Moriguchi, M. Takagi, K. Matsumoto, K. Miyazono and Y. Gotoh, Induction of apoptosis by ASK1, a mammalian MAPKKK that activates SAPK/JNK and p38 signaling pathways, *Science*, 1997, **275**, 90–94.
154. S. G. Cho, Y. H. Lee, H. S. Park, K. Ryoo, K. W. Kang, J. Park, S. J. Eom, M. J. Kim, T. S. Chang, S. Y. Choi, J. Shim, Y. Kim, M. S. Dong, M. J. Lee, S. G. Kim, H. Ichijo and E. J. Choi, Glutathione S-transferase mu modulates the stress-activated signals by suppressing apoptosis signal-regulating kinase 1, *J. Biol. Chem.*, 2001, **276**, 12749–12755.
155. M. Subramanyam, N. Takahashi, Y. Hasegawa, T. Mohri and Y. Okada, Inhibition of protein kinase Akt1 by apoptosis signal-regulating kinase-1 (ASK1) is involved in apoptotic inhibition of regulatory volume increase, *J. Biol. Chem.*, 2010, **285**, 6109–6117.
156. H. Nishitoh, A. Matsuzawa, K. Tobiume, K. Saegusa, K. Takeda, K. Inoue, S. Hori, A. Kakizuka and H. Ichijo, ASK1 is essential for endoplasmic reticulum stress-induced neuronal cell death triggered by expanded polyglutamine repeats, *Genes Dev.*, 2002, **16**, 1345–1355.
157. V. N. Ivanov and Z. Ronai, p38 protects human melanoma cells from UV-induced apoptosis through down-regulation of NF-κB activity and Fas expression, *Oncogene*, 2000, **19**, 3003–3012.
158. K. W. Yoon, J. H. Cho, J. K. Lee, Y. H. Kang, J. S. Chae, Y. M. Kim, J. Kim, E. K. Kim, S. E. Kim, J. H. Baik, U. P. Naik, S. G. Cho and E. J. Choi, CIB1 functions as a Ca^{2+}-sensitive modulator of stress-induced signaling by targeting ASK1, *Proc. Natl. Acad. Sci. U. S. A.*, 2009, **106**, 17389–17394.
159. K. Takeda, T. Noguchi, I. Naguro and H. Ichijo, Apoptosis signal-regulating kinase 1 in stress and immune response, *Annu. Rev. Pharmacol. Toxicol.*, 2008, **48**, 199–225.
160. J. J. Song, J. G. Rhee, M. Suntharalingam, S. A. Walsh, D. R. Spitz and Y. J. Lee, Role of glutaredoxin in metabolic oxidative stress. Glutaredoxin as a sensor of oxidative stress mediated by H_2O_2, *J. Biol. Chem.*, 2002, **277**, 46566–46575.
161. S. Dorion, H. Lambert and J. Landry, Activation of the p38 signaling pathway by heat shock involves the dissociation of glutathione S-transferase Mu from Ask1, *J. Biol. Chem.*, 2002, **277**, 30792–30797.
162. J. Waak, S. S. Weber, K. Gorner, C. Schall, H. Ichijo, T. Stehle and P. J. Kahle, Oxidizable residues mediating protein stability and cytoprotective interaction of DJ-1 with apoptosis signal-regulating kinase 1, *J. Biol. Chem.*, 2009, **284**, 14245–14257.
163. S. Y. Kim, T. J. Kim and K. Y. Lee, A novel function of peroxiredoxin 1 (Prx-1) in apoptosis signal-regulating kinase 1 (ASK1)-mediated signaling pathway, *FEBS Lett.*, 2008, **582**, 1913–1918.
164. L. Zhang, J. Chen and H. Fu, Suppression of apoptosis signal-regulating kinase 1-induced cell death by 14-3-3 proteins, *Proc. Natl. Acad. Sci. U. S. A.*, 1999, **96**, 8511–8515.

165. K. Morita, M. Saitoh, K. Tobiume, H. Matsuura, S. Enomoto, H. Nishitoh and H. Ichijo, Negative feedback regulation of ASK1 by protein phosphatase 5 (PP5) in response to oxidative stress, *EMBO J.*, 2001, **20**, 6028–6036.
166. H. S. Park, S. G. Cho, C. K. Kim, H. S. Hwang, K. T. Noh, M. S. Kim, S. H. Huh, M. J. Kim, K. Ryoo, E. K. Kim, W. J. Kang, J. S. Lee, J. S. Seo, Y. G. Ko, S. Kim and E. J. Choi, Heat shock protein hsp72 is a negative regulator of apoptosis signal-regulating kinase 1, *Mol. Cell. Biol.*, 2002, **22**, 7721–7730.
167. A. H. Kim, G. Khursigara, X. Sun, T. F. Franke and M. V. Chao, Akt phosphorylates and negatively regulates apoptosis signal-regulating kinase 1, *Mol. Cell. Biol.*, 2001, **21**, 893–901.
168. P. J. Nadeau, S. J. Charette, M. B. Toledano and J. Landry, Disulfide bond-mediated multimerization of Ask1 and its reduction by thioredoxin-1 regulate H(2)O(2)-induced c-Jun NH(2)-terminal kinase activation and apoptosis, *Mol. Biol. Cell*, 2007, **18**, 3903–3913.
169. G. Fujino, T. Noguchi, A. Matsuzawa, S. Yamauchi, M. Saitoh, K. Takeda and H. Ichijo, Thioredoxin and TRAF family proteins regulate reactive oxygen species-dependent activation of ASK1 through reciprocal modulation of the N-terminal homophilic interaction of ASK1, *Mol. Cell. Biol.*, 2007, **27**, 8152–8163.
170. R. Zhang, X. He, W. Liu, M. Lu, J. T. Hsieh and W. Min, AIP1 mediates TNF-α-induced ASK1 activation by facilitating dissociation of ASK1 from its inhibitor 14-3-3, *J. Clin. Invest.*, 2003, **111**, 1933–1943.
171. H. Y. Chang, H. Nishitoh, X. Yang, H. Ichijo and D. Baltimore, Activation of apoptosis signal-regulating kinase 1 (ASK1) by the adapter protein Daxx, *Science*, 1998, **281**, 1860–1863.
172. P. J. Nadeau, S. J. Charette and J. Landry, REDOX reaction at ASK1-Cys250 is essential for activation of JNK and induction of apoptosis, *Mol. Biol. Cell*, 2009, **20**, 3628–3637.
173. A. Holmgren, Thioredoxin, *Annu. Rev. Biochem.*, 1985, **54**, 237–271.
174. J. Lu and A. Holmgren, The thioredoxin antioxidant system, *Free Radical Biol. Med.*, 2014, **66**, 75–87.
175. Y. Liu and W. Min, Thioredoxin promotes ASK1 ubiquitination and degradation to inhibit ASK1-mediated apoptosis in a redox activity-independent manner, *Circ. Res.*, 2002, **90**, 1259–1266.
176. A. Lippoldt, C. A. Padilla, H. Gerst, B. Andbjer, E. Richter, A. Holmgren and K. Fuxe, Localization of thioredoxin in the rat brain and functional implications, *J. Neurosci.*, 1995, **15**, 6747–6756.
177. J. R. Godoy, M. Funke, W. Ackermann, P. Haunhorst, S. Oesteritz, F. Capani, H. P. Elsasser and C. H. Lillig, Redox atlas of the mouse. Immunohistochemical detection of glutaredoxin-, peroxiredoxin-, and thioredoxin-family proteins in various tissues of the laboratory mouse, *Biochim. Biophys. Acta*, 2011, **1810**, 2–92.
178. J. Bai, H. Nakamura, Y. W. Kwon, M. Tanito, S. Ueda, T. Tanaka, I. Hattori, S. Ban, T. Momoi, Y. Kitao, S. Ogawa and J. Yodoi, Does thioredoxin-1

prevent mitochondria- and endoplasmic reticulum-mediated neurotoxicity of 1-methyl-4-phenyl-1,2,3,6-tetrahydropyridine? *Antioxid. Redox Signaling*, 2007, **9**, 603–608.

179. L. Arodin, A. Miranda-Vizuete, P. Swoboda and A. P. Fernandes, Protective effects of the thioredoxin and glutaredoxin systems in dopamine-induced cell death, *Free Radical Biol. Med.*, 2014, **73**, 328–336.
180. M. M. Gallogly, D. W. Starke and J. J. Mieyal, Mechanistic and kinetic details of catalysis of thiol-disulfide exchange by glutaredoxins and potential mechanisms of regulation, *Antioxid. Redox Signaling*, 2009, **11**, 1059–1081.
181. S. A. Gravina and J. J. Mieyal, Thioltransferase is a specific glutathionyl mixed disulfide oxidoreductase, *Biochemistry*, 1993, **32**, 3368–3376.
182. S. Balijepalli, P. S. Tirumalai, K. V. Swamy, M. R. Boyd, J. J. Mieyal and V. Ravindranath, Rat brain thioltransferase: regional distribution, immunological characterization, and localization by fluorescent *in situ* hybridization, *J. Neurochem.*, 1999, **72**, 1170–1178.
183. R. S. Kenchappa, L. Diwakar, J. Annepu and V. Ravindranath, Estrogen and neuroprotection: higher constitutive expression of glutaredoxin in female mice offers protection against MPTP-mediated neurodegeneration, *FASEB J.*, 2004, **18**, 1102–1104.
184. H. Rodriguez-Rocha, A. Garcia Garcia, L. Zavala-Flores, S. Li, N. Madayiputhiya and R. Franco, Glutaredoxin 1 protects dopaminergic cells by increased protein glutathionylation in experimental Parkinson's disease, *Antioxid. Redox Signaling*, 2012, **17**, 1676–1693.
185. W. M. Johnson, C. Yao, S. L. Siedlak, W. Wang, X. Zhu, G. A. Caldwell, A. L. Wilson-Delfosse, J. J. Mieyal and S. G. Chen, Glutaredoxin deficiency exacerbates neurodegeneration in C. elegans models of Parkinson's disease, *Hum. Mol. Genet.*, 2015, **24**, 1322–1335.
186. J. S. Mo, J. Jung, J. H. Yoon, J. A. Hong, M. Y. Kim, E. J. Ann, M. S. Seo, Y. H. Choi and H. S. Park, DJ-1 modulates the p38 mitogen-activated protein kinase pathway through physical interaction with apoptosis signal-regulating kinase 1, *J. Cell. Biochem.*, 2010, **110**, 229–237.
187. Y. O. Akazawa, Y. Saito, T. Hamakubo, Y. Masuo, Y. Yoshida, K. Nishio, M. Shichiri, T. Miyasaka, H. Iwanari, Y. Mochizuki, T. Kodama, N. Noguchi and E. Niki, Elevation of oxidized DJ-1 in the brain and erythrocytes of Parkinson disease model animals, *Neurosci. Lett.*, 2010, **483**, 201–205.
188. T. Kinumi, J. Kimata, T. Taira, H. Ariga and E. Niki, Cysteine-106 of DJ-1 is the most sensitive cysteine residue to hydrogen peroxide-mediated oxidation *in vivo* in human umbilical vein endothelial cells, *Biochem. Biophys. Res. Commun.*, 2004, **317**, 722–728.
189. M. A. Wilson, The role of cysteine oxidation in DJ-1 function and dysfunction, *Antioxid. Redox Signaling*, 2011, **15**, 111–122.
190. R. M. Canet-Aviles, M. A. Wilson, D. W. Miller, R. Ahmad, C. McLendon, S. Bandyopadhyay, M. J. Baptista, D. Ringe, G. A. Petsko and M. R. Cookson, The Parkinson's disease protein DJ-1 is neuroprotective due to cysteine-sulfinic acid-driven mitochondrial localization, *Proc. Natl. Acad. Sci. U. S. A.*, 2004, **101**, 9103–9108.

191. J. Y. Im, K. W. Lee, E. Junn and M. M. Mouradian, DJ-1 protects against oxidative damage by regulating the thioredoxin/ASK1 complex, *Neurosci. Res.*, 2010, **67**, 203–208.
192. R. Bandopadhyay, A. E. Kingsbury, M. R. Cookson, A. R. Reid, I. M. Evans, A. D. Hope, A. M. Pittman, T. Lashley, R. Canet-Aviles, D. W. Miller, C. McLendon, C. Strand, A. J. Leonard, P. M. Abou-Sleiman, D. G. Healy, H. Ariga, N. W. Wood, R. de Silva, T. Revesz, J. A. Hardy and A. J. Lees, The expression of DJ-1 (PARK7) in normal human CNS and idiopathic Parkinson's disease, *Brain*, 2004, **127**, 420–430.
193. N. Kotaria, U. Hinz, S. Zechel and O. von Bohlen Und Halbach, Localization of DJ-1 protein in the murine brain, *Cell Tissue Res.*, 2005, **322**, 503–507.
194. H. Z. Chae, K. Robison, L. B. Poole, G. Church, G. Storz and S. G. Rhee, Cloning and sequencing of thiol-specific antioxidant from mammalian brain: alkyl hydroperoxide reductase and thiol-specific antioxidant define a large family of antioxidant enzymes, *Proc. Natl. Acad. Sci. U. S. A.*, 1994, **91**, 7017–7021.
195. L. B. Poole, The catalytic mechanism of peroxiredoxins, *Subcell. Biochem.*, 2007, **44**, 61–81.
196. R. M. Jarvis, S. M. Hughes and E. C. Ledgerwood, Peroxiredoxin 1 functions as a signal peroxidase to receive, transduce, and transmit peroxide signals in mammalian cells, *Free Radical Biol. Med.*, 2012, **53**, 1522–1530.
197. J. Goemaere and B. Knoops, Peroxiredoxin distribution in the mouse brain with emphasis on neuronal populations affected in neurodegenerative disorders, *J. Comp. Neurol.*, 2012, **520**, 258–280.
198. J. Fang, T. Nakamura, D. H. Cho, Z. Gu and S. A. Lipton, S-nitrosylation of peroxiredoxin 2 promotes oxidative stress-induced neuronal cell death in Parkinson's disease, *Proc. Natl. Acad. Sci. U. S. A.*, 2007, **104**, 18742–18747.
199. K. Krapfenbauer, E. Engidawork, N. Cairns, M. Fountoulakis and G. Lubec, Aberrant expression of peroxiredoxin subtypes in neurodegenerative disorders, *Brain Res.*, 2003, **967**, 152–160.
200. G. Boka, P. Anglade, D. Wallach, F. Javoy-Agid, Y. Agid and E. C. Hirsch, Immunocytochemical analysis of tumor necrosis factor and its receptors in Parkinson's disease, *Neurosci. Lett.*, 1994, **172**, 151–154.
201. M. Mogi, A. Togari, T. Kondo, Y. Mizuno, O. Komure, S. Kuno, H. Ichinose and T. Nagatsu, Caspase activities and tumor necrosis factor receptor R1 (p55) level are elevated in the substantia nigra from parkinsonian brain, *J. Neural Transm.*, 2000, **107**, 335–341.
202. Y. Gotoh and J. A. Cooper, Reactive oxygen species- and dimerization-induced activation of apoptosis signal-regulating kinase 1 in tumor necrosis factor-α signal transduction, *J. Biol. Chem.*, 1998, **273**, 17477–17482.
203. H. Liu, H. Nishitoh, H. Ichijo and J. M. Kyriakis, Activation of apoptosis signal-regulating kinase 1 (ASK1) by tumor necrosis factor receptor-associated factor 2 requires prior dissociation of the ASK1 inhibitor thioredoxin, *Mol. Cell. Biol.*, 2000, **20**, 2198–2208.

204. K. Tobiume, A. Matsuzawa, T. Takahashi, H. Nishitoh, K. Morita, K. Takeda, O. Minowa, K. Miyazono, T. Noda and H. Ichijo, ASK1 is required for sustained activations of JNK/p38 MAP kinases and apoptosis, *EMBO Rep.*, 2001, **2**, 222–228.
205. H. Nishitoh, M. Saitoh, Y. Mochida, K. Takeda, H. Nakano, M. Rothe, K. Miyazono and H. Ichijo, ASK1 is essential for JNK/SAPK activation by TRAF2, *Mol. Cell*, 1998, **2**, 389–395.
206. S. Shiizaki, I. Naguro and H. Ichijo, Activation mechanisms of ASK1 in response to various stresses and its significance in intracellular signaling, *Adv. Biol. Regul.*, 2014, **53**, 135–144.
207. E. C. Hirsch, S. Hunot, P. Damier and B. Faucheux, Glial cells and inflammation in Parkinson's disease: a role in neurodegeneration? *Ann. Neurol.*, 1998, **44**, S115–S120.
208. K. Sriram, J. M. Matheson, S. A. Benkovic, D. B. Miller, M. I. Luster and J. P. O'Callaghan, Mice deficient in TNF receptors are protected against dopaminergic neurotoxicity: implications for Parkinson's disease, *FASEB J.*, 2002, **16**, 1474–1476.
209. K. Sriram, J. M. Matheson, S. A. Benkovic, D. B. Miller, M. I. Luster and J. P. O'Callaghan, Deficiency of TNF receptors suppresses microglial activation and alters the susceptibility of brain regions to MPTP-induced neurotoxicity: role of TNF-α, *FASEB J.*, 2006, **20**, 670–682.
210. E. Rousselet, J. Callebert, K. Parain, C. Joubert, S. Hunot, A. Hartmann, C. Jacque, F. Perez-Diaz, C. Cohen-Salmon, J. M. Launay and E. C. Hirsch, Role of TNF-α receptors in mice intoxicated with the parkinsonian toxin MPTP, *Exp. Neurol.*, 2002, **177**, 183–192.
211. B. Ferger, A. Leng, A. Mura, B. Hengerer and J. Feldon, Genetic ablation of tumor necrosis factor-alpha (TNF-α) and pharmacological inhibition of TNF-synthesis attenuates MPTP toxicity in mouse striatum, *J. Neurochem.*, 2004, **89**, 822–833.
212. C. Zhao, Z. Ling, M. B. Newman, A. Bhatia and P. M. Carvey, TNF-α knockout and minocycline treatment attenuates blood-brain barrier leakage in MPTP-treated mice, *Neurobiol. Dis.*, 2007, **26**, 36–46.
213. C. J. Barnum, X. Chen, J. Chung, J. Chang, M. Williams, N. Grigoryan, R. J. Tesi and M. G. Tansey, Peripheral administration of the selective inhibitor of soluble Tumor necrosis factor (TNF) XPro(R)1595 attenuates nigral cell loss and glial activation in 6-OHDA hemiparkinsonian rats, *J. Parkinson's Dis.*, 2014, **4**, 349–360.
214. M. K. McCoy, K. A. Ruhn, T. N. Martinez, F. E. McAlpine, A. Blesch and M. G. Tansey, Intranigral lentiviral delivery of dominant-negative TNF attenuates neurodegeneration and behavioral deficits in hemiparkinsonian rats, *Mol. Ther.*, 2008, **16**, 1572–1579.
215. J. J. Song and Y. J. Lee, Role of the ASK1-SEK1-JNK1-HIPK1 signal in Daxx trafficking and ASK1 oligomerization, *J. Biol. Chem.*, 2003, **278**, 47245–47252.
216. U. Saeed, A. Ray, R. K. Valli, A. M. Kumar and V. Ravindranath, DJ-1 loss by glutaredoxin but not glutathione depletion triggers Daxx translocation and cell death, *Antioxid. Redox Signaling*, 2010, **13**, 127–144.

217. E. Junn, H. Taniguchi, B. S. Jeong, X. Zhao, H. Ichijo and M. M. Mouradian, Interaction of DJ-1 with Daxx inhibits apoptosis signal-regulating kinase 1 activity and cell death, *Proc. Natl. Acad. Sci. U. S. A.*, 2005, **102**, 9691–9696.
218. S. Karunakaran, U. Saeed, M. Mishra, R. K. Valli, S. D. Joshi, D. P. Meka, P. Seth and V. Ravindranath, Selective activation of p38 mitogen-activated protein kinase in dopaminergic neurons of substantia nigra leads to nuclear translocation of p53 in 1-methyl-4-phenyl-1,2,3,6-tetrahydropyridine-treated mice, *J. Neurosci.*, 2008, **28**, 12500–12509.
219. K. W. Lee, X. Zhao, J. Y. Im, H. Grosso, W. H. Jang, T. W. Chan, P. K. Sonsalla, D. C. German, H. Ichijo, E. Junn and M. M. Mouradian, Apoptosis signal-regulating kinase 1 mediates MPTP toxicity and regulates glial activation, *PLoS One*, 2012, **7**, e29935.
220. M. Niso-Santano, R. A. Gonzalez-Polo, J. M. Bravo-San Pedro, R. Gomez-Sanchez, I. Lastres-Becker, M. A. Ortiz-Ortiz, G. Soler, J. M. Moran, A. Cuadrado and J. M. Fuentes, Activation of apoptosis signal-regulating kinase 1 is a key factor in paraquat-induced cell death: modulation by the Nrf2/Trx axis, *Free Radical Biol. Med.*, 2010, **48**, 1370–1381.
221. K. W. Lee, J. M. Woo, J. Y. Im, E. S. Park, L. He, H. Ichijo, E. Junn and M. M. Mouradian, Apoptosis signal-regulating kinase 1 modulates the phenotype of α-synuclein transgenic mice, *Neurobiol. Aging*, 2015, **36**, 519–526.
222. A. Cuadrado and A. R. Nebreda, Mechanisms and functions of p38 MAPK signaling, *Biochem. J.*, 2010, **429**, 403–417.
223. Y. T. Ip and R. J. Davis, Signal transduction by the c-Jun N-terminal kinase (JNK)–from inflammation to development, *Curr. Opin. Cell Biol.*, 1998, **10**, 205–219.
224. Y. Shen, R. Luche, B. Wei, M. L. Gordon, C. D. Diltz and N. K. Tonks, Activation of the Jnk signaling pathway by a dual-specificity phosphatase, JSP-1, *Proc. Natl. Acad. Sci. U. S. A.*, 2001, **98**, 13613–13618.
225. A. M. Bode and Z. Dong, The functional contrariety of JNK, *Mol. Carcinog.*, 2007, **46**, 591–598.
226. V. Waetzig and T. Herdegen, Context-specific inhibition of JNKs: overcoming the dilemma of protection and damage, *Trends Pharmacol. Sci.*, 2005, **26**, 455–461.
227. J. Wang, L. Chen, C. I. Ko, L. Zhang, A. Puga and Y. Xia, Distinct signaling properties of mitogen-activated protein kinase kinases 4 (MKK4) and 7 (MKK7) in embryonic stem cell (ESC) differentiation, *J. Biol. Chem.*, 2012, **287**, 2787–2797.
228. J. Yamauchi, Y. Kaziro and H. Itoh, Differential regulation of mitogen-activated protein kinase kinase 4 (MKK4) and 7 (MKK7) by signaling from G protein βγ subunit in human embryonal kidney 293 cells, *J. Biol. Chem.*, 1999, **274**, 1957–1965.
229. C. R. Weston and R. J. Davis, The JNK signal transduction pathway, *Curr. Opin. Genet. Dev.*, 2002, **12**, 14–21.

230. Y. J. Kim, W. S. Lee, C. Ip, H. Z. Chae, E. M. Park and Y. M. Park, Prx1 suppresses radiation-induced c-Jun NH_2-terminal kinase signaling in lung cancer cells through interaction with the glutathione S-transferase Pi/c-Jun NH_2-terminal kinase complex, *Cancer Res.*, 2006, **66**, 7136–7142.
231. H. Klintworth, K. Newhouse, T. Li, W. S. Choi, R. Faigle and Z. Xia, Activation of c-Jun N-terminal protein kinase is a common mechanism underlying paraquat- and rotenone-induced dopaminergic cell apoptosis, *Toxicol. Sci.*, 2007, **97**, 149–162.
232. J. Peng, X. O. Mao, F. F. Stevenson, M. Hsu and J. K. Andersen, The herbicide paraquat induces dopaminergic nigral apoptosis through sustained activation of the JNK pathway, *J. Biol. Chem.*, 2004, **279**, 32626–32632.
233. M. S. Saporito, B. A. Thomas and R. W. Scott, MPTP activates c-Jun NH_2-terminal kinase (JNK) and its upstream regulatory kinase MKK4 in nigrostriatal neurons *in vivo*, *J. Neurochem.*, 2000, **75**, 1200–1208.
234. L. H. Wang, C. G. Besirli and E. M. Johnson Jr, Mixed-lineage kinases: a target for the prevention of neurodegeneration, *Annu. Rev. Pharmacol. Toxicol.*, 2004, **44**, 451–474.
235. I. Ferrer, R. Blanco, M. Carmona, B. Puig, M. Barrachina, C. Gomez and S. Ambrosio, Active, phosphorylation-dependent mitogen-activated protein kinase (MAPK/ERK), stress-activated protein kinase/c-Jun N-terminal kinase (SAPK/JNK), and p38 kinase expression in Parkinson's disease and Dementia with Lewy bodies, *J. Neural Transm.*, 2001, **108**, 1383–1396.
236. W. Duan, X. Zhu, B. Ladenheim, Q. S. Yu, Z. Guo, J. Oyler, R. G. Cutler, J. L. Cadet, N. H. Greig and M. P. Mattson, p53 inhibitors preserve dopamine neurons and motor function in experimental parkinsonism, *Ann. Neurol.*, 2002, **52**, 597–606.
237. C. Perier, J. Bove, D. C. Wu, B. Dehay, D. K. Choi, V. Jackson-Lewis, S. Rathke-Hartlieb, P. Bouillet, A. Strasser, J. B. Schulz, S. Przedborski and M. Vila, Two molecular pathways initiate mitochondria-dependent dopaminergic neurodegeneration in experimental Parkinson's disease, *Proc. Natl. Acad. Sci. U. S. A.*, 2007, **104**, 8161–8166.
238. Parkinson's Study Group, The safety and tolerability of a mixed lineage kinase inhibitor (CEP-1347) in PD, *Neurology*, 2004, **62**, 330–332.
239. J. Raingeaud, S. Gupta, J. S. Rogers, M. Dickens, J. Han, R. J. Ulevitch and R. J. Davis, Pro-inflammatory cytokines and environmental stress cause p38 mitogen-activated protein kinase activation by dual phosphorylation on tyrosine and threonine, *J. Biol. Chem.*, 1995, **270**, 7420–7426.
240. Y. Jiang, C. Chen, Z. Li, W. Guo, J. A. Gegner, S. Lin and J. Han, Characterization of the structure and function of a new mitogen-activated protein kinase (p38β), *J. Biol. Chem.*, 1996, **271**, 17920–17926.
241. C. Lechner, M. A. Zahalka, J. F. Giot, N. P. Moller and A. Ullrich, ERK6, a mitogen-activated protein kinase involved in C2C12 myoblast differentiation, *Proc. Natl. Acad. Sci. U. S. A.*, 1996, **93**, 4355–4359.

242. Z. Li, Y. Jiang, R. J. Ulevitch and J. Han, The primary structure of p38γ: a new member of p38 group of MAP kinases, *Biochem. Biophys. Res. Commun.*, 1996, **228**, 334–340.
243. Y. Jiang, H. Gram, M. Zhao, L. New, J. Gu, L. Feng, F. Di Padova, R. J. Ulevitch and J. Han, Characterization of the structure and function of the fourth member of p38 group mitogen-activated protein kinases, p38δ, *J. Biol. Chem.*, 1997, **272**, 30122–30128.
244. S. K. Hanks and T. Hunter, Protein kinases 6. The eukaryotic protein kinase superfamily: kinase (catalytic) domain structure and classification, *FASEB J.*, 1995, **9**, 576–596.
245. Y. Jiang, Z. Li, E. M. Schwarz, A. Lin, K. Guan, R. J. Ulevitch and J. Han, Structure-function studies of p38 mitogen-activated protein kinase. Loop 12 influences substrate specificity and autophosphorylation, but not upstream kinase selection, *J. Biol. Chem.*, 1997, **272**, 11096–11102.
246. C. G. Parker, J. Hunt, K. Diener, M. McGinley, B. Soriano, G. A. Keesler, J. Bray, Z. Yao, X. S. Wang, T. Kohno and H. S. Lichenstein, Identification of stathmin as a novel substrate for p38δ, *Biochem. Biophys. Res. Commun.*, 1998, **249**, 791–796.
247. G. A. Keesler, J. Bray, J. Hunt, D. A. Johnson, T. Gleason, Z. Yao, S. W. Wang, C. Parker, H. Yamane, C. Cole and H. S. Lichenstein, Purification and activation of recombinant p38 isoforms α, β, γ, and δ, *Protein Expression Purif.*, 1998, **14**, 221–228.
248. F. Wu, Z. Wang, J. H. Gu, J. B. Ge, Z. Q. Liang and Z. H. Qin, p38(MAPK)/p53-Mediated Bax induction contributes to neurons degeneration in rotenone-induced cellular and rat models of Parkinson's disease, *Neurochem. Int.*, 2013, **63**, 133–140.
249. M. Karin and T. Hunter, Transcriptional control by protein phosphorylation: signal transmission from the cell surface to the nucleus, *Curr. Biol.*, 1995, **5**, 747–757.
250. N. Sakai, T. Wada, K. Furuichi, Y. Iwata, K. Yoshimoto, K. Kitagawa, S. Kokubo, M. Kobayashi, S. Takeda, H. Kida, K. Kobayashi, N. Mukaida, K. Matsushima and H. Yokoyama, p38 MAPK phosphorylation and NF-κB activation in human crescentic glomerulonephritis, *Nephrol., Dial., Transplant.*, 2002, **17**, 998–1004.
251. D. V. Bulavin, S. Saito, M. C. Hollander, K. Sakaguchi, C. W. Anderson, E. Appella and A. J. Fornace Jr, Phosphorylation of human p53 by p38 kinase coordinates N-terminal phosphorylation and apoptosis in response to UV radiation, *EMBO J.*, 1999, **18**, 6845–6854.
252. J. L. Perfettini, M. Castedo, R. Nardacci, F. Ciccosanti, P. Boya, T. Roumier, N. Larochette, M. Piacentini and G. Kroemer, Essential role of p53 phosphorylation by p38 MAPK in apoptosis induction by the HIV-1 envelope, *J. Exp. Med.*, 2005, **201**, 279–289.
253. P. A. Trimmer, T. S. Smith, A. B. Jung and J. P. Bennett Jr, Dopamine neurons from transgenic mice with a knockout of the p53 gene resist MPTP neurotoxicity, *Neurodegeneration*, 1996, **5**, 233–239.

254. E. H. Heiss, Y. D. Schilder and V. M. Dirsch, Chronic treatment with resveratrol induces redox stress- and ataxia telangiectasia-mutated (ATM)-dependent senescence in p53-positive cancer cells, *J. Biol. Chem.*, 2007, **282**, 26759–26766.
255. G. Melino, F. Bernassola, M. Ranalli, K. Yee, W. X. Zong, M. Corazzari, R. A. Knight, D. R. Green, C. Thompson and K. H. Vousden, p73 Induces apoptosis *via* PUMA transactivation and Bax mitochondrial translocation, *J. Biol. Chem.*, 2004, **279**, 8076–8083.
256. M. Lukosz, S. Jakob, N. Büchner, T. C. Zschauer, J. Altschmied and J. Haendeler, Nuclear redox signaling, *Antioxid. Redox Signaling*, 2010, **12**, 713–742.
257. T. K. Rudolph and B. A. Freeman, Transduction of redox signaling by electrophile-protein reactions, *Sci. Signaling*, 2009, **2**, re7.
258. N. K. Tonks, Protein tyrosine phosphatases: from genes, to function, to disease, *Nat. Rev. Mol. Cell Biol.*, 2006, **7**, 833–846.
259. R. P. Guttmann, Redox regulation of cysteine-dependent enzymes, *J. Anim. Sci.*, 2010, **88**, 1297–1306.
260. R. J. Mailloux, X. Jin and W. G. Willmore, Redox regulation of mitochondrial function with emphasis on cysteine oxidation reactions, *Redox Biol.*, 2014, **2**, 123–139.
261. T. R. Hurd, R. Requejo, A. Filipovska, S. Brown, T. A. Prime, A. J. Robinson, I. M. Fearnley and M. P. Murphy, Redox regulation of mitochondrial function with emphasis on cysteine oxidation reactions, *J. Biol. Chem.*, 2008, **283**, 24801–24815.
262. V. Petronilli, P. Costantini, L. Scorrano, R. Colonna, S. Passamonti and P. Bernardi, The voltage sensor of the mitochondrial permeability transition pore is tuned by the oxidation-reduction state of vicinal thiols. Increase of the gating potential by oxidants and its reversal by reducing agents, *J. Biol. Chem.*, 1994, **269**, 16638–16642.
263. A. P. Halestrap and C. Brenner, The adenine nucleotide translocase: a central component of the mitochondrial permeability transition pore and key player in cell death, *Curr. Med. Chem.*, 2003, **10**, 1507–1525.
264. M. W. Dodson and M. Guo, Pink1, Parkin, DJ-1 and mitochondrial dysfunction in Parkinson's disease, *Curr. Opin. Neurobiol.*, 2007, **17**, 331–337.
265. C. Henchcliffe and M. F. Beal, Mitochondrial biology and oxidative stress in Parkinson disease pathogenesis, *Nat. Clin. Pract. Neurol.*, 2008, **4**, 600–609.
266. M. Varcin, E. Bentea, Y. Michotte and S. Sarre, Oxidative stress in genetic mouse models of Parkinson's disease, *Oxid. Med. Cell. Longevity*, 2012, **2012**, 624925.
267. Y. Zhang, V. L. Dawson and T. M. Dawson, Oxidative stress and genetics in the pathogenesis of Parkinson's disease, *Neurobiol. Dis.*, 2000, **7**, 240–250.

CHAPTER 9

Iron Metabolism in Parkinson's Disease

GUOFEN GAO, LIN-HAO YOU AND YAN-ZHONG CHANG*

Laboratory of Molecular Iron Metabolism, College of Life Science, Hebei Normal University, Shijiazhuang, Hebei Province, 050024, China
*E-mail: chang7676@163.com

9.1 Brain Iron Homeostasis

Iron (Fe) is an essential trace element in the human body and plays critical roles in several biologically important processes, including O_2 transport and storage, electron transport, energy metabolism, and antioxidant and DNA synthesis.[1–3] The balance of Fe metabolism in the brain is also very important for maintaining the normal functions of the central nervous system (CNS) where it is also involved in neurotransmitter synthesis and nerve myelination.[4,5] Mounting evidence has indicated that alterations in brain Fe metabolism may trigger a cascade of deleterious events, which are tightly associated with the etiology of neurodegenerative diseases.[6,7] Parkinson's disease (PD) is a common neurodegenerative disease characterized by the loss of dopaminergic (DAergic) neurons in the *substantia nigra* (SN) and formation of filamentous intraneuronal inclusions.[8] Previous researches have reported that Fe is heavily accumulated in the DAergic neurons of the SN in PD patients[9–11] and changes in the expression of key Fe transporters have also been observed in PD patients or animal models.[12–14] Excessive Fe deposition in neurons is expected to generate massive reactive oxygen species (ROS)

Issues in Toxicology No. 34
Oxidative Stress and Redox Signalling in Parkinson's Disease
Edited by Rodrigo Franco, Jonathan A. Doorn and Jean-Christophe Rochet

Published by the Royal Society of Chemistry, www.rsc.org

through the Fenton reaction and increase the level of oxidative stress. These ROS would subsequently damage the structure of cell membranes, proteins and nucleic acids, leading to apoptosis and necrosis of DAergic neurons. Thus, Fe might be one of the causes of DAergic neuronal death in the SN during the progression of PD.

9.1.1 Brain Iron Transport and Distribution

The human body contains in average 3–5 g of Fe. Dietary Fe is absorbed predominately in the duodenum and enters blood circulation in the small intestine. Once in blood circulation, ferric iron (Fe^{3+}) binds to apotransferrin and forms transferrin (Tf). Tf is the major vehicle for Fe transport in the body, and carries Fe to other cells and tissues through the circulation. At the target cell, Tf binds to transferrin receptors (TfR) on the cell membrane, and the TfR-Tf-Fe complex is then endocytosed into the cell, where Fe is released. Free Fe either enters mitochondrion for utilization in metabolic processes, such as synthesis of hemoglobin and Fe-sulfur (S) cluster, or it is incorporated into the cytosolic Fe-storage protein, ferritin, which serves as a cellular store of Fe.

The brain is one of the most abundant Fe-containing tissues. Iron needs to pass the blood–brain barrier (BBB) in order to enter the brain. Tf-Fe in the blood circulation is uptaken at the surface of cerebral capillary endothelia, mainly through the classic TfR-mediated endocytosis.[15] Ferrous iron (Fe^{2+}) can also enter the BBB *via* a proton (H^+) driven transporter, divalent metal transporter 1 (DMT1).[16,17] In endothelia, Fe is released and transported across the abluminal membrane of the barriers into the cerebral compartment. This process likely involves Fe exporter ferroportin 1 (FPN1) on the abluminal membrane, but the exact mechanism remains to be explored (Figure 9.1).[2,6,18]

The elemental Fe released into the brain interstitial fluid either binds to brain Tf, becoming available for neurons and neuroglia expressing TfR,[19] or enters neurons as free Fe *via* DMT1 or the trivalent cation-specific transporter (TCT1) for ferrous and Fe^{3+}, respectively.[20–22] The export of Fe from neurons or neuroglia is also through FPN1 (Figure 9.1). The excess Fe in the brain can be exported back to the interstitial fluid, and be released into the cerebrospinal fluid in the brain ventricles through bulk flow.[15,18] The apical microvilli of choroidal epithelia then capture free Fe by TfR or DMT1 and transport it back to the blood circulation.[6]

Iron distribution in different brain regions is not even. Perl's staining has shown that Fe concentration in the extrapyramidal system is most abundant, especially in the basal ganglia region;[23,24] the Fe concentrations in globus pallidus, SN, red nucleus and cerebellar dentate gyrus are also high; in brain cortex, Fe concentration is relatively low,[23] whereas the white matter and medulla oblongata have the lowest Fe levels.[25] Iron widely distributes in both neuron and glial cells, including astrocytes, microglia and oligodendroglia, in which the oligodendroglia are found to have the highest Fe concentration.[24,26]

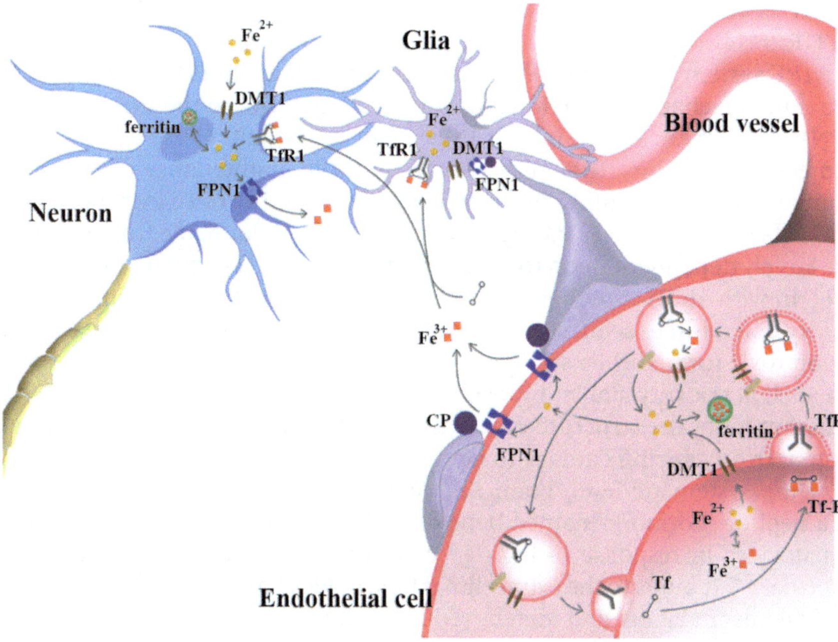

Figure 9.1 Schematic view of brain iron transport. Tf-Fe in the blood circulation is uptaken at the surface of cerebral capillary endothelia, mainly through the classic TfR-mediated endocytosis. Free Fe^{2+} also enters the brain barrier by DMT1. In endothelia, Fe is released and transported across the abluminal membrane of the BBB into the cerebral compartment *via* the Fe exporter FPN1. Iron released into the brain interstitial fluid either binds to brain Tf, becoming available for neurons and neuroglia expressing TfR or enters neurons as free Fe *via* the metal transporter DMT1. The export of Fe from neurons or neuroglia is also through FPN1. Most intracellular Fe is stored in ferritin.

9.1.2 Regulation of Brain Iron Homeostasis

Iron homeostasis in brain is regulated precisely. At the cellular level, Fe homeostasis is mainly controlled by TfR, DMT1 and FPN1. It has been reported that the uneven distribution of TfR in cerebral endothelia is responsible for the differences of Fe concentrations in different brain regions.[27] Iron concentrations are high in the striatum and the hippocampus where higher TfR density and Fe uptake rate are also observed,[27] but are low in the cortex and the brain stem.[23,28] Similar to the Fe regulation at the periphery, Fe homeostasis in brain is tightly regulated by iron regulatory proteins (IRPs) IRP1 and IRP2.[29] When the brain cellular Fe concentration is low, the active center of IRPs binds to the stem-loop structure of the Fe-responsive element (IRE) located at the 3′-untranslated region (UTR) of TfR mRNA. This binding stabilizes TfR mRNA and increases its cellular expression level, thereby increasing Fe uptake. When the Fe concentration is high, the active center of IRP is occupied by four Fe-S, which block the binding of IRP to the IRE of TfR, resulting in low TfR translation level and reduced Fe uptake.[30] The IRP/IRE system also regulates the stability of DMT1 with IRE (+IRE), through the same mechanism as TfR. However, binding of IRP to the IRE (5′ UTR) of FPN1 and ferritin would inhibit their translation, and cause lower protein expression.[29,30] Thus, IRPs play a key role in the maintenance of cellular Fe homeostasis by regulating the expression and stability of Fe transporters.

Inside the cells, the Fe storage levels and the cellular liable iron level (LIP) are largely dependent upon the availability of ferritin. Ferritin is a ubiquitous protein with an Fe core that can accommodate up to 4500 Fe atoms.[31,32] The ability of ferritin to sequester Fe provides its dual function, Fe segregation in a non-toxic form and Fe storage.[31,33] However, the storage of Fe in the ferritin shell is not very stable. In response to certain severe conditions, such as inflammation, the structure of ferritin shell may be affected with Fe release from ferritin causing damage to neurons.[34]

At the systematic level, brain Fe homeostasis may involve the regulation of a peptide 'hormone' named hepcidin.[35,36] Hepcidin is mainly produced by hepatocytes in response to high Fe concentration, inflammatory stimuli or hypoxia.[37,38] It binds to the extracellular loop of FPN1 and causes its internalization and degradation, and thereby reduces cellular Fe efflux.[39–41] Several recent studies reported the identification of hepcidin producing cells in the brain and investigated hepcidin's functions under normal and pathological conditions.[35,36,42,43] Zechel *et al.* showed that hepcidin is widely expressed in different brain areas, including the cortex, hippocampus, thalamus, cerebellum, spinal cord, and so on, in both neurons and astrocytes.[42] Increased hepcidin expression was detected in the choroid plexus of the brain in response to peripheral inflammation.[43] Studies have also found that hepcidin mRNA levels in different brain regions are increased with aging and injection of hepcidin into the lateral cerebral ventricle decreased FPN1 levels and resulted in brain Fe overload.[36] These findings implied the important regulatory role of hepcidin on brain Fe metabolism, though the cellular mechanisms remain to be elucidated.

9.2 Iron Metabolism and Parkinson's Disease

Evidence accumulated thus far has shown that Fe metabolism is closely related to the production of oxidative stress and the pathogenesis of neurodegenerative diseases.[3] It has been revealed that in brain Fe levels increase with aging, notably in the areas of SN and *locus coeruleus*, and the elevated levels of brain Fe are inversely related to functional impairment of cognitive ability in neurodegenerative diseases and in elderly individuals.[44] PD, as one of the most common neurodegenerative diseases has been found to be closely associated with abnormal Fe metabolism. Iron heavily accumulates in DAergic neurons in SN of parkinsonian brain and MPTP-induced PD model,[9–11] accompanied with significant changes in Fe transporters and regulators, which increases ROS generation and oxidative damage, and finally leads to apoptosis of DAergic neurons.

9.2.1 Symptoms of Parkinson's Disease

Parkinson's disease is one of the most common neurodegenerative diseases that is characterized by the loss of DAergic neurons in the SN *pars compacta* (SNpc), and finally leads to motor dysfunctions, with major clinical symptoms of rigidity, resting tremor, bradykinesia and postural instability.[8] These motor symptoms are collectively called parkinsonism. It more commonly occurs in older people, with most cases found after the age of 50. In addition to motor dysfunctions, neuropsychiatric disturbances are also commonly observed in parkinsonian individuals. These disorders include speech, cognition, behavior and mood, *etc.*

The cause of PD is unclear. Only 10% of the cases are thought to have a genetic origin. Many risks and protective factors have been investigated, and an increased risk of PD was found in people who were exposed to notoriously unhealthy environments, such as neurotoxins.[45–47] The pathology of the disease is characterized by the accumulation of alpha-synuclein (α-synuclein) into protein inclusions called Lewy bodies in neurons, which lead to insufficient formation and activity of dopamine (DA) produced in neurons of the SNpc.[48] Lewy bodies are the universal pathological hallmark of the idiopathic disorder, although the distribution of Lewy bodies varies from one individual to another in the parkinsonian brain.[49,50] The expression and anatomical distribution of the Lewy bodies is directly related to the degree of the clinical symptoms of each individual. The diagnosis of typical cases is mainly based on symptoms, while neuroimaging of the brain is also used to confirm PD. The SNpc and several other brain regions show neuron death, whereas the remaining nerve cells contain Lewy bodies.

9.2.2 Iron Accumulation Accelerates the Symptomatology of Parkinson's Disease

Midbrain is the main place that DAergic neurons locate and release DA. Dopaminergic neurons play an essential role in regulating the normal functions of brain, including motor behavior, motive, study and memory, emotion, addiction and stress. Under physiological conditions, Fe acts as a cofactor for tyrosine hydroxylase (TH), an enzyme catalyzing the rate-limiting step in DA synthesis. *In vitro* studies have shown that TH activity is dose-dependently stimulated by Fe.[51] Insufficient Fe supply to the brain will affect the production of DA. Studies have shown that Fe shortage could cause a decrease in DA receptors, which can lead to a marked loss of learning ability if Fe deficiency occurs in the early stages of CNS development.[52]

However, excess Fe is a severe damaging factor to brain function. Similar as in other neurodegenerative diseases, Fe induces the generation of ROS and oxidative stress, playing indispensable roles in the pathogenesis of PD. It has been reported that Fc content is 35% higher in the SN of PD patient than in normal people.[10] People with a higher dietary non-heme Fe intake usually have higher probability of developing PD.[53] However, there is no convincing experimental evidence to answer the question of whether Fe overload causes DAergic neuronal death in the SN of PD patients or if Fe toxicity is simply a byproduct of PD progression. Nevertheless, several studies have shown that the increase in brain Fe can induce parkinsonism, or accelerate PD symptoms.

Since 1993, people have used the infusion of Fe^{3+} into the SNpc to create a model of dose-related, progressive parkinsonism including a reduction in DAergic activity.[54] It has also been reported that hereditary haemochromatosis patients develop parkinsonian syndromes that most likely are caused by abnormal Fe deposition in the basal ganglia and cerebellum.[55,56] Further studies have shown that certain mutations in the HFE gene, encoding the human hemochromatosis protein, are associated with neurodegenerative disorders, including PD.[57] You *et al.* recently injected directly ferric citrate (FAC) into the cerebral ventricle of mice.[58] They observed that there was a clear aggravated DAergic cell death and behavioral PD symptoms of mice.[58] In addition, the commonly used PD animal models are associated with Fe accumulation in the SN, such as 6-hydroxydopamine (6-OHDA)-induced PD model[59,60] and the 1-methyl-4-phenyl-1,2,3,6-tetrahydropyridine (MPTP)-treated PD model.[58,61,62] Both 6-OHDA and MPTP can selectively destroy DAergic neurons in the brain, likely by producing ROS such as superoxide radical ($O_2^{\bullet-}$), and involving the participation of Fe *via* the Fenton reaction.[63] 6-OHDA can induce Fe release from ferritin, and released free Fe further accelerates the progression of neuronal loss by generating more ROS.

Recently, it has been found that injection of the proteasome inhibitor lactacystin that impairs the function of the ubiquitin-proteasome system (UPS) can also cause significant DAergic neuron degeneration, which is accompanied by excessive Fe accumulation in the SN of rodent brains.[64] Further *in vitro* studies found that the profound increases in labile Fe occur prior to neuronal cell death. They proposed that increase labile Fe participates in ROS production and aggregation of ubiquitin-conjugated proteins, which then leads to cell injury and death.[64]

9.2.3 Mechanism of Iron Accumulation in the *Substantia Nigra* of Parkinson's Disease Brains

Research has reported that Fe is heavily accumulated in the brain of PD patients.[9–11] The SN is the main pathological area in PD, and the ratio of Fe content to tissue dry weight is extraordinarily high, and in some cases, it is even higher than that of liver.[52] Iron selectively accumulated in SN of midbrain in PD patients is found to be associated with changes in expression of Fe-transporters and Fe storage proteins.

9.2.3.1 Changes in Iron-Uptake Proteins

Several studies have indicated that increased expression of TfR is observed in PD, indicating that the Tf/TfR-dependent Fe transport system plays an important role in PD pathogenesis, which causes an increased accumulation of Fe in brain cells.[13,65] Intracerebroventricular injection (ICV) of Fe can induce Fe accumulation in the caudate putamen and SN.[58] Injected FE can cross the BBB from the cerebrospinal fluid and induce extracellular Fe overload by passively increasing Fe uptake. A study by Kaur *et al.* showed that glutathione (GSH) depletion in DAergic neurons in the early stages of PD results in an increased expression in TfR1 and an elevation in labile Fe pool.[65] Researchers have also found increased expression of TfR1 in 1-methyl-4-phenylpyridinium (MPP^+, the active metabolite of MPTP used as a PD model) treated neuronal cells.[55] Therefore the Tf/TfR1 Fe transport system may play an important role in the Fe deposition of PD patients. Besides TfR1, recent studies have also found that TfR2-mediated Fe transport in mitochondria may also relate to the Fe deposition process of PD. Rotenone, a pesticide that interferes with the electron-transport chain (ETC) in mitochondria, is also used to induce oxidative damage and death of DAergic neurons. Rotenone treatment induces Fe deposition in mitochondria, which correlates with an increased expression of TfR2 in the mitochondrial membrane.[66] Further studies have shown that rotenone not only induces TfR2 expression, but also increases the binding activity between Tf and TfR2, leading to the accumulation of Tf-bound Fe in mitochondria. TfR2 is selectively expressed in the mitochondria of human DAergic neurons, and PD patients have Tf accumulation and oxidative damage in mitochondria, which may indicate that the Tf/TfR2 Fe transport system also plays an important role in PD pathogenesis.

The alteration of another Fe transporter, DMT1, which transports non-Tf-bound Fe into cells, was also observed in PD. Both in the MPTP-treated PD mouse model and in the 6-OHDA-treated rat model an increased expression of DMT1 in SN region was found, which was concomitant with Fe accumulation, oxidative stress, and DAergic cell loss.[67,68] A recent study further confirmed this phenomenon with PD patient. With Perls' staining, Salazar *et al.* found that neuromelanin-containing DAergic neurons in the mesencephalon of PD patients exhibited significantly higher DMT1 labeling than that of control subjects.[12] Besides, they also found that a mutation in DMT1 that impairs Fe transport protected rodents against parkinsonism induced by MPTP and 6-OHDA.[12] The up-regulation of DMT1 expression was found to be IRE/IRP dependent,[12] further attesting its role in Fe accumulation in PD. Over-expression of DMT1 in MES23.5 DAergic cells dramatically increases intracellular Fe accumulation, oxidative stress and ultimately cell apoptosis.[69] These studies all support a critical role for DMT1 in Fe-mediated neurodegeneration in PD.

9.2.3.2 Changes in Iron-Release Proteins

Cellular Fe homeostasis depends on the balance of two processes, Fe-uptake and release. As previously stated, Fe-uptake proteins transport more Fe to brain cells in PD, which is associated with Fe deposition in SN. Meanwhile, it has also been found that Fe efflux is also changed in PD. FPN1 is the only known Fe exporter on the cell membrane. Under normal conditions, FPN1 exports excess Fe out of cells to keep a constant amount of Fe inside the cell, maintaining intracellular Fe homeostasis. Studies have shown that, in the 6-OHDA PD model, FPN1 expression in the SN was reduced and was accompanied with Fe accumulation,[14] which indicates FPN1 is also involved in PD pathogenesis.

However, it remains unclear whether the abnormal expression of DMT1, TfR, and FPN1 causes the accumulation of Fe in SN, or if the accumulation of Fe in SN leads to changes in the expression of these Fe transporters. Both DMT1 (+IRE) and FPN1 mRNA levels have the classic IRE at 3′-UTR and 5′-UTR, respectively, and their expression is regulated by IRPs. IRPs expression responds to small changes in intracellular Fe, and form IRP/IRE complex. The IRP/IRE system stabilizes DMT1 (+IRE) mRNA and increases DMT1 expression, while it also decreases FPN1 expression. Therefore, if in the cells that contain high amounts of Fe, IRP cannot bind to IRE, the expression of DMT1 will be down-regulated, and the expression of FPN1 will be enhanced, but this is in contrast to what it is observed in PD. Thus, these observations support the hypothesis that the changes in DMT1 and FPN1 expression cause Fe accumulation in the SN of PD, not the opposite way.

Ceruloplasmin (CP) is a ferroxidase enzyme that is involved in Fe metabolism by oxidation of Fe^{2+} into Fe^{3+}.[70] CP stabilizes FPN1 and facilitates Fe export from endothelial cells and other cell types. Research has demonstrated that in the CP-knockout ($CP^{-/-}$) mice there is a significant increase in Fe content in the cerebellum and brainstem.[71] In $CP^{-/-}$ mice, Fe release is clearly decreased due to CP deficiency. It has also been reported that the expression of DMT1 (-IRE) is increased in the cerebellum of these mice and that TfR1 is elevated in the endothelial cells of the blood–brain barrier, increasing Fe uptake in the presence of Fe accumulation in neural cells of aged $CP^{-/-}$ mice.[72] These data indicate that $CP^{-/-}$ mice accumulate intracellular Fe in the brain. Another study has shown that $CP^{-/-}$ mice present increased vulnerability to rotenone-induced neurotoxicity.[73] A study with PD patients previously showed that there is a significant correlation between lower serum CP levels and younger age of onset in PD,[74] indicating low CP levels might be causative in PD pathogenesis.

9.2.3.3 Changes in Iron-Storage Proteins

In addition to Fe uptake and release proteins, some Fe storage/binding proteins may also contribute to Fe accumulation in PD. Lactoferrin (Lf), a globular glycoprotein with a molecular mass of about 80 kDa, is an Fe-containing multifunctional protein of the Tf family, which functions mainly in host defense against infection and severe inflammation. It has been reported that Lf accumulates in the brain during neurodegenerative disorders.[75] Quantitative immunohistochemical study with normal individuals and patients affected with PD has shown that higher immunolabeling levels of Lf are found in the surviving neurons of the SN and ventral tegmental area of PD brains when compared to control cases.[75] A study by Faucheux *et al.*

demonstrated that expression of lactoferrin receptors (LfR) is increased in the mesencephalon of patients with PD,[76] and also found that LfR is not only expressed in neuronal cell membrane, but also in the membrane of microvascular endothelial cells. These data may suggest that LfR probably has similar functions as TfR, playing important roles in Fe uptake at the BBB. All these results suggest that Lf may participate actively in the mechanism of neuronal degeneration in PD.

The abnormal expression of other Fe-containing proteins, such as ferritin, mitochondrial ferritin and neuromelanin (NM), are also involved in the neuronal degeneration in PD. Ferritin, the primary intracellular Fe-storage protein, incorporates free Fe in its 24 subunit shell and keeps Fe in a soluble and non-toxic form. Although Fe heavily accumulates in the SN of PD patients, studies have found that the L-ferritin levels decreased in SN of PD.[77] Further studies found that there is a difference in the structure of ferritin from PD brains, which results in low Fe-storage capacity of ferritin and Fe release from ferritin shell,[77] thereby the availability of Fe is elevated. Another study showed that increases in free Fe levels in the SN of PD went beyond the maximum capacity of ferritin binding, which may involve the regulation of ferritin expression by IRP.[78] A sustained Fe IRP1 binding activity was determined in the SN of patients of PD, which associates with a lack of up-regulation of ferritin expression.[78] In addition, researchers have found that the expression of H-ferritin is decreased during the normal aging process, which further disrupts the normal Fe-storage system and exacerbates the progress of PD.[79]

Neuromelanin is a dark pigment in the brain that is structurally related to melanin. NM is mainly expressed in large quantities in the DAergic cells of the SN.[80,81] NM can chelate Fe. Recent studies have investigated its role in protecting neurons in the SN from Fe-induced oxidative stress, delaying PD symptom appearance and slowing disease progression.[82] It has been shown that patients with PD have 50% the amount of NM in the SN as compared to similar individuals of the same age without PD.[79,82] The death of NM containing neurons in the SNpc, and *locus coeruleus* visualized *in vivo* with MRI has been linked to PD, and also NM.[83] Released Fe from NM can promote oxidative damage in mitochondria, and subsequently the dysfunction of mitochondria would lead to a higher amount of DAergic cell death. These studies propose the protective role of NM in PD pathogenesis, which was consistent with the finding that PD and degenerative brain disorders such as dementia are more common among white-skinned populations, who have lower amounts of NM, as compared to darker-skinned people.[84]

9.2.3.4 *Changes in the Distribution of Intracellular Iron*

Mitochondrial ferritin (FtMt) is a novel Fe-storage protein structurally and functionally similar to the cytosolic H-ferritin. FtMt protects mitochondria from Fe-induced oxidative damage presumably through sequestration of potentially harmful excess free Fe. Studies have shown that FtMt plays a protective role in the pathogenesis of neurodegenerative diseases.[85]

Altered expression of FtMt was observed in the cerebellum of Ireb2$^{-/-}$ mice, which presents a profound dysregulation of neuronal Fe metabolism and associates with the neurodegeneration.[60] Overexpression of FtMt significantly protected neuronal cells from the 6-OHDA-induced neuronal cell death and alleviated parkinsonian phenotype.[60] These results indicate the potentially important role that FtMt plays in the pathological events of PD.

9.2.4 Iron and the Aggregation of α-Synuclein

α-synuclein is a protein with 140 amino acids, which is abundant in the human brain. It mainly locates at presynaptic terminals of neurons, and may play important roles in regulating the release of DA and maintaining a supply of synaptic vesicles in presynaptic terminals. It is predominantly a neuronal protein expressed in the neocortex, hippocampus, SN, thalamus, and cerebellum.[86] Under normal conditions, wild-type α-synuclein forms a stably folded tetramer that resists aggregation.[87] Nevertheless, in some pathological conditions α-synuclein aggregates to form insoluble fibrils characterized by Lewy bodies, such as in PD.[88] The aggregation mechanism of α-synuclein is uncertain, but has been proposed to be associated to an altered conformation related to pathogenic aggregation. Mutations in the α-synuclein gene have been identified in families with PD.[88] Aggregated α-synuclein can interact with the cell membrane, form pore-like structures, and depolarize the membrane potential, finally disrupting normal functions of the cell, leading to DAergic cell death and PD.[88]

Recent evidence indicates that Fe accumulation in the SN of PD may induce the aggregation of α-synuclein,[89,90] and then cause apoptosis of DAergic cells, although the detailed mechanism is unclear. A recent study found that the 5′-UTR of α-synuclein mRNA contains a 46-nucleotide RNA loop that is very similar with the IRE region in H-ferritin.[91] This loop is predicted to be an IRE element on α-synuclein, which responds to the IRP regulation at high Fe concentration and leads to α-synuclein aggregation, while isoforms of α-synuclein protein lacking IRE element cannot form aggregates.[91] Therefore, it could be inferred that this IRE element plays an important role in α-synuclein aggregation. In this case, at high concentration of Fe, less IRP binds to the IRE of α-synuclein mRNA, inhibition of α-synuclein mRNA translation is reduced, and the increased α-synuclein expression may lead to the formation of protein aggregates due to incorrect protein folding. Therefore, Fe-induced α-synuclein aggregation most likely is an important cause of DAergic neuronal death.

9.2.5 Iron, ROS and Apoptosis of Dopaminergic Neurons in the *Substantia Nigra* of Parkinson's Disease

At physiological conditions, ROS regulate cell growth, gene expression, antimicrobial, and inflammatory responses. This basal amount of free radicals can be cleared rapidly by the body to prevent its damaging effects after accumulation. However, deposition of Fe in brain cells induces oxidative stress by ROS production, especially *via* generation of ˙OH. Increased Fe^{2+} levels can lead to the production of the highly reactive ˙OH *via* the Fenton reaction. Increased Fe levels can also generate peroxyl/alkoxyl radicals due to Fe^{2+}-dependent lipid peroxidation.[92] Iron-induced oxidative stress can be very destructive because of a positive-feedback loop can be developed from the release of more free Fe from the Fe-containing proteins, such as ferritin, heme proteins and Fe–S clusters in the mitochondrial respiratory chain. As a result, the toxic effect of Fe-induced oxidative stress is exacerbated (Figure 9.2).

Excess ROS production can damage cellular macromolecules including proteins, lipids, chromosomal and mitochondrial DNA, and affect protein expression and finally trigger the intrinsic apoptotic pathway (Figure 9.2). p53 is a cellular stress sensor that is phosphorylated and stabilized in response to stress.[93] Activated p53 initiates apoptosis by transcriptionally activating the pro-apoptotic B-cell CLL/lymphoma 2 (Bcl-2) family members and repressing antiapoptotic Bcl-2 proteins and cellular inhibitor of apoptosis protein (cIAPs). Bcl-2 family proteins can regulate the release of cytochrome C from mitochondria and formation of the apoptosome, a catalytic multiprotein platform that activates caspase 9. Activated caspase 9 then cleaves caspase 3 resulting in downstream events that finally lead to cell death.[93] It has been reported that oxidative stress levels in PD patients

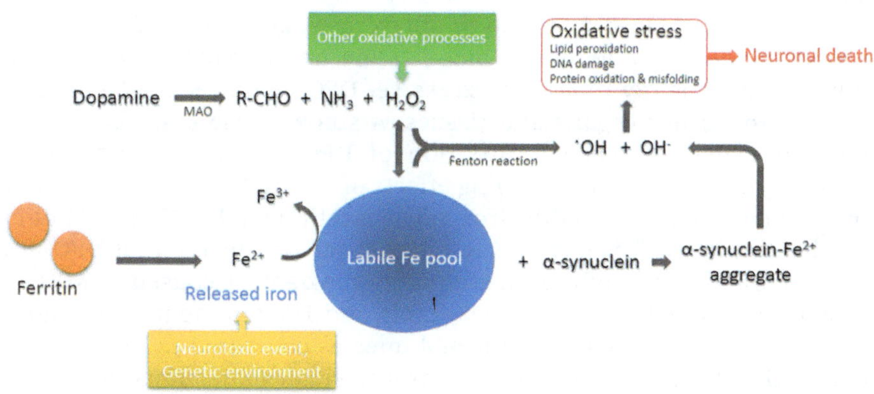

Figure 9.2 Iron-induced neurodegeneration in Parkinson's disease. Iron release from ferritin, the generation of reactive ˙OH and induction of oxidative stress in DAergic neurons within the SNpc. PD-related neurotoxins induce accumulation of Fe at sites where they initiate neurodegeneration. Iron is thought to enter a labile Fe pool, where it is accessible to participate in the Fenton reaction with hydrogen peroxide (H_2O_2), which is generated by DA metabolism *via* MAO (monoamine oxidase), and other oxidative processes. The Fenton reaction produces ˙OH. The resulting effect is oxidative stress as a consequence of increased membrane lipid peroxidation, DNA damage, protein oxidation and misfolding. Labile Fe can also cause aggregation of α-synuclein and amyloid-β (Aβ) to form toxic aggregates, which, in turn, can initiate ˙OH generation, causing oxidative stress.

are increased, and Fe accumulation in the SN of PD is believed to induce more oxidative damage and directly participate in the neurodegeneration in PD.[63,94–96]

9.3 Iron-Related Therapeutic Approaches for PD

Therapeutic approaches against PD have been studied for many years. Some medications (such as levodopa and DA agonists), surgery (such as deep-brain stimulation) and multidisciplinary management have been used to relieve from PD symptoms.[97,98] However, as the pathogenic mechanism of PD remains unclear, there is no efficient prevention or treatment method for PD so far. Since dysregulation of Fe metabolism and increased oxidative stress play pivotal roles in the neurotoxicity and neurodegeneration in PD, many current studies are focused on exploring the functions Fe-chelators, antioxidant, and other Fe-related therapeutic approaches.

The Fe-chelation therapy could be a viable neuroprotective approach for treatment of PD.[3] Chelation has the potential to prevent Fe-induced ROS, oxidative stress and aggregation of α-synuclein, and the limited *in vitro* and *in vivo* neuroprotective studies that have been carried out so far seem to support this idea.

Desferrioxamine (DFO) has been used for a long time in the clinic as an Fe-chelating agent to remove excess Fe from the body. It is a bacterial siderophore produced by the Actinobacteria *Streptomyces pilosus*, which acts by binding free Fe in the bloodstream and enhancing its elimination in the urine. By removing excess Fe, DFO could reduce Fe-induced damage to various organs and tissues. A study by Xie *et al.* has shown that intracerebroventricular injection of DFO before the injection of 6-OHDA could block the damaging effects of 6-OHDA, preventing the loss of DA and its metabolites dihydroxyphenylacetic acid (DOPAC) and homovanilic acid (HVA).[99] You *et al.* showed that the pre-injection of DFO into the cerebral ventricle of mouse attenuates the damage caused by MPTP.[58] Less Fe accumulation in SN, ROS levels and DAergic neuron apoptosis are observed in the DFO pre-injected mice as compared to mice treated only with MPTP.[58] Besides intracerebroventricular injections, intranasal administration of DFO has also been tried in the rat PD model that was established by overexpressing α-synuclein.[100] By analyzing the behavior of animals and the histopathological changes in the brain, it has been shown that DFO decreases the number of pathological α-synuclein formations and improves the motor behavior in rats, but it did not seem to protect them from DAergic cell death. These results indicate that intranasal administration of DFO could slow PD progression, but since DFO has poor penetration across the BBB, its application in the treatment of PD still needs to be improved.

The effects of two synthesized, brain-permeable Fe chelators, named VK-28 [5-[[4-(2-hydroxyethyl)-1-piperazinyl]methyl]-8-quinolinol] and M30 [5-(*N*-methyl-*N*-propargyaminomethyl)-8-hydroxyquinoline], on the treatment of PD have been studied by Youdim's group in the past few years with three different PD animal models, induced by 6-OHDA, MPTP or lactacystin.[101–103] In the 6-OHDA-induced PD rat model, VK28 was injected either intraventricularly or intraperitoneally. They found that intraventricularly injection of VK28 at a low dose was able to completely protect against 6-OHDA-induced striatal DAergic lesion, whereas the intraperitoneally injection of VK28 at a higher dose showed 68% protection.[102] In the lactacystin induced PD mouse model, VK28 or M30 was administered intraperitoneally. They found that both VK28 and M30 significantly improved the behavioral performance and attenuated lactacystin-induced DAergic neuron loss, possibly *via* preventing proteasomal inhibition, Fe accumulation and microglial activation in the SN.[103] In addition, M30 restored the level of anti-apoptotic protein Bcl-2, indicating that apoptotic signals were inhibited.[103] Furthermore, preliminary studies in animals[102,104,105] have not shown any interference with neurotransmitter metabolism or nicotinamide adenine dinucleotide (NADH) oxidoreductase (complex I) activity, but the use of Fe chelators still need to be carefully controlled to avoid toxic effects.[106] These findings suggest that brain-permeable Fe chelator can efficiently improve DAergic neuron survival even with different PD animal models, indicating that Fe chelation might be a promising therapeutic target for PD.

Antioxidant supplementation is another commonly used therapeutic approach in the treatment of oxidative stress-related diseases, such as cancer, stroke and neurodegenerative diseases. Antioxidants are molecules that can inhibit the oxidation of other molecules. Antioxidants terminate free-radical chain reactions by removing free-radical intermediates, and inhibit other oxidation reactions, while they become oxidized. Therefore antioxidants are often reducing agents such as thiols, ascorbic acid, or polyphenols.[107] Since oxidation is often catalyzed by metals, some compounds contribute to antioxidant defense by chelating metals and preventing them from catalyzing redox reactions such as the Fenton reaction. A study by Dowling *et al.* determined the ability of isoflavone metal agents in chelating copper (Cu^{2+}) and Fe^{3+}.[108] Another study by Vlachodimitropoulou *et al.* showed the function of quercetin as an efficient Fe chelator and investigated its mechanism of reducing intracellular labile Fe and preventing the production of $^{\bullet}OH$.[109] The activity of oak leaf extracts was also studied, and results showed that oak leaf extracts work as topical antioxidants by scavenging free radicals and chelating Fe.[110]

The functions of some other compounds and other approaches in the treatment of PD have also been studied. Epigallocatechin gallate (EGCG) and curcumin polyphenols are neuroprotective in animal models of PD, most likely due to their antioxidant, Fe chelating and anti-inflammatory activities.[111,112] Lycopene has functions as an antioxidant, anti-cancer and anti-aging agent, and prevents cardiovascular disease, but the chemical structure of lycopene is unstable. A recent study by Tian *et al.* showed that lycopene liposomes could increase and prolong the antioxidative effect of lycopene and reduce lipid peroxidation obviously compared with lycopene in rats.[113]

9.4 Conclusions

PD, as one of the most common neurodegenerative diseases, has been found to be closely associated with abnormal Fe metabolism. Mounting evidence has indicated that Fe is heavily deposited in DAergic neurons in SN of parkinsonian brain, and may participate in the progression of PD pathogenesis. Excessive Fe deposition in neurons is expected to generate massive ROS through the Fenton reaction, which increases the level of oxidative stress and subsequently leads to apoptosis and necrosis of DAergic neurons. Extensive studies have been performed by researchers on the alterations of Fe metabolism in PD models. Although the exact role of Fe in PD pathogenesis remains unclear, the findings accumulated thus far have been very useful for exploring novel approaches for treatment of PD. The Fe-chelation therapy has been carried out with PD animal models and found to be a viable neuroprotective approach to relieve PD symptoms. Other Fe-related therapeutic targets against the changes in the expression of Fe transport proteins or Fe regulation proteins in PD models are also promising approaches as shown by many current studies. Further studies on the precise mechanisms that lead to the abnormal Fe metabolism in PD would be greatly beneficial to the therapy of PD.

References

1. T. Ganz and E. Nemeth, Regulation of iron acquisition and iron distribution in mammals, *Biochim. Biophys. Acta*, 2006, **1763**, 690–699.
2. T. Moos, T. Rosengren Nielsen, T. Skjorringe and E. H. Morgan, Iron trafficking inside the brain, *J. Neurochem.*, 2007, **103**, 1730–1740.
3. L. Zecca, M. B. Youdim, P. Riederer, J. R. Connor and R. R. Crichton, Iron, brain ageing and neurodegenerative disorders, *Nat. Rev. Neurosci.*, 2004, **5**, 863–873.
4. W. van Gelder, M. I. Huijskes-Heins, M. I. Cleton-Soeteman, J. P. van Dijk and H. G. van Eijk, Iron uptake in blood-brain barrier endothelial cells cultured in iron-depleted and iron-enriched media, *J. Neurochem.*, 1998, **71**, 1134–1140.
5. R. B. Mounsey and P. Teismann, Chelators in the treatment of iron accumulation in Parkinson's disease, *Int. J. Cell Biol.*, 2012, **2012**, 983245.
6. E. Mills, X. P. Dong, F. Wang and H. Xu, Mechanisms of brain iron transport: insight into neurodegeneration and CNS disorders, *Future Med. Chem.*, 2010, **2**, 51–64.
7. R. J. Ward, F. A. Zucca, J. H. Duyn, R. R. Crichton and L. Zecca, The role of iron in brain ageing and neurodegenerative disorders, *Lancet Neurol.*, 2014, **13**, 1045–1060.
8. J. Parkinson, An essay on the shaking palsy. 1817, *J. Neuropsychiatry Clin. Neurosci.*, 2002, **14**, 223–236, discussion 222.
9. D. T. Dexter, F. R. Wells, F. Agid, Y. Agid, A. J. Lees, P. Jenner and C. D. Marsden, Increased nigral iron content in postmortem parkinsonian brain, *Lancet*, 1987, **2**, 1219–1220.

10. E. Sofic, P. Riederer, H. Heinsen, H. Beckmann, G. P. Reynolds, G. Hebenstreit and M. B. Youdim, Increased iron (III) and total iron content in post mortem substantia nigra of parkinsonian brain, *J. Neural Transm.*, 1988, **74**, 199–205.
11. P. Jenner and C. W. Olanow, Understanding cell death in Parkinson's disease, *Ann. Neurol.*, 1998, **44**, S72–S84.
12. J. Salazar, N. Mena, S. Hunot, A. Prigent, D. Alvarez-Fischer, M. Arredondo, C. Duyckaerts, V. Sazdovitch, L. Zhao, L. M. Garrick, M. T. Nunez, M. D. Garrick, R. Raisman-Vozari and E. C. Hirsch, Divalent metal transporter 1 (DMT1) contributes to neurodegeneration in animal models of Parkinson's disease, *Proc. Natl. Acad. Sci. U. S. A.*, 2008, **105**, 18578–18583.
13. S. V. Kalivendi, S. Kotamraju, S. Cunningham, T. Shang, C. J. Hillard and B. Kalyanaraman, 1-Methyl-4-phenylpyridinium (MPP+)-induced apoptosis and mitochondrial oxidant generation: role of transferrin-receptor-dependent iron and hydrogen peroxide, *Biochem. J.*, 2003, **371**, 151–164.
14. T. Wang and E. A. Craig, Binding of yeast frataxin to the scaffold for Fe-S cluster biogenesis, Isu, *J. Biol. Chem.*, 2008, **283**, 12674–12679.
15. M. W. Bradbury, Transport of iron in the blood-brain-cerebrospinal fluid system, *J. Neurochem.*, 1997, **69**, 443–454.
16. A. J. Siddappa, R. B. Rao, J. D. Wobken, E. A. Leibold, J. R. Connor and M. K. Georgieff, Developmental changes in the expression of iron regulatory proteins and iron transport proteins in the perinatal rat brain, *J. Neurosci. Res.*, 2002, **68**, 761–775.
17. T. Skjorringe, L. B. Moller and T. Moos, Impairment of interrelated iron- and copper homeostatic mechanisms in brain contributes to the pathogenesis of neurodegenerative disorders, *Front. Pharmacol.*, 2012, **3**, 169.
18. W. Zheng and A. D. Monnot, Regulation of brain iron and copper homeostasis by brain barrier systems: implication in neurodegenerative diseases, *Pharmacol. Ther.*, 2012, **133**, 177–188.
19. J. Han, J. R. Day, J. R. Connor and J. L. Beard, Gene expression of transferrin and transferrin receptor in brains of control *vs.* iron-deficient rats, *Nutr. Neurosci.*, 2003, **6**, 1–10.
20. Z. K. Attieh, C. K. Mukhopadhyay, V. Seshadri, N. A. Tripoulas and P. L. Fox, Ceruloplasmin ferroxidase activity stimulates cellular iron uptake by a trivalent cation-specific transport mechanism, *J. Biol. Chem.*, 1999, **274**, 1116–1123.
21. Z. M. Qian, Y. M. Pu, Q. Wang, Y. Ke, Y. D. Yao, W. F. Chen, X. Shen, Y. M. Feng and P. L. Tang, Cerebellar granule cells acquire transferrin-free iron by a carrier-mediated process, *Neuroscience*, 1999, **92**, 577–582.
22. Y. Ke and Z. M. Qian, Brain iron metabolism: neurobiology and neurochemistry, *Prog. Neurobiol.*, 2007, **83**, 149–173.
23. C. M. Morris, J. M. Candy, A. E. Oakley, C. A. Bloxham and J. A. Edwardson, Histochemical distribution of non-haem iron in the human brain, *Acta Anat.*, 1992, **144**, 235–257.

24. T. Moos, D. Trinder and E. H. Morgan, Cellular distribution of ferric iron, ferritin, transferrin and divalent metal transporter 1 (DMT1) in substantia nigra and basal ganglia of normal and beta2-microglobulin deficient mouse brain, *Cell Mol. Biol. (Noisy-le-grand)*, 2000, **46**, 549–561.
25. B. Drayer, P. Burger, R. Darwin, S. Riederer, R. Herfkens and G. A. Johnson, MRI of brain iron, *AJR, Am. J. Roentgenol.*, 1986, **147**, 103–110.
26. J. R. Connor, S. L. Menzies, S. M. St Martin and E. J. Mufson, Cellular distribution of transferrin, ferritin, and iron in normal and aged human brains, *J. Neurosci. Res.*, 1990, **27**, 595–611.
27. R. Deane, W. Zheng and B. V. Zlokovic, Brain capillary endothelium and choroid plexus epithelium regulate transport of transferrin-bound and free iron into the rat brain, *J. Neurochem.*, 2004, **88**, 813–820.
28. N. Sugawara, T. Ikeda, C. Sugawara, Y. Kohgo, J. Kato and N. Takeichi, Regional distribution of copper, zinc and iron in the brain in Long-Evans Cinnamon (LEC) rats with a new mutation causing hereditary hepatitis, *Brain Res.*, 1992, **588**, 287–290.
29. T. A. Rouault, The role of iron regulatory proteins in mammalian iron homeostasis and disease, *Nat. Chem. Biol.*, 2006, **2**, 406–414.
30. R. Leipuviene and E. C. Theil, The family of iron responsive RNA structures regulated by changes in cellular iron and oxygen, *Cell. Mol. Life Sci.*, 2007, **64**, 2945–2955.
31. P. M. Harrison and P. Arosio, The ferritins: molecular properties, iron storage function and cellular regulation, *Biochim. Biophys. Acta*, 1996, **1275**, 161–203.
32. E. C. Theil, Ferritin: structure, gene regulation, and cellular function in animals, plants, and microorganisms, *Annu. Rev. Biochem.*, 1987, **56**, 289–315.
33. F. M. Torti and S. V. Torti, Regulation of ferritin genes and protein, *Blood*, 2002, **99**, 3505–3516.
34. P. Biemond, A. J. Swaak, H. G. van Eijk and J. F. Koster, Superoxide dependent iron release from ferritin in inflammatory diseases, *Free Radic. Biol. Med.*, 1988, **4**, 185–198.
35. R. R. Crichton, D. T. Dexter and R. J. Ward, Brain iron metabolism and its perturbation in neurological diseases, *J. Neural Transm.*, 2011, **118**, 301–314.
36. S. M. Wang, L. J. Fu, X. L. Duan, D. R. Crooks, P. Yu, Z. M. Qian, X. J. Di, J. Li, T. A. Rouault and Y. Z. Chang, Role of hepcidin in murine brain iron metabolism, *Cell. Mol. Life Sci.*, 2010, **67**, 123–133.
37. C. H. Park, E. V. Valore, A. J. Waring and T. Ganz, Hepcidin, a urinary antimicrobial peptide synthesized in the liver, *J. Biol. Chem.*, 2001, **276**, 7806–7810.
38. G. Nicolas, C. Chauvet, L. Viatte, J. L. Danan, X. Bigard, I. Devaux, C. Beaumont, A. Kahn and S. Vaulont, The gene encoding the iron regulatory peptide hepcidin is regulated by anemia, hypoxia, and inflammation, *J. Clin. Invest.*, 2002, **110**, 1037–1044.

39. G. Ramey, J. C. Deschemin, B. Durel, F. Canonne-Hergaux, G. Nicolas and S. Vaulont, Hepcidin targets ferroportin for degradation in hepatocytes, *Haematologica*, 2010, **95**, 501–504.
40. G. J. Anderson and F. Wang, Essential but toxic: controlling the flux of iron in the body, *Clin. Exp. Pharmacol. Physiol.*, 2012, **39**, 719–724.
41. E. Nemeth, M. S. Tuttle, J. Powelson, M. B. Vaughn, A. Donovan, D. M. Ward, T. Ganz and J. Kaplan, Hepcidin regulates cellular iron efflux by binding to ferroportin and inducing its internalization, *Science*, 2004, **306**, 2090–2093.
42. S. Zechel, K. Huber-Wittmer and O. von Bohlen und Halbach, Distribution of the iron-regulating protein hepcidin in the murine central nervous system, *J. Neurosci. Res.*, 2006, **84**, 790–800.
43. F. Marques, A. M. Falcao, J. C. Sousa, G. Coppola, D. Geschwind, N. Sousa, M. Correia-Neves and J. A. Palha, Altered iron metabolism is part of the choroid plexus response to peripheral inflammation, *Endocrinology*, 2009, **150**, 2822–2828.
44. J. Hagemeier, J. J. Geurts and R. Zivadinov, Brain iron accumulation in aging and neurodegenerative disorders, *Expert Rev. Neurother.*, 2012, **12**, 1467–1480.
45. J. M. Gorell, C. C. Johnson, B. A. Rybicki, E. L. Peterson, G. X. Kortsha, G. G. Brown and R. J. Richardson, Occupational exposure to manganese, copper, lead, iron, mercury and zinc and the risk of Parkinson's disease, *Neurotoxicology*, 1999, **20**, 239–247.
46. K. M. Powers, T. Smith-Weller, G. M. Franklin, W. T. Longstreth Jr, P. D. Swanson and H. Checkoway, Parkinson's disease risks associated with dietary iron, manganese, and other nutrient intakes, *Neurology*, 2003, **60**, 1761–1766.
47. J. Peng, L. Peng, F. F. Stevenson, S. R. Doctrow and J. K. Andersen, Iron and paraquat as synergistic environmental risk factors in sporadic Parkinson's disease accelerate age-related neurodegeneration, *J. Neurosci.*, 2007, **27**, 6914–6922.
48. M. H. Polymeropoulos, C. Lavedan, E. Leroy, S. E. Ide, A. Dehejia, A. Dutra, B. Pike, H. Root, J. Rubenstein, R. Boyer, E. S. Stenroos, S. Chandrasekharappa, A. Athanassiadou, T. Papapetropoulos, W. G. Johnson, A. M. Lazzarini, R. C. Duvoisin, G. Di Iorio, L. I. Golbe and R. L. Nussbaum, Mutation in the alpha-synuclein gene identified in families with Parkinson's disease, *Science*, 1997, **276**, 2045–2047.
49. C. A. Davie, A review of Parkinson's disease, *Br. Med. Bull.*, 2008, **86**, 109–127.
50. W. J. Schulz-Schaeffer, The synaptic pathology of alpha-synuclein aggregation in dementia with Lewy bodies, Parkinson's disease and Parkinson's disease dementia, *Acta Neuropathol.*, 2010, **120**, 131–143.
51. W. D. Rausch, Y. Hirata, T. Nagatsu, P. Riederer and K. Jellinger, Tyrosine hydroxylase activity in caudate nucleus from Parkinson's disease: effects of iron and phosphorylating agents, *J. Neurochem.*, 1988, **50**, 202–208.

52. D. Berg, Ultrasound in the (premotor) diagnosis of Parkinson's disease, *Parkinsonism Relat. Disord.*, 2007, **13**(suppl. 3), S429–S433.
53. G. Logroscino, X. Gao, H. Chen, A. Wing and A. Ascherio, Dietary iron intake and risk of Parkinson's disease, *Am. J. Epidemiol.*, 2008, **168**, 1381–1388.
54. G. J. Sengstock, C. W. Olanow, R. A. Menzies, A. J. Dunn and G. W. Arendash, Infusion of iron into the rat substantia nigra: nigral pathology and dose-dependent loss of striatal dopaminergic markers, *J. Neurosci. Res.*, 1993, **35**, 67–82.
55. J. E. Nielsen, L. N. Jensen and K. Krabbe, Hereditary haemochromatosis: a case of iron accumulation in the basal ganglia associated with a parkinsonian syndrome, *J. Neurol. Neurosurg. Psychiatry*, 1995, **59**, 318–321.
56. D. J. Costello, S. L. Walsh, H. J. Harrington and C. H. Walsh, Concurrent hereditary haemochromatosis and idiopathic Parkinson's disease: a case report series, *J. Neurol. Neurosurg. Psychiatry*, 2004, **75**, 631–633.
57. W. Nandar and J. R. Connor, HFE gene variants affect iron in the brain, *J. Nutr.*, 2011, **141**, 729S–739S.
58. L. H. You, F. Li, L. Wang, S. E. Zhao, S. M. Wang, L. L. Zhang, L. H. Zhang, X. L. Duan, P. Yu and Y. Z. Chang, Brain iron accumulation exacerbates the pathogenesis of MPTP-induced Parkinson's disease, *Neuroscience*, 2015, **284**, 234–246.
59. Y. He, T. Lee and S. K. Leong, Time course of dopaminergic cell death and changes in iron, ferritin and transferrin levels in the rat substantia nigra after 6-hydroxydopamine (6-OHDA) lesioning, *Free Radical Res.*, 1999, **31**, 103–112.
60. Z. H. Shi, G. Nie, X. L. Duan, T. Rouault, W. S. Wu, B. Ning, N. Zhang, Y. Z. Chang and B. L. Zhao, Neuroprotective mechanism of mitochondrial ferritin on 6-hydroxydopamine-induced dopaminergic cell damage: implication for neuroprotection in Parkinson's disease, *Antioxid. Redox Signaling*, 2010, **13**, 783–796.
61. H. Mochizuki, H. Imai, K. Endo, K. Yokomizo, Y. Murata, N. Hattori and Y. Mizuno, Iron accumulation in the substantia nigra of 1-methyl-4-phenyl-1,2,3,6-tetrahydropyridine (MPTP)-induced hemiparkinsonian monkeys, *Neurosci. Lett.*, 1994, **168**, 251–253.
62. M. B. Youdim, E. Grunblatt, Y. Levites, G. Maor and S. Mandel, Early and late molecular events in neurodegeneration and neuroprotection in Parkinson's disease MPTP model as assessed by cDNA microarray; the role of iron, *Neurotoxic. Res.*, 2002, **4**, 679–689.
63. D. Blum, S. Torch, N. Lambeng, M. Nissou, A. L. Benabid, R. Sadoul and J. M. Verna, Molecular pathways involved in the neurotoxicity of 6-OHDA, dopamine and MPTP: contribution to the apoptotic theory in Parkinson's disease, *Prog. Neurobiol.*, 2001, **65**, 135–172.
64. W. Le, Role of iron in UPS impairment model of Parkinson's disease, *Parkinsonism Relat. Disord.*, 2014, **20**(suppl. 1), S158–S161.

65. D. Kaur, D. Lee, S. Ragapolan and J. K. Andersen, Glutathione depletion in immortalized midbrain-derived dopaminergic neurons results in increases in the labile iron pool: implications for Parkinson's disease, *Free Radic. Biol. Med.*, 2009, **46**, 593–598.
66. P. G. Mastroberardino, E. K. Hoffman, M. P. Horowitz, R. Betarbet, G. Taylor, D. Cheng, H. M. Na, C. A. Gutekunst, M. Gearing, J. Q. Trojanowski, M. Anderson, C. T. Chu, J. Peng and J. T. Greenamyre, A novel transferrin/TfR2-mediated mitochondrial iron transport system is disrupted in Parkinson's disease, *Neurobiol. Dis.*, 2009, **34**, 417–431.
67. H. Jiang, Z. M. Qian and J. X. Xie, [Increased DMT1 expression and iron content in MPTP-treated C57BL/6 mice], *Sheng Li Xue Bao*, 2003, **55**, 571–576.
68. H. Jiang, N. Song, H. Xu, S. Zhang, J. Wang and J. Xie, Up-regulation of divalent metal transporter 1 in 6-hydroxydopamine intoxication is IRE/IRP dependent, *Cell Res.*, 2010, **20**, 345–356.
69. H. M. Xu, H. Jiang, J. Wang, B. Luo and J. X. Xie, Over-expressed human divalent metal transporter 1 is involved in iron accumulation in MES23.5 cells, *Neurochem. Int.*, 2008, **52**, 1044–1051.
70. I. De Domenico, D. M. Ward, M. C. di Patti, S. Y. Jeong, S. David, G. Musci and J. Kaplan, Ferroxidase activity is required for the stability of cell surface ferroportin in cells expressing GPI-ceruloplasmin, *EMBO J.*, 2007, **26**, 2823–2831.
71. B. N. Patel, R. J. Dunn, S. Y. Jeong, Q. Zhu, J. P. Julien and S. David, Ceruloplasmin regulates iron levels in the CNS and prevents free radical injury, *J. Neurosci.*, 2002, **22**, 6578–6586.
72. S. Y. Jeong and S. David, Age-related changes in iron homeostasis and cell death in the cerebellum of ceruloplasmin-deficient mice, *J. Neurosci.*, 2006, **26**, 9810–9819.
73. K. Kaneko, A. Hineno, K. Yoshida and S. Ikeda, Increased vulnerability to rotenone-induced neurotoxicity in ceruloplasmin-deficient mice, *Neurosci. Lett.*, 2008, **446**, 56–58.
74. K. J. Bharucha, J. K. Friedman, A. S. Vincent and E. D. Ross, Lower serum ceruloplasmin levels correlate with younger age of onset in Parkinson's disease, *J. Neurol.*, 2008, **255**, 1957–1962.
75. B. Leveugle, B. A. Faucheux, C. Bouras, N. Nillesse, G. Spik, E. C. Hirsch, Y. Agid and P. R. Hof, Cellular distribution of the iron-binding protein lactotransferrin in the mesencephalon of Parkinson's disease cases, *Acta Neuropathol.*, 1996, **91**, 566–572.
76. B. A. Faucheux, N. Nillesse, P. Damier, G. Spik, A. Mouatt-Prigent, A. Pierce, B. Leveugle, N. Kubis, J. J. Hauw and Y. Agid, *et al.*, Expression of lactoferrin receptors is increased in the mesencephalon of patients with Parkinson disease, *Proc. Natl. Acad. Sci. U. S. A.*, 1995, **92**, 9603–9607.
77. D. Koziorowski, A. Friedman, P. Arosio, P. Santambrogio and D. Dziewulska, ELISA reveals a difference in the structure of substantia nigra ferritin in Parkinson's disease and incidental Lewy body compared to control, *Parkinsonism Relat. Disord.*, 2007, **13**, 214–218.

78. B. A. Faucheux, M. E. Martin, C. Beaumont, S. Hunot, J. J. Hauw, Y. Agid and E. C. Hirsch, Lack of up-regulation of ferritin is associated with sustained iron regulatory protein-1 binding activity in the substantia nigra of patients with Parkinson's disease, *J. Neurochem.*, 2002, **83**, 320–330.
79. M. Fasano, B. Bergamasco and L. Lopiano, Modifications of the iron-neuromelanin system in Parkinson's disease, *J. Neurochem.*, 2006, **96**, 909–916.
80. L. Zecca, A. Stroppolo, A. Gatti, D. Tampellini, M. Toscani, M. Gallorini, G. Giaveri, P. Arosio, P. Santambrogio, R. G. Fariello, E. Karatekin, M. H. Kleinman, N. Turro, O. Hornykiewicz and F. A. Zucca, The role of iron and copper molecules in the neuronal vulnerability of locus coeruleus and substantia nigra during aging, *Proc. Natl. Acad. Sci. U. S. A.*, 2004, **101**, 9843–9848.
81. L. Zecca, C. Bellei, P. Costi, A. Albertini, E. Monzani, L. Casella, M. Gallorini, L. Bergamaschi, A. Moscatelli, N. J. Turro, M. Eisner, P. R. Crippa, S. Ito, K. Wakamatsu, W. D. Bush, W. C. Ward, J. D. Simon and F. A. Zucca, New melanic pigments in the human brain that accumulate in aging and block environmental toxic metals, *Proc. Natl. Acad. Sci. U. S. A.*, 2008, **105**, 17567–17572.
82. L. Zecca, F. A. Zucca, A. Albertini, E. Rizzio and R. G. Fariello, A proposed dual role of neuromelanin in the pathogenesis of Parkinson's disease, *Neurology*, 2006, **67**, S8–S11.
83. M. Sasaki, E. Shibata, K. Tohyama, J. Takahashi, K. Otsuka, K. Tsuchiya, S. Takahashi, S. Ehara, Y. Terayama and A. Sakai, Neuromelanin magnetic resonance imaging of locus ceruleus and substantia nigra in Parkinson's disease, *Neuroreport*, 2006, **17**, 1215–1218.
84. K. Stepien, A. Dzierzega-Lecznar and I. Tam, [The role of neuromelanin in Parkinson's disease–new concepts], *Wiad. Lek.*, 2007, **60**, 563–569.
85. G. Gao and Y. Z. Chang, Mitochondrial ferritin in the regulation of brain iron homeostasis and neurodegenerative diseases, *Front. Pharmacol.*, 2014, **5**, 19.
86. P. J. McLean, H. Kawamata, S. Ribich and B. T. Hyman, Membrane association and protein conformation of alpha-synuclein in intact neurons. Effect of Parkinson's disease-linked mutations, *J. Biol. Chem.*, 2000, **275**, 8812–8816.
87. S. Byard, A. Abraham, P. J. T. Boulton, R. K. Harris and P. Hodgkinson, A multi-technique approach to the study of structural stability and desolvation of two unusual channel hydrate solvates of finasteride, *J. Pharm. Sci.*, 2012, **101**, 176–186.
88. N. M. Bonini and B. I. Giasson, Snaring the function of alpha-synuclein, *Cell*, 2005, **123**, 359–361.
89. V. N. Uversky, Neuropathology, biochemistry, and biophysics of alpha-synuclein aggregation, *J. Neurochem.*, 2007, **103**, 17–37.
90. M. Kostka, T. Hogen, K. M. Danzer, J. Levin, M. Habeck, A. Wirth, R. Wagner, C. G. Glabe, S. Finger, U. Heinzelmann, P. Garidel, W. Duan, C. A. Ross, H. Kretzschmar and A. Giese, Single particle characterization

of iron-induced pore-forming alpha-synuclein oligomers, *J. Biol. Chem.*, 2008, **283**, 10992–11003.

91. A. L. Friedlich, R. E. Tanzi and J. T. Rogers, The 5′-untranslated region of Parkinson's disease alpha-synuclein messengerRNA contains a predicted iron responsive element, *Mol. Psychiatry*, 2007, **12**, 222–223.
92. E. Pollitt, Early iron deficiency anemia and later mental retardation, *Am. J. Clin. Nutr.*, 1999, **69**, 4–5.
93. S. Haupt, M. Berger, Z. Goldberg and Y. Haupt, Apoptosis–the p53 network, *J. Cell Sci.*, 2003, **116**, 4077–4085.
94. D. Berg and H. Hochstrasser, Iron metabolism in Parkinsonian syndromes, *Mov. Disord.*, 2006, **21**, 1299–1310.
95. A. I. Bush, Metals and neuroscience, *Curr. Opin. Chem. Biol.*, 2000, **4**, 184–191.
96. K. Hanrott, L. Gudmunsen, M. J. O'Neill and S. Wonnacott, 6-hydroxydopamine-induced apoptosis is mediated *via* extracellular auto-oxidation and caspase 3-dependent activation of protein kinase Cdelta, *J. Biol. Chem.*, 2006, **281**, 5373–5382.
97. S. B. Dunnett and A. Bjorklund, Prospects for new restorative and neuroprotective treatments in Parkinson's disease, *Nature*, 1999, **399**, A32–A39.
98. P. A. LeWitt and D. Nyholm, New developments in levodopa therapy, *Neurology*, 2004, **62**, S9–S16.
99. X. Junxia, J. Hong, C. Wenfang and Q. Ming, Dopamine release rather than content in the caudate putamen is associated with behavioral changes in the iron rat model of Parkinson's disease, *Exp. Neurol.*, 2003, **182**, 483–489.
100. F. Febbraro, K. J. Andersen, V. Sanchez-Guajardo, N. Tentillier and M. Romero-Ramos, Chronic intranasal deferoxamine ameliorates motor defects and pathology in the alpha-synuclein rAAV Parkinson's model, *Exp. Neurol.*, 2013, **247**, 45–58.
101. S. Gal, H. Zheng, M. Fridkin and M. B. Youdim, Novel multifunctional neuroprotective iron chelator-monoamine oxidase inhibitor drugs for neurodegenerative diseases. *In vivo* selective brain monoamine oxidase inhibition and prevention of MPTP-induced striatal dopamine depletion, *J. Neurochem.*, 2005, **95**, 79–88.
102. D. B. Shachar, N. Kahana, V. Kampel, A. Warshawsky and M. B. Youdim, Neuroprotection by a novel brain permeable iron chelator, VK-28, against 6-hydroxydopamine lession in rats, *Neuropharmacology*, 2004, **46**, 254–263.
103. W. Zhu, W. Xie, T. Pan, P. Xu, M. Fridkin, H. Zheng, J. Jankovic, M. B. Youdim and W. Le, Prevention and restoration of lactacystin-induced nigrostriatal dopamine neuron degeneration by novel brain-permeable iron chelators, *FASEB J.*, 2007, **21**, 3835–3844.
104. M. B. Youdim, G. Stephenson and D. Ben Shachar, Ironing iron out in Parkinson's disease and other neurodegenerative diseases with iron chelators: a lesson from 6-hydroxydopamine and iron chelators, desferal and VK-28, *Ann. N. Y. Acad. Sci.*, 2004, **1012**, 306–325.

105. D. Kaur, F. Yantiri, S. Rajagopalan, J. Kumar, J. Q. Mo, R. Boonplueang, V. Viswanath, R. Jacobs, L. Yang, M. F. Beal, D. DiMonte, I. Volitaskis, L. Ellerby, R. A. Cherny, A. I. Bush and J. K. Andersen, Genetic or pharmacological iron chelation prevents MPTP-induced neurotoxicity *in vivo*: a novel therapy for Parkinson's disease, *Neuron*, 2003, **37**, 899–909.
106. D. T. Dexter, R. J. Ward, A. Florence, P. Jenner and R. R. Crichton, Effects of desferrithiocin and its derivatives on peripheral iron and striatal dopamine and 5-hydroxytryptamine metabolism in the ferrocene-loaded rat, *Biochem. Pharmacol.*, 1999, **58**, 151–155.
107. H. Sies, Oxidative stress: oxidants and antioxidants, *Exp. Physiol.*, 1997, **82**, 291–295.
108. S. Dowling, F. Regan and H. Hughes, The characterisation of structural and antioxidant properties of isoflavone metal chelates, *J. Inorg. Biochem.*, 2010, **104**, 1091–1098.
109. E. Vlachodimitropoulou, P. A. Sharp and R. J. Naftalin, Quercetin-iron chelates are transported *via* glucose transporters, *Free Radic. Biol. Med.*, 2011, **50**, 934–944.
110. I. F. Almeida, P. Valentao, P. B. Andrade, R. M. Seabra, T. M. Pereira, M. H. Amaral, P. C. Costa and M. F. Bahia, Oak leaf extract as topical antioxidant: free radical scavenging and iron chelating activities and *in vivo* skin irritation potential, *Biofactors*, 2008, **33**, 267–279.
111. S. Mandel, O. Weinreb, T. Amit and M. B. Youdim, Cell signaling pathways in the neuroprotective actions of the green tea polyphenol (-)-epigallocatechin-3-gallate: implications for neurodegenerative diseases, *J. Neurochem.*, 2004, **88**, 1555–1569.
112. L. Baum and A. Ng, Curcumin interaction with copper and iron suggests one possible mechanism of action in Alzheimer's disease animal models, *J. Alzheimers Dis.*, 2004, **6**, 367–377, discussion 443–369.
113. Y. Y. Tian, L. Ge, X. L. Duan, Z. Q. Gao and Y. Z. Chang, [Lycopene liposomes: lycopene release *in vitro* and pharmaceutical behaviors and antioxidation *in vivo*], *Yaoxue Xuebao*, 2007, **42**, 1107–1111.

CHAPTER 10

Protein Oxidation, Quality-Control Mechanisms and Parkinson's Disease

PABLO HERNANDEZ-FRANCO, ANNADURAI ANANDHAN AND RODRIGO FRANCO*

Redox Biology Center & School of Veterinary Medicine and Biomedical Sciences, University of Nebraska-Lincoln, Lincoln, NE 68583, USA
*E-mail: rfrancocruz2@unl.edu

10.1 Introduction

Proteostasis is defined as the integrated mechanisms within cells that control protein biogenesis, folding, trafficking and degradation.[1] The integrity of the proteome is essential for cellular homeostasis, function and survival and is continuously challenged under both physiological and pathological conditions. Cells have evolved a complex and hierarchical array of processes called protein quality-control mechanisms to ensure protein integrity that include chaperones and protein sorting/segregation and degradation pathways. Protein quality control starts even before protein synthesis and continues throughout their 'lifespan'. The diversity of mechanisms involved in quality control are in direct relationship to the high degree of heterogeneity in protein structure, lifespan, abundance and distribution within the proteome.[2,3]

Accumulation of misfolded protein aggregates is a hallmark in Parkinson's disease (PD) and several other neurodegenerative disorders.[4] The loss of

Issues in Toxicology No. 34
Oxidative Stress and Redox Signalling in Parkinson's Disease
Edited by Rodrigo Franco, Jonathan A. Doorn and Jean-Christophe Rochet

Published by the Royal Society of Chemistry, www.rsc.org

neuromelanin-containing dopaminergic neurons in the *substantia nigra* (A9) is the specific hallmark in PD. PD diagnosis made at autopsy, is also based on the presence of intraneuronal, eosinophilic inclusions called Lewy bodies (LBs). LBs are protein inclusions or clumps found throughout the brain of PD patients. Alterations in protein quality-control mechanisms involved in protein folding and clearance of misfolded protein aggregates have been observed during aging, the major risk factor for most neurodegenerative disorders including PD. In addition to aging, the etiology of PD involves genetic and environmental risk factors that have been shown to promote protein misfolding and aggregation as well by oxidative damage and energy failure (Figure 10.1).[5]

In this chapter, we will review the proposed mechanisms by which PD risk factors (aging, genetics and environmental exposures) promote protein misfolding and aggregation and impair protein quality-control mechanisms. Special emphasis will be placed on the role of oxidative stress in the dysfunction of the chaperone network, the ubiquitin-proteasome (UPS) and the autophagosome-lysosome system in PD.

10.2 Misfolded Protein Aggregation and Accumulation in PD

Histopathological analyses of postmortem brain samples identified the presence of intraneuronal fibrillar aggregates referred to as LBs as a hallmark in PD. LBs are not exclusive of PD and can also be found in other LB

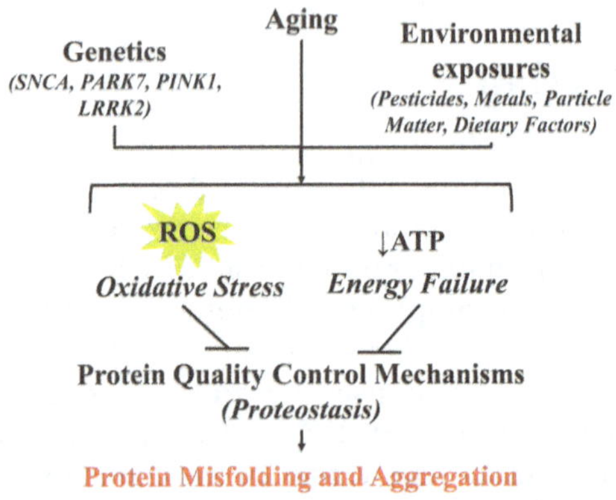

Figure 10.1 PD-risk factors impair protein quality-control mechanisms. Proteostasis is defined as the integrated mechanisms within cells that control protein biogenesis, folding, trafficking and degradation. Cells have evolved a complex and hierarchical array of processes called protein quality-control mechanisms to ensure protein integrity that include chaperones and protein sorting/segregation and degradation pathways. The presence of intraneuronal, eosinophilic inclusions called LBs is a hallmark of PD, which is linked to alterations in protein quality-control mechanisms involved in protein folding and clearance of misfolded protein aggregates PD risk factors aging, genetic alterations and environmental exposures have been shown to promote protein misfolding and aggregation in association with oxidative damage and energy failure.

diseases such as dementia with LBs (DLB), incidental LB disease (ILBD), Alzheimer's disease with predominantly limbic LBs (ADLB), lysosomal storage disorders (Gaucher disease), multiple system atrophy and neurodegeneration with brain iron accumulation. Proteomic analyses of LBs have identified over 100 protein in LBs from which alpha (α)-synuclein is the major component. While initial studies proposed that LB inclusion-mediated toxicity was involved in dopaminergic cell loss, recent studies have proposed that LB formation might be a protective mechanism against the toxicity of misfolded intermediates.[6]

10.2.1 α-Synuclein: Mutations and Misfolding

α-synuclein is a 140 amino acid protein encoded by the SNCA gen located in the long (q) arm of chromosome 4 at position 21. α-synuclein does not have a defined structure and is considered a natively unfolded and intrinsically disordered protein lacking well-organized secondary structure. However, upon binding to negatively charged lipids or on prolonged incubation periods it forms α-helical and β-rich structures, respectively.[7] Natively unfolded proteins require interacting partners in order take on a specific tertiary conformation. Thus, it is very likely that α-synuclein structure under physiological conditions is determined by the presence of molecular interacting partners. α-synuclein can form multimers, and it has been proposed that at physiological levels and under normal conditions α-synuclein exists as a tetramer (Figure 10.2). α-synuclein is composed of three distinct regions (Figure 10.3): (1) an amino terminus (residues 1–60) with apolipoprotein lipid-binding motifs predicted to form amphiphilic helices, (2) a central hydrophobic region (61–95) called non-Aβ component (NAC), and (3) a carboxyl terminus that is highly negatively charged, and is unstructured. α-synuclein is localized abundantly in presynaptic terminals in neurons (1% of total cytosolic protein in the nervous system). The physiological function of α-synuclein is still unclear, but research suggests a role in regulating synaptic neurotransmission, vesicle traffic, and vesicular neurotransmitter storage. However, our lack of a clear understanding about the exact physiological role of α-synuclein does not allow us to determine whether its pathological role is linked to a toxic gain of function, or an impairment in its physiological role.[8]

Only about 10% of PD cases are linked to hereditably monogenic mutations. Mutations in SNCA cause early-onset autosomal-dominant PD. Missense mutations (A30P, E46K, H50Q, G51D, A53E and A53T, see Figure 10.3) as well as duplications and triplications have been described to be linked to familial PD so far. Duplications have also been reported in sporadic cases. Gene dosage is directly associated with an earlier onset, more severe phenotype, and faster disease progression. SNCA mutants E46K, H50Q and A53T are more prompt to generate stable β sheets that oligomerize and progressively form protofibrils, and insoluble fibrils (Figure 10.2).[9] Of importance, H50, which is also subject to oxidative modification (*see section below*), has a key role in protecting α-synuclein from aggregation.[10] Importantly, oligomers or protofibrils intermediates act as seeds that accelerate the conversion

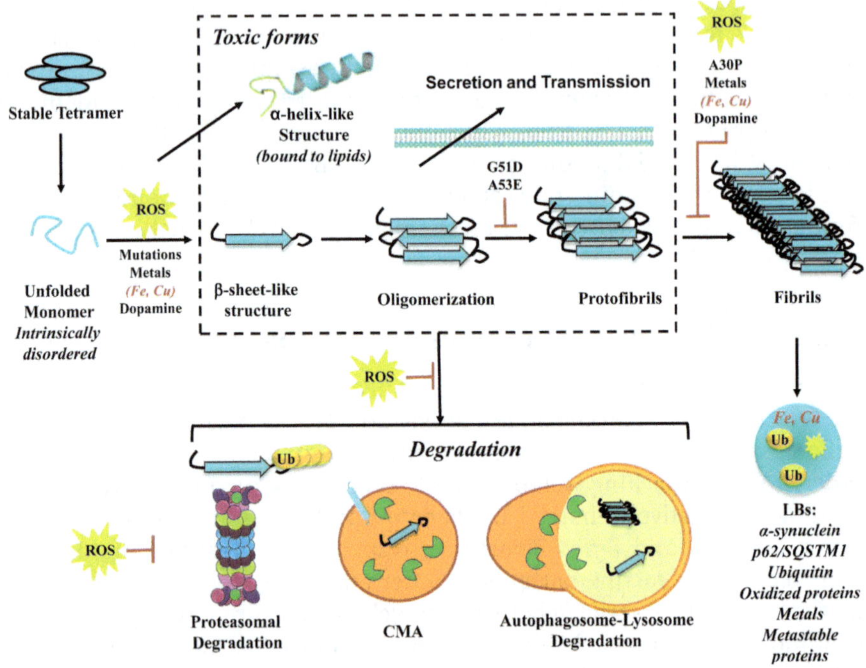

Figure 10.2 α-synuclein aggregation and degradation. α-synuclein is considered a natively unfolded and intrinsically disordered protein. However, upon binding to negatively charged lipids or on prolonged incubation periods it forms α-helical and β-rich structures, respectively. α-synuclein can form multimers, and it has been proposed that under normal conditions α-synuclein exists as a tetramer. SNCA mutations (E46K, H50Q and A53T), oxidation, interaction with metals and increased levels of α-synuclein prompt the generation of stable β sheets that oligomerize and progressively form protofibrils. However, G51D, A53E, and A30P mutations and other oxidative modifications also reduce α-synuclein fibrillation or aggregation, suggesting that early oligomeric forms are the toxic species of α-synuclein. The UPS and the autophagosome-lysosome system are complementary pathways, where alterations in the rate of one system are reported to modify those of the other. Accordingly, clearance of misfolded/aggregated α-synuclein has been shown to be mediated by both autophagy (macroautophagy and CMA) and the UPS. Importantly, components within the autophagy and UPS are sensitive to inactivation by oxidative stress.

of α-synuclein into aggregates and because every intermediate product is more stable than the next one this process becomes largely irreversible. Furthermore, α-synuclein aggregation is also promoted by increased α-synuclein levels, post-translational modifications including oxidation, and interaction with metals. However, the fact that G51D, A53E, and A30P mutations reduce α-synuclein fibrillation or aggregation suggests that: (1) aggregation is not involved in the pathological consequences of SNCA genetic modifications; (2) early oligomeric forms are the toxic species of α-synuclein; (3) a likely loss

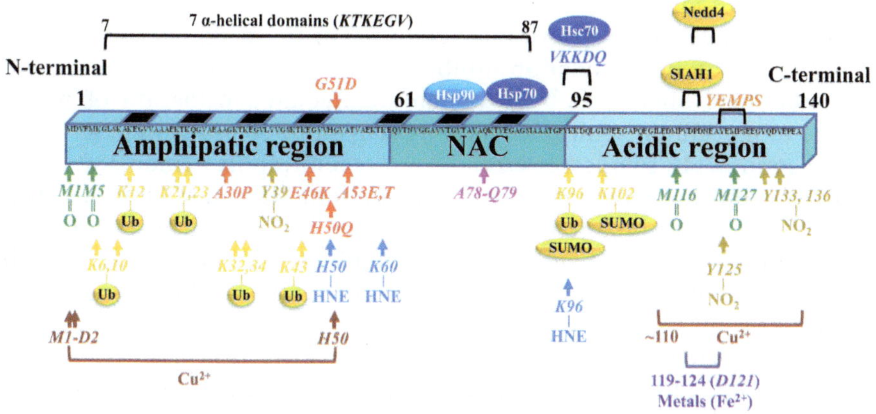

Figure 10.3 α-synuclein modifications that regulate its aggregation and degradation. α-synuclein is a 140 amino acid protein composed of three regions: (1) an amino terminus (residues 1–60), (2) a central hydrophobic region (61–95) called non-Aβ component (NAC), and (3) a carboxyl terminus that is highly negatively charged, and unstructured. The figure highlights mutations (in red), oxidative modifications (tyrosine nitration [dark yellow], methionine sulfoxide [green], 4-HNE adducts [blue] and the oxidized catechol binding region [orange]), binding regions for metals (Cu in brown, Fe in purple), and ubiquitin-like modifications (Ub and SUMO, in yellow). Binding regions of E3-ligases and Hsps are also indicated, as well as the cleaved region site for MMP3 (pink).

of function is associated with α-synuclein toxicity; or (4) that multiple mechanisms might be involved in the toxicity of α-synuclein (Figure 10.2).[11]

10.2.1.1 *Redox Regulation of α-Synuclein Aggregation*

Oxidative stress has been proposed to promote α-synuclein aggregation and toxicity.[12] α-synuclein has been shown to be modified by tyrosine nitration,[13–17] dityrosine crosslinking[18,19] and dopamine quinones (byproducts from the oxidation of dopamine),[20,21] which differentially affects its ability to oligomerize, form stable fibrils and protein inclusions.[22] Both oxidative stress and dopamine-related catecholamines inhibit protofibril-to-fibril conversion by oxidation or adduct formation of α-synuclein.[23–25] Oxidized GSSG also interacts with α-synuclein to promote amyloid formation.[26]

Specific regions and amino acids in α-synuclein have been described as target for redox modifications (Figure 10.3). Y39, Y125, Y133 and Y136 have been reported to be nitrated in response to oxidizing agents or by the overexpression of monoamine oxidase (MAO) that catalyses the oxidation of catecholamines (dopamine).[14,22,27] Importantly, Y39 or Y125 seem to be also involved in dityrsoine crosslinking of α-synuclein monomers.[19,28–30] Tyrosine nitration not only regulates α-synuclein oligomerization and fibrillogenesis, but also its interaction with lipid bilayers.[29,31] Oxidized catechols have been reported to interact non-covalently with α-synuclein residues 125-YEMPS-129 in the C-terminal region or in the neighboring region of Y39.[25,32] Methionine oxidation inhibits fibrillation of α-synuclein stabilizing non-toxic oligomers.[22,33,34] Dopamine-mediated oxidation of methionine to methionine sulfoxide (M(O)) seems to preferably occur at M127, but also in M116 to a lesser extent.[35,36] A recent report demonstrates that from the four methionines subject to oxidation, M1 and M5 but not M116 and M127 can be repaired by methionine sulfoxide reductases (MSRs).[37] Oxidation (M1 and M5) and dopamine modification of α-synuclein inhibits its degradation by the proteasome or chaperone-mediated autophagy, respectively.[38]

Lipid peroxidation byproducts also target α-synuclein. The highly reactive aldehyde, 4-hydroxy-2-nonenal (HNE) generates α-synculein adducts *via* covalent modification (Michael addition) of H50, and to a lesser extent K60 and K96 modulating its oligomerization and fibrillation.[39–42] 4-oxo-2-nonenal (ONE) and acrolein also modify α-synuclein and enhance its oligomerization.[42–44] Similarly, the highly cytotoxic aldehyde 3,4-dihydroxyphenylacetaldehyde (DOPAL), a byproduct from the oxidation of dopamine by MAO, has been reported to modify α-synuclein *via* Schiff-base and Michael-addition adduct formation through modification of lysines within the α-helical domains (Figure 10.3).[45] Oxidative byproducts of cholesterol have also been reported to accelerate α-synuclein aggregation.[46]

Oxidative glycation reactions between the amino groups of proteins and reducing sugars result in highly reactive advanced glycation end products (AGEs). AGEs cause protease-resistant crosslinking of peptides and proteins, leading to protein aggregation. As such, AGEs also promote α-synuclein crosslinking and inclusion body formation.[47] Dicarbonyl compounds methyglyoxal and glyoxal generated by cell metabolism, glucose autoxidation and lipid peroxidation can generate AGEs, and have also been found to promote α-synuclein oligomerization while reducing its fibrillization.[48]

Redox-active metal ions iron (Fe) and copper (Cu) are known to bind α-synuclein and promote its fibrillogenesis and aggregation, and this binding, and more importantly, the metal-α-synuclein toxic interaction is modified by the presence of mutations.[49–52] Interestingly, metal-catalyzed oxidation of α-synuclein has been reported to inhibit the formation of α-synuclein filaments with the increased accumulation of oligomers and protofibrils (Figure 10.2).[23,53–55] In addition, metal interaction with α-synuclein has also been shown to promote the formation of pore-like oligomers.[56] Fe and other divalent metals interact with α-synuclein with low affinity (mM) at the 119-DPDNEA-124 domain involving D121, D122 and N123 residues primarily (Figure 10.3).[57] In contrast to other metals, Cu can bind to α-synuclein at low micromolar range (0.1 μM) being the metal that most effectively promotes α-synuclein oligomerization.[58] α-synuclein has two high-affinity binding sites for Cu at the 1-MD-2 and H50 at physiological pH.[57–61] A low-binding site has also been described at the C-terminal domain (Figure 10.3).[60–62]

10.3 Protein Quality-Control Mechanisms in PD

Proteins are continually at risk of damage, misfolding and aggregation. Protein folding and maintenance in their functional form is mediated by an elaborate network of quality-control mechanisms that involve quality control checkpoints, degradation or repair machineries, and stress responses that are activated to counteract the deleterious effects of protein misfolding.[2] We will next summarize the dysfunction in protein quality-control mechanisms associated with PD.

10.3.1 Protein Synthesis

Proteostasis starts at the level of the synthesis of nascent polypeptide chains in the ribosome where quality-control mechanisms participate in the process of eukaryotic translation to ensure efficiency, quality and accuracy of polypeptide synthesis. While only one in every 10 000 amino acids is estimated to be misincorporated in polypeptide chains, given the size of the proteome tens of millions of aberrant proteins are likely to be produced. The mRNA is an important 'quality' control checkpoint for translation efficiency. Translational errors have important effects destabilizing protein structure and folding. Importantly, mutations in genes involved in quality control of nascent polypeptide synthesis such as tRNA synthetases that disrupt translational fidelity increasing the rate of misincorporation of amino acids, or ubiquitin ligases that impair cotranslational degradation of anomalous polypeptides, lead to neurodegeneration.[63,64] The correct attachment of amino acids to each tRNA species that is required for protein synthesis is an energy-dependent process carried out by aminoacyl-tRNA synthetases. Additionally, aminoacyl-tRNA synthetases also hydrolyze (edit) an incorrectly attached amino acid, and oxidative stress induces protein mistranslation and degradation by impairment of aminoacyl-tRNA synthetase editing.[65] Oxidative stress also diverts tRNA synthetases to the nucleus to protect against oxidative damage.[66] Moreover, oxidized mRNA, protein mistranslation and subsequent degradation have been recently recognized as important contributors to neurodegeneration.[67–72]

Modulation of translational rates is also important in facilitating the binding of chaperones that prevent aggregate formation. As such, protein aggregates contain proteins as nascent unstructured polypeptides suggesting that chaperones failed to bind them and found them in time to avoid aggregation. LRRK2 stimulates mRNA translation by phosphorylation of eukaryotic initiation factor 4E (eIF4E)-binding protein (4E-BP), which releases it from its inhibitory binding to eIF4E enabling cap-dependent translation initiation.[73] In addition, the pathogenic LRRK2 G2019S has been recently shown to stimulate both cap-dependent and cap-independent mRNA translation and induces a bulk increase in protein synthesis *via* phosphorylation of the ribosomal protein s15.[74] LRRK2 has also been shown to negatively regulate micro-RNA-mediated translational repression.[75] PINK1 and Parkin have also been shown to regulate the translation of respiratory chain mRNA components through displacement of translational repressors.[76] Another study reported that PINK1 promotes cap-independent internal ribosomal entry site (IRES)-mediated translation.[77] Interestingly, α-synuclein 5′-UTR harbors an IRES, and oxidative stress stimulated SNCA protein translation *via* enhanced IRES activity.[78] Inhibition of protein translation exerts a protective effect against the toxicity of PINK1 mutants.[79] The eukaryotic translation initiation factor 4-gamma (EIF4G1) is a scaffold protein for the binding of eIF4A and eIF4E to form the mRNA cap complex, required for initiating cap-dependent translation, and mutations in EIF4G1 have also been linked to familial PD.[80] However, the effects of altered protein translation in protein aggregate formation have not been clearly delineated.

10.3.2 Protein Folding, Unfolding and Disaggregation by Chaperones

Chaperones interact with other proteins and aid in their folding or assembly without being part of its final structure. Chaperones are usually part of protein machineries that assist various processes within protein quality control including folding, unfolding or disaggregation of misfolded proteins or aggregates, which are subsequently refolded, or targeted for degradation. Chaperones also assist macromolecular complex assembly and protein traffic across cellular compartments. Chaperones act on a wide variety of substrates, whose activity regulates the folding or unfolding of non-native polypeptides, or mediates the protection of nascent polypeptide subunits during their assembly process. Heat shock proteins (Hsps) are ATP-dependent chaperones classified in different families according to their molecular weight. Chaperones do not usually work alone but as chaperoning machines formed by various members of the same family.[81,82]

Hsp families have different functions. In association with various cofactors, Hsp70s (17 genes in humans) direct substrates for unfolding, disaggregation, refolding or degradation as well as translocation across organelle membranes. Its activity depends on dynamic interactions between the ATPase and substrate-binding domains, and their interaction with co-chaperones such as the Hsp40 proteins (J proteins or DnaJs) and nucleotide exchange factors (NEFs). Hsp40s (49 members in humans) stimulate Hsp70 ATPase and are thought to act as the substrate recruiter for Hsp70.[81–84] With 5 genes in humans, Hsp90s are highly abundant and ubiquitous chaperones with diverse biological roles and a multitude of interacting partners that regulate their activity. Hsp90s binds non-native polypeptides preventing their aggregation. They also participate at a late stage of folding of substrates important in cellular signalling and development, and they also target substrates for proteolysis. Hsp90s function is modulated by co-chaperones, substrates and post-translational modifications. Sixteen Hsp60 (CCT or chaperonins) genes have been described in humans. Hsp60s participate in *de novo* protein folding and assembly as well as protein refolding.[81,82] Small Hsps (10 in humans) are ATP-independent chaperones that function as holdases to buffer protein aggregation primarily during stress conditions (Figure 10.4).[81,82]

Protein folding can begin cotranslationally as soon as the N-terminal part of the polypeptide emerges from the ribosome. The majority of proteins must fold into defined 3D structures that represent a compromise between

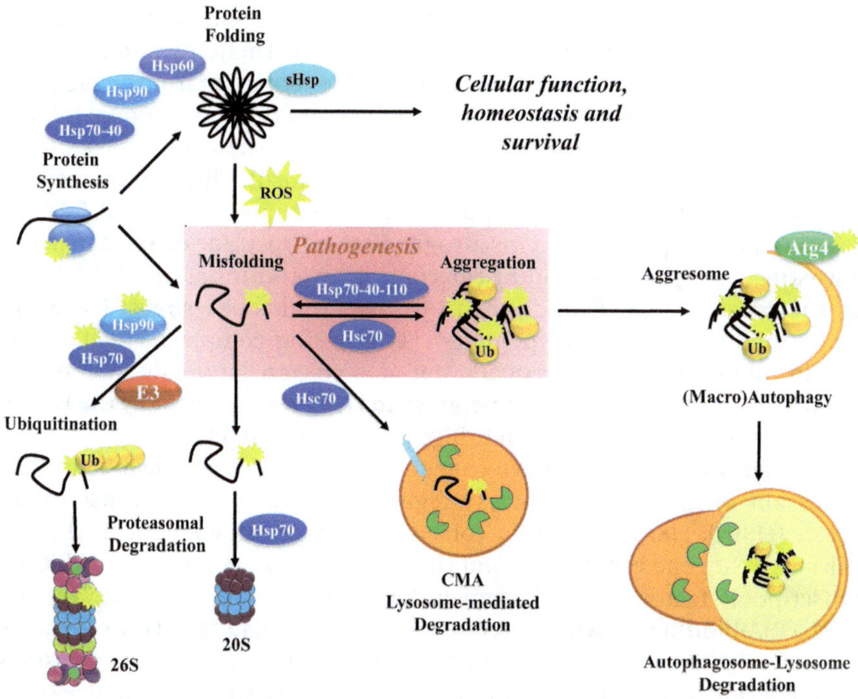

Figure 10.4 Protein-degradation pathways involved in the clearance of misfolded oxidized proteins and aggregates. As complementary pathways both the UPS and autophagy are involved in the clearance of misfolded/oxidized proteins and aggregates. Oxidized proteins can be degraded by ubiquitin-dependent and independent pathways *via* the 26S or 20S proteasome. Energy (ATP)/ubiquitin-independent degradation by the 20S proteasome seems to mediate the degradation of the majority of oxidized proteins. Covalent crosslinks, disulfide bonds, hydrophobic interactions, and heavily oxidized stable protein aggregates are thought to be unsuitable for proteasomal degradation and thus are considered to be degraded largely by macroautophagy (autophagy), which also mediates the clearance of oxidized organelles (mitophagy). However, the Hsp70-Hsp40-Hsp110 chaperone machinery has been reported to mediate ATP-dependent disassembly of amyloids fibril and aggregates that can be degraded by the proteasome. A dysfunction in these protein quality-control mechanisms is directly linked to the pathogenesis of PD. Oxidative stress (yellow stars) is not only involved in the generation of oxidized protein aggregates but it also impairs different aspects of protein quality control including protein-degradation pathways, protein synthesis and folding (regulated by Hsp70-Hsp40, Hsp90, Hsp60 and sHsp chaperone machineries). Protein misfolding and aggregation target enzymes required for normal cell function and homeostasis.

thermodynamic stability and the conformational flexibility required for proper functionality. As a consequence, protein structure can be marginally stable even under physiological conditions (metastable) and thus highly susceptible to misfolding and aggregation. Around 30% of proteins, such as α-synuclein, are classified as intrinsically unstructured or disordered and contain regions thought to adopt a structure that depends on its interaction with binding partners. During translation ribosome-binding chaperones (nascent-chain-associated complex (NACA), MPP11 [DnaJC12] and Hsp70L1) interact first with the nascent polypeptides to prevent them from engaging in unfavorable intra- and intermolecular interactions.[85] Cotranslational folding is followed by the heat shock proteins Hsp40 and Hsp70, which also mediate folding of overall unfolded proteins in the cytosol. Thus, an increase in the levels of unfolded proteins would also affect co-translational folding. Hsp70-Hsp40 chaperone systems also mediate polypeptide chain transfer to downstream chaperones. Folding can continue post-translationally by downstream transfer of polypeptide chains to chaperonins or Hsp90s. The chaperonin tailless complex polypeptide-1 (TCP-1) ring complex (TRiC) assists protein folding post-translationally but also co-translationally in cooperation with Hsp70. Transfer of polypeptides to Hsp90 is mediated by the Hsp90 organizing protein (HOP), which bridges Hsp70 and Hsp90.[82]

Oxidative stress is well known to regulate the chaperone machinery. The expression of chaperone genes is regulated by heat-shock transcription factors (HSFs). HSF1 binds to a consensus heat-shock element (HSE) and activates HSP gene expression, and this transcription is assisted by dFOXO transcription factor.[86] Initial studies demonstrated the sensitivity of the HSF1 DNA-binding activity to reduction/oxidative insults.[87] HSF1 remains in an inert state as inactive monomers bound to chaperones/co-chaperones including HSP90, HSP70, HSP40 and others. Oxidative stress can induce HSF1 activation by recruitment of the chaperones that repress HSF1 to oxidized (misfolded) proteins, thus allowing HSF1 translocation to the nucleus. In addition, HSF1 can be directly activated by oxidative stress *via* the oxidation of two redox-sensitive cysteines in its DNA-binding domain.[88,89] Hsp90 and Hsp40 protect the activity of the proteasome from being inactivated by oxidative stress.[90,91] Hsp70 and sHsps have also been reported to protect against oxidative stress by regulating the levels/activity of antioxidant systems.[92–94] Work by Jacob demonstrated that bacterial Hsp33 contains redox-sensitive cysteines whose oxidation activates its chaperone function.[95] Hsp90 is S-nitros(yl)ated at the C-terminal domain (C597), which affects its chaperone function by reducing its ATPase activity.[96,97] In addition, Hsp90 was found to be inactivated by cleavage upon oxidative stress.[98] Inducible Hsp72 (HspA1A), but not constitutive Hsc70 (HspA8), is also inactivated by oxidation of C267 and C306.[99] Interestingly, a recent report demonstrated that one of the mechanisms by which sHSP27 protects against oxidative stress and apoptosis is by binding to the 3′UTR of the mRNA of the proapoptotic BH3-only protein Bim, which represses its translation.[100]

Hsp70 binds prefibrillar α-synuclein species in the NAC region and inhibits fibril formation, aggregation and toxicity (Figure 10.3).[101–110] HspA8/Hsc70 also binds and inhibits the toxicity of α-synuclein fibers.[111,112] Hsp70 disaggregase-associated chaperone components (HspA8/Hsc70, Hsp40/DNAJB1 and Hsp110/NEF) have been recently reported to mediate ATP-dependent disassembly of α-synuclein amyloids *via* fibril fragmentation and depolymerization (Figure 10.4).[113] HspA8/Hsc70 and its co-chaperone DnaJC5 were recently identified to be involved in the release of α-synuclein to the extracellular space.[114] In the extracellular space Hsp70 reduces α-synuclein oligomerization and toxicity.[115] Conversely, the unfolding/refolding activities of Hsp70/Hsp40 are inhibited by α-synuclein oligomers.[116] α-synuclein induced co-aggregation with Hsp70 is prevented by the co-chaperone Hip (ST13).[117] Interestingly, while the ATPase activity of Hsp70 is required for preventing α-synuclein toxicity, it is not necessary to mediate α-synuclein degradation.[118]

Hsp90 was reported to be the predominant Hsp co-localized with α-synuclein in LBs, Lewy neurites, and glial cell inclusions in PD brains as well as transgenic mouse models.[119] PD-related genes LRRK2, Parkin, PINK1, DJ-1 and α-synuclein are substrates for Hsp90 (https://www.picard.ch/downloads/Hsp90interactors.pdf). Hsp90 inhibits the aggregation of soluble oligomers of α-synuclein.[120] Hsp90 has also been reported to bind monomeric α-synuclein by also interacting with the NAC region (Figure 10.3), and regulate amyloidogenesis in an ATP-dependent manner.[121]

Small Hsps alphaB-crystallin, Hsp27, Hsp20, HspB8, and HspB2B3 bind to α-synuclein and inhibit α-synuclein aggregation.[122–128] The levels of Hsp27 are decreased in the *substantia nigra*.[129] Hsp27 displaces α-synuclein-Cu interaction and the concomitant generation of ROS and amyloidogenesis.[130] Hsp27 overexpression was reported to be ineffective in preventing α-synuclein toxicity.[131] However, a different study reported that Hsp27 prevents apoptotic cell death induced by both WT and mutant forms, while Hsp70 preferentially protects against WT α-synuclein.[132] HSF1 activation also protects against α-synuclein *via* Hsp70.[133] Hsp90 inhibitors are known to induce a stress response that upregulates Hsp70 *via* HSF1, and are reported to protect against α-synuclein-induced neurodegeneration.[134–137] α-synuclein impairs HSF1 function by promoting its degradation *via* Nedd4.[138]

10.3.3 Protein Degradation Pathways

As in many other disorders linked to the aggregation of misfolded proteins, neurodegeneration in PD is linked to the dysfunction of the two major protein-degradation pathways: the UPS and the autophagosome-lysosome system.[5] However, the exact mechanisms by which PD-related risk factors (aging, genetics and environmental toxicants) contribute to the dysfunction in these protein-degradation processes are unclear. Clearance of misfolded/aggregated α-synuclein has been shown to be mediated by both autophagy and the UPS.[139–141] The UPS and the autophagosome-lysosome system are complementary pathways, where alterations in the rate of one system are reported to modify those of the other. For example, impairment of the UPS triggers autophagy.[142–144] We next review the contribution of both the UPS and the autophagosome-lysosome system to the degradation of α-synuclein, oxidized proteins and the mechanisms by which oxidative stress impairs these homeostatic mechanisms.

10.3.3.1 The Ubiquitin Proteasome System (UPS)

A number of proteins have been identified as major components of LBs including α-synuclein and ubiquitin.[6,145] The selective conjugation of ubiquitin to proteins regulates a number of processes that include protein sorting, signalling and degradation. Ubiquitin has seven lysine residues (K6, K11, K27, K29, K33, K48 and K63) and a M1 that are used for (poly-)ubiquitination. Protein ubiquitination (or ubiquitylation) is mediated by the E1 (ubiquitin activating enzyme)-E2 (ubiquitin conjugating enzyme)-E3 (ubiquitin ligase) system. Mono-, di-, tri-, and polyubiquitinated forms of α-synuclein are found in LBs.[146] α-synuclein has 15 K residues from which 9 of them have been reported to be ubiquitinated *in vitro* or *in vivo* (Figure 10.3), and 3 (K12, K21 and K23) were found to be ubiquitinated in phosphorylated α-synuclein from LNs.[146,147] Ubiquitination of α-synuclein modulates both its degradation by the proteasome and the formation of protein inclusions LBs.[148–151] Lysine 48 polyubiquitination-chains of α-synuclein target it to proteasomal degradation.[152] However, some reports suggest that α-synuclein monomers protofibrils, filaments and oligomers/aggregates, as well as mutations inhibit the proteasome.[106,153–156] Interestingly, ubiquitin-independent α-synuclein degradation *via* the 20S proteasome has also been reported.[157]

Different E3 ubiquitin-ligases target α-synuclein. Seven in absentia homologs (SIAH1 and 2) mono- and di-ubiquitinate α-synuclein in K10, K12, K21, K23, K34, K43 and K96, which promotes its aggregation and increases its accumulation when protein degradation pathways are impaired. α-synuclein is reported to have a VXP motif (118–120 amino acids) that fit the core consensus sequence for binding to the substrate-recognition site of SIAH1 (Figure 10.3).[158,159] Knockdown of the deubiquitinase USP9X (Ubiquitin Specific Peptidase 9, X-Linked) increases the levels of monoubiquitinated α-synuclein and facilitates protein inclusion formation.[160] The E6 associated protein (E6-AP) and the carboxyl terminus of Hsp70-interacting protein (CHIP) promote the degradation of WT and mutant α-synuclein.[161–165] In addition, the tumor necrosis factor-receptor associated factor 6 (TRAF6) promotes atypical ubiquitin linkage formation (K6, K27 and K29) in WT and mutant A53T α-synuclein but, while in misfolded mutant L166P DJ-1 this promotes its accumulation in protein aggregates, this effect has not been corroborated for α-synuclein.[166] On the other hand, the ubiquitin ligase Nedd4 (neural precursor cell expressed developmentally down-regulated protein 4) binds to the carboxyl terminus of α-synuclein (residues 120–133, Figure 10.3) and attaches K63-linked ubiquitin chains primarily at K21 and K96 to target α-synuclein to the endosomal-lysosomal pathway.[167] Other ubiquitin-like protein modifications regulate α-synuclein. Recently, K96 and K102 α-synuclein sumoylation was reported to regulate its aggregation.[168–171] Synphilin-1 is a cytosolic protein that interacts with α-synuclein and regulates its aggregation, while also being found in LBs, and both SIAH1 and Parkin ubiquitinate synphilin-1.[172,173]

What is the relevance of all these ubiquitin modifications? The fact that α-synuclein can be ubiquitinated in different lysines, by different ligases, in different manners (*via* assembly of different Ub chains) might suggest a very complex interaction between ubiquitin modifications and the existence of a ubiquitin code for the homeostatic regulation of α-synuclein.[174] This question has started to become addressed by the generation of synthetic ubiquitinated α-synuclein chains. Recent reports have clearly demonstrated that K48 polyubiquitinated chains of α-synuclein at K12 (di- and tetra-) are resistant to deubiquitinases and are avidly degraded by the proteasome, while K48 monoubiquitination of α-synuclein stabilizes its monomeric form.[149] Monoubiquitination of N-terminal K6, K12, K21 and K23 promotes degradation of α-synuclein, in contrast with the lack of significant turnover of monoubiquitinated K32, K34, and K96 forms.[175] Another report demonstrated that ubiquitination of α-synuclein in the core region K32, K34 and K43 inhibits fiber formation.[148] It is important to also consider that an interplay between different modifications must also be considered. For example, a number of studies have demonstrated that α-synuclein phosphorylation regulates its modification by ubiquitin-like proteins, and some of the lysines that are ubiquitinated are also reported to be targeted by oxidative modifications (Figure 10.3).[39–42,146,176]

Accumulation of ubiquitin-bound proteins including α-synuclein in PD has been proposed to be associated with an impairment in proteasomal activity.[5] Indeed, a dysfunction in the proteasome is found in the *substantia nigra* in PD.[177] Pharmacological inhibition of proteasomal activity has been reported to induce selective dopaminergic cell death in the *substantia nigra*,[178–184] but contradictory observations also exist.[185–187] These contradictory findings might be explained by a number of reasons such as the lack of complete inhibition of proteasomal activity by the agents used, their secondary effects on the activity of other proteases,[188] a secondary effect on the regulation of signalling cascades,[189–192] or a compensatory response by other protein quality-control mechanisms.[143,193] More recently, genetic targeting of the 19S proteasome cap has been shown to induce the formation of ubiquitinated inclusions but no cell death in dopaminergic and neuronal cell cultures.[194] However, conditional *in vivo* knockout of an essential subunit of the 19S regulatory particle, PSMC1 (Rpt2/S4), causes neurodegeneration and the formation of intraneuronal LB-like inclusions in neurons of the nigrostriatal pathway.[195]

The levels of endogenous ubiquitin-conjugates depend on the balance between (1) the rate of ubiquitin conjugation determined by the availability of ubiquitin, the activity of the E1-E2-E3 system and ATP levels; and (2) the rate of turnover of ubiquitin-conjugates (degradation/deubiquitination), which depends on the activity of the proteasome, autophagy and deubiquitinating enzymes.[196] Several components of the UPS can present different sensitivities to oxidative damage/modulation (Figure 10.2).[197–204] Deubiquitinating enzymes are inhibited by oxidation of their catalytic cysteine residue.[205,206] The proteasome has been reported to be more susceptible to oxidative inhibition than ubiquitin-conjugating enzymes.[197,207,208] Accordingly, mild to moderate oxidative stress upregulates ubiquitin levels and the activity of the ubiquitin-conjugating system promoting the formation of Ub-conjugates, and reducing the activity of the proteasome. In contrast, extensive but not lethal oxidative stress reduces the formation of ubiquitin-conjugates by inactivating ubiquitin-conjugating enzymes promoting the accumulation/aggregation of damaged/abnormal proteins.[207] However, this observation might depend on the oxidative insult and the potency/duration of oxidative stress. We have observed that the PD-related pesticide paraquat, that generates ROS *via* its redox cycling properties, induces a depletion of ubiquitin prior to the inhibition of proteasome activity, which might be linked to an impairment in ubiquitin protein synthesis.[209] Other studies have demonstrated that nitrosylation of Parkin in PD-related models inhibits its ubiquitin ligase activity.[200] These observations might explain the presence of ubiquitin-negative protein inclusions in PD brains.[210,211]

An increased accumulation of byproducts of protein oxidation is found in PD brains.[212] A large variety of oxidative protein modifications can be induced by reactive oxygen/nitrogen species, or byproducts of oxidative stress. Oxidized proteins can form oligomeric complexes resulting in the formation of protein aggregates. Irreversibly oxidized proteins such as protein carbonyls have to be degraded in order to maintain proper cellular homeostasis. Ubiquitin-dependent and independent degradation of oxidized proteins by the 26S or 20S proteasome has been reported (Figure 10.4). Energy (ATP)/ubiquitin-independent degradation by the 20S proteasome seems to mediate the degradation of the majority of oxidized proteins.[213] Interestingly, Hsp70 has been shown to mediate the dissociation and reassociation of the 26S proteasome during adaptation to oxidative stress. This mechanism provides cells with an immediate capacity to degrade oxidatively damaged proteins by the 20S proteasome while preserving the 19S regulators in a bound complex.[214] A recent report demonstrates that chaperones and proteins involved in energy metabolism, cytoskeleton/intermediate filaments, and protein translation/ribosome biogenesis are specifically ubiquitinated in response to oxidative stress.[215] However, covalent crosslinks, disulfide bonds, hydrophobic interactions, and heavily oxidized stable protein aggregates are thought to be unsuitable for proteasomal degradation.

The Arg/N-end rule pathway of the UPS is a proteolytic system defined by the presence of destabilizing N-terminal residues/sequences (degrons) in short-lived proteins. Primary destabilizing amino acids (arginine, histidine and lysine) specify degrons recognized by E3 ubiquitin ligases (UBR) that direct proteins to degradation *via* the 26S proteasome. Secondary destabilizing residues (aspartate, glutamate or oxidized cysteines [sulfinylated or sulfonylated]) are processed by arginyl-transfer RNA transferases (ATE) to conjugate a destabilizing primary arginine residue. Tertiary destabilizing residues (glutamine, asparagines and reduced cysteines) require chemical modifications to generate secondary destabilizing amino acids. Exposure of N-terminal destabilizing amino acids is mediated by methionine-aminopeptidases (to remove the initial methionine), or by endoprotease cleavage.[216] N-terminal secondary sulfinylated or sulfonylated cysteine degrons have been proposed to be formed by hydrogen peroxide and nitric oxide.[217,218] However, recent studies suggest that cysteamine dioxygenase might mediate this process.[219] LBs contain both full-length and cleaved fragments of α-synuclein that are thought to be generated by proteases such as calpains and metalloproteinase-3 (MMP3). MMP3 cleaves α-synuclein between A78 and Q79 yielding N-terminal and C-terminal fragments that are more prone to aggregation compared to the full-length protein (Figure 10.3). Q79 C-terminal α-synuclein soluble fragments were recently found to be short-lived peptides targeted by the Arg/N-end rule pathway.[220]

10.3.3.2 The Autophagosome–Lysosome System

Three major types of autophagy processes are recognized: autophagy (macroautophagy), microautophagy, and chaperone-mediated autophagy. Autophagy is responsible for the breakdown of proteins and organelles in the cell, and it is considered a necessary homeostatic mechanism for cell survival. Autophagy starts with the formation of a phagophore, which matures into a double-membrane autophagosome (Figure 10.4). The phagophore has been hypothesized to be generated *de novo* from intracellular membrane sources that include the endoplasmic reticulum (ER), the outer mitochondrial membrane, and the plasma membrane. The autophagosome then fuses with a lysosome, forming an autolysosome where cellular cargo is degraded by lysosomal hydrolyses.[221] Autophagy has been suggested to play a major role in the removal of oxidized protein aggregates by their incomplete degradation within the lysosomal compartment that results in the formation of polymerized lipofuscin-like aggregates consisting of oxidized polypeptides (Figure 10.4).[222,223] Impairment of autophagy in the nervous system, leads to neurodegeneration accompanied by polyubiquitinated protein aggregation in cytoplasmic inclusion bodies.[224,225] A disruption in autophagic pathways has also been linked to PD pathogenesis,[5,226–228] where morphological features of autophagy have been observed in *postmortem* human samples of PD patients.[229] Conditional disruption of autophagy in dopaminergic cells *in vivo* leads to the accumulation of ubiquitinated protein aggregates.[230,231] Autophagy mediates the clearance of α-synuclein aggregates,[232] and some studies suggest that this mechanisms is complementary to the UPS.[140,141,233,234] Importantly, α-synuclein aggregates are themselves resistant to degradation, while at the same time they impair autophagy.[235,236] Pharmacological or genetic upregulation of autophagy protects against dopaminergic neurodegeneration induced by α-synuclein.[237–239]

Autophagy was initially considered as a non-selective bulk degradation pathway. However, several forms of selective autophagy have been recently identified to participate in the clearance of proteins, lipids or organelles. Autophagic adaptors p62, NBR1 (neighbor of Brca1 gene), NDP52 (nuclear dot protein 52 kDa), Optineurin and TOLLIP (toll interacting protein) selectively target ubiquitinated proteins and aggregates for degradation by autophagy.[240] Interestingly, silencing of the ubiquitin-binding protein p62/SQSTM1 (sequestosome-1), which targets proteins for selective degradation *via* the autophagosome-lysosome pathway, enhances the accumulation of oxidized proteins,[241] supporting a role for protein ubiquitination in the clearance of oxidized proteins by autophagy.[242] p62 and NBR1 have also been found accumulated in LBs,[243,244] and α-synuclein inclusions and oxidized proteins can be degraded through the p62-dependent autophagy clearance (Figure 10.4).[245,246]

Chaperone-mediated autophagy (CMA) specializes in degrading cytosolic proteins initially identified by a chaperone that subsequently delivers them to the lysosomes. Protein degradation by CMA is mediated by a specific motif (KFERQ-like) recognized by Hsc70 (HspA8/Hsc70) (Figure 10.4). Interaction of the targeted protein with LAMP-2A (lysosome-associated membrane protein type 2A) mediates its unfolding and translocation into the lysosomal lumen.[247] CMA has been demonstrated to be involved in the turnover of oxidized proteins (Figure 10.4).[248] High levels of several components of the lysosomal translocation complex, such as the chaperoning activities of Hsp70 and HspA8/Hsc70, as well as transcriptional up-regulation of LAMP-2, are induced under oxidative stress.[248,249] Impairment of CMA up-regulates macroautophagy in order to maintain normal rates of long-lived protein degradation. Interestingly, up-regulation of autophagy is unable to compensate for the increased sensitivity of CMA-deficient cells to oxidative stress.[250] CMA also contributes to the clearance of α-synuclein *via* the interaction with 95-VKKDQ-99 amino acids (Figures 10.2 and 10.3),[251–253] but pathogenic mutants are reported to inhibit it.[139] Overexpression of LAMP-2 ameliorates α-synuclein-induced degeneration *in vivo*.[254] Impairment of lysosomal protein hydrolases such as cathepsin D and beta-glucocerebrosidase (GBA1), whose mutations are lined to familial PD, lead to the increased accumulation of α-synuclein aggregates.[255,256]

Oxidative stress modulates autophagy flux rates (from the induction of the phagophore, to the clearance of autophagolysosomes). But the effects that oxidative stress has on autophagy and the role of autophagy determining cell fate seem to be cell type and stimuli specific. We and others have shown that oxidative stress induced by PD-related toxicants induces the accumulation of autophagosomes, which can be attributed to either an increase or a decrease in autophagy flux.[257–261] Discrepancy in this matter is likely related to the time point or stage of cell stress/demise at which autophagy flux is being determined. Induction of autophagy by oxidative stress can be linked to a protective response against cellular damage that can both exert protective or deleterious effects in the cell (when excessive degradation of cellular components occurs).[228,262] While early or mild oxidative stress induced by PD-related insults is expected to trigger an adaptive-protective induction of autophagy

(Figure 10.4). Prolonged or excessive oxidative damage will certainly lead to impairment of autophagy flux. Because autophagy regulates a number of processes such as cell metabolism (by degradation of cellular components and promoting the recycling and salvage of cellular nutrients),[263] turnover of cellular organelles (mitochondria)[264] and signalling (such as the antioxidant response by Nrf2),[265] it is cumbersome to draw a clear picture of the contribution of autophagy-mediated degradation of oxidized/misfolded aggregates to PD neurodegeneration, just by pharmacological or genetic manipulation of autophagy rates. In general terms, however, autophagy seems to exert a protective role against dopaminergic cell death in PD.[209,258,266]

Autophagy signalling cascades have been reported to be modulated by redox modifications. Reversible conjugation of the Atg8 (including LC3) family of proteins to the autophagosomal membrane is mediated by the Atg4 family of cysteine proteases. Atg4 cleave Atg8 allowing its conjugation to PE. Then Atg4 cleaves Atg8 (LC3)-PE releasing it from the membrane in a cycle where Atg4 must be inactivated to ensure the conjugation of Atg8(LC3) to the autophagosomal membrane. After the autophagosome fuses with the lysosome, Atg4 is reactivated in order to delipidate and recycle Atg8 (LC3). Mitochondrial hydrogen peroxide oxidizes and inactivates Atg4 following the initial cleavage of LC3, ensuring the structural integrity of the mature autophagosome.[267] More recently, a thiol/disulfide exchange reaction between Atg4 and thioredoxin has also been reported to regulate Atg4 activation.[268]

10.3.4 Protein Quality Control in Organelles

10.3.4.1 Endoplasmic Reticulum Stress and the Unfolded Protein Response

Close to one-third of all newly synthesized proteins are targeted and translocated to the ER where they interact with chaperones and undergo cotranslational modifications that regulate their folding. The oxidizing environment in the ER favors the formation of disulfide bonds that stabilize a protein. Protein disulfide isomerases (PDIs) catalyze the formation and isomerization of disulfide bonds by thiol-disulfide exchange. The resultant reduced PDI is recycled back to oxidized PDI by the flavin adenine dinucleoatide (FAD)-linked sulfhydryl oxidase, Ero1, with the concomitant formation of hydrogen peroxide.[269] Oxidative protein folding in the ER is a significant source of ROS. Interestingly, the toxicity of complex I inhibitors was reported to be decreased by inhibition of PDI.[270] In contrast, in PD brains, PDI is found to be inactivated by nitros(yl)ation.[271]

The homeostasis of the ER can be altered by a number of conditions including luminal calcium depletion, oxidative stress, and mutations in proteins that traffic through the secretory pathway leading to the disruption of protein folding in the ER. ER-associated degradation (ERAD) mediates the degradation of misfolded proteins by dislocation of selected substrates across the ER membrane to be further degraded by the UPS.[272] Accumulation of ERAD substrates is found in PD-derived neurons, suggesting a dysfunction in protein quality control in PD.[273]

Accumulation of misfolded/unfolded proteins induces ER stress and activates the unfolded protein response (UPR). Activation of the UPR is mediated by three main stress sensors located at the ER membrane: (1) the inositol requiring kinase 1α (IRE1α); (2) the cyclic AMP-dependent activating transcription factor-6 (ATF6α); and (3) the protein kinase RNA-like ER kinase (PERK). The ER chaperone glucose regulated protein 78 (HspA5/Grp [glucose regulated protein]78/BiP [binding immunoglobulin protein]) binds and inhibits signalling by the three stress sensors. Upon accumulation of misfolded proteins in the ER disrupts HspA5/Grp78 association with the transducers that become activated to regulate misfolded protein load *via* the induction of downstream signalling cascades that control protein folding and clearance. IRE1α is a serine–threonine kinase and endoribonuclease that processes the mRNA encoding the transcription factor X-Box binding protein-1 (XBP1) into a more stable and active form (XBP1s). IRE1α can also degrade mRNAs of proteins predicted to be difficult to fold. ATF6α is a transcription factor that traffics to the Golgi and undergoes subsequent proteolytic processing to release active ATF6α. Upon activation, PERK phosphorylates the eukaryotic initiation factor 2α (eIF2α) resulting in attenuation of general protein translation and a decrease in misfolded protein overload. eIF2α phosphorylation also prompts the translation of the activating transcription factor 4 (ATF4). XBP1, ATF6α and ATF4 regulate the transcription of genes involved in promoting ER protein-folding capacity and ERAD. Under chronic ER stress, ATF4 also regulates the expression of pro-apoptotic CCAAT/-enhancer-binding protein homologous protein (CHOP), growth arrest and DNA damage-inducible protein 34 (GADD34) and Bcl-2 proteins. GADD34 also re-establishes protein translation by inducing the dephosphorylation of eIF2α *via* protein phosphatase 1C (PP1C).[274,275]

An increase in HspA5/Grp78 and phosphorylated PERK and eIF2α was found in LB dementias and PD brains.[276,277] α-synuclein is found accumulated in ER microsomes in PD brains and transgenic mouse models.[278] α-synuclein increases the levels of ER chaperones Hsp90B1/Grp94/endoplasmin and HspA5/Grp78 and induces the activation of the UPR *via* PERK and ATF6α signalling pathways. α-synuclein has also been reported to induce alternative splicing of XBP1 mRNA, activation of pro-aoptotic CHOP, and increase levels of ATF4. However, α-synuclein-induced UPR was found to be abnormal as it had deficient eIF2α activation.[278–280] HspA5/Grp78 binds α-synuclein and protects against dopaminergic cell death.[280,281] In contrast, oxidation and inactivation of HspA5/Grp78 protects against ER stress.[282] PD-related insults induce the UPR.[283,284] PERK and eIF2α protect against dopaminergic cell death-induced 6-hydroxydopamine (6-OHDA) and paraquat, respectively.[261,285] In contrast, CHOP upregulation mediates dopaminergic cell death induced by 6-OHDA and complex I inhibition.[286] ATF6α and XBP1 are also reported to exert protection against complex I inhibition, 6-OHDA and α-synuclein.[287–290]

XBP1 silencing itself triggers chronic ER stress and dopaminergic neuron degeneration in adult mice.[288] Importantly, α-synuclein has also been reported to impair ATF6α signalling,[291] while complex I inhibition reduces XBP1 mRNA splicing *via* nitric oxide production.[292] On the other hand, Parkin is upregulated by ATF4 upon activation of the UPR.[293] Conversely, Parkin modulates DJ-1 by regulation of XBP1 levels.[294]

10.3.4.2 Mitochondria Protein Quality Control

The central role of mitochondrial dysfunction was recognized very early in the PD field, and this topic is reviewed in depth in Chapter 3 of this book. We will next focus on addressing the role of mitochondrial dynamics and protein quality-control pathways in PD. Mitochondrial dysfunction in PD has been linked to the indirect alteration of mitochondrial dynamics (fusion and fission, turnover, biogenesis, and transport) and protein quality-control mechanisms (proteostasis).

Mitochondria are dynamic organelles that undergo continuous events of biogenesis, remodeling and turnover. Fusion and fission are opposing processes working in concert to maintain the homeostasis of mitochondria and their physiological function. Fusion enables content to be mixed between neighboring mitochondria and has been proposed to rescue 'moderately' dysfunctional mitochondria. Fusion is mediated by oligomerization of dynamin GTPases mitofusins 1 and 2 (Mfns) at the outer membrane tethering adjacent mitochondria together, and subsequently fusing the inner membranes by the optic atrophy GTPase Opa1. Fission represents a quality-control mechanism to transform damaged elongated mitochondria into a form suitable for engulfment by mitochondrial autophagy or mitophagy. Fission requires recruitment of the GTPase Drp1 *via* mitochondrial surface receptors (mitochondrial fission 1 protein [Fis1], mitochondrial fission factor [Mff] and mitochondrial dynamics proteins 49 and 51 [MiD49, MiD51]) for assembly of the fission machinery that subsequently triggers membrane scission.[295] Complex I inhibition promotes Drp1-mediated fission and mitophagy. Conversely, inhibition of fission protects against dopaminergic cell death induced by Complex I inhibitors.[296] Mitophagy has been shown to be regulated by PINK1 and the translocation of Parkin from the cytoplasm to 'defective' mitochondria. Fusion arrest by polyubiquitination and proteasomal degradation of mitofusins (Mfn1 and Mfn2) by Parkin is required for mitophagy, while polyubiquitination of the voltage-dependent anion channel 1 (VDAC1) recruits the ubiquitin binding protein p62 that also interacts with the autophagosomal protein LC3/GABARAP. Loss of the E3-ubiquitin ligase Parkin and PINK1 has been shown to promote Drp1-dependent fission, loss of mitochondrial membrane potential and energy failure.[297] In addition, recent reports have demonstrated that Parkin and PINK1 mediate the formation of mitochondria-derived vesicles for the degradation of oxidized and damaged proteins *via* the lysosome.[298,299]

Mitochondrial protein quality control (MQC) is a set of evolutionary conserved mechanisms involved in the recognition and correction of the mitochondrial proteome. Mitochondrial proteome accounts for 900–1500 polypeptides from which ~500 proteins are localized in the matrix, ~60 in the inner intermembrane space (IMS), while most of them are embedded in the inner mitochondrial membrane (IM). Only 13 mitochondrial proteins are derived from mtDNA thus, the vast majority of proteins are imported into the organelle, and in many cases large complexes are assembled inside the matrix. The mitochondrial proteome is highly susceptible to oxidation due to the intrinsic generation of ROS in this organelle.[300,301] Mitochondrial chaperones are central players in MQC. The mitochondrial ATP-dependent chaperone HspA9/Grp75/mtHsp70/mortalin is essential for the import and folding of nuclear-encoded mitochondrial proteins, and for the degradation of misfolded/damaged proteins. Reduced levels of HspA9/mtHsp70 have been found in the brains of PD patients.[302,303] In addition, mutations in the ATPase (R126W) and substrate-binding domains (P509S) were identified in PD patients.[304,305] Proteostatic stress induced by these pathogenic HspA9/mtHsp70 mutants is rescued by Parkin and PINK1-mediated mitophagy.[306,307] HspA9/mtHsp70 protects against A53T α-synuclein toxicity,[308] while is itself targeted by dopamine oxidation.[309] mtHsp40/DnaJ3/Tid1 (tumorous imaginal disc 1) cochaperone was found to be degraded in response to 6-OHDA toxicity.[310] On the other hand, phosphorylation of mtHsp90/TRAP1 (TNF receptor-associated protein 1) by PINK1 protects against oxidative stress.[311] mtHsp90/TRAP1 also reduces α-synuclein toxicity.[312] Importantly, mitochondrial dysfunction induced by mtHsp90/Trap1 deficiency or PINK1 mutants is reciprocally rescued by PINK1 or mtHsp90/Trap1 overexpression, respectively.[313,314]

Mitochondrial proteases are classified in ATP-dependent (AAA^+, ATPase associated with diverse cellular activities) and -independent proteases. AAA^+s include ClpXP and LonP1 proteases in the matrix involved in the degradation of oxidized or aggregated polypeptides. AAA^+ proteases localized to the IM include the matrix (m)-AAA protease whose active site is facing the matrix and the intermembrane space (i)-AAA peptidase that faces the IMS. The ATP-independent proteases include: (a) processing peptidases (MPP [matrix metalloproteinase protease], Oct1, Icp55, Cym1, and IMP) involved in the removal and/or degradation of mitochondrial targeting sequences and thus proper biogenesis, sorting, and stabilization of matrix-and IM-targeted proteins; (b) soluble peptidases (Neurolysin, Atp23, and Omi/HtrA2 [HtrA Serine Peptidase 2]) that contribute to MQC in the IMS; and (c) IM-bound proteases (PARL [presenilin-associated rhomboid-like protein], Oma1) that mediate both quality control of IM proteins and regulate mitochondrial dynamics.[300,301]

Accumulation of misfolded proteins in the mitochondria is sensed by PINK1 to induce Parkin-mediated mitophagy.[315] Omi/HtrA2 is found aggregated in LBs[316] and mutations (G399S, A141S, P143A and R404W) have been associated with PD.[317–320] PINK1 interacts and phosphorylates Omi/HtrA2.[321] Omi/HtrA2 triggers the accumulation of unfolded proteins in the mitochondria and the induction of pro-apoptotic CHOP.[322] Omi/HtrA2 also cleaves ubiquitin (Ub) carboxyl-terminal hydrolase L1 (UCH-L1), another PD-related protein.[323] Importantly, Omi/HtrA2 is a well-known pro-apoptotic inducer upon its release from the mitochondria, and its overexpression exerts a toxic effect.[324,325] PARL degrades PINK1 and this process is modulated by pathogenic PINK1 mutations.[326–328] As such, PINK1 cleavage has been proposed to regulate mitophagy.[329] Recently, a misense mutation in PARL (S77N) has been linked to PD.[330] LonP1, MMP, m-AAA and ClpXP are also involved in PINK1 cleavage or degradation.[315,331] PINK1 cleavage by MPP was found to be coupled to its import and reduced accumulation at the mitochondrial surface,[331] while PINK1-cleaved forms were reported to be further degraded by LonP1 in the mitochondrial matrix.[332]

10.4 Conclusions and Perspectives

In this chapter, we have revised the alterations in protein quality-control mechanisms (protein synthesis, folding, and turnover) induced by oxidative stress and PD-risk factors. We have made special emphasis regarding the role of the dysfunction in these homeostatic mechanisms in the synthesis, accumulation, misfolding and aggregation of α-synuclein. In addition, we have also reviewed the mechanisms involved in protein quality control in the ER and mitochondria. The work summarized here clearly highlights the well-established role of proteostatic mechanisms in the pathogenesis of PD.

Acknowledgements

Research in Dr Franco's lab is supported by the National Institutes of Health Grant P20RR17675, Centers of Biomedical Research Excellence (COBRE), the Scientist Development Grant of the American Heart Association (12SDG12090015), and the Office of Research of the University of Nebraska-Lincoln. R.M.F. was supported by grant DBI-1461240 from NSF. P.H.-F. was supported by a Postdoctoral Fellowship (238703) from the National Council of Science and Technology (CONACYT, Mexico).

References

1. E. T. Powers and W. E. Balch, Diversity in the origins of proteostasis networks–a driver for protein function in evolution, *Nat. Rev. Mol. Cell Biol.*, 2013, **14**, 237–248.
2. S. Wolff, J. S. Weissman and A. Dillin, Differential scales of protein quality control, *Cell*, 2014, **157**, 52–64.
3. B. Chen, M. Retzlaff, T. Roos and J. Frydman, Cellular strategies of protein quality control, *Cold Spring Harbor Perspect. Biol.*, 2011, **3**, a004374.
4. D. Chhangani and A. Mishra, Protein quality control system in neurodegeneration: a healing company hard to beat but failure is fatal, *Mol. Neurobiol.*, 2013, **48**, 141–156.

5. C. Cook, C. Stetler and L. Petrucelli, Disruption of protein quality control in Parkinson's disease, *Cold Spring Harbor Perspect. Med.*, 2012, **2**, a009423.
6. K. Wakabayashi, K. Tanji, S. Odagiri, Y. Miki, F. Mori and H. Takahashi, The Lewy body in Parkinson's disease and related neurodegenerative disorders, *Mol. Neurobiol.*, 2013, **47**, 495–508.
7. A. Deleersnijder, M. Gerard, Z. Debyser and V. Baekelandt, The remarkable conformational plasticity of alpha-synuclein: blessing or curse? *Trends Mol. Med.*, 2013, **19**, 368–377.
8. N. M. Kanaan and F. P. Manfredsson, Loss of functional alpha-synuclein: a toxic event in Parkinson's disease? *J. Parkinson's Dis.*, 2012, **2**, 249–267.
9. C. Klein and A. Westenberger, Genetics of Parkinson's disease, *Cold Spring Harbor Perspect. Med.*, 2012, **2**, a008888.
10. Y. C. Chi, G. S. Armstrong, D. N. Jones, E. Z. Eisenmesser and C. W. Liu, Residue histidine 50 plays a key role in protecting alpha-synuclein from aggregation at physiological pH, *J. Biol. Chem.*, 2014, **289**, 15474–15481.
11. D. E. Mor, S. E. Ugras, M. J. Daniels and H. Ischiropoulos, Dynamic structural flexibility of alpha-synuclein, *Neurobiol. Dis.*, 2016, **88**, 66–74.
12. S. Schildknecht, H. R. Gerding, C. Karreman, M. Drescher, H. A. Lashuel, T. F. Outeiro, D. A. Di Monte and M. Leist, Oxidative and nitrative alpha-synuclein modifications and proteostatic stress: implications for disease mechanisms and interventions in synucleinopathies, *J. Neurochem.*, 2013, **125**, 491–511.
13. H. M. Gao, P. T. Kotzbauer, K. Uryu, S. Leight, J. Q. Trojanowski and V. M. Lee, Neuroinflammation and oxidation/nitration of alpha-synuclein linked to dopaminergic neurodegeneration, *J. Neurosci.*, 2008, **28**, 7687–7698.
14. E. H. Norris, B. I. Giasson, H. Ischiropoulos and V. M. Lee, Effects of oxidative and nitrative challenges on alpha-synuclein fibrillogenesis involve distinct mechanisms of protein modifications, *J. Biol. Chem.*, 2003, **278**, 27230–27240.
15. G. Yamin, V. N. Uversky and A. L. Fink, Nitration inhibits fibrillation of human alpha-synuclein *in vitro* by formation of soluble oligomers, *FEBS Lett.*, 2003, **542**, 147–152.
16. S. Przedborski, Q. Chen, M. Vila, B. I. Giasson, R. Djaldatti, S. Vukosavic, J. M. Souza, V. Jackson-Lewis, V. M. Lee and H. Ischiropoulos, Oxidative post-translational modifications of alpha-synuclein in the 1-methyl-4-phenyl-1,2,3,6-tetrahydropyridine (MPTP) mouse model of Parkinson's disease, *J. Neurochem.*, 2001, **76**, 637–640.
17. B. I. Giasson, J. E. Duda, I. V. Murray, Q. Chen, J. M. Souza, H. I. Hurtig, H. Ischiropoulos, J. Q. Trojanowski and V. M. Lee, Oxidative damage linked to neurodegeneration by selective alpha-synuclein nitration in synucleinopathy lesions, *Science*, 2000, **290**, 985–989.
18. C. Andrekopoulos, H. Zhang, J. Joseph, S. Kalivendi and B. Kalyanaraman, Bicarbonate enhances alpha-synuclein oligomerization and nitration: intermediacy of carbonate radical anion and nitrogen dioxide radical, *Biochem. J.*, 2004, **378**, 435–447.

19. J. M. Souza, B. I. Giasson, Q. Chen, V. M. Lee and H. Ischiropoulos, Dityrosine cross-linking promotes formation of stable alpha -synuclein polymers. Implication of nitrative and oxidative stress in the pathogenesis of neurodegenerative synucleinopathies, *J. Biol. Chem.*, 2000, **275**, 18344–18349.
20. M. Bisaglia, S. Mammi and L. Bubacco, Kinetic and structural analysis of the early oxidation products of dopamine: analysis of the interactions with alpha-synuclein, *J. Biol. Chem.*, 2007, **282**, 15597–15605.
21. H. T. Li, D. H. Lin, X. Y. Luo, F. Zhang, L. N. Ji, H. N. Du, G. Q. Song, J. Hu, J. W. Zhou and H. Y. Hu, Inhibition of alpha-synuclein fibrillization by dopamine analogs *via* reaction with the amino groups of alpha-synuclein. Implication for dopaminergic neurodegeneration, *FEBS J.*, 2005, **272**, 3661–3672.
22. W. Xiang, J. C. Schlachetzki, S. Helling, J. C. Bussmann, M. Berlinghof, T. E. Schaffer, K. Marcus, J. Winkler, J. Klucken and C. M. Becker, Oxidative stress-induced posttranslational modifications of alpha-synuclein: specific modification of alpha-synuclein by 4-hydroxy-2-nonenal increases dopaminergic toxicity, *Mol. Cell. Neurosci.*, 2013, **54**, 71–83.
23. N. B. Cole, D. D. Murphy, J. Lebowitz, L. Di Noto, R. L. Levine and R. L. Nussbaum, Metal-catalyzed oxidation of alpha-synuclein: helping to define the relationship between oligomers, protofibrils, and filaments, *J. Biol. Chem.*, 2005, **280**, 9678–9690.
24. K. A. Conway, J. C. Rochet, R. M. Bieganski and P. T. Lansbury Jr, Kinetic stabilization of the alpha-synuclein protofibril by a dopamine-alpha-synuclein adduct, *Science*, 2001, **294**, 1346–1349.
25. J. R. Mazzulli, M. Armakola, M. Dumoulin, I. Parastatidis and H. Ischiropoulos, Cellular oligomerization of alpha-synuclein is determined by the interaction of oxidized catechols with a C-terminal sequence, *J. Biol. Chem.*, 2007, **282**, 31621–31630.
26. S. R. Paik, D. Lee, H. J. Cho, E. N. Lee and C. S. Chang, Oxidized glutathione stimulated the amyloid formation of alpha-synuclein, *FEBS Lett.*, 2003, **537**, 63–67.
27. S. R. Danielson, J. M. Held, B. Schilling, M. Oo, B. W. Gibson and J. K. Andersen, Preferentially increased nitration of alpha-synuclein at tyrosine-39 in a cellular oxidative model of Parkinson's disease, *Anal. Chem.*, 2009, **81**, 7823–7828.
28. C. D. Borsarelli, L. J. Falomir-Lockhart, V. Ostatna, J. A. Fauerbach, H. H. Hsiao, H. Urlaub, E. Palecek, E. A. Jares-Erijman and T. M. Jovin, Biophysical properties and cellular toxicity of covalent crosslinked oligomers of alpha-synuclein formed by photoinduced side-chain tyrosyl radicals, *Free Radical Biol. Med.*, 2012, **53**, 1004–1015.
29. R. Burai, N. Ait-Bouziad, A. Chiki and H. A. Lashuel, Elucidating the Role of Site-Specific Nitration of alpha-Synuclein in the Pathogenesis of Parkinson's Disease *via* Protein Semisynthesis and Mutagenesis, *J. Am. Chem. Soc.*, 2015, **137**, 5041–5052.
30. T. Takahashi, H. Yamashita, T. Nakamura, Y. Nagano and S. Nakamura, Tyrosine 125 of alpha-synuclein plays a critical role for dimerization following nitrative stress, *Brain Res.*, 2002, **938**, 73–80.

31. E. Sevcsik, A. J. Trexler, J. M. Dunn and E. Rhoades, Allostery in a disordered protein: oxidative modifications to alpha-synuclein act distally to regulate membrane binding, *J. Am. Chem. Soc.*, 2011, **133**, 7152–7158.
32. W. Zhou, A. Gallagher, D. P. Hong, C. Long, A. L. Fink and V. N. Uversky, At low concentrations, 3,4-dihydroxyphenylacetic acid (DOPAC) binds non-covalently to alpha-synuclein and prevents its fibrillation, *J. Mol. Biol.*, 2009, **388**, 597–610.
33. V. N. Uversky, G. Yamin, P. O. Souillac, J. Goers, C. B. Glaser and A. L. Fink, Methionine oxidation inhibits fibrillation of human alpha-synuclein *in vitro*, *FEBS Lett.*, 2002, **517**, 239–244.
34. W. Zhou, C. Long, S. H. Reaney, D. A. Di Monte, A. L. Fink and V. N. Uversky, Methionine oxidation stabilizes non-toxic oligomers of alpha-synuclein through strengthening the auto-inhibitory intra-molecular long-range interactions, *Biochim. Biophys. Acta*, 2010, **1802**, 322–330.
35. K. Nakaso, N. Tajima, S. Ito, M. Teraoka, A. Yamashita, Y. Horikoshi, D. Kikuchi, S. Mochida, K. Nakashima and T. Matsura, Dopamine-mediated oxidation of methionine 127 in alpha-synuclein causes cytotoxicity and oligomerization of alpha-synuclein, *PLoS One*, 2013, **8**, e55068.
36. S. L. Leong, C. L. Pham, D. Galatis, M. T. Fodero-Tavoletti, K. Perez, A. F. Hill, C. L. Masters, F. E. Ali, K. J. Barnham and R. Cappai, Formation of dopamine-mediated alpha-synuclein-soluble oligomers requires methionine oxidation, *Free Radical Biol. Med.*, 2009, **46**, 1328–1337.
37. A. Binolfi, A. Limatola, S. Verzini, J. Kosten, F. X. Theillet, H. M. Rose, B. Bekei, M. Stuiver, M. van Rossum and P. Selenko, Intracellular repair of oxidation-damaged alpha-synuclein fails to target C-terminal modification sites, *Nat. Commun.*, 2016, 7, 10251.
38. B. Alvarez-Castelao, M. Goethals, J. Vandekerckhove and J. G. Castano, Mechanism of cleavage of alpha-synuclein by the 20S proteasome and modulation of its degradation by the RedOx state of the N-terminal methionines, *Biochim. Biophys. Acta*, 2014, **1843**, 352–365.
39. Z. Qin, D. Hu, S. Han, S. H. Reaney, D. A. Di Monte and A. L. Fink, Effect of 4-hydroxy-2-nonenal modification on alpha-synuclein aggregation, *J. Biol. Chem.*, 2007, **282**, 5862–5870.
40. A. Trostchansky, S. Lind, R. Hodara, T. Oe, I. A. Blair, H. Ischiropoulos, H. Rubbo and J. M. Souza, Interaction with phospholipids modulates alpha-synuclein nitration and lipid-protein adduct formation, *Biochem. J.*, 2006, **393**, 343–349.
41. E. J. Bae, D. H. Ho, E. Park, J. W. Jung, K. Cho, J. H. Hong, H. J. Lee, K. P. Kim and S. J. Lee, Lipid peroxidation product 4-hydroxy-2-nonenal promotes seeding-capable oligomer formation and cell-to-cell transfer of alpha-synuclein, *Antioxid. Redox Signaling*, 2013, **18**, 770–783.
42. T. Nasstrom, T. Fagerqvist, M. Barbu, M. Karlsson, F. Nikolajeff, A. Kasrayan, M. Ekberg, L. Lannfelt, M. Ingelsson and J. Bergstrom, The lipid peroxidation products 4-oxo-2-nonenal and 4-hydroxy-2-nonenal promote the formation of alpha-synuclein oligomers with distinct biochemical, morphological, and functional properties, *Free Radical Biol. Med.*, 2011, **50**, 428–437.

43. M. Shamoto-Nagai, W. Maruyama, Y. Hashizume, M. Yoshida, T. Osawa, P. Riederer and M. Naoi, In parkinsonian substantia nigra, alpha-synuclein is modified by acrolein, a lipid-peroxidation product, and accumulates in the dopamine neurons with inhibition of proteasome activity, *J. Neural Transm.*, 2007, **114**, 1559–1567.
44. T. Nasstrom, T. Wahlberg, M. Karlsson, F. Nikolajeff, L. Lannfelt, M. Ingelsson and J. Bergstrom, The lipid peroxidation metabolite 4-oxo-2-nonenal cross-links alpha-synuclein causing rapid formation of stable oligomers, *Biochem. Biophys. Res. Commun.*, 2009, **378**, 872–876.
45. C. Follmer, E. Coelho-Cerqueira, D. Y. Yatabe-Franco, G. D. Araujo, A. S. Pinheiro, G. B. Domont and D. Eliezer, Oligomerization and Membrane-binding Properties of Covalent Adducts Formed by the Interaction of alpha-Synuclein with the Toxic Dopamine Metabolite 3,4-Dihydroxyphenylacetaldehyde (DOPAL), *J. Biol. Chem.*, 2015, **290**, 27660–27679.
46. D. A. Bosco, D. M. Fowler, Q. Zhang, J. Nieva, E. T. Powers, P. Wentworth Jr, R. A. Lerner and J. W. Kelly, Elevated levels of oxidized cholesterol metabolites in Lewy body disease brains accelerate alpha-synuclein fibrilization, *Nat. Chem. Biol.*, 2006, **2**, 249–253.
47. S. Shaikh and L. F. Nicholson, Advanced glycation end products induce *in vitro* cross-linking of alpha-synuclein and accelerate the process of intracellular inclusion body formation, *J. Neurosci. Res.*, 2008, **86**, 2071–2082.
48. D. Lee, C. W. Park, S. R. Paik and K. Y. Choi, The modification of alpha-synuclein by dicarbonyl compounds inhibits its fibril-forming process, *Biochim. Biophys. Acta*, 2009, **1794**, 421–430.
49. Bharathi and K. S. Rao, Molecular understanding of copper and iron interaction with alpha-synuclein by fluorescence analysis, *J. Mol. Neurosci.*, 2008, **35**, 273–281.
50. J. Levin, T. Hogen, A. S. Hillmer, B. Bader, F. Schmidt, F. Kamp, H. A. Kretzschmar, K. Botzel and A. Giese, Generation of ferric iron links oxidative stress to alpha-synuclein oligomer formation, *J. Parkinson's Dis.*, 2011, **1**, 205–216.
51. N. Ostrerova-Golts, L. Petrucelli, J. Hardy, J. M. Lee, M. Farer and B. Wolozin, The A53T alpha-synuclein mutation increases iron-dependent aggregation and toxicity, *J. Neurosci.*, 2000, **20**, 6048–6054.
52. Z. J. Zhu, K. C. Wu, W. H. Yung, Z. M. Qian and Y. Ke, Differential interaction between iron and mutant alpha-synuclein causes distinctive Parkinsonian phenotypes in Drosophila, *Biochim. Biophys. Acta*, 2016, **1862**, 518–525.
53. G. Meloni and M. Vasak, Redox activity of alpha-synuclein-Cu is silenced by Zn(7)-metallothionein-3, *Free Radical Biol. Med.*, 2011, **50**, 1471–1479.
54. T. Kowalik-Jankowska, A. Rajewska, E. Jankowska, K. Wisniewska and Z. Grzonka, Products of Cu(II)-catalyzed oxidation of the N-terminal fragments of alpha-synuclein in the presence of hydrogen peroxide, *J. Inorg. Biochem.*, 2006, **100**, 1623–1631.

55. V. N. Uversky, J. Li and A. L. Fink, Metal-triggered structural transformations, aggregation, and fibrillation of human alpha-synuclein. A possible molecular NK between Parkinson's disease and heavy metal exposure, *J. Biol. Chem.*, 2001, **276**, 44284–44296.
56. M. Kostka, T. Hogen, K. M. Danzer, J. Levin, M. Habeck, A. Wirth, R. Wagner, C. G. Glabe, S. Finger, U. Heinzelmann, P. Garidel, W. Duan, C. A. Ross, H. Kretzschmar and A. Giese, Single particle characterization of iron-induced pore-forming alpha-synuclein oligomers, *J. Biol. Chem.*, 2008, **283**, 10992–11003.
57. A. Binolfi, R. M. Rasia, C. W. Bertoncini, M. Ceolin, M. Zweckstetter, C. Griesinger, T. M. Jovin and C. O. Fernandez, Interaction of alpha-synuclein with divalent metal ions reveals key differences: a link between structure, binding specificity and fibrillation enhancement, *J. Am. Chem. Soc.*, 2006, **128**, 9893–9901.
58. A. Binolfi, G. R. Lamberto, R. Duran, L. Quintanar, C. W. Bertoncini, J. M. Souza, C. Cervenansky, M. Zweckstetter, C. Griesinger and C. O. Fernandez, Site-specific interactions of Cu(II) with alpha and beta-synuclein: bridging the molecular gap between metal binding and aggregation, *J. Am. Chem. Soc.*, 2008, **130**, 11801–11812.
59. Y. H. Sung, C. Rospigliosi and D. Eliezer, NMR mapping of copper binding sites in alpha-synuclein, *Biochim. Biophys. Acta*, 2006, **1764**, 5–12.
60. R. M. Rasia, C. W. Bertoncini, D. Marsh, W. Hoyer, D. Cherny, M. Zweckstetter, C. Griesinger, T. M. Jovin and C. O. Fernandez, Structural characterization of copper(II) binding to alpha-synuclein: Insights into the bioinorganic chemistry of Parkinson's disease, *Proc. Natl. Acad. Sci. U. S. A.*, 2005, **102**, 4294–4299.
61. R. De Ricco, D. Valensin, S. Dell'Acqua, L. Casella, P. Dorlet, P. Faller and C. Hureau, Remote His50 Acts as a Coordination Switch in the High-Affinity N-Terminal Centered Copper(II) Site of alpha-Synuclein, *Inorg. Chem.*, 2015, **54**, 4744–4751.
62. S. R. Paik, H. J. Shin and J. H. Lee, Metal-catalyzed oxidation of alpha-synuclein in the presence of Copper(II) and hydrogen peroxide, *Arch. Biochem. Biophys.*, 2000, **378**, 269–277.
63. J. W. Lee, K. Beebe, L. A. Nangle, J. Jang, C. M. Longo-Guess, S. A. Cook, M. T. Davisson, J. P. Sundberg, P. Schimmel and S. L. Ackerman, iting-defective tRNA synthetase causes protein misfolding and neurodegeneration, *Nature*, 2006, **443**, 50–55.
64. J. Chu, N. A. Hong, C. A. Masuda, B. V. Jenkins, K. A. Nelms, C. C. Goodnow, R. J. Glynne, H. Wu, E. Masliah, C. A. Joazeiro and S. A. Kay, A mouse forward genetics screen identifies LISTERIN as an E3 ubiquitin ligase involved in neurodegeneration, *Proc. Natl. Acad. Sci. U. S. A.*, 2009, **106**, 2097–2103.
65. J. Ling and D. Soll, Severe oxidative stress induces protein mistranslation through impairment of an aminoacyl-tRNA synthetase editing site, *Proc. Natl. Acad. Sci. U. S. A.*, 2010, **107**, 4028–4033.
66. N. Wei, Y. Shi, L. N. Truong, K. M. Fisch, T. Xu, E. Gardiner, G. Fu, Y. S. Hsu, S. Kishi, A. I. Su, X. Wu and X. L. Yang, Oxidative stress diverts

tRNA synthetase to nucleus for protection against DNA damage, *Mol. Cell*, 2014, **56**, 323–332.

67. M. Tanaka, P. B. Chock and E. R. Stadtman, Oxidized messenger RNA induces translation errors, *Proc. Natl. Acad. Sci. U. S. A.*, 2007, **104**, 66–71.
68. X. Shan, H. Tashiro and C. L. Lin, The identification and characterization of oxidized RNAs in Alzheimer's disease, *J. Neurosci.*, 2003, **23**, 4913–4921.
69. X. Shan, Y. Chang and C. L. Lin, Messenger RNA oxidation is an early event preceding cell death and causes reduced protein expression, *FASEB J.*, 2007, **21**, 2753–2764.
70. Y. Chang, Q. Kong, X. Shan, G. Tian, H. Ilieva, D. W. Cleveland, J. D. Rothstein, D. R. Borchelt, P. C. Wong and C. L. Lin, Messenger RNA oxidation occurs early in disease pathogenesis and promotes motor neuron degeneration in ALS, *PLoS One*, 2008, **3**, e2849.
71. C. L. Simms, B. H. Hudson, J. W. Mosior, A. S. Rangwala and H. S. Zaher, An active role for the ribosome in determining the fate of oxidized mRNA, *Cell Rep.*, 2014, **9**, 1256–1264.
72. Q. Ding, E. Dimayuga and J. N. Keller, Oxidative stress alters neuronal RNA- and protein-synthesis: Implications for neural viability, *Free Radical Res.*, 2007, **41**, 903–910.
73. Y. Imai, S. Gehrke, H. Q. Wang, R. Takahashi, K. Hasegawa, E. Oota and B. Lu, Phosphorylation of 4E-BP by LRRK2 affects the maintenance of dopaminergic neurons in Drosophila, *EMBO J.*, 2008, **27**, 2432–2443.
74. I. Martin, J. W. Kim, B. D. Lee, H. C. Kang, J. C. Xu, H. Jia, J. Stankowski, M. S. Kim, J. Zhong, M. Kumar, S. A. Andrabi, Y. Xiong, D. W. Dickson, Z. K. Wszolek, A. Pandey, T. M. Dawson and V. L. Dawson, Ribosomal protein s15 phosphorylation mediates LRRK2 neurodegeneration in Parkinson's disease, *Cell*, 2014, **157**, 472–485.
75. S. Gehrke, Y. Imai, N. Sokol and B. Lu, Pathogenic LRRK2 negatively regulates microRNA-mediated translational repression, *Nature*, 2010, **466**, 637–641.
76. S. Gehrke, Z. Wu, M. Klinkenberg, Y. Sun, G. Auburger, S. Guo and B. Lu, PINK1 and Parkin control localized translation of respiratory chain component mRNAs on mitochondria outer membrane, *Cell Metab.*, 2015, **21**, 95–108.
77. W. Lin, N. L. Wadlington, L. Chen, X. Zhuang, J. R. Brorson and U. J. Kang, Loss of PINK1 attenuates HIF-1alpha induction by preventing 4E-BP1-dependent switch in protein translation under hypoxia, *J. Neurosci.*, 2014, **34**, 3079–3089.
78. P. Koukouraki and E. Doxakis, Constitutive translation of human alpha-synuclein is mediated by the 5'-untranslated region, *Open Biol.*, 2016, **6**, 160022.
79. S. Liu and B. Lu, Reduction of protein translation and activation of autophagy protect against PINK1 pathogenesis in Drosophila melanogaster, *PLoS Genet.*, 2010, **6**, e1001237.

80. M. C. Chartier-Harlin, J. C. Dachsel, C. Vilarino-Guell, S. J. Lincoln, F. Lepretre, M. M. Hulihan, J. Kachergus, A. J. Milnerwood, L. Tapia, M. S. Song, E. Le Rhun, E. Mutez, L. Larvor, A. Duflot, C. Vanbesien-Mailliot, A. Kreisler, O. A. Ross, K. Nishioka, A. I. Soto-Ortolaza, S. A. Cobb, H. L. Melrose, B. Behrouz, B. H. Keeling, J. A. Bacon, E. Hentati, L. Williams, A. Yanagiya, N. Sonenberg, P. J. Lockhart, A. C. Zubair, R. J. Uitti, J. O. Aasly, A. Krygowska-Wajs, G. Opala, Z. K. Wszolek, R. Frigerio, D. M. Maraganore, D. Gosal, T. Lynch, M. Hutchinson, A. R. Bentivoglio, E. M. Valente, W. C. Nichols, N. Pankratz, T. Foroud, R. A. Gibson, F. Hentati, D. W. Dickson, A. Destee and M. J. Farrer, Translation initiator EIF4G1 mutations in familial Parkinson disease, *Am. J. Hum. Genet.*, 2011, **89**, 398–406.
81. H. Saibil, Chaperone machines for protein folding, unfolding and disaggregation, *Nat. Rev. Mol. Cell Biol.*, 2013, **14**, 630–642.
82. Y. E. Kim, M. S. Hipp, A. Bracher, M. Hayer-Hartl and F. U. Hartl, Molecular chaperone functions in protein folding and proteostasis, *Annu. Rev. Biochem.*, 2013, **82**, 323–355.
83. S. M. Doyle, O. Genest and S. Wickner, Protein rescue from aggregates by powerful molecular chaperone machines, *Nat. Rev. Mol. Cell Biol.*, 2013, **14**, 617–629.
84. C. Koutras and J. E. Braun, J protein mutations and resulting proteostasis collapse, *Front. Cell. Neurosci.*, 2014, **8**, 191.
85. S. Preissler and E. Deuerling, Ribosome-associated chaperones as key players in proteostasis, *Trends Biochem. Sci.*, 2012, **37**, 274–283.
86. M. R. Donovan and M. T. Marr 2nd, dFOXO Activates Large and Small Heat Shock Protein Genes in Response to Oxidative Stress to Maintain Proteostasis in Drosophila, *J. Biol. Chem.*, 2016, **291**, 19042–19050.
87. L. E. Huang, H. Zhang, S. W. Bae and A. Y. Liu, Thiol reducing reagents inhibit the heat shock response. Involvement of a redox mechanism in the heat shock signal transduction pathway, *J. Biol. Chem.*, 1994, **269**, 30718–30725.
88. S. G. Ahn and D. J. Thiele, Redox regulation of mammalian heat shock factor 1 is essential for Hsp gene activation and protection from stress, *Genes Dev.*, 2003, **17**, 516–528.
89. K. Niforou, C. Cheimonidou and I. P. Trougakos, Molecular chaperones and proteostasis regulation during redox imbalance, *Redox Biol.*, 2014, **2**, 323–332.
90. M. Conconi, I. Petropoulos, I. Emod, E. Turlin, F. Biville and B. Friguet, Protection from oxidative inactivation of the 20S proteasome by heat-shock protein 90, *Biochem. J.*, 1998, **333**(pt 2), 407–415.
91. S. A. Kim, S. Chang, J. H. Yoon and S. G. Ahn, TAT-Hsp40 inhibits oxidative stress-mediated cytotoxicity *via* the inhibition of Hsp70 ubiquitination, *FEBS Lett.*, 2008, **582**, 734–740.
92. Y. J. Lee, H. N. Cho, D. I. Jeoung, J. W. Soh, C. K. Cho, S. Bae, H. Y. Chung, S. J. Lee and Y. S. Lee, HSP25 overexpression attenuates oxidative stress-induced apoptosis: roles of ERK1/2 signaling and manganese superoxide dismutase, *Free Radical Biol. Med.*, 2004, **36**, 429–444.

93. A. P. Arrigo, S. Virot, S. Chaufour, W. Firdaus, C. Kretz-Remy and C. Diaz-Latoud, Hsp27 consolidates intracellular redox homeostasis by upholding glutathione in its reduced form and by decreasing iron intracellular levels, *Antioxid. Redox Signaling*, 2005, **7**, 414–422.
94. S. Guo, W. Wharton, P. Moseley and H. Shi, Heat shock protein 70 regulates cellular redox status by modulating glutathione-related enzyme activities, *Cell Stress Chaperones*, 2007, **12**, 245–254.
95. U. Jakob, W. Muse, M. Eser and J. C. Bardwell, Chaperone activity with a redox switch, *Cell*, 1999, **96**, 341–352.
96. A. Martinez-Ruiz, L. Villanueva, C. Gonzalez de Orduna, D. Lopez-Ferrer, M. A. Higueras, C. Tarin, I. Rodriguez-Crespo, J. Vazquez and S. Lamas, S-nitrosylation of Hsp90 promotes the inhibition of its ATPase and endothelial nitric oxide synthase regulatory activities, *Proc. Natl. Acad. Sci. U. S. A.*, 2005, **102**, 8525–8530.
97. M. Retzlaff, M. Stahl, H. C. Eberl, S. Lagleder, J. Beck, H. Kessler and J. Buchner, Hsp90 is regulated by a switch point in the C-terminal domain, *EMBO Rep.*, 2009, **10**, 1147–1153.
98. J. P. Castro, S. Reeg, V. Botelho, H. Almeida and T. Grune, HSP90 cleavage associates with oxidized proteins accumulation after oxidative stress, *Free Radical Biol. Med.*, 2014, **75**(suppl. 1), S24–S25.
99. Y. Miyata, J. N. Rauch, U. K. Jinwal, A. D. Thompson, S. Srinivasan, C. A. Dickey and J. E. Gestwicki, Cysteine reactivity distinguishes redox sensing by the heat-inducible and constitutive forms of heat shock protein 70, *Chem. Biol.*, 2012, **19**, 1391–1399.
100. D. Davila, E. M. Jimenez-Mateos, C. M. Mooney, G. Velasco, D. C. Henshall and J. H. Prehn, Hsp27 binding to the 3'UTR of bim mRNA prevents neuronal death during oxidative stress-induced injury: a novel cytoprotective mechanism, *Mol. Biol. Cell*, 2014, **25**, 3413–3423.
101. M. M. Dedmon, J. Christodoulou, M. R. Wilson and C. M. Dobson, Heat shock protein 70 inhibits alpha-synuclein fibril formation *via* preferential binding to prefibrillar species, *J. Biol. Chem.*, 2005, **280**, 14733–14740.
102. T. R. Flower, L. S. Chesnokova, C. A. Froelich, C. Dixon and S. N. Witt, Heat shock prevents alpha-synuclein-induced apoptosis in a yeast model of Parkinson's disease, *J. Mol. Biol.*, 2005, **351**, 1081–1100.
103. C. Huang, H. Cheng, S. Hao, H. Zhou, X. Zhang, J. Gao, Q. H. Sun, H. Hu and C. C. Wang, Heat shock protein 70 inhibits alpha-synuclein fibril formation *via* interactions with diverse intermediates, *J. Mol. Biol.*, 2006, **364**, 323–336.
104. K. Kilpatrick, J. A. Novoa, T. Hancock, C. J. Guerriero, P. Wipf, J. L. Brodsky and L. Segatori, Chemical induction of Hsp70 reduces alpha-synuclein aggregation in neuroglioma cells, *ACS Chem. Biol.*, 2013, **8**, 1460–1468.
105. J. Klucken, Y. Shin, E. Masliah, B. T. Hyman and P. J. McLean, Hsp70 Reduces alpha-Synuclein Aggregation and Toxicity, *J. Biol. Chem.*, 2004, **279**, 25497–25502.

106. E. Lindersson, R. Beedholm, P. Hojrup, T. Moos, W. Gai, K. B. Hendil and P. H. Jensen, Proteasomal inhibition by alpha-synuclein filaments and oligomers, *J. Biol. Chem.*, 2004, **279**, 12924–12934.
107. K. C. Luk, I. P. Mills, J. Q. Trojanowski and V. M. Lee, Interactions between Hsp70 and the hydrophobic core of alpha-synuclein inhibit fibril assembly, *Biochemistry*, 2008, **47**, 12614–12625.
108. F. Opazo, A. Krenz, S. Heermann, J. B. Schulz and B. H. Falkenburger, Accumulation and clearance of alpha-synuclein aggregates demonstrated by time-lapse imaging, *J. Neurochem.*, 2008, **106**, 529–540.
109. F. Yu, H. Xu, M. Zhuo, L. Sun, A. Dong and X. Liu, Impairment of redox state and dopamine level induced by alpha-synuclein aggregation and the prevention effect of hsp70, *Biochem. Biophys. Res. Commun.*, 2005, **331**, 278–284.
110. Y. Zhou, G. Gu, D. R. Goodlett, T. Zhang, C. Pan, T. J. Montine, K. S. Montine, R. H. Aebersold and J. Zhang, Analysis of alpha-synuclein-associated proteins by quantitative proteomics, *J. Biol. Chem.*, 2004, **279**, 39155–39164.
111. S. Pemberton, K. Madiona, L. Pieri, M. Kabani, L. Bousset and R. Melki, Hsc70 protein interaction with soluble and fibrillar alpha-synuclein, *J. Biol. Chem.*, 2011, **286**, 34690–34699.
112. V. Redeker, S. Pemberton, W. Bienvenut, L. Bousset and R. Melki, Identification of protein interfaces between alpha-synuclein, the principal component of Lewy bodies in Parkinson disease, and the molecular chaperones human Hsc70 and the yeast Ssa1p, *J. Biol. Chem.*, 2012, **287**, 32630–32639.
113. X. Gao, M. Carroni, C. Nussbaum-Krammer, A. Mogk, N. B. Nillegoda, A. Szlachcic, D. L. Guilbride, H. R. Saibil, M. P. Mayer and B. Bukau, Human Hsp70 Disaggregase Reverses Parkinson's-Linked alpha-Synuclein Amyloid Fibrils, *Mol. Cell*, 2015, **59**, 781–793.
114. S. N. Fontaine, D. Zheng, J. J. Sabbagh, M. D. Martin, D. Chaput, A. Darling, J. H. Trotter, A. R. Stothert, B. A. Nordhues, A. Lussier, J. Baker, L. Shelton, M. Kahn, L. J. Blair, S. M. Stevens Jr and C. A. Dickey, DnaJ/Hsc70 chaperone complexes control the extracellular release of neurodegenerative-associated proteins, *EMBO J.*, 2016, **35**, 1537–1549.
115. K. M. Danzer, W. P. Ruf, P. Putcha, D. Joyner, T. Hashimoto, C. Glabe, B. T. Hyman and P. J. McLean, Heat-shock protein 70 modulates toxic extracellular alpha-synuclein oligomers and rescues trans-synaptic toxicity, *FASEB J.*, 2011, **25**, 326–336.
116. M. P. Hinault, A. F. Cuendet, R. U. Mattoo, M. Mensi, G. Dietler, H. A. Lashuel and P. Goloubinoff, Stable alpha-synuclein oligomers strongly inhibit chaperone activity of the Hsp70 system by weak interactions with J-domain co-chaperones, *J. Biol. Chem.*, 2010, **285**, 38173–38182.
117. C. Roodveldt, C. W. Bertoncini, A. Andersson, A. T. van der Goot, S. T. Hsu, R. Fernandez-Montesinos, J. de Jong, T. J. van Ham, E. A. Nollen, D. Pozo, J. Christodoulou and C. M. Dobson, Chaperone proteostasis

in Parkinson's disease: stabilization of the Hsp70/alpha-synuclein complex by Hip, *EMBO J.*, 2009, **28**, 3758–3770.
118. J. Klucken, Y. Shin, B. T. Hyman and P. J. McLean, A single amino acid substitution differentiates Hsp70-dependent effects on alpha-synuclein degradation and toxicity, *Biochem. Biophys. Res. Commun.*, 2004, **325**, 367–373.
119. K. Uryu, C. Richter-Landsberg, W. Welch, E. Sun, O. Goldbaum, E. H. Norris, C. T. Pham, I. Yazawa, K. Hilburger, M. Micsenyi, B. I. Giasson, N. M. Bonini, V. M. Lee and J. Q. Trojanowski, Convergence of heat shock protein 90 with ubiquitin in filamentous alpha-synuclein inclusions of alpha-synucleinopathies, *Am. J. Pathol.*, 2006, **168**, 947–961.
120. S. Daturpalli, C. A. Waudby, S. Meehan and S. E. Jackson, Hsp90 inhibits alpha-synuclein aggregation by interacting with soluble oligomers, *J. Mol. Biol.*, 2013, **425**, 4614–4628.
121. S. F. Falsone, A. J. Kungl, A. Rek, R. Cappai and K. Zangger, The molecular chaperone Hsp90 modulates intermediate steps of amyloid assembly of the Parkinson-related protein alpha-synuclein, *J. Biol. Chem.*, 2009, **284**, 31190–31199.
122. I. B. Bruinsma, K. A. Bruggink, K. Kinast, A. A. Versleijen, I. M. Segers-Nolten, V. Subramaniam, H. B. Kuiperij, W. Boelens, R. M. de Waal and M. M. Verbeek, Inhibition of alpha-synuclein aggregation by small heat shock proteins, *Proteins*, 2011, **79**, 2956–2967.
123. T. F. Outeiro, J. Klucken, K. E. Strathearn, F. Liu, P. Nguyen, J. C. Rochet, B. T. Hyman and P. J. McLean, Small heat shock proteins protect against alpha-synuclein-induced toxicity and aggregation, *Biochem. Biophys. Res. Commun.*, 2006, **351**, 631–638.
124. A. Rekas, C. G. Adda, J. Andrew Aquilina, K. J. Barnham, M. Sunde, D. Galatis, N. A. Williamson, C. L. Masters, R. F. Anders, C. V. Robinson, R. Cappai and J. A. Carver, Interaction of the molecular chaperone alphaB-crystallin with alpha-synuclein: effects on amyloid fibril formation and chaperone activity, *J. Mol. Biol.*, 2004, **340**, 1167–1183.
125. C. A. Waudby, T. P. Knowles, G. L. Devlin, J. N. Skepper, H. Ecroyd, J. A. Carver, M. E. Welland, J. Christodoulou, C. M. Dobson and S. Meehan, The interaction of alphaB-crystallin with mature alpha-synuclein amyloid fibrils inhibits their elongation, *Biophys. J.*, 2010, **98**, 843–851.
126. J. Wang, E. Martin, V. Gonzales, D. R. Borchelt and M. K. Lee, Differential regulation of small heat shock proteins in transgenic mouse models of neurodegenerative diseases, *Neurobiol. Aging*, 2008, **29**, 586–597.
127. P. J. McLean, H. Kawamata, S. Shariff, J. Hewett, N. Sharma, K. Ueda, X. O. Breakefield and B. T. Hyman, TorsinA and heat shock proteins act as molecular chaperones: suppression of alpha-synuclein aggregation, *J. Neurochem.*, 2002, **83**, 846–854.
128. D. Cox, J. A. Carver and H. Ecroyd, Preventing alpha-synuclein aggregation: the role of the small heat-shock molecular chaperone proteins, *Biochim. Biophys. Acta*, 2014, **1842**, 1830–1843.

129. S. Chen and I. R. Brown, Neuronal expression of constitutive heat shock proteins: implications for neurodegenerative diseases, *Cell Stress Chaperones*, 2007, **12**, 51–58.
130. A. Asthana, M. Bollapalli, R. Tangirala, R. Bakthisaran and C. Mohan Rao, Hsp27 suppresses the Cu(2+)-induced amyloidogenicity, redox activity, and cytotoxicity of alpha-synuclein by metal ion stripping, *Free Radical Biol. Med.*, 2014, **72**, 176–190.
131. T. C. Moloney, R. Hyland, D. O'Toole, A. Paucard, D. Kirik, A. O'Doherty, A. M. Gorman and E. Dowd, Heat shock protein 70 reduces alpha-synuclein-induced predegenerative neuronal dystrophy in the alpha-synuclein viral gene transfer rat model of Parkinson's disease, *CNS Neurosci. Ther.*, 2014, **20**, 50–58.
132. A. Zourlidou, M. D. Payne Smith and D. S. Latchman, HSP27 but not HSP70 has a potent protective effect against alpha-synuclein-induced cell death in mammalian neuronal cells, *J. Neurochem.*, 2004, **88**, 1439–1448.
133. X. Liangliang, H. Yonghui, E. Shunmei, G. Shoufang, Z. Wei and Z. Jiangying, Dominant-positive HSF1 decreases alpha-synuclein level and alpha-synuclein-induced toxicity, *Mol. Biol. Rep.*, 2010, **37**, 1875–1881.
134. N. R. McFarland, H. Dimant, L. Kibuuka, D. Ebrahimi-Fakhari, C. A. Desjardins, K. M. Danzer, M. Danzer, Z. Fan, M. A. Schwarzschild, W. Hirst and P. J. McLean, Chronic treatment with novel small molecule Hsp90 inhibitors rescues striatal dopamine levels but not alpha-synuclein-induced neuronal cell loss, *PLoS One*, 2014, **9**, e86048.
135. P. J. McLean, J. Klucken, Y. Shin and B. T. Hyman, Geldanamycin induces Hsp70 and prevents alpha-synuclein aggregation and toxicity *in vitro*, *Biochem. Biophys. Res. Commun.*, 2004, **321**, 665–669.
136. P. Putcha, K. M. Danzer, L. R. Kranich, A. Scott, M. Silinski, S. Mabbett, C. D. Hicks, J. M. Veal, P. M. Steed, B. T. Hyman and P. J. McLean, Brain-permeable small-molecule inhibitors of Hsp90 prevent alpha-synuclein oligomer formation and rescue alpha-synuclein-induced toxicity, *J. Pharmacol. Exp. Ther.*, 2010, **332**, 849–857.
137. R. Xiong, W. Zhou, D. Siegel, R. R. Kitson, C. R. Freed, C. J. Moody and D. Ross, A Novel Hsp90 Inhibitor Activates Compensatory Heat Shock Protein Responses and Autophagy and Alleviates Mutant A53T alpha-Synuclein Toxicity, *Mol. Pharmacol.*, 2015, **88**, 1045–1054.
138. E. Kim, B. Wang, N. Sastry, E. Masliah, P. T. Nelson, H. Cai and F. F. Liao, NEDD4-mediated HSF1 degradation underlies alpha-synucleinopathy, *Hum. Mol. Genet.*, 2016, **25**, 211–222.
139. A. M. Cuervo, L. Stefanis, R. Fredenburg, P. T. Lansbury and D. Sulzer, Impaired degradation of mutant alpha-synuclein by chaperone-mediated autophagy, *Science*, 2004, **305**, 1292–1295.
140. D. Ebrahimi-Fakhari, I. Cantuti-Castelvetri, Z. Fan, E. Rockenstein, E. Masliah, B. T. Hyman, P. J. McLean and V. K. Unni, Distinct roles *in vivo* for the ubiquitin-proteasome system and the autophagy-lysosomal pathway in the degradation of alpha-synuclein, *J. Neurosci.*, 2011, **31**, 14508–14520.

141. J. L. Webb, B. Ravikumar, J. Atkins, J. N. Skepper and D. C. Rubinsztein, Alpha-Synuclein is degraded by both autophagy and the proteasome, *J. Biol. Chem.*, 2003, **278**, 25009–25013.
142. V. I. Korolchuk, A. Mansilla, F. M. Menzies and D. C. Rubinsztein, Autophagy inhibition compromises degradation of ubiquitin-proteasome pathway substrates, *Mol. Cell*, 2009, **33**, 517–527.
143. D. C. Rubinsztein, Autophagy induction rescues toxicity mediated by proteasome inhibition, *Neuron*, 2007, **54**, 854–856.
144. V. I. Korolchuk, F. M. Menzies and D. C. Rubinsztein, Mechanisms of cross-talk between the ubiquitin-proteasome and autophagy-lysosome systems, *FEBS Lett.*, 2010, **584**, 1393–1398.
145. S. Kuzuhara, H. Mori, N. Izumiyama, M. Yoshimura and Y. Ihara, Lewy bodies are ubiquitinated. A light and electron microscopic immunocytochemical study, *Acta Neuropathol.*, 1988, **75**, 345–353.
146. J. P. Anderson, D. E. Walker, J. M. Goldstein, R. de Laat, K. Banducci, R. J. Caccavello, R. Barbour, J. Huang, K. Kling, M. Lee, L. Diep, P. S. Keim, X. Shen, T. Chataway, M. G. Schlossmacher, P. Seubert, D. Schenk, S. Sinha, W. P. Gai and T. J. Chilcote, Phosphorylation of Ser-129 is the dominant pathological modification of alpha-synuclein in familial and sporadic Lewy body disease, *J. Biol. Chem.*, 2006, **281**, 29739–29752.
147. T. Nonaka, T. Iwatsubo and M. Hasegawa, Ubiquitination of alpha-synuclein, *Biochemistry*, 2005, **44**, 361–368.
148. F. Meier, T. Abeywardana, A. Dhall, N. P. Marotta, J. Varkey, R. Langen, C. Chatterjee and M. R. Pratt, Semisynthetic, site-specific ubiquitin modification of alpha-synuclein reveals differential effects on aggregation, *J. Am. Chem. Soc.*, 2012, **134**, 5468–5471.
149. M. Haj-Yahya, B. Fauvet, Y. Herman-Bachinsky, M. Hejjaoui, S. N. Bavikar, S. V. Karthikeyan, A. Ciechanover, H. A. Lashuel and A. Brik, Synthetic polyubiquitinated alpha-Synuclein reveals important insights into the roles of the ubiquitin chain in regulating its pathophysiology, *Proc. Natl. Acad. Sci. U. S. A.*, 2013, **110**, 17726–17731.
150. C. Liu, E. Fei, N. Jia, H. Wang, R. Tao, A. Iwata, N. Nukina, J. Zhou and G. Wang, Assembly of lysine 63-linked ubiquitin conjugates by phosphorylated alpha-synuclein implies Lewy body biogenesis, *J. Biol. Chem.*, 2007, **282**, 14558–14566.
151. M. C. Bennett, J. F. Bishop, Y. Leng, P. B. Chock, T. N. Chase and M. M. Mouradian, Degradation of alpha-synuclein by proteasome, *J. Biol. Chem.*, 1999, **274**, 33855–33858.
152. F. K. Lee, A. K. Wong, Y. W. Lee, O. W. Wan, H. Y. Chan and K. K. Chung, The role of ubiquitin linkages on alpha-synuclein induced-toxicity in a Drosophila model of Parkinson's disease, *J. Neurochem.*, 2009, **110**, 208–219.
153. N. Y. Zhang, Z. Tang and C. W. Liu, alpha-Synuclein protofibrils inhibit 26 S proteasome-mediated protein degradation: understanding the cytotoxicity of protein protofibrils in neurodegenerative disease pathogenesis, *J. Biol. Chem.*, 2008, **283**, 20288–20298.

154. Y. Tanaka, S. Engelender, S. Igarashi, R. K. Rao, T. Wanner, R. E. Tanzi, A. Sawa, V. L. Dawson, T. M. Dawson and C. A. Ross, Inducible expression of mutant alpha-synuclein decreases proteasome activity and increases sensitivity to mitochondria-dependent apoptosis, *Hum. Mol. Genet.*, 2001, **10**, 919–926.
155. T. Nonaka and M. Hasegawa, A cellular model to monitor proteasome dysfunction by alpha-synuclein, *Biochemistry*, 2009, **48**, 8014–8022.
156. H. Snyder, K. Mensah, C. Theisler, J. Lee, A. Matouschek and B. Wolozin, Aggregated and monomeric alpha-synuclein bind to the S6' proteasomal protein and inhibit proteasomal function, *J. Biol. Chem.*, 2003, **278**, 11753–11759.
157. G. K. Tofaris, R. Layfield and M. G. Spillantini, alpha-synuclein metabolism and aggregation is linked to ubiquitin-independent degradation by the proteasome, *FEBS Lett.*, 2001, **509**, 22–26.
158. J. T. Lee, T. C. Wheeler, L. Li and L. S. Chin, Ubiquitination of alpha-synuclein by Siah-1 promotes alpha-synuclein aggregation and apoptotic cell death, *Hum. Mol. Genet.*, 2008, **17**, 906–917.
159. R. Rott, R. Szargel, J. Haskin, V. Shani, A. Shainskaya, I. Manov, E. Liani, E. Avraham and S. Engelender, Monoubiquitylation of alpha-synuclein by seven in absentia homolog (SIAH) promotes its aggregation in dopaminergic cells, *J. Biol. Chem.*, 2008, **283**, 3316–3328.
160. R. Rott, R. Szargel, J. Haskin, R. Bandopadhyay, A. J. Lees, V. Shani and S. Engelender, alpha-Synuclein fate is determined by USP9X-regulated monoubiquitination, *Proc. Natl. Acad. Sci. U. S. A.*, 2011, **108**, 18666–18671.
161. S. A. Mulherkar, J. Sharma and N. R. Jana, The ubiquitin ligase E6-AP promotes degradation of alpha-synuclein, *J. Neurochem.*, 2009, **110**, 1955–1964.
162. L. V. Kalia, S. K. Kalia, H. Chau, A. M. Lozano, B. T. Hyman and P. J. McLean, Ubiquitinylation of alpha-synuclein by carboxyl terminus Hsp70-interacting protein (CHIP) is regulated by Bcl-2-associated athanogene 5 (BAG5), *PLoS One*, 2011, **6**, e14695.
163. H. Dimant, L. Zhu, L. N. Kibuuka, Z. Fan, B. T. Hyman and P. J. McLean, Direct visualization of CHIP-mediated degradation of alpha-synuclein *in vivo*: implications for PD therapeutics, *PLoS One*, 2014, **9**, e92098.
164. Y. Shin, J. Klucken, C. Patterson, B. T. Hyman and P. J. McLean, The co-chaperone carboxyl terminus of Hsp70-interacting protein (CHIP) mediates alpha-synuclein degradation decisions between proteasomal and lysosomal pathways, *J. Biol. Chem.*, 2005, **280**, 23727–23734.
165. J. E. Tetzlaff, P. Putcha, T. F. Outeiro, A. Ivanov, O. Berezovska, B. T. Hyman and P. J. McLean, CHIP targets toxic alpha-Synuclein oligomers for degradation, *J. Biol. Chem.*, 2008, **283**, 17962–17968.
166. S. Zucchelli, M. Codrich, F. Marcuzzi, M. Pinto, S. Vilotti, M. Biagioli, I. Ferrer and S. Gustincich, TRAF6 promotes atypical ubiquitination of mutant DJ-1 and alpha-synuclein and is localized to Lewy bodies in sporadic Parkinson's disease brains, *Hum. Mol. Genet.*, 2010, **19**, 3759–3770.

167. G. K. Tofaris, H. T. Kim, R. Hourez, J. W. Jung, K. P. Kim and A. L. Goldberg, Ubiquitin ligase Nedd4 promotes alpha-synuclein degradation by the endosomal-lysosomal pathway, *Proc. Natl. Acad. Sci. U. S. A.*, 2011, **108**, 17004–17009.
168. P. Krumova, E. Meulmeester, M. Garrido, M. Tirard, H. H. Hsiao, G. Bossis, H. Urlaub, M. Zweckstetter, S. Kugler, F. Melchior, M. Bahr and J. H. Weishaupt, Sumoylation inhibits alpha-synuclein aggregation and toxicity, *J. Cell Biol.*, 2011, **194**, 49–60.
169. V. Dorval and P. E. Fraser, Small ubiquitin-like modifier (SUMO) modification of natively unfolded proteins tau and alpha-synuclein, *J. Biol. Chem.*, 2006, **281**, 9919–9924.
170. Y. M. Kim, W. H. Jang, M. M. Quezado, Y. Oh, K. C. Chung, E. Junn and M. M. Mouradian, Proteasome inhibition induces alpha-synuclein SUMOylation and aggregate formation, *J. Neurol. Sci.*, 2011, **307**, 157–161.
171. T. Abeywardana and M. R. Pratt, Extent of inhibition of alpha-synuclein aggregation *in vitro* by SUMOylation is conjugation site- and SUMO isoform-selective, *Biochemistry*, 2015, **54**, 959–961.
172. Y. Nagano, H. Yamashita, T. Takahashi, S. Kishida, T. Nakamura, E. Iseki, N. Hattori, Y. Mizuno, A. Kikuchi and M. Matsumoto, Siah-1 facilitates ubiquitination and degradation of synphilin-1, *J. Biol. Chem.*, 2003, **278**, 51504–51514.
173. K. K. Chung, Y. Zhang, K. L. Lim, Y. Tanaka, H. Huang, J. Gao, C. A. Ross, V. L. Dawson and T. M. Dawson, Parkin ubiquitinates the alpha-synuclein-interacting protein, synphilin-1: implications for Lewy-body formation in Parkinson disease, *Nat. Med.*, 2001, **7**, 1144–1150.
174. R. Yau and M. Rape, The increasing complexity of the ubiquitin code, *Nat. Cell Biol.*, 2016, **18**, 579–586.
175. T. Abeywardana, Y. H. Lin, R. Rott, S. Engelender and M. R. Pratt, Site-specific differences in proteasome-dependent degradation of monoubiquitinated alpha-synuclein, *Chem. Biol.*, 2013, **20**, 1207–1213.
176. H. Shahpasandzadeh, B. Popova, A. Kleinknecht, P. E. Fraser, T. F. Outeiro and G. H. Braus, Interplay between sumoylation and phosphorylation for protection against alpha-synuclein inclusions, *J. Biol. Chem.*, 2014, **289**, 31224–31240.
177. K. S. McNaught and P. Jenner, Proteasomal function is impaired in substantia nigra in Parkinson's disease, *Neurosci. Lett.*, 2001, **297**, 191–194.
178. F. Fornai, P. Lenzi, M. Gesi, M. Ferrucci, G. Lazzeri, C. L. Busceti, R. Ruffoli, P. Soldani, S. Ruggieri, M. G. Alessandri and A. Paparelli, Fine structure and biochemical mechanisms underlying nigrostriatal inclusions and cell death after proteasome inhibition, *J. Neurosci.*, 2003, **23**, 8955–8966.
179. K. S. McNaught, D. P. Perl, A. L. Brownell and C. W. Olanow, Systemic exposure to proteasome inhibitors causes a progressive model of Parkinson's disease, *Ann. Neurol.*, 2004, **56**, 149–162.

180. H. J. Rideout, I. C. Lang-Rollin, M. Savalle and L. Stefanis, Dopaminergic neurons in rat ventral midbrain cultures undergo selective apoptosis and form inclusions, but do not up-regulate iHSP70, following proteasomal inhibition, *J. Neurochem.*, 2005, **93**, 1304–1313.
181. A. H. Schapira, M. W. Cleeter, J. R. Muddle, J. M. Workman, J. M. Cooper and R. H. King, Proteasomal inhibition causes loss of nigral tyrosine hydroxylase neurons, *Ann. Neurol.*, 2006, **60**, 253–255.
182. F. Sun, V. Anantharam, D. Zhang, C. Latchoumycandane, A. Kanthasamy and A. G. Kanthasamy, Proteasome inhibitor MG-132 induces dopaminergic degeneration in cell culture and animal models, *Neurotoxicology*, 2006, **27**, 807–815.
183. W. Xie, X. Li, C. Li, W. Zhu, J. Jankovic and W. Le, Proteasome inhibition modeling nigral neuron degeneration in Parkinson's disease, *J. Neurochem.*, 2010, **115**, 188–199.
184. B. Y. Zeng, S. Bukhatwa, A. Hikima, S. Rose and P. Jenner, Reproducible nigral cell loss after systemic proteasomal inhibitor administration to rats, *Ann. Neurol.*, 2006, **60**, 248–252.
185. N. Kadoguchi, H. Kimoto, R. Yano, H. Kato and T. Araki, Failure of acute administration with proteasome inhibitor to provide a model of Parkinson's disease in mice, *Metab. Brain Dis.*, 2008, **23**, 147–154.
186. J. H. Kordower, N. M. Kanaan, Y. Chu, R. Suresh Babu, J. Stansell 3rd, B. T. Terpstra, C. E. Sortwell, K. Steece-Collier and T. J. Collier, Failure of proteasome inhibitor administration to provide a model of Parkinson's disease in rats and monkeys, *Ann. Neurol.*, 2006, **60**, 264–268.
187. A. B. Manning-Bog, S. H. Reaney, V. P. Chou, L. C. Johnston, A. L. McCormack, J. Johnston, J. W. Langston and D. A. Di Monte, Lack of nigrostriatal pathology in a rat model of proteasome inhibition, *Ann. Neurol.*, 2006, **60**, 256–260.
188. A. F. Kisselev and A. L. Goldberg, Proteasome inhibitors: from research tools to drug candidates, *Chem. Biol.*, 2001, **8**, 739–758.
189. J. Bai, H. Nakamura, S. Ueda, Y. W. Kwon, T. Tanaka, S. Ban and J. Yodoi, Proteasome-dependent degradation of cyclin D1 in 1-methyl-4-phenylpyridinium ion (MPP+)-induced cell cycle arrest, *J. Biol. Chem.*, 2004, **279**, 38710–38714.
190. H. Sawada, R. Kohno, T. Kihara, Y. Izumi, N. Sakka, M. Ibi, M. Nakanishi, T. Nakamizo, K. Yamakawa, H. Shibasaki, N. Yamamoto, A. Akaike, M. Inden, Y. Kitamura, T. Taniguchi and S. Shimohama, Proteasome mediates dopaminergic neuronal degeneration, and its inhibition causes alpha-synuclein inclusions, *J. Biol. Chem.*, 2004, **279**, 10710–10719.
191. N. Yamamoto, H. Sawada, Y. Izumi, T. Kume, H. Katsuki, S. Shimohama and A. Akaike, Proteasome inhibition induces glutathione synthesis and protects cells from oxidative stress: relevance to Parkinson disease, *J. Biol. Chem.*, 2007, **282**, 4364–4372.
192. M. Inden, J. Kondo, Y. Kitamura, K. Takata, K. Nishimura, T. Taniguchi, H. Sawada and S. Shimohama, Proteasome inhibitors protect against

degeneration of nigral dopaminergic neurons in hemiparkinsonian rats, *J. Pharmacol. Sci.*, 2005, **97**, 203–211.

193. S. Kageyama, Y. S. Sou, T. Uemura, S. Kametaka, T. Saito, R. Ishimura, T. Kouno, L. Bedford, R. J. Mayer, M. S. Lee, M. Yamamoto, S. Waguri, K. Tanaka and M. Komatsu, Proteasome dysfunction activates autophagy and the Keap1-Nrf2 pathway, *J. Biol. Chem.*, 2014, **289**, 24944–24955.
194. A. Droggiti, C. C. Ho, L. Stefanis, W. T. Dauer and H. J. Rideout, Targeted disruption of neuronal 19S proteasome subunits induces the formation of ubiquitinated inclusions in the absence of cell death, *J. Neurochem.*, 2011, **119**, 630–643.
195. L. Bedford, D. Hay, A. Devoy, S. Paine, D. G. Powe, R. Seth, T. Gray, I. Topham, K. Fone, N. Rezvani, M. Mee, T. Soane, R. Layfield, P. W. Sheppard, T. Ebendal, D. Usoskin, J. Lowe and R. J. Mayer, Depletion of 26S proteasomes in mouse brain neurons causes neurodegeneration and Lewy-like inclusions resembling human pale bodies, *J. Neurosci.*, 2008, **28**, 8189–8198.
196. M. J. Clague and S. Urbe, Ubiquitin: same molecule, different degradation pathways, *Cell*, 2010, **143**, 682–685.
197. X. Zhang, J. Zhou, A. F. Fernandes, J. R. Sparrow, P. Pereira, A. Taylor and F. Shang, The proteasome: a target of oxidative damage in cultured human retina pigment epithelial cells, *Invest. Ophthalmol. Visual Sci.*, 2008, **49**, 3622–3630.
198. J. Jahngen-Hodge, M. S. Obin, X. Gong, F. Shang, T. R. Nowell Jr, J. Gong, H. Abasi, J. Blumberg and A. Taylor, Regulation of ubiquitin-conjugating enzymes by glutathione following oxidative stress, *J. Biol. Chem.*, 1997, **272**, 28218–28226.
199. M. Obin, F. Shang, X. Gong, G. Handelman, J. Blumberg and A. Taylor, Redox regulation of ubiquitin-conjugating enzymes: mechanistic insights using the thiol-specific oxidant diamide, *FASEB J.*, 1998, **12**, 561–569.
200. D. Yao, Z. Gu, T. Nakamura, Z. Q. Shi, Y. Ma, B. Gaston, L. A. Palmer, E. M. Rockenstein, Z. Zhang, E. Masliah, T. Uehara and S. A. Lipton, Nitrosative stress linked to sporadic Parkinson's disease: S-nitrosylation of parkin regulates its E3 ubiquitin ligase activity, *Proc. Natl. Acad. Sci. U. S. A.*, 2004, **101**, 10810–10814.
201. T. Ishii, T. Sakurai, H. Usami and K. Uchida, Oxidative modification of proteasome: identification of an oxidation-sensitive subunit in 26 S proteasome, *Biochemistry*, 2005, **44**, 13893–13901.
202. M. Caballero, P. B. Liton, D. L. Epstein and P. Gonzalez, Proteasome inhibition by chronic oxidative stress in human trabecular meshwork cells, *Biochem. Biophys. Res. Commun.*, 2003, **308**, 346–352.
203. K. Okada, C. Wangpoengtrakul, T. Osawa, S. Toyokuni, K. Tanaka and K. Uchida, 4-Hydroxy-2-nonenal-mediated impairment of intracellular proteolysis during oxidative stress. Identification of proteasomes as target molecules, *J. Biol. Chem.*, 1999, **274**, 23787–23793.

204. Q. Ding and J. N. Keller, Proteasome inhibition in oxidative stress neurotoxicity: implications for heat shock proteins, *J. Neurochem.*, 2001, **77**, 1010–1017.
205. X. M. Cotto-Rios, M. Bekes, J. Chapman, B. Ueberheide and T. T. Huang, Deubiquitinases as a signaling target of oxidative stress, *Cell Rep.*, 2012, **2**, 1475–1484.
206. J. G. Lee, K. Baek, N. Soetandyo and Y. Ye, Reversible inactivation of deubiquitinases by reactive oxygen species *in vitro* and in cells, *Nat. Commun.*, 2013, **4**, 1568.
207. F. Shang and A. Taylor, Ubiquitin-proteasome pathway and cellular responses to oxidative stress, *Free Radical Biol. Med.*, 2011, **51**, 5–16.
208. F. Shang, X. Gong and A. Taylor, Activity of ubiquitin-dependent pathway in response to oxidative stress. Ubiquitin-activating enzyme is transiently up-regulated, *J. Biol. Chem.*, 1997, **272**, 23086–23093.
209. J. Navarro-Yepes, A. Anandhan, E. Bradley, I. Bohovych, B. Yarabe, A. de Jong, H. Ovaa, Y. Zhou, O. Khalimonchuk, B. Quintanilla-Vega and R. Franco, Inhibition of Protein Ubiquitination by Paraquat and 1-Methyl-4-Phenylpyridinium Impairs Ubiquitin-Dependent Protein Degradation Pathways, *Mol. Neurobiol.*, 2016, **53**, 5229–5251.
210. M. G. Spillantini, R. A. Crowther, R. Jakes, M. Hasegawa and M. Goedert, alpha-Synuclein in filamentous inclusions of Lewy bodies from Parkinson's disease and dementia with lewy bodies, *Proc. Natl. Acad. Sci. U. S. A.*, 1998, **95**, 6469–6473.
211. E. Gomez-Tortosa, K. Newell, M. C. Irizarry, J. L. Sanders and B. T. Hyman, alpha-Synuclein immunoreactivity in dementia with Lewy bodies: morphological staging and comparison with ubiquitin immunostaining, *Acta Neuropathol.*, 2000, **99**, 352–357.
212. Z. I. Alam, S. E. Daniel, A. J. Lees, D. C. Marsden, P. Jenner and B. Halliwell, A generalised increase in protein carbonyls in the brain in Parkinson's but not incidental Lewy body disease, *J. Neurochem.*, 1997, **69**, 1326–1329.
213. R. Raynes, L. C. Pomatto and K. J. Davies, Degradation of oxidized proteins by the proteasome: Distinguishing between the 20S, 26S, and immunoproteasome proteolytic pathways, *Mol. Aspects Med.*, 2016, **50**, 41–55.
214. T. Grune, B. Catalgol, A. Licht, G. Ermak, A. M. Pickering, J. K. Ngo and K. J. Davies, HSP70 mediates dissociation and reassociation of the 26S proteasome during adaptation to oxidative stress, *Free Radical Biol. Med.*, 2011, **51**, 1355–1364.
215. M. Kastle, S. Reeg, A. Rogowska-Wrzesinska and T. Grune, Chaperones, but not oxidized proteins, are ubiquitinated after oxidative stress, *Free Radical Biol. Med.*, 2012, **53**, 1468–1477.
216. T. Tasaki, S. M. Sriram, K. S. Park and Y. T. Kwon, The N-end rule pathway, *Annu. Rev. Biochem.*, 2012, **81**, 261–289.
217. R. G. Hu, J. Sheng, X. Qi, Z. Xu, T. T. Takahashi and A. Varshavsky, The N-end rule pathway as a nitric oxide sensor controlling the levels of multiple regulators, *Nature*, 2005, **437**, 981–986.

218. K. G. Reddie and K. S. Carroll, Expanding the functional diversity of proteins through cysteine oxidation, *Curr. Opin. Chem. Biol.*, 2008, **12**, 746–754.
219. D. A. Weits, B. Giuntoli, M. Kosmacz, S. Parlanti, H. M. Hubberten, H. Riegler, R. Hoefgen, P. Perata, J. T. van Dongen and F. Licausi, Plant cysteine oxidases control the oxygen-dependent branch of the N-end-rule pathway, *Nat. Commun.*, 2014, **5**, 3425.
220. C. S. Brower, K. I. Piatkov and A. Varshavsky, Neurodegeneration-associated protein fragments as short-lived substrates of the N-end rule pathway, *Mol. Cell*, 2013, **50**, 161–171.
221. Z. Yang and D. J. Klionsky, Mammalian autophagy: core molecular machinery and signaling regulation, *Curr. Opin. Cell Biol.*, 2010, **22**, 124–131.
222. R. A. Dunlop, U. T. Brunk and K. J. Rodgers, Oxidized proteins: mechanisms of removal and consequences of accumulation, *IUBMB Life*, 2009, **61**, 522–527.
223. N. Chondrogianni, I. Petropoulos, S. Grimm, K. Georgila, B. Catalgol, B. Friguet, T. Grune and E. S. Gonos, Protein damage, repair and proteolysis, *Mol. Aspects Med.*, 2014, **35**, 1–71.
224. M. Komatsu, S. Waguri, T. Chiba, S. Murata, J. Iwata, I. Tanida, T. Ueno, M. Koike, Y. Uchiyama, E. Kominami and K. Tanaka, Loss of autophagy in the central nervous system causes neurodegeneration in mice, *Nature*, 2006, **441**, 880–884.
225. T. Hara, K. Nakamura, M. Matsui, A. Yamamoto, Y. Nakahara, R. Suzuki-Migishima, M. Yokoyama, K. Mishima, I. Saito, H. Okano and N. Mizushima, Suppression of basal autophagy in neural cells causes neurodegenerative disease in mice, *Nature*, 2006, **441**, 885–889.
226. K. Tanaka and N. Matsuda, Proteostasis and neurodegeneration: the roles of proteasomal degradation and autophagy, *Biochim. Biophys. Acta*, 2014, **1843**, 197–204.
227. A. M. Choi, S. W. Ryter and B. Levine, Autophagy in human health and disease, *N. Engl. J. Med.*, 2013, **368**, 651–662.
228. J. Navarro-Yepes, M. Burns, A. Anandhan, O. Khalimonchuk, L. M. del Razo, B. Quintanilla-Vega, A. Pappa, M. I. Panayiotidis and R. Franco, Oxidative stress, redox signaling, and autophagy: cell death *versus* survival, *Antioxid. Redox Signaling*, 2014, **21**, 66–85.
229. P. Anglade, S. Vyas, F. Javoy-Agid, M. T. Herrero, P. P. Michel, J. Marquez, A. Mouatt-Prigent, M. Ruberg, E. C. Hirsch and Y. Agid, Apoptosis and autophagy in nigral neurons of patients with Parkinson's disease, *Histol. Histopathol.*, 1997, **12**, 25–31.
230. I. Ahmed, Y. Liang, S. Schools, V. L. Dawson, T. M. Dawson and J. M. Savitt, Development and characterization of a new Parkinson's disease model resulting from impaired autophagy, *J. Neurosci.*, 2012, **32**, 16503–16509.
231. L. G. Friedman, M. L. Lachenmayer, J. Wang, L. He, S. M. Poulose, M. Komatsu, G. R. Holstein and Z. Yue, Disrupted autophagy leads to dopaminergic axon and dendrite degeneration and promotes presynaptic

accumulation of alpha-synuclein and LRRK2 in the brain, *J. Neurosci.*, 2012, **32**, 7585–7593.

232. D. Petroi, B. Popova, N. Taheri-Talesh, S. Irniger, H. Shahpasandzadeh, M. Zweckstetter, T. F. Outeiro and G. H. Braus, Aggregate clearance of alpha-synuclein in Saccharomyces cerevisiae depends more on autophagosome and vacuole function than on the proteasome, *J. Biol. Chem.*, 2012, **287**, 27567–27579.
233. K. Ancolio, C. Alves da Costa, K. Ueda and F. Checler, Alpha-synuclein and the Parkinson's disease-related mutant Ala53Thr-alpha-synuclein do not undergo proteasomal degradation in HEK293 and neuronal cells, *Neurosci. Lett.*, 2000, **285**, 79–82.
234. S. Batelli, E. Peverelli, S. Rodilossi, G. Forloni and D. Albani, Macroautophagy and the proteasome are differently involved in the degradation of alpha-synuclein wild type and mutated A30P in an *in vitro* inducible model (PC12/TetOn), *Neuroscience*, 2011, **195**, 128–137.
235. S. A. Tanik, C. E. Schultheiss, L. A. Volpicelli-Daley, K. R. Brunden and V. M. Lee, Lewy body-like alpha-synuclein aggregates resist degradation and impair macroautophagy, *J. Biol. Chem.*, 2013, **288**, 15194–15210.
236. A. R. Winslow, C. W. Chen, S. Corrochano, A. Acevedo-Arozena, D. E. Gordon, A. A. Peden, M. Lichtenberg, F. M. Menzies, B. Ravikumar, S. Imarisio, S. Brown, C. J. O'Kane and D. C. Rubinsztein, alpha-Synuclein impairs macroautophagy: implications for Parkinson's disease, *J. Cell Biol.*, 2010, **190**, 1023–1037.
237. Q. He, J. B. Koprich, Y. Wang, W. B. Yu, B. G. Xiao, J. M. Brotchie and J. Wang, Treatment with Trehalose Prevents Behavioral and Neurochemical Deficits Produced in an AAV alpha-Synuclein Rat Model of Parkinson's Disease, *Mol. Neurobiol.*, 2016, **53**, 2258–2268.
238. M. Decressac, B. Mattsson, P. Weikop, M. Lundblad, J. Jakobsson and A. Bjorklund, TFEB-mediated autophagy rescues midbrain dopamine neurons from alpha-synuclein toxicity, *Proc. Natl. Acad. Sci. U. S. A.*, 2013, **110**, E1817–E1826.
239. B. Spencer, R. Potkar, M. Trejo, E. Rockenstein, C. Patrick, R. Gindi, A. Adame, T. Wyss-Coray and E. Masliah, Beclin 1 gene transfer activates autophagy and ameliorates the neurodegenerative pathology in alpha-synuclein models of Parkinson's and Lewy body diseases, *J. Neurosci.*, 2009, **29**, 13578–13588.
240. A. Khaminets, C. Behl and I. Dikic, Ubiquitin-Dependent And Independent Signals In Selective Autophagy, *Trends Cell Biol.*, 2016, **26**, 6–16.
241. L. Wang, M. Cano and J. T. Handa, p62 provides dual cytoprotection against oxidative stress in the retinal pigment epithelium, *Biochim. Biophys. Acta*, 2014, **1843**, 1248–1258.
242. C. Marques, P. Pereira, A. Taylor, J. N. Liang, V. N. Reddy, L. I. Szweda and F. Shang, Ubiquitin-dependent lysosomal degradation of the HNE-modified proteins in lens epithelial cells, *FASEB J.*, 2004, **18**, 1424–1426.

243. K. Zatloukal, C. Stumptner, A. Fuchsbichler, H. Heid, M. Schnoelzer, L. Kenner, R. Kleinert, M. Prinz, A. Aguzzi and H. Denk, p62 Is a common component of cytoplasmic inclusions in protein aggregation diseases, *Am. J. Pathol.*, 2002, **160**, 255–263.
244. S. Odagiri, K. Tanji, F. Mori, A. Kakita, H. Takahashi and K. Wakabayashi, Autophagic adapter protein NBR1 is localized in Lewy bodies and glial cytoplasmic inclusions and is involved in aggregate formation in alpha-synucleinopathy, *Acta Neuropathol.*, 2012, **124**, 173–186.
245. Y. Watanabe, H. Tatebe, K. Taguchi, Y. Endo, T. Tokuda, T. Mizuno, M. Nakagawa and M. Tanaka, p62/SQSTM1-dependent autophagy of Lewy body-like alpha-synuclein inclusions, *PLoS One*, 2012, **7**, e52868.
246. K. Tanji, S. Odagiri, Y. Miki, A. Maruyama, Y. Nikaido, J. Mimura, F. Mori, E. Warabi, T. Yanagawa, S. Ueno, K. Itoh and K. Wakabayashi, p62 Deficiency Enhances alpha-Synuclein Pathology in Mice, *Brain Pathol.*, 2015, **25**, 552–564.
247. M. Xilouri and L. Stefanis, Chaperone mediated autophagy to the rescue: A new-fangled target for the treatment of neurodegenerative diseases, *Mol. Cell. Neurosci.*, 2015, **66**, 29–36.
248. R. Kiffin, C. Christian, E. Knecht and A. M. Cuervo, Activation of chaperone-mediated autophagy during oxidative stress, *Mol. Biol. Cell*, 2004, **15**, 4829–4840.
249. M. K. Callahan, D. Chaillot, C. Jacquin, P. R. Clark and A. Menoret, Differential acquisition of antigenic peptides by Hsp70 and Hsc70 under oxidative conditions, *J. Biol. Chem.*, 2002, **277**, 33604–33609.
250. A. C. Massey, S. Kaushik, G. Sovak, R. Kiffin and A. M. Cuervo, Consequences of the selective blockage of chaperone-mediated autophagy, *Proc. Natl. Acad. Sci. U. S. A.*, 2006, **103**, 5805–5810.
251. K. A. Malkus and H. Ischiropoulos, Regional deficiencies in chaperone-mediated autophagy underlie alpha-synuclein aggregation and neurodegeneration, *Neurobiol. Dis.*, 2012, **46**, 732–744.
252. S. K. Mak, A. L. McCormack, A. B. Manning-Bog, A. M. Cuervo and D. A. Di Monte, Lysosomal degradation of alpha-synuclein *in vivo*, *J. Biol. Chem.*, 2010, **285**, 13621–13629.
253. T. Vogiatzi, M. Xilouri, K. Vekrellis and L. Stefanis, Wild type alpha-synuclein is degraded by chaperone-mediated autophagy and macroautophagy in neuronal cells, *J. Biol. Chem.*, 2008, **283**, 23542–23556.
254. M. Xilouri, O. R. Brekk, N. Landeck, P. M. Pitychoutis, T. Papasilekas, Z. Papadopoulou-Daifoti, D. Kirik and L. Stefanis, Boosting chaperone-mediated autophagy *in vivo* mitigates alpha-synuclein-induced neurodegeneration, *Brain*, 2013, **136**, 2130–2146.
255. E. J. Bae, N. Y. Yang, C. Lee, S. Kim, H. J. Lee and S. J. Lee, Haploinsufficiency of cathepsin D leads to lysosomal dysfunction and promotes cell-to-cell transmission of alpha-synuclein aggregates, *Cell Death Dis.*, 2015, **6**, e1901.

256. E. J. Bae, N. Y. Yang, C. Lee, H. J. Lee, S. Kim, S. P. Sardi and S. J. Lee, Loss of glucocerebrosidase 1 activity causes lysosomal dysfunction and alpha-synuclein aggregation, *Exp. Mol. Med.*, 2015, **47**, e188.
257. B. J. Mader, V. N. Pivtoraiko, H. M. Flippo, B. J. Klocke, K. A. Roth, L. R. Mangieri and J. J. Shacka, Rotenone inhibits autophagic flux prior to inducing cell death, *ACS Chem. Neurosci.*, 2012, **3**, 1063–1072.
258. A. Garcia-Garcia, A. Anandhan, M. Burns, H. Chen, Y. Zhou and R. Franco, Impairment of Atg5-dependent autophagic flux promotes paraquat- and MPP(+)-induced apoptosis but not rotenone or 6-hydroxydopamine toxicity, *Toxicol. Sci.*, 2013, **136**, 166–182.
259. M. Miyara, Y. Kotake, W. Tokunaga, S. Sanoh and S. Ohta, Mild MPP+ exposure impairs autophagic degradation through a novel lysosomal acidity-independent mechanism, *J. Neurochem.*, 2016, **139**, 294–308.
260. J. H. Zhu, C. Horbinski, F. Guo, S. Watkins, Y. Uchiyama and C. T. Chu, Regulation of autophagy by extracellular signal-regulated protein kinases during 1-methyl-4-phenylpyridinium-induced cell death, *Am. J. Pathol.*, 2007, **170**, 75–86.
261. M. Niso-Santano, J. M. Bravo-San Pedro, R. Gomez-Sanchez, V. Climent, G. Soler, J. M. Fuentes and R. A. Gonzalez-Polo, ASK1 overexpression accelerates paraquat-induced autophagy *via* endoplasmic reticulum stress, *Toxicol. Sci.*, 2011, **119**, 156–168.
262. G. Filomeni, D. De Zio and F. Cecconi, Oxidative stress and autophagy: the clash between damage and metabolic needs, *Cell Death Differ.*, 2015, **22**, 377–388.
263. J. Kaur and J. Debnath, Autophagy at the crossroads of catabolism and anabolism, *Nat. Rev. Mol. Cell Biol.*, 2015, **16**, 461–472.
264. R. L. de Vries and S. Przedborski, Mitophagy and Parkinson's disease: be eaten to stay healthy, *Mol. Cell. Neurosci.*, 2013, **55**, 37–43.
265. T. Jiang, B. Harder, M. Rojo de la Vega, P. K. Wong, E. Chapman and D. D. Zhang, p62 links autophagy and Nrf2 signaling, *Free Radical Biol. Med.*, 2015, **88**, 199–204.
266. M. A. Lynch-Day, K. Mao, K. Wang, M. Zhao and D. J. Klionsky, The role of autophagy in Parkinson's disease, *Cold Spring Harbor Perspect. Med.*, 2012, **2**, a009357.
267. R. Scherz-Shouval, E. Shvets, E. Fass, H. Shorer, L. Gil and Z. Elazar, Reactive oxygen species are essential for autophagy and specifically regulate the activity of Atg4, *EMBO J.*, 2007, **26**, 1749–1760.
268. M. E. Perez-Perez, M. Zaffagnini, C. H. Marchand, J. L. Crespo and S. D. Lemaire, The yeast autophagy protease Atg4 is regulated by thioredoxin, *Autophagy*, 2014, **10**, 1953–1964.
269. D. A. Hudson, S. A. Gannon and C. Thorpe, Oxidative protein folding: from thiol-disulfide exchange reactions to the redox poise of the endoplasmic reticulum, *Free Radical Biol. Med.*, 2015, **80**, 171–182.
270. S. Lehtonen, M. Jaronen, P. Vehvilainen, M. Lakso, M. Rudgalvyte, V. Keksa-Goldsteine, G. Wong, M. J. Courtney, J. Koistinaho and G. Goldsteins, Inhibition of Excessive Oxidative Protein Folding Is Protective

in MPP Toxicity-Induced Parkinson's Disease Models, *Antioxid. Redox Signaling*, 2016, **25**, 485–497.

271. T. Uehara, T. Nakamura, D. Yao, Z. Q. Shi, Z. Gu, Y. Ma, E. Masliah, Y. Nomura and S. A. Lipton, S-nitrosylated protein-disulphide isomerase links protein misfolding to neurodegeneration, *Nature*, 2006, **441**, 513–517.
272. J. A. Olzmann, R. R. Kopito and J. C. Christianson, The mammalian endoplasmic reticulum-associated degradation system, *Cold Spring Harbor Perspect. Biol.*, 2013, **5**, a013185.
273. C. Y. Chung, V. Khurana, P. K. Auluck, D. F. Tardiff, J. R. Mazzulli, F. Soldner, V. Baru, Y. Lou, Y. Freyzon, S. Cho, A. E. Mungenast, J. Muffat, M. Mitalipova, M. D. Pluth, N. T. Jui, B. Schule, S. J. Lippard, L. H. Tsai, D. Krainc, S. L. Buchwald, R. Jaenisch and S. Lindquist, Identification and rescue of alpha-synuclein toxicity in Parkinson patient-derived neurons, *Science*, 2013, **342**, 983–987.
274. G. Mercado, V. Castillo, P. Soto and A. Sidhu, ER stress and Parkinson's disease: Pathological inputs that converge into the secretory pathway, *Brain Res.*, 2016, **1648**, 626–632.
275. G. Mercado, P. Valdes and C. Hetz, An ERcentric view of Parkinson's disease, *Trends Mol. Med.*, 2013, **19**, 165–175.
276. J. H. Baek, D. Whitfield, D. Howlett, P. Francis, E. Bereczki, C. Ballard, T. Hortobagyi, J. Attems and D. Aarsland, Unfolded protein response is activated in Lewy body dementias, *Neuropathol. Appl. Neurobiol.*, 2016, **42**, 352–365.
277. J. J. Hoozemans, E. S. van Haastert, P. Eikelenboom, R. A. de Vos, J. M. Rozemuller and W. Scheper, Activation of the unfolded protein response in Parkinson's disease, *Biochem. Biophys. Res. Commun.*, 2007, **354**, 707–711.
278. E. Colla, P. Coune, Y. Liu, O. Pletnikova, J. C. Troncoso, T. Iwatsubo, B. L. Schneider and M. K. Lee, Endoplasmic reticulum stress is important for the manifestations of alpha-synucleinopathy *in vivo*, *J. Neurosci.*, 2012, **32**, 3306–3320.
279. A. Bellucci, L. Navarria, M. Zaltieri, E. Falarti, S. Bodei, S. Sigala, L. Battistin, M. Spillantini, C. Missale and P. Spano, Induction of the unfolded protein response by alpha-synuclein in experimental models of Parkinson's disease, *J. Neurochem.*, 2011, **116**, 588–605.
280. M. S. Gorbatyuk, A. Shabashvili, W. Chen, C. Meyers, L. F. Sullivan, M. Salganik, J. H. Lin, A. S. Lewin, N. Muzyczka and O. S. Gorbatyuk, Glucose regulated protein 78 diminishes alpha-synuclein neurotoxicity in a rat model of Parkinson disease, *Mol. Ther.*, 2012, **20**, 1327–1337.
281. M. Salganik, V. G. Sergeyev, V. Shinde, C. A. Meyers, M. S. Gorbatyuk, J. H. Lin, S. Zolotukhin and O. S. Gorbatyuk, The loss of glucose-regulated protein 78 (GRP78) during normal aging or from siRNA knockdown augments human alpha-synuclein (alpha-syn) toxicity to rat nigral neurons, *Neurobiol. Aging*, 2015, **36**, 2213–2223.

282. J. Wang, K. A. Pareja, C. A. Kaiser and C. S. Sevier, Redox signaling *via* the molecular chaperone BiP protects cells against endoplasmic reticulum-derived oxidative stress, *eLife*, 2014, **3**, e03496.
283. W. A. Holtz and K. L. O'Malley, Parkinsonian mimetics induce aspects of unfolded protein response in death of dopaminergic neurons, *J. Biol. Chem.*, 2003, **278**, 19367–19377.
284. W. A. Holtz, J. M. Turetzky, Y. J. Jong and K. L. O'Malley, Oxidative stress-triggered unfolded protein response is upstream of intrinsic cell death evoked by parkinsonian mimetics, *J. Neurochem.*, 2006, **99**, 54–69.
285. E. J. Ryu, H. P. Harding, J. M. Angelastro, O. V. Vitolo, D. Ron and L. A. Greene, Endoplasmic reticulum stress and the unfolded protein response in cellular models of Parkinson's disease, *J. Neurosci.*, 2002, **22**, 10690–10698.
286. R. M. Silva, V. Ries, T. F. Oo, O. Yarygina, V. Jackson-Lewis, E. J. Ryu, P. D. Lu, S. J. Marciniak, D. Ron, S. Przedborski, N. Kholodilov, L. A. Greene and R. E. Burke, CHOP/GADD153 is a mediator of apoptotic death in substantia nigra dopamine neurons in an *in vivo* neurotoxin model of parkinsonism, *J. Neurochem.*, 2005, **95**, 974–986.
287. N. Egawa, K. Yamamoto, H. Inoue, R. Hikawa, K. Nishi, K. Mori and R. Takahashi, The endoplasmic reticulum stress sensor, ATF6alpha, protects against neurotoxin-induced dopaminergic neuronal death, *J. Biol. Chem.*, 2011, **286**, 7947–7957.
288. P. Valdes, G. Mercado, R. L. Vidal, C. Molina, G. Parsons, F. A. Court, A. Martinez, D. Galleguillos, D. Armentano, B. L. Schneider and C. Hetz, Control of dopaminergic neuron survival by the unfolded protein response transcription factor XBP1, *Proc. Natl. Acad. Sci. U. S. A.*, 2014, **111**, 6804–6809.
289. A. Ray, S. Zhang, C. Rentas, K. A. Caldwell and G. A. Caldwell, RTCB-1 mediates neuroprotection *via* XBP-1 mRNA splicing in the unfolded protein response pathway, *J. Neurosci.*, 2014, **34**, 16076–16085.
290. L. Si, T. Xu, F. Wang, Q. Liu and M. Cui, X-box-binding protein 1-modified neural stem cells for treatment of Parkinson's disease, *Neural Regener. Res.*, 2012, **7**, 736–740.
291. J. J. Credle, P. A. Forcelli, M. Delannoy, A. W. Oaks, E. Permaul, D. L. Berry, V. Duka, J. Wills and A. Sidhu, alpha-Synuclein-mediated inhibition of ATF6 processing into COPII vesicles disrupts UPR signaling in Parkinson's disease, *Neurobiol. Dis.*, 2015, **76**, 112–125.
292. R. Nakato, Y. Ohkubo, A. Konishi, M. Shibata, Y. Kaneko, T. Iwawaki, T. Nakamura, S. A. Lipton and T. Uehara, Regulation of the unfolded protein response *via* S-nitrosylation of sensors of endoplasmic reticulum stress, *Sci. Rep.*, 2015, **5**, 14812.
293. L. Bouman, A. Schlierf, A. K. Lutz, J. Shan, A. Deinlein, J. Kast, Z. Galehdar, V. Palmisano, N. Patenge, D. Berg, T. Gasser, R. Augustin, D. Trumbach, I. Irrcher, D. S. Park, W. Wurst, M. S. Kilberg, J. Tatzelt and K. F. Winklhofer, Parkin is transcriptionally regulated by ATF4: evidence for an interconnection between mitochondrial stress and ER stress, *Cell Death Differ.*, 2011, **18**, 769–782.

294. E. Duplan, E. Giaime, J. Viotti, J. Sevalle, O. Corti, A. Brice, H. Ariga, L. Qi, F. Checler and C. Alves da Costa, ER-stress-associated functional link between Parkin and DJ-1 *via* a transcriptional cascade involving the tumor suppressor p53 and the spliced X-box binding protein XBP-1, *J. Cell Sci.*, 2013, **126**, 2124–2133.
295. C. Perier and M. Vila, Mitochondrial biology and Parkinson's disease, *Cold Spring Harbor Perspect. Med.*, 2012, **2**, a009332.
296. V. S. Van Laar and S. B. Berman, The interplay of neuronal mitochondrial dynamics and bioenergetics: implications for Parkinson's disease, *Neurobiol. Dis.*, 2013, **51**, 43–55.
297. A. K. Lutz, N. Exner, M. E. Fett, J. S. Schlehe, K. Kloos, K. Lammermann, B. Brunner, A. Kurz-Drexler, F. Vogel, A. S. Reichert, L. Bouman, D. Vogt-Weisenhorn, W. Wurst, J. Tatzelt, C. Haass and K. F. Winklhofer, Loss of parkin or PINK1 function increases Drp1-dependent mitochondrial fragmentation, *J. Biol. Chem.*, 2009, **284**, 22938–22951.
298. G. L. McLelland, V. Soubannier, C. X. Chen, H. M. McBride and E. A. Fon, Parkin and PINK1 function in a vesicular trafficking pathway regulating mitochondrial quality control, *EMBO J.*, 2014, **33**, 282–295.
299. A. Tanaka, M. M. Cleland, S. Xu, D. P. Narendra, D. F. Suen, M. Karbowski and R. J. Youle, Proteasome and p97 mediate mitophagy and degradation of mitofusins induced by Parkin, *J. Cell Biol.*, 2010, **191**, 1367–1380.
300. I. Bohovych, S. S. Chan and O. Khalimonchuk, Mitochondrial protein quality control: the mechanisms guarding mitochondrial health, *Antioxid. Redox Signaling*, 2015, **22**, 977–994.
301. H. B. Suliman and C. A. Piantadosi, Mitochondrial Quality Control as a Therapeutic Target, *Pharmacol. Rev.*, 2016, **68**, 20–48.
302. J. Jin, C. Hulette, Y. Wang, T. Zhang, C. Pan, R. Wadhwa and J. Zhang, Proteomic identification of a stress protein, mortalin/mthsp70/GRP75: relevance to Parkinson disease, *Mol. Cell. Proteomics*, 2006, **5**, 1193–1204.
303. M. Shi, J. Jin, Y. Wang, R. P. Beyer, E. Kitsou, R. L. Albin, M. Gearing, C. Pan and J. Zhang, Mortalin: a protein associated with progression of Parkinson disease? *J. Neuropathol. Exp. Neurol.*, 2008, **67**, 117–124.
304. L. F. Burbulla, C. Schelling, H. Kato, D. Rapaport, D. Woitalla, C. Schiesling, C. Schulte, M. Sharma, T. Illig, P. Bauer, S. Jung, A. Nordheim, L. Schols, O. Riess and R. Kruger, Dissecting the role of the mitochondrial chaperone mortalin in Parkinson's disease: functional impact of disease-related variants on mitochondrial homeostasis, *Hum. Mol. Genet.*, 2010, **19**, 4437–4452.
305. A. V. Goswami, M. Samaddar, D. Sinha, J. Purushotham and P. D'Silva, Enhanced J-protein interaction and compromised protein stability of mtHsp70 variants lead to mitochondrial dysfunction in Parkinson's disease, *Hum. Mol. Genet.*, 2012, **21**, 3317–3332.
306. L. F. Burbulla, J. C. Fitzgerald, K. Stegen, J. Westermeier, A. K. Thost, H. Kato, D. Mokranjac, J. Sauerwald, L. M. Martins, D. Woitalla, D. Rapaport, O. Riess, T. Proikas-Cezanne, T. M. Rasse and R. Kruger, Mitochondrial proteolytic stress induced by loss of mortalin function is rescued by Parkin and PINK1, *Cell Death Dis.*, 2014, **5**, e1180.

307. H. Yang, X. Zhou, X. Liu, L. Yang, Q. Chen, D. Zhao, J. Zuo and W. Liu, Mitochondrial dysfunction induced by knockdown of mortalin is rescued by Parkin, *Biochem. Biophys. Res. Commun.*, 2011, **410**, 114–120.
308. F. T. Liu, Y. Chen, Y. J. Yang, L. Yang, M. Yu, J. Zhao, J. J. Wu, F. Huang, W. Liu, Z. T. Ding and J. Wang, Involvement of mortalin/GRP75/mthsp70 in the mitochondrial impairments induced by A53T mutant alpha-synuclein, *Brain Res.*, 2015, **1604**, 52–61.
309. V. S. Van Laar, A. J. Mishizen, M. Cascio and T. G. Hastings, Proteomic identification of dopamine-conjugated proteins from isolated rat brain mitochondria and SH-SY5Y cells, *Neurobiol. Dis.*, 2009, **34**, 487–500.
310. J. Proft, J. Faraji, J. C. Robbins, F. C. Zucchi, X. Zhao, G. A. Metz and J. E. Braun, Identification of bilateral changes in TID1 expression in the 6-OHDA rat model of Parkinson's disease, *PLoS One*, 2011, **6**, e26045.
311. J. W. Pridgeon, J. A. Olzmann, L. S. Chin and L. Li, PINK1 protects against oxidative stress by phosphorylating mitochondrial chaperone TRAP1, *PLoS Biol.*, 2007, **5**, e172.
312. E. K. Butler, A. Voigt, A. K. Lutz, J. P. Toegel, E. Gerhardt, P. Karsten, B. Falkenburger, A. Reinartz, K. F. Winklhofer and J. B. Schulz, The mitochondrial chaperone protein TRAP1 mitigates alpha-Synuclein toxicity, *PLoS Genet.*, 2012, **8**, e1002488.
313. A. C. Costa, S. H. Loh and L. M. Martins, Drosophila Trap1 protects against mitochondrial dysfunction in a PINK1/parkin model of Parkinson's disease, *Cell Death Dis.*, 2013, **4**, e467.
314. L. Zhang, P. Karsten, S. Hamm, J. H. Pogson, A. K. Muller-Rischart, N. Exner, C. Haass, A. J. Whitworth, K. F. Winklhofer, J. B. Schulz and A. Voigt, TRAP1 rescues PINK1 loss-of-function phenotypes, *Hum. Mol. Genet.*, 2013, **22**, 2829–2841.
315. S. M. Jin and R. J. Youle, The accumulation of misfolded proteins in the mitochondrial matrix is sensed by PINK1 to induce PARK2/Parkin-mediated mitophagy of polarized mitochondria, *Autophagy*, 2013, **9**, 1750–1757.
316. Y. Kawamoto, Y. Kobayashi, Y. Suzuki, H. Inoue, H. Tomimoto, I. Akiguchi, H. Budka, L. M. Martins, J. Downward and R. Takahashi, Accumulation of HtrA2/Omi in neuronal and glial inclusions in brains with alpha-synucleinopathies, *J. Neuropathol. Exp. Neurol.*, 2008, **67**, 984–993.
317. K. M. Strauss, L. M. Martins, H. Plun-Favreau, F. P. Marx, S. Kautzmann, D. Berg, T. Gasser, Z. Wszolek, T. Muller, A. Bornemann, H. Wolburg, J. Downward, O. Riess, J. B. Schulz and R. Kruger, Loss of function mutations in the gene encoding Omi/HtrA2 in Parkinson's disease, *Hum. Mol. Genet.*, 2005, **14**, 2099–2111.
318. V. Bogaerts, K. Nuytemans, J. Reumers, P. Pals, S. Engelborghs, B. Pickut, E. Corsmit, K. Peeters, J. Schymkowitz, P. P. De Deyn, P. Cras, F. Rousseau, J. Theuns and C. Van Broeckhoven, Genetic variability in the mitochondrial serine protease HTRA2 contributes to risk for Parkinson disease, *Hum. Mutat.*, 2008, **29**, 832–840.

319. C. H. Lin, M. L. Chen, G. S. Chen, C. H. Tai and R. M. Wu, Novel variant Pro143Ala in HTRA2 contributes to Parkinson's disease by inducing hyperphosphorylation of HTRA2 protein in mitochondria, *Hum. Genet.*, 2011, **130**, 817–827.
320. H. Unal Gulsuner, S. Gulsuner, F. N. Mercan, O. E. Onat, T. Walsh, H. Shahin, M. K. Lee, O. Dogu, T. Kansu, H. Topaloglu, B. Elibol, C. Akbostanci, M. C. King, T. Ozcelik and A. B. Tekinay, Mitochondrial serine protease HTRA2 p.G399S in a kindred with essential tremor and Parkinson disease, *Proc. Natl. Acad. Sci. U. S. A.*, 2014, **111**, 18285–18290.
321. E. S. Alnemri, HtrA2 and Parkinson's disease: think PINK? *Nat. Cell Biol.*, 2007, **9**, 1227–1229.
322. N. Moisoi, K. Klupsch, V. Fedele, P. East, S. Sharma, A. Renton, H. Plun-Favreau, R. E. Edwards, P. Teismann, M. D. Esposti, A. D. Morrison, N. W. Wood, J. Downward and L. M. Martins, Mitochondrial dysfunction triggered by loss of HtrA2 results in the activation of a brain-specific transcriptional stress response, *Cell Death Differ.*, 2009, **16**, 449–464.
323. D. W. Park, M. K. Nam and H. Rhim, The serine protease HtrA2 cleaves UCH-L1 and inhibits its hydrolase activity: implication in the UCH-L1-mediated cell death, *Biochem. Biophys. Res. Commun.*, 2011, **415**, 24–29.
324. N. Casadei, P. Sood, T. Ulrich, P. Fallier-Becker, N. Kieper, S. Helling, C. May, E. Glaab, J. Chen, S. Nuber, K. Marcus, D. Rapaport, T. Ott, O. Riess, R. Kruger and J. C. Fitzgerald, Mitochondrial defects and neurodegeneration in mice overexpressing wild-type or G399S mutant HtrA2, *Hum. Mol. Genet.*, 2016, **25**, 459–471.
325. J. K. Jiang, X. Ma, Q. Y. Wu, W. B. Qian, N. Wang, S. S. Shi, J. L. Han, J. Y. Zhao, S. Y. Jiang and C. H. Wan, Upregulation of mitochondrial protease HtrA2/Omi contributes to manganese-induced neuronal apoptosis in rat brain striatum, *Neuroscience*, 2014, **268**, 169–179.
326. C. Meissner, H. Lorenz, A. Weihofen, D. J. Selkoe and M. K. Lemberg, The mitochondrial intramembrane protease PARL cleaves human Pink1 to regulate Pink1 trafficking, *J. Neurochem.*, 2011, **117**, 856–867.
327. S. M. Jin, M. Lazarou, C. Wang, L. A. Kane, D. P. Narendra and R. J. Youle, Mitochondrial membrane potential regulates PINK1 import and proteolytic destabilization by PARL, *J. Cell Biol.*, 2010, **191**, 933–942.
328. E. Deas, H. Plun-Favreau, S. Gandhi, H. Desmond, S. Kjaer, S. H. Loh, A. E. Renton, R. J. Harvey, A. J. Whitworth, L. M. Martins, A. Y. Abramov and N. W. Wood, PINK1 cleavage at position A103 by the mitochondrial protease PARL, *Hum. Mol. Genet.*, 2011, **20**, 867–879.
329. C. Meissner, H. Lorenz, B. Hehn and M. K. Lemberg, Intramembrane protease PARL defines a negative regulator of PINK1- and PARK2/Parkin-dependent mitophagy, *Autophagy*, 2015, **11**, 1484–1498.
330. G. Shi, J. R. Lee, D. A. Grimes, L. Racacho, D. Ye, H. Yang, O. A. Ross, M. Farrer, G. A. McQuibban and D. E. Bulman, Functional alteration of PARL contributes to mitochondrial dysregulation in Parkinson's disease, *Hum. Mol. Genet.*, 2011, **20**, 1966–1974.

331. A. W. Greene, K. Grenier, M. A. Aguileta, S. Muise, R. Farazifard, M. E. Haque, H. M. McBride, D. S. Park and E. A. Fon, Mitochondrial processing peptidase regulates PINK1 processing, import and Parkin recruitment, *EMBO Rep.*, 2012, **13**, 378–385.
332. R. E. Thomas, L. A. Andrews, J. L. Burman, W. Y. Lin and L. J. Pallanck, PINK1-Parkin pathway activity is regulated by degradation of PINK1 in the mitochondrial matrix, *PLoS Genet.*, 2014, **10**, e1004279.

CHAPTER 11

At the Intersection Between Mitochondrial Dysfunction and Lysosomal Autophagy: Role of PD-Related Neurotoxins and Gene Products

JOSEPHAT M. ASIAGO[†a], TREVOR B. DOYLE[†a,b], VARTIKA MISHRA[†a], AURÉLIE DE RUS JACQUET[a] AND JEAN-CHRISTOPHE ROCHET*[a,b]

[a]Department of Medicinal Chemistry and Molecular Pharmacology, Purdue University, West Lafayette, IN 47907, USA; [b]Purdue Institute for Integrative Neuroscience, Purdue University, West Lafayette, IN 47907, USA
*E-mail: jrochet@purdue.edu

11.1 Introduction

Parkinson's disease (PD) is a complex neurodegenerative disorder that was first described in detail by Dr James Parkinson in 1817 in 'An essay on the shaking palsy'. In this report, Dr Parkinson presented the cases of six

† These authors contributed equally.

Issues in Toxicology No. 34
Oxidative Stress and Redox Signalling in Parkinson's Disease
Edited by Rodrigo Franco, Jonathan A. Doorn and Jean-Christophe Rochet

Published by the Royal Society of Chemistry, www.rsc.org

patients whom he identified in London, three of whom he examined personally.[1] Later, non-tremulous forms of PD were identified by Charcot, who also specified that the disease was associated with a 'slowness of movement'.[2] Today, PD is recognized as involving characteristic motor symptoms that include tremor at rest, rigidity, akinesia (or bradykinesia), and postural instability, collectively represented by the acronym '*TRAP*'.[2] PD patients also present with an array of non-motor symptoms, including olfactory dysfunction,[3] gastrointestinal deficits,[4] autonomic failure,[5] sleep disorders,[6] depression,[7] compulsive behaviors,[8,9] and dementia.[10]

The motor symptoms of PD are largely attributed to a pronounced loss of dopaminergic neurons in the *substantia nigra pars compacta* (SNpc) region of the midbrain.[11] Another neuropathological hallmark of postmortem PD brains is the presence in various brain regions of Lewy bodies, cytosolic inclusions enriched with fibrillar forms of the presynaptic protein α-synuclein (aSyn).[12–14] Braak and colleagues[15,16] have described an evolution of aSyn neuropathology in the brains of PD patients that spans six stages identified based on which regions of the brain show evidence of Lewy body accumulation: stages 1 and 2, medulla oblongata and pontine tegmentum; stage 3, SNpc; and stages 4–6, mesocortex and neocortex.

Familial forms of PD are associated with a mutation in a single gene and account for approximately 5–10% of total cases.[17] The remaining 90–95% of PD cases, referred to as 'idiopathic' or 'sporadic', have no known genetic cause. Evidence suggests that the risk of idiopathic PD is modulated by a combination of environmental influences (*e.g.* exposure to herbicides, pesticides, or organic solvents; living in a rural environment; well-water consumption; occupational exposures associated with welding or mining)[18–24] and genetic risk factors,[25,26] including common allelic variants that only weakly increase the probability of developing PD.[17]

A major risk factor of PD is aging. Whereas PD affects approximately 0.3% of the total U.S. population, the incidence of the disease increases with age such that PD patients comprise 2% of the population over 65 years of age[27] and 5% of the population aged 85 years or greater.[28] With medical and technological advances, the life expectancy of humans continues to increase, resulting in a progressive aging of the population. In 2010, there were 630 000 diagnosed PD cases in the USA, and total costs associated with these cases, including medical and non-medical expenses, were estimated to be over $14 billion.[29] As our population continues to age, the number of people suffering from PD is expected to double by 2040,[29] thus imposing a tremendous economic burden on our health care system. Currently, PD patients are treated with drugs that temporarily help manage their symptoms, including the dopamine precursor L-3,4-dihydroxyphenylalanine (L-DOPA or levodopa) and dopamine agonists such as ropinirole, pramipexole, and rotigotine.[30] Importantly, however, these drugs have no impact on the progress of the disease, and ultimately their effectiveness becomes limited by debilitating side effects.

A significant hindrance for developing disease-modifying therapies for PD has been an incomplete understanding of the underlying pathogenic mechanisms. Accordingly, a major effort in the PD field has been directed towards elucidating the molecular underpinnings of the disease. Multiple lines of evidence suggest that defects in two cellular phenomena play a key role in neurodegeneration in PD: mitochondrial respiration and macroautophagy. In this chapter, we review neuropathological, toxicological, and genetic findings supporting the premise that mitochondrial dysfunction and autophagic impairment are involved in PD pathobiology. Moreover, we examine evidence suggesting that defects in mitochondrial function can provoke autophagic impairment in PD, and *vice versa*. Finally, we discuss how insights into the roles of mitochondrial and autophagic dysfunction in PD can set the stage for developing therapies to alleviate neurotoxicity in the brains of patients.

11.2 Neuropathological Evidence for Mitochondrial Deficits and Autophagic Impairment in PD

11.2.1 Evidence of Mitochondrial Deficits in Postmortem PD Brains

Mitochondria are often referred to as 'powerhouses of the cell'[31] because of their central role in modulating cellular metabolism and bioenergetics. Multiple lines of evidence suggest that mitochondrial dysfunction contributes to neurodegeneration in PD. Postmortem analysis of the SNpc and frontal cortex regions in the brains of PD patients has revealed a decreased activity of mitochondrial complex I, but not other complexes in the electron-transport chain (ETC).[32–34] In addition, complex I subunits in PD brains are characterized by increased oxidative damage in the form of protein carbonyls.[35] Evidence of impaired complex I activity has also been observed in peripheral tissues of PD patients, including platelets[36] and skeletal muscle.[37] Not only does a loss of complex I activity produce a decrease in mitochondrial ATP, but it also results in a 'leakage' of electrons from the ETC, leading to the formation of reactive oxygen species (ROS) that damage mitochondrial DNA, lipids, and proteins, thereby causing further mitochondrial impairment.[38–41] Consistent with this idea, extensive oxidative damage has been observed in the brains and peripheral tissues of PD patients.[42–45]

Oxidative stress occurs under conditions that lead to an imbalance between ROS production and antioxidant activity.[18] Dopaminergic neurons contain high levels of ROS due to the presence of dopamine. Similar to other catecholamines, dopamine undergoes auto-oxidation to form the *o*-quinone form, a reaction that involves the generation of superoxide radicals.[46] In addition, dopamine is readily metabolized by tyrosinase and monoamine oxidase, two ROS-generating enzymes.[18,47] Free radicals formed as a result of dopamine auto-oxidation or metabolism can damage mitochondrial electron-transport subunits, thereby leading to further oxidative stress associated with mitochondrial dysfunction as part of a vicious cycle.[48] Because dopaminergic neurons have higher basal levels of oxidative stress than other types of neurons, they are more susceptible to pathogenic mechanisms that further up-regulate ROS, including mitochondrial dysfunction.

11.2.2 Evidence of Autophagic Impairment in Postmortem PD Brains

There are three types of autophagy in mammalian cells: (i) microautophagy, a process by which the lysosomal membrane surrounds and engulfs cytoplasmic components for degradation; (ii) chaperone-mediated autophagy (CMA), involving the targeting of proteins with a specific sequence motif for uptake into lysosomes with the assistance of the molecular chaperone, Hsc70; and (iii) macroautophagy, a multi-step process that plays a key role in the clearance of long-lived proteins, including aggregated forms of aSyn, and the removal of dysfunctional or damaged organelles such as mitochondria.[49,50] Macroautophagy (the focus of this chapter) involves the sequestration of organelles and cytoplasmic components *via* the formation of a double-membrane-bound organelle termed an autophagosome.[50,51] Autophagosomes fuse with lysosomes to form autolysosomes, wherein the intralumenal components are degraded by lysosomal hydrolases (Figure 11.1). Macroautophagy is regulated by various protein complexes, including (i) the Atg1/unc-51-like kinase (ULK) complex, which is negatively regulated by the target of rapamycin (mTOR) macromolecular complex 1 (mTORC1) and plays a key role in initiating autophagosome formation; (ii) the beclin-1 complex, which mediates autophagosome nucleation by stimulating the formation of phosphatidylinositol-3-phosphate (PI3P); and (iii) microtubule-associated-protein-1 light chain 3 (LC3), which is involved in autophagosome maturation.[50–52] LC3 is modified *via* a ubiquitination-like system to form LC3-I, which in turn is modified to a membrane-bound form, LC3-II, by conjugation to phosphatidylethanolamine *via* a reaction mediated by the Atg5-Atg12 complex. Importantly, LC3-II localizes to the autophagosome and autolysosome and is thus a characteristic marker of these vesicular structures[53] (Figure 11.1). Macroautophagy is a critical cellular process for the removal of aggregated proteins that cannot be degraded *via* the ubiquitin-proteasome system (UPS) and dysfunctional organelles, including mitochondria having undergone a loss of membrane potential (the removal of mitochondria *via* macroautophagy is referred to as mitophagy).[54–56] Proteins and organelles targeted for removal by macroautophagy are typically ubiquitinated, and as a result of this modification they are directed to the autophagosomal membrane by adaptor proteins that bind to both LC3 and ubiquitin, including p62/SQSTM1 and optineurin.[57,58] Through the remainder of this chapter, macroautophagy is referred to as 'autophagy', and p62/SQSTM1 is referred to as 'p62'.

A number of neuropathological studies have revealed evidence of impaired lysosomal clearance in the brains of sporadic or familial PD patients and patients with dementia with Lewy bodies (DLB), a related synucleinopathy disorder.[59–66] Immunohistochemical data revealed that levels of lysosome-associated membrane protein 1 (LAMP1) and the lysosomal enzyme

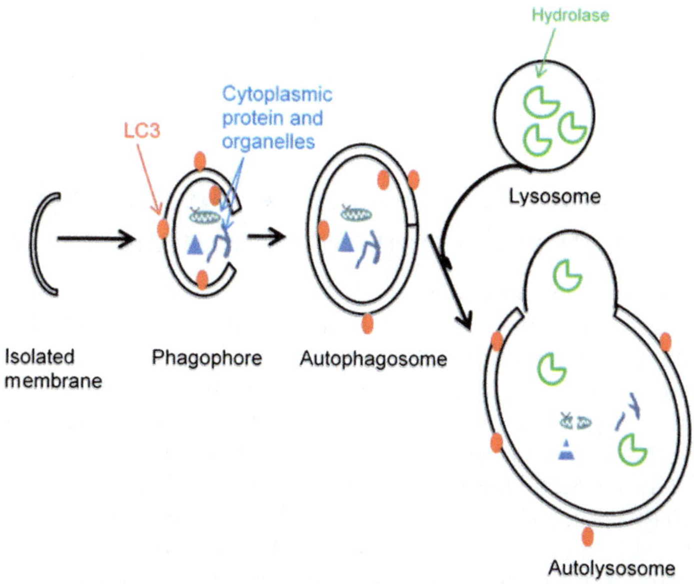

Figure 11.1 Diagram illustrating key steps involved in macroautophagy. Macroautophagy is initiated by the formation of a cup-shaped isolated double membrane that engulfs cytoplasmic proteins and organelles. The autophagic marker protein LC3-II is bound to the membrane of the phagophore and autophagosomes and is also present in autolysosomes, which are formed as a result of autophagosome–lysosome fusion. Cytoplasmic proteins and organelles in the autolysosome are degraded by lysosomal hydrolases.

cathepsin D are decreased in neurons of the SNpc (and in particular in nigral neurons containing aSyn inclusions) in postmortem PD brains.[59,61] Another study revealed evidence of decreased levels of the autophagy regulatory protein Atg7, increased mTOR levels, and the accumulation of morphologically abnormal autophagosomes in the brains of DLB patients.[60] Several groups have reported that the levels of various autophagy-related proteins are altered in the brains of PD or DLB patients, including ULK-1, ULK-2, beclin-1, and LC3.[60–62,65] The fact that lysosomal clearance appears to be disrupted in the brains of patients with synucleinopathy disorders implies that autophagic impairment could contribute to PD pathogenesis *via* the accumulation of protein aggregates and dysfunctional organelles. A limitation of studies aimed at characterizing autophagy in the postmortem brains of PD patients is that an increase in autophagosome levels can be interpreted in two very different ways: namely, it could reflect an increased production of autophagosomes associated with an aberrant upregulation of the autophagic pathway, or it could reflect an inhibition of autophagosome clearance associated with the disruption of a late step on the pathway (*e.g.* autophagosome–lysosome fusion).[67] This caveat has been addressed in cellular and animal models of PD by determining the effects of blocking autophagic clearance with a pharmacological agent (*e.g.* bafilomycin, an inhibitor of autophagosome–lysosome fusion, or chloroquine, an inhibitor of acid-dependent, autolysosomal protein degradation) on autophagosome accumulation triggered by PD-related insults.[68–71]

11.3 Toxicological Evidence for Mitochondrial Deficits and Autophagic Impairment in PD

Epidemiological evidence suggests that exposure to environmental toxins is linked to an increased risk of PD. In a rigorous case-controlled study, the pesticide rotenone and the herbicide paraquat (PQ) were found to be linked to an increased risk of PD (odds ratio, 2.5 for each).[72] Combined exposure to PQ and the fungicide maneb, which is used in the same geographical areas as PQ,[73] has also been associated with an increased incidence of parkinsonism.[74,75] The synthetic neurotoxin 1-methyl-4-phenyl-1,2,3,6-tetrahydropyridine (MPTP) has been linked to PD as a result of the discovery that individuals exposed to this agent develop a severe PD-like phenotype.[76] Another important neurotoxin in the PD field is 6-hydroxydopamine (6-OHDA). In contrast to rotenone, PQ, and MPTP, 6-OHDA has not been linked to human PD, but it has been used extensively to model dopaminergic neurodegeneration in *in vivo* models of the disease, particularly in rats.[77–80]

All of the agents outlined above have been used experimentally to develop cellular and animal models of PD in the laboratory (described in more detail below). Typically, these models consist of neuronal cells, primary mesencephalic cultures, or rodents exposed to rotenone, PQ (with or without maneb), MPP^+ or MPTP (in the case of cellular and animal models, respectively), or 6-OHDA. These studies have revealed that each of these agents elicits cytotoxicity *via* mechanisms involving mitochondrial dysfunction, oxidative stress, and autophagic impairment.

11.3.1 Rotenone

Rotenone is a natural isoflavonoid that diffuses across the cell membrane and inhibits complex I, a proton-translocating NADH:quinone oxidoreductase enzyme of the mitochondrial ETC.[81,82] Specifically, rotenone interferes with quinone reduction at the hydrophobic site of complex I.[83,84] Rotenone-mediated inhibition of this step results in (i) a build-up of superoxide, leading to oxidative stress; (ii) a disruption of oxidative phosphorylation, leading to a decrease in cellular ATP levels; and (iii) a collapse of the mitochondrial membrane potential (MMP).[38–41,85,86]

Systemic exposure of rats to chronic, low doses of rotenone produces motor deficits and neuropathological features similar to those observed in human PD patients, including the selective degeneration of striatal nerve terminals and nigral dopaminergic neurons and the presence of Lewy-like inclusions enriched with aSyn.[43,87,88] A number of studies in cell culture have also established a link between rotenone exposure and PD-related features, including aggregation of aSyn, an increase in ubiquitination, and caspase-induced cytotoxicity in cells chronically exposed to rotenone.[38,89,90] Overexpression of the single-subunit nicotinamide adenine dinucleotide (reduced) dehydrogenase (NDI1) from *Saccharomyces cerevisiae*, a protein that can functionally complement the entire multisubunit mammalian complex I but is resistant to rotenone-mediated inhibition, interfered with rotenone toxicity in SK-N-MC neuroblastoma cells[38] and with nigral neurodegeneration in rotenone-treated rats,[91] suggesting that rotenone elicits neurotoxicity by disrupting complex I activity. Evidence suggests that cell death in cellular and animal models of rotenone intoxication is primarily a result of oxidative stress rather than ATP depletion.[38,90] Dopaminergic neurons (in contrast to GABAergic neurons) undergo preferential degeneration in primary mesencephalic cultures exposed to rotenone,[92] suggesting that dopaminergic neurons are particularly susceptible to the toxic effects of mitochondrial inhibition and ROS production.

Multiple lines of evidence suggest that autophagic impairment plays a key role in rotenone-mediated neurotoxicity. Autophagosome levels have been shown to be increased in the SNpc region of rats exposed to rotenone either systemically[93] or *via* stereotaxic infusion.[94] A study of the effects of rotenone on autophagic flux in differentiated SH-SY5Y cells revealed that the toxicant induced up-regulation of the autophagosome marker LC3-II without affecting the increase in LC3-II levels elicited by bafilomycin, suggesting that rotenone disrupts autophagosome clearance in this cellular model.[95] In another study, rotenone was found to induce a decrease in LC3-II levels in primary cortical neurons cultured in the absence or presence of bafilomycin or chloroquine, suggesting that the toxicant interferes with the induction of autophagy in this model.[96] Rotenone toxicity is alleviated or amplified in neuronal cells co-treated with the autophagy inducer rapamycin or the autophagy inhibitor 3-methyladenine, respectively,[96–98] suggesting that autophagic impairment contributes to rotenone-mediated neurodegeneration. Consistent with this idea, a number of other agents have been shown to attenuate rotenone neurotoxicity in SH-SY5Y cells by stimulating autophagy, including celastrol,[99] deferoxamine,[100] erythropoietin,[101] kaempferol,[102] and lithium.[103]

Several mechanisms have been proposed to account for the inhibitory effects of rotenone on autophagy, including (i) a build-up of ROS, leading to the Akt-mediated inhibition of beclin-1 complex formation,[104] an increase in lysosomal pH or membrane permeability,[93,104] and/or a disruption of the microtubule network required for autophagosome trafficking to lysosomes;[105] (ii) a decrease in cellular ATP levels, leading to a defect in ATP-dependent lysosomal acidification mediated by vacuolar ATPase;[95] and (iii) perturbation of autophagic gene expression,[106] including up- and down-regulation of the genes encoding mTOR and beclin-1, respectively.[101]

In contrast to its inhibitory effects on autophagy, rotenone has been found to activate mitophagy.[96] Evidence suggests that rotenone induces the redistribution of cardiolipin from the inner mitochondrial membrane (IMM) to the outer mitochondrial membrane (OMM).[107,108] Because cardiolipin in the OMM is bound by LC3, and because LC3 is involved in autophagosome formation and cargo recognition, the externalization of cardiolipin serves as a signal for the elimination of damaged mitochondria *via* mitophagy.[107]

11.3.2 PQ and Maneb

PQ (1, 1′-dimethyl-4, 4′-bipyridinium dichloride) is a redox-reactive heterocycle that triggers oxidative stress through its conversion from the dicationic form (PQ^{2+}) to the monocation (PQ^{+}) *via* a reaction cycle catalyzed by NADPH oxidase and nitric oxide synthase (NOS) that also involves the conversion of O_2 to superoxide.[109–111] Maneb, a polymer formed from a complex of manganese with ethylene bis(dithiocarbamate) (Mn-EBDC), has been shown to trigger a build-up of ROS.[112] Maneb was also found to interfere with the UPS in one study[113] but not another.[114]

PQ has been shown to trigger selective dopaminergic cell death and neurite retraction in primary midbrain cultures.[110,115,116] Moreover, several groups have reported that rodents exposed to PQ or a combination of PQ and maneb develop features reminiscent of PD. Mice receiving intraperitoneal injections of PQ showed evidence of motor deficits,[117] selective nigral dopaminergic neurotoxicity,[118,119] and aSyn aggregation.[120] Mice injected with PQ were also reported to undergo a loss of dopaminergic neurons accompanied by an increase in lipid peroxidation and tyrosine nitration in the SNpc,[121] suggesting that PQ-mediated oxidative stress contributed significantly to nigral degeneration in these animals. In another study, mice treated with a combination of PQ and maneb exhibited more pronounced nigral dopaminergic cell death compared to animals exposed to either agent individually.[73,122] In contrast to these findings, Smeyne *et al.*[123] recently reported that mice exposed to PQ at doses near the maximally tolerated (nonlethal) amount showed no evidence of nigral dopaminergic cell death or striatal terminal degeneration, apparently because only low levels of the toxicant penetrated the mouse BBB. The discrepancy between these findings and those of earlier studies suggesting that PQ elicits nigrostriatal toxicity may be explained by differences in experimental design (*e.g.* blinding, protocol for tissue processing) or the degree to which the mice may have been exposed to inflammatory insults prior to PQ exposure.

Data reported by a number of groups suggest that PQ (or a combination of PQ and maneb) elicits mitochondrial dysfunction in preclinical models. Mitochondria isolated from the striatum of rats exposed to PQ *via* a chronic treatment (10 mg kg^{-1}, weekly ip injection, 4 weeks)[124] or an acute treatment (25 mg kg^{-1} ip, 24 h)[125] were characterized by reduced activity of NADH-cytochrome c reductase (complexes I–III) and cytochrome c oxidase (complex IV) activities and a decrease in mitochondrial O_2 consumption rates. Additional evidence suggests that complex III (ubiquinol:ferricytochrome c oxidoreductase) plays an important role in PQ-mediated ROS generation.[126,127]

Proteomic analysis of nigrostriatal tissues of mice exposed to a combination of PQ and maneb revealed altered expression levels of various mitochondrial proteins, including subunits of ETC complexes,[128] and the major constituent of maneb, Mn-EBDC, was found to inhibit mitochondrial complex III.[129]

An important question raised by the observations outlined above is whether PQ directly inhibits respiratory chain complexes (similar to rotenone or MPP^+), or whether it carries out its inhibitory effect *via* an indirect mechanism involving the redox cycling-mediated production of ROS that in turn cause oxidative damage to ETC proteins. The results of one study indicate that PQ has a substantially lower affinity for (and inhibitory activity against) complex I in isolated mitochondria compared to rotenone or MPP^+.[130] Moreover, PQ exhibited no inhibitory effect on mitochondrial O_2 consumption immediately after being added to N27 dopaminergic cells, in contrast to rotenone and MPP^+.[131] These findings imply that PQ interferes with mitochondrial respiration *via* an indirect mechanism, potentially involving ROS-mediated damage to ETC components. Consistent with this idea, mitochondria isolated from the striatum of PQ-treated rats show evidence of enhanced ROS (*e.g.* superoxide) accumulation, lipid peroxidation, and cardiolipin oxidation/depletion compared to mitochondria from untreated animals.[124,125,132] Additional studies revealed that PQ triggered an increase in cytosolic ROS that preceded an increase in mitochondrial ROS and a decrease in the MMP in the PC12 neuronal cell line,[133] although another group reported that PQ induced ROS formation and IMM depolarization in isolated mitochondria.[134] Collectively, these findings suggest that PQ-mediated mitochondrial dysfunction in the brain results from oxidative damage triggered by ROS formed either in the cytosol or in mitochondria, presumably *via* a redox cycling process.

Autophagic perturbations have been reported to contribute to neurodegeneration elicited by PQ. In one study, mTOR was found to be up-regulated (consistent with a defect in autophagic flux) in the striata of mice treated with PQ or PQ plus maneb.[114] PQ has also been shown to induce an increase in autophagosome levels (detected *via* microscopy or by measuring LC3-II levels) in SK-N-SH and SH-SY5Y neuroblastoma cells.[135,136] Evidence from studies in neuronal cell culture models suggests that PQ interferes with autophagic flux *via* mechanisms related to its ability to trigger oxidative stress, including the disruption of lysosomal acidification (presumably as a result of lysosomal membrane permeabilization (LMP))[137] and the depletion of cytosolic ubiquitin coupled with p62 inactivation.[138] Additional findings suggest that PQ induces misfolding of the PD-related gene product parkin,[139] an E3 ubiquitin ligase that plays a critical role in mitophagy (see below). Neuronal cells in which autophagy is inhibited with 3-methyladenine or *via* overexpression of a dominant-negative form of the regulatory protein Atg5 are sensitized to PQ toxicity,[135,136] whereas kaempferol, an upregulator of autophagy, has been found to protect against PQ-induced neuronal cell death.[102] Together, these results imply that autophagic disruption is involved in PQ-mediated neurodegeneration.

11.3.3 MPTP or MPP^+

In the late 1970s and early 1980s, a number of individuals who had self-injected synthetic preparations of the meperidine analog, 1-methyl-4-phenyl-4-propionoxypiperidine (MPPP), were found to present with severe PD-like symptoms.[76,140] The compound responsible for the patients' symptoms was identified as MPTP. Additional studies revealed that MPTP crosses the blood–brain barrier (BBB) and is converted to MPP^+ in astrocytes by monoamine oxidase B. MPP^+ then causes selective toxicity to dopaminergic neurons as a result of its uptake by the dopamine transporter.[141,142] The discovery of a link between MPTP and a PD-like syndrome in humans has led to a wealth of studies of PD pathogenic mechanisms involving the use of cells treated with MPP^+ and rodents (usually mice) or non-human primates exposed to MPTP as cellular and animal models, respectively.[78,143,144] Although Lewy-like inclusions are typically not observed in the brains of mice subjected to MPTP treatment,[145,146] inclusions that stained positive for aSyn and ubiquitin were formed in the SNpc region of mice subjected to a chronic MPTP treatment *via* continuous infusion.[147]

Evidence reported over the past three decades suggests that MPP^+ elicits neurotoxicity by inhibiting complex I *via* interactions with the same binding site as that targeted by rotenone,[148] resulting in a disruption of mitochondrial electron transport and a build-up of ROS.[149–151] Consistent with this idea, over-expression of yeast NDI1 was found to alleviate toxicity elicited by MPP^+ in SK-N-MC cells, and NDI1 over-expression from a viral vector injected in mouse SNpc attenuated dopaminergic cell death and behavioral deficits associated with MPTP intoxication.[152,153] In addition to the canonical mechanism of complex I inhibition, other mechanisms of MPTP/MPP^+ neurotoxicity have been proposed, including (i) the disruption of overall mitochondrial transport as a result of oxidative stress, leading to energy deficits at axon terminals in dopaminergic neurons;[154] and (ii) the displacement of dopamine from dopaminergic vesicles, leading to an increase in oxidative stress resulting from dopamine autoxidation in the cytosol.[155]

Multiple observations suggest that a disruption of autophagy is involved in toxicity elicited by MPP^+. Autophagosome levels were shown to be elevated in the SNpc region of rats infused intranigrally with MPP^+[156] and in MPP^+-treated SK-N-SH cells, SH-SY5Y cells, MN9D dopaminergic neuronal cells, and primary dopaminergic neurons.[135,157–159] Autophagosome levels were not increased in SK-N-SH cells co-treated with MPP^+ and chloroquine *versus* chloroquine alone, suggesting that MPP^+ caused an impairment of autophagic flux in this cellular model.[135] In addition, MPP^+ was shown *via* microscopy to induce an increase in levels of punctate forms of the autophagy substrate p62, consistent with a defect in autophagic flux, in MN9D cells and primary dopaminergic neurons.[157,158] The autophagy inducer rapamycin has been shown to alleviate MPTP-mediated neurodegeneration in mice.[160] In addition, neuronal cells in which autophagy is down-regulated *via* the over-expression of a dominant-negative form of Atg5 are sensitized to

MPP^+ toxicity,[135] whereas agents that promote autophagy, including kaempferol[102] and valproate,[161] protect against MPP^+-induced neuronal cell death. Moreover, a combination of valproate and lithium was found to attenuate behavioral deficits and nigral dopaminergic cell death associated with MPTP intoxication in mice.[162]

Data from studies involving MPTP-treated mice and BE-M17 neuroblastoma cells exposed to MPP^+ suggest that MPP^+ interferes with autophagic flux in dopaminergic neurons by eliciting a decrease in neuronal lysosome content, apparently *via* a mechanism involving LMP mediated by ROS or the pro-apoptotic, channel-forming protein Bax.[61,163] In addition to causing a decrease in lysosomal content, LMP is predicted to contribute to neurodegeneration by triggering a release of lysosomal proteases into the neuronal cytosol.[61] In another study, autophagy was found to be inhibited *via* a mechanism involving the inactivation of cathepsin D in the absence of changes in lysosomal acidity in SH-SY5Y cells subjected to a mild MPP^+ exposure.[164] Like PQ, MPP^+ has been shown to elicit ubiquitin depletion together with p62 inactivation[138] and parkin misfolding and aggregation[139,165] in neuronal cell-culture models.

Despite evidence that MPTP/MPP^+ elicits neurotoxicity by inhibiting autophagic flux as outlined above, other findings suggest that MPTP/MPP^+ in fact induces an overstimulation of autophagy, and in turn this effect plays a role in MPTP/MPP^+-mediated neurodegeneration.[156,159,166–168] In one study, mice infused intranigrally with an siRNA targeting Atg7 prior to intranigral infusion with MPP^+ were characterized by reduced levels of activated caspase-9 and increased levels of TH in the SNpc compared to animals infused with MPP+ alone.[156] In another study, mice with a conditional knockout of Atg7 in midbrain dopaminergic neurons were found to be resistant to MPTP-induced nigral dopaminergic cell death and striatal degeneration, although the knockout mice had markedly reduced numbers of nigral dopaminergic neurons and striatal terminals compared to wild-type animals.[166] Studies in SH-SY5Y cells have led to the elucidation of mechanisms that could account for MPTP/MPP^+-mediated autophagy up-regulation, including: (i) LC3 dephosphorylation, based on evidence that phosphorylated LC3 has a reduced propensity to be recruited to autophagosomes;[167] and (ii) activation of extracellular signal-regulated protein kinase (ERK), a kinase involved in stimulating autophagy and mitophagy *via* a beclin-1-independent process.[159,168]

It is unclear why MPTP/MPP^+ was found to inhibit autophagic flux in some studies but to stimulate flux in others, or why different groups found that MPTP/MPP^+-mediated neurotoxicity was exacerbated or alleviated by down-regulating autophagy. As one possibility, the results of these studies may be confounded by the fact that MPP^+ can have different effects on autophagy during early *versus* late stages of intoxication.[156,161] Moreover, the effects of MPTP or MPP^+ on autophagy, and the degree to which autophagy is protective as opposed to neurotoxic in cells or animals exposed to MPTP/MPP^+, could vary significantly as a function of differences in the magnitude of autophagy induction or the availability of feedback mechanisms to constrain autophagic flux among different experimental models.[159,168]

11.3.4 6-OHDA

In contrast to rotenone, PQ, maneb, and MPTP, 6-OHDA lacks the ability to penetrate the BBB. Accordingly, it is typically administered into the brains of rats, mice, or monkeys *via* stereotaxic injection. Rats lesioned unilaterally in the SNpc, medial forebrain bundle (MFB), or striatum with 6-OHDA exhibit a classic motor phenotype that consists of ipsilateral or contralateral rotations relative to the side of the lesion in response to the administration of amphetamine or apomorphine, respectively.[169–172] Because of its structural similarity to dopamine, 6-OHDA is taken up into catecholaminergic neurons by DAT, and this uptake in turn accounts for the toxin's ability to trigger preferential destruction of catecholaminergic neurons and their terminals, including dopaminergic neurons and terminals of the nigrostriatal pathway. Data from multiple groups suggest that 6-OHDA elicits dopaminergic cell death *via* the generation of quinones or ROS that cause oxidative damage.[77,78,80] Similar to the MPTP model of PD, the brains of 6-OHDA-treated rats do not show the presence of Lewy-like aSyn inclusions.[77,78,80,169,172]

Evidence from various studies suggests that 6-OHDA triggers mitochondrial impairment *via* multiple mechanisms. A number of groups have reported that 6-OHDA disrupts mitochondrial respiration by inhibiting complex I of the ETC in rat SNpc[173,174] or in isolated brain mitochondria.[175,176] 6-OHDA has also been found to impair mitochondrial respiration in the N-2a neuroblastoma cell line by maintaining cytochrome c in a reduced state (*via* the reduction of Fe(III) to Fe(II)).[177] In contrast, another group reported that 6-OHDA has only a modest inhibitory effect on mitochondrial O_2 consumption (less pronounced than that of rotenone or MPP^+) in differentiated SH-SY5Y cells, implying that 6-OHDA-mediated oxidative stress occurs *via* mechanisms unrelated to defects in mitochondrial oxidative phosphorylation in this cell line.[178,179]

Additional findings from studies in SH-SY5Y cells suggest that 6-OHDA can cause mitochondrial dysfunction by promoting the translocation of the mitochondrial fission protein DrpI from the cytosol to the mitochondrial surface, thereby triggering an excess of mitochondrial fragmentation.[180–182] The stimulatory effect of 6-OHDA on the mitochondrial translocation of DrpI was found to be attenuated by the antioxidant molecule MitoQ, suggesting that this process depends on 6-OHDA-mediated oxidative stress.[181] Another study revealed that 6-OHDA induces defects in mitochondrial axonal transport in mouse midbrain cultures, apparently *via* a mechanism involving a buildup of ROS.[183] Transgenic mice overexpressing the mitochondrial enzyme manganese superoxide dismutase (MnSOD) exhibit a less pronounced striatal injury and less extensive nigral dopaminergic cell death following intrastriatal 6-OHDA injection compared to non-transgenic animals, suggesting that mitochondrial superoxide plays a key role in 6-OHDA-mediated nigrostriatal degeneration.[184]

Multiple lines of evidence suggest that autophagy is up-regulated in the brains of rodents exposed to 6-OHDA. In one study, levels of LC3-positive puncta likely corresponding to autophagosomes were found to be increased in nigral dopaminergic neurons of rats injected in the SNpc with 6-OHDA compared to control animals.[185] Other groups have reported that rats injected in the MFB with 6-OHDA exhibit (i) an increase in midbrain levels of LC3, together with a decrease in midbrain levels of the autophagy substrate protein p62;[186] and (ii) a transient up-regulation of beclin-1 in the striatum, hippocampus, and cortex.[187] SK-N-SH cells treated with 6-OHDA show increased levels of autophagic flux (reflected by an increase in LC3-II levels in cells incubated in the presence 6-OHDA plus chloroquine *versus* chloroquine alone), consistent with the idea that 6-OHDA stimulates rather than inhibits autophagy.[135] Data from studies in dopaminergic neuronal cells suggest that 6-OHDA activates autophagy by triggering a build-up of ROS, which can in turn antagonize the autophagy repressor mTOR by activating the AMPK/Raptor signalling pathway[188] or by inducing an increase in levels of Tnfaip8 l1/Oxi-β, a protein that stabilizes the mTOR repressor TSC2.[189]

Rats injected intranigrally with 6-OHDA exhibit a less pronounced degree of nigral neurodegeneration when pretreated with the autophagy inhibitor 3-methyladenine compared to rats injected with 6-OHDA alone.[185] Moreover, expression of constitutively active forms of the kinase Akt and the GTPase Rheb, two mTOR activators, results in a decrease in retrograde axonal degeneration in mice injected intrastriatally with 6-OHDA.[190] These observations suggest that the up-regulation of autophagy observed in the brains of 6-OHDA-treated rats contributes to neurodegeneration in this model.

Like rotenone (see above), 6-OHDA induces the redistribution of cardiolipin from the IMM to the OMM, thus promoting the elimination of damaged mitochondria *via* mitophagy.[107] The induction of autophagy in SH-SY5Y cells exposed to 6-OHDA is inhibited upon cotreatment of the cells with a small-molecule inhibitor of the mitochondrial fission protein DrpI, suggesting that mitochondrial fission promotes mitophagy, potentially because one of the daughter mitochondria produced as a result of each fission event has a disrupted membrane potential and is therefore targeted for mitophagy.[182,191] These findings imply that 6-OHDA stimulates not only mitochondrial fission, but also mitophagy as a result of its stimulatory effect on the mitochondrial translocation of DrpI (see above). Finally, data reported by Chu and colleagues[192,193] suggest that 6-OHDA can also promote mitophagy in SH-SY5Y cells *via* a mechanism involving the activation of a mitochondrial pool of ERK1/2.

11.4 Genetic Evidence for Mitochondrial Deficits and Autophagic Impairment in PD

At least 20 chromosomal loci, referred to as PARK loci, have been identified as being involved in monogenic forms of PD, and these loci are classified chronologically (PARK 1–20) according to when they were identified.[25,194] A number of these genes encode proteins that regulate mitochondrial function and autophagy, including aSyn, parkin, PINK1, DJ-1, and ATP13A2 (reviewed below). Another PD-related protein with modulatory effects on mitochondrial function and autophagy, LRRK2, has been discussed extensively in several reviews[195–197] and is not considered further here because of space limitations.

11.4.1 aSyn

Missense mutations in the aSyn gene (encoding the substitutions A30P, E46K, H50Q, G51D, A53E, and A53T) have been linked to familial PD[198–205] (Figure 11.2). Moreover, duplication and triplication mutations that increase the copy number of the wild-type aSyn gene have been identified.[206,207] Additional data from genome-wide association studies (GWAS) have demonstrated a strong link between variation in the SNCA gene encoding aSyn and the risk of both familial and sporadic PD.[208,209] aSyn is a 140-amino acid residue protein that localizes to presynaptic terminals, where it is present as both cytosolic and membrane-bound forms.[210] The aSyn sequence is subdivided into three domains: (i) an N-terminal domain (residues 1–85), consisting of seven imperfect repeats, six of which contain the conserved hexamer sequence KTK(E/Q)GV (Figure 11.2), which plays a critical role in the association of aSyn with phospholipid membranes;[211,212] (ii) a central hydrophobic domain, also referred to as the non-amyloid-β-component of Alzheimer's disease (NAC) domain (residues 61–95), consisting primarily of hydrophobic residues that initiate aggregation;[213] and (iii) the carboxy-terminal (C-terminal) region (residues 96–140), enriched with negatively charged residues (aspartate, glutamate) and proline residues (Figure 11.2). Although the binding of aSyn to phospholipid membranes is presumably essential for the

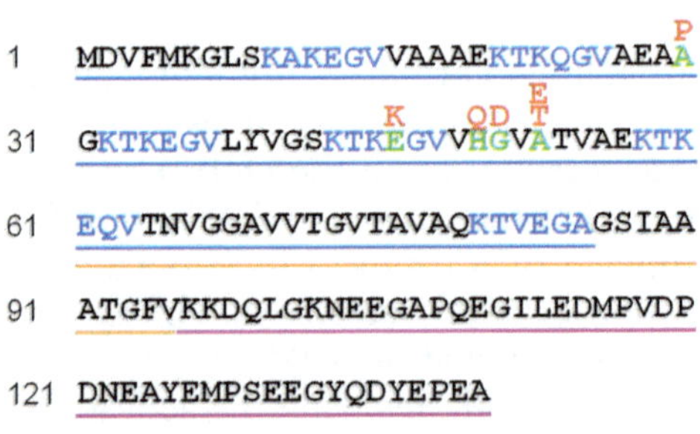

Figure 11.2 Amino acid sequence of aSyn shown in one-letter code. The N-terminal repeat region spanning residues 1–85 is underlined in blue. The central hydrophobic domain spanning residues 61–95 is underlined in orange. The unfolded C-terminal region enriched with proline, aspartate, and glutamate residues spanning residues 96–140 is underlined in violet. Amino acid substitutions in familial PD mutants (A30P, E46K, H50Q, G51D, A53E, and A53T) are indicated in red. Adapted from *Bio-nanoimaging: Protein Misfolding and Aggregation,* Griggs, A. M., *et al.*, Edited by Vladimir Uversky and Yuri Lyubchenko, Role of aberrant α-synuclein–membrane interactions in Parkinson's disease, pp. 443–452, Copyright 2014, with permission from Elsevier.

protein's normal function related to the regulation of synaptic vesicle release and neurotransmission,[211,214–219] aSyn is also known to form aggregates at membrane surfaces.[220–226]

A number of aSyn substitution mutants, including E46K,[227–229] H50Q,[230–233] and A53T,[234] have been shown to form amyloid-like fibrils more rapidly than wild-type aSyn. In contrast, the neurotoxic effects of other familial PD mutants (*e.g.* A30P) appear to correlate better with the formation of pre-fibrillar intermediates (oligomers or protofibrils) rather than accelerated fibrillization.[234,235] Some aSyn substitutions appear to favor the protein's oligomerization at the surface of phospholipid vesicles (*e.g.* A30P, G51D), and evidence suggests that membrane-induced aSyn aggregation plays a key role in aSyn neurotoxicity, potentially *via* a mechanism involving the permeabilization of dopaminergic vesicles.[225,226,236–238] Increases in aSyn expression resulting from duplication or triplication of the SNCA gene,[206,207] as well as the more modest upregulation of aSyn associated with aging[239] or the presence of certain polymorphisms in the 5′ promoter or 3′ enhancer regions of SNCA,[240–243] are predicted to favor aSyn aggregation *via* mass action in both the absence and presence of membranes.[226,244]

Multiple lines of evidence suggest that aSyn causes mitochondrial structural and functional impairments that play a role in the protein's neurotoxicity.[245,246] In one study, the brains of A53T transgenic mice were found to be characterized by extensive mitochondrial structural anomalies and damage to mitochondrial DNA.[247] A number of groups have reported that aSyn binds the OMM,[248–251] whereas another study revealed that aSyn is imported to the IMM *via* a mechanism that depends on the presence of a cryptic targeting sequence spanning residues 1–32.[252] aSyn expression has been shown to trigger mitochondrial fragmentation, presumably as a result of the protein's interactions with the OMM.[253,254] Additional findings suggest that aSyn binds to mitochondria-associated ER membranes, which play a role in modulating mitochondrial Ca^{2+} import and mitochondrial dynamics, and aSyn knockdown or over-expression of the wild-type protein or PD-linked mutants was found to cause a disruption of ER-mitochondria contacts, a loss of mitochondrial Ca^{2+} homeostasis, and increased mitochondrial fragmentation.[255,256] Post-translationally modified forms of aSyn have been shown to bind to the outer membrane translocase protein TOM20, leading to defects in mitochondrial protein import and electron transport, a build-up of ROS, and a disruption of the MMP.[257]

aSyn over-expression has been shown to result in a loss of complex I activity in cell-culture models and in the brains of transgenic mice, apparently as a result of inhibitory effects of the normal form of the protein as opposed to pathogenic species.[252–254,256,258] The inhibitory effect of aSyn on complex I implies that aSyn expression should sensitize neurons to the toxic effects of complex I inhibitors. Consistent with this idea, aSyn knockout mice and rats depleted of aSyn *via* intranigral injection of shRNA-encoding virus are resistant to neurotoxicity elicited by MPTP and rotenone, respectively.[259,260]

Data from *in vivo* studies suggest that at high expression levels, aSyn is cleared *via* autophagy in transgenic mouse brains, whereas at lower expression levels (*e.g.* in non-transgenic mice), aSyn is eliminated *via* proteasomal degradation.[261] aSyn transgenic mice show evidence of autophagosome accumulation and Atg7 downregulation, suggesting that aSyn overexpression leads to a disruption of autophagy.[60] Mechanisms by which aSyn has been reported to impair autophagic flux include: (i) antagonism of Rab1a, leading to a disruption of the formation of autophagosome precursors termed 'omegasomes';[262] (ii) inhibition of a step that occurs downstream of autophagosome formation (*e.g.* autophagosome–lysosome fusion);[263] (iii) binding of the protein high mobility group box 1 (HMGB1), leading to the disruption of an interaction between HMGB1 and beclin-1 that plays a key role in autophagic induction;[264] (iv) inhibition of the kinase JNK1, resulting in a decrease in Bcl-2 phosphorylation that in turn favors Bcl-2/beclin-1 complex formation and disfavors the initiation of autophagy;[265] and (v) disruption of Rab1a-mediated trafficking of lysosomal hydrolases, leading to deficits in lysosomal degradation.[266] Because aSyn is eliminated *via* autophagy under conditions that favor its accumulation and aggregation, aSyn-mediated disruption of autophagy is predicted to cause a further accumulation of aSyn, leading to the formation of toxic aSyn oligomeric species involved in neurodegeneration. Stimulation of autophagy *via* rapamycin infusion into the lateral ventricle or lentivirus-mediated overexpression of beclin-1 or Atg7 in the cortex and hippocampus attenuates neuropathology in aSyn transgenic mice,[60,71] suggesting that aSyn-mediated autophagic impairment contributes to neurotoxicity in this model.

11.4.2 Parkin/PINK1

Genetic analyses of autosomal recessive and sporadic cases of PD have led to the identification of mutations in genes encoding two enzymes, PINK1 and parkin.[267–270] Mutations in the parkin gene are the most common mutations associated with early-onset, autosomal-recessive PD.[25] Evidence suggests that parkin overexpression can rescue PINK1 knock-out, whereas PINK1 overexpression fails to rescue parkin knock-out.[271–273] These observations imply that PINK1 and parkin function on a shared pathway, with PINK1 acting upstream of parkin.

PINK1 is a 62.8 kDa serine/threonine kinase with an N-terminal mitochondrial localization sequence (MLS). The protein is rapidly degraded by the UPS in cells with normally functioning mitochondria but is stabilized on the OMM in response to mitochondrial stress.[274–281] When targeted to impaired mitochondria with a disrupted membrane potential, PINK1 escapes degradation and instead undergoes homodimerization and autophosphorylation. These steps increase the kinase activity of PINK1, and further promote parkin translocation to the mitochondria.

Parkin is a 51.5 kDa E3 ubiquitin ligase that naturally exists in a non-catalytic, autoinhibited state.[275,282] The protein consists of an N-terminal 'ubiquitin-like' (Ubl) domain that shares homology with ubiquitin and a C-terminal domain composed of three RING (*r*eally *i*nteresting *n*ew *g*ene) finger domains (RING0, RING1, RING2), two of which (RING1 and RING2) are separated by an IBR (*in*-*b*etween *R*ING finger) domain.[283] The RING-IBR-RING motif mediates the recruitment of E2 ubiquitin conjugating enzymes and protein substrates targeted for ubiquitination. PINK1 activates latent parkin *via* two steps (Figure 11.3). Direct phosphorylation of parkin on serine 65 in the ubiquitin-like domain induces conformational changes that reverse the natural autoinhibited state of parkin and expose residues necessary for the catalytic transfer of ubiquitin onto a specific substrate.[284,285] Introduction of the phosphomimetic substitution S65E was insufficient to fully activate parkin ubiquitination, indicating that further steps may be necessary to fully activate catalysis.[285] PINK1 has also been shown to phosphorylate serine 65 of ubiquitin, and phosphoubiquitin is in turn hypothesized

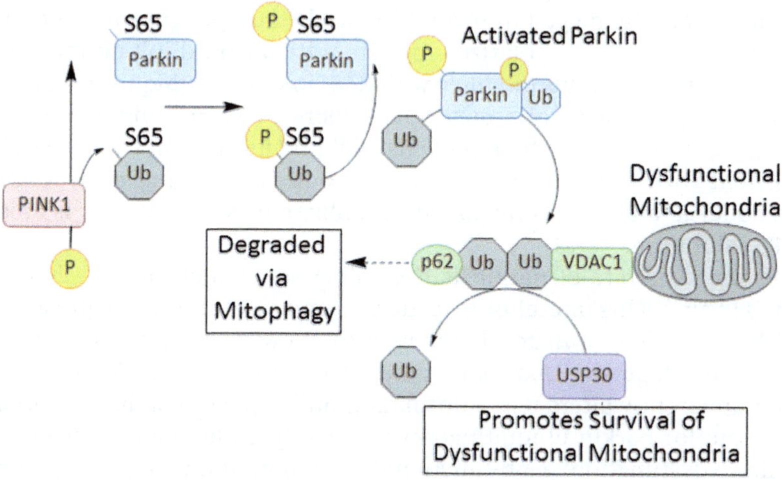

Figure 11.3 Model illustrating key events involved in PINK1- and parkin-mediated mitophagy. In response to mitochondrial stress, PINK1 translocates from the cytosol to the OMM where it phosphorylates both parkin and ubiquitin on serine residue 65. The ubiquitin ligase activity of parkin is usually repressed through structural autoinhibition but is induced by PINK1 *via* a two-step phosphorylation-dependent mechanism. First, the phosphorylation of parkin by PINK1 reverses the naturally autoinhibited state. This is followed by the interaction of parkin with phosphoubiquitin, resulting in allosteric activation of the enzyme. Once activated, parkin catalyzes the transfer of ubiquitin onto specific substrates on dysfunctional mitochondria, thereby targeting the mitochondria for clearance *via* mitophagy. The intermediary protein p62 serves as a link to facilitate the interaction of the ubiquitin tag with the phagophore (developing autophagosome). USP30 catalyzes the removal of ubiquitin from targeted mitochondrial proteins, promoting their retention as part of a mechanism that functionally opposes PINK1- and parkin-mediated mitophagy.

to bind allosterically to parkin, disrupting hydrophobic interactions and promoting the exposure of the catalytic site for full activation of parkin (Figure 11.3).[280] These findings suggest that PINK1-mediated parkin activation occurs *via* a two-step phosphorylation-dependent mechanism involving the direct phosphorylation of parkin and allosteric modulation of parkin by phosphoubiquitin.

Data from studies in immortalized cells and primary cortical neurons suggest that PINK1 and parkin function together to regulate the elimination of damaged mitochondria (*e.g.* mitochondria having undergone oxidative damage or a loss of membrane potential) through mitophagy.[275,276,286,287] Upon activation, parkin catalyzes the transfer of ubiquitin onto mitochondrial substrates, targeting them for mitophagy. The ubiquitination of mitochondrial outer membrane proteins with K27- or K63-linked polyubiquitin chains serves as a 'tag' for the elimination of damaged mitochondria because ubiquitinated mitochondrial proteins are recognized and bound by adaptor proteins such as p62 and optineurin, which simultaneously interact with LC3 and thus direct the tagged mitochondria to the autophagosome membrane (Figure 11.3).[286,288,289] Depolarized mitochondria also exhibit an increase in outer membrane proteins modified with K48-linked polyubiquitin chains, the classic linkage topology of proteasome substrates, suggesting that the UPS-mediated degradation of these proteins could be involved in mitophagy.[290,291] Consistent with this idea, the 26S proteasome is recruited to depolarized mitochondria, and parkin-mediated mitophagy in SH-SY5Y cells is ablated by UPS inhibition.[290]

One key substrate of parkin is the voltage-dependent anion channel 1 (VDAC1).[286,292] This metabolite transporter is commonly expressed on the OMM and is known to oligomerize to allow passage of cytochrome c to mediate apoptosis. Upon stress-induced mitochondrial depolarization, VDAC1 apparently alters its conformation and exposes residues to become a substrate for parkin ubiquitination exclusively under conditions of mitochondrial dysfunction. Non-functional parkin mutants fail to polyubiquitinate VDAC1 during mitochondrial stress,[286] further supporting that ubiquitination of VDAC1 is dependent on parkin activity. Knockdown of VDAC1 expression with siRNA inhibits intracellular clearance of depolarized mitochondria, indicating that VDAC1 is a necessary substrate for the process of mitophagy.[286] Parkin-mediated VDAC1 ubiquitination may serve as a time-dependent, anti-apoptotic mechanism that promotes rapid mitophagy of dysfunctional mitochondria, while simultaneously preventing cell death.

The deubiquitinase USP30 mechanistically opposes PINK1/parkin-mediated mitophagy by removing the ubiquitin tag that targets dysfunctional mitochondria to the autophagosome.[293–296] USP30 is localized on the OMM and can remove the ubiquitin tag of parkin targets, promoting the retention of mitochondria in the cytoplasm and inhibiting clearance. Catalytically inactive USP30 fails to prevent parkin-mediated mitochondrial clearance, and knockdown of endogenous USP30 dramatically enhances mitophagy of dysfunctional mitochondria.[294,296] In a parkin-deficient *Drosophila* model of

PD, USP30 knockdown rescued behavioral motor deficits and further protected flies from PQ-induced neurotoxicity.[294]

The stimulation of parkin through depletion of the MMP has been shown to decrease levels of intracellular USP30, whereas non-functional parkin mutants did not alter USP30 expression.[294] Further investigation confirmed that USP30 is a substrate of parkin, and can be ubiquitinated during mitochondrial dysfunction, targeting the enzyme for proteolysis.[294] These results suggest that not only is parkin tagging mitochondria with ubiquitin for clearance, but it also attenuates a catalytic process that opposes mitophagy. USP30 inhibitors have the potential to be effective therapies to prevent oxidative damage by promoting the clearance of dysfunctional mitochondria.

Additional findings suggest that PINK1 and parkin interact with proteins involved in mitochondrial fusion and fission, and these interactions play a key role in regulating the interplay (or 'crosstalk') between mitochondrial dynamics and mitophagy.[244,297–299] The preservation of healthy mitochondrial networks is fundamental to proper cellular function and maintaining the integrity of cellular signalling pathways. Therefore, depolarized or dysfunctional mitochondria must be quickly isolated and degraded promptly to prevent damage to both the greater mitochondrial network, as well as the cell itself. The mitochondrial fusion/fission machinery is part of an elegant quality control system that has evolved for the identification and removal of dysfunctional mitochondria. The mitochondrial Rho small GTPase Miro functions as a kinesin-mitochondria adapter protein, facilitating transport within the cell and promoting mitochondrial interactions necessary for fusion and fission to occur.[300,301] Mitochondrial fusion has been shown to be critically dependent upon mitofusins 1 and 2 (Mfn1 and Mfn2), which facilitate the interaction and merging of OMMs.[302,303] In contrast, mitochondrial fission depends on Drp1 polymerization to complete the division of the dividing organelle by pinching off the two halves.[300,302,304] Parkin and PINK1 have been shown to promote either fission or fusion in different mammalian cell-culture models, and defects in these regulatory processes could lead to a disruption of mitochondrial quality control and a build-up of dysfunctional mitochondria (reviewed in ref. 244).

Chemical depolarization of Mfn1-deficient cells was reported to induce parkin recruitment to mitochondria.[305] However, Mfn2 null cells failed to recruit parkin under identical treatment conditions, an effect that could be reversed by Mfn2 expression in the deficient cells. PINK1-dependent phosphorylation of Mfn2 is a necessary molecular tag to recruit parkin to the OMM, leading to the ubiquitination of Mfn1 and Mfn2.[303,305–308] In turn, ubiquitinated Mfn1 or Mfn2 is degraded by the proteasome, apparently as a mechanism to ensure that depolarized mitochondria are segregated and do not escape mitophagy by undergoing fusion with functional mitochondria.[290,291,309,310]

The mitochondrial fission protein Drp1 is a necessary component for mitochondrial fragmentation, a functional step preceding mitophagy.[299,311] Drp1 has been shown to interact with the second ring finger domain of parkin, which catalyzes the transfer of ubiquitin to Drp1 in cells.[304]

Rather than promoting the recruitment of mitophagy machinery, the ubiquitination of Drp1 apparently targets the protein for proteasomal degradation.[304] The selective degradation of Drp1 may serve to further isolate dysfunctional mitochondria.

It has been proposed that Miro ubiquitination and subsequent proteolytic degradation may effectively isolate damaged mitochondrial units by preventing their transport throughout the cell.[301,312] Indeed, mitochondrial membrane depolarization induces rapid and robust PINK1 and parkin-mediated ubiquitination of Miro on the mitochondrial outer membrane. However, the slow degradation kinetics of Miro support a model in which the ubiquitination of Miro recruits mitophagy machinery (*e.g.* p62) to mitochondria, rather than stalling mitochondrial transport by the removal of Miro *via* proteolysis.[313,314]

11.4.3 DJ-1

Deletions and point mutations resulting in a loss of functional DJ-1 protein are involved in autosomal recessive PD.[315] In an initial study, a six-exon deletion and a missense mutation resulting in the substitution of proline for leucine at residue 166 (L166P) were identified in a Dutch family and an Italian family, respectively.[315] Since then, additional deletions and point mutations in the DJ-1 gene have been linked to cases of familial PD.[316–320] Other studies revealed a mutation in the DJ-1 promoter.[319] DJ-1 is a 189-residue protein that exists as a homodimer under physiological conditions.[321–324] Based on its structure, DJ-1 has been classified as a member of the ThiJ/PfpI superfamily of proteins.[325–330] Although DJ-1 contains a conserved cysteine residue at position 106 (C106) in a pocket that resembles the active sites of proteases in the ThiJ/PfpI superfamily, this residue is not part of a catalytic triad and, therefore, is not thought to be involved in proteolytic activity.[325] Nevertheless, C106 is proposed to be a functionally important residue of DJ-1, perhaps as an oxidative stress-sensing residue given its high reactivity with pro-oxidants such as H_2O_2.[331–333] DJ-1 undergoes oxidation to the sulfinic acid form more rapidly than the other two DJ-1 cysteine residues, C46, located at the dimer interface, and C53, a relatively exposed residue. An appropriately located glutamate residue at position 18 (E18) lowers the sidechain pK_a of C106[334] and thus favors C106 oxidation to the sulfinic acid state, which has been proposed to be the active form of DJ-1 involved in enhancing mitochondrial function.[335,336] DJ-1 with C106 oxidized to the sulfinic acid is referred to as the '2O' form of the protein.[324,332]

Over-oxidized forms of DJ-1 have been identified in the brains of PD and Alzheimer's disease patients.[337] With age, oxidation of DJ-1 in rats, flies and humans has been found to increase, and this also affects the ability of the protein to suppress oxidative stress.[338] Over-oxidation of DJ-1 at residue C106 destabilizes the structure and can compromise the function of DJ-1.[323,332] Oxidation at methionine residues of DJ-1 causes a loss of stability of the protein, making it more susceptible to unfolding and aggregation.[324] Thus, DJ-1 dysfunction resulting from destabilizing oxidative modifications could also play a role in more common sporadic forms of PD.[323,337,338]

Evidence suggests that DJ-1 alleviates toxicity elicited by oxidative insults. The over-expression of wild-type DJ-1, but not familial mutant forms of the protein, protects neuronal cells from ROS,[339] whereas down-regulation of DJ-1 sensitizes cells to oxidative stress.[340] DJ-1 has also been shown to protect neurons from rotenone-induced oxidative stress and cell death.[92,341] DJ-1 protects against toxicity elicited by H_2O_2, 6-OHDA, and rotenone by up-regulating the antioxidant molecule glutathione, whereas it induces an increase in levels of the heat shock protein 70 (HSP 70) in neuronal cultures expressing mutant aSyn.[92,342] These results suggest that DJ-1 protects against various PD-related insults, and the mechanism of DJ-1 neuroprotection depends on the type of stress.

Multiple lines of evidence suggest that DJ-1 can localize to mitochondria in mammalian cell lines. In one set of studies, wild-type DJ-1 (but not the C106A mutant) was found to undergo oxidation to the 2O form, translocate to the OMM, and carry out a cytoprotective effect in M17 cells exposed to oxidative insults, including rotenone and PQ.[335,336] These data suggest that the oxidation of cysteine 106 is part of an oxidative stress 'sensing' mechanism, promoting relocalization of DJ-1 from the cytosol to the OMM. In contrast, another group reported that DJ-1 translocated from the cytosol to the OMM in SK-N-BE(2)C dopaminergic neuroblastoma cells exposed to H_2O_2, but in this case the relocalization of the protein was not affected by the C106S substitution.[343] A variant of DJ-1 targeted to the OMM in SK-N-BE(2)C cells had a greater protective effect against cytotoxicity elicited by H_2O_2 compared to the wild-type protein.[343] Other investigators observed that a fraction of DJ-1 was present in the mitochondrial matrix and intermembrane space under basal conditions, and the degree of DJ-1 localization to mitochondria was not influenced by oxidative stress.[344] Data obtained using a novel split-GFP technique revealed that wild-type DJ-1 was localized to the OMM and to a lesser extent the mitochondrial matrix in HeLa cells under basal conditions.[345] Because DJ-1 lacks a canonical MLS, the molecular mechanism by which it is transported into the interior of mitochondria is unclear, although one group reported that a 12-residue segment in the N-terminal region could be involved in the protein's mitochondrial localization.[346]

The data outlined above imply that DJ-1 carries out at least part of its neuroprotective function in mitochondria, and, therefore, that a loss of DJ-1 function could lead to mitochondrial impairment. Consistent with this idea, mitochondria isolated from the brains of DJ-1 knockout mice were found to have higher ROS levels compared to mitochondria from wild-type animals.[333] DJ-1$^{-/-}$ mouse embryonic fibroblasts (MEFs) were characterized by elevated mitochondrial ROS levels, decreased respiration rates, and a reduced MMP compared to wild-type cells.[347] An increase in mitochondrial ROS levels and a reduction in MMP was also observed by another group in DJ-1$^{-/-}$ MEFs, although mitochondria isolated from the mutant fibroblasts or from the brains of knockout mice were found to have normal respiration rates.[348]

A study of SN4741 dopaminergic neuronal cells depleted of DJ-1 revealed a disruption of complex I assembly and deficits in mitochondrial electron transport, O_2 consumption, and MMP.[349] Moreover, mitochondria were found to undergo enhanced fission in DJ-1 knockout MEFs and in fibroblasts from familial PD patients with a DJ-1 mutation,[347] as well as in M17 cells overexpressing DJ-1 familial mutants.[350] Data from two additional studies indicated that DJ-1$^{-/-}$ MEFs and primary cortical neurons and M17 cells depleted of DJ-1 *via* shRNA-mediated knockdown exhibited increased mitochondrial fragmentation, and this defect was ameliorated by treating the cells with a membrane-permeable antioxidant or by expressing PINK1 or parkin.[351,352] Another group reported that the recruitment of parkin to depolarized mitochondria in primary cortical neurons and MEFs subjected to oxidative stress was attenuated by expressing wild-type DJ-1 but not the mutant form, C106A.[287] These findings imply that DJ-1 is involved in modulating mitochondrial morphology *via* a pathway that involves the suppression of oxidative stress and occurs in parallel to PINK1/parkin-mediated mitophagy.[287,351–353]

Dopaminergic neurons in the SNpc selectively express the L-type Ca(v)1.3 channel, and expression of this channel alters neuronal firing rates, producing a steady pacemaking rhythm.[354] Here, the term 'pacemaking' refers to the fact that SNpc dopaminergic neurons slowly and regularly fire broad action potentials lasting several milliseconds. Calcium entry through the L-type Ca(v)1.3 channel results in intracellular Ca^{2+} accumulation that may play a role in modulating normal physiological functions specific to SNpc dopaminergic neurons.[354] At the same time, the persistent Ca^{2+} influx associated with pacemaking imposes a metabolic burden on mitochondria involved in buffering Ca^{2+} uptake in SNpc dopaminergic neurons, resulting in potentially harmful mitochondrial dysfunction and oxidative stress.[355,356] Deletion of DJ-1 enhances oxidative stress associated with Ca(v)1.3-mediated Ca^{2+} entry into SNpc dopaminergic neurons, apparently *via* a mechanism involving down-regulation of the mitochondrial uncoupling proteins UCP4 and UCP5, which in turn are involved in attenuating mitochondrial ROS production.[357]

Several observations suggest that DJ-1 can modulate lysosomal autophagy, although different results have been obtained from studies involving different cellular models. One group reported that DJ-1 knockout MEFs showed evidence under basal conditions of impaired lysosomal activity, reduced autophagosome levels (based on LC3-II levels determined *via* Western blotting), reduced colocalization between mitochondria and lysosomes, and an increase in mitochondrial mass compared to fibroblasts from wild-type animals.[347] The impaired autophagic clearance of mitochondria in DJ-1$^{-/-}$ MEFs observed in this study correlated with a decrease in the activation of mitochondrial ERK1/2, suggesting that a loss of DJ-1 function disrupts ERK1/2-mediated autophagy induction. Consistent with these findings, DJ-1 was found to induce autophagy and alleviate neurodegeneration *via* an ERK-dependent mechanism when expressed from an rAAV vector in the SNpc region of rotenone-treated rats.[358] In contrast, results obtained by two other groups suggested that DJ-1 dysfunction led to a *stimulation* of autophagy.

In one study, M17 cells depleted of DJ-1 *via* stable, shRNA-mediated knockdown were found to have elevated autophagosome levels, an effect that was amplified by treating the cells with bafilomycin.[352] In another study, DJ-1 knockout MEFs were found to have reduced basal levels of the autophagy substrate protein p62, and p62 levels were restored upon treating the cells with bafilomycin.[351]

11.4.4 ATP13A2

ATP13A2 is a P-type ATPase encoded by the PARK9 gene, located between the PARK6 and PARK7 loci.[359,360] Mutations in the PARK9 gene are associated with Kufor–Rakeb syndrome (KRS), a rare form of autosomal recessive, early-onset atypical PD with pyramidal degeneration and dementia.[360] There are three alternatively spliced ATP13A2 isoforms expressed in humans.[361] Isoform 1 encodes a protein with 1180 amino acids consisting of 10 transmembrane domains.[362] Isoform 2 encodes a protein of 1175 amino acids.[361] The loss of 5 residues in isoform 2 compared to isoform 1 is the result of an in-frame deletion near the N-terminus.[361] Isoform 3 encodes a protein with 1158 amino acid residues. Compared to isoform 1, this variant has an in-frame deletion of 117 bases, generating a highly diverse C-terminus.[361] Variant 3 is retained in the endoplasmic reticulum (ER) where it is rapidly degraded.[361] Data from qRT-PCR analyses indicate that ATP13A2 is predominantly expressed in the ventral midbrain in human brain,[360] consistent with the pronounced nigral degeneration in the brains of KRS patients. Moreover, ATP13A2 levels are reduced in nigral dopaminergic neurons of sporadic PD patients compared to non-diseased individuals, and the protein has been detected in Lewy bodies.[363] These observations suggest that ATP13A2 dysfunction could be involved in the pathogenesis of PD as well as KRS.

A number of groups have reported that ATP13A2 is localized to lysosomes and early and late endosomes when expressed as an epitope-tagged form or a GFP fusion in mammalian cell lines and in cultured neurons.[360,363–368] Endogenous ATP13A2 was found to be present in multivesicular bodies (MVBs) (a late endosomal compartment) and to amphisomes, structures that result from the fusion of MVBs with autophagosomes, but was not associated with lysosomes.[368] P-type ATPases are characterized by the ability to pump ions across biological membranes *via* a process involving ATP hydrolysis,[362] suggesting that ATP13A2 could play a role in the ATP-dependent transport of cations across endosomal or lysosomal membranes. In support of this idea, deletion of the ATP13A2 homolog in *S. cerevisiae* sensitizes yeast cells to Mn^{2+} toxicity.[369,370] Mammalian cells depleted of ATP13A2 exhibit enhanced sensitivity to Zn^{2+}-mediated cytotoxicity, whereas neuronal cells over-expressing ATP13A2 show increased resistance to the toxic effects of Zn^{2+}, Mn^{2+}, and Cd^{2+}.[368,371–373] Collectively, these data suggest that gene–environment interactions (*i.e.* a combination of the effects of ATP13A2 dysfunction and stress associated with elevated levels of metal ions) play a key role in neurodegeneration in the brains of patients with PARK9 mutations.[369]

The fact that ATP13A2 is localized to endolysosomal compartments has led to the suggestion that the protein could play a role in lysosomal protein degradation.[360] Consistent with this hypothesis, several groups reported that fibroblasts from KRS patients with *PARK9* gene mutations and neuronal cells depleted of ATP13A2 *via* shRNA-mediated knockdown showed evidence of autophagosome and autolysosome accumulation, lysosomal dysfunction (including incomplete processing of the lysosomal proteases cathepsin D and cathepsin B, decreased cathepsin activity, elevated lysosomal pH), defective autophagosome–lysosome fusion, and impaired degradation of lysosomal protein substrates.[363,366,367] Another group found that ATP132A2 knockdown led to a disruption of basal autophagic flux and a suppression of rapamycin-mediated autophagic induction in SH-SY5Y cells or primary cortical neurons.[374] Moreover, the cortices of ATP13A2 knockout mice were characterized by (i) a build-up of lysosomal proteins and lipids and lipofuscin, a lipid-derived autofluorescent pigment; (ii) the presence of ubiquitinated protein aggregates; and (iii) defects in cathepsin D trafficking and maturation.[375,376]

Several groups have reported that ATP13A2 silencing in neuronal cells leads to an increase in basal aSyn levels and a decrease in cell viability.[363,373,377] In one study of ATP13A2 knockout mice, levels of Triton X-100-insoluble aSyn were found to be elevated in the hippocampus, a region of the brain that also showed a marked accumulation of lipofuscin.[375] Additional studies have revealed that ATP13A2 enhances the removal of aSyn from the cell by promoting the formation of extracellular vesicles named exosomes, and ATP13A2-mediated Zn^{2+} import into MVBs may facilitate this clearance process by promoting lipid clustering favoring invagination of the MVB membrane, MVB-autophagosome fusion, or exosome release at the cell membrane.[368,378] Collectively, these findings suggest that ATP13A2 alleviates aSyn neurotoxicity by stimulating the protein's clearance from the cell *via* autophagy or exosome-mediated exocytosis, thus preventing the accumulation of neurotoxic aSyn aggregates.[368,378,379] Consistent with this idea, the over-expression of human ATP13A2 was found to alleviate aSyn-mediated neurodegeneration in a transgenic *C. elegans* model and in rat primary midbrain cultures.[369] In addition, down-regulation of endogenous aSyn was found to alleviate neurotoxicity resulting from ATP13A2 silencing in mouse primary cortical cultures.[366]

Contrary to these findings, studies of a line of ATP13A2 knockout mice (different from that referred to above) revealed no evidence of aSyn neuropathology even at 18 months of age, and mice with a double ATP13A2/aSyn knockout or that over-expressed human A53T aSyn on an ATP13A2 knockout background had similar neuropathological features compared to ATP13A2 knockout mice with a normal level of aSyn expression.[376] Moreover, rats injected intranigrally with AAV encoding aSyn and ATP13A2 were found to have similar neuropathological and behavioral features as rats injected with aSyn virus alone, suggesting that ATP13A2 fails to alleviate aSyn-mediated neurodegeneration in this model.[380] The discrepancies between these results and data from other studies suggesting that ATP13A2 modulates aSyn accumulation and neurotoxicity (see above) likely reflect differences in the nature of the models used to explore ATP13A2/aSyn interactions. Ultimately, achieving insights into whether ATP13A2 dysfunction leads to aSyn neuropathology in KRS will require examining the postmortem brains of patients with PARK9 mutations.

Several lines of evidence suggest that ATP13A2 can ameliorate mitochondrial dysfunction. Two groups reported that ATP13A2 expression led to enhanced viability of human neuroblastoma cells exposed to the complex I inhibitors, rotenone and MPP^+.[371,381] A study of human fibroblasts from KRS patients with ATP13A2 mutations revealed enhanced mitochondrial fragmentation, reduced mitochondrial DNA integrity, decreased ATP production, and elevated maximum respiration rates, and the respiratory chain defect was rescued by co-expressing wild-type ATP13A2.[382] Additional findings indicated that olfactory neurospheres from patients with ATP13A2 mutations were sensitized to Zn^{2+}-mediated mitochondrial defects, including mitochondrial fragmentation and depolarization and a decrease in ATP synthesis rates, and these toxic effects were ameliorated in cells over-expressing wild-type ATP13A2 or treated with an agent that promotes mitochondrial fusion.[372] Other studies revealed that ATP13A2 knockdown induced mitochondrial fragmentation and an increase in mitochondrial mass in SH-SY5Y cells or rat primary cortical neurons[365,374] and triggered a build-up of mitochondrially-derived ROS and an increase in O_2 consumption rates in SH-SY5Y cells.[374]

11.5 Interrelationships Between Autophagic Impairment and Mitochondrial Dysfunction

Data from multiple studies in PD models have led to an improved understanding of interrelationships between autophagic impairment and mitochondrial dysfunction, and how the interplay between these two neurotoxic phenomena is involved in PD pathogenesis. As autophagy plays a central role in the elimination of dysfunctional mitochondria, it is hypothesized that autophagic impairment triggers a build-up of defective mitochondria in the cell, in turn leading to increased ROS accumulation.[54,179,383,384] In support of this idea, siRNA-mediated knockdown of Atg7 leads to an increase in mitochondrial mass, O_2 consumption rates, and mitochondrial H_2O_2 in SH-SY5Y cells.[374] Moreover, MEFs treated with bafilomycin A1 show evidence of a decrease in MMP, presumably because of the accumulation of dysfunctional mitochondria associated with autophagic impairment.[347] Conversely, a build-up of dysfunctional mitochondria that generate ROS is predicted to trigger autophagic impairment by producing an increased lysosomal burden[385] and by causing oxidative damage to endolysosomal components (*e.g.* ROS-mediated LMP).[61,386]

Based on these considerations, one can propose a model to describe how environmental and genetic perturbations can simultaneously disrupt mitochondrial function and autophagy, leading to neurodegeneration in the brains

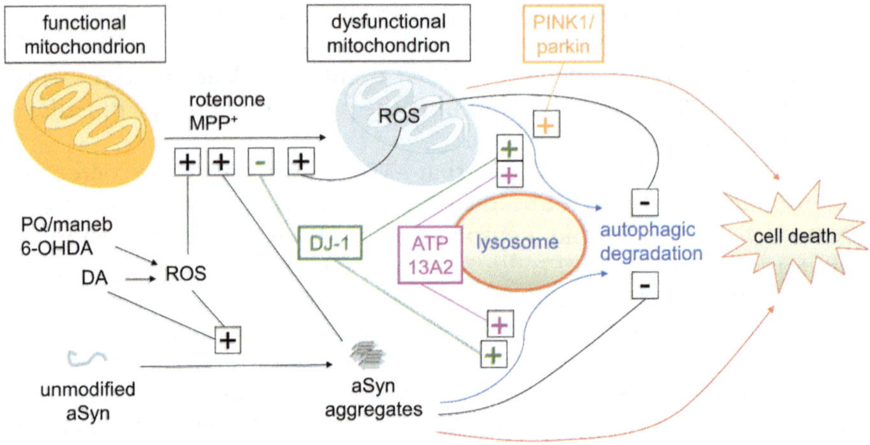

Figure 11.4 Model illustrating environmental and genetic perturbations of the interplay between mitochondrial function and autophagy in PD. See Section 11.4.5 for details.

of PD patients (Figure 11.4). Environmental toxins cause mitochondrial dysfunction either *via* direct inhibition of electron transport (rotenone, MPTP/ MPP^+) or by causing a build-up of cytosolic ROS that damage mitochondrial components (PQ, 6-OHDA). In turn, dysfunctional mitochondria produce high levels of ROS, leading to a vicious cycle of oxidative stress and mitochondrial impairment. A build-up of cytosolic ROS (including ROS produced and released by dysfunctional mitochondria) is predicted to favor aSyn aggregation *via* a mechanism involving the post-translational modification of the protein, and aSyn aggregates can elicit further mitochondrial impairment. Mitochondrial functional defects are attenuated by (i) parkin and PINK1, through their involvement in mitophagy; (ii) DJ-1, through its ability to enhance mitochondrial function, suppress mitochondrial ROS formation, and stimulate autophagy; and (iii) ATP13A2, through its ability to promote autophagic clearance. Moreover, a build-up of potentially neurotoxic aSyn aggregates is alleviated by DJ-1 and ATP13A2 as a result of their stimulatory effects on autophagy. Under conditions of pronounced mitochondrial dysfunction or aSyn aggregation, the autophagic machinery becomes overwhelmed by an excessive clearance burden or impaired as a result of oxidative damage. In turn, this level of dysfunction of cellular quality control systems leads to a vicious cycle involving a further build-up of defective mitochondria and aggregated aSyn, ROS accumulation, and lysosomal dysfunction, ultimately resulting in neurodegeneration.

11.6 Summary and Future Directions

The research findings discussed in this chapter represent important advances in our understanding of the role of mitochondrial dysfunction and autophagic impairment in PD over the past two decades. Neuropathological evidence indicates that reduced complex I activity and impaired lysosomal clearance are characteristic features of the postmortem brains of PD patients. Moreover, data from studies of various PD-related neurotoxins (rotenone, PQ/maneb, MPTP/MPP^+, and 6-OHDA) and of gene products involved in familial PD (aSyn, parkin, PINK1, DJ-1, ATP13A2) strongly support the idea that defects in mitochondrial function and autophagy play a central role in PD pathogenesis. The results of these studies have also revealed a close interrelationship between mitochondrial dysfunction and autophagic impairment and yielded insights into how the interplay between these two toxic phenomena contributes to neurodegeneration in PD.

A goal for future research will be to determine the effects of other recently identified PD-related neurotoxins (*e.g.* heterocyclic amines[387]) and gene products (*e.g.* VPS35 [388–390]) on mitochondrial function and autophagic clearance in PD models. Moreover, it will be important to determine which results from studies in cellular models of PD also hold true *in vivo*. In this regard, an examination of the effects of PINK1 and parkin deficiency in motor neurons of the intact *Drosophila* larval nervous system revealed that (i) a loss of PINK1 or parkin function triggers alterations in the morphology of mitochondria in cell bodies, but not in axons, in contrast to morphological changes observed in both compartments of PINK1- or parkin-deficient neurons *in vitro*; and (ii) parkin-mediated mitophagy appears to be a rare event in mature motor neurons *in vivo*.[391,392] This example underscores the point that discrepancies between *in vitro versus in vivo* data must be resolved, and (by extension) the extent to which observations from preclinical models relate to pathogenic mechanisms in the brains of human patients must be determined.

Insights from the studies examined in this chapter could lay the foundation for developing new therapies aimed at ameliorating mitochondrial dysfunction and autophagic impairment in the brains of PD patients. A number of strategies for alleviating mitochondrial defects are under investigation, including promoting mitochondrial biogenesis,[393] stimulation of PINK1 activity,[394] and inhibition of L-type Ca(v)1.3 channel activity.[395] In addition, various approaches for rescuing autophagic impairment are being explored, including inhibition of mTOR,[61,263] stimulation of transcription factor EB (TFEB, an activator of lysosomal biogenesis),[61,396] interfering with LMP *via* Bax channel inhibition,[163] and promoting lysosomal repair by enhancing the function of apolipoprotein D.[137] Given that over-activation of autophagy has been found to play a role in neurodegeneration in some cellular and animal models of PD,[156,159,166–168] there could in fact be risks associated with therapies aimed at stimulating autophagic flux. Accordingly, an important goal of future research will be to identify conditions in human brain that justify administering pro-autophagy therapies to PD patients. Evidence that autophagy is impaired in neuroblastoma cells with reduced metabolic activity implies that administering drugs that stimulate mitochondrial function and autophagy to PD patients with metabolic deficits in the brain, potentially identified *via* imaging or biomarker analysis,[397] could be a viable therapeutic strategy.[263] Ultimately, pharmacological agents that ameliorate mitochondrial and autophagic impairment could have a highly favorable impact by lowering the risk of PD and/or slowing disease progression in carefully selected patient cohorts.

References

1. J. Parkinson, An essay on the shaking palsy. 1817, *J. Neuropsychiatry Clin. Neurosci.*, 2002, **14**(2), 223–236, discussion 2, PubMed PMID: 11983801. Epub 2002/05/02. eng.
2. J. Jankovic, Parkinson's disease: clinical features and diagnosis, *J. Neurol., Neurosurg. Psychiatry*, 2008, **79**(4), 368–376, PubMed PMID: 18344392. Epub 2008/03/18. eng.
3. M. M. Ponsen, D. Stoffers, J. Booij, B. L. van Eck-Smit, E. Wolters and H. W. Berendse, Idiopathic hyposmia as a preclinical sign of Parkinson's disease, *Ann. Neurol.*, 2004, **56**(2), 173–181, PubMed PMID: 15293269. Epub 2004/08/05. eng.
4. A. Fasano, N. P. Visanji, L. W. Liu, A. E. Lang and R. F. Pfeiffer, Gastrointestinal dysfunction in Parkinson's disease, *Lancet Neurol.*, 2015, **14**(6), 625–639, PubMed PMID: 25987282. Epub 2015/05/20. eng.
5. L. M. Allcock, K. Ullyart, R. A. Kenny and D. J. Burn, Frequency of orthostatic hypotension in a community based cohort of patients with Parkinson's disease, *J. Neurol., Neurosurg. Psychiatry*, 2004, **75**(10), 1470–1471, PubMed PMID: 15377699. Pubmed Central PMCID: 1738761. Epub 2004/09/21. eng.
6. M. D. Gjerstad, G. Alves, T. Wentzel-Larsen, D. Aarsland and J. P. Larsen, Excessive daytime sleepiness in Parkinson disease: is it the drugs or the disease? *Neurology*, 2006, **67**(5), 853–858, PubMed PMID: 16966550. Epub 2006/09/13. eng.
7. D. Aarsland, K. Bronnick, U. Ehrt, P. P. De Deyn, S. Tekin and M. Emre, *et al.*, Neuropsychiatric symptoms in patients with Parkinson's disease and dementia: frequency, profile and associated care giver stress, *J. Neurol., Neurosurg. Psychiatry*, 2007, **78**(1), 36–42, PubMed PMID: 16820421. Pubmed Central PMCID: 2117797. Epub 2006/07/06. eng.
8. R. D. Palmiter, Is dopamine a physiologically relevant mediator of feeding behavior? *Trends Neurosci.*, 2007, **30**(8), 375–381, PubMed PMID: 17604133. Epub 2007/07/03. eng.
9. J. M. Miyasaki, K. Al Hassan, A. E. Lang and V. Voon, Punding prevalence in Parkinson's disease, *Mov. Disord.*, 2007, **22**(8), 1179–1181, PubMed PMID: 17230464. Epub 2007/01/19. eng.
10. D. Aarsland, K. Andersen, J. P. Larsen, A. Lolk, H. Nielsen and P. Kragh-Sorensen, Risk of dementia in Parkinson's disease: a community-based, prospective study, *Neurology*, 2001, **56**(6), 730–736, PubMed PMID: 11274306. Epub 2001/03/29. eng.
11. J. M. Shulman, P. L. De Jager and M. B. Feany, Parkinson's disease: genetics and pathogenesis, *Annu. Rev Pathol*, 2011, **6**, 193–222, PubMed PMID: 21034221. Epub 2010/11/03. eng.
12. M. S. Pollanen, D. W. Dickson and C. Bergeron, Pathology and biology of the Lewy body, *J. Neuropathol. Exp. Neurol.*, 1993, **52**(3), 183–191, PubMed PMID: 7684074. Epub 1993/05/01. eng.

13. M. Baba, S. Nakajo, P. H. Tu, T. Tomita, K. Nakaya and V. M. Lee, *et al.*, Aggregation of alpha-synuclein in Lewy bodies of sporadic Parkinson's disease and dementia with Lewy bodies, *Am. J. Pathol.*, 1998, **152**(4), 879–884, PubMed PMID: 9546347. Pubmed Central PMCID: 1858234. Epub 1998/04/18. eng.
14. M. G. Spillantini, M. L. Schmidt, V. M. Lee, J. Q. Trojanowski, R. Jakes and M. Goedert, Alpha-synuclein in Lewy bodies, *Nature*, 1997, **388**(6645), 839–840, PubMed PMID: 9278044. Epub 1997/08/28. eng.
15. H. Braak, K. Del Tredici, U. Rub, R. A. de Vos, E. N. Jansen Steur and E. Braak, Staging of brain pathology related to sporadic Parkinson's disease, *Neurobiol. Aging*, 2003, **24**(2), 197–211, PubMed PMID: 12498954. Epub 2002/12/25. eng.
16. H. Braak, E. Ghebremedhin, U. Rub, H. Bratzke and K. Del Tredici, Stages in the development of Parkinson's disease-related pathology, *Cell Tissue Res.*, 2004, **318**(1), 121–134, PubMed PMID: 15338272. Epub 2004/09/01. eng.
17. T. Gasser, Usefulness of Genetic Testing in PD and PD Trials: A Balanced Review, *J. Parkinsons Dis.*, 2015, **5**(2), 209–215, PubMed PMID: 25624421. Pubmed Central PMCID: PMC4923738. Epub 2015/01/28. eng.
18. O. Hwang, Role of oxidative stress in Parkinson's disease, *Exp. Neurobiol.*, 2013, **22**(1), 11–17, PubMed PMID: 23585717. Pubmed Central PMCID: 3620453. Epub 2013/04/16. eng.
19. A. H. Schapira and P. Jenner, Etiology and pathogenesis of Parkinson's disease, *Mov. Disord.*, 2011, **26**(6), 1049–1055, PubMed PMID: 21626550. Epub 2011/06/01. eng.
20. J. A. Obeso, M. C. Rodriguez-Oroz, C. G. Goetz, C. Marin, J. H. Kordower and M. Rodriguez, *et al.*, Missing pieces in the Parkinson's disease puzzle, *Nat. Med.*, 2010, **16**(6), 653–661, PubMed PMID: 20495568. Epub 2010/05/25. eng.
21. A. Elbaz, J. Clavel, P. J. Rathouz, F. Moisan, J. P. Galanaud and B. Delemotte, *et al.*, Professional exposure to pesticides and Parkinson disease, *Ann. Neurol.*, 2009, **66**(4), 494–504, PubMed PMID: 19847896. Epub 2009/10/23. eng.
22. J. M. Gorell, C. C. Johnson, B. A. Rybicki, E. L. Peterson and R. J. Richardson, The risk of Parkinson's disease with exposure to pesticides, farming, well water, and rural living, *Neurology*, 1998, **50**(5), 1346–1350, PubMed PMID: 9595985. Epub 1998/05/22. eng.
23. A. S. Dhillon, G. L. Tarbutton, J. L. Levin, G. M. Plotkin, L. K. Lowry and J. T. Nalbone, *et al.*, Pesticide/environmental exposures and Parkinson's disease in East Texas, *J. Agromedicine*, 2008, **13**(1), 37–48, PubMed PMID: 19042691. Epub 2008/12/02. eng.
24. K. M. Semchuk, E. J. Love and R. G. Lee, Parkinson's disease and exposure to agricultural work and pesticide chemicals, *Neurology*, 1992, **42**(7), 1328–1335, PubMed PMID: 1620342. Epub 1992/07/01. eng.

25. D. G. Hernandez, X. Reed and A. B. Singleton, Genetics in Parkinson disease: Mendelian *versus* non-Mendelian inheritance, *J. Neurochem.*, 2016, **139**(suppl. 1), 59–74, PubMed PMID: 27090875. Pubmed Central PMCID: PMC5155439. Epub 2016/04/20. eng.
26. J. Trinh and M. Farrer, Advances in the genetics of Parkinson disease, *Nat. Rev. Neurol.*, 2013, **9**(8), 445–454, PubMed PMID: 23857047. Epub 2013/07/17. eng.
27. A. E. Lang and A. M. Lozano, Parkinson's disease. First of two parts, *N. Engl. J. Med.*, 1998, **339**(15), 1044–1053, PubMed PMID: 9761807. Epub 1998/10/08. eng.
28. L. M. de Lau, P. C. Giesbergen, M. C. de Rijk, A. Hofman, P. J. Koudstaal and M. M. Breteler, Incidence of parkinsonism and Parkinson disease in a general population: the Rotterdam Study, *Neurology*, 2004, **63**(7), 1240–1244, PubMed PMID: 15477545. Epub 2004/10/13. eng.
29. S. L. Kowal, T. M. Dall, R. Chakrabarti, M. V. Storm and A. Jain, The current and projected economic burden of Parkinson's disease in the United States, *Mov. Disord.*, 2013, **28**(3), 311–318, PubMed PMID: 23436720. Epub 2013/02/26. eng.
30. J. Jankovic and W. Poewe, Therapies in Parkinson's disease, *Curr. Opin. Neurol.*, 2012, **25**(4), 433–447, PubMed PMID: 22691758. Epub 2012/06/14. eng.
31. P. Siekevitz, Powerhouse of the cell, *Sci. Am.*, 1957, **197**(1), 131–140.
32. A. H. Schapira, J. M. Cooper, D. Dexter, P. Jenner, J. B. Clark and C. D. Marsden, Mitochondrial complex I deficiency in Parkinson's disease, *Lancet*, 1989, **333**(8649), 1269, PubMed PMID: 2566813. Epub 1989/06/03. eng.
33. W. D. Parker Jr, J. K. Parks and R. H. Swerdlow, Complex I deficiency in Parkinson's disease frontal cortex, *Brain Res.*, 2008, **1189**, 215–218, PubMed PMID: 18061150. Pubmed Central PMCID: 2295283. Epub 2007/12/07. eng.
34. J. T. Greenamyre, T. B. Sherer, R. Betarbet and A. V. Panov, Complex I and Parkinson's disease, *IUBMB Life*, 2001, **52**(3–5), 135–141, PubMed PMID: 11798025. Epub 2002/01/19. eng.
35. P. M. Keeney, J. Xie, R. A. Capaldi and J. P. Bennett Jr, Parkinson's disease brain mitochondrial complex I has oxidatively damaged subunits and is functionally impaired and misassembled, *J. Neurosci.*, 2006, **26**(19), 5256–5264, PubMed PMID: 16687518. Epub 2006/05/12. eng.
36. H. Yoshino, Y. Nakagawa-Hattori, T. Kondo and Y. Mizuno, Mitochondrial complex I and II activities of lymphocytes and platelets in Parkinson's disease, *J. Neural Transm.: Parkinson's Dis. Dementia Sect.*, 1992, **4**(1), 27–34, PubMed PMID: 1347219. Epub 1992/01/01. eng.
37. O. Blin, C. Desnuelle, O. Rascol, M. Borg, H. Peyro Saint Paul and J. P. Azulay, *et al.*, Mitochondrial respiratory failure in skeletal muscle from patients with Parkinson's disease and multiple system atrophy, *J. Neurol. Sci.*, 1994, **125**(1), 95–101, PubMed PMID: 7964895. Epub 1994/08/01. eng.

38. T. B. Sherer, R. Betarbet, C. M. Testa, B. B. Seo, J. R. Richardson and J. H. Kim, *et al.*, Mechanism of toxicity in rotenone models of Parkinson's disease, *J. Neurosci.*, 2003, **23**(34), 10756–10764, PubMed PMID: 14645467. Epub 2003/12/03. eng.
39. L. Tretter, I. Sipos and V. Adam-Vizi, Initiation of neuronal damage by complex I deficiency and oxidative stress in Parkinson's disease, *Neurochem. Res.*, 2004, **29**(3), 569–577, PubMed PMID: 15038604. Epub 2004/03/25. eng.
40. M. T. Lin and M. F. Beal, Mitochondrial dysfunction and oxidative stress in neurodegenerative diseases, *Nature*, 2006, **443**(7113), 787–795, PubMed PMID: 17051205. Epub 2006/10/20. eng.
41. R. K. Chaturvedi and M. Flint Beal, Mitochondrial diseases of the brain, *Free Radical Biol. Med.*, 2013, **63**, 1–29, PubMed PMID: 23567191. Epub 2013/04/10. eng.
42. Z. I. Alam, A. Jenner, S. E. Daniel, A. J. Lees, N. Cairns and C. D. Marsden, *et al.*, Oxidative DNA damage in the parkinsonian brain: an apparent selective increase in 8-hydroxyguanine levels in substantia nigra, *J. Neurochem.*, 1997, **69**(3), 1196–1203, PubMed PMID: 9282943. Epub 1997/09/01. eng.
43. R. Betarbet, T. B. Sherer, G. MacKenzie, M. Garcia-Osuna, A. V. Panov and J. T. Greenamyre, Chronic systemic pesticide exposure reproduces features of Parkinson's disease, *Nat. Neurosci.*, 2000, **3**(12), 1301–1306, PubMed PMID: 11100151. Epub 2000/12/02. eng.
44. J. Zhang, G. Perry, M. A. Smith, D. Robertson, S. J. Olson and D. G. Graham, *et al.*, Parkinson's disease is associated with oxidative damage to cytoplasmic DNA and RNA in substantia nigra neurons, *Am. J. Pathol.*, 1999, **154**(5), 1423–1429, PubMed PMID: 10329595. Pubmed Central PMCID: 1866598. Epub 1999/05/18. eng.
45. L. H. Sanders and J. Timothy Greenamyre, Oxidative damage to macromolecules in human Parkinson disease and the rotenone model, *Free Radical Biol. Med.*, 2013, **62**, 111–120, PubMed PMID: 23328732. Pubmed Central PMCID: 3677955. Epub 2013/01/19. eng.
46. D. G. Graham, S. M. Tiffany, W. R. Bell Jr and W. F. Gutknecht, Autoxidation *versus* covalent binding of quinones as the mechanism of toxicity of dopamine, 6-hydroxydopamine, and related compounds toward C1300 neuroblastoma cells *in vitro*, *Mol. Pharmacol.*, 1978, **14**(4), 644–653, PubMed PMID: 567274. Epub 1978/07/01. eng.
47. P. Munoz, S. Huenchuguala, I. Paris and J. Segura-Aguilar, Dopamine oxidation and autophagy, *Parkinson's Dis.*, 2012, **2012**, 920953, PubMed PMID: 22966478. Pubmed Central PMCID: 3433151. Epub 2012/09/12. eng.
48. S. B. Berman and T. G. Hastings, Dopamine oxidation alters mitochondrial respiration and induces permeability transition in brain mitochondria: implications for Parkinson's disease, *J. Neurochem.*, 1999, **73**(3), 1127–1137, PubMed PMID: 10461904. Epub 1999/08/26. eng.

49. S. Kaushik and A. M. Cuervo, Chaperone-mediated autophagy: a unique way to enter the lysosome world, *Trends Cell Biol.*, 2012, **22**(8), 407–417, PubMed PMID: 22748206. Pubmed Central PMCID: PMC3408550. Epub 2012/07/04. eng.
50. K. R. Parzych and D. J. Klionsky, An overview of autophagy: morphology, mechanism, and regulation, *Antioxid. Redox Signaling*, 2014, **20**(3), 460–473, PubMed PMID: 23725295. Pubmed Central PMCID: PMC3894687. Epub 2013/06/04. eng.
51. B. Ravikumar, S. Sarkar, J. E. Davies, M. Futter, M. Garcia-Arencibia and Z. W. Green-Thompson, *et al.*, Regulation of mammalian autophagy in physiology and pathophysiology, *Physiol. Rev.*, 2010, **90**(4), 1383–1435, PubMed PMID: 20959619. Epub 2010/10/21. eng.
52. J. Navarro-Yepes, M. Burns, A. Anandhan, O. Khalimonchuk, L. M. del Razo and B. Quintanilla-Vega, *et al.*, Oxidative stress, redox signaling, and autophagy: cell death *versus* survival, *Antioxid. Redox Signaling*, 2014, **21**(1), 66–85, PubMed PMID: 24483238. Pubmed Central PMCID: PMC4048575. Epub 2014/02/04. eng.
53. Y. Kabeya, N. Mizushima, T. Uero, A. Yamamoto, T. Kirisako and T. Noda, *et al.*, LC3, a mammalian homologue of yeast Apg8p, is localized in autophagosome membranes after processing, *EMBO J.*, 2000, **19**(21), 5720–5728, PubMed PMID: ISI:000165235000013. English.
54. J. Lee, S. Giordano and J. Zhang, Autophagy, mitochondria and oxidative stress: cross-talk and redox signalling, *Biochem. J.*, 2012, **441**(2), 523–540, PubMed PMID: 22187934. Pubmed Central PMCID: PMC3258656. Epub 2011/12/23. eng.
55. B. J. Ryan, S. Hoek, E. A. Fon and R. Wade-Martins, Mitochondrial dysfunction and mitophagy in Parkinson's: from familial to sporadic disease, *Trends Biochem. Sci.*, 2015, **40**(4), 200–210, PubMed PMID: 25757399. Epub 2015/03/12. eng.
56. R. J. Youle and D. P. Narendra, Mechanisms of mitophagy, *Nat. Rev. Mol. Cell Biol.*, 2011, **12**(1), 9–14, PubMed PMID: 21179058. Pubmed Central PMCID: PMC4780047. Epub 2010/12/24. eng.
57. M. Lippai and P. Low, The role of the selective adaptor p62 and ubiquitin-like proteins in autophagy, *BioMed Res. Int.*, 2014, **2014**, 832704, PubMed PMID: 25013806. Pubmed Central PMCID: PMC4075091. Epub 2014/07/12. eng.
58. A. Stolz, A. Ernst and I. Dikic, Cargo recognition and trafficking in selective autophagy, *Nat. Cell Biol.*, 2014, **16**(6), 495–501, PubMed PMID: 24875736. Epub 2014/05/31. eng.
59. Y. Chu, H. Dodiya, P. Aebischer, C. W. Olanow and J. H. Kordower, Alterations in lysosomal and proteasomal markers in Parkinson's disease: relationship to alpha-synuclein inclusions, *Neurobiol. Dis.*, 2009, **35**(3), 385–398, PubMed PMID: 19505575. Epub 2009/06/10. eng.
60. L. Crews, B. Spencer, P. Desplats, C. Patrick, A. Paulino and E. Rockenstein, *et al.*, Selective molecular alterations in the autophagy pathway in patients with Lewy body disease and in models of alpha-synucleinopathy,

PLoS One, 2010, **5**(2), e9313, PubMed PMID: 20174468. Pubmed Central PMCID: PMC2824828. Epub 2010/02/23. eng.

61. B. Dehay, J. Bove, N. Rodriguez-Muela, C. Perier, A. Recasens and P. Boya, *et al.*, Pathogenic lysosomal depletion in Parkinson's disease, *J. Neurosci.*, 2010, **30**(37), 12535–12544, PubMed PMID: 20844148. Epub 2010/09/17. eng.
62. K. Tanji, F. Mori, A. Kakita, H. Takahashi and K. Wakabayashi, Alteration of autophagosomal proteins (LC3, GABARAP and GATE-16) in Lewy body disease, *Neurobiol. Dis.*, 2011, **43**(3), 690–697, PubMed PMID: 21684337. Epub 2011/06/21. eng.
63. G. K. Tofaris, Lysosome-dependent pathways as a unifying theme in Parkinson's disease, *Mov. Disord.*, 2012, **27**(11), 1364–1369, PubMed PMID: 22927213. Epub 2012/08/29. eng.
64. K. Wakabayashi, K. Tanji, S. Odagiri, Y. Miki, F. Mori and H. Takahashi, The Lewy body in Parkinson's disease and related neurodegenerative disorders, *Mol. Neurobiol.*, 2013, **47**(2), 495–508, PubMed PMID: 22622968. Epub 2012/05/25. eng.
65. Y. Miki, K. Tanji, F. Mori, J. Utsumi, H. Sasaki and A. Kakita, *et al.*, Alteration of Upstream Autophagy-Related Proteins (ULK1, ULK2, Beclin1, VPS34 and AMBRA1) in Lewy Body Disease, *Brain Pathol.*, 2016, **26**(3), 359–370, PubMed PMID: 26260450. Epub 2015/08/12. eng.
66. D. Toulorge, A. H. Schapira and R. Hajj, Molecular changes in the postmortem parkinsonian brain, *J. Neurochem.*, 2016, **139**(suppl. 1), 27–58, PubMed PMID: 27381749. Epub 2016/07/07. eng.
67. R. K. Dagda, T. Das Banerjee and E. Janda, How Parkinsonian toxins dysregulate the autophagy machinery, *Int. J. Mol. Sci.*, 2013, **14**(11), 22163–22189, PubMed PMID: 24217228. Pubmed Central PMCID: PMC3856058. Epub 2013/11/13. eng.
68. D. C. Rubinsztein, A. M. Cuervo, B. Ravikumar, S. Sarkar, V. Korolchuk and S. Kaushik, *et al.*, In search of an "autophagomometer", *Autophagy*, 2009, **5**(5), 585–589, PubMed PMID: 19411822. Epub 2009/05/05. eng.
69. N. Mizushima, T. Yoshimori and B. Levine, Methods in mammalian autophagy research, *Cell*, 2010, **140**(3), 313–326, PubMed PMID: 20144757. Pubmed Central PMCID: 2852113. Epub 2010/02/11. eng.
70. J. Klucken, A. M. Poehler, D. Ebrahimi-Fakhari, J. Schneider, S. Nuber and E. Rockenstein, *et al.*, Alpha-synuclein aggregation involves a bafilomycin A 1-sensitive autophagy pathway, *Autophagy*, 2012, **8**(5), 754–766, PubMed PMID: 22647715. Pubmed Central PMCID: PMC3378419. Epub 2012/06/01. eng.
71. B. Spencer, R. Potkar, M. Trejo, E. Rockenstein, C. Patrick and R. Gindi, *et al.*, Beclin 1 gene transfer activates autophagy and ameliorates the neurodegenerative pathology in alpha-synuclein models of Parkinson's and Lewy body diseases, *J. Neurosci.*, 2009, **29**(43), 13578–13588, PubMed PMID: 19864570. Pubmed Central PMCID: PMC2812014. Epub 2009/10/30. eng.

72. C. M. Tanner, F. Kamel, G. W. Ross, J. A. Hoppin, S. M. Goldman and M. Korell, *et al.*, Rotenone, Paraquat and Parkinson's Disease, *Environ. Health Perspect.*, 2011, **119**(6), 866–872, PubMed PMID: 21269927. Epub 2011/01/29. Eng.
73. M. Thiruchelvam, E. K. Richfield, R. B. Baggs, A. W. Tank and D. A. Cory-Slechta, The nigrostriatal dopaminergic system as a preferential target of repeated exposures to combined paraquat and maneb: implications for Parkinson's disease, *J. Neurosci.*, 2000, **20**(24), 9207–9214, PubMed PMID: 11124998. Epub 2000/01/11. eng.
74. S. Costello, M. Cockburn, J. Bronstein, X. Zhang and B. Ritz, Parkinson's disease and residential exposure to maneb and paraquat from agricultural applications in the central valley of California, *Am. J. Epidemiol.*, 2009, **169**(8), 919–926, PubMed PMID: 19270050. Pubmed Central PMCID: PMC2727231. Epub 2009/03/10. eng.
75. A. Wang, S. Costello, M. Cockburn, X. Zhang, J. Bronstein and B. Ritz, Parkinson's disease risk from ambient exposure to pesticides, *Eur. J. Epidemiol.*, 2011, **26**(7), 547–555, PubMed PMID: 21505849. Pubmed Central PMCID: PMC3643971. Epub 2011/04/21. eng.
76. J. W. Langston, P. Ballard, J. W. Tetrud and I. Irwin, Chronic Parkinsonism in humans due to a product of meperidine-analog synthesis, *Science*, 1983, **219**(4587), 979–980, PubMed PMID: 6823561. Epub 1983/02/25. eng.
77. J. Blesa, S. Phani, V. Jackson-Lewis and S. Przedborski, Classic and new animal models of Parkinson's disease, *J. Biomed. Biotechnol.*, 2012, **2012**, 845618, PubMed PMID: 22536024. Pubmed Central PMCID: PMC3321500. Epub 2012/04/27. eng.
78. J. Bove and C. Perier, Neurotoxin-based models of Parkinson's disease, *Neuroscience*, 2012, **211**, 51–76, PubMed PMID: 22108613. Epub 2011/11/24. eng.
79. J. R. Cannon and J. T. Greenamyre, Neurotoxic *in vivo* models of Parkinson's disease recent advances, *Prog. Brain Res.*, 2010, **184**, 17–33, PubMed PMID: 20887868. Epub 2010/10/05. eng.
80. G. E. Meredith, P. K. Sonsalla and M. F. Chesselet, Animal models of Parkinson's disease progression, *Acta Neuropathol.*, 2008, **115**(4), 385–398, PubMed PMID: 18273623. Pubmed Central PMCID: PMC2562277. Epub 2008/02/15. eng.
81. J. Hirst, Towards the molecular mechanism of respiratory complex I, *Biochem. J.*, 2009, **425**(2), 327–339, PubMed PMID: 20025615. Epub 2009/12/23. eng.
82. V. Zickermann, S. Kerscher, K. Zwicker, M. A. Tocilescu, M. Radermacher and U. Brandt, Architecture of complex I and its implications for electron transfer and proton pumping, *Biochim. Biophys. Acta*, 2009, **1787**(6), 574–583, PubMed PMID: 19366614. Pubmed Central PMCID: PMC2699368. Epub 2009/04/16. eng.
83. E. Darrouzet, J. P. Issartel, J. Lunardi and A. Dupuis, The 49-kDa subunit of NADH-ubiquinone oxidoreductase (Complex I) is involved in the binding of piericidin and rotenone, two quinone-related inhibitors, *FEBS Lett.*, 1998, **431**(1), 34–38, PubMed PMID: 9684860. Epub 1998/07/31. eng.

84. M. A. Tocilescu, V. Zickermann, K. Zwicker and U. Brandt, Quinone binding and reduction by respiratory complex I., *Biochim. Biophys. Acta*, 2010, **1797**(12), 1883–1890, PubMed PMID: 20493164. Epub 2010/05/25. eng.
85. S. E. Stephans, G. W. Miller, A. I. Levey and J. T. Greenamyre, Acute mitochondrial and chronic toxicological effects of 1-methyl-4-phenylpyridinium in human neuroblastoma cells, *Neurotoxicology*, 2002, **23**(4–5), 569–580, PubMed PMID: 12428729. Epub 2002/11/14. eng.
86. C. M. Testa, T. B. Sherer and J. T. Greenamyre, Rotenone induces oxidative stress and dopaminergic neuron damage in organotypic substantia nigra cultures, *Brain Res. Mol. Brain Res.*, 2005, **134**(1), 109–118, PubMed PMID: 15790535. Epub 2005/03/26. eng.
87. M. Alam and W. J. Schmidt, Rotenone destroys dopaminergic neurons and induces parkinsonian symptoms in rats, *Behav. Brain Res.*, 2002, **136**(1), 317–324, PubMed PMID: 12385818. Epub 2002/10/19. eng.
88. J. R. Cannon, V. Tapias, H. M. Na, A. S. Honick, R. E. Drolet and J. T. Greenamyre, A highly reproducible rotenone model of Parkinson's disease, *Neurobiol. Dis.*, 2009, **34**(2), 279–290, PubMed PMID: 19385059. Pubmed Central PMCID: 2757935. Epub 2009/04/23. eng.
89. T. B. Sherer, R. Betarbet, A. K. Stout, S. Lund, M. Baptista and A. V. Panov, *et al.*, An *in vitro* model of Parkinson's disease: linking mitochondrial impairment to altered alpha-synuclein metabolism and oxidative damage, *J. Neurosci.*, 2002, **22**(16), 7006–7015, PubMed PMID: 12177198. Epub 2002/08/15. eng.
90. M. Watabe and T. Nakaki, ATP depletion does not account for apoptosis induced by inhibition of mitochondrial electron transport chain in human dopaminergic cells, *Neuropharmacology*, 2007, **52**(2), 536–541, PubMed PMID: 17027047. Epub 2006/10/10. eng.
91. M. Marella, B. B. Seo, E. Nakamaru-Ogiso, J. T. Greenamyre, A. Matsuno-Yagi and T. Yagi, Protection by the NDI1 gene against neurodegeneration in a rotenone rat model of Parkinson's disease, *PLoS One*, 2008, **3**(1), e1433, PubMed PMID: 18197244. Pubmed Central PMCID: PMC2175531. Epub 2008/01/17. eng.
92. F. Liu, J. L. Nguyen, J. D. Hulleman, L. Li and J.-C. Rochet, Mechanisms of DJ-1 neuroprotection in a cellular model of Parkinson's disease, *J. Neurochem.*, 2008, **105**, 2435–2453.
93. F. Wu, H. D. Xu, J. J. Guan, Y. S. Hou, J. H. Gu and X. C. Zhen, *et al.*, Rotenone impairs autophagic flux and lysosomal functions in Parkinson's disease, *Neuroscience*, 2015, **284**, 900–911, PubMed PMID: 25446361. Epub 2014/12/03. eng.
94. N. Xiong, J. Xiong, M. Jia, L. Liu, X. Zhang and Z. Chen, *et al.*, The role of autophagy in Parkinson's disease: rotenone-based modeling, *Behav. Brain Funct.*, 2013, **9**, 13, PubMed PMID: 23497442. Pubmed Central PMCID: PMC3606411. Epub 2013/03/19. eng.
95. B. J. Mader, V. N. Pivtoraiko, H. M. Flippo, B. J. Klocke, K. A. Roth and L. R. Mangieri, *et al.*, Rotenone inhibits autophagic flux prior to inducing cell death, *ACS Chem. Neurosci.*, 2012, **3**(12), 1063–1072, PubMed PMID: 23259041. Pubmed Central PMCID: PMC3526971. Epub 2012/12/22. eng.

96. S. Giordano, M. Dodson, S. Ravi, M. Redmann, X. Ouyang and V. M. Darley Usmar, *et al.*, Bioenergetic adaptation in response to autophagy regulators during rotenone exposure, *J. Neurochem.*, 2014, **131**(5), 625–633, PubMed PMID: 25081478. Pubmed Central PMCID: PMC4454361. Epub 2014/08/02. eng.
97. T. Pan, P. Rawal, Y. Wu, W. Xie, J. Jankovic and W. Le, Rapamycin protects against rotenone-induced apoptosis through autophagy induction, *Neuroscience*, 2009, **164**(2), 541–551, PubMed PMID: 19682553. Epub 2009/08/18. eng.
98. K. Radad, R. Moldzio and W. D. Rausch, Rapamycin protects dopaminergic neurons against rotenone-induced cell death in primary mesencephalic cell culture, *Folia Neuropathol.*, 2015, **53**(3), 250–261, PubMed PMID: 26443316. Epub 2015/10/08. eng.
99. Y. N. Deng, J. Shi, J. Liu and Q. M. Qu, Celastrol protects human neuroblastoma SH-SY5Y cells from rotenone-induced injury through induction of autophagy, *Neurochem. Int.*, 2013, **63**(1), 1–9, PubMed PMID: 23619395. Epub 2013/04/27. eng.
100. Y. Wu, X. Li, W. Xie, J. Jankovic, W. Le and T. Pan, Neuroprotection of deferoxamine on rotenone-induced injury *via* accumulation of HIF-1 alpha and induction of autophagy in SH-SY5Y cells, *Neurochem. Int.*, 2010, **57**(3), 198–205, PubMed PMID: 20546814. Epub 2010/06/16. eng.
101. W. Jang, H. J. Kim, H. Li, K. D. Jo, M. K. Lee and H. O. Yang, The Neuroprotective Effect of Erythropoietin on Rotenone-Induced Neurotoxicity in SH-SY5Y Cells Through the Induction of Autophagy, *Mol. Neurobiol.*, 2016, **53**(6), 3812–3821, PubMed PMID: 26156288. Epub 2015/07/15. eng.
102. G. Filomeni, I. Graziani, D. De Zio, L. Dini, D. Centonze and G. Rotilio, *et al.*, Neuroprotection of kaempferol by autophagy in models of rotenone-mediated acute toxicity: possible implications for Parkinson's disease, *Neurobiol. Aging*, 2012, **33**(4), 767–785, PubMed PMID: 20594614. Epub 2010/07/03. eng.
103. L. Hou, N. Xiong, L. Liu, J. Huang, C. Han and G. Zhang, *et al.*, Lithium protects dopaminergic cells from rotenone toxicity *via* autophagy enhancement, *BMC Neurosci.*, 2015, **16**, 82, PubMed PMID: 26608648. Pubmed Central PMCID: PMC4658766. Epub 2015/11/27. eng.
104. R. Pal, L. Bajaj, J. Sharma, M. Palmieri, A. Di Ronza and P. Lotfi, *et al.*, NADPH oxidase promotes Parkinsonian phenotypes by impairing autophagic flux in an mTORC1-independent fashion in a cellular model of Parkinson's disease, *Sci. Rep.*, 2016, **6**, 22866, PubMed PMID: 26960433. Pubmed Central PMCID: PMC4785399. Epub 2016/03/11. eng.
105. L. Bonet-Ponce, S. Saez-Atienzar, C. da Casa, J. Sancho-Pelluz, J. M. Barcia and N. Martinez-Gil, *et al.*, Rotenone Induces the Formation of 4-Hydroxynonenal Aggresomes. Role of ROS-Mediated Tubulin Hyperacetylation and Autophagic Flux Disruption, *Mol. Neurobiol.*, 2016, **53**(9), 6194–6208, PubMed PMID: 26558631. Epub 2016/11/01. eng.

106. Y. W. Yap, M. J. Chen, Z. F. Peng, J. Manikandan, J. M. Ng and R. M. Llanos, *et al.*, Gene expression profiling of rotenone-mediated cortical neuronal death: evidence for inhibition of ubiquitin-proteasome system and autophagy-lysosomal pathway, and dysfunction of mitochondrial and calcium signaling, *Neurochem. Int.*, 2013, **62**(5), 653–663, PubMed PMID: 23186747. Epub 2012/11/29. eng.
107. C. T. Chu, J. Ji, R. K. Dagda, J. F. Jiang, Y. Y. Tyurina and A. A. Kapralov, *et al.*, Cardiolipin externalization to the outer mitochondrial membrane acts as an elimination signal for mitophagy in neuronal cells, *Nat. Cell Biol.*, 2013, **15**(10), 1197–1205, PubMed PMID: 24036476. Pubmed Central PMCID: PMC3806088. Epub 2013/09/17. eng.
108. C. T. Chu, H. Bayir and V. E. Kagan, LC3 binds externalized cardiolipin on injured mitochondria to signal mitophagy in neurons: implications for Parkinson disease, *Autophagy*, 2014, **10**(2), 376–378, PubMed PMID: 24351649. Epub 2013/12/20. eng.
109. B. J. Day, M. Patel, L. Calavetta, L. Y. Chang and J. S. Stamler, A mechanism of paraquat toxicity involving nitric oxide synthase, *Proc. Natl. Acad. Sci. U. S. A.*, 1999, **96**(22), 12760–12765, PubMed PMID: 10535996. Pubmed Central PMCID: 23088. Epub 1999/10/27. eng.
110. D. Bonneh-Barkay, W. J. Langston and D. A. Di Monte, Toxicity of redox cycling pesticides in primary mesencephalic cultures, *Antioxid. Redox Signaling*, 2005, 7(5–6), 649–653, PubMed PMID: 15890009. Epub 2005/05/14. eng.
111. D. Bonneh-Barkay, S. H. Reaney, W. J. Langston and D. A. Di Monte, Redox cycling of the herbicide paraquat in microglial cultures, *Brain Res. Mol. Brain Res.*, 2005, **134**(1), 52–56, PubMed PMID: 15790529. Epub 2005/03/26. eng.
112. V. A. Fitsanakis, V. Amarnath, J. T. Moore, K. S. Montine, J. Zhang and T. J. Montine, Catalysis of catechol oxidation by metal-dithiocarbamate complexes in pesticides, *Free Radical Biol. Med.*, 2002, **33**(12), 1714–1723, PubMed PMID: 12488139. Epub 2002/12/19. eng.
113. Y. Zhou, F. S. Shie, P. Piccardo, T. J. Montine and J. Zhang, Proteasomal inhibition induced by manganese ethylene-bis-dithiocarbamate: relevance to Parkinson's disease, *Neuroscience*, 2004, **128**(2), 281–291, PubMed PMID: 15350641. Epub 2004/09/08. eng.
114. J. Wills, J. Credle, A. W. Oaks, V. Duka, J. H. Lee and J. Jones, *et al.*, Paraquat, but not maneb, induces synucleinopathy and tauopathy in striata of mice through inhibition of proteasomal and autophagic pathways, *PLoS One*, 2012, 7(1), e30745, PubMed PMID: 22292029. Pubmed Central PMCID: PMC3264632. Epub 2012/02/01. eng.
115. J. Peng, F. F. Stevenson, S. R. Doctrow and J. K. Andersen, Superoxide dismutase/catalase mimetics are neuroprotective against selective paraquat-mediated dopaminergic neuron death in the substantial nigra: implications for Parkinson disease, *J. Biol. Chem.*, 2005, **280**(32), 29194–29198, PubMed PMID: 15946937. Epub 2005/06/11. eng.

116. A. de Rus Jacquet, M. A. Tambe, S. Y. Ma, G. P. McCabe, J. H. Vest and J. C. Rochet, Pikuni-Blackfeet traditional medicine: Neuroprotective activities of medicinal plants used to treat Parkinson's disease-related symptoms, *J. Ethnopharmacol.*, 2017, **206**, 393–407.
117. A. I. Brooks, C. A. Chadwick, H. A. Gelbard, D. A. Cory-Slechta and H. J. Federoff, Paraquat elicited neurobehavioral syndrome caused by dopaminergic neuron loss, *Brain Res.*, 1999, **823**(1–2), 1–10, PubMed PMID: 10095006. Epub 1999/03/30. eng.
118. A. L. McCormack, M. Thiruchelvam, A. B. Manning-Bog, C. Thiffault, J. W. Langston and D. A. Cory-Slechta, *et al.*, Environmental risk factors and Parkinson's disease: selective degeneration of nigral dopaminergic neurons caused by the herbicide paraquat, *Neurobiol. Dis.*, 2002, **10**(2), 119–127, PubMed PMID: 12127150.
119. M. Thiruchelvam, A. McCormack, E. K. Richfield, R. B. Baggs, A. W. Tank and D. A. Di Monte, *et al.*, Age-related irreversible progressive nigrostriatal dopaminergic neurotoxicity in the paraquat and maneb model of the Parkinson's disease phenotype, *Eur. J. Neurosci.*, 2003, **18**(3), 589–600, PubMed PMID: 12911755. Epub 2003/08/13. eng.
120. A. B. Manning-Bog, A. L. McCormack, J. Li, V. N. Uversky, A. L. Fink and D. A. Di Monte, The herbicide paraquat causes up-regulation and aggregation of alpha-synuclein in mice: paraquat and alpha-synuclein, *J. Biol. Chem.*, 2002, **277**(3), 1641–1644, PubMed PMID: 11707429. Epub 2001/11/15. eng.
121. A. L. McCormack, J. G. Atienza, L. C. Johnston, J. K. Andersen, S. Vu and D. A. Di Monte, Role of oxidative stress in paraquat-induced dopaminergic cell degeneration, *J. Neurochem.*, 2005, **93**(4), 1030–1037, PubMed PMID: 15857406. Epub 2005/04/29. eng.
122. M. Thiruchelvam, B. J. Brockel, E. K. Richfield, R. B. Baggs and D. A. Cory-Slechta, Potentiated and preferential effects of combined paraquat and maneb on nigrostriatal dopamine systems: environmental risk factors for Parkinson's disease? *Brain Res.*, 2000, **873**(2), 225–234, PubMed PMID: 10930548. Epub 2000/08/10. eng.
123. R. J. Smeyne, C. B. Breckenridge, M. Beck, Y. Jiao, M. T. Butt and J. C. Wolf, *et al.*, Assessment of the Effects of MPTP and Paraquat on Dopaminergic Neurons and Microglia in the Substantia Nigra Pars Compacta of C57BL/6 Mice, *PLoS One*, 2016, **11**(10), e0164094, PubMed PMID: 27788145. Pubmed Central PMCID: PMC5082881. Epub 2016/10/28. eng.
124. A. Czerniczyniec, A. G. Karadayian, J. Bustamante, R. A. Cutrera and S. Lores-Arnaiz, Paraquat induces behavioral changes and cortical and striatal mitochondrial dysfunction, *Free Radical Biol. Med.*, 2011, **51**(7), 1428–1436, PubMed PMID: 21802509. Epub 2011/08/02. eng.
125. A. Czerniczyniec, E. M. Lanza, A. G. Karadayian, J. Bustamante and S. Lores-Arnaiz, Impairment of striatal mitochondrial function by acute paraquat poisoning, *J. Bioenerg. Biomembr.*, 2015, **47**(5), 395–408, PubMed PMID: 26350412. Epub 2015/09/10. eng.

126. P. R. Castello, D. A. Drechsel and M. Patel, Mitochondria are a major source of paraquat-induced reactive oxygen species production in the brain, *J. Biol. Chem.*, 2007, **282**(19), 14186–14193, PubMed PMID: 17389593. Pubmed Central PMCID: PMC3088512. Epub 2007/03/29. eng.
127. D. A. Drechsel and M. Patel, Differential contribution of the mitochondrial respiratory chain complexes to reactive oxygen species production by redox cycling agents implicated in parkinsonism, *Toxicol. Sci.*, 2009, **112**(2), 427–434, PubMed PMID: 19767442. Pubmed Central PMCID: PMC2777080. Epub 2009/09/22. eng.
128. A. Dixit, G. Srivastava, D. Verma, M. Mishra, P. K. Singh and O. Prakash, *et al.*, Minocycline, levodopa and MnTMPyP induced changes in the mitochondrial proteome profile of MPTP and maneb and paraquat mice models of Parkinson's disease, *Biochim. Biophys. Acta*, 2013, **1832**(8), 1227–1240, PubMed PMID: 23562983. Epub 2013/04/09. eng.
129. J. Zhang, V. A. Fitsanakis, G. Gu, D. Jing, M. Ao and V. Amarnath, *et al.*, Manganese ethylene-bis-dithiocarbamate and selective dopaminergic neurodegeneration in rat: a link through mitochondrial dysfunction, *J. Neurochem.*, 2003, **84**(2), 336–346, PubMed PMID: 12558996. Epub 2003/02/01. eng.
130. J. R. Richardson, Y. Quan, T. B. Sherer, J. T. Greenamyre and G. W. Miller, Paraquat neurotoxicity is distinct from that of MPTP and rotenone, *Toxicol. Sci.*, 2005, **88**(1), 193–201, PubMed PMID: 16141438. Epub 2005/09/06. eng.
131. B. P. Dranka, J. Zielonka, A. G. Kanthasamy and B. Kalyanaraman, Alterations in bioenergetic function induced by Parkinson's disease mimetic compounds: lack of correlation with superoxide generation, *J. Neurochem.*, 2012, **122**(5), 941–951, PubMed PMID: 22708893. Pubmed Central PMCID: PMC3423581. Epub 2012/06/20. eng.
132. A. Czerniczyniec, S. Lores-Arnaiz and J. Bustamante, Mitochondrial susceptibility in a model of paraquat neurotoxicity, *Free Radical Res.*, 2013, **47**(8), 614–623, PubMed PMID: 23692144. Epub 2013/05/23. eng.
133. C. L. Huang, C. C. Chao, Y. C. Lee, M. K. Lu, J. J. Cheng and Y. C. Yang, *et al.*, Paraquat Induces Cell Death Through Impairing Mitochondrial Membrane Permeability, *Mol. Neurobiol.*, 2016, **53**(4), 2169–2188, PubMed PMID: 25947082. Epub 2015/05/08. eng.
134. S. McCarthy, M. Somayajulu, M. Sikorska, H. Borowy-Borowski and S. Pandey, Paraquat induces oxidative stress and neuronal cell death; neuroprotection by water-soluble Coenzyme Q10, *Toxicol. Appl. Pharmacol.*, 2004, **201**(1), 21–31, PubMed PMID: 15519605. Epub 2004/11/03. eng.
135. A. Garcia-Garcia, A. Anandhan, M. Burns, H. Chen, Y. Zhou and R. Franco, Impairment of Atg5-dependent autophagic flux promotes paraquat- and MPP(+)-induced apoptosis but not rotenone or 6-hydroxydopamine toxicity, *Toxicol. Sci.*, 2013, **136**(1), 166–182, PubMed PMID: 23997112. Pubmed Central PMCID: PMC3829573. Epub 2013/09/03. eng.
136. R. A. Gonzalez-Polo, M. Niso-Santano, M. A. Ortiz-Ortiz, A. Gomez-Martin, J. M. Moran and L. Garcia-Rubio, *et al.*, Inhibition of paraquat-induced

autophagy accelerates the apoptotic cell death in neuroblastoma SH-SY5Y cells, *Toxicol. Sci.*, 2007, **97**(2), 448–458, PubMed PMID: 17341480. Epub 2007/03/08. eng.

137. R. Pascua-Maestro, S. Diez-Hermano, C. Lillo, M. D. Ganfornina and D. Sanchez, Protecting cells by protecting their vulnerable lysosomes: Identification of a new mechanism for preserving lysosomal functional integrity upon oxidative stress, *PLoS Genet.*, 2017, **13**(2), e1006603, PubMed PMID: 28182653. Epub 2017/02/10. eng.
138. J. Navarro-Yepes, A. Anandhan, E. Bradley, I. Bohovych, B. Yarabe and A. de Jong, *et al.*, Inhibition of Protein Ubiquitination by Paraquat and 1-Methyl-4-Phenylpyridinium Impairs Ubiquitin-Dependent Protein Degradation Pathways, *Mol. Neurobiol.*, 2016, **53**(8), 5229–5251, PubMed PMID: 26409479. Pubmed Central PMCID: PMC4842169. Epub 2015/09/28. eng.
139. C. Wang, H. S. Ko, B. Thomas, F. Tsang, K. C. Chew and S. P. Tay, *et al.*, Stress-induced alterations in parkin solubility promote parkin aggregation and compromise parkin's protective function, *Hum. Mol. Genet.*, 2005, **14**(24), 3885–3897, PubMed PMID: 16278233. Epub 2005/11/10. eng.
140. G. C. Davis, A. C. Williams, S. P. Markey, M. H. Ebert, E. D. Caine and C. M. Reichert, *et al.*, Chronic Parkinsonism secondary to intravenous injection of meperidine analogues, *Psychiatry Res.*, 1979, **1**(3), 249–254, PubMed PMID: 298352. Epub 1979/12/01. eng.
141. R. R. Gainetdinov, F. Fumagalli, S. R. Jones and M. G. Caron, Dopamine transporter is required for *in vivo* MPTP neurotoxicity: evidence from mice lacking the transporter, *J. Neurochem.*, 1997, **69**(3), 1322–1325, PubMed PMID: 9282960. Epub 1997/09/01. eng.
142. J. A. Javitch, R. J. D'Amato, S. M. Strittmatter and S. H. Snyder, Parkinsonism-inducing neurotoxin, *N*-methyl-4-phenyl-1,2,3,6-tetrahydropyridine: uptake of the metabolite *N*-methyl-4-phenylpyridine by dopamine neurons explains selective toxicity, *Proc. Natl. Acad. Sci. U. S. A.*, 1985, **82**(7), 2173–2177, PubMed PMID: 3872460. Pubmed Central PMCID: PMC397515. Epub 1985/04/01. eng.
143. J. Bove, D. Prou, C. Perier and S. Przedborski, Toxin-induced models of Parkinson's disease, *NeuroRx*, 2005, **2**(3), 484–494, PubMed PMID: 16389312. Pubmed Central PMCID: PMC1144492. Epub 2006/01/04. eng.
144. M. W. Jakowec and G. M. Petzinger, 1-methyl-4-phenyl-1,2,3,6-tetrahydropyridine-lesioned model of parkinson's disease, with emphasis on mice and nonhuman primates, *Comp. Med.*, 2004, **54**(5), 497–513, PubMed PMID: 15575363. Epub 2004/12/04. eng.
145. G. Halliday, M. T. Herrero, K. Murphy, H. McCann, F. Ros-Bernal and C. Barcia, *et al.*, No Lewy pathology in monkeys with over 10 years of severe MPTP Parkinsonism, *Mov. Disord.*, 2009, **24**(10), 1519–1523, PubMed PMID: 19526568. Epub 2009/06/16. eng.

146. M. Shimoji, L. Zhang, A. S. Mandir, V. L. Dawson and T. M. Dawson, Absence of inclusion body formation in the MPTP mouse model of Parkinson's disease, *Brain Res. Mol. Brain Res.*, 2005, **134**(1), 103–108, PubMed PMID: 15790534. Epub 2005/03/26. eng.
147. F. Fornai, O. M. Schluter, P. Lenzi, M. Gesi, R. Ruffoli and M. Ferrucci, *et al.*, Parkinson-like syndrome induced by continuous MPTP infusion: convergent roles of the ubiquitin-proteasome system and alpha-synuclein, *Proc. Natl. Acad. Sci. U. S. A.*, 2005, **102**(9), 3413–3418, PubMed PMID: 15716361. Pubmed Central PMCID: PMC552938. Epub 2005/02/18. eng.
148. R. R. Ramsay, M. J. Krueger, S. K. Youngster, M. R. Gluck, J. E. Casida and T. P. Singer, Interaction of 1-methyl-4-phenylpyridinium ion (MPP+) and its analogs with the rotenone/piericidin binding site of NADH dehydrogenase, *J. Neurochem.*, 1991, **56**(4), 1184–1190, PubMed PMID: 2002336. Epub 1991/04/01. eng.
149. Y. Mizuno, N. Sone and T. Saitoh, Effects of 1-methyl-4-phenyl-1,2,3,6-tetrahydropyridine and 1-methyl-4-phenylpyridinium ion on activities of the enzymes in the electron transport system in mouse brain, *J. Neurochem.*, 1987, **48**(6), 1787–1793, PubMed PMID: 3106573. Epub 1987/06/01. eng.
150. W. J. Nicklas, S. K. Youngster, M. V. Kindt and R. E. Heikkila, MPTP, MPP+ and mitochondrial function, *Life Sci.*, 1987, **40**(8), 721–729, PubMed PMID: 3100899. Epub 1987/02/23. eng.
151. H. Rodriguez-Rocha, A. Garcia-Garcia, C. Pickett, S. Li, J. Jones and H. Chen, *et al.*, Compartmentalized oxidative stress in dopaminergic cell death induced by pesticides and complex I inhibitors: distinct roles of superoxide anion and superoxide dismutases, *Free Radical Biol. Med.*, 2013, **61**, 370–383, PubMed PMID: 23602909. Pubmed Central PMCID: PMC3883883. Epub 2013/04/23. eng.
152. B. B. Seo, E. Nakamaru-Ogiso, T. R. Flotte, A. Matsuno-Yagi and T. Yagi, *In vivo* complementation of complex I by the yeast Ndi1 enzyme. Possible application for treatment of Parkinson disease, *J. Biol. Chem.*, 2006, **281**(20), 14250–14255, PubMed PMID: 16543240. Epub 2006/03/18. eng.
153. J. R. Richardson, W. M. Caudle, T. S. Guillot, J. L. Watson, E. Nakamaru-Ogiso and B. B. Seo, *et al.*, Obligatory role for complex I inhibition in the dopaminergic neurotoxicity of 1-methyl-4-phenyl-1,2,3,6-tetrahydropyridine (MPTP), *Toxicol. Sci.*, 2007, **95**(1), 196–204, PubMed PMID: 17038483. Epub 2006/10/14. eng.
154. J. S. Kim-Han, J. A. Antenor-Dorsey and K. L. O'Malley, The parkinsonian mimetic, MPP+, specifically impairs mitochondrial transport in dopamine axons, *J. Neurosci.*, 2011, **31**(19), 7212–7221, PubMed PMID: 21562285. Pubmed Central PMCID: PMC3140916. Epub 2011/05/13. eng.
155. J. Lotharius and K. L. O'Malley, The parkinsonism-inducing drug 1-methyl-4-phenylpyridinium triggers intracellular dopamine oxidation. A novel mechanism of toxicity, *J. Biol. Chem.*, 2000, **275**(49), 38581–38588, PubMed PMID: 10969076. Epub 2000/09/02. eng.

156. K. C. Hung, H. J. Huang, M. W. Lin, Y. P. Lei and A. M. Lin, Roles of autophagy in MPP+-induced neurotoxicity *in vivo*: the involvement of mitochondria and alpha-synuclein aggregation, *PLoS One*, 2014, **9**(3), e91074, PubMed PMID: 24646838. Pubmed Central PMCID: PMC3960112. Epub 2014/03/22. eng.
157. J. Lim, H. W. Kim, M. B. Youdim, I. J. Rhyu, K. M. Choe and Y. J. Oh, Binding preference of p62 towards LC3-ll during dopaminergic neurotoxin-induced impairment of autophagic flux, *Autophagy*, 2011, **7**(1), 51–60, PubMed PMID: 21045561. Epub 2010/11/04. eng.
158. J. Lim, Y. Lee, S. Jung, M. B. Youdim and Y. J. Oh, Impaired autophagic flux is critically involved in drug-induced dopaminergic neuronal death, *Parkinsonism Relat. Disord.*, 2014, **20**(suppl. 1), S162–S166, PubMed PMID: 24262172. Epub 2013/11/23. eng.
159. J. H. Zhu, C. Horbinski, F. Guo, S. Watkins, Y. Uchiyama and C. T. Chu, Regulation of autophagy by extracellular signal-regulated protein kinases during 1-methyl-4-phenylpyridinium-induced cell death, *Am. J. Pathol.*, 2007, **170**(1), 75–86, PubMed PMID: 17200184. Pubmed Central PMCID: PMC1762689. Epub 2007/01/04. eng.
160. K. Liu, N. Shi, Y. Sun, T. Zhang and X. Sun, Therapeutic effects of rapamycin on MPTP-induced Parkinsonism in mice, *Neurochem. Res.*, 2013, **38**(1), 201–207, PubMed PMID: 23117422. Epub 2012/11/03. eng.
161. Y. Zhang, J. Y. Wu, L. H. Weng, X. X. Li, L. J. Yu and Y. Xu, Valproic acid protects against MPP+-mediated neurotoxicity in SH-SY5Y Cells through autophagy, *Neurosci. Lett.*, 2017, **638**, 60–68, PubMed PMID: 27956235. Epub 2016/12/14. eng.
162. X. Z. Li, X. P. Chen, K. Zhao, L. M. Bai, H. Zhang and X. P. Zhou, Therapeutic effects of valproate combined with lithium carbonate on MPTP-induced parkinsonism in mice: possible mediation through enhanced autophagy, *Int. J. Neurosci.*, 2013, **123**(2), 73–79, PubMed PMID: 22978383. Epub 2012/09/18. eng.
163. J. Bove, M. Martinez-Vicente, B. Dehay, C. Perier, A. Recasens and A. Bombrun, *et al.*, BAX channel activity mediates lysosomal disruption linked to Parkinson disease, *Autophagy*, 2014, **10**(5), 889–900, PubMed PMID: 24686337. Pubmed Central PMCID: PMC5119069. Epub 2014/04/02. eng.
164. M. Miyara, Y. Kotake, W. Tokunaga, S. Sanoh and S. Ohta, Mild MPP+ exposure impairs autophagic degradation through a novel lysosomal acidity-independent mechanism, *J. Neurochem.*, 2016, **139**(2), 294–308, PubMed PMID: 27309572. Epub 2016/06/17. eng.
165. F. Meng, D. Yao, Y. Shi, J. Kabakoff, W. Wu and J. Reicher, *et al.*, Oxidation of the cysteine-rich regions of parkin perturbs its E3 ligase activity and contributes to protein aggregation, *Mol. Neurodegener.*, 2011, **6**, 34, PubMed PMID: 21595948. Pubmed Central PMCID: PMC3120712. Epub 2011/05/21. eng.
166. X. Y. Niu, H. J. Huang, J. B. Zhang, C. Zhang, W. G. Chen and C. Y. Sun, *et al.*, Deletion of autophagy-related gene 7 in dopaminergic neurons

prevents their loss induced by MPTP, *Neuroscience*, 2016, **339**, 22–31, PubMed PMID: 27693472. Epub 2016/11/05. eng.
167. S. J. Cherra 3rd, S. M. Kulich, G. Uechi, M. Balasubramani, J. Mountzouris and B. W. Day, *et al.*, Regulation of the autophagy protein LC3 by phosphorylation, *J. Cell Biol.*, 2010, **190**(4), 533–539, PubMed PMID: 20713600. Pubmed Central PMCID: PMC2928022. Epub 2010/08/18. eng.
168. C. T. Chu, J. Zhu and R. Dagda, Beclin 1-independent pathway of damage-induced mitophagy and autophagic stress: implications for neurodegeneration and cell death, *Autophagy*, 2007, **3**(6), 663–666, PubMed PMID: 17622797. Pubmed Central PMCID: PMC2779565. Epub 2007/07/12. eng.
169. M. Decressac, B. Mattsson and A. Bjorklund, Comparison of the behavioural and histological characteristics of the 6-OHDA and alpha-synuclein rat models of Parkinson's disease, *Exp. Neurol.*, 2012, **235**(1), 306–315, PubMed PMID: 22394547. Epub 2012/03/08. eng.
170. F. Hefti, E. Melamed, B. J. Sahakian and R. J. Wurtman, Circling behavior in rats with partial, unilateral nigro-striatal lesions: effect of amphetamine, apomorphine, and DOPA, *Pharmacol., Biochem. Behav.*, 1980, **12**(2), 185–188, PubMed PMID: 7189592. Epub 1980/02/01. eng.
171. R. E. Johnson, T. Schallert and J. B. Becker, Akinesia and postural abnormality after unilateral dopamine depletion, *Behav. Brain Res.*, 1999, **104**(1–2), 189–196, PubMed PMID: 11125738. Epub 2000/12/28. eng.
172. H. S. Lindgren, M. J. Lelos and S. B. Dunnett, Do alpha-synuclein vector injections provide a better model of Parkinson's disease than the classic 6-hydroxydopamine model? *Exp. Neurol.*, 2012, **237**(1), 36–42, PubMed PMID: 22727767. Epub 2012/06/26. eng.
173. F. Dabbeni-Sala, S. Di Santo, D. Franceschini, S. D. Skaper and P. Giusti, Melatonin protects against 6-OHDA-induced neurotoxicity in rats: a role for mitochondrial complex I activity, *FASEB J.*, 2001, **15**(1), 164–170, PubMed PMID: 11149904. Epub 2001/01/10. eng.
174. A. Kupsch, W. Schmidt, Z. Gizatullina, G. Debska-Vielhaber, J. Voges and F. Striggow, *et al.*, 6-Hydroxydopamine impairs mitochondrial function in the rat model of Parkinson's disease: respirometric, histological, and behavioral analyses, *J. Neural Transm. (Vienna)*, 2014, **121**(10), 1245–1257, PubMed PMID: 24627045. Epub 2014/03/15. eng.
175. Y. Glinka, K. F. Tipton and M. B. Youdim, Nature of inhibition of mitochondrial respiratory complex I by 6-Hydroxydopamine, *J. Neurochem.*, 1996, **66**(5), 2004–2010, PubMed PMID: 8780029. Epub 1996/05/01. eng.
176. Y. Y. Glinka and M. B. Youdim, Inhibition of mitochondrial complexes I and IV by 6-hydroxydopamine, *Eur. J. Pharmacol.*, 1995, **292**(3–4), 329–332, PubMed PMID: 7796873. Epub 1995/03/16. eng.
177. E. A. Mazzio, R. R. Reams and K. F. Soliman, The role of oxidative stress, impaired glycolysis and mitochondrial respiratory redox failure in the cytotoxic effects of 6-hydroxydopamine *in vitro*, *Brain Res.*, 2004, **1004**(1–2), 29–44, PubMed PMID: 15033417. Epub 2004/03/23. eng.

178. S. Giordano, J. Lee, V. M. Darley-Usmar and J. Zhang, Distinct effects of rotenone, 1-methyl-4-phenylpyridinium and 6-hydroxydopamine on cellular bioenergetics and cell death, *PLoS One*, 2012, **7**(9), e44610, PubMed PMID: 22970265. Pubmed Central PMCID: PMC3435291. Epub 2012/09/13. eng.
179. S. Giordano, V. Darley-Usmar and J. Zhang, Autophagy as an essential cellular antioxidant pathway in neurodegenerative disease, *Redox Biol.*, 2014, **2**, 82–90, PubMed PMID: 24494187. Pubmed Central PMCID: PMC3909266. Epub 2014/02/05. eng.
180. M. Gomez-Lazaro, N. A. Bonekamp, M. F. Galindo, J. Jordan and M. Schrader, 6-Hydroxydopamine (6-OHDA) induces Drp1-dependent mitochondrial fragmentation in SH-SY5Y cells, *Free Radical Biol. Med.*, 2008, **44**(11), 1960–1969, PubMed PMID: 18395527. Epub 2008/04/09. eng.
181. M. E. Solesio, T. A. Prime, A. Logan, M. P. Murphy, M. Del Mar Arroyo-Jimenez and J. Jordan, *et al.*, The mitochondria-targeted anti-oxidant MitoQ reduces aspects of mitochondrial fission in the 6-OHDA cell model of Parkinson's disease, *Biochim. Biophys. Acta*, 2013, **1832**(1), 174–182, PubMed PMID: 22846607. Epub 2012/08/01. eng.
182. M. E. Solesio, S. Saez-Atienzar, J. Jordan and M. F. Galindo, Characterization of mitophagy in the 6-hydoxydopamine Parkinson's disease model, *Toxicol. Sci.*, 2012, **129**(2), 411–420, PubMed PMID: 22821850. Epub 2012/07/24. eng.
183. X. Lu, J. S. Kim-Han, S. Harmon, S. E. Sakiyama-Elbert and K. L. O'Malley, The Parkinsonian mimetic, 6-OHDA, impairs axonal transport in dopaminergic axons, *Mol. Neurodegener.*, 2014, **9**, 17, PubMed PMID: 24885281. Pubmed Central PMCID: PMC4016665. Epub 2014/06/03. eng.
184. J. Callio, T. D. Oury and C. T. Chu, Manganese superoxide dismutase protects against 6-hydroxydopamine injury in mouse brains, *J. Biol. Chem.*, 2005, **280**(18), 18536–18542, PubMed PMID: 15755737. Pubmed Central PMCID: PMC1885201. Epub 2005/03/10. eng.
185. L. Li, X. Wang, X. Fei, L. Xia, Z. Qin and Z. Liang, Parkinson's disease involves autophagy and abnormal distribution of cathepsin L, *Neurosci. Lett.*, 2011, **489**(1), 62–67, PubMed PMID: 21134415. Epub 2010/12/08. eng.
186. L. Huang, M. Deng, Y. He, S. Lu, S. Liu and Y. Fang, beta-asarone increases MEF2D and TH levels and reduces alpha-synuclein level in 6-OHDA-induced rats *via* regulating the HSP70/MAPK/MEF2D/Beclin-1 pathway: Chaperone-mediated autophagy activation, macroautophagy inhibition and HSP70 up-expression, *Behav. Brain Res.*, 2016, **313**, 370–379, PubMed PMID: 27444243. Epub 2016/07/23. eng.
187. S. Zhang, Z. F. Xue, L. P. Huang, R. M. Fang, Y. P. He and L. Li, *et al.*, Dynamic expressions of Beclin 1 and tyrosine hydroxylase in different areas of 6-hydroxydopamine-induced Parkinsonian rats, *Cell. Mol. Neurobiol.*, 2013, **33**(7), 973–981, PubMed PMID: 23868341. Epub 2013/07/23. eng.

188. K. Arsikin, T. Kravic-Stevovic, M. Jovanovic, B. Ristic, G. Tovilovic and N. Zogovic, *et al.*, Autophagy-dependent and -independent involvement of AMP-activated protein kinase in 6-hydroxydopamine toxicity to SH-SY5Y neuroblastoma cells, *Biochim. Biophys. Acta*, 2012, **1822**(11), 1826–1836, PubMed PMID: 22917563. Epub 2012/08/25. eng.
189. J. Y. Ha, J. S. Kim, Y. H. Kang, E. Bok, Y. S. Kim and J. H. Son, Tnfaip8 l1/Oxi-beta binds to FBXW5, increasing autophagy through activation of TSC2 in a Parkinson's disease model, *J. Neurochem.*, 2014, **129**(3), 527–538, PubMed PMID: 24444419. Epub 2014/01/22. eng.
190. H. C. Cheng, S. R. Kim, T. F. Oo, T. Kareva, O. Yarygina and M. Rzhetskaya, *et al.*, Akt suppresses retrograde degeneration of dopaminergic axons by inhibition of macroautophagy, *J. Neurosci.*, 2011, **31**(6), 2125–2135, PubMed PMID: 21307249. Pubmed Central PMCID: PMC3075005. Epub 2011/02/11. eng.
191. M. F. Galindo, M. E. Solesio, S. Atienzar-Aroca, M. J. Zamora and J. Jordan Bueso, Mitochondrial dynamics and mitophagy in the 6-hydroxydopamine preclinical model of Parkinson's disease, *Parkinson's Dis.*, 2012, **2012**, 131058, PubMed PMID: 22966477. Pubmed Central PMCID: PMC3431121. Epub 2012/09/12. eng.
192. R. K. Dagda, J. Zhu, S. M. Kulich and C. T. Chu, Mitochondrially localized ERK2 regulates mitophagy and autophagic cell stress: implications for Parkinson's disease, *Autophagy*, 2008, **4**(6), 770–782, PubMed PMID: 18594198. Pubmed Central PMCID: PMC2574804. Epub 2008/07/03. eng.
193. S. M. Kulich, C. Horbinski, M. Patel and C. T. Chu, 6-Hydroxydopamine induces mitochondrial ERK activation, *Free Radical Biol. Med.*, 2007, **43**(3), 372–383, PubMed PMID: 17602953. Pubmed Central PMCID: PMC2023873. Epub 2007/07/03. eng.
194. R. Inzelberg, Y. Samuels, E. Azizi, N. Qutob, L. Inzelberg and E. Domany, *et al.*, Parkinson disease (PARK) genes are somatically mutated in cutaneous melanoma, *Neurol. Genet.*, 2016, **2**(3), e70, PubMed PMID: 27123489. Pubmed Central PMCID: PMC4832432. Epub 2016/04/29. eng.
195. A. R. Esteves and S. M. Cardoso, LRRK2 at the Crossroad Between Autophagy and Microtubule Trafficking: Insights into Parkinson's Disease, *The Neuroscientist*, 2017, **23**(1), 16–26, PubMed PMID: 26740081. Epub 2016/01/08. eng.
196. D. A. Roosen and M. R. Cookson, LRRK2 at the interface of autophagosomes, endosomes and lysosomes, *Mol. Neurodegener.*, 2016, **11**(1), 73, PubMed PMID: 27927216. Pubmed Central PMCID: PMC5142374. Epub 2016/12/09. eng.
197. C. Manzoni, The LRRK2-macroautophagy axis and its relevance to Parkinson's disease, *Biochem. Soc. Trans.*, 2017, **45**(1), 155–162, PubMed PMID: 28202669. Pubmed Central PMCID: PMC5310720. Epub 2017/02/17. eng.
198. M. H. Polymeropoulos, C. Lavedan, E. Leroy, S. E. Ide, A. Dehejia and A. Dutra, *et al.*, Mutation in the alpha-synuclein gene identified in families with Parkinson's disease, *Science*, 1997, **276**(5321), 2045–2047, PubMed PMID: 9197268. Epub 1997/06/27. eng.

199. R. Kruger, W. Kuhn, T. Muller, D. Woitalla, M. Graeber and S. Kosel, *et al.*, Ala30Pro mutation in the gene encoding alpha-synuclein in Parkinson's disease, *Nat. Genet.*, 1998, **18**(2), 106–108, PubMed PMID: 9462735. Epub 1998/02/14. eng.
200. J. J. Zarranz, J. Alegre, J. C. Gomez-Esteban, E. Lezcano, R. Ros and I. Ampuero, *et al.*, The new mutation, E46K, of alpha-synuclein causes Parkinson and Lewy body dementia, *Ann. Neurol.*, 2004, **55**(2), 164–173, PubMed PMID: 14755719. Epub 2004/02/03. eng.
201. A. P. Kiely, Y. T. Asi, E. Kara, P. Limousin, H. Ling and P. Lewis, *et al.*, alpha-Synucleinopathy associated with G51D SNCA mutation: a link between Parkinson's disease and multiple system atrophy? *Acta Neuropathol.*, 2013, **125**(5), 753–769, PubMed PMID: 23404372. Pubmed Central PMCID: 3681325. Epub 2013/02/14. eng.
202. S. Lesage, M. Anheim, F. Letournel, L. Bousset, A. Honore and N. Rozas, *et al.*, G51D alpha-synuclein mutation causes a novel Parkinsonian-pyramidal syndrome, *Ann. Neurol.*, 2013, **73**(4), 459–471, PubMed PMID: 23526723. Epub 2013/03/26. eng.
203. P. Pasanen, L. Myllykangas, M. Siitonen, A. Raunio, S. Kaakkola and J. Lyytinen, *et al.*, A novel alpha-synuclein mutation A53E associated with atypical multiple system atrophy and Parkinson's disease-type pathology, *Neurobiol. Aging*, 2014, **35**(9), 2180 e1–2180 e5, PubMed PMID: 24746362.
204. C. Proukakis, C. G. Dudzik, T. Brier, D. S. Mackay, J. M. Cooper and G. L. Millhauser, *et al.*, A novel alpha-synuclein missense mutation in Parkinson disease, *Neurology*, 2013, **80**(11), 1062–1064, PubMed PMID: 23427326. Epub 2013/02/22. eng.
205. S. Petrucci, M. Ginevrino and E. M. Valente, Phenotypic spectrum of alpha-synuclein mutations: New insights from patients and cellular models, *Parkinsonism Relat. Disord.*, 2016, **22**(suppl. 1), S16–S20, PubMed PMID: 26341711. Epub 2015/09/06. eng.
206. M. C. Chartier-Harlin, J. Kachergus, C. Roumier, V. Mouroux, X. Douay and S. Lincoln, *et al.*, Alpha-synuclein locus duplication as a cause of familial Parkinson's disease, *Lancet*, 2004, **364**(9440), 1167–1169, PubMed PMID: 15451224. Epub 2004/09/29. eng.
207. A. B. Singleton, M. Farrer, J. Johnson, A. Singleton, S. Hague and J. Kachergus, *et al.*, alpha-Synuclein locus triplication causes Parkinson's disease, *Science*, 2003, **302**(5646), 841, PubMed PMID: 14593171. Epub 2003/11/01. eng.
208. N. Pankratz, J. B. Wilk, J. C. Latourelle, A. L. DeStefano, C. Halter and E. W. Pugh, *et al.*, Genomewide association study for susceptibility genes contributing to familial Parkinson disease, *Hum. Genet.*, 2009, **124**(6), 593–605, PubMed PMID: 18985386. Pubmed Central PMCID: PMC2627511. Epub 2008/11/06. eng.
209. J. Simon-Sanchez, C. Schulte, J. M. Bras, M. Sharma, J. R. Gibbs and D. Berg, *et al.*, Genome-wide association study reveals genetic risk underlying Parkinson's disease, *Nat. Genet.*, 2009, **41**(12), 1308–1312, PubMed PMID: 19915575. Pubmed Central PMCID: PMC2787725. Epub 2009/11/17. eng.

210. J. M. George, The synucleins, *Genome Biol.*, 2002, **3**(1), reviews3002.1-reviews3002.6, PubMed PMID: 11806835. Pubmed Central PMCID: 150459. Epub 2002/01/25. eng.
211. W. S. Davidson, A. Jonas, D. F. Clayton and J. M. George, Stabilization of alpha-synuclein secondary structure upon binding to synthetic membranes, *J. Biol. Chem.*, 1998, **273**(16), 9443–9449, PubMed PMID: 9545270. Epub 1998/05/23. eng.
212. R. J. Perrin, W. S. Woods, D. F. Clayton and J. M. George, Interaction of human alpha-Synuclein and Parkinson's disease variants with phospholipids. Structural analysis using site-directed mutagenesis, *J. Biol. Chem.*, 2000, **275**(44), 34393–34398, PubMed PMID: 10952980. Epub 2000/08/23. eng.
213. B. I. Giasson, I. V. Murray, J. Q. Trojanowski and V. M. Lee, A hydrophobic stretch of 12 amino acid residues in the middle of alpha-synuclein is essential for filament assembly, *J. Biol. Chem.*, 2001, **276**(4), 2380–2386, PubMed PMID: 11060312. Epub 2000/11/04. eng.
214. P. H. Jensen, M. S. Nielsen, R. Jakes, C. G. Dotti and M. Goedert, Binding of α-synuclein to brain vesicles is abolished by familial Parkinson's disease mutation, *J. Biol. Chem.*, 1998, **273**(41), 26292–26294.
215. D. D. Murphy, S. M. Rueter, J. Q. Trojanowski and V. M. Lee, Synucleins are developmentally expressed, and alpha-synuclein regulates the size of the presynaptic vesicular pool in primary hippocampal neurons, *J. Neurosci.*, 2000, **20**(9), 3214–3220, PubMed PMID: 10777786. Epub 2000/04/25. eng.
216. D. E. Cabin, K. Shimazu, D. Murphy, N. B. Cole, W. Gottschalk and K. L. McIlwain, *et al.*, Synaptic vesicle depletion correlates with attenuated synaptic responses to prolonged repetitive stimulation in mice lacking alpha-synuclein, *J. Neurosci.*, 2002, **22**(20), 8797–8807, PubMed PMID: 12388586. Epub 2002/10/22. eng.
217. K. E. Larsen, Y. Schmitz, M. D. Troyer, E. Mosharov, P. Dietrich and A. Z. Quazi, *et al.*, Alpha-synuclein overexpression in PC12 and chromaffin cells impairs catecholamine release by interfering with a late step in exocytosis, *J. Neurosci.*, 2006, **26**(46), 11915–11922, PubMed PMID: 17108165. Epub 2006/11/17. eng.
218. L. L. Venda, S. J. Cragg, V. L. Buchman and R. Wade-Martins, alpha-Synuclein and dopamine at the crossroads of Parkinson's disease, *Trends Neurosci.*, 2010, **33**(12), 559–568, PubMed PMID: 20961626. Pubmed Central PMCID: 3631137.
219. J. Diao, J. Burre, S. Vivona, D. J. Cipriano, M. Sharma and M. Kyoung, *et al.*, Native alpha-synuclein induces clustering of synaptic-vesicle mimics *via* binding to phospholipids and synaptobrevin-2/VAMP2, *eLife*, 2013, **2**, e00592, PubMed PMID: 23638301. Pubmed Central PMCID: PMC3639508. Epub 2013/05/03. eng.
220. C. Galvagnion, A. K. Buell, G. Meisl, T. C. Michaels, M. Vendruscolo and T. P. Knowles, *et al.*, Lipid vesicles trigger alpha-synuclein aggregation by stimulating primary nucleation, *Nat. Chem. Biol.*, 2015, **11**(3), 229–234, PubMed PMID: 25643172. Epub 2015/02/03. eng.

221. F. Haque, A. P. Pandey, L. R. Cambrea, J. C. Rochet and J. S. Hovis, Adsorption of alpha-synuclein on lipid bilayers: modulating the structure and stability of protein assemblies, *J. Phys. Chem. B*, 2010, **114**(11), 4070–4081, PubMed PMID: 20187615. Pubmed Central PMCID: 2855901. Epub 2010/03/02. eng.
222. E. Jo, A. A. Darabie, K. Han, A. Tandon, P. E. Fraser and J. McLaurin, alpha-Synuclein-synaptosomal membrane interactions: implications for fibrillogenesis, *Eur. J. Biochem.*, 2004, **271**(15), 3180–3189, PubMed PMID: 15265037.
223. H.-J. Lee, C. Choi and S.-J. Lee, Membrane-bound α-synuclein has a high aggregation propensity and the ability to seed the aggregation of the cytosolic form, *J. Biol. Chem.*, 2002, **277**, 671–678.
224. A. P. Pandey, F. Haque, J. C. Rochet and J. S. Hovis, Clustering of alpha-synuclein on supported lipid bilayers: role of anionic lipid, protein, and divalent ion concentration, *Biophys. J.*, 2009, **96**(2), 540–551, PubMed PMID: 19167303.
225. D. Ysselstein, B. Dehay, I. M. Costantino, G. P. McCabe, M. P. Frosch and J. M. George, *et al.*, Endosulfine-alpha inhibits membrane-induced alpha-synuclein aggregation and protects against alpha-synuclein neurotoxicity, *Acta Neuropathol. Commun.*, 2017, **5**(1), 3, PubMed PMID: 28069058. Pubmed Central PMCID: PMC5223451. Epub 2017/01/11. eng.
226. D. Ysselstein, M. Joshi, V. Mishra, A. M. Griggs, J. M. Asiago and G. P. McCabe, *et al.*, Effects of impaired membrane interactions on alpha-synuclein aggregation and neurotoxicity, *Neurobiol. Dis.*, 2015, **79**, 150–163, PubMed PMID: 25931201. Pubmed Central PMCID: PMC4458405. Epub 2015/05/02. eng.
227. W. Choi, S. Zibaee, R. Jakes, L. C. Serpell, B. Davletov and R. A. Crowther, *et al.*, Mutation E46K increases phospholipid binding and assembly into filaments of human alpha-synuclein, *FEBS Lett.*, 2004, **576**(3), 363–368, PubMed PMID: 15498564.
228. R. A. Fredenburg, C. Rospigliosi, R. K. Meray, J. C. Kessler, H. A. Lashuel and D. Eliezer, *et al.*, The impact of the E46K mutation on the properties of alpha-synuclein in its monomeric and oligomeric states, *Biochemistry*, 2007, **46**(24), 7107–7118, PubMed PMID: 17530780.
229. E. A. Greenbaum, C. L. Graves, A. J. Mishizen-Eberz, M. A. Lupoli, D. R. Lynch and S. W. Englander, *et al.*, The E46K mutation in alpha-synuclein increases amyloid fibril formation, *J. Biol. Chem.*, 2005, **280**(9), 7800–7807, PubMed PMID: 15632170.
230. D. Ghosh, M. Mondal, G. M. Mohite, P. K. Singh, P. Ranjan and A. Anoop, *et al.*, The Parkinson's Disease-Associated H50Q Mutation Accelerates alpha-Synuclein Aggregation *in vitro*, *Biochemistry*, 2013, **52**(40), 6925–6927, PubMed PMID: 24047453. Epub 2013/09/21. eng.
231. O. Khalaf, B. Fauvet, A. Oueslati, I. Dikiy, A. L. Mahul-Mellier and F. S. Ruggeri, *et al.*, The H50Q Mutation Enhances alpha-Synuclein

Aggregation, Secretion, and Toxicity, *J. Biol. Chem.*, 2014, **289**(32), 21856–21876, PubMed PMID: 24936070. Pubmed Central PMCID: 4139205. Epub 2014/06/18. eng.

232. R. Porcari, C. Proukakis, C. A. Waudby, B. Bolognesi, P. P. Mangione and J. F. Paton, *et al.*, The H50Q mutation induces a 10-fold decrease in the solubility of alpha-synuclein, *J. Biol. Chem.*, 2015, **290**(4), 2395–2404, PubMed PMID: 25505181. Pubmed Central PMCID: PMC4303689. Epub 2014/12/17. eng.
233. N. J. Rutherford, B. D. Moore, T. E. Golde and B. I. Giasson, Divergent effects of the H50Q and G51D SNCA mutations on the aggregation of alpha-synuclein, *J. Neurochem.*, 2014, **131**(6), 859–867, PubMed PMID: 24984882. Epub 2014/07/06. Eng.
234. K. A. Conway, S. J. Lee, J. C. Rochet, T. T. Ding, R. E. Williamson and P. T. Lansbury Jr, Acceleration of oligomerization, not fibrillization, is a shared property of both alpha-synuclein mutations linked to early-onset Parkinson's disease: implications for pathogenesis and therapy, *Proc. Natl. Acad. Sci. U. S. A.*, 2000, **97**(2), 571–576, PubMed PMID: 10639120. Pubmed Central PMCID: 15371. Epub 2000/01/19. eng.
235. J. Li, V. N. Uversky and A. L. Fink, Effect of familial Parkinson's disease point mutations A30P and A53T on the structural properties, aggregation, and fibrillation of human α-synuclein, *Biochemistry*, 2001, **40**, 11604–11613.
236. G. Comellas, L. R. Lemkau, D. H. Zhou, J. M. George and C. M. Rienstra, Structural intermediates during alpha-synuclein fibrillogenesis on phospholipid vesicles, *J. Am. Chem. Soc.*, 2012, **134**(11), 5090–5099, PubMed PMID: 22352310. Pubmed Central PMCID: 3331674. Epub 2012/02/23. eng.
237. J. H. Lee, C. S. Hong, S. Lee, J. E. Yang, Y. I. Park and D. Lee, *et al.*, Radiating amyloid fibril formation on the surface of lipid membranes through unit-assembly of oligomeric species of alpha-synuclein, *PLoS One*, 2012, 7(10), e47580, PubMed PMID: 23077644. Pubmed Central PMCID: 3471876. Epub 2012/10/19. eng.
238. N. P. Reynolds, A. Soragni, M. Rabe, D. Verdes, E. Liverani and S. Handschin, *et al.*, Mechanism of membrane interaction and disruption by alpha-synuclein, *J. Am. Chem. Soc.*, 2011, **133**(48), 19366–19375, PubMed PMID: 21978222. Epub 2011/10/08. eng.
239. C. W. Olanow and P. Brundin, Parkinson's disease and alpha synuclein: is Parkinson's disease a prion-like disorder? *Mov. Disord.*, 2013, **28**(1), 31–40, PubMed PMID: 23390095. Epub 2013/02/08. eng.
240. O. Chiba-Falek, J. A. Kowalak, M. E. Smulson and R. L. Nussbaum, Regulation of alpha-synuclein expression by poly (ADP ribose) polymerase-1 (PARP-1) binding to the NACP-Rep1 polymorphic site upstream of the SNCA gene, *Am. J. Hum. Genet.*, 2005, **76**(3), 478–492, PubMed PMID: 15672325. Pubmed Central PMCID: PMC1196399. Epub 2005/01/27. eng.

241. O. Chiba-Falek, J. W. Touchman and R. L. Nussbaum, Functional analysis of intra-allelic variation at NACP-Rep1 in the alpha-synuclein gene, *Hum. Genet.*, 2003, **113**(5), 426–431, PubMed PMID: 12923682. Epub 2003/08/19. eng.
242. K. D. Cronin, D. Ge, P. Manninger, C. Linnertz, A. Rossoshek and B. M. Orrison, *et al.*, Expansion of the Parkinson disease-associated SNCA-Rep1 allele upregulates human alpha-synuclein in transgenic mouse brain, *Hum. Mol. Genet.*, 2009, **18**(17), 3274–3285, PubMed PMID: 19498036. Pubmed Central PMCID: PMC2722989. Epub 2009/06/06. eng.
243. F. Soldner, Y. Stelzer, C. S. Shivalila, B. J. Abraham, J. C. Latourelle and M. I. Barrasa, *et al.*, Parkinson-associated risk variant in distal enhancer of alpha-synuclein modulates target gene expression, *Nature*, 2016, **533**(7601), 95–99, PubMed PMID: 27096366. Pubmed Central PMCID: PMC5042324. Epub 2016/04/21. eng.
244. J. C. Rochet, B. A. Hay and M. Guo, Molecular insights into Parkinson's disease, *Prog. Mol. Biol. Transl. Sci.*, 2012, **107**, 125–188, PubMed PMID: 22482450. Epub 2012/04/10. eng.
245. K. Nakamura, alpha-Synuclein and mitochondria: partners in crime? *Neurotherapeutics*, 2013, **10**(3), 391–399, PubMed PMID: 23512373. Pubmed Central PMCID: PMC3701775. Epub 2013/03/21. eng.
246. A. Bose and M. F. Beal, Mitochondrial dysfunction in Parkinson's disease, *J. Neurochem.*, 2016, **139**(suppl. 1), 216–231, PubMed PMID: 27546335. Epub 2016/08/23. eng.
247. L. J. Martin, Y. Pan, A. C. Price, W. Sterling, N. G. Copeland and N. A. Jenkins, *et al.*, Parkinson's disease alpha-synuclein transgenic mice develop neuronal mitochondrial degeneration and cell death, *J. Neurosci.*, 2006, **26**(1), 41–50, PubMed PMID: 16399671.
248. N. B. Cole, D. Dieuliis, P. Leo, D. C. Mitchell and R. L. Nussbaum, Mitochondrial translocation of alpha-synuclein is promoted by intracellular acidification, *Exp. Cell Res.*, 2008, **314**(10), 2076–2089, PubMed PMID: 18440504.
249. K. Nakamura, V. M. Nemani, E. K. Wallender, K. Kaehlcke, M. Ott and R. H. Edwards, Optical reporters for the conformation of alpha-synuclein reveal a specific interaction with mitochondria, *J. Neurosci.*, 2008, **28**(47), 12305–12317, PubMed PMID: 19020024.
250. M. S. Parihar, A. Parihar, M. Fujita, M. Hashimoto and P. Ghafourifar, Mitochondrial association of alpha-synuclein causes oxidative stress, *Cell. Mol. Life Sci.*, 2008, **65**(7–8), 1272–1284, PubMed PMID: 18322646.
251. M. S. Parihar, A. Parihar, M. Fujita, M. Hashimoto and P. Ghafourifar, Alpha-synuclein overexpression and aggregation exacerbates impairment of mitochondrial functions by augmenting oxidative stress in human neuroblastoma cells, *Int. J. Biochem. Cell Biol.*, 2009, **41**(10), 2015–2024, PubMed PMID: 19460457.
252. L. Devi, V. Raghavendran, B. M. Prabhu, N. G. Avadhani and H. K. Anandatheerthavarada, Mitochondrial import and accumulation of

alpha-synuclein impair complex I in human dopaminergic neuronal cultures and Parkinson disease brain, *J. Biol. Chem.*, 2008, **283**(14), 9089–9100, PubMed PMID: 18245082.

253. F. Kamp, N. Exner, A. K. Lutz, N. Wender, J. Hegermann and B. Brunner, *et al.*, Inhibition of mitochondrial fusion by alpha-synuclein is rescued by PINK1, Parkin and DJ-1, *Embo Journal*, 2010, **29**(20), 3571–3589, PubMed PMID: 20842103. Pubmed Central PMCID: 2964170. Epub 2010/09/16. eng.
254. K. Nakamura, V. M. Nemani, F. Azarbal, G. Skibinski, J. M. Levy and K. Egami, *et al.*, Direct Membrane Association Drives Mitochondrial Fission by the Parkinson Disease-associated Protein {alpha}-Synuclein, *J. Biol. Chem.*, 2011, **286**(23), 20710–20726, PubMed PMID: 21489994. Pubmed Central PMCID: 3121472. Epub 2011/04/15. eng.
255. T. Cali, D. Ottolini, A. Negro and M. Brini, alpha-Synuclein controls mitochondrial calcium homeostasis by enhancing endoplasmic reticulum-mitochondria interactions, *J. Biol. Chem.*, 2012, **287**(22), 17914–17929, PubMed PMID: 22453917. Pubmed Central PMCID: 3365710. Epub 2012/03/29. eng.
256. C. Guardia-Laguarta, E. Area-Gomez, C. Rub, Y. Liu, J. Magrane and D. Becker, *et al.*, alpha-Synuclein is localized to mitochondria-associated ER membranes, *J. Neurosci.*, 2014, **34**(1), 249–259, PubMed PMID: 24381286. Pubmed Central PMCID: PMC3866487. Epub 2014/01/02. eng.
257. R. Di Maio, P. J. Barrett, E. K. Hoffman, C. W. Barrett, A. Zharikov and A. Borah, *et al.*, alpha-Synuclein binds to TOM20 and inhibits mitochondrial protein import in Parkinson's disease, *Sci. Transl. Med.*, 2016, **8**(342), 342ra78, PubMed PMID: 27280685. Epub 2016/06/10. eng.
258. V. Loeb, E. Yakunin, A. Saada and R. Sharon, The transgenic overexpression of alpha-synuclein and not its related pathology associates with complex I inhibition, *J. Biol. Chem.*, 2010, **285**(10), 7334–7343, PubMed PMID: 20053987. Pubmed Central PMCID: PMC2844181. Epub 2010/01/08. eng.
259. W. Dauer, N. Kholodilov, M. Vila, A. C. Trillat, R. Goodchild and K. E. Larsen, *et al.*, Resistance of alpha-synuclein null mice to the parkinsonian neurotoxin MPTP, *Proc. Natl. Acad. Sci. U. S. A.*, 2002, **99**(22), 14524–14529, PubMed PMID: 12376616. Pubmed Central PMCID: PMC137916. Epub 2002/10/12. eng.
260. A. D. Zharikov, J. R. Cannon, V. Tapias, Q. Bai, M. P. Horowitz and V. Shah, *et al.*, shRNA targeting alpha-synuclein prevents neurodegeneration in a Parkinson's disease model, *J. Clin. Invest.*, 2015, **121**(7), 2721–2735, PubMed PMID: 26075822. Epub 2015/06/16. Eng.
261. D. Ebrahimi-Fakhari, I. Cantuti-Castelvetri, Z. Fan, E. Rockenstein, E. Masliah and B. T. Hyman, *et al.*, Distinct roles *in vivo* for the ubiquitin-proteasome system and the autophagy-lysosomal pathway in the degradation of alpha-synuclein, *J. Neurosci.*, 2011, **31**(41), 14508–14520, PubMed PMID: 21994367. Pubmed Central PMCID: 3587176. Epub 2011/10/14. eng.

262. A. R. Winslow, C. W. Chen, S. Corrochano, A. Acevedo-Arozena, D. E. Gordon and A. A. Peden, *et al.*, alpha-Synuclein impairs macroautophagy: implications for Parkinson's disease, *J. Cell Biol.*, 2010, **190**(6), 1023–1037, PubMed PMID: 20855506. Pubmed Central PMCID: 3101586. Epub 2010/09/22. eng.
263. W. H. Yu, B. Dorado, H. Y. Figueroa, L. Wang, E. Planel and M. R. Cookson, *et al.*, Metabolic activity determines efficacy of macroautophagic clearance of pathological oligomeric alpha-synuclein, *Am. J. Pathol.*, 2009, **175**(2), 736–747, PubMed PMID: 19628769. Pubmed Central PMCID: 2716969. Epub 2009/07/25. eng.
264. J. X. Song, J. H. Lu, L. F. Liu, L. L. Chen, S. S. Durairajan and Z. Yue, *et al.*, HMGB1 is involved in autophagy inhibition caused by SNCA/alpha-synuclein overexpression: a process modulated by the natural autophagy inducer corynoxine B, *Autophagy*, 2014, **10**(1), 144–154, PubMed PMID: 24178442. Pubmed Central PMCID: PMC4389868. Epub 2013/11/02. eng.
265. J. Q. Yan, Y. H. Yuan, Y. N. Gao, J. Y. Huang, K. L. Ma and Y. Gao, *et al.*, Overexpression of human E46K mutant alpha-synuclein impairs macroautophagy *via* inactivation of JNK1-Bcl-2 pathway, *Mol. Neurobiol.*, 2014, **50**(2), 685–701, PubMed PMID: 24833599. Epub 2014/05/17. eng.
266. J. R. Mazzulli, F. Zunke, O. Isacson, L. Studer and D. Krainc, alpha-Synuclein-induced lysosomal dysfunction occurs through disruptions in protein trafficking in human midbrain synucleinopathy models, *Proc. Natl. Acad. Sci. U. S. A.*, 2016, **113**(7), 1931–1936, PubMed PMID: 26839413. Pubmed Central PMCID: PMC4763774. Epub 2016/02/04. eng.
267. T. Gasser, Mendelian forms of Parkinson's disease, *Biochim. Biophys. Acta*, 2009, **1792**(7), 587–596, PubMed PMID: 19168133.
268. T. Kitada, S. Asakawa, N. Hattori, H. Matsumine, Y. Yamamura and S. Minoshima, *et al.*, Mutations in the parkin gene cause autosomal recessive juvenile parkinsonism, *Nature*, 1998, **392**(6676), 605–608, PubMed PMID: 9560156.
269. C. B. Lucking, A. Durr, V. Bonifati, J. Vaughan, G. De Michele and T. Gasser, *et al.*, Association between early-onset Parkinson's disease and mutations in the parkin gene, *N. Engl. J. Med.*, 2000, **342**(21), 1560–1567, PubMed PMID: 10824074.
270. C. Klein, A. Grunewald and K. Hedrich, Early-onset parkinsonism associated with PINK1 mutations: frequency, genotypes, and phenotypes, *Neurology*, 2006, **66**(7), 1129–1130, author reply -30. PubMed PMID: 16606941.
271. A. J. Whitworth and L. J. Pallanck, The PINK1/Parkin pathway: a mitochondrial quality control system? *J. Bioenerg. Biomembr.*, 2009, **41**(6), 499–503, PubMed PMID: 19967438.
272. I. E. Clark, M. W. Dodson, C. Jiang, J. H. Cao, J. R. Huh and J. H. Seol, *et al.*, Drosophila pink1 is required for mitochondrial function and interacts genetically with parkin, *Nature*, 2006, **441**(7097), 1162–1166, PubMed PMID: 16672981.

273. Y. Yang, S. Gehrke, Y. Imai, Z. Huang, Y. Ouyang and J. W. Wang, *et al.*, Mitochondrial pathology and muscle and dopaminergic neuron degeneration caused by inactivation of Drosophila Pink1 is rescued by Parkin, *Proc. Natl. Acad. Sci. U. S. A.*, 2006, **103**(28), 10793–10798, PubMed PMID: 16818890. Pubmed Central PMCID: PMC1502310.
274. R. D. Mills, C. H. Sim, S. S. Mok, T. D. Mulhern, J. G. Culvenor and H. C. Cheng, Biochemical aspects of the neuroprotective mechanism of PTEN-induced kinase-1 (PINK1), *J. Neurochem.*, 2008, **105**(1), 18–33, PubMed PMID: 18221368.
275. N. Matsuda, S. Sato, K. Shiba, K. Okatsu, K. Saisho and C. A. Gautier, *et al.*, PINK1 stabilized by mitochondrial depolarization recruits Parkin to damaged mitochondria and activates latent Parkin for mitophagy, *J. Cell Biol.*, 2010, **189**(2), 211–221, PubMed PMID: 20404107. Pubmed Central PMCID: 2856912. Epub 2010/04/21. eng.
276. D. P. Narendra, S. M. Jin, A. Tanaka, D. F. Suen, C. A. Gautier and J. Shen, *et al.*, PINK1 is selectively stabilized on impaired mitochondria to activate Parkin, *PLoS Biol.*, 2010, **8**(1), e1000298, PubMed PMID: 20126261. Pubmed Central PMCID: 2811155. Epub 2010/02/04. eng.
277. K. Okatsu, T. Oka, M. Iguchi, K. Imamura, H. Kosako and N. Tani, *et al.*, PINK1 autophosphorylation upon membrane potential dissipation is essential for Parkin recruitment to damaged mitochondria, *Nat. Commun.*, 2012, **3**, 1016, PubMed PMID: 22910362. Pubmed Central PMCID: PMC3432468.
278. L. Aerts, K. Craessaerts, B. De Strooper and V. A. Morais, PINK1 kinase catalytic activity is regulated by phosphorylation on serines 228 and 402, *J. Biol. Chem.*, 2015, **290**(5), 2798–2811, PubMed PMID: 25527497. Pubmed Central PMCID: PMC4317039.
279. K. Okatsu, M. Uno, F. Koyano, E. Go, M. Kimura and T. Oka, *et al.*, A dimeric PINK1-containing complex on depolarized mitochondria stimulates Parkin recruitment, *J. Biol. Chem.*, 2013, **288**(51), 36372–36384, PubMed PMID: 24189060. Pubmed Central PMCID: PMC3868751.
280. F. Koyano, K. Okatsu, H. Kosako, Y. Tamura, E. Go and M. Kimura, *et al.*, Ubiquitin is phosphorylated by PINK1 to activate parkin, *Nature*, 2014, **510**(7503), 162–166, PubMed PMID: 24784582.
281. C. Vives-Bauza, C. Zhou, Y. Huang, M. Cui, R. L. de Vries and J. Kim, *et al.*, PINK1-dependent recruitment of Parkin to mitochondria in mitophagy, *Proc. Natl. Acad. Sci. U. S. A.*, 2010, **107**(1), 378–383, PubMed PMID: 19966284. Pubmed Central PMCID: PMC2806779.
282. T. Wauer and D. Komander, Structure of the human Parkin ligase domain in an autoinhibited state, *EMBO J.*, 2013, **32**(15), 2099–2112, PubMed PMID: 23727886. Pubmed Central PMCID: PMC3730226.
283. J. F. Trempe, V. Sauve, K. Grenier, M. Seirafi, M. Y. Tang and M. Menade, *et al.*, Structure of parkin reveals mechanisms for ubiquitin ligase activation, *Science*, 2013, **340**(6139), 1451–1455, PubMed PMID: 23661642. Epub 2013/05/11. eng.

284. C. Kondapalli, A. Kazlauskaite, N. Zhang, H. I. Woodroof, D. G. Campbell and R. Gourlay, *et al.*, PINK1 is activated by mitochondrial membrane potential depolarization and stimulates Parkin E3 ligase activity by phosphorylating Serine 65, *Open Biol.*, 2012, **2**(5), 120080, PubMed PMID: 22724072. Pubmed Central PMCID: PMC3376738.
285. M. Iguchi, Y. Kujuro, K. Okatsu, F. Koyano, H. Kosako and M. Kimura, *et al.*, Parkin-catalyzed ubiquitin-ester transfer is triggered by PINK1-dependent phosphorylation, *J. Biol. Chem.*, 2013, **288**(30), 22019–22032, PubMed PMID: 23754282. Pubmed Central PMCID: PMC3724655.
286. S. Geisler, K. M. Holmstrom, D. Skujat, F. C. Fiesel, O. C. Rothfuss and P. J. Kahle, *et al.*, PINK1/Parkin-mediated mitophagy is dependent on VDAC1 and p62/SQSTM1, *Nat. Cell Biol.*, 2010, **12**(2), 119–131, PubMed PMID: 20098416. Epub 2010/01/26. eng.
287. A. P. Joselin, S. J. Hewitt, S. M. Callaghan, R. H. Kim, Y. H. Chung and T. W. Mak, *et al.*, ROS-dependent regulation of Parkin and DJ-1 localization during oxidative stress in neurons, *Hum. Mol. Genet.*, 2012, **21**(22), 4888–4903, PubMed PMID: 22872702. Epub 2012/08/09. eng.
288. M. Lazarou, D. A. Sliter, L. A. Kane, S. A. Sarraf, C. Wang and J. L. Burman, *et al.*, The ubiquitin kinase PINK1 recruits autophagy receptors to induce mitophagy, *Nature*, 2015, **524**(7565), 309–314, PubMed PMID: 26266977. Pubmed Central PMCID: PMC5018156.
289. S. Pankiv, T. H. Clausen, T. Lamark, A. Brech, J. A. Bruun and H. Outzen, *et al.*, p62/SQSTM1 binds directly to Atg8/LC3 to facilitate degradation of ubiquitinated protein aggregates by autophagy, *J. Biol. Chem.*, 2007, **282**(33), 24131–24145, PubMed PMID: 17580304.
290. N. C. Chan, A. M. Salazar, A. H. Pham, M. J. Sweredoski, N. J. Kolawa and R. L. Graham, *et al.*, Broad activation of the ubiquitin-proteasome system by Parkin is critical for mitophagy, *Hum. Mol. Genet.*, 2011, **20**(9), 1726–1737, PubMed PMID: 21296869. Pubmed Central PMCID: PMC3071670. Epub 2011/02/08. eng.
291. A. Schreiber and M. Peter, Substrate recognition in selective autophagy and the ubiquitin-proteasome system, *Biochim. Biophys. Acta*, 2014, **1843**(1), 163–181, PubMed PMID: 23545414. Epub 2013/04/03. eng.
292. S. Abu-Hamad, N. Arbel, D. Calo, L. Arzoine, A. Israelson and N. Keinan, *et al.*, The VDAC1 N-terminus is essential both for apoptosis and the protective effect of anti-apoptotic proteins, *J. Cell Sci.*, 2009, **122**(pt 11), 1906–1916, PubMed PMID: 19461077.
293. J. R. Liang, A. Martinez, J. D. Lane, U. Mayor, M. J. Clague and S. Urbe, USP30 deubiquitylates mitochondrial Parkin substrates and restricts apoptotic cell death, *EMBO Rep.*, 2015, **16**(5), 618–627, PubMed PMID: 25739811. Pubmed Central PMCID: PMC4428036.
294. B. Bingol, J. S. Tea, L. Phu, M. Reichelt, C. E. Bakalarski and Q. Song, *et al.*, The mitochondrial deubiquitinase USP30 opposes parkin-mediated mitophagy, *Nature*, 2014, **510**(7505), 370–375, PubMed PMID: 24896179.
295. Y. Wang, M. Serricchio, M. Jauregui, R. Shanbhag, T. Stoltz and C. T. Di Paolo, *et al.*, Deubiquitinating enzymes regulate PARK2-mediated

mitophagy, *Autophagy*, 2015, **11**(4), 595–606, PubMed PMID: 25915564. Pubmed Central PMCID: PMC4502823.

296. N. Nakamura and S. Hirose, Regulation of mitochondrial morphology by USP30, a deubiquitinating enzyme present in the mitochondrial outer membrane, *Mol. Biol. Cell*, 2008, **19**(5), 1903–1911, PubMed PMID: 18287522. Pubmed Central PMCID: PMC2366858.
297. F. Mouton-Liger, M. Jacoupy, J. C. Corvol and O. Corti, PINK1/Parkin-dependent mitochondrial surveillance: from pleiotropy to Parkinson's disease, *Front. Mol. Neurosci.*, 2017, **10**:120, PubMed PMID: 28507507. Pubmed Central PMCID: PMC5410576.
298. O. S. Shirihai, M. Song and G. W. Dorn 2nd, How mitochondrial dynamism orchestrates mitophagy, *Circ. Res.*, 2015, **116**(11), 1835–1849, PubMed PMID: 25999423. Pubmed Central PMCID: PMC4443843. Epub 2015/05/23. eng.
299. G. Twig and O. S. Shirihai, The interplay between mitochondrial dynamics and mitophagy, *Antioxid. Redox Signaling*, 2011, **14**(10), 1939–1951, PubMed PMID: 21128700. Pubmed Central PMCID: PMC3078508. Epub 2010/12/07. eng.
300. D. T. Chang and I. J. Reynolds, Mitochondrial trafficking and morphology in healthy and injured neurons, *Prog Neurobiol.*, 2006, **80**(5), 241–268, PubMed PMID: 17188795.
301. S. Liu, T. Sawada, S. Lee, W. Yu, G. Silverio and P. Alapatt, *et al.*, Parkinson's disease-associated kinase PINK1 regulates Miro protein level and axonal transport of mitochondria, *PLoS Genet.*, 2012, **8**(3), e1002537, PubMed PMID: 22396657. Pubmed Central PMCID: PMC3291531.
302. B. Westermann, Bioenergetic role of mitochondrial fusion and fission, *Biochim. Biophys. Acta*, 2012, **1817**(10), 1833–1838, PubMed PMID: 22409868.
303. E. Ziviani, R. N. Tao and A. J. Whitworth, Drosophila parkin requires PINK1 for mitochondrial translocation and ubiquitinates mitofusin, *Proc. Natl. Acad. Sci. U. S. A.*, 2010, **107**(11), 5018–5023, PubMed PMID: 20194754. Pubmed Central PMCID: PMC2841909.
304. H. Wang, P. Song, L. Du, W. Tian, W. Yue and M. Liu, *et al.*, Parkin ubiquitinates Drp1 for proteasome-dependent degradation: implication of dysregulated mitochondrial dynamics in Parkinson disease, *J. Biol. Chem.*, 2011, **286**(13), 11649–11658, PubMed PMID: 21292769. Pubmed Central PMCID: PMC3064217.
305. Y. Chen and G. W. Dorn 2nd, PINK1-phosphorylated mitofusin 2 is a Parkin receptor for culling damaged mitochondria, *Science*, 2013, **340**(6131), 471–475, PubMed PMID: 23620051. Pubmed Central PMCID: PMC3774525.
306. L. Pallanck, Mitophagy: mitofusin recruits a mitochondrial killer, *Curr. Biol.*, 2013, **23**(13), R570–R572, PubMed PMID: 23845246.
307. A. C. Poole, R. E. Thomas, S. Yu, E. S. Vincow and L. Pallanck, The mitochondrial fusion-promoting factor mitofusin is a substrate of the PINK1/parkin pathway, *PLoS One*, 2010, **5**(4), e10054, PubMed PMID: 20383334. Pubmed Central PMCID: PMC2850930.

308. A. Nardin, E. Schrepfer and E. Ziviani, Counteracting PINK/Parkin Deficiency in the Activation of Mitophagy: A Potential Therapeutic Intervention for Parkinson's Disease, *Curr. Neuropharmacol.*, 2016, **14**(3), 250–259, PubMed PMID: 26517048. Pubmed Central PMCID: PMC4857628.
309. A. Tanaka, M. M. Cleland, S. Xu, D. P. Narendra, D. F. Suen and M. Karbowski, *et al.*, Proteasome and p97 mediate mitophagy and degradation of mitofusins induced by Parkin, *J. Cell Biol.*, 2010, **191**(7), 1367–1380, PubMed PMID: 21173115. Pubmed Central PMCID: PMC3010068. Epub 2010/12/22. eng.
310. L. Glauser, S. Sonnay, K. Stafa and D. J. Moore, Parkin promotes the ubiquitination and degradation of the mitochondrial fusion factor mitofusin 1, *J. Neurochem.*, 2011, **118**(4), 636–645, PubMed PMID: 21615408. Epub 2011/05/28. eng.
311. R. K. Dagda, S. J. Cherra 3rd, S. M. Kulich, A. Tandon, D. Park and C. T. Chu, Loss of PINK1 function promotes mitophagy through effects on oxidative stress and mitochondrial fission, *J. Biol. Chem.*, 2009, **284**(20), 13843–13855, PubMed PMID: 19279012. Pubmed Central PMCID: PMC2679485.
312. X. Wang, D. Winter, G. Ashrafi, J. Schlehe, Y. L. Wong and D. Selkoe, *et al.*, PINK1 and Parkin target Miro for phosphorylation and degradation to arrest mitochondrial motility, *Cell*, 2011, **147**(4), 893–906, PubMed PMID: 22078885. Pubmed Central PMCID: PMC3261796.
313. N. Birsa, R. Norkett, T. Wauer, T. E. Mevissen, H. C. Wu and T. Foltynie, *et al.*, Lysine 27 ubiquitination of the mitochondrial transport protein Miro is dependent on serine 65 of the Parkin ubiquitin ligase, *J. Biol. Chem.*, 2014, **289**(21), 14569–14582, PubMed PMID: 24671417. Pubmed Central PMCID: PMC4031514.
314. A. Ordureau, J. M. Heo, D. M. Duda, J. A. Paulo, J. L. Olszewski and D. Yanishevski, *et al.*, Defining roles of PARKIN and ubiquitin phosphorylation by PINK1 in mitochondrial quality control using a ubiquitin replacement strategy, *Proc. Natl. Acad. Sci. U. S. A.*, 2015, **112**(21), 6637–6642, PubMed PMID: 25969509. Pubmed Central PMCID: PMC4450373.
315. V. Bonifati, P. Rizzu, M. J. van Baren, O. Schaap, G. J. Breedveld and E. Krieger, *et al.*, Mutations in the DJ-1 gene associated with autosomal recessive early-onset parkinsonism, *Science*, 2003, **299**(5604), 256–259, PubMed PMID: 12446870.
316. P. J. Kahle, J. Waak and T. Gasser, DJ-1 and prevention of oxidative stress in Parkinson's disease and other age-related disorders, *Free Radical Biol. Med.*, 2009, **47**(10), 1354–1361, PubMed PMID: 19686841. Epub 2009/08/19. eng.
317. P. M. Abou-Sleiman, D. G. Healy, N. Quinn, A. J. Lees and N. W. Wood, The role of pathogenic DJ-1 mutations in Parkinson's disease, *Ann. Neurol.*, 2003, **54**(3), 283–286, PubMed PMID: 12953260. Epub 2003/09/04. eng.

318. R. Hering, K. M. Strauss, X. Tao, A. Bauer, D. Woitalla and E. M. Mietz, *et al.*, Novel homozygous p.E64D mutation in DJ1 in early onset Parkinson disease (PARK7), *Hum. Mutat.*, 2004, **24**(4), 321–329, PubMed PMID: 15365989. Epub 2004/09/15. eng.
319. G. Annesi, G. Savettieri, P. Pugliese, M. D'Amelio, P. Tarantino and P. Ragonese, *et al.*, DJ-1 mutations and parkinsonism-dementia-amyotrophic lateral sclerosis complex, *Ann. Neurol.*, 2005, **58**(5), 803–807, PubMed PMID: 16240358. Epub 2005/10/22. eng.
320. L. N. Clark, S. Afridi, H. Mejia-Santana, J. Harris, E. D. Louis and L. J. Cote, *et al.*, Analysis of an early-onset Parkinson's disease cohort for DJ-1 mutations, *Mov. Disord.*, 2004, **19**(7), 796–800, PubMed PMID: 15254937. Epub 2004/07/16. eng.
321. J. A. Olzmann, K. Brown, K. D. Wilkinson, H. D. Rees, Q. Huai and H. Ke, *et al.*, Familial Parkinson's disease-associated L166P mutation disrupts DJ-1 protein folding and function, *J. Biol. Chem.*, 2004, **279**(9), 8506–8515, PubMed PMID: 14665635.
322. S. Baulac, M. J. LaVoie, J. Strahle, M. G. Schlossmacher and W. Xia, Dimerization of Parkinson's disease-causing DJ-1 and formation of high molecular weight complexes in human brain, *Mol. Cell. Neurosci.*, 2004, **27**(3), 236–246, PubMed PMID: 15519239. Epub 2004/11/03. eng.
323. J. D. Hulleman, H. Mirzaei, E. Guigard, K. L. Taylor, S. S. Ray and C. M. Kay, *et al.*, Destabilization of DJ-1 by familial substitution and oxidative modifications: implications for Parkinson's disease, *Biochemistry*, 2007, **46**(19), 5776–5789, PubMed PMID: 17451229. Epub 2007/04/25. eng.
324. A. G. Madian, J. Hindupur, J. D. Hulleman, N. Diaz-Maldonado, V. R. Mishra and E. Guigard, *et al.*, Effect of single amino acid substitution on oxidative modifications of the Parkinson's disease-related protein, DJ-1, *Mol. Cell. Proteomics*, 2012, **11**(2), M111.010892, PubMed PMID: 22104028. Pubmed Central PMCID: 3277756. Epub 2011/11/23. eng.
325. M. A. Wilson, J. L. Collins, Y. Hod, D. Ringe and G. A. Petsko, The 1.1-A resolution crystal structure of DJ-1, the protein mutated in autosomal recessive early onset Parkinson's disease, *Proc. Natl. Acad. Sci. U. S. A.*, 2003, **100**(16), 9256–9261, PubMed PMID: 12855764.
326. X. Tao and L. Tong, Crystal structure of human DJ-1, a protein associated with early onset Parkinson's disease, *J. Biol. Chem.*, 2003, **278**(33), 31372–31379, PubMed PMID: 12761214. Epub 2003/05/23. eng.
327. K. Honbou, N. N. Suzuki, M. Horiuchi, T. Niki, T. Taira and H. Ariga, *et al.*, The crystal structure of DJ-1, a protein related to male fertility and Parkinson's disease, *J. Biol. Chem.*, 2003, **278**(33), 31380–31384, PubMed PMID: 12796482. Epub 2003/06/11. eng.
328. Q. Huai, Y. Sun, H. Wang, L. S. Chin, L. Li and H. Robinson, *et al.*, Crystal structure of DJ-1/RS and implication on familial Parkinson's disease, *FEBS Lett.*, 2003, **549**(1–3), 171–175, PubMed PMID: 12914946. Epub 2003/08/14. eng.

329. S. J. Lee, S. J. Kim, I. K. Kim, J. Ko, C. S. Jeong and G. H. Kim, *et al.*, Crystal structures of human DJ-1 and Escherichia coli Hsp31, which share an evolutionarily conserved domain, *J. Biol. Chem.*, 2003, **278**(45), 44552–44559, PubMed PMID: 12939276. Epub 2003/08/27. eng.
330. G. Malgieri and D. Eliezer, Structural effects of Parkinson's disease linked DJ-1 mutations, *Protein Sci.*, 2008, **17**(5), 855–868, PubMed PMID: 18436956. Pubmed Central PMCID: 2327288. Epub 2008/04/26. eng.
331. T. Kinumi, J. Kimata, T. Taira, H. Ariga and E. Niki, Cysteine-106 of DJ-1 is the most sensitive cysteine residue to hydrogen peroxide-mediated oxidation *in vivo* in human umbilical vein endothelial cells, *Biochem. Biophys. Res. Commun.*, 2004, **317**(3), 722–728, PubMed PMID: 15081400. Epub 2004/04/15. eng.
332. W. Zhou, M. Zhu, M. A. Wilson, G. A. Petsko and A. L. Fink, The oxidation state of DJ-1 regulates its chaperone activity toward alpha-synuclein, *J. Mol. Biol.*, 2006, **356**(4), 1036–1048, PubMed PMID: 16403519.
333. E. Andres-Mateos, C. Perier, L. Zhang, B. Blanchard-Fillion, T. M. Greco and B. Thomas, *et al.*, DJ-1 gene deletion reveals that DJ-1 is an atypical peroxiredoxin-like peroxidase, *Proc. Natl. Acad. Sci. U. S. A.*, 2007, **104**(37), 14807–14812, PubMed PMID: 17766438. Pubmed Central PMCID: 1976193. Epub 2007/09/04. eng.
334. A. C. Witt, M. Lakshminarasimhan, B. C. Remington, S. Hasim, E. Pozharski and M. A. Wilson, Cysteine pKa depression by a protonated glutamic acid in human DJ-1, *Biochemistry*, 2008, **47**(28), 7430–7440, PubMed PMID: 18570440. Pubmed Central PMCID: 2760839. Epub 2008/06/24. eng.
335. R. M. Canet-Aviles, M. A. Wilson, D. W. Miller, R. Ahmad, C. McLendon and S. Bandyopadhyay, *et al.*, The Parkinson's disease protein DJ-1 is neuroprotective due to cysteine-sulfinic acid-driven mitochondrial localization, *Proc. Natl. Acad. Sci. U. S. A.*, 2004, **101**(24), 9103–9108, PubMed PMID: 15181200.
336. J. Blackinton, M. Lakshminarasimhan, K. J. Thomas, R. Ahmad, E. Greggio and A. S. Raza, *et al.*, Formation of a stabilized cysteine sulfinic acid is critical for the mitochondrial function of the parkinsonism protein DJ-1, *J. Biol. Chem.*, 2009, **284**(10), 6476–6485, PubMed PMID: 19124468. Pubmed Central PMCID: 2649108. Epub 2009/01/07. eng.
337. J. Choi, M. C. Sullards, J. A. Olzmann, H. D. Rees, S. T. Weintraub and D. E. Bostwick, *et al.*, Oxidative damage of DJ-1 is linked to sporadic Parkinson and Alzheimer diseases, *J. Biol. Chem.*, 2006, **281**(16), 10816–10824, PubMed PMID: 16517609.
338. M. C. Meulener, K. Xu, L. Thomson, H. Ischiropoulos and N. M. Bonini, Mutational analysis of DJ-1 in Drosophila implicates functional inactivation by oxidative damage and aging, *Proc. Natl. Acad. Sci. U. S. A.*, 2006, **103**(33), 12517–12522, PubMed PMID: 16894167. Pubmed Central PMCID: 1533799. Epub 2006/08/09. eng.

339. K. Takahashi-Niki, T. Niki, T. Taira, S. M. Iguchi-Ariga and H. Ariga, Reduced anti-oxidative stress activities of DJ-1 mutants found in Parkinson's disease patients, *Biochem. Biophys. Res. Commun.*, 2004, **320**(2), 389–397, PubMed PMID: 15219840. Epub 2004/06/29. eng.
340. C. Martinat, S. Shendelman, A. Jonason, T. Leete, M. F. Beal and L. Yang, *et al.*, Sensitivity to oxidative stress in DJ-1-deficient dopamine neurons: an ES- derived cell model of primary Parkinsonism, *PLoS Biol.*, 2004, **2**(11), e327, PubMed PMID: 15502868. Pubmed Central PMCID: 521171. Epub 2004/10/27. eng.
341. R. H. Kim, P. D. Smith, H. Aleyasin, S. Hayley, M. P. Mount and S. Pownall, *et al.*, Hypersensitivity of DJ-1-deficient mice to 1-methyl-4-phenyl-1,2,3,6-tetrahydropyrindine (MPTP) and oxidative stress, *Proc. Natl. Acad. Sci. U. S. A.*, 2005, **102**, 5215–5220, PubMed PMID: 15784737.
342. W. Zhou and C. R. Freed, DJ-1 up-regulates glutathione synthesis during oxidative stress and inhibits A53T alpha-synuclein toxicity, *J. Biol. Chem.*, 2005, **280**(52), 43150–43158, PubMed PMID: 16227205. Epub 2005/10/18. eng.
343. E. Junn, W. H. Jang, X. Zhao, B. S. Jeong and M. M. Mouradian, Mitochondrial localization of DJ-1 leads to enhanced neuroprotection, *J. Neurosci. Res.*, 2009, **87**(1), 123–129, PubMed PMID: 18711745. Pubmed Central PMCID: 2752655. Epub 2008/08/20. eng.
344. L. Zhang, M. Shimoji, B. Thomas, D. J. Moore, S. W. Yu and N. I. Marupudi, *et al.*, Mitochondrial localization of the Parkinson's disease related protein DJ-1: implications for pathogenesis, *Hum. Mol. Genet.*, 2005, **14**(14), 2063–2073, PubMed PMID: 15944198. Epub 2005/06/10. eng.
345. T. Cali, D. Ottolini, M. E. Soriano and M. Brini, A new split-GFP-based probe reveals DJ-1 translocation into the mitochondrial matrix to sustain ATP synthesis upon nutrient deprivation, *Hum. Mol. Genet.*, 2015, **24**(4), 1045–1060, PubMed PMID: 25305074. Epub 2014/10/12. eng.
346. C. Maita, H. Maita, S. M. Iguchi-Ariga and H. Ariga, Monomer DJ-1 and its N-terminal sequence are necessary for mitochondrial localization of DJ-1 mutants, *PLoS One*, 2013, **8**(1), e54087, PubMed PMID: 23326576. Pubmed Central PMCID: 3542337. Epub 2013/01/18. eng.
347. G. Krebiehl, S. Ruckerbauer, L. F. Burbulla, N. Kieper, B. Maurer and J. Waak, *et al.*, Reduced basal autophagy and impaired mitochondrial dynamics due to loss of Parkinson's disease-associated protein DJ-1, *PLoS One*, 2010, **5**(2), e9367, PubMed PMID: 20186336. Pubmed Central PMCID: 2826413. Epub 2010/02/27. eng.
348. E. Giaime, H. Yamaguchi, C. A. Gautier, T. Kitada and J. Shen, Loss of DJ-1 does not affect mitochondrial respiration but increases ROS production and mitochondrial permeability transition pore opening, *PLoS One*, 2012, **7**(7), e40501, PubMed PMID: 22792356. Pubmed Central PMCID: PMC3392228. Epub 2012/07/14. eng.
349. J. Y. Heo, J. H. Park, S. J. Kim, K. S. Seo, J. S. Han and S. H. Lee, *et al.*, DJ-1 null dopaminergic neuronal cells exhibit defects in mitochondrial

function and structure: involvement of mitochondrial complex I assembly, *PLoS One*, 2012, **7**(3), e32629, PubMed PMID: 22403686. Pubmed Central PMCID: PMC3293835. Epub 2012/03/10. eng.

350. X. Wang, T. G. Petrie, Y. Liu, J. Liu, H. Fujioka and X. Zhu, Parkinson's disease-associated DJ-1 mutations impair mitochondrial dynamics and cause mitochondrial dysfunction, *J. Neurochem.*, 2012, **121**(5), 830–839, PubMed PMID: 22428580. Pubmed Central PMCID: PMC3740560. Epub 2012/03/21. eng.
351. I. Irrcher, H. Aleyasin, E. L. Seifert, S. J. Hewitt, S. Chhabra and M. Phillips, *et al.*, Loss of the Parkinson's disease-linked gene DJ-1 perturbs mitochondrial dynamics, *Hum. Mol. Genet.*, 2010, **19**(19), 3734–3746, PubMed PMID: 20639397. Epub 2010/07/20. eng.
352. K. J. Thomas, M. K. McCoy, J. Blackinton, A. Beilina, M. van der Brug and A. Sandebring, *et al.*, DJ-1 acts in parallel to the PINK1/parkin pathway to control mitochondrial function and autophagy, *Hum. Mol. Genet.*, 2011, **20**(1), 40–50, PubMed PMID: 20940149. Pubmed Central PMCID: 3000675. Epub 2010/10/14. eng.
353. L. Y. Hao, B. I. Giasson and N. M. Bonini, DJ-1 is critical for mitochondrial function and rescues PINK1 loss of function, *Proc. Natl. Acad. Sci. U. S. A.*, 2010, **107**(21), 9747–9752, PubMed PMID: 20457924. Pubmed Central PMCID: 2906840. Epub 2010/05/12. eng.
354. E. Dragicevic, J. Schiemann and B. Liss, Dopamine midbrain neurons in health and Parkinson's disease: emerging roles of voltage-gated calcium channels and ATP-sensitive potassium channels, *Neuroscience*, 2015, **284**, 798–814, PubMed PMID: 25450964. Epub 2014/12/03. eng.
355. D. J. Surmeier, J. N. Guzman, J. Sanchez and P. T. Schumacker, Physiological phenotype and vulnerability in Parkinson's disease, *Cold Spring Harbor Perspect. Med.*, 2012, **2**(7), a009290, PubMed PMID: 22762023. Pubmed Central PMCID: PMC3385938. Epub 2012/07/05. eng.
356. D. I. Dryanovski, J. N. Guzman, Z. Xie, D. J. Galteri, L. A. Volpicelli-Daley and V. M. Lee, *et al.*, Calcium entry and alpha-synuclein inclusions elevate dendritic mitochondrial oxidant stress in dopaminergic neurons, *J. Neurosci.*, 2013, **33**(24), 10154–10164, PubMed PMID: 23761910. Pubmed Central PMCID: PMC3682382. Epub 2013/06/14. eng.
357. J. N. Guzman, J. Sanchez-Padilla, D. Wokosin, J. Kondapalli, E. Ilijic and P. T. Schumacker, *et al.*, Oxidant stress evoked by pacemaking in dopaminergic neurons is attenuated by DJ-1, *Nature*, 2010, **468**(7324), 696–700, PubMed PMID: 21068725. Epub 2010/11/12. eng.
358. H. Gao, W. Yang, Z. Qi, L. Lu, C. Duan and C. Zhao, *et al.*, DJ-1 protects dopaminergic neurons against rotenone-induced apoptosis by enhancing ERK-dependent mitophagy, *J. Mol. Biol.*, 2012, **423**(2), 232–248, PubMed PMID: 22898350. Epub 2012/08/18. eng.
359. D. J. Hampshire, E. Roberts, Y. Crow, J. Bond, A. Mubaidin and A. L. Wriekat, *et al.*, Kufor-Rakeb syndrome, pallido-pyramidal degeneration with supranuclear upgaze paresis and dementia, maps to 1p36, *J. Med. Genet.*, 2001, **38**(10), 680–682, PubMed PMID: 11584046. Pubmed Central PMCID: 1734748. Epub 2001/10/05. eng.

360. A. Ramirez, A. Heimbach, J. Grundemann, B. Stiller, D. Hampshire and L. P. Cid, *et al.*, Hereditary parkinsonism with dementia is caused by mutations in ATP13A2, encoding a lysosomal type 5 P-type ATPase, *Nat. Genet.*, 2006, **38**(10), 1184–1191, PubMed PMID: 16964263.
361. J. Ugolino, S. Fang, C. Kubisch and M. J. Monteiro, Mutant Atp13a2 proteins involved in parkinsonism are degraded by ER-associated degradation and sensitize cells to ER-stress induced cell death, *Hum. Mol. Genet.*, 2011, **20**(18), 3565–3577, PubMed PMID: 21665991. Pubmed Central PMCID: 3159557. Epub 2011/06/15. eng.
362. W. Kuhlbrandt, Biology, structure and mechanism of P-type ATPases, *Nat. Rev. Mol. Cell Biol.*, 2004, **5**(4), 282–295, PubMed PMID: 15071553.
363. B. Dehay, A. Ramirez, M. Martinez-Vicente, C. Perier, M. H. Canron and E. Doudnikoff, *et al.*, Loss of P-type ATPase ATP13A2/PARK9 function induces general lysosomal deficiency and leads to Parkinson disease neurodegeneration, *Proc. Natl. Acad. Sci. U. S. A.*, 2012, **109**(24), 9611–9616, PubMed PMID: 22647602. Epub 2012/06/01. eng.
364. A. Podhajska, A. Musso, A. Trancikova, K. Stafa, R. Moser and S. Sonnay, *et al.*, Common pathogenic effects of missense mutations in the P-type ATPase ATP13A2 (PARK9) associated with early-onset parkinsonism, *PLoS One*, 2012, **7**(6), e39942, PubMed PMID: 22768177. Pubmed Central PMCID: 3386943. Epub 2012/07/07. eng.
365. D. Ramonet, A. Podhajska, K. Stafa, S. Sonnay, A. Trancikova and E. Tsika, *et al.*, PARK9-associated ATP13A2 localizes to intracellular acidic vesicles and regulates cation homeostasis and neuronal integrity, *Hum. Mol. Genet.*, 2012, **21**(8), 1725–1743, PubMed PMID: 22186024. Pubmed Central PMCID: 3465694. Epub 2011/12/22. eng.
366. M. Usenovic, E. Tresse, J. R. Mazzulli, J. P. Taylor and D. Krainc, Deficiency of ATP13A2 leads to lysosomal dysfunction, alpha-synuclein accumulation, and neurotoxicity, *J. Neurosci.*, 2012, **32**(12), 4240–4246, PubMed PMID: 22442086. Pubmed Central PMCID: PMC3462811. Epub 2012/03/24. eng.
367. H. Matsui, F. Sato, S. Sato, M. Koike, Y. Taruno and S. Saiki, *et al.*, ATP13A2 deficiency induces a decrease in cathepsin D activity, fingerprint-like inclusion body formation, and selective degeneration of dopaminergic neurons, *FEBS Lett.*, 2013, **587**(9), 1316–1325, PubMed PMID: 23499937. Epub 2013/03/19. eng.
368. S. M. Kong, B. K. Chan, J. S. Park, K. J. Hill, J. B. Aitken and L. Cottle, *et al.*, Parkinson's disease-linked human PARK9/ATP13A2 maintains zinc homeostasis and promotes alpha-Synuclein externalization *via* exosomes, *Hum. Mol. Genet.*, 2014, **23**(11), 2816–2833, PubMed PMID: 24603074. Epub 2014/03/08. eng.
369. A. D. Gitler, A. Chesi, M. L. Geddie, K. E. Strathearn, S. Hamamichi and K. J. Hill, *et al.*, Alpha-synuclein is part of a diverse and highly conserved interaction network that includes PARK9 and manganese toxicity, *Nat. Genet.*, 2009, **41**(3), 308–315, PubMed PMID: 19182805.
370. A. Chesi, A. Kilaru, X. Fang, A. A. Cooper and A. D. Gitler, The role of the Parkinson's disease gene PARK9 in essential cellular pathways and

the manganese homeostasis network in yeast, *PLoS One*, 2012, **7**(3), e34178, PubMed PMID: 22457822. Pubmed Central PMCID: 3311584. Epub 2012/03/30. eng.

371. J. P. Covy, E. A. Waxman and B. I. Giasson, Characterization of cellular protective effects of ATP13A2/PARK9 expression and alterations resulting from pathogenic mutants, *J. Neurosci. Res.*, 2012, **90**(12), 2306–2316, PubMed PMID: 22847264. Epub 2012/08/01. eng.
372. J. S. Park, B. Koentjoro, D. Veivers, A. Mackay-Sim and C. M. Sue, Parkinson's disease-associated human ATP13A2 (PARK9) deficiency causes zinc dyshomeostasis and mitochondrial dysfunction, *Hum. Mol. Genet.*, 2014, **23**(11), 2802–2815, PubMed PMID: 24399444. Pubmed Central PMCID: 4014187. Epub 2014/01/09. eng.
373. T. Tsunemi and D. Krainc, Zn2+ dyshomeostasis caused by loss of ATP13A2/PARK9 leads to lysosomal dysfunction and alpha-synuclein accumulation, *Hum. Mol. Genet.*, 2014, **23**(11), 2791–2801, PubMed PMID: 24334770. Pubmed Central PMCID: 4014186. Epub 2013/12/18. eng.
374. A. M. Gusdon, J. Zhu, B. Van Houten and C. T. Chu, ATP13A2 regulates mitochondrial bioenergetics through macroautophagy, *Neurobiol. Dis.*, 2012, **45**(3), 962–972, PubMed PMID: 22198378. Pubmed Central PMCID: 3291101. Epub 2011/12/27. eng.
375. P. J. Schultheis, S. M. Fleming, A. K. Clippinger, J. Lewis, T. Tsunemi and B. Giasson, *et al.*, Atp13a2-deficient mice exhibit neuronal ceroid lipofuscinosis, limited alpha-synuclein accumulation and age-dependent sensorimotor deficits, *Hum. Mol. Genet.*, 2013, **22**(10), 2067–2082, PubMed PMID: 23393156. Pubmed Central PMCID: PMC3633373. Epub 2013/02/09. eng.
376. L. R. Kett, B. Stiller, M. M. Bernath, I. Tasset, J. Blesa and V. Jackson-Lewis, *et al.*, alpha-Synuclein-independent histopathological and motor deficits in mice lacking the endolysosomal Parkinsonism protein Atp13a2, *J. Neurosci.*, 2015, **35**(14), 5724–5742, PubMed PMID: 25855184. Pubmed Central PMCID: PMC4388928. Epub 2015/04/10. eng.
377. M. Usenovic and D. Krainc, Lysosomal dysfunction in neurodegeneration: the role of ATP13A2/PARK9, *Autophagy*, 2012, **8**(6), 987–988, PubMed PMID: 22561922. Pubmed Central PMCID: 3427268. Epub 2012/05/09. eng.
378. T. Tsunemi, K. Hamada and D. Krainc, ATP13A2/PARK9 regulates secretion of exosomes and alpha-synuclein, *J. Neurosci.*, 2014, **34**(46), 15281–15287, PubMed PMID: 25392495. Pubmed Central PMCID: PMC4228131. Epub 2014/11/14. eng.
379. B. Dehay, M. Martinez-Vicente, G. A. Caldwell, K. A. Caldwell, Z. Yue and M. R. Cookson, *et al.*, Lysosomal impairment in Parkinson's disease, *Mov. Disord*, 2013, **28**(6), 725–732, PubMed PMID: 23580333. Epub 2013/04/13. eng.
380. G. Daniel, A. Musso, E. Tsika, A. Fiser, L. Glauser and O. Pletnikova, *et al.*, alpha-Synuclein-induced dopaminergic neurodegeneration in a rat model of Parkinson's disease occurs independent of ATP13A2 (PARK9),

Neurobiol. Dis., 2015, **73**, 229–243, PubMed PMID: 25461191. Epub 2014/12/03. eng.

381. T. Holemans, D. M. Sorensen, S. van Veen, S. Martin, D. Hermans and G. C. Kemmer, *et al.*, A lipid switch unlocks Parkinson's disease-associated ATP13A2, *Proc. Natl. Acad. Sci. U. S. A.*, 2015, **112**(29), 9040–9045, PubMed PMID: 26134396. Pubmed Central PMCID: PMC4517210. Epub 2015/07/03. eng.
382. A. Grunewald, B. Arns, P. Seibler, A. Rakovic, A. Munchau and A. Ramirez, *et al.*, ATP13A2 mutations impair mitochondrial function in fibroblasts from patients with Kufor-Rakeb syndrome, *Neurobiol. Aging*, 2012, **33**(8), 1843 e1–1843 e7, PubMed PMID: 22296644. Epub 2012/02/03. eng.
383. D. R. Green, L. Galluzzi and G. Kroemer, Mitochondria and the autophagy-inflammation-cell death axis in organismal aging, *Science*, 2011, **333**(6046), 1109–1112, PubMed PMID: 21868666. Pubmed Central PMCID: PMC3405151. Epub 2011/08/27. eng.
384. B. Wang, N. Abraham, G. Gao and Q. Yang, Dysregulation of autophagy and mitochondrial function in Parkinson's disease, *Transl. Neurodegener.*, 2016, **5**, 19, PubMed PMID: 27822367. Pubmed Central PMCID: PMC5087125. Epub 2016/11/09. eng.
385. S. M. Cardoso, C. F. Pereira, P. I. Moreira, D. M. Arduino, A. R. Esteves and C. R. Oliveira, Mitochondrial control of autophagic lysosomal pathway in Alzheimer's disease, *Exp. Neurol.*, 2010, **223**(2), 294–298, PubMed PMID: 19559703. Epub 2009/06/30. eng.
386. J. Demers-Lamarche, G. Guillebaud, M. Tlili, K. Todkar, N. Belanger and M. Grondin, *et al.*, Loss of Mitochondrial Function Impairs Lysosomes, *J. Biol. Chem.*, 2016, **291**(19), 10263–10276, PubMed PMID: 26987902. Pubmed Central PMCID: PMC4858975. Epub 2016/03/19. eng.
387. A. M. Griggs, Z. S. Agim, V. R. Mishra, M. A. Tambe, A. E. Director-Myska and K. W. Turteltaub, *et al.*, 2-Amino-1-methyl-6-phenylimidazo[4,5-b] pyridine (PhIP) is Selectively Toxic to Primary Dopaminergic Neurons *In vitro*, *Toxicol. Sci.*, 2014, **140**(1), 179–189, PubMed PMID: 24718704.
388. F. L. Tang, W. Liu, J. X. Hu, J. R. Erion, J. Ye and L. Mei, *et al.*, VPS35 Deficiency or Mutation Causes Dopaminergic Neuronal Loss by Impairing Mitochondrial Fusion and Function, *Cell Rep.*, 2015, **12**(10), 1631–1643, PubMed PMID: 26321632. Pubmed Central PMCID: PMC4565770. Epub 2015/09/01. eng.
389. W. Wang, X. Wang, H. Fujioka, C. Hoppel, A. L. Whone and M. A. Caldwell, *et al.*, Parkinson's disease-associated mutant VPS35 causes mitochondrial dysfunction by recycling DLP1 complexes, *Nat. Med.*, 2016, **22**(1), 54–63, PubMed PMID: 26618722. Pubmed Central PMCID: PMC4826611. Epub 2015/12/01. eng.
390. E. Zavodszky, M. N. Seaman, K. Moreau, M. Jimenez-Sanchez, S. Y. Breusegem and M. E. Harbour, *et al.*, Mutation in VPS35 associated with Parkinson's disease impairs WASH complex association and inhibits autophagy, *Nat. Commun.*, 2014, **5**, 3828, PubMed PMID: 24819384. Pubmed Central PMCID: PMC4024763. Epub 2014/05/14. eng.

391. S. Devireddy, A. Liu, T. Lampe and P. J. Hollenbeck, The Organization of Mitochondrial Quality Control and Life Cycle in the Nervous System *In vivo* in the Absence of PINK1, *J. Neurosci.*, 2015, **35**(25), 9391–9401, PubMed PMID: 26109662. Pubmed Central PMCID: PMC4478254. Epub 2015/06/26. eng.
392. H. Sung, L. C. Tandarich, K. Nguyen and P. J. Hollenbeck, Compartmentalized Regulation of Parkin-Mediated Mitochondrial Quality Control in the Drosophila Nervous System *In vivo*, *J. Neurosci.*, 2016, **36**(28), 7375–7391, PubMed PMID: 27413149. Pubmed Central PMCID: PMC4945662. Epub 2016/07/15. eng.
393. T. Valero, Mitochondrial biogenesis: pharmacological approaches, *Curr. Pharm. Des.*, 2014, **20**(35), 5507–5509, PubMed PMID: 24606795. Epub 2014/03/13. eng.
394. N. T. Hertz, A. Berthet, M. L. Sos, K. S. Thorn, A. L. Burlingame and K. Nakamura, *et al.*, A neo-substrate that amplifies catalytic activity of parkinson's-disease-related kinase PINK1, *Cell*, 2013, **154**(4), 737–747, PubMed PMID: 23953109. Pubmed Central PMCID: PMC3950538. Epub 2013/08/21. eng.
395. D. J. Surmeier, P. T. Schumacker, J. D. Guzman, E. Ilijic, B. Yang and E. Zampese, Calcium and Parkinson's disease, *Biochem. Biophys. Res. Commun.*, 2017, **483**(4), 1013–1019, PubMed PMID: 27590583. Epub 2016/09/04. eng.
396. T. E. Moors, J. J. Hoozemans, A. Ingrassia, T. Beccari, L. Parnetti and M. C. Chartier-Harlin, *et al.*, Therapeutic potential of autophagy-enhancing agents in Parkinson's disease, *Mol. Neurodegener.*, 2017, **12**(1), 11, PubMed PMID: 28122627. Pubmed Central PMCID: PMC5267440. Epub 2017/01/27. eng.
397. H. Okazawa, M. Ikawa, T. Tsujikawa, Y. Kiyono and M. Yoneda, Brain imaging for oxidative stress and mitochondrial dysfunction in neurodegenerative diseases, *Q J. Nucl. Med. Mol. Imaging*, 2014, **58**(4), 387–397, PubMed PMID: 25366710. Epub 2014/11/05. eng.

CHAPTER 12

Genes, Aging, and Parkinson's Disease

CHIARA MILANESE[a,b] AND PIER G. MASTROBERARDINO*[a]

[a]Department of Genetics, Erasmus University Medical Center Rotterdam, 3015 CN Rotterdam, The Netherlands; [b]Fondazione Ri.Med, Palermo 90133, Italy
*E-mail: p.g.mastroberardino@erasmusmc.nl

12.1 Introduction

Aging is characterized by a progressive deterioration of physiological functions and concomitant increased risk of morbidity and mortality. Aging constitutes a major risk factor for most complex diseases, including neurodegenerative disorders and Parkinson's disease (PD). In advanced economies, lifespan extension has increased dramatically, the population is rapidly aging, and in the near future healthcare and general costs for the growing proportion of elders will substantially impact society. Unraveling the mechanisms promoting healthy aging to devise effective intervention to delay or avoid the onset of age-relate pathologies therefore constitutes a matter of primary social and economic importance.

The timing and the dynamics of lifespan increase observed during the last century suggest that the principal factors responsible for such improvement are not genetic (*i.e.* not caused by acquisition of advantageous genetic variants) because the pace of lifespan extension has been too rapid to be compatible with the evolutionary timescale. Nonetheless, multiple elements indicate

Issues in Toxicology No. 34
Oxidative Stress and Redox Signalling in Parkinson's Disease
Edited by Rodrigo Franco, Jonathan A. Doorn and Jean-Christophe Rochet

Published by the Royal Society of Chemistry, www.rsc.org

that genes influence lifespan and understanding the extent of genetic contribution is fundamental to decipher this phenomenon and to identify potential targets to mitigate the noxious effects of aging.

A discussion on genetics of aging cannot ignore the fundamental problem posed by the heterogeneity of the aging phenotype. While a long lifespan is a well-recognizable trait, healthy aging is definitively more complex and can be defined in several ways, such as the preservation of certain functions such as cognition or the absence of morbidity in advanced stages of life. Decoding the genetic landscape favoring healthy aging is therefore challenging because of the extreme heterogeneity in the phenotype.

The exact role of aging on the etiopathogenesis of PD and of other neurodegenerative disorders is still debated. There is consensus, however, that promoting successful aging, which is associated with a desirable general health status and, eventually, extended lifespan, would be highly desirable in view of the aging role as a main risk factor for complex diseases. An overall good health status and longevity are to some extent correlated because it has been observed that long-lived individuals tend to be healthy for most of their lives. Longitudinal studies following the same cohort of cases in time, in fact, have shown that a lower presence in midlife of risk factor for conditions such as cardiovascular diseases predicts longer and healthier lifespan.[1–4] The progression rate and the severity of aging are influenced by complex interactions between genes and environment, including lifestyle. In this section we will analyze the genetic and environmental factors so far known to impact aging and discuss their relevance for the control of cellular oxido–reductive homeostasis and PD.

12.1.1 Do Genes Influence Lifespan?

The aging phenotype is heterogeneous and is produced by the concomitant action of numerous biological factors. It is therefore extremely challenging to determine the features of genetic landscapes favoring longevity and/or healthy aging.

A first level of information on aging genetics can be inferred by epidemiological studies designed to determine whether a prominent trait of aging, for instance longevity, is inheritable and, if so, to what extent. Here, the influence of genes on aging emerges from several investigations that approached the problem from different perspectives. Analyses in large cohorts of twins demonstrated that the lifespan of homozygous pairs was more similar than that observed for same-sex heterozygous siblings[5,6] and these investigations estimated longevity heritability to be around 25%. Investigations in cohorts from the general population (*i.e.* population-based studies) led to results that are comparable with those gathered in twins. For instance, studies on centenarians have shown that siblings and offspring of exceptionally long-lived individuals have higher chances to live longer than average;[7–10] here, the contribution to heritability is slightly lower than that observed in twins and has been estimated to be around 20%.

When looking at these figures, however, it should be considered that the followed phenotype – longevity – is multi-dimensional because it derives from a combination of factors; healthy aging, for instance, may be promoted by favorable environmental and societal conditions. Therefore, selecting a population exclusively on the basis of its members' lifespan can lead to loss of information about potential genetic information because of confounding factors. Additional investigative strategies following alternative and more specific aging-related phenotypes, which in turn contribute to longevity, might provide better insights into the effects of genes on aging. Here, examples come from studies monitoring traits such as weakness, defined as hand-grip strength, or lower extremity function display much higher degrees of heritability.[11,12] Overall, these researches provided preliminary evidence that the duration of life may be influenced by genetic factors. Although it has been suggested that genetic influence of longevity varies among different ethnicities,[13] further investigations are required to properly address this issue, which is intrinsically biased by the different life-styles among these groups as well as inequalities in their socioeconomic conditions and life expectancy.[14]

12.1.2 Which Genes Influence Lifespan?

Provided that epidemiological studies ascertained the influence of genes on aging, the next logical step would be to identify the specific genes underlying lifespan extension or early aging. Traditionally, these studies are performed through linkage analysis, in which statistical methods are used to determine the association between genetic markers (*e.g.* a certain genomic region) and a specific trait. In its more sophisticated form, linkage analysis involves genome-wide genotyping for single nucleotide polymorphisms (SNPs) using microarray technology[15] or next-generation DNA sequencing[16] (genome-wide association studies or GWAS). Linkage studies, however, present some intrinsic difficulties: sample size is often an issue, as very large populations are required to reach adequate power and, seeking association with a heterogeneous, multi-component trait such as longevity, only aggravates the problem.[17] Linkage studies, in fact, are rather suitable to detect variants that exert strong effects on the studied trait. Because general wisdom assumes that genetically human aging is produced by a combination of multiple *loci*, each with modest effect, linkage studies seem to be further penalized in this specific field. Genotyping of SNPs implies additional challenges as SNPs number in the human genome is of the order of 10^6 and might even exceed 10^7.[18,19] Under these circumstances, association studies on aging have traditionally suffered from a significant proportion of false-positive data. Careful validation, possibly through meta-analysis of pooled data from different studies, is imperative.[20]

On the basis of these technical and conceptual hurdles, it is not surprising that the results from linkage studies to discover the genetic basis of longevity failed to meet expectations. More encouraging results have been obtained in studies exploring association between a specific aging trait and candidate genes, in cases and control groups. In contrast to genome-wide studies, this approach focuses on *loci* that, on the basis of previous biological knowledge (*e.g.* derived from animal models), constitute presumptive candidates to influence aging. Because the strategy is circumscribed to determined *loci*, it has more power than linkage analysis to detect variants with modest effects. It should, however, be noted that case–control studies focused on aging involve the comparison between elderly and young cohorts and that this design poses distinctive methodological challenges concerning selection of appropriate controls groups. For instance, an insidious assumption is that the frequency of the alleles influencing lifespan is equal in the two initial populations of young and elders, or, in other words, that the populations of young and elders are not stratified. The frequency of the studied allele, therefore, should be comparable in a cohort of centenarians born in the first decade of the twentieth century (assuming that the study is performed in the second decade of the twenty-first century) and in a group of putative controls in their seventies, born in the third decade of the same century. This supposition, however, may be inaccurate because of several factors. Migration fluxes in the considered population, for instance, may change significantly the frequency of a certain allele and – because of substantial geographic mobility in certain periods of the past century – two populations born only a few decades apart may exhibit significant difference in allelic frequencies. Obviously, genetic studies are designed to distinguish between different ethnic groups. It should be also noted, however, that such distinction rarely occurs within the Caucasians ethnicity and thus groups such as Germans, English, and Italians are treated as one, despite plausible differences in allelic frequencies. The issue of initial allelic frequency of genes eventually influencing lifespan constitutes a serious confounding factor leading to spurious results. An excellent article by Lewis and Brunner[21] provides additional details on this and other biases undermining association studies comparing young controls and elder cases, and also invites caution when considering and interpreting data from this kind of analysis.

Given the complexity and the challenges of genetic studies on longevity, it is not surprising that an exiguous number of *loci* associated with extended lifespan have been identified thus far. Although several reports have indicated numerous genes that, in principle, might be associated with longevity, most of these candidates have not been confirmed in subsequent independent investigations and therefore their relevance for longevity remains unclear. In fact, only two genes have been consistently associated with longevity in multiple studies: Apolipoprotein E (*APOE*) and the Forkhead-box O (*FOXO*) transcription factor 3A (*FOXO3A*),[22–24] also reviewed in ref. 25,26. Importantly, contribution of these genes to longevity has been recently confirmed in a large multi-center study (the Cohorts for Heart and Aging Research in Genomic Epidemiology Consortium) that performed meta-analysis of data from a very large population of longevity cases and relative controls.[27] Given the study's considerable sample size, which is generally regarded as the critical ingredient for successful GWAS, it is intriguing that no genes other than *APOE* or *FOXO3A* emerged as candidates with significant association. The authors ascribe such an outcome to the intrinsic heterogeneity of longevity as a trait and the involvement of rare allelic variants; alternatively, regulation of longevity might depend upon different combinations of genetic variants, which would have modest effects individually, but interact synergistically when simultaneously present in a genotype. In all such cases, associations would not be captured by GWAS.

In the present chapter, we will focus the discussion on APOE or FOXO3A, also highlighting their relevance for PD and redox homeostasis; for those *loci* with uncertain association, the reader should refer to other excellent reviews.[17,25,26]

12.2 Apolipoprotein E

Apolipoprotein E (APOE) is a 34 kDa protein involved in lipid transport and delivery between the different tissues of the organism.[28] Although its function is not limited to the central nervous system (CNS), for the purposes of this chapter we will emphasize its role in the brain, referring the reader to other excellent reviews for a more general discussion (*e.g.* ref. 28). A seminal and classic article by Brown and Goldstein, who pioneered studies on receptor-mediated cholesterol transport, still represent an excellent overview of this biological process and is highly recommended for further reading.[29] APOE is particularly important in cholesterol transport and reuptake from the blood and it was originally described as a component of the very low density lipoproteins (VLDL) and the chilomicrons. These particles are spherical macromolecular complexes instrumental in transporting hydrophobic lipids otherwise insoluble in the bloodstream. The designation VDLD derives from the initial characterization of this family of particles on the basis of their density inferred by ultracentrifugation analysis.[30] Structurally, lipoproteins consist of a lipidic core wrapped in a phospholipid monolayer, which in turn embeds the apolipoproteins. The lipidic core, essentially a lipid droplet, is composed by triacylglycerols and cholesterol, and the latter is esterified in its single polar hydroxyl residue to further ease solubilization by the phospholipid monolayer and therefore transport. Humans express at least ten apolipoproteins, including APOE, which differ in their molecular features and functions.[30]

In the brain, APOE is primarily synthesized in astrocytes[31] – and to a much lesser extent by microglia[32] – and is essential for glial-mediated delivery of cholesterol to neurons. Cell-culture experiments indicate that APOE is secreted by astrocytes as a lipoprotein complex of discoidal shape, containing little lipid. *In vivo*, however, in samples from human cerebrospinal fluid (CSF), APOE complexes have been reported to form spherical particles containing esterified cholesterol in its core.[33] Whether this discrepancy is due to artifacts in cell-culture condition or rather reflects a maturation process of the particle from secretion to release into the CSF, which would also lead to enrichment of the lipid core, is unclear. It is, however, ascertained that association with lipids is vital for APOE stability as demonstrated by the sharp reduction in its levels in the brain of mice lacking the ATP-binding cassette, sub-family A, member 1(Abca1) enzyme, which specifically lipidates APOE.[34,35]

Despite the fact that neurons can autonomously produce cholesterol, glial supply remains crucial to generate and maintain synaptic connections.[36–38] APOE is the principal apolipoprotein in the brain as also evidenced in studies on knockout mice, which show reduced brain functional connectivity as determined by resting-state functional MRI, decreased immunoreactivity for the post-synaptic marker PSD-05, and reduced cerebral blood flow.[39]

APOE is a polymorphic gene encoding for a protein regulating lipid homeostasis. In humans, APOE exists in three allelic forms at a single gene locus on chromosome 19, which are denominated APOE2, APOE3, and APOE4 (also known as ε2, ε3, ε4)[40] with respective frequency of 8.4%, 77.9%, and 13.7% in the global population. The assortment of these alleles in the genotype exerts a moderate impact on both common diseases and longevity. The ε4 allele is in fact a moderate risk factor for Alzheimer's disease (AD) and cardiovascular diseases[28] and is also significantly less presented in the genotype of long-lived individuals, and in centenarians in particular.[22,23] On the contrary, the ε2 allele might exert a protective function against AD and cardiovascular disorders and is enriched in centenarians.[22,23] The precise reasons underlying these effects are poorly understood. It is, however, clear that the polymorphisms in ε2, ε3, and ε4 modify basic biological properties of the respectively coded proteins: APOE2, APOE3, and APOE4.

12.2.1 LDL Receptors

The mechanism used by apolipoproteins to mediate lipid internalization is receptor-mediated endocytosis. In mammals, the LDL receptor (LDLR) family encompasses at least seven different genes and APOE can bind to most, if not all of its members.[41,42] Some members, for instance LRP1 and megalin, are essential for the proper development of the nervous system, as evidenced by embryonic lethality of knock-out mice.[43,44] The receptor–ligand complex is internalized in clathrin-coated endocytic vesicles that eventually fuse with lysosomes, whose enzymes hydrolyze the esterified cholesterol allowing its release in the cytosol. The LDLR receptor escapes degradation and is recycled on the cell surface to participate to the next LDL uptake cycle.[42] Importantly, LDLR function is not limited to vesicle transport *via* endocytosis as they also actively participate to cellular signal transduction,[42] which is mediated by specific interactors with diverse functions binding to the cytosolic domain of the receptor. On the basis of actual knowledge, the majority of genes encoding for LDLR interactors mediating signalling lack elements that might link them to PD or even to dopaminergic (DAergic) neurons function. In some cases, which are described in the next paragraphs, functional evidence might suggest potential involvement. Data in support of this conjecture, however, are lacking thus far.

The LDLR interacting protein Disabled-1 (DAB1) constitutes the intracellular effector of Reelin (RLN), an extracellular matrix glycoprotein essential for neuronal migration and positioning during development, which activates a pathway that is also relevant for the development of DAergic neurons.[45,46] The pathway is also important in the adult brain and its deregulation has been implicated in AD and Creutzfeld–Jacob disease.[47–49] RLN interferes with aggregation of amyloid-beta[50] and accumulates in amyloid-like deposits during normal aging.[51] No reports describe investigations on the role of the RLN/DAB1 in PD thus far. RLN involvement in DAergic neuron development and its abnormalities in normal aging, however, might indicate that this pathway could also be modified in PD.

Post-synaptic density protein 95 (PSD-95) is a scaffolding protein that stabilizes glutamate receptor, and *N*-methyl-D-aspartic acid (NMDA) receptors in particular in synapses. More recently it has been shown that PSD-95 also interacts with D1 and D2 dopamine (DA) receptors and, accordingly, mice with targeted deletion of PSD-95 show motor impairments, striatal degeneration, and altered DA-glutamate interplay.[52] Additionally, PSD-95 modulates D1 dopamine receptor trafficking and influences L-DOPA dyskinesia, which is the major undesired side effect in DA replacing palliative therapy in PD.[53]

CAPON regulates neuronal nitric oxide synthase (nNOS),[54] which has been implicated in neurodegeneration in some PD models.[55,56] Additionally, CAPON might participate in iron (Fe) homeostasis regulation,[57] which is severely deranged in the *substantia nigra pars compacta* (SNpc) of PD patients.[58]

A further nexus between LDLR and PD might derive from the ability of its members, and of LRP1 in particular, to bind proteins that might have an implication in the disease such as lactoferrin[59–61] and alpha2-macroglobulin.[62]

12.2.2 Effects of Polymorphisms on APOE Function

APOE isoforms present primary sequence differences at the level of aminoacidic residues 112, 158, or both. Both sites can present either a cysteine (Cys) or an arginine (Arg) residue that are differently assorted in the isoforms: Cys112, Cys158 in APOE2, Cys112, Arg158 in APOE3, Arg112, Arg158 in APOE4. The mechanistic details underling the detrimental effects of the E4 allele are dimly understood, yet it is clear that these substitutions alter ApoE structure and biological properties, as reviewed by Zhong and Weisgraber.[63]

Amino acid substitutions in APOE3 and APOE4 directly influence the structural properties of the protein. In its mutant forms, APOE better succeeds in maintaining relatively stable equilibrium states that are intermediates between the unfolded and the native configuration. These folding intermediates – called molten globules – exhibit distinctive properties different from those of the protein in its native state, which might result in detrimental consequences.[64] For instance, in the molten globule state APOE4 exposes otherwise buried proteolytic sites[65] and is therefore more susceptible to proteolysis. A higher prevalence of molten globules, which occurs by virtue of their increased stability in APOE3 and APOE4 variants, might therefore underlie detrimental processes in genotypes carrying these polymorphisms. Additional conformational changes concern the tertiary structure of the native protein state; an Arg substitution at residue 112 leads to a conformational shift that allows a novel interaction between Arg61 and glutamate (Glu)255, which compacts the protein conformation. Conversely, when a Cys is present at position 112, Arg61 is shielded by two helices in the structure and is therefore not available for polar interactions with Glu255. Cysteine to Arg substitution at position 158 sharply reduces LDL-receptor binding activity. The same does not occur, however, for the Cys to Arg substitution at residue 112. It is also unknown if the detrimental consequences associated with APOE3 and APOE4 isoforms, which both carry an Arg at residue 158, are related to the low LDL-receptor binding activity.[66]

Furthermore, substitutions in APOE variants also alter the protein stability, particularly at its N-terminal, and render APOE4 less stable than APOE3, which is in turn less stable than APOE2. Instability has important pathogenic implications because it renders the isoform more prone to aggregation, at least *in vitro*.[67,68] ApoE can also promote aggregation of other proteins, most notably amyloid-beta[69] and in patients APOE4 expression levels correlate with the extent of amyloid deposition.[70]

Cysteine residues in positions 112 and 158 can engage in disulfide bonds forming homodimers with reduced LDL-receptor binding activity. Homodimers, obviously, cannot form in the APOE4 form where Cys are substituted. A significant portion of APOE2 and APOE3 circulating in plasma is in the form of homodimers (about 50%),[71] which constitute an even larger proportion in the CSF.[72] There are only limited studies investigating how oxidative stress intrinsic to aging or neurodegenerative diseases impacts Cys residues of APOE and the eventual functional consequences. *S*-nitros(yl)ation of APOE2 and APOE3 after interaction with NOS1 has been recently described.[73] Studies on other modifications such as *S*-glutathionylation or the formation irreversible oxidative modifications (*i.e.* sulfenic, sulfinic, or sulfonic acids) have not been performed thus far, at least to the best of our knowledge.

12.2.3 APOE and Parkinson's Disease

Traditionally, APOE polymorphisms have been associated with higher risk of AD. However, more recent evidence indicates that APOE may also be relevant for PD, and particularly for some aspect of its pathology. Genetic case-control association studies have shown that APOE4 increases the risk of dementia in synucleopathies (*i.e.* in PD, PD with dementia, and dementia with Lewy bodies).[74] Additionally, an exhaustive, multi-centric study on a relatively large cohort of PD case revealed that, in patients suffering with dementia, APOE4 allele is associated with more severe deficit in multiple cognitive domains.[75] Interestingly, in PD patients without dementia, APOE4 association was limited to the cognitive domain of semantic verbal fluency, whose deterioration is typical of early AD rather than PD and is rather is attributable to the derangement of the temporal cortex. This evidence suggests that APOE influences processes that are predominant in AD pathology, but are to some extent shared also by PD. It should also be emphasized, however, that the APOE genotype did not correlate with measures of AD neuropathological changes in studies on PD autopsy specimens[76] and thus additional studies will be necessary to fully understand whether the APOE4 genotype effectively increases the load of AD-like neuropathological changes in PD. The molecular mechanisms underlying APOE effects in PD are obscure. The process seems to be unrelated to amyloid-beta processing because, in PD patients, the APOE genotype did not correlate with the brain levels of amyloid-beta.[77] Increase in both APOE and LRP1 expression has been observed in neuromelanin (NM) containing DAergic neurons of the SNpc[78] and led to the hypothesis that perturbation in lipid metabolism might participate to deterioration. APOE cascade is also activated in PD animal models, as demonstrated in 1-methyl-4-phenyl-1,2,3,6-tetrahydropyridine (MPTP)-treated mice.[79]

12.2.4 APOE Oxidative Stress

APOE genotype also influences redox tolerance and homeostasis, which are central to PD etiopathogenesis.[80–82] APOE can protect cells from hydrogen peroxide (H_2O_2) toxicity, it can scavenge reactive oxygen species (ROS), and it can restrain copper (Cu)-catalyzed lipoprotein oxidation,[83] as reviewed in ref. 84. These properties are enhanced in E2 and E3 isoforms, when compared with E4 (E2>E3>E4). Accordingly, AD patients carrying the E4 polymorphism exhibit signs of redox imbalance such as lipid peroxidation in the brain[85] and increased ROS in the bloodstream.[86,87] APOE also protects synaptosomes from amyloid-beta-induced oxidation in an allele-dependent way (E2>E3>E4).[88,89] Additionally, APOE interferes with the metabolism of metals. Initial experiments used metal-chelate affinity chromatography to demonstrate APOE capacity to interact with different metal cations and determined that the protein has strong affinity for Cu, and to a lesser extent, for ferric (Fe^{3+}), ferrous (Fe^{2+}), and zinc (Zn) cations.[83] The exact binding metal site remains to be identified, but it has been speculated that the four-helix bundle at the protein N-terminus is structurally compatible with metal coordination.[83] The latter, however, does not differ among the different isoforms.[83] Whether these factors participate to PD etiopathogenesis is unknown, but it is tempting to speculate that APOE genotype might contribute to perturbation of Fe metabolism, which is central to PD pathogenesis.[58]

12.3 FOXO3A and FOXO Family

FOXO3A belongs to the sub-family of the FOXO transcription factors, which are homologs of *C. elegans* DAF-16.[90,91] Members of the FOXO family are highly conserved, especially at the level of the 110-amino acid DNA binding domain, and are found in species ranging from yeast to humans. FOXOs play a central role in multiple biological processes, including cell cycle and differentiation,[90,91] DNA-damage repair,[92,93] metabolism,[94–96] apoptosis,[97,98] oxidative stress resistance,[99,100] and longevity.[101–104] Four members of the FOXO family, FOXO1A, FOXO3A, FOXO4 and FOXO6 have previously been described[105] in mammals and are ubiquitously expressed with the exception of FOXO6, which is mainly expressed in the CNS and liver.[96,106,107] They all share the common characteristic of being regulated by two evolutionally conserved signalling pathways: the canonical insulin/IGF1/PI3K and AKT signalling pathways[108–110] negatively regulating FOXOs activity, and the oxidative stress JNK-mediated signalling activating the members of the FOXO family.[111,112] In addition, FOXOs activities are fine tuned by several other signalling pathways that include acetylation, ubiquination, methylation, adenosine monophosphate-activated protein kinase (AMPK) regulation, and thiol oxidation.[112,113]

Because the multiple binding sites located within their structure, FOXOs can function both as transcriptional activators or repressors, probably depending on the range of associated cofactors. These transcriptional factors have overlapping gene expression patterns both during development and in the adult. In addition, FOXO1, FOXO3A and FOXO4 bind to the same DNA target sequence and are capable of regulating the same target genes. FOXOs members show functional redundancy, as demonstrated by deletion studies in mice. Besides FOXO1 null mutation, which induces lethality due to incomplete vascular development, FOXO3A null mutants were found to be viable, although showing hematological abnormalities together with widespread organ inflammation[114] and abnormal ovarian follicular development in females. FOXO4 null mutants show no phenotype.[115]

12.3.1 FOXO3A Biological Functions

As previously mentioned, the members of the FOXO family and FOXO3A in particular, have been involved in a vast range of cellular processes. Several reports have shown that overexpression of the constitutively active nuclear form of FOXO3A induces cell-cycle arrest in the G1 phase by enhancing the expression of the cyclin-dependent kinase inhibitor p27.[116] Expression of the native form of FOXO3A has been shown to directly activate transcription of the p130 gene, which can induce cells to enter in a quiescent state[117] and to directly activate the DNA-damage response *via* the expression of the growth arrest and DNA-damage-inducible protein alpha (*GADD45a*) gene by inducing a delay in the G2-M phase in CCl39 fibroblasts exposed to ultraviolet (UV) irradiation.[118] FOXO3A is also directly involved in apoptosis induction, since it was found to activate the intrinsic apoptotic pathway through the modulation of genes belonging to the B-cell CLL/lymphoma 2 (Bcl-2) family such as Bcl-2-like protein 11 (Bim),[119] p53 upregulated modulator of apoptosis (PUMA),[120] TNF-related apoptosis-inducing ligand (TRAIL),[121] and Fas ligand.[122]

One of the most recent discoveries about FOXO3A activities is related to its significant role in modulation of mitochondria energy metabolism and in the regulation of response to oxidative stress. The effects of FOXO3A on mitochondrial functions were studied on colon cancer cell lines, where activation of FOXO3A resulted in a substantial down-regulation of mitochondrial genes and changes in the levels of mitochondrial proteins such as mitochondrial import receptor subunit TOM20 homolog (TOM20) and cytochrome C oxidase1 (COX 1). FOXO3A activation also caused a decrease in mitochondrial DNA copy number and a reduction in the mitochondrial respiration activity that was mediated by a FOXO3A-dependent inhibition of c-myc.[123] Controversially, in mouse fibroblasts and skeletal myotubes subjected to glucose restriction, FOXO3A accumulates in mitochondria in an AMPK-dependent manner. Upon sirtuin 3 (SIRT3) activation, FOXO3A mediates mitochondrial polymerase recruitment to DNA with consequent transcription of the core catalytic subunits of the electron transport chain (ETC).[124]

FOXO3A has been revealed to have a prominent role in protecting cells from elevated oxidative stress levels. The activation of FOXO3A *via* the AKT pathway protects quiescent cells from oxidative stress by directly increasing the expression of the manganese superoxide dismutase (MnSOD) gene.[125] As in many transcription factors, reversible acetylation, which increases transactivation by increasing the affinity for the DNA, fine tunes the FOXO3A function.[126] Here, an important role is played by SIRTs, a class of histone deacetylases dependent from nicotinamide adenine dinucleotide (NAD^+) able to suppress transcription on the genome-wide scale.[127] Because they are regulated by NAD^+, SIRTs constitute a functional nexus between metabolism and gene expression regulation and have been implicated in multiple biological processes including the pathogenesis of neurodegenerative disorders and aging.[127,128] SIRT3-dependent activation of FOXO3A increases the expression of MnSOD and catalase (CAT) in different cell types,[129–131] thus increasing cell survival. SIRT3-mediated deacetylation of FOXO3A positively regulates the expression of dynamin-1-like protein (DRP1), mitochondrial fission 1 protein (FIS1) and mitofusin-2 (MFN2) to coordinate mitochondrial fission and fusion in human endothelial cells.[132] It has also been demonstrated that resveratrol, a polyphenolic compound contained in red fruits and wine, mediates its potent antioxidant effect by activating the SIRT3/FOXO3A pathway, which in turn leads to up-regulation of mitochondrial genes such as ATP synthase 6 (ATP6), COX1, cytochrome b (CYTB) and NADH dehydrogenase 2 and 5 (ND2 and ND5) and leads to reduction of mitochondrial ROS reduction production.[128] Reperfusion after hypoxia, a condition leading to massive ROS production, affects cellular localization of FOXO3A by inhibition of the AKT pathway and induction of a rapid nuclear translocation of this transcription factor.[133] Detrimental effects of hypoxia/reperfusion in rat cardiomyocytes can be mitigated by acetylcholine treatment. Mechanistically, this intervention strongly reduces oxidative stress by inducing MnSOD expression through FOXO3A/peroxisome proliferator-activated receptor gamma coactivator 1-alpha (PGC-1α) signalling pathway, and up-regulating Cu/Zn superoxide dismutase (CuZnSOD) in the cytoplasm, thus protecting against oxidative injury and cellular apoptosis.[134] Finally, it has been demonstrated that, in hearts from rat subjected to ischemia/reperfusion, resveratrol administration activates the SIRT3/FOXO3A axis promoting the activation of the PINK1/Parkin pathway potentiating the mitochondrial fission leading to mitophagy in rat cardiomyocites.[135]

12.3.2 FOXO3A and Protein Homeostasis

The role of FOXO members in the regulation of protein homeostasis and protein degradation has been exhaustively studied in muscle tissues and increasing evidence shows that FOXOs are involved in both the autophagy–lysosomal and in the ubiquitin–proteasome pathways.[136–139]

In myotubes, FOXO3A stimulates lysosomal proteolysis by inducing the expression of many autophagy-related genes involved in various steps of the process, including microtubule-associated protein 1 light chain 3 beta (LC3B), GABARAPL1 (GEC1), Beclin 1 (BECN1), phosphatidylinositol 3-kinase 3 (VPS34), unc-51 like autophagy activating kinase 2 (ULK2), and the ubiquitin-like protein ATG12 (APG12L).[140] In mouse skeletal muscle, stimulation of the lysosomal proteolytic pathway leads to atrophy[141] and FOXO3A contributes to this process by activating the Akt1 pathway up-regulating the expression of Bcl-2/adenovirus E1B 19 kDa protein-interacting protein 3 (BNIP3), a Bcl-2 related protein that are involved in the modulation of autophagy in multiple cell systems.[142–145] FOXO3A has been involved in mitophagy as well. In muscles, FOXO3A activates the expression of MUL1, a mitochondrial E3 ubiquitin ligase promoting fragmentation, depolarization, and clearance of mitochondria through the autophagy-lysosome pathway targeting the MFN2 protein in skeletal muscle *in vivo* and in C2C12 myotubes.[146] In the previous section, we have already described the FOXO3A-dependent activation of the PINK1-Parkin pathway in rat cardiomyocites, with consequent increase in the mitochondria fission process and the selective degradation of dysfunctional mitochondria.[135] Additionally, Tseng and colleagues showed that SIRT3-related deacetylation of FOXO3A regulates the expression of the primary mitophagy mediators BNIP3, Bcl-2/adenovirus E1B 19 kDa protein-interacting protein 3-like (BNIP3L or NIX) and LC3 in endothelial cells.[132]

In addition to autophagy and mitophagy, FOXO3A is also involved in the ubiquitin–proteasome system (UPS), a protein quality-control process responsible for the degradation of cytosolic proteins[147] and the removal of damaged proteins and protein aggregates.[148] In skeletal muscle atrophy, FOXO3A directly regulates the expression of the muscle specific E3 ubiquitin ligase MURF1 upon activation mediated by AMPK-α2.[149] FOXO3A can also mediate the crosstalk signalling between the intracellular kinases PI3K/AKT and mitogen-activated protein kinase (MAPK) kinase (MEK)/extracellular signal-regulated kinase (Erk) pathways to coordinate the transcription levels of ATROGIN-1 (an E3 ubiquitin ligase also known as FBOX32) and Ubiquitin C in glucocorticoid-induced skeletal muscle atrophy.[133]

12.3.3 FOXO3A and Parkinson's Disease

Overall, these data indicate that FOXO3A is at the crossroad of pathways and mechanisms of great relevance for PD etiopathogenesis, including mitochondrial function, redox homeostasis, PINK1 and Parkin biology, protein quality control, and Erk signalling, to mention a few.[80,82,94,150,151] On these premises, the paucity of studies to better evaluate a potential role of FOXO3A in PD is rather surprising; at the time this chapter was written, a PubMed search using the keywords 'FOXO3A' and 'Parkinson' retrieved only six items. Recent studies, however, show that activation of the FOXO3A pathway may have neuroprotective effects in PD signalling models *in vitro* and *in vivo*[152] and, accordingly, that SIRT2 mediated FOXO3A deacetylation, and subsequent suppression of its activity, exacerbates MPTP toxicity in mice.[153] These experimental data substantiate the hypothesis that FOXO3A might participate in PD pathogenesis and pave the way for future and more detailed investigations.

12.4 Role of Other Genes Emerged from Animal Studies in Aging

As in human studies, exploring the genetics of aging in laboratory animal models presents its peculiar challenges. An investigator should always question whether the aging phenotype elicited by a mutation in a single gene is representative of natural aging, or if it rather reflects general deterioration of the organism.[154,155] Additionally, the extent of correlation of a mutation in a single gene with longevity should be carefully pondered.[154,155] Bearing in mind these caveats, studies performed in different laboratory model organisms suggested that several candidate genes might play a relevant role in increasing lifespan and in promoting healthy aging.

The identified genetic mutants have been classified into three different clusters based on their effects on life extension: (1) long-lived mutants that can extend longevity; (2) short-term mutants that are prone to develop aging-related disorders and significantly reduce lifespan, and (3) mutants that have no direct effect on longevity but can be used to study in detail molecular mechanisms and relevant pathways involved in aging processes.[154]

Among them, it is worth highlighting the very well described insulin/IGF-signalling (IIS) system, which has been identified for the first time in *Caenorhabditis elegans*. Mutations in the ISS genes Daf-2 and age-1, and in the FOXO homolog Daf-16 are associated with extensions of lifespan, increased resistance to oxidative stress and reduced oxidative damage.[156–159] Similarly, in *Drosophila melanogaster* and in the long-lived Ames and Snell Dwarf mouse strains, depressed insulin signalling results in increased life extension.[160,161] Accordingly, over-expression of the *Klotho* gene to repress intracellular signalling between insulin and insulin-like growth factor receptor 1 (IGF-1R) in mice leads to lifespan extension.[162] Complete ablation of IGF-1R as well as of the insulin receptor (IR), is highly deleterious because it leads to insulin resistance and insulin-related disorders such as obesity and diabetes, which are all aging-related conditions that adversely impact lifespan in humans and animal models.[163] Controversially, selective IGF-1R ablation and impairment of the IIS pathway in the CNS decrease amyloid-beta accumulation and prevent mortality in an AD mouse model.[164] In humans, down-regulation of the IIS has been observed in postmortem brain of patient of sporadic AD;[165] however, the identification in centenarians of reduced circulating IGF-1 levels in association to normal insulin production is beneficial for survival in old age,[166–168] suggesting that modulation of the IIS system could reduce the occurrence of age-related diseases, therefore improving the quality of natural aging and extend lifespan.

Biological deterioration typical of aging has also been ascribed to accumulation of macromolecular damage in biological molecules caused by byproducts of metabolism such as ROS. This theory was originally proposed by Denham Harman in 1956, is commonly known as the free-radical theory of aging, and has been held as a plausible model since its original conception.[169] Accumulation in time of ROS-mediated chemical modifications to DNA (*i.e.* DNA damage) could be particularly harmful because it may corrupt the genetic information and therefore the cell function at its core.[170] Consistently, mutations in certain biological pathways responsible for repairing DNA lesions are associated with a phenotype that closely resembles aging, even though it progresses at a much faster pace and is consistently described as accelerated aging. In humans, these defects result in devastating progeroid conditions such as the Cockayne syndrome. Consensus on the extent of affinity between natural and accelerated aging, however, is not unanimous and the topic is still the object of a passionate debate between scholars.[171] The nucleotide excision repair (NER) pathway is a versatile mechanism that is able to amend several types of DNA lesions and may contribute to the pathogenesis of neurodegenerative diseases.[172,173] Mutations in genes involved in the NER system, and in transcription coupled nucleotide excision repair (TC-NER) branch in particular, are often associated with the occurrence of age-related phenotype and aging-related disorders early in life and to a remarkable reduction in lifespan, suggesting that mutations in genes involved in the DNA-repair pathways could play a relevant role in accelerating the aging processes.[174] Disorders such as Cockayne syndrome, xeroderma pigmentosus (XP), or trichothiodystrophy (TTD) are characterized by the progressive appearance of premature aging phenotypes in the first decade of life.[175] In particular, CS, XP and TTD patients develop defects typically associated with aging such as hearing loss, cataract, cachexia, oxidative damage and progressive neurological degeneration in early life and exhibit a severely reduced lifespan.[172,174,176] Several organism models have been developed to dissect in detail the correlation between defects in the NER system, the manifestation of multiple age-related defects, and lifespan.[177–182] These models have also been used to study aging-related disorders,[183–185] despite the fact that the rapidly developing phenotype is in contrast with the chronic, slow progressing one observed in natural aging and in neurodegenerative conditions such as PD.[155]

A further set of evidences connecting DNA damage and aging stems from evidence obtained in a mouse model harboring a defective form of the mitochondrial DNA (mtDNA) polymerase (polymerase gamma or POLG). This enzyme has a 3′-5′ exonuclease activity that is essential to proofread mtDNA and that was ablated in mutant mice. Consequently, this strain is characterized by pronounced accumulation of mutations in mtDNA, was accordingly named the '*mtDNA-mutator*' mouse, and displays premature aging and reduced lifespan.[186,187] This model is interesting also because it establishes a further link between the aging, DNA repair, and the free-radical theory of aging. An implication of this theory is in fact that ROS production, which occurs especially during mitochondrial respiration, may induce mutations in mtDNA, which in turn may impair mitochondrial activity to set off a vicious circle that ultimately leads to further increase in ROS production and thus faster progression of aging.[169] This theory indeed fits with findings showing that specimens from aged individuals often exhibit fewer and structurally abnormal mitochondria,[188] increased oxidative damage,[189] and declined mitochondrial respiration activity.[190,191] In line with these evidences there are observations that long-lived animals show reduced oxidative damage and increased resistance to oxidative stress.[192,193] However, only a minor increase of oxidative damage and normal ROS production levels were observed in the '*mtDNA mutator*' POL-G mutants, weakening the nexus between mt-DNA oxidative damage and premature aging.[187] Similarly, mice deficient in oxoguanine DNA glycosylase (OGG1), a gene responsible for the oxidative DNA lesion 8-hydroxy-2-deoxyguanosine (8-OHdG) removal, show increased mt-DNA oxidation levels compared to wild-type mice, but no signs of increased oxidative stress and regular mitochondrial activity.[194]

Exposure to sources of oxidative stress can certainly lead to aging; a classical example comes from the skin, where UV-exposure constitutes a major contributor to the aging process.[195,196] However, the free-radical theory of aging has been challenged by some recent studies in genetically modified mouse models. In fact, genetic manipulation of antioxidant enzymes failed to conclusively demonstrate direct connections between overexpression of antioxidant genes and longevity as well as between increased level of oxidative stress and reduction in lifespan.[197,198] Null mutants for glutathione peroxidases 1 (GPx1) and MnSOD genes, or heterozygous mutants for glutathione peroxidase 4 (GPx4), for instance, showed no difference in lifespan compared to wild-type animals, despite increased levels of oxidative damage and elevated sensitivity to oxidative stress.[199] Deletion of the CuZnSOD gene showed a 30% reduction of life extension, which was, however, attributable to an increased incidence of hepatocellular carcinoma rather than decreased antioxidant capabilities.[200] Mutation in the thioredoxin 2 gene (Trx2) was associated with a diminished mitochondrial activity, with an increased production in ROS, and with a slight decrease in lifespan (7%).[198,201] Overexpression of MnSOD, CuZnSOD, GPx4, and catalase increased resistance to oxidative stress, yet led to no improvements in longevity.[198] Catalase is normally expressed in peroxisomes and mediates dismutation of H_2O_2 to O_2 and H_2O.

Interestingly, redirection of catalase expression *via* a targeting presequence to the mitochondria (MCAT mice), which normally do not contain this enzyme, resulted in significant delay of age-related anomalies such as cardiac dysfunction and cataract, as well as significant lifespan extension.[202] Subsequent studies further substantiated the original findings.[203–205] Overall, the results in MCAT mice reinforce the concepts that excessive oxidation might indeed favor the aging process and reveal that targeting antioxidant intervention to critical subcellular compartments is imperative for success. Nonetheless, factors other than abatement of oxidative stress might contribute to MCAT mice longevity, as also suggested in the original publication by the authors[202] who speculated that chronic reduction of H_2O_2-mediated intracellular signalling might also participate to the phenotype. Altogether, these findings indicate that further studies are required to unravel the mechanisms integrating redox tolerance, DNA repair, and aging.

12.4.1 Are Aging-Modifying Genes Discovered in Laboratory Animals Relevant for PD?

All the processes mentioned in the previous section have been implicated in PD pathogenesis, at least to some extent, and might therefore be relevant to explain the relationship between aging and this disorder.

The involvement of both exogenous and endogenous oxidative stress in PD etiopathology has been extensively documented.[206–208] Nonetheless, we are still unable to manipulate the cellular redox environment, as evidenced by the disappointing outcome of numerous clinical trials based on antioxidant compounds.[209,210] Failure to operate effective redox control likely reflects the complexity of the network controlling the redox environment, which cannot be maintained at its homeostasis by simply providing electron donors (*i.e.* antioxidant molecules). Manipulation should probably be attempted on different levels, for instance targeting also the signalling pathways that participate in the redox control in a given cell type (*e.g.* DAergic neurons). Future studies will be necessary to address these issues and understand the role of age-related redox alteration on PD pathogenesis.

A role for DNA damage and repair in PD is gradually emerging. Leucine-rich repeat kinase 2 (LRRK2) mutations, for instance, have been found to cause mtDNA-damage. The latter also specifically increases in DAergic neurons of PD patients and PD animal models.[211,212] DNA quality control is an issue of particular relevance for neurons, which are post-mitotic cells and must preserve their DNA lifelong. It is tempting to speculate that DAergic neurons, which are intrinsically more oxidized than other neuronal population even in normal conditions,[80,213] might suffer from a progressively higher burden of DNA-damage accumulation and, in time, might become sensitive to environmental hazards related to PD. Some studies to address this possibility have been already performed,[214] but additional experiments are certainly required to produce conclusive evidence.

Also, the growth hormone and the insulin pathways have been studied in PD. Although no alterations in the growth hormone (GH)/IGF-1 axis have been reported in PD patients, IGF-1 provides protection against DAergic degeneration in PD cellular and rodent models.[215–217] GH production, however, is stimulated by several drugs currently used to mitigate PD symptoms, such as levodopa.[218,219] A potential involvement of GH in the principal levodopa undesired side effect, dyskinesia, has not been explored thus far. Finally, a recent review summarizes the findings suggesting that several pathways deregulated in PD are also altered in diabetes and therefore establishes a mechanistic nexus between these diseases.[220]

The evidence and concepts discussed in this chapter suggest that multiple factors involved in natural aging might constitute potential modifiers of PD pathogenesis. Surprisingly, studies to explore this possibility are relatively uncommon, therefore leaving ample room for future investigations.

References

1. D. F. Terry, M. J. Pencina, R. S. Vasan, J. M. Murabito, P. A. Wolf, M. K. Hayes, D. Levy, R. B. D'Agostino and E. J. Benjamin, Cardiovascular risk factors predictive for survival and morbidity-free survival in the oldest-old Framingham Heart Study participants, *J. Am. Geriatr. Soc.*, 2005, **53**, 1944–1950.
2. B. J. Willcox, Q. He, R. Chen, K. Yano, K. H. Masaki, J. S. Grove, T. A. Donlon, D. C. Willcox and J. D. Curb, Midlife risk factors and healthy survival in men, *JAMA*, 2006, **296**, 2343–2350.
3. A. Dutta, W. Henley, I. Lang, D. Llewellyn, J. Guralnik, R. B. Wallace and D. Melzer, Predictors of extraordinary survival in the Iowa established populations for epidemiologic study of the elderly: cohort follow-up to "extinction", *J. Am. Geriatr. Soc.*, 2011, **59**, 963–971.
4. L. B. Yates, L. Djousse, T. Kurth, J. E. Buring and J. M. Gaziano, Exceptional longevity in men: modifiable factors associated with survival and function to age 90 years, *Arch. Intern. Med.*, 2008, **168**, 284–290.
5. M. McGue, J. W. Vaupel, N. Holm and B. Harvald, Longevity is moderately heritable in a sample of Danish twins born 1870–1880, *J. Gerontol.*, 1993, **48**, B237–B244.
6. A. M. Herskind, M. McGue, N. V. Holm, T. I. Sorensen, B. Harvald and J. W. Vaupel, The heritability of human longevity: a population-based study of 2872 Danish twin pairs born 1870–1900, *Hum. Genet.*, 1996, **97**, 319–323.
7. T. Perls, M. Shea-Drinkwater, J. Bowen-Flynn, S. B. Ridge, S. Kang, E. Joyce, M. Daly, S. J. Brewster, L. Kunkel and A. A. Puca, Exceptional familial clustering for extreme longevity in humans, *J. Am. Geriatr. Soc.*, 2000, **48**, 1483–1485.

8. T. T. Perls, J. Wilmoth, R. Levenson, M. Drinkwater, M. Cohen, H. Bogan, E. Joyce, S. Brewster, L. Kunkel and A. Puca, Life-long sustained mortality advantage of siblings of centenarians, *Proc. Natl. Acad. Sci. U. S. A.*, 2002, **99**, 8442–8447.
9. B. J. Willcox, D. C. Willcox, Q. He, J. D. Curb and M. Suzuki, Siblings of Okinawan centenarians share lifelong mortality advantages, *J. Gerontol., Ser. A*, 2006, **61**, 345–354.
10. D. F. Terry, M. A. Wilcox, M. A. McCormick, J. Y. Pennington, E. A. Schoenhofen, S. L. Andersen and T. T. Perls, Lower all-cause, cardiovascular, and cancer mortality in centenarians' offspring, *J. Am. Geriatr. Soc.*, 2004, **52**, 2074–2076.
11. D. Carmelli and T. Reed, Stability and change in genetic and environmental influences on hand-grip strength in older male twins, *J. Appl. Physiol. (1985)*, 2000, **89**, 1879–1883.
12. D. Carmelli, M. Kelly-Hayes, P. A. Wolf, G. E. Swan, L. M. Jack, T. Reed and J. M. Guralnik, The contribution of genetic influences to measures of lower-extremity function in older male twins, *J. Gerontol., Ser. A*, 2000, **55**, B49–B53.
13. J. H. Lee, A. Flaquer, R. Costa, H. Andrews, P. Cross, R. Lantigua, N. Schupf, M. X. Tang and R. Mayeux, Genetic influences on life span and survival among elderly African-Americans, Caribbean Hispanics, and Caucasians, *Am. J. Med. Genet., Part A*, 2004, **128A**, 159–164.
14. R. S. Levine, J. E. Foster, R. E. Fullilove, M. T. Fullilove, N. C. Briggs, P. C. Hull, B. A. Husaini and C. H. Hennekens, Black-white inequalities in mortality and life expectancy, 1933-1999: implications for healthy people 2010, *Public Health Rep.*, 2001, **116**, 474–483.
15. M. M. Shi, Technologies for individual genotyping: detection of genetic polymorphisms in drug targets and disease genes, *Am. J. PharmacoGenomics*, 2002, **2**, 197–205.
16. H. Kilpinen and J. C. Barrett, How next-generation sequencing is transforming complex disease genetics, *Trends Genet.*, 2013, **29**, 23–30.
17. K. Christensen, T. E. Johnson and J. W. Vaupel, The quest for genetic determinants of human longevity: challenges and insights, *Nat. Rev. Genet.*, 2006, **7**, 436–448.
18. C. International HapMap, A haplotype map of the human genome, *Nature*, 2005, **437**, 1299–1320.
19. C. Genomes Project, G. R. Abecasis, D. Altshuler, A. Auton, L. D. Brooks, R. M. Durbin, R. A. Gibbs, M. E. Hurles and G. A. McVean, A map of human genome variation from population-scale sequencing, *Nature*, 2010, **467**, 1061–1073.
20. J. P. Ioannidis, Genetic associations: false or true? *Trends Mol. Med.*, 2003, **9**, 135–138.
21. S. J. Lewis and E. J. Brunner, Methodological problems in genetic association studies of longevity–the apolipoprotein E gene as an example, *Int. J. Epidemiol.*, 2004, **33**, 962–970.

22. F. Schachter, L. Faure-Delanef, F. Guenot, H. Rouger, P. Froguel, L. Lesueur-Ginot and D. Cohen, Genetic associations with human longevity at the APOE and ACE loci, *Nat. Genet.*, 1994, **6**, 29–32.
23. K. Kervinen, M. J. Savolainen, J. Salokannel, A. Hynninen, J. Heikkinen, C. Ehnholm, M. J. Koistinen and Y. A. Kesaniemi, Apolipoprotein E and B polymorphisms–longevity factors assessed in nonagenarians, *Atherosclerosis*, 1994, **105**, 89–95.
24. B. J. Willcox, T. A. Donlon, Q. He, R. Chen, J. S. Grove, K. Yano, K. H. Masaki, D. C. Willcox, B. Rodriguez and J. D. Curb, FOXO3A genotype is strongly associated with human longevity, *Proc. Natl. Acad. Sci. U. S. A.*, 2008, **105**, 13987–13992.
25. H. E. Wheeler and S. K. Kim, Genetics and genomics of human ageing, *Philos. Trans. R. Soc. London, Ser. B*, 2011, **366**, 43–50.
26. J. M. Murabito, R. Yuan and K. L. Lunetta, The search for longevity and healthy aging genes: insights from epidemiological studies and samples of long-lived individuals, *J. Gerontol., Ser. A*, 2012, **67**, 470–479.
27. L. Broer, A. S. Buchman, J. Deelen, D. S. Evans, J. D. Faul, K. L. Lunetta, P. Sebastiani, J. A. Smith, A. V. Smith, T. Tanaka, L. Yu, A. M. Arnold, T. Aspelund, E. J. Benjamin, P. L. De Jager, G. Eirkisdottir, D. A. Evans, M. E. Garcia, A. Hofman, R. C. Kaplan, S. L. Kardia, D. P. Kiel, B. A. Oostra, E. S. Orwoll, N. Parimi, B. M. Psaty, F. Rivadeneira, J. I. Rotter, S. Seshadri, A. Singleton, H. Tiemeier, A. G. Uitterlinden, W. Zhao, S. Bandinelli, D. A. Bennett, L. Ferrucci, V. Gudnason, T. B. Harris, D. Karasik, L. J. Launer, T. T. Perls, P. E. Slagboom, G. J. Tranah, D. R. Weir, A. B. Newman, C. M. van Duijn and J. M. Murabito, GWAS of longevity in CHARGE consortium confirms APOE and FOXO3 candidacy, *J. Gerontol., Ser. A*, 2015, **70**, 110–118.
28. R. W. Mahley and S. C. Rall Jr, Apolipoprotein E: far more than a lipid transport protein, *Annu. Rev. Genomics Hum. Genet.*, 2000, **1**, 507–537.
29. M. S. Brown and J. L. Goldstein, A receptor-mediated pathway for cholesterol homeostasis, *Science*, 1986, **232**, 34–47.
30. D. L. Nelson, M. M. Cox and A. L. Lehninger, *Lehninger Principles of Biochemistry*, W.H. Freeman, New York, 4th edn, 2005.
31. R. E. Pitas, J. K. Boyles, S. H. Lee, D. Foss and R. W. Mahley, Astrocytes synthesize apolipoprotein E and metabolize apolipoprotein E-containing lipoproteins, *Biochim. Biophys. Acta*, 1987, **917**, 148–161.
32. T. Uchihara, C. Duyckaerts, Y. He, K. Kobayashi, D. Seilhean, P. Amouyel and J. J. Hauw, ApoE immunoreactivity and microglial cells in Alzheimer's disease brain, *Neurosci. Lett.*, 1995, **195**, 5–8.
33. M. J. LaDu, S. M. Gilligan, J. R. Lukens, V. G. Cabana, C. A. Reardon, L. J. Van Eldik and D. M. Holtzman, Nascent astrocyte particles differ from lipoproteins in CSF, *J. Neurochem.*, 1998, **70**, 2070–2081.
34. V. Hirsch-Reinshagen, S. Zhou, B. L. Burgess, L. Bernier, S. A. McIsaac, J. Y. Chan, G. H. Tansley, J. S. Cohn, M. R. Hayden and C. L. Wellington, Deficiency of ABCA1 impairs apolipoprotein E metabolism in brain, *J. Biol. Chem.*, 2004, **279**, 41197–41207.

35. S. E. Wahrle, H. Jiang, M. Parsadanian, J. Legleiter, X. Han, J. D. Fryer, T. Kowalewski and D. M. Holtzman, ABCA1 is required for normal central nervous system ApoE levels and for lipidation of astrocyte-secreted apoE, *J. Biol. Chem.*, 2004, **279**, 40987–40993.
36. D. H. Mauch, K. Nagler, S. Schumacher, C. Goritz, E. C. Muller, A. Otto and F. W. Pfrieger, CNS synaptogenesis promoted by glia-derived cholesterol, *Science*, 2001, **294**, 1354–1357.
37. F. W. Pfrieger, Cholesterol homeostasis and function in neurons of the central nervous system, *Cell. Mol. Life Sci.*, 2003, **60**, 1158–1171.
38. C. C. Liu, T. Kanekiyo, H. Xu and G. Bu, Apolipoprotein E and Alzheimer disease: risk, mechanisms and therapy, *Nat. Rev. Neurol.*, 2013, **9**, 106–118.
39. V. Zerbi, M. Wiesmann, T. L. Emmerzaal, D. Jansen, M. Van Beek, M. P. Mutsaers, C. F. Beckmann, A. Heerschap and A. J. Kiliaan, Resting-state functional connectivity changes in aging apoE4 and apoE-KO mice, *J. Neurosci.*, 2014, **34**, 13963–13975.
40. M. V. Starkov, L. F. Shashkina and M. I. Remezova, [A morphohistochemical study of the embryotropic action of steroids with androgenic properties], *Arkh. Patol.*, 1975, **37**, 45–50.
41. M. Dieckmann, M. F. Dietrich and J. Herz, Lipoprotein receptors–an evolutionarily ancient multifunctional receptor family, *Biol. Chem.*, 2010, **391**, 1341–1363.
42. J. Herz and H. H. Bock, Lipoprotein receptors in the nervous system, *Annu. Rev. Biochem.*, 2002, **71**, 405–434.
43. J. Herz, D. E. Clouthier and R. E. Hammer, LDL receptor-related protein internalizes and degrades uPA-PAI-1 complexes and is essential for embryo implantation, *Cell*, 1992, **71**, 411–421.
44. J. Herz, D. E. Couthier and R. E. Hammer, Correction: LDL receptor-related protein internalizes and degrades uPA-PAI-1 complexes and is essential for embryo implantation, *Cell*, 1993, **73**, 428.
45. W. Y. Kang, S. S. Kim, S. K. Cho, S. Kim, H. Suh-Kim and Y. D. Lee, Migratory defect of mesencephalic dopaminergic neurons in developing reeler mice, *Anat. Cell Biol.*, 2010, **43**, 241–251.
46. A. Sharaf, H. H. Bock, B. Spittau, E. Bouche and K. Krieglstein, ApoER2 and VLDLr are required for mediating reelin signalling pathway for normal migration and positioning of mesencephalic dopaminergic neurons, *PLoS One*, 2013, **8**, e71091.
47. A. Herring, A. Donath, K. M. Steiner, M. P. Widera, S. Hamzehian, D. Kanakis, K. Kolble, A. ElAli, D. M. Hermann, W. Paulus and K. Keyvani, Reelin depletion is an early phenomenon of Alzheimer's pathology, *J. Alzheimers Dis.*, 2012, **30**, 963–979.
48. T. Muller, C. Loosse, A. Schrotter, A. Schnabel, S. Helling, R. Egensperger and K. Marcus, The AICD interacting protein DAB1 is up-regulated in Alzheimer frontal cortex brain samples and causes deregulation of proteins involved in gene expression changes, *Curr. Alzheimer Res.*, 2011, **8**, 573–582.

49. R. Gavin, I. Ferrer and J. A. del Rio, Involvement of Dab1 in APP processing and beta-amyloid deposition in sporadic Creutzfeldt-Jakob patients, *Neurobiol. Dis.*, 2010, **37**, 324–329.
50. L. Pujadas, D. Rossi, R. Andres, C. M. Teixeira, B. Serra-Vidal, A. Parcerisas, R. Maldonado, E. Giralt, N. Carulla and E. Soriano, Reelin delays amyloid-beta fibril formation and rescues cognitive deficits in a model of Alzheimer's disease, *Nat. Commun.*, 2014, **5**, 3443.
51. I. Knuesel, M. Nyffeler, C. Mormede, M. Muhia, U. Meyer, S. Pietropaolo, B. K. Yee, C. R. Pryce, F. M. LaFerla, A. Marighetto and J. Feldon, Age-related accumulation of Reelin in amyloid-like deposits, *Neurobiol. Aging*, 2009, **30**, 697–716.
52. J. Zhang, T. Saur, A. N. Duke, S. G. Grant, D. M. Platt, J. K. Rowlett, O. Isacson and W. D. Yao, Motor impairments, striatal degeneration, and altered dopamine-glutamate interplay in mice lacking PSD-95, *J. Neurogenet.*, 2014, **28**, 98–111.
53. G. Porras, A. Berthet, B. Dehay, Q. Li, L. Ladepeche, E. Normand, S. Dovero, A. Martinez, E. Doudnikoff, M. L. Martin-Negrier, Q. Chuan, B. Bloch, D. Choquet, E. Boue-Grabot, L. Groc and E. Bezard, PSD-95 expression controls L-DOPA dyskinesia through dopamine D1 receptor trafficking, *J. Clin. Invest.*, 2012, **122**, 3977–3989.
54. S. R. Jaffrey, A. M. Snowman, M. J. Eliasson, N. A. Cohen and S. H. Snyder, CAPON: a protein associated with neuronal nitric oxide synthase that regulates its interactions with PSD95, *Neuron*, 1998, **20**, 115–124.
55. I. Joniec, A. Ciesielska, I. Kurkowska-Jastrzebska, A. Przybylkowski, A. Czlonkowska and A. Czlonkowski, Age- and sex-differences in the nitric oxide synthase expression and dopamine concentration in the murine model of Parkinson's disease induced by 1-methyl-4-phenyl-1,2,3,6-tetrahydropyridine, *Brain Res.*, 2009, **1261**, 7–19.
56. T. Shang, S. Kotamraju, H. Zhao, S. V. Kalivendi, C. J. Hillard and B. Kalyanaraman, Sepiapterin attenuates 1-methyl-4-phenylpyridinium-induced apoptosis in neuroblastoma cells transfected with neuronal NOS: role of tetrahydrobiopterin, nitric oxide, and proteasome activation, *Free Radical Biol. Med.*, 2005, **39**, 1059–1074.
57. J. H. Cheah, S. F. Kim, L. D. Hester, K. W. Clancy, S. E. Patterson 3rd, V. Papadopoulos and S. H. Snyder, NMDA receptor-nitric oxide transmission mediates neuronal iron homeostasis *via* the GTPase Dexras1, *Neuron*, 2006, **51**, 431–440.
58. P. G. Mastroberardino, E. K. Hoffman, M. P. Horowitz, R. Betarbet, G. Taylor, D. Cheng, H. M. Na, C. A. Gutekunst, M. Gearing, J. Q. Trojanowski, M. Anderson, C. T. Chu, J. Peng and J. T. Greenamyre, A novel transferrin/TfR2-mediated mitochondrial iron transport system is disrupted in Parkinson's disease, *Neurobiol. Dis.*, 2009, **34**, 417–431.
59. E. Rousseau, P. P. Michel and E. C. Hirsch, The iron-binding protein lactoferrin protects vulnerable dopamine neurons from degeneration by preserving mitochondrial calcium homeostasis, *Mol. Pharmacol.*, 2013, **84**, 888–898.

60. B. A. Faucheux, N. Nillesse, P. Damier, G. Spik, A. Mouatt-Prigent, A. Pierce, B. Leveugle, N. Kubis, J. J. Hauw and Y. Agid, *et al.*, Expression of lactoferrin receptors is increased in the mesencephalon of patients with Parkinson disease, *Proc. Natl. Acad. Sci. U. S. A.*, 1995, **92**, 9603–9607.
61. E. C. Hirsch, Iron transport in Parkinson's disease, *Parkinsonism Relat. Disord.*, 2009, **15**(suppl. 3), S209–S211.
62. R. Kruger, A. M. Menezes-Saecker, L. Schols, W. Kuhn, T. Muller, D. Woitalla, D. Berg, K. Berger, H. Przuntek, J. T. Epplen and O. Riess, Genetic analysis of the alpha2-macroglobulin gene in early- and late-onset Parkinson's disease, *NeuroReport*, 2000, **11**, 2439–2442.
63. N. Zhong and K. H. Weisgraber, Understanding the association of apolipoprotein E4 with Alzheimer disease: clues from its structure, *J. Biol. Chem.*, 2009, **284**, 6027–6031.
64. O. B. Ptitsyn, Molten globule and protein folding, *Adv. Protein Chem.*, 1995, **47**, 83–229.
65. J. A. Morrow, D. M. Hatters, B. Lu, P. Hochtl, K. A. Oberg, B. Rupp and K. H. Weisgraber, Apolipoprotein E4 forms a molten globule. A potential basis for its association with disease, *J. Biol. Chem.*, 2002, **277**, 50380–50385.
66. K. H. Weisgraber, Apolipoprotein E: structure-function relationships, *Adv. Protein Chem.*, 1994, **45**, 249–302.
67. D. M. Hatters, N. Zhong, E. Rutenber and K. H. Weisgraber, Amino-terminal domain stability mediates apolipoprotein E aggregation into neurotoxic fibrils, *J. Mol. Biol.*, 2006, **361**, 932–944.
68. T. Wisniewski, E. M. Castano, A. Golabek, T. Vogel and B. Frangione, Acceleration of Alzheimer's fibril formation by apolipoprotein E *in vitro*, *Am. J. Pathol.*, 1994, **145**, 1030–1035.
69. D. A. Sanan, K. H. Weisgraber, S. J. Russell, R. W. Mahley, D. Huang, A. Saunders, D. Schmechel, T. Wisniewski, B. Frangione and A. D. Roses, *et al.*, Apolipoprotein E associates with beta amyloid peptide of Alzheimer's disease to form novel monofibrils. Isoform apoE4 associates more efficiently than apoE3, *J. Clin. Invest.*, 1994, **94**, 860–869.
70. D. E. Schmechel, A. M. Saunders, W. J. Strittmatter, B. J. Crain, C. M. Hulette, S. H. Joo, M. A. Pericak-Vance, D. Goldgaber and A. D. Roses, Increased amyloid beta-peptide deposition in cerebral cortex as a consequence of apolipoprotein E genotype in late-onset Alzheimer disease, *Proc. Natl. Acad. Sci. U. S. A.*, 1993, **90**, 9649–9653.
71. K. H. Weisgraber and L. H. Shinto, Identification of the disulfide-linked homodimer of apolipoprotein E3 in plasma. Impact on receptor binding activity, *J. Biol. Chem.*, 1991, **266**, 12029–12034.
72. K. S. Montine, C. N. Bassett, J. J. Ou, W. R. Markesbery, L. L. Swift and T. J. Montine, Apolipoprotein E allelic influence on human cerebrospinal fluid apolipoproteins, *J. Lipid Res.*, 1998, **39**, 2443–2451.
73. A. J. Abrams, A. Farooq and G. Wang, S-nitrosylation of ApoE in Alzheimer's disease, *Biochemistry*, 2011, **50**, 3405–3407.

74. D. Tsuang, J. B. Leverenz, O. L. Lopez, R. L. Hamilton, D. A. Bennett, J. A. Schneider, A. S. Buchman, E. B. Larson, P. K. Crane, J. A. Kaye, P. Kramer, R. Woltjer, J. Q. Trojanowski, D. Weintraub, A. S. Chen-Plotkin, D. J. Irwin, J. Rick, G. D. Schellenberg, G. S. Watson, W. Kukull, P. T. Nelson, G. A. Jicha, J. H. Neltner, D. Galasko, E. Masliah, J. F. Quinn, K. A. Chung, D. Yearout, I. F. Mata, J. Y. Wan, K. L. Edwards, T. J. Montine and C. P. Zabetian, APOE epsilon4 increases risk for dementia in pure synucleinopathies, *JAMA Neurol.*, 2013, **70**, 223–228.
75. I. F. Mata, J. B. Leverenz, D. Weintraub, J. Q. Trojanowski, H. I. Hurtig, V. M. Van Deerlin, B. Ritz, R. Rausch, S. L. Rhodes, S. A. Factor, C. Wood-Siverio, J. F. Quinn, K. A. Chung, A. L. Peterson, A. J. Espay, F. J. Revilla, J. Devoto, S. C. Hu, B. A. Cholerton, J. Y. Wan, T. J. Montine, K. L. Edwards and C. P. Zabetian, APOE, MAPT, and SNCA genes and cognitive performance in Parkinson disease, *JAMA Neurol.*, 2014, **71**, 1405–1412.
76. D. J. Irwin, M. T. White, J. B. Toledo, S. X. Xie, J. L. Robinson, V. Van Deerlin, V. M. Lee, J. B. Leverenz, T. J. Montine, J. E. Duda, H. I. Hurtig and J. Q. Trojanowski, Neuropathologic substrates of Parkinson disease dementia, *Ann. Neurol.*, 2012, **72**, 587–598.
77. S. N. Gomperts, J. J. Locascio, D. Rentz, A. Santarlasci, M. Marquie, K. A. Johnson and J. H. Growdon, Amyloid is linked to cognitive decline in patients with Parkinson disease without dementia, *Neurology*, 2013, **80**, 85–91.
78. M. M. Wilhelmus, J. G. Bol, E. S. Van Haastert, A. J. Rozemuller, G. Bu, B. Drukarch and J. J. Hoozemans, Apolipoprotein E and LRP1 Increase Early in Parkinson's Disease Pathogenesis, *Am. J. Pathol.*, 2011, **179**, 2152–2156.
79. D. Domenger, D. Dea, L. Theroux, L. Moquin, A. Gratton and J. Poirier, The MPTP neurotoxic lesion model of Parkinson's disease activates the apolipoprotein E cascade in the mouse brain, *Exp. Neurol.*, 2012, **233**, 513–522.
80. M. P. Horowitz, C. Milanese, R. Di Maio, X. Hu, L. M. Montero, L. H. Sanders, V. Tapias, S. Sepe, W. A. van Cappellen, E. A. Burton, J. T. Greenamyre and P. G. Mastroberardino, Single-cell redox imaging demonstrates a distinctive response of dopaminergic neurons to oxidative insults, *Antioxid. Redox Signaling*, 2011, **15**, 855–871.
81. P. G. Mastroberardino, A. L. Orr, X. Hu, H. M. Na and J. T. Greenamyre, A FRET-based method to study protein thiol oxidation in histological preparations, *Free Radical Biol. Med.*, 2008, **45**, 971–981.
82. M. F. Beal, Mitochondria, oxidative damage, and inflammation in Parkinson's disease, *Ann. N. Y. Acad. Sci.*, 2003, **991**, 120–131.
83. M. Miyata and J. D. Smith, Apolipoprotein E allele-specific antioxidant activity and effects on cytotoxicity by oxidative insults and beta-amyloid peptides, *Nat. Genet.*, 1996, **14**, 55–61.
84. L. Jofre-Monseny, A. M. Minihane and G. Rimbach, Impact of apoE genotype on oxidative stress, inflammation and disease risk, *Mol. Nutr. Food Res.*, 2008, **52**, 131–145.

85. D. A. Butterfield, A. Castegna, C. M. Lauderback and J. Drake, Evidence that amyloid beta-peptide-induced lipid peroxidation and its sequelae in Alzheimer's disease brain contribute to neuronal death, *Neurobiol. Aging*, 2002, **23**, 655–664.
86. Y. Ihara, T. Hayabara, K. Sasaki, Y. Fujisawa, R. Kawada, T. Yamamoto, Y. Nakashima, S. Yoshimune, M. Kawai, M. Kibata and S. Kuroda, Free radicals and superoxide dismutase in blood of patients with Alzheimer's disease and vascular dementia, *J. Neurol. Sci.*, 1997, **153**, 76–81.
87. Y. Ihara, T. Hayabara, K. Sasaki, R. Kawada, Y. Nakashima and S. Kuroda, Relationship between oxidative stress and apoE phenotype in Alzheimer's disease, *Acta Neurol. Scand.*, 2000, **102**, 346–349.
88. C. M. Lauderback, J. Kanski, J. M. Hackett, N. Maeda, M. S. Kindy and D. A. Butterfield, Apolipoprotein E modulates Alzheimer's Abeta(1-42)-induced oxidative damage to synaptosomes in an allele-specific manner, *Brain Res.*, 2002, **924**, 90–97.
89. C. M. Lauderback, J. M. Hackett, J. N. Keller, S. Varadarajan, L. Szweda, M. Kindy, W. R. Markesbery and D. A. Butterfield, Vulnerability of synaptosomes from apoE knock-out mice to structural and oxidative modifications induced by A beta(1-40): implications for Alzheimer's disease, *Biochemistry*, 2001, **40**, 2548–2554.
90. K. L. Clark, E. D. Halay, E. Lai and S. K. Burley, Co-crystal structure of the HNF-3/fork head DNA-recognition motif resembles histone H5, *Nature*, 1993, **364**, 412–420.
91. T. Obsil and V. Obsilova, Structural basis for DNA recognition by FOXO proteins, *Biochim. Biophys. Acta*, 2011, **1813**, 1946–1953.
92. A. Klagge, C. Weidinger, K. Krause, B. Jessnitzer, M. Gutknecht and D. Fuhrer, The role of FOXO3 in DNA damage response in thyrocytes, *Endocr.-Relat. Cancer*, 2011, **18**, 555–564.
93. M. M. Mueller, L. Castells-Roca, V. Babu, M. A. Ermolaeva, R. U. Muller, P. Frommolt, A. B. Williams, S. Greiss, J. I. Schneider, T. Benzing, B. Schermer and B. Schumacher, DAF-16/FOXO and EGL-27/GATA promote developmental growth in response to persistent somatic DNA damage, *Nat. Cell Biol.*, 2014, **16**, 1168–1179.
94. D. N. Hauser and T. G. Hastings, Mitochondrial dysfunction and oxidative stress in Parkinson's disease and monogenic parkinsonism, *Neurobiol. Dis.*, 2013, **51**, 35–42.
95. M. Safra, R. Fickentscher, M. Levi-Ferber, Y. M. Danino, A. Haviv-Chesner, M. Hansen, T. Juven-Gershon, M. Weiss and S. Henis-Korenblit, The FOXO transcription factor DAF-16 bypasses ire-1 requirement to promote endoplasmic reticulum homeostasis, *Cell Metab.*, 2014, **20**, 870–881.
96. K. E. van der Vos and P. J. Coffer, The extending network of FOXO transcriptional target genes, *Antioxid. Redox Signaling*, 2011, **14**, 579–592.
97. X. Zhao, Y. Liu, L. Du, L. He, B. Ni, J. Hu, D. Zhu and Q. Chen, Threonine 32 (Thr32) of FoxO3 is critical for TGF-beta-induced apoptosis *via* Bim in hepatocarcinoma cells, *Protein Cell*, 2014, **6**, 127–138.

98. Z. Fu and D. J. Tindall, FOXOs, cancer and regulation of apoptosis, *Oncogene*, 2008, **27**, 2312–2319.
99. N. P. Gomez-Crisostomo, E. Rodriguez Martinez and S. Rivas-Arancibia, Oxidative stress activates the transcription factors FoxO 1a and FoxO 3a in the hippocampus of rats exposed to low doses of ozone, *Oxid. Med. Cell. Longevity*, 2014, **2014**, 805764.
100. Y. Akasaki, O. Alvarez-Garcia, M. Saito, B. Carames, Y. Iwamoto and M. K. Lotz, FoxO transcription factors support oxidative stress resistance in human chondrocytes, *Arthritis Rheumatol.*, 2014, **66**, 3349–3358.
101. C. Y. Ewald, J. N. Landis, J. P. Abate, C. T. Murphy and T. K. Blackwell, Dauer-independent insulin/IGF-1-signalling implicates collagen remodelling in longevity, *Nature*, 2015, **519**, 97–101.
102. K. Warnhoff, J. T. Murphy, S. Kumar, D. L. Schneider, M. Peterson, S. Hsu, J. Guthrie, J. D. Robertson and K. Kornfeld, The DAF-16 FOXO transcription factor regulates natc-1 to modulate stress resistance in Caenorhabditis elegans, linking insulin/IGF-1 signaling to protein N-terminal acetylation, *PLoS Genet.*, 2014, **10**, e1004703.
103. R. S. Nho and P. Hergert, FoxO3a and disease progression, *World J. Biol. Chem.*, 2014, **5**, 346–354.
104. D. H. Kim, M. H. Park, E. K. Lee, Y. J. Choi, K. W. Chung, K. M. Moon, M. J. Kim, H. J. An, J. W. Park, N. D. Kim, B. P. Yu and H. Y. Chung, The roles of FoxOs in modulation of aging by calorie restriction, *Biogerontology*, 2015, **16**, 1–14.
105. H. Huang and D. J. Tindall, Dynamic FoxO transcription factors, *J. Cell Sci.*, 2007, **120**, 2479–2487.
106. D. A. Salih, A. J. Rashid, D. Colas, L. de la Torre-Ubieta, R. P. Zhu, A. A. Morgan, E. E. Santo, D. Ucar, K. Devarajan, C. J. Cole, D. V. Madison, M. Shamloo, A. J. Butte, A. Bonni, S. A. Josselyn and A. Brunet, FoxO6 regulates memory consolidation and synaptic function, *Genes Dev.*, 2012, **26**, 2780–2801.
107. M. F. Hoekman, F. M. Jacobs, M. P. Smidt and J. P. Burbach, Spatial and temporal expression of FoxO transcription factors in the developing and adult murine brain, *Gene Expr. Patterns*, 2006, **6**, 134–140.
108. X. Huo, S. Liu, T. Shao, H. Hua, Q. Kong, J. Wang, T. Luo and Y. Jiang, GSK3 protein positively regulates type I insulin-like growth factor receptor through forkhead transcription factors FOXO1/3/4, *J. Biol. Chem.*, 2014, **289**, 24759–24770.
109. K. C. Arden and W. H. Biggs 3rd, Regulation of the FoxO family of transcription factors by phosphatidylinositol-3 kinase-activated signaling, *Arch. Biochem. Biophys.*, 2002, **403**, 292–298.
110. G. Tzivion, M. Dobson and G. Ramakrishnan, FoxO transcription factors; Regulation by AKT and 14-3-3 proteins, *Biochim. Biophys. Acta*, 2011, **1813**, 1938–1945.
111. M. A. Essers, S. Weijzen, A. M. dc Vries-Smits, I. Saarloos, N. D. de Ruiter, J. L. Bos and B. M. Burgering, FOXO transcription factor activation by oxidative stress mediated by the small GTPase Ral and JNK, *EMBO J.*, 2004, **23**, 4802–4812.

112. A. Eijkelenboom and B. M. Burgering, FOXOs: signalling integrators for homeostasis maintenance, *Nat. Rev. Mol. Cell Biol.*, 2013, **14**, 83–97.
113. Y. Wang, Y. Zhou and D. T. Graves, FOXO transcription factors: their clinical significance and regulation, *BioMed Res. Int.*, 2014, **2014**, 925350.
114. F. Seiler, J. Hellberg, P. M. Lepper, A. Kamyschnikow, C. Herr, M. Bischoff, F. Langer, H. J. Schafers, F. Lammert, M. D. Menger, R. Bals and C. Beisswenger, FOXO transcription factors regulate innate immune mechanisms in respiratory epithelial cells, *J. Immunol.*, 2013, **190**, 1603–1613.
115. T. Hosaka, W. H. Biggs 3rd, D. Tieu, A. D. Boyer, N. M. Varki, W. K. Cavenee and K. C. Arden, Disruption of forkhead transcription factor (FOXO) family members in mice reveals their functional diversification, *Proc. Natl. Acad. Sci. U. S. A.*, 2004, **101**, 2975–2980.
116. R. H. Medema, G. J. Kops, J. L. Bos and B. M. Burgering, AFX-like Forkhead transcription factors mediate cell-cycle regulation by Ras and PKB through p27kip1, *Nature*, 2000, **404**, 782–787.
117. G. J. Kops, R. H. Medema, J. Glassford, M. A. Essers, P. F. Dijkers, P. J. Coffer, E. W. Lam and B. M. Burgering, Control of cell cycle exit and entry by protein kinase B-regulated forkhead transcription factors, *Mol. Cell. Biol.*, 2002, **22**, 2025–2036.
118. H. Tran, A. Brunet, J. M. Grenier, S. R. Datta, A. J. Fornace Jr, P. S. DiStefano, L. W. Chiang and M. E. Greenberg, DNA repair pathway stimulated by the forkhead transcription factor FOXO3a through the Gadd45 protein, *Science*, 2002, **296**, 530–534.
119. J. Wang, S. Liu, Y. Yin, M. Li, B. Wang, L. Yang and Y. Jiang, FOXO3-mediated up-regulation of Bim contributes to rhein-induced cancer cell apoptosis, *Apoptosis*, 2015, **20**, 399–409.
120. R. Akhter, P. Sanphui and S. C. Biswas, The essential role of p53-up-regulated modulator of apoptosis (Puma) and its regulation by FoxO3a transcription factor in beta-amyloid-induced neuron death, *J. Biol. Chem.*, 2014, **289**, 10812–10822.
121. Y. Sakoe, K. Sakoe, K. Kirito, K. Ozawa and N. Komatsu, FOXO3A as a key molecule for all-trans retinoic acid-induced granulocytic differentiation and apoptosis in acute promyelocytic leukemia, *Blood*, 2010, **115**, 3787–3795.
122. G. Marfe, M. Tafani, F. Fiorito, U. Pagnini, G. Iovane and L. De Martino, Involvement of FOXO transcription factors, TRAIL-FasL/Fas, and sirtuin proteins family in canine coronavirus type II-induced apoptosis, *PLoS One*, 2011, **6**, e27313.
123. E. C. Ferber, B. Peck, O. Delpuech, G. P. Bell, P. East and A. Schulze, FOXO3a regulates reactive oxygen metabolism by inhibiting mitochondrial gene expression, *Cell Death Differ.*, 2012, **19**, 968–979.
124. A. Peserico, F. Chiacchiera, V. Grossi, A. Matrone, D. Latorre, M. Simonatto, A. Fusella, J. G. Ryall, L. W. Finley, M. C. Haigis, G. Villani, P. L. Puri, V. Sartorelli and C. Simone, A novel AMPK-dependent FoxO3A-SIRT3 intramitochondrial complex sensing glucose levels, *Cell. Mol. Life Sci.*, 2013, **70**, 2015–2029.

125. G. J. Kops, T. B. Dansen, P. E. Polderman, I. Saarloos, K. W. Wirtz, P. J. Coffer, T. T. Huang, J. L. Bos, R. H. Medema and B. M. Burgering, Forkhead transcription factor FOXO3a protects quiescent cells from oxidative stress, *Nature*, 2002, **419**, 316–321.
126. H. Daitoku, J. Sakamaki and A. Fukamizu, Regulation of FoxO transcription factors by acetylation and protein-protein interactions, *Biochim. Biophys. Acta*, 2011, **1813**, 1954–1960.
127. F. Zhang, S. Wang, L. Gan, P. S. Vosler, Y. Gao, M. J. Zigmond and J. Chen, Protective effects and mechanisms of sirtuins in the nervous system, *Prog. Neurobiol.*, 2011, **95**, 373–395.
128. X. Zhou, M. Chen, X. Zeng, J. Yang, H. Deng, L. Yi and M. T. Mi, Resveratrol regulates mitochondrial reactive oxygen species homeostasis through Sirt3 signaling pathway in human vascular endothelial cells, *Cell Death Dis.*, 2014, **5**, e1576.
129. E. L. Greer and A. Brunet, FOXO transcription factors at the interface between longevity and tumor suppression, *Oncogene*, 2005, **24**, 7410–7425.
130. X. Kong, R. Wang, Y. Xue, X. Liu, H. Zhang, Y. Chen, F. Fang and Y. Chang, Sirtuin 3, a new target of PGC-1alpha, plays an important role in the suppression of ROS and mitochondrial biogenesis, *PLoS One*, 2010, **5**, e11707.
131. N. R. Sundaresan, M. Gupta, G. Kim, S. B. Rajamohan, A. Isbatan and M. P. Gupta, Sirt3 blocks the cardiac hypertrophic response by augmenting Foxo3a-dependent antioxidant defense mechanisms in mice, *J. Clin. Invest.*, 2009, **119**, 2758–2771.
132. A. H. Tseng, S. S. Shieh and D. L. Wang, SIRT3 deacetylates FOXO3 to protect mitochondria against oxidative damage, *Free Radical Biol. Med.*, 2013, **63**, 222–234.
133. B. Zheng, S. Ohkawa, H. Li, T. K. Roberts-Wilson and S. R. Price, FOXO3a mediates signaling crosstalk that coordinates ubiquitin and atrogin-1/MAFbx expression during glucocorticoid-induced skeletal muscle atrophy, *FASEB J.*, 2010, **24**, 2660–2669.
134. L. Sun, W. J. Zang, H. Wang, M. Zhao, X. J. Yu, X. He, Y. Miao and J. Zhou, Acetylcholine Promotes ROS Detoxification Against Hypoxia/reoxygenation-Induced Oxidative Stress Through FoxO3a/PGC-1alpha Dependent Superoxide Dismutase, *Cell. Physiol. Biochem.*, 2014, **34**, 1614–1625.
135. S. Das, G. Mitrovsky, H. R. Vasanthi and D. K. Das, Antiaging properties of a grape-derived antioxidant are regulated by mitochondrial balance of fusion and fission leading to mitophagy triggered by a signaling network of Sirt1-Sirt3-Foxo3-PINK1-PARKIN, *Oxid. Med. Cell. Longevity*, 2014, **2014**, 345105.
136. M. M. Desgeorges, D. Freyssenet, S. Chanon, J. Castells, P. Pugniere, D. Bechet, A. Peinnequin, X. Devillard and A. Defour, Post-transcriptional regulation of autophagy in C2C12 myotubes following starvation and nutrient restoration, *Int. J. Biochem. Cell Biol.*, 2014, **54**, 208–216.

137. A. M. Sanchez, H. Bernardi, G. Py and R. B. Candau, Autophagy is essential to support skeletal muscle plasticity in response to endurance exercise, *Am. J. Physiol.: Regul., Integr. Comp. Physiol.*, 2014, **307**, R956–R969.
138. H. Tang, K. Inoki, M. Lee, E. Wright, A. Khuong, A. Khuong, S. Sugiarto, M. Garner, J. Paik, R. A. DePinho, D. Goldman, K. L. Guan and J. B. Shrager, mTORC1 promotes denervation-induced muscle atrophy through a mechanism involving the activation of FoxO and E3 ubiquitin ligases, *Sci. Signaling*, 2014, **7**, ra18.
139. A. E. Webb and A. Brunet, FOXO transcription factors: key regulators of cellular quality control, *Trends Biochem. Sci.*, 2014, **39**, 159–169.
140. J. Zhao, J. J. Brault, A. Schild, P. Cao, M. Sandri, S. Schiaffino, S. H. Lecker and A. L. Goldberg, FoxO3 coordinately activates protein degradation by the autophagic/lysosomal and proteasomal pathways in atrophying muscle cells, *Cell Metab.*, 2007, **6**, 472–483.
141. D. Bechet, A. Tassa, D. Taillandier, L. Combaret and D. Attaix, Lysosomal proteolysis in skeletal muscle, *Int. J. Biochem. Cell Biol.*, 2005, **37**, 2098–2114.
142. D. J. Cao, N. Jiang, A. Blagg, J. L. Johnstone, R. Gondalia, M. Oh, X. Luo, K. C. Yang, J. M. Shelton, B. A. Rothermel, T. G. Gillette, G. W. Dorn and J. A. Hill, Mechanical unloading activates FoxO3 to trigger Bnip3-dependent cardiomyocyte atrophy, *J. Am. Heart Assoc.*, 2013, **2**, e000016.
143. C. Mammucari, G. Milan, V. Romanello, E. Masiero, R. Rudolf, P. Del Piccolo, S. J. Burden, R. Di Lisi, C. Sandri, J. Zhao, A. L. Goldberg, S. Schiaffino and M. Sandri, FoxO3 controls autophagy in skeletal muscle *in vivo*, *Cell Metab.*, 2007, **6**, 458–471.
144. C. Mammucari, S. Schiaffino and M. Sandri, Downstream of Akt: FoxO3 and mTOR in the regulation of autophagy in skeletal muscle, *Autophagy*, 2008, **4**, 524–526.
145. K. Tracy and K. F. Macleod, Regulation of mitochondrial integrity, autophagy and cell survival by BNIP3, *Autophagy*, 2007, **3**, 616–619.
146. S. Lokireddy, I. W. Wijesoma, S. Teng, S. Bonala, P. D. Gluckman, C. McFarlane, M. Sharma and R. Kambadur, The ubiquitin ligase Mul1 induces mitophagy in skeletal muscle in response to muscle-wasting stimuli, *Cell Metab.*, 2012, **16**, 613–624.
147. P. Low, The role of ubiquitin-proteasome system in ageing, *Gen. Comp. Endocrinol.*, 2011, **172**, 39–43.
148. N. P. Dantuma and L. C. Bott, The ubiquitin-proteasome system in neurodegenerative diseases: precipitating factor, yet part of the solution, *Front. Mol. Neurosci.*, 2014, **7**, 70.
149. A. Jaitovich, M. Angulo, E. Lecuona, L. A. Dada, L. C. Welch, Y. Cheng, G. Gusarova, E. Ceco, C. Liu, M. Shigemura, E. Barreiro, C. Patterson, G. A. Nader and J. I. Sznajder, High CO2 Levels Cause Skeletal Muscle Atrophy *via* AMP-activated Kinase (AMPK), FoxO3a Protein, and Muscle-specific Ring Finger Protein 1 (MuRF1), *J. Biol. Chem.*, 2015, **290**, 9183–9194.

150. E. Lin, J. E. Cavanaugh, R. K. Leak, R. G. Perez and M. J. Zigmond, Rapid activation of ERK by 6-hydroxydopamine promotes survival of dopaminergic cells, *J. Neurosci. Res.*, 2008, **86**, 108–117.
151. A. M. Pickrell and R. J. Youle, The roles of PINK1, parkin, and mitochondrial fidelity in Parkinson's disease, *Neuron*, 2015, **85**, 257–273.
152. Y. Jia, S. J. Mo, Q. Q. Feng, M. L. Zhan, L. S. OuYang, J. C. Chen, Y. X. Ma, J. J. Wu and W. L. Lei, EPO-dependent activation of PI3K/Akt/FoxO3a signalling mediates neuroprotection in *in vitro* and *in vivo* models of Parkinson's disease, *J. Mol. Neurosci.*, 2014, **53**, 117–124.
153. L. Liu, A. Arun, L. Ellis, C. Peritore and G. Donmez, SIRT2 enhances 1-methyl-4-phenyl-1,2,3,6-tetrahydropyridine (MPTP)-induced nigrostriatal damage *via* apoptotic pathway, *Front. Aging Neurosci.*, 2014, **6**, 184.
154. S. Hekimi, How genetic analysis tests theories of animal aging, *Nat. Genet.*, 2006, **38**, 985–991.
155. R. A. Miller, Genes against aging, *J. Gerontol., Ser. A*, 2012, **67**, 495–502.
156. C. T. Murphy and P. J. Hu, Insulin/insulin-like growth factor signaling in C. elegans, *WormBook*, 2013, **26**, 1–43.
157. N. Ishii, S. Goto and P. S. Hartman, Protein oxidation during aging of the nematode Caenorhabditis elegans, *Free Radical Biol. Med.*, 2002, **33**, 1021–1025.
158. S. Yanase, K. Yasuda and N. Ishii, Adaptive responses to oxidative damage in three mutants of Caenorhabditis elegans (age-1, mev-1 and daf-16) that affect life span, *Mech. Ageing Dev.*, 2002, **123**, 1579–1587.
159. K. Yasuda, H. Adachi, Y. Fujiwara and N. Ishii, Protein carbonyl accumulation in aging dauer formation-defective (daf) mutants of Caenorhabditis elegans, *J. Gerontol., Ser. A*, 1999, **54**, B47–B51, discussion B52–43.
160. L. Partridge, N. Alic, I. Bjedov and M. D. Piper, Ageing in Drosophila: the role of the insulin/Igf and TOR signalling network, *Exp. Gerontol.*, 2011, **46**, 376–381.
161. A. Lorenzini, A. B. Salmon, C. Lerner, C. Torres, Y. Ikeno, S. Motch, R. McCarter and C. Sell, Mice producing reduced levels of insulin-like growth factor type 1 display an increase in maximum, but not mean, life span, *J. Gerontol., Ser. A*, 2014, **69**, 410–419.
162. H. Kurosu, M. Yamamoto, J. D. Clark, J. V. Pastor, A. Nandi, P. Gurnani, O. P. McGuinness, H. Chikuda, M. Yamaguchi, H. Kawaguchi, I. Shimomura, Y. Takayama, J. Herz, C. R. Kahn, K. P. Rosenblatt and M. Kuro-o, Suppression of aging in mice by the hormone Klotho, *Science*, 2005, **309**, 1829–1833.
163. M. Haluzik, S. Yakar, O. Gavrilova, J. Setser, Y. Boisclair and D. LeRoith, Insulin resistance in the liver-specific IGF-1 gene-deleted mouse is abrogated by deletion of the acid-labile subunit of the IGF-binding protein-3 complex: relative roles of growth hormone and IGF-1 in insulin resistance, *Diabetes*, 2003, **52**, 2483–2489.
164. S. Freude, M. M. Hettich, C. Schumann, O. Stohr, L. Koch, C. Kohler, M. Udelhoven, U. Leeser, M. Muller, N. Kubota, T. Kadowaki, W. Krone,

H. Schroder, J. C. Bruning and M. Schubert, Neuronal IGF-1 resistance reduces Abeta accumulation and protects against premature death in a model of Alzheimer's disease, *FASEB J.*, 2009, **23**, 3315–3324.

165. L. Frolich, D. Blum-Degen, H. G. Bernstein, S. Engelsberger, J. Humrich, S. Laufer, D. Muschner, A. Thalheimer, A. Turk, S. Hoyer, R. Zochling, K. W. Boissl, K. Jellinger and P. Riederer, Brain insulin and insulin receptors in aging and sporadic Alzheimer's disease, *J. Neural Transm.*, 1998, **105**, 423–438.
166. D. van Heemst, M. Beekman, S. P. Mooijaart, B. T. Heijmans, B. W. Brandt, B. J. Zwaan, P. E. Slagboom and R. G. Westendorp, Reduced insulin/IGF-1 signalling and human longevity, *Aging Cell*, 2005, **4**, 79–85.
167. M. Bonafe, M. Barbieri, F. Marchegiani, F. Olivieri, E. Ragno, C. Giampieri, E. Mugianesi, M. Centurelli, C. Franceschi and G. Paolisso, Polymorphic variants of insulin-like growth factor I (IGF-I) receptor and phosphoinositide 3-kinase genes affect IGF-I plasma levels and human longevity: cues for an evolutionarily conserved mechanism of life span control, *J. Clin. Endocrinol. Metab.*, 2003, **88**, 3299–3304.
168. M. Barbieri, M. Bonafe, C. Franceschi and G. Paolisso, Insulin/IGF-I-signaling pathway: an evolutionarily conserved mechanism of longevity from yeast to humans, *Am. J. Physiol.: Endocrinol. Metab.*, 2003, **285**, E1064–E1071.
169. D. Harman, Aging: a theory based on free radical and radiation chemistry, *J. Gerontol.*, 1956, **11**, 298–300.
170. J. H. Hoeijmakers, DNA damage, aging, and cancer, *N. Engl. J. Med.*, 2009, **361**, 1475–1485.
171. R. A. Miller, 'Accelerated aging': a primrose path to insight? *Aging Cell*, 2004, **3**, 47–51.
172. J. de Boer and J. H. Hoeijmakers, Nucleotide excision repair and human syndromes, *Carcinogenesis*, 2000, **21**, 453–460.
173. S. Sepe, C. Payan-Gomez, C. Milanese, J. H. Hoeijmakers and P. G. Mastroberardino, Nucleotide excision repair in chronic neurodegenerative diseases, *DNA Repair*, 2013, **12**, 568–577.
174. J. A. Marteijn, H. Lans, W. Vermeulen and J. H. Hoeijmakers, Understanding nucleotide excision repair and its roles in cancer and ageing, *Nat. Rev. Mol. Cell Biol.*, 2014, **15**, 465–481.
175. A. R. Lehmann, DNA repair-deficient diseases, xeroderma pigmentosum, Cockayne syndrome and trichothiodystrophy, *Biochimie*, 2003, **85**, 1101–1111.
176. K. H. Kraemer, N. J. Patronas, R. Schiffmann, B. P. Brooks, D. Tamura and J. J. DiGiovanna, Xeroderma pigmentosum, trichothiodystrophy and Cockayne syndrome: a complex genotype-phenotype relationship, *Neuroscience*, 2007, **145**, 1388–1396.
177. S. Barnhoorn, L. M. Uittenboogaard, D. Jaarsma, W. P. Vermeij, M. Tresini, M. Weymaere, H. Menoni, R. M. Brandt, M. C. de Waard, S. M. Botter, A. H. Sarker, N. G. Jaspers, G. T. van der Horst, P. K. Cooper, J. H. Hoeijmakers and I. van der Pluijm, Cell-autonomous progeroid changes

in conditional mouse models for repair endonuclease XPG deficiency, *PLoS Genet.*, 2014, **10**, e1004686.

178. D. Jaarsma, I. van der Pluijm, M. C. de Waard, E. D. Haasdijk, R. Brandt, M. Vermeij, Y. Rijksen, A. Maas, H. van Steeg, J. H. Hoeijmakers and G. T. van der Horst, Age-related neuronal degeneration: complementary roles of nucleotide excision repair and transcription-coupled repair in preventing neuropathology, *PLoS Genet.*, 2011, 7, e1002405.
179. A. Ahmad, A. R. Robinson, A. Duensing, E. van Drunen, H. B. Beverloo, D. B. Weisberg, P. Hasty, J. H. Hoeijmakers and L. J. Niedernhofer, ERCC1-XPF endonuclease facilitates DNA double-strand break repair, *Mol. Cell. Biol.*, 2008, **28**, 5082–5092.
180. G. Weeda, I. Donker, J. de Wit, H. Morreau, R. Janssens, C. J. Vissers, A. Nigg, H. van Steeg, D. Bootsma and J. H. Hoeijmakers, Disruption of mouse ERCC1 results in a novel repair syndrome with growth failure, nuclear abnormalities and senescence, *Curr. Biol.*, 1997, **7**, 427–439.
181. J. W. Astin, N. J. O'Neil and P. E. Kuwabara, Nucleotide excision repair and the degradation of RNA pol II by the Caenorhabditis elegans XPA and Rsp5 orthologues, RAD-3 and WWP-1, *DNA Repair (Amst)*, 2008, 7, 267–280.
182. H. Lans and W. Vermeulen, Nucleotide Excision Repair in Caenorhabditis elegans, *Mol. Biol. Int.*, 2011, **2011**, 542795.
183. D. Jaarsma, I. Van Der Pluijm, M. De Waard, E. Haasdijk, R. Brandt, M. Vermeij, Y. Rijksen, A. Maas, H. Van Steeg, J. H. Hoeijmakers and G. T. van der Horst, Age-related neuronal degeneration: complementary roles of nucleotide excision repair and transcription-coupled repair in preventing neuropathology, *PLoS Genet.*, 2011, **7**, e1002405.
184. I. Karakasilioti, I. Kamileri, G. Chatzinikolaou, T. Kosteas, E. Vergadi, A. R. Robinson, I. Tsamardinos, T. A. Rozgaja, S. Siakouli, C. Tsatsanis, L. J. Niedernhofer and G. A. Garinis, DNA damage triggers a chronic autoinflammatory response, leading to fat depletion in NER progeria, *Cell Metab.*, 2013, **18**, 403–415.
185. K. E. Diderich, C. Nicolaije, M. Priemel, J. H. Waarsing, J. S. Day, R. M. Brandt, A. F. Schilling, S. M. Botter, H. Weinans, G. T. van der Horst, J. H. Hoeijmakers and J. P. van Leeuwen, Bone fragility and decline in stem cells in prematurely aging DNA repair deficient trichothiodystrophy mice, *Age (Dordr.)*, 2012, **34**, 845–861.
186. A. Trifunovic, A. Wredenberg, M. Falkenberg, J. N. Spelbrink, A. T. Rovio, C. E. Bruder, Y. M. Bohlooly, S. Gidlof, A. Oldfors, R. Wibom, J. Tornell, H. T. Jacobs and N. G. Larsson, Premature ageing in mice expressing defective mitochondrial DNA polymerase, *Nature*, 2004, **429**, 417–423.
187. A. Trifunovic, A. Hansson, A. Wredenberg, A. T. Rovio, E. Dufour, I. Khvorostov, J. N. Spelbrink, R. Wibom, H. T. Jacobs and N. G. Larsson, Somatic mtDNA mutations cause aging phenotypes without affecting reactive oxygen species production, *Proc. Natl. Acad. Sci. U. S. A.*, 2005, **102**, 17993–17998.

188. H. Frenzel and J. Feimann, Age-dependent structural changes in the myocardium of rats. A quantitative light- and electron-microscopic study on the right and left chamber wall, *Mech. Ageing Dev.*, 1984, **27**, 29–41.
189. K. D. Jacob, N. Noren Hooten, A. R. Trzeciak and M. K. Evans, Markers of oxidant stress that are clinically relevant in aging and age-related disease, *Mech. Ageing Dev.*, 2013, **134**, 139–157.
190. D. A. Cottrell and D. M. Turnbull, Mitochondria and ageing, *Curr. Opin. Clin. Nutr. Metab. Care*, 2000, **3**, 473–478.
191. A. Trifunovic and N. G. Larsson, Mitochondrial dysfunction as a cause of ageing, *J. Intern. Med.*, 2008, **263**, 167–178.
192. M. Olahova and E. A. Veal, A peroxiredoxin, PRDX-2, is required for insulin secretion and insulin/IIS-dependent regulation of stress resistance and longevity, *Aging Cell*, 2015, **14**, 558–568.
193. D. Ahn, D. S. Cha, E. B. Lee, B. J. Kim, S. Y. Lee, H. Jeon, M. S. Ahn, H. W. Lim, H. Y. Lee and D. K. Kim, The Longevity Properties of 1,2,3,4,6-Penta-O-Galloyl-beta-D-Glucose from Curcuma longa in Caenorhabditis elegans, *Biomol. Ther. (Seoul)*, 2013, **21**, 442–446.
194. J. A. Stuart, B. M. Bourque, N. C. de Souza-Pinto and V. A. Bohr, No evidence of mitochondrial respiratory dysfunction in OGG1-null mice deficient in removal of 8-oxodeoxyguanine from mitochondrial DNA, *Free Radical Biol. Med.*, 2005, **38**, 737–745.
195. D. R. Bickers and M. Athar, Oxidative stress in the pathogenesis of skin disease, *J. Invest. Dermatol.*, 2006, **126**, 2565–2575.
196. V. E. Kagan, E. R. Kisin, K. Kawai, B. F. Serinkan, A. N. Osipov, E. A. Serbinova, I. Wolinsky and A. A. Shvedova, Toward mechanism-based antioxidant interventions: lessons from natural antioxidants, *Ann. N. Y. Acad. Sci.*, 2002, **959**, 188–198.
197. V. I. Perez, A. Bokov, H. Van Remmen, J. Mele, Q. Ran, Y. Ikeno and A. Richardson, Is the oxidative stress theory of aging dead? *Biochim. Biophys. Acta*, 2009, **1790**, 1005–1014.
198. V. I. Perez, H. Van Remmen, A. Bokov, C. J. Epstein, J. Vijg and A. Richardson, The overexpression of major antioxidant enzymes does not extend the lifespan of mice, *Aging Cell*, 2009, **8**, 73–75.
199. Y. C. Jang, V. I. Perez, W. Song, M. S. Lustgarten, A. B. Salmon, J. Mele, W. Qi, Y. Liu, H. Liang, A. Chaudhuri, Y. Ikeno, C. J. Epstein, H. Van Remmen and A. Richardson, Overexpression of Mn superoxide dismutase does not increase life span in mice, *J. Gerontol., Ser. A*, 2009, **64**, 1114–1125.
200. S. Elchuri, T. D. Oberley, W. Qi, R. S. Eisenstein, L. Jackson Roberts, H. Van Remmen, C. J. Epstein and T. T. Huang, CuZnSOD deficiency leads to persistent and widespread oxidative damage and hepatocarcinogenesis later in life, *Oncogene*, 2005, **24**, 367–380.
201. V. I. Perez, C. M. Lew, L. A. Cortez, C. R. Webb, M. Rodriguez, Y. Liu, W. Qi, Y. Li, A. Chaudhuri, H. Van Remmen, A. Richardson and Y. Ikeno, Thioredoxin 2 haploinsufficiency in mice results in impaired

mitochondrial function and increased oxidative stress, *Free Radical Biol. Med.*, 2008, **44**, 882–892.
202. S. E. Schriner, N. J. Linford, G. M. Martin, P. Treuting, C. E. Ogburn, M. Emond, P. E. Coskun, W. Ladiges, N. Wolf, H. Van Remmen, D. C. Wallace and P. S. Rabinovitch, Extension of murine life span by overexpression of catalase targeted to mitochondria, *Science*, 2005, **308**, 1909–1911.
203. D. F. Dai, L. F. Santana, M. Vermulst, D. M. Tomazela, M. J. Emond, M. J. MacCoss, K. Gollahon, G. M. Martin, L. A. Loeb, W. C. Ladiges and P. S. Rabinovitch, Overexpression of catalase targeted to mitochondria attenuates murine cardiac aging, *Circulation*, 2009, **119**, 2789–2797.
204. P. M. Treuting, N. J. Linford, S. E. Knoblaugh, M. J. Emond, J. F. Morton, G. M. Martin, P. S. Rabinovitch and W. C. Ladiges, Reduction of age-associated pathology in old mice by overexpression of catalase in mitochondria, *J. Gerontol. A Biol. Sci. Med. Sci.*, 2008, **63**, 813–822.
205. A. Umanskaya, G. Santulli, W. Xie, D. C. Andersson, S. R. Reiken and A. R. Marks, Genetically enhancing mitochondrial antioxidant activity improves muscle function in aging, *Proc. Natl. Acad. Sci. U. S. A.*, 2014, **111**, 15250–15255.
206. R. Betarbet, T. B. Sherer, G. MacKenzie, M. Garcia-Osuna, A. V. Panov and J. T. Greenamyre, Chronic systemic pesticide exposure reproduces features of Parkinson's disease, *Nat. Neurosci.*, 2000, **3**, 1301–1306.
207. D. A. Di Monte, M. Lavasani and A. B. Manning-Bog, Environmental factors in Parkinson's disease, *Neurotoxicology*, 2002, **23**, 487–502.
208. C. M. Tanner, F. Kamel, G. W. Ross, J. A. Hoppin, S. M. Goldman, M. Korell, C. Marras, G. S. Bhudhikanok, M. Kasten, A. R. Chade, K. Comyns, M. B. Richards, C. Meng, B. Priestley, H. H. Fernandez, F. Cambi, D. M. Umbach, A. Blair, D. P. Sandler and J. W. Langston, Rotenone, paraquat, and Parkinson's disease, *Environ. Health Perspect.*, 2011, **119**, 866–872.
209. D. Athauda and T. Foltynie, The ongoing pursuit of neuroprotective therapies in Parkinson disease, *Nat. Rev. Neurol.*, 2015, **11**, 25–40.
210. B. J. Snow, F. L. Rolfe, M. M. Lockhart, C. M. Frampton, J. D. O'Sullivan, V. Fung, R. A. Smith, M. P. Murphy, K. M. Taylor and G. Protect Study, A double-blind, placebo-controlled study to assess the mitochondria-targeted antioxidant MitoQ as a disease-modifying therapy in Parkinson's disease, *Mov. Disord.*, 2010, **25**, 1670–1674.
211. L. H. Sanders, J. McCoy, X. Hu, P. G. Mastroberardino, B. C. Dickinson, C. J. Chang, C. T. Chu, B. Van Houten and J. T. Greenamyre, Mitochondrial DNA damage: molecular marker of vulnerable nigral neurons in Parkinson's disease, *Neurobiol. Dis.*, 2014, **70**, 214–223.
212. L. H. Sanders, J. Laganiere, O. Cooper, S. K. Mak, B. J. Vu, Y. A. Huang, D. E. Paschon, M. Vangipuram, R. Sundararajan, F. D. Urnov, J. W. Langston, P. D. Gregory, H. S. Zhang, J. T. Greenamyre, O. Isacson and B. Schule, LRRK2 mutations cause mitochondrial DNA damage in iPSC-derived neural cells from Parkinson's disease patients: reversal by gene correction, *Neurobiol. Dis.*, 2014, **62**, 381–386.

213. J. N. Guzman, J. Sanchez-Padilla, D. Wokosin, J. Kondapalli, E. Ilijic, P. T. Schumacker and D. J. Surmeier, Oxidant stress evoked by pacemaking in dopaminergic neurons is attenuated by DJ-1, *Nature*, 2010, **468**, 696–700.
214. F. Cardozo-Pelaez, M. Sanchez-Contreras and A. B. Nevin, Ogg1 null mice exhibit age-associated loss of the nigrostriatal pathway and increased sensitivity to MPTP, *Neurochem. Int.*, 2012, **61**, 721–730.
215. J. Guan, R. Krishnamurthi, H. J. Waldvogel, R. L. Faull, R. Clark and P. Gluckman, N-terminal tripeptide of IGF-1 (GPE) prevents the loss of TH positive neurons after 6-OHDA induced nigral lesion in rats, *Brain Res.*, 2000, **859**, 286–292.
216. A. Quesada, B. Y. Lee and P. E. Micevych, PI3 kinase/Akt activation mediates estrogen and IGF-1 nigral DA neuronal neuroprotection against a unilateral rat model of Parkinson's disease, *Dev. Neurobiol.*, 2008, **68**, 632–644.
217. D. Offen, B. Shtaif, D. Hadad, A. Weizman, E. Melamed and I. Gil-Ad, Protective effect of insulin-like-growth-factor-1 against dopamine-induced neurotoxicity in human and rodent neuronal cultures: possible implications for Parkinson's disease, *Neurosci. Lett.*, 2001, **316**, 129–132.
218. S. Schaefer, T. Vogt, T. Nowak, P. H. Kann and K. b. German, Pituitary function and the somatotrophic system in patients with idiopathic Parkinson's disease under chronic dopaminergic therapy, *J. Neuroendocrinol.*, 2008, **20**, 104–109.
219. H. P. Vogel and R. Ketsche, Effect of hypoglycaemia, TRH and levodopa on plasma growth hormone, prolactin, thyrotropin and cortisol in Parkinson's disease before and during therapy, *J. Neurol.*, 1986, **233**, 149–152.
220. J. A. Santiago and J. A. Potashkin, Shared dysregulated pathways lead to Parkinson's disease and diabetes, *Trends Mol. Med.*, 2013, **19**, 176–186.

CHAPTER 13

Biomarkers of Oxidative Stress in Parkinson's Disease

EMILIO FERNÁNDEZ

Laboratory of Molecular Neurology and Neurophysiology (BIO127), Department of Medical Physiology and Biophysics, School of Medicine, Universidad de Sevilla, Sevilla, Spain
*E-mail: efespejo@us.es

13.1 Biomarkers

Biomarkers are indicators of either normal biological and pathogenic processes or pharmacological responses to a therapeutic intervention, according to the Biomarkers Definitions Working Group.[1] It is evident that biomarkers must help to diagnose disease. Regarding Parkinson's disease (PD), biomarkers must either differentiate susceptible individuals from healthy subjects before onset of symptoms (sensitivity) or to identify true idiopathic Parkinson's Disease from imitators once symptoms are present (specificity).

First, a good sensitivity biomarker is needed because neuroprotective agents should be given as soon as the pathology is detected, during the premotor phase. The premotor phase is known to start 4–5 years before the motor symptoms are referred by the patient, according to imaging and clinical studies.[2–4] A premotor phase is defined through a range of symptoms such as rapid eye movement behavior disorder, olfactory loss, and autonomic dysfunction including constipation and loss of cardiac reflexes. It is well known that motor symptoms develop once loss of *substantia nigra* (SN) neurons is

Issues in Toxicology No. 34
Oxidative Stress and Redox Signalling in Parkinson's Disease
Edited by Rodrigo Franco, Jonathan A. Doorn and Jean-Christophe Rochet

Published by the Royal Society of Chemistry, www.rsc.org

higher than 50%, a fact that points to an insidious onset of the disease that could be delayed or prevented by neuroprotective agents. Many authors suggest that, if the decline rate of dopaminergic (DAergic) function is lower than expected, there would be a preclinical phase as well.[4] The preclinical phase is devoid of clinical signs and symptoms but PD-specific pathology would be already present.[5] Recent data suggest that pathology can originate outside the central nervous system (CNS) long before the SN is affected, such as the mesenteric plexus or neuronal tissue of the heart.[6,7]

Secondly, once diagnosis of PD is established, a specificity biomarker should be strongly correlated to a surrogate end-point to give an objective evolution of the disease. These biomarkers are also called surrogate biomarkers because they substitute for a clinical end-point, defined as a variable that reflects how a patient feels, functions or survives. A surrogate biomarker should predict clinical benefit or harm, or lack of benefit.[8] Biomarkers of disease progression after diagnosis are also important for the development of disease-modifying medications. These surrogate biomarkers are quite difficult to find given the variability of symptoms and evolution of each patient, and at least they should reflect disease progression without being masked by age, sex, medication, other coexisting diseases, or any factors other than progression of PD.[9]

An ideal biomarker should have several characteristics: (a) it should be strongly correlated with PD without the influence of intermediate variables, (b) treatment has no direct effect on the biomarker, (c) measures of the biomarker are reproducible in different times or health centers, (d) small changes in PD evolution are detected, and finally (e) it should be inexpensive and non-invasive.

At present the only biomarker accepted is DaTSCAN, an imaging biomarker based on single-photon emission computed tomography (SPECT).[10] Biochemical biomarkers are not available. Two SPECT radioligands based on ioflupane (^{123}I) can be used, and ^{123}I-β-CIT and ^{123}I-FP-CIT SPECT have high power to discriminate PD from normal aging. It is well known that deterioration of motor function (bradykinesia, rigidity, daily activity performance) parallels to functional presynaptic DAergic imaging as assessed by DaTSCAN.[11] The main clinical application of DaTSCAN imaging is the differentiation of parkinsonism from essential tremor or drug-induced, vascular or psychogenic parkinsonism. DaTSCAN sensitivity and specificity are excellent, with values ranging from 95% to 100%.[10]

13.2 Oxidative Stress

Oxidative stress is defined as an imbalance between the production of reactive oxidative species (ROS) and antioxidant mechanisms, and it is considered an important pathogenic mechanism in Parkinson's disease and other neurodegenerative diseases.[12–15] Reactive oxidative species include ROS, reactive nitrogen species (RNS), and free halogenation radicals. In fact, all reactive species are interconnected.

Regarding ROS-induced oxidative stress, the generation of free radicals and ROS is part of the normal cellular metabolism. Superoxide anion radicals ($O_2^{\bullet-}$) that are produced in the electronic-transport chain (ETC), are the most important ROS. Logically, the first line of defense of the organism is to minimize the production of $O_2^{\bullet-}$, and metabolic O_2 is converted directly into H_2O thanks to the action of cytochromes of the ETC. However, around 0.1–1% of electrons transferred to O_2 can yield $O_2^{\bullet-}$. Free radicals are normally scavenged by cells, through conversion of $O_2^{\bullet-}$ to hydrogen peroxide (H_2O_2) under the control of $O_2^{\bullet-}$ dismutases (SODs), and then the formation of H_2O from the reaction of H_2O_2 with reduced glutathione (GSH), under the control of GSH peroxidase (GPx). The H_2O_2 produced from $O_2^{\bullet-}$ is also eliminated through catalases and several peroxidases. Catalases carry out dismutation reactions, and most catalases are heme-containing enzymes. Peroxidases use several reductants to reduce H_2O_2 to two molecules of H_2O.

Another type of oxidative stress linked to overproduction of ROS is halogenative stress. This type of stress leads to excess of several molecules such as advanced oxidation protein products (AOPP), a group of proteins byproducts with strong halogenation. These protein products originate as a result of the action of free radicals such as chloramines and hypochlorous acid (HOCl), and they are known to be augmented in several diseases.[16] This halogenative process could also modify free amino acids such as tyrosine (Tyr) leading to the formation of free 3-chlorotyrosine, 3-bromotyrosine, or 3-iodotyrosine.

Oxidation of methionine (Met) residues of proteins is another type of oxidation related to excess of H_2O_2, chloramines and HOCl.[17,18] Methionine is oxidized to methionine sulfoxide (MetSO),[18] and this oxidation process is known to alter the normal functionality of several proteins that are related to PD.[19] Glycation is another modification of proteins that is highly related to oxidative processes, and it induces the release of H_2O_2.[20,21] Among other effects, glycation leads to the production of advanced glycation end-products (AGEs) that are formed after the reaction of amino groups of proteins with reducing sugars such as glucose or trioses.[20,22]

Nitrative stress and *S*-nitros(yl)ation stress are two types of oxidative stress where oxidative modifications of molecules, mostly proteins, are secondary to excess of nitric oxide ($^{\bullet}NO$), a molecule that can react with ROS giving RNS such as peroxynitrite ($ONOO^-$). Regarding nitrative stress, nitration of Tyr residues leading to 3-nitrotyrosine (3-NT) proteins or free 3-NT is the most prominent change, and this type of oxidative stress has been associated with human pathologies and neurodegenerative diseases.[23,24] In Parkinson's disease and Lewy body (LB) dementia, 3-NT is demonstrated to represent a pathogenic mechanism, and nitrated Tyr residues of diverse PD-related proteins such as neurofilaments or α-synuclein (α-syn) are detected inside LBs.[25] Biochemically, nitrated proteins and amines can be formed through different pathways. As explained, the formation of $ONOO^-$ through $^{\bullet}NO$ and $O_2^{\bullet-}$ is the main pathway.[26] As regards *S*-nitros(yl)ation, it is a reaction transferring a nitroso group to a Cys thiol/sulfhydryl (P-SH) to form an *S*-nitrosothiol derivative (P-SNO),[27] and this type of oxidative modification has been associated with alteration of protein function, and human pathologies and neurodegenerative diseases as well.[28,29] In PD, many proteins that are localized in LBs are known to be *S*-nitros(yl)ated such as α-syn, Parkin, and protein disulfide isomerase (PDI).[27]

13.3 Candidate Biomarkers for ROS-Induced Stress

Oxidative stress mediated by excess of $O_2^{\bullet-}$ and H_2O_2 (or peroxidation stress) can be detected in brain tissue, blood and cerebrospinal fluid (CSF) of PD patients. Many studies indicate that SNpc tissue is subjected to oxidative stress because this tissue shows enhancement of several oxidative markers such as lipid-peroxidation products,[30,31] 8-hydroxyguanosine (8-OHdG, marker of oxidative stress to DNA),[32] protein carbonyls (markers of protein oxidation),[33] advanced lipoxidation end-products (ALEs), and advanced glycation end products (AGEs).[34] Since SNpc tissue is not accessible for obtaining biomarkers in patients, only PD-related biomarkers localized in peripheral fluids such as blood or CSF are of interest. As regards blood, there have been detected increases of 8-OHdG, malondialdehyde (MDA), and vitamin E, and reduction of the antioxidant enzyme GPx.[35,36] Interestingly, there is a significant correlation between leucocyte 8-OHdG level and disease severity in PD.[36] MDA is formed from ROS-degraded polyunsaturated lipids.[37] This compound is a reactive aldehyde that causes toxic stress in cells and forms covalent protein adducts or ALEs.

Regarding CSF, Hong *et al.*[38] have pointed out that DJ-1, reported to act as an antioxidant peroxidase, is reduced in the initial phases of PD. These authors also demonstrated that both DJ-1 and α-syn levels are decreased in Parkinson's patients *versus* controls or Alzheimer's patients.[38] Buhmann *et al.*[39] detected enhanced lipid peroxidation in the CSF, and Ilic *et al.*[35] reported high levels of MDA. Other biomarkers for oxidative stress such as the oxidized forms of coenzyme-Q and 8-OHdG have been detected in CSF of PD patients as well.[40] CSF levels of 8-OHdG correlate with the duration of the illness but not the severity of PD.[40] The antioxidant activity of superoxide dismutase 1 (SOD1 or CuZnSOD) is reduced in the CSF of PD, together with ferroxidase activity and cerulopasmin.[41–43] These latter two molecules are known to be neuroprotectant because they limit the generation of toxic hydroxyl radicals ($^{\bullet}OH$).[44] Regarding antioxidative biomarkers of the GSH system, several studies have indicated that levels of GSH are diminished in CSF and brain tissue of PD,[45–47] and GPx has been reported to be either augmented or reduced in the CSF.[47–50]

All mentioned molecules could be viewed as peripheral biomarkers for PD, but they do not have a good diagnostic value because they are disturbed in other neurodegenerative diseases. For instance, SOD1 is also reduced in CSF of patients with Alzheimer's disease, Huntington's disease, and amyotrophic lateral sclerosis.[42] AGEs are also implicated in Alzheimer's disease.[51]

Hence, their specificity is low, and searching for more sensitive and specific biomarkers is an active field of research. In this context, several authors have proposed that a combination of biomarkers would be better to account for a pathogenic heterogeneity of PD.[52] A 'multiple biomarker' for PD remains to be discerned.

Altered proteins or adducted protein products related to PD could be more specific biomarkers. In addition, the use of oxidative modification of proteins as a biomarker may have some advantages in comparison with other non-protein biomarkers, because of their greater stability, early formation, and longer lifespan.[53] Thus, ROS-induced stress can lead to misfolding of PD-related proteins such as α-syn and Parkin, not only in brain tissue but also in peripheral fluids. Misfolded α-syn (as abnormal β-sheets instead of normal α-helices) tends to form protofibrils that would precipitate forming fibrils, which in turn constitute the core of the LBs. Protofibrils have been shown to be neurotoxic, pointing to the fact that protofibrils rather than fibrils could be the deleterious species of α-syn.[54,55] Association of α-syn with oxidized lipid metabolites can lead to mitochondrial dysfunction that in turn leads to DAergic neuron death.[55] For instance, acrolein, a lipid-peroxidation product, may initiate vicious cycles of modification and aggregation of α-syn, and impairment of the proteolysis system, leading to neuronal death in PD.[56] ROS-induced stress can induce specific structural changes in α-syn leading to covalent aggregation.[57] These changes depend of Tyr, and oxidation of Tyr39 (tyrosine residue at the amine terminus of the molecule) is essential for formation of fibrils.[58]

13.3.1 Halogenation

Characterizing protein biomarkers for halogenative oxidative stress is of great interest for biomedicine. As explained, the use of oxidative modification of proteins as biomarkers may have some advantages in comparison with non-protein biomarkers.[53] Molecules that are markers for halogenative stress are AOPP,[59] which are known to be augmented in serum of PD patients.[60] These products originate as a result of the action of free radicals such as chloramines and HOCl on proteins and may act as inflammatory mediators.[59,61,62] Other hypohalous acids like hypobromous acid (HOBr) and hypoiodous acid (HOI) can be involved as well.[63] Peroxidases such as myeloperoxidase (MPO) catalyze these reactions leading to hypohalous acids from halogens and excess of H_2O_2 and, as a consequence, halogenated proteins and amines are formed. Proteins are mostly halogenated at Tyr and lysine (Lys) residues, and AOPP can conjugate with human serum albumin (HSA) giving AOPP-HSA conjugates. AOPP-HSA conjugates may act as inflammatory mediators triggering the oxidative 'ignition' of neutrophiles, monocytes and T-lymphocytes.[64,65] Enzymes of the GSH system are related to these end-products because AOPP accumulation coexists with decreased GPx level in uremic patients.[66,67] AOPP are associated with neuroinflammation, and they could be linked to initial inflammatory processes in PD, where immunological and inflammatory responses in the vicinity of DAergic neurons have been detected.[68–71] It has been postulated that, once the neurodegeneration process begins, activation of the immune system and neuroinflammation could be related to an immunological response against aberrant products of oxidation.[70] Recently, the presence of halogenative stress in serum and, to a lesser extent, CSF of parkinsonian patients leading to excess of AOPP have been reported.[60] Serum concentration of these protein products is progressively reduced over time, and levodopa treatment contributes to this reduction.[60] Serum levels of AOPP could be a prognostic marker of duration of PD (Figure 13.1). However, specificity of AOPP as a PD biomarker is low because these products also accumulate in other diseases such as diabetes mellitus, uremic syndrome, atherosclerosis, systemic sclerosis and acquired immune deficiency syndrome (AIDS).[66,71–76] Again, the need of 'multiple biomarker' for enhancing specificity and sensitivity is clear.

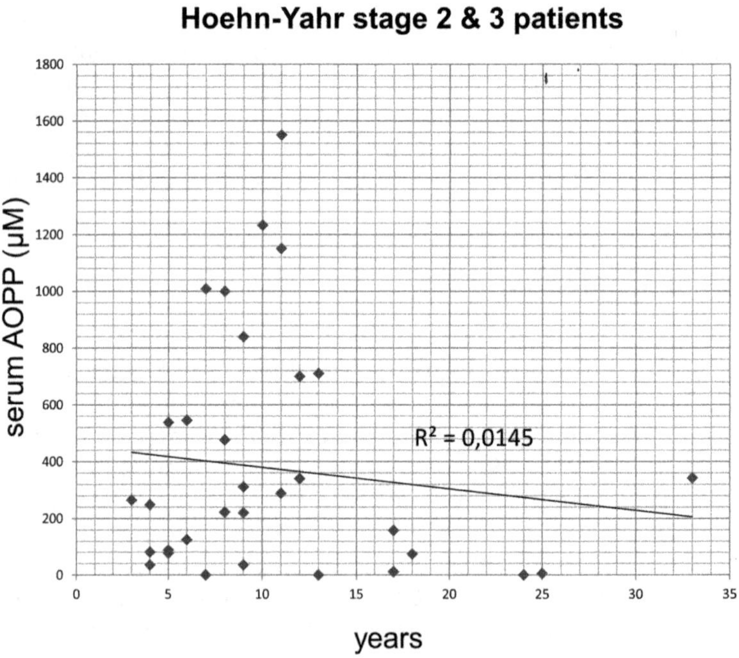

Figure 13.1 Linear regression between serum levels of AOPP, and the corresponding duration of PD in years. Patients (n = 34) are at stage 2 or 3 of the Hoehn–Yahr scale. It is observed that those patients with a duration of disease longer than 13 years present low serum AOPP levels. Serum level of AOPP could be a prognostic marker of PD duration.[60] The Hoehn–Yahr scale is a commonly used system for describing how the symptoms of PD progress (from 1, unilateral involvement only, to 5, confinement to bed or wheelchair unless aided) (Copyright © 2013, Mary Ann Liebert, Inc.).

13.3.2 Methionine Oxidation

Oxidation of Met residues of proteins is caused by excess of H_2O_2, chloramines and HOCl.[17,18] Peroxynitrite, which is formed after the reaction of $^{\bullet}NO$ with $O_2^{\bullet-}$ can be involved as well.[17] Methionine is oxidized to MetSO.[18] Levels of MetSO are controlled by Met sulfoxide reductases (Msr), whose deficiency also induces Met oxidation. The methionine oxidation process also alters the functionality of several proteins that are related to PD.[19] Among PD-related proteins, α-syn is easily oxidized in the Met127 residue, a process where dopamine (DA) plays a significant role.[19]

13.3.3 Glycation and Carbonylation

Another type of AOPP, AGEs, also accumulate in neurodegenerative diseases such as PD and Alzheimer's disease.[20,77–79] AGEs such as pentosidine and pyrraline are formed after the reaction of amino groups of proteins with reducing sugars such as glucose or trioses.[20,22] Accumulation of AGEs is toxic because it can modify proteins favoring protein aggregation, and it damages extracellular matrix.[20,80] AGEs formation seems to be part of very early processes in the course of PD,[20,78,81] and receptors for AGEs (RAGE) are expressed in the brain of PD patients.[82] The presence of AGEs is related to crosslinking of PD-related proteins such as α-syn.[20] This protein can be glycated at fifteen Lys residues, and glycated α-syn is colocalized with AGEs in LBs. However, the localization of glycated α-syn in peripheral fluids of PD has not been reported so far.

Enzymes of the GSH system are related to the formation of these end-products because GSH levels are reduced leading to increased cellular damage by protein glycation and formation of crosslinking of cytoskeletal proteins.[79,83] However, these GSH-mediated effects seem not to occur in blood of PD patients because molecules encompassing the GSH system are not affected in blood regardless of the presence of oxidative stress.[84] AGEs are known to exacerbate oxidative stress in cells because they induce the release of H_2O_2.[20,85]

The formation of AGEs can be induced by carbonylation stress too, another type of protein oxidation.[86] Carbonylation is a process that forms ketones or aldehydes that react by 2,4-dinitrophenyl hydrazine to form hydrazones. Carbonylation stress can alter protein functionality irreversibly, because enzymatic reversal of the protein-carbonyl modifications has not been detected.[86] Protein carbonyls are good markers for protein oxidation,[33] but their specificity is low because they are augmented in several neurodegenerative diseases.[86]

13.4 Candidate Biomarkers for RNS-Induced Stress

As explained, excess of $^{\bullet}NO$ can lead to either nitrative or *S*-nitros(yl)ation stress. Oxidative modifications of proteins after nitration and *S*-nitros(yl)ation affects at Tyr residues and Cys residues of proteins, respectively.

13.4.1 Nitration

Nitration of Tyr residues of proteins, giving rise to nitrated proteins, is an oxidative modification of proteins secondary to excess of $^{\bullet}NO$ during nitrative stress. Another effect of nitrative stress is the nitration of free tyrosine leading to 3-NT, a free nitrosoamine. 3-nitrotyrosine proteins and amines are related to several human diseases, and they can play a significant role in PD.[23,24] Of note is that 3-NT residues are detected in proteins forming LBs, mostly neurofilaments and α-synuclein.[25,26] 3-nitrotyrosine proteins can be generated through several pathways. The main one is formation of $ONOO^-$ after the reaction of $^{\bullet}NO$ with $O_2^{\bullet-}$. Peroxynitrite promotes protein nitration likely through the formation of nitrite radicals (NO_2^-), and carbonyl radicals. These carbonyl radicals transform tyrosyl residues of proteins into tyrosyl radicals, and NO_2^- reacts with tyrosyl residues to form nitrotyrosyl residues in proteins.[87] Peroxynitrite is an important nitrative factor of PD-related proteins such as manganese superoxide dismutase (MnSOD or SOD2), tyrosine-hydroxylase (TH), and α-syn,[88–91] and $^{\bullet}NO$ excess is detected in blood and SN tissue from EP patients.[91–94] Other pathways leading to protein nitration are dependent on peroxidases such as MPO or the oxidation of nitrites forming nitrogen dioxide ($^{\bullet}NO_2$), which can nitrate proteins as well.[93,94]

The pathogenic role of nitrated proteins or free nitrosoamines is not well known. Tyrosine nitration can either stimulate or diminish the activity of a protein, or protein functionality can be either changed or inhibited. All of these mechanisms have been associated with human pathologies.[23] In addition, nitration can activate the immune system and the production of 3-NT antibodies, leading to glial and phagocyte overstimulation too.[70,95] In this context, PD has been associated with activation of the immune system, and neurodegeneration could be an aberrant immune reaction to oxidation and nitration products. As regards free nitrosoamines, 3-NT is known to be neurotoxic in animal models of parkinsonism,[96] although its presence has not been detected in serum or CSF of parkinsonian patients.[97]

Among nitrated proteins, nitrated MnSOD is present in LBs, and nitrated MnSOD levels are enhanced in the CSF of a group of PD patients.[90] Its functional role is not well known, but downregulation of MnSOD leads to mitochondrial vacuolization and oxidized lipid deposits in mice,[98] phenomena that are observed in PD. Furthermore, the recombinant MnSOD with 3-NT in position 34 retains only 3% of enzymatic activity, leading to a failure of dismutation of $O_2^{\bullet-}$ to H_2O_2.[99] Tyrosine hydroxylase is another enzyme that can be selectively nitrated, a reaction that could alter normal function of this enzyme, which is critical for the biosynthesis of DA. In animal models of PD, nitration of TH leads to inhibition of TH activity,[100,101] but it is not clear how its function is altered in humans suffering from PD. Some authors have proposed that TH inhibition could play a role in PD accelerating DA depletion.[102]

13.4.1.1 *Nitrated Species of Serum Proteins Such as α-Synuclein as Possible Biomarkers*

Nitrated α-synuclein (3-NT-α-syn) is another candidate biomarker for nitrative stress. This protein is the main component of LBs, and it has been associated to the pathogenesis of PD.[103–105] Several nitrative agents are generated by DA metabolism and they can induce α-synuclein aggregation.[106] α-synuclein inclusions in PD brains can be labeled with anti-3-NT antibodies, indicating that this protein is a target of nitration.[107] This protein contains 140 amino-acids with four Tyr residues (Tyr39, Tyr125, Tyr133, Tyr136),[108,109] which are readily nitrated, and it is known that nitrated amino acids can modify the functional activity of the protein. Thus, excess nitration of Tyr residues of α-syn reduces its binding to vesicles and its degradation rate.[110] Nitration of Tyr39 and Tyr125/136 residues increases α-syn aggregation,[111] a fact thought to be important in the pathogenesis of PD because the conversion of soluble α-syn into insoluble aggregated form is a crucial factor in this disease.[112] Finally, nitration of α-syn also leads to a neurotoxic form of the protein that can kill culture DAergic neurons.[113,114]

It is believed that parkinsonian nitration of α-syn mostly occurs in brain tissue, because α-syn is easily detected in neurons and extracellular fluid.[115] However, α-syn is also present in blood.[116] Blood α-syn can be nitrated as well, and it might participate in brain aggregation. As explained, excess nitration of proteins can modify their functional activity, producing new properties such as cytotoxicity and pro-aggregation.[110–115] In this context, in Fernandez's lab it has been further demonstrated that there are signs of nitrative stress in serum of PD patients, leading to augmentation of nitrated proteins levels.[97] In addition, new bands of nitrated proteins are observed in the immunoblots of patients' serum, as shown in Figure 13.2. The identification of these nitrated proteins deserves further investigation because they could represent new biomarkers.[97] Among these proteins, 3-NT-α-syn is detected in serum, and nitration of α-syn is enhanced as evaluated with immunoblot based on antibodies aimed at the nitrated Tyr125 and Tyr136 residues. Interestingly, ~57 kDa bands of 3-NT-α-syn are mostly observed after immunoblots (unpublished results), and these bands seem to correspond to oligomeric conformations of the molecule,[117–119] likely tetramers or trimers, containing unfolded α-syn monomers forming a globular protein.[116] Other authors have reported the presence of monomers and dimers of α-syn and 3-NT-αSYN, and each 'natively unfolded' monomer of α-syn has a molecular mass of ~14 kDa.[120] The high structural plasticity of α-syn is well known, which allows it to achieve ensembles of different molecular sizes.[118] In Fernandez's lab, it has been observed that filtered PD serum (with Amicon® 50K filters) is devoid of albumin, and only contains 'free' low-mass proteins such as species of 3-NT-α-syn. This filtered serum, not the complete one, induces large proteinaceous inclusions in culture neurons of SN, and inclusions express 3-NT-α-syn (Figure 13.3). The intraneuronal presence of proteinaceous inclusions expressing 3-NT-α-syn is a hallmark of PD.[25] These findings suggest that PD serum without albumin contains 'particular'

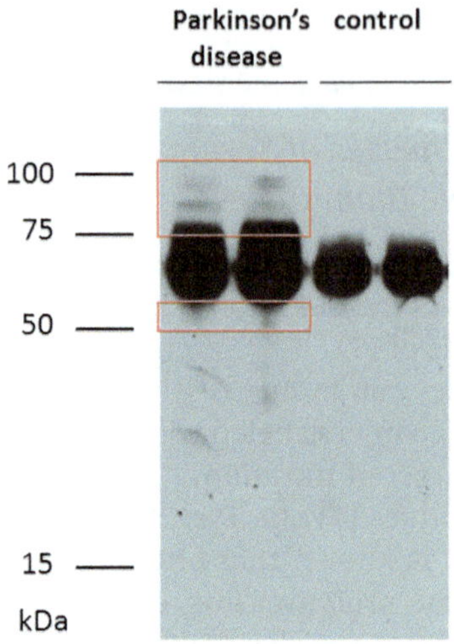

Figure 13.2 Immunoblot analysis of complete serum of early-parkinsonian patients and healthy controls after 3-NT staining. There are several nitrated proteins with molecular mass at 75–100 kDa and 50–60 kDa (red squares) that are characteristic of patients, and they are not present in the serum of controls. The identification of these nitrated proteins in serum of parkinsonian patients deserves further investigation. Abbrev.: early-PD, early Parkinson's disease.

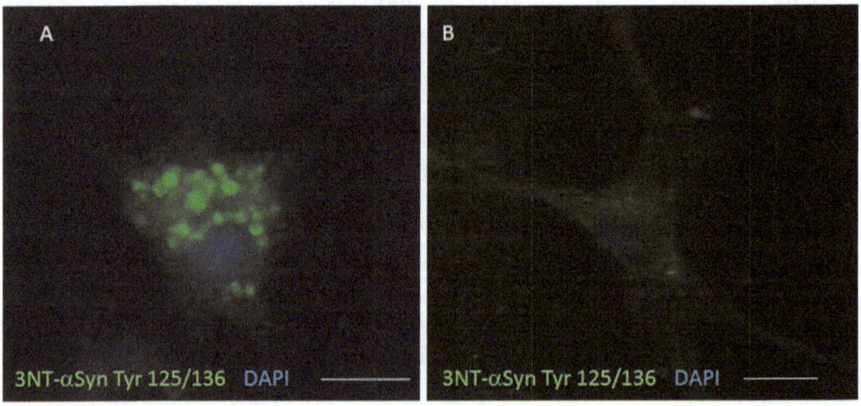

Figure 13.3 Immunocytochemical features of SN neurons after exposure to parkinsonian serum or control medium. A. Culture *substantia nigra* neuron after 24-h exposure to a culture medium of filtered serum of early-parkinsonian patients. Serum was filtered with 50K filters. Numerous large inclusions positive to 3-NT-α-syn (3NT-αSYN, green color) are observed inside the cytoplasm. The filtered serum contains 'albumin-free' proteins. These findings suggest that PD serum without albumin contains 'particular' low-mass molecules that induce 3-NT-α-syn-positive inclusions in SN neurons, a hallmark of PD.[25] These 'particular' species could represent new biomarkers for PD. B. Cultured SN neuron after 24 h exposure to control medium. The normal signal of 3-NT-α-syn can be observed. Expression of the nitrated protein is low, with a diffuse and perinuclear appearance. No large inclusions are detected. 3-NT-α-syn was stained with monoclonal antibody against nitrated α-syn with nitration of Tyr125 and 136 residues or 3-NT-α-syn Tyr125/136 (mouse monoclonal anti-nitro-α/β-synuclein antibody, nSYN12, Millipore, USA). Abbrev.: PD, Parkinson's disease. Bar = 5 μm.

low-mass molecules that induce 3-NT-α-syn-positive inclusions in SN neurons. These 'particular' species could represent new biomarkers for PD, and they deserve further investigation.

13.4.1.2 Nitration, Halogenation and the Thyroid Gland: A New Field for Searching Biomarkers for Parkinson's Disease

The thyroid gland has been linked to PD, because some researchers have found more prevalence of some thyroid dysfunctions among PD patients or they have detected sympathetic denervation of the thyroid gland in PD patients, a hallmark of the disease because it also occurs in other organs such as heart and rectum.[121–124] Recently, Fernandez's research group has studied if the thyroid gland could be involved in idiopathic PD and nitrative stress.[125] The findings indicate that the levels of the enzyme thyroperoxydase (TPO) are elevated in serum of early-disease patients, and TPO levels are negatively correlated with serum nitrated proteins levels (Figure 13.4). These changes are not influenced by the type of medication of patients. The elevation of TPO could represent a novel indicator of dysfunction of the thyroid gland in early PD. This enzyme is a peroxidase, and peroxidase excess is expected to be positively correlated with nitrative stress,[125] even though TPO levels were found to be negatively correlated with serum 3-NT protein levels. These observations led to the hypothesis that the thyroid gland and thyroperoxidase participate on nitration of serum proteins, and they could influence parkinsonian nitrative stress. In addition, TPO is critical for iodination of tyrosine residues on thyroglobulin for the production of thyroid hormones. In other words, TPO induces the

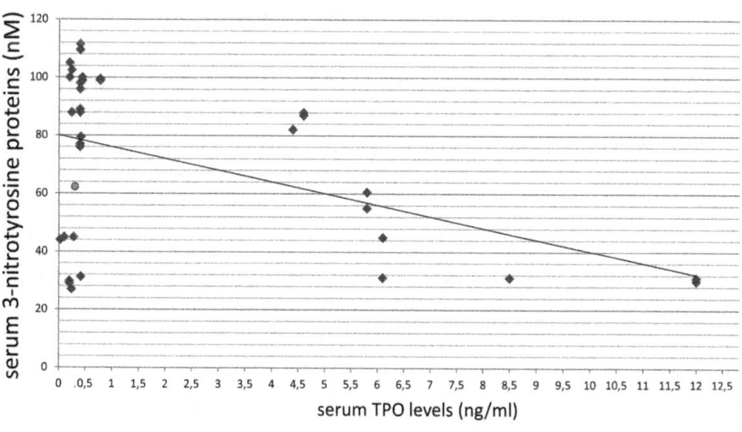

Figure 13.4 Correlation between serum levels of thyroperoxidase (TPO) and nitrated proteins in early-disease patients (Hoehn–Yahr stages 1 and 2, $n = 34$). Correlation was found to be significant ($R = -0.461$, $p < 0.006$; Pearson's test). The mean value in control subjects is represented as a gray dot. The elevation of TPO could represent a novel indicator of dysfunction of the thyroid gland in early Parkinson's disease, and TPO could be a biomarker.[125] (Copyright © 2014, Mary Ann Liebert, Inc.).

synthesis of halogenated proteins. In this context, the AOPP family of proteins also includes iodinated products,[62,63] and it is worth noting that AOPP levels significantly correlate with TPO levels in patients, further supporting a role for the thyroid gland in PD (unpublished results).

13.4.2 *S*-nitros(yl)ation

As explained before, *S*-nitros(yl)ation it is a reaction transferring a nitroso group to a Cys thiol/sulfhydryl (P-SH) to form an *S*-nitrosothiol derivative (P-SNO).[27] This type of oxidative stress has been associated with alteration of protein function, and human pathologies and neurodegenerative diseases as well.[28,29] *S*-nitros(yl)ated proteins are potential biomarkers for PD, and their presence in peripheral fluids deserves further investigation.

S-nitros(yl)ated protein disulfide isomerase (PDI-SNO) is a promising biomarker because of its pathological activity and its presence in PD brains.[126] It could be a factor leading to accumulation of misfolded proteins in PD, although its specificity is low because PDI-SNO has also been detected in AD and ALS.[126–128] Normally, PDI attenuates abnormal protein accumulation and ubiquitination, affording neuroprotection. Interestingly, PDI-SNO exerts an opposite effect to that of 'wild-type' PDI, and it mediates, not attenuates, aggregation of PD-related proteins such as α-syn and synphilin-1.[126,129,130] It is quite interesting that PDI is also localized in blood, on the surface of platelets (cell-surface PDI or csPDI).[131] On platelets, PDI denitrosates *S*-nitrosothiols releasing ˙NO that attenuates platelet activation. This action could be impaired in PD if csPDI is *S*-nitros(yl)ated in this disease. The presence of PDI in red blood cells has also been pointed out.[132,133]

The protein Parkin is found to be *S*-nitrosylated in PD as well.[134] Parkin is an E3 ubiquitin lygase involved in the ubiquitination of proteins, and this activity is inhibited by *S*-nitros(yl)ation. The inhibition of parkin activity by *S*-nitros(yl)ation could contribute to the degenerative process by impairing the ubiquitination of parkin substrates,[134] although its role in human PD is not well known. Parkin is also a transcriptional repressor of p53,[135] but Parkin-SNO enhances expression of the p53 gene leading to cell death.[136] This functional change could also play a role in PD. It is worth recalling that recessive mutations in the *PARK2* gene encoding Parkin are responsible for many juvenile and early-onset cases of parkinsonism.[137]

Peroxiredoxins (Prx) are enzymes that are also *S*-nitros(yl)ated in PD. Prx are antioxidant enzymes that exert a neuroprotective role through the reduction of H_2O_2.[138] Among Prx, Prx2 and Prx6 are enhanced in several neurodegenerative diseases, and they are *S*-nitros(yl)ated in PD.[139,140] *S*-nitros(yl)ation targets the Cys51 and Cys127 residues of the molecule, and inhibits its neuroprotective function. More interestingly, it seems that there is a differential up-regulation of peroxiredoxins in AD and PD, suggestive of the diversity and distinct functional roles of these proteins. For instance, it has been proposed that up-regulation of Prx6 may be used to discriminate PD from AD.[139]

Finally, *S*-nitros(yl)ated glyceraldehide-3-phosphate dehydrogenase (GAPDH-SNO) is another potential biomarker in PD. GAPDH is an enzyme involved in the process of glycolysis, but GAPDH-SNO is known to be neurotoxic and it could participate in cell death in PD.[141] GAPDH is *S*-nitros(yl)ated at Cys150 and interacts with seven in absentia homolog 1 (SIAH1), an E3-ubiquitin-ligase, forming the GAPDH-SNO/SIAH1 protein complex. This complex is neurotoxic, and it induces cell death.[142]

13.5 Conclusions

A good sensitivity biomarker for PD is urgently needed because neuroprotective agents could be of benefit for patients if given as soon as the pathology is detected. Oxidative stress is considered as an important pathogenic mechanism in PD. New biomarkers could be found analyzing molecules that are specifically altered by oxidative insults, and they should be better searched in peripheral fluids such as blood and CSF because they are easily accessible, in contrast to neural tissue. Molecular changes are caused by oxidative reactions induced by reactive species, including peroxidation, halogenation, glycation, carbonylation, Met oxidation, nitration and *S*-nitros(yl)ation. Several biomarkers related to these oxidative reactions have been proposed or are under study (see Table 13.1). A combination of biomarkers ('multiple biomarker') has been proposed to better account for the pathogenic heterogeneity of PD. The discovery of these new biomarkers could improve the knowledge and treatment of PD.

Table 13.1 Candidates of biomarkers for Parkinson's disease and oxidative reactions.[a]

Oxidative Modification	Potential Biomarkers
Peroxidation	DJ-1[38]
	Malondialdehyde (MDA)[35,36]
	Superoxide dismutase 1 (SOD1)[41]
	Ferroxidase[41–43]
	Glutathione (GSH)[47]
	Glutathione peroxidase (GPx)[47–50]
	Acrolein[56]
Halogenation	Advanced oxidation protein products (AOPP)[60]
	3-chlorotyrosine[60,102]
	3-bromotyrosine[60]
	3-iodotyrosine[60]
Methionine oxidation	Met-oxidized α-synuclein[19]
Glycation	Advanced glycation end-products (AGEs)[20,34,78,81]
	Glycated α-synuclein[20]
Nitration	Superoxide dismutase 2 (SOD2)[90]
	Tyrosine hydroxylase (TH)[100,101]
	Nitrated α-synuclein (3-NT-α-syn)[97]
	3-nitrotyrosine (3-NT)[23,24]
S-Nitros(yl)ation	Protein disulfide isomerase (PDI)[126]
	Parkin[134]
	Peroxiredoxins (Prx)[139,140]
	Glyceraldehide-3-phosphate dehydrogenase (GAPDH)[141]

[a]These biomarkers should be found in easily accessible peripheral fluids such as blood and CSF of patients.

References

1. Biomarkers Definitions Working Group, Biomarkers and surrogate endpoints: preferred definitions and conceptual framework, *Clin. Pharmacol. Ther.*, 2001, **69**, 89–95.
2. J. M. Fearnley and A. J. Lees, Ageing and Parkinson's disease: substantia nigra regional selectivity, *Brain*, 1991, **114**, 2283–2301.
3. P. K. Morrish, G. V. Sawle and D. J. Brooks, An [18F]dopa-PET and clinical study of the rate of progression in Parkinson's disease, *Brain*, 1996, **119**, 585–591.
4. J. Schwarz, A. Storch, W. Koch, O. Pogarell, P. E. Radau and K. Tatsch, Loss of dopamine transporter binding in Parkinson's disease follows a single exponential rather than linear decline, *J. Nucl. Med.*, 2004, **45**(10), 1694–1697.
5. M. B. Stern, A. Lang and W. Poewe, Toward a redefinition of Parkinson's disease, *Mov. Disord.*, 2012, **27**(1), 54–60.
6. E. C. Wolters and H. Braak, Parkinson's disease: premotor clinico-pathological correlations, *J. Neural Transm., Suppl.*, 2006, **70**, 309–319.
7. A. Bloch, A. Probst, H. Bissig, H. Adams and M. Tolnay, Alpha-synuclein pathology of the spinal and peripheral autonomic nervous system in neurologically unimpaired elderly subjects, *Neuropathol. Appl. Neurobiol.*, 2006, **32**(3), 284–295.
8. A. W. Michell, S. J. Lewis, T. Foltynie and R. A. Barker, Biomarkers and Parkinson's disease, *Brain*, 2004, **127**, 1693–1705.
9. K. Marek, D. Jennings, G. Tamagnan and J. Seibyl, Biomarkers for Parkinson's disease: tools to assess Parkinson's disease onset and progression, *Ann. Neurol.*, 2008, **64**(suppl. 2), S111–S121.
10. A. Varrone and C. Halldin, Molecular imaging of the dopamine transporter, *J. Nucl. Med.*, 2010, **51**, 1331–1334.
11. W. Maetzler, I. Liepelt and D. Berg, Progression of Parkinson's disease in the clinical phase: potential markers, *Lancet Neurol.*, 2009, **8**, 1158–1171.
12. S. Fahn and G. Cohen, The oxidant stress hypothesis in Parkinson's disease: evidence supporting it, *Ann. Neurol.*, 1992, **32**, 804–812.
13. P. Jenner, Oxidative stress in Parkinson's disease, *Ann. Neurol.*, 2003, **53**(suppl. 3), S26–S36.
14. A. H. Schapira, C. W. Olanow, J. T. Greenamyre and E. Bezard, Slowing of neurodegeneration in Parkinson's disease and Huntington's disease: future therapeutic perspectives, *Lancet*, 2014, **384**, 545–555.
15. V. Dias, E. Junn and M. M. Mouradian, The role of oxidative stress in Parkinson's disease, *J. Parkinsons. Dis.*, 2013, **3**(4), 461–491.
16. V. Witko-Sarsat, T. Nguyen-Khoa, P. Jungers, T. B. Drüeke and B. Descamps-Latscha, Advanced oxidation protein products as a novel molecular basis of oxidative stress in uraemia, *Nephrol., Dial., Transplant.*, 1999, **14**, 76–78.
17. C. B. Glaser, G. Yamin, V. N. Uversky and A. L. Fink, Methionine oxidation, alpha-synuclein and Parkinson's disease, *Biochim. Biophys. Acta*, 2005, **1703**(2), 157–169.

18. W. Vogt, Oxidation of methionyl residues in proteins: tools, targets, and reversal, *Free Radic. Biol. Med.*, 1995, **18**(1), 93–105.
19. K. Nakaso, N. Tajima, S. Ito, M. Teraoka, A. Yamashita, Y. Horikoshi, D. Kikuchi, S. Mochida, K. Nakashima and T. Matsura, Dopamine-mediated oxidation of methionine-127 in α-synuclein causes cytotoxicity and oligomerization of α-synuclein, *PLoS One*, 2013, **8**(2), e55068.
20. G. Münch, H. J. Lüth, A. Wong, T. Arendt, E. Hirsch, R. Ravid and P. Riederer, Crosslinking of alpha-synuclein by advanced glycation end-products–an early pathophysiological step in Lewy body formation? *J. Chem. Neuroanat.*, 2000, **20**(3–4), 253–257.
21. M. A. Smith, P. L. Richey, S. Taneda, R. K. Kutty, L. M. Sayre, V. M. Monnier and G. Perry, Advanced Maillard reaction end products, free radicals, and protein oxidation in Alzheimer's disease, *Ann. N. Y. Acad. Sci.*, 1994, **738**, 447–454.
22. N. A. Ansari and A. R. Moinuddin, Glycated lysine residues: a marker for non-enzymatic protein glycation in age-related diseases, *Dis. Markers*, 2011, **30**(6), 317–324.
23. H. Ischiropoulos, Protein tyrosine nitration – An update, *Arch. Biochem. Biophys.*, 2009, **484**, 117–121.
24. H. Ischiropoulos and J. S. Beckman, Oxidative stress and nitration in neurodegeneration: cause, effect, or association? *J. Clin. Invest.*, 2003, **111**(2), 163–169.
25. B. I. Giasson, J. E. Duda, I. V. Murray, Q. Chen, J. M. Souza, H. I. Hurtig, H. Ischiropoulos, J. Q. Trojanowski and V. M. Lee, Oxidative damage linked to neurodegeneration by selective alpha-synuclein nitration in synucleinopathy lesions, *Science*, 2000, **290**(5493), 985–989.
26. P. F. Good, A. Hsu, P. Werner, D. P. Perl and C. W. Olanow, Protein nitration in Parkinson's disease, *J. Neuropathol. Exp. Neurol.*, 1998, **57**, 338–342.
27. T. Nakamura and S. A. Lipton, Redox modulation by *S*-nitros(yl)ation contributes to protein misfolding, mitochondrial dynamics, and neuronal synaptic damage in neurodegenerative diseases, *Cell Death Differ.*, 2011, **18**(9), 1478–1486.
28. S. A. Lipton, Y. B. Choi, Z. H. Pan, S. Z. Lei, H. S. Chen, N. J. Sucher, J. Loscalzo, D. J. Singel and J. S. Stamler, A redox-based mechanism for the neuroprotective and neurodestructive effects of nitric oxide and related nitroso-compounds, *Nature*, 1993, **364**(6438), 626–632.
29. J. S. Stamler, Redox signaling: nitrosylation and related target interactions of nitric oxide, *Cell*, 1994, **78**(6), 931–936.
30. D. Dexter, C. Carter, F. Agid, Y. Agid, A. J. Lees, P. Jenner and C. D. Marsden, Lipid peroxidation as cause of nigral cell death in Parkinson's disease, *Lancet*, 1986, **2**(8507), 639–640.
31. D. T. Dexter, C. J. Carter, F. R. Wells, F. Javoy-Agid, Y. Agid, A. Lees, P. Jenner and C. D. Marsden, Basal lipid peroxidation in substantia nigra is increased in Parkinson's disease, *J. Neurochem.*, 1989, **52**, 381–389.

32. Z. I. Alam, A. Jenner, S. E. Daniel, A. J. Lees, N. Cairns, C. D. Marsden, P. Jenner and B. Halliwell, Oxidative DNA damage in the parkinsonian brain: an apparent selective increase in 8-hydroxyguanine levels in substantia nigra, *J. Neurochem.*, 1997, **69**(3), 1196–1203.
33. E. Floor and M. G. Wetzel, Increased protein oxidation in human substantia nigra pars compacta in comparison with basal ganglia and prefrontal cortex measured with an improved dinitrophenylhydrazine assay, *J. Neurochem.*, 1998, **70**, 268–275.
34. R. Castellani, M. A. Smith, P. L. Richey and G. Perry, Glycoxidation and oxidative stress in Parkinson disease and diffuse Lewy body disease, *Brain Res.*, 1996, **737**(1–2), 195–200.
35. T. V. Ilic, M. Jovanovic, A. Jovicic and M. Tomovic, Oxidative stress indicators are elevated in *de novo* Parkinson's disease patients, *Funct. Neurol.*, 1999, **14**, 141–147.
36. C. M. Chen, J. L. Liu, Y. R. Wu, Y. C. Chen, H. S. Cheng, M. L. Cheng and D. T. Chiu, Increased oxidative damage in peripheral blood correlates with severity of Parkinson's disease, *Neurobiol. Dis.*, 2009, **33**(3), 429–435.
37. W. A. Pryor and J. P. Stanley, A suggested mechanism for the production of malondialdehyde during the autoxidation of polyunsaturated fatty acids. Nonenzymatic production of prostaglandin endoperoxides during autoxidation, *J. Org. Chem.*, 1975, **40**, 3615–3617.
38. Z. Hong, M. Shi, K. A. Chung, J. F. Quinn, E. R. Peskind, D. Galasko, J. Jankovic, C. P. Zabetian, J. B. Leverenz, G. Baird, T. J. Montine, A. M. Hancock, H. Hwang, C. Pan, J. Bradner, U. J. Kang, P. H. Jensen and J. Zhang, DJ-1 and alpha-synuclein in human cerebrospinal fluid as biomarkers of Parkinson's disease, *Brain*, 2010, **133**(pt 3), 713–726.
39. C. Buhmann, S. Arlt, A. Kontush, T. Möller-Bertram, S. Sperber, M. Oechsner, H. J. Stuerenburg and U. Beisiegel, Plasma and CSF markers of oxidative stress are increased in Parkinson's disease and influenced by antiparkinsonian medication, *Neurobiol. Dis.*, 2004, **15**(1), 160–170.
40. C. Isobe, T. Abe and Y. Terayama, L-Dopa therapy increases homocysteine concentration in cerebrospinal fluid from patients with Parkinson's disease, *J. Clin. Neurosci.*, 2010, **17**(6), 717–721.
41. M. C. Boll, J. Sotelo, E. Otero, M. Alcaraz-Zubeldia and C. Rios, Reduced ferroxidase activity in the cerebrospinal fluid from patients with Parkinson's disease, *Neurosci. Lett.*, 1999, **265**(3), 155–158.
42. M. C. Boll, M. Alcaraz-Zubeldia, S. Montes and C. Rios, Free copper, ferroxidase and SOD1 activities, lipid peroxidation and NO(x) content in the CSF. A different marker profile in four neurodegenerative diseases, *Neurochem. Res.*, 2008, **33**(9), 1717–1723.
43. F. Abdi, J. F. Quinn, J. Jankovic, M. McIntosh, J. B. Leverenz, E. Peskind, R. Nixon, J. Nutt, K. Chung, C. Zabetian, A. Samii, M. Lin, S. Hattan, C. Pan, Y. Wang, J. Jin, D. Zhu, G. J. Li, Y. Liu, D. Waichunas, T. J. Montine and J. Zhang, Detection of biomarkers with a multiplex quantitative proteomic platform in cerebrospinal fluid of patients with neurodegenerative disorders, *J. Alzheimers Dis.*, 2006, **9**(3), 293–348.

44. D. A. Loeffler, A. J. DeMaggio, P. L. Juneau, C. M. Brickman, G. A. Mashour, J. H. Finkelman, N. Pomara and P. A. LeWitt, Ceruloplasmin is increased in cerebrospinal fluid in Alzheimer's disease but not Parkinson's disease, *Alzheimer Dis. Assoc. Disord.*, 1994, **8**(3), 190–197.
45. T. L. Perry, D. V. Godin and S. Hansen, Parkinson's disease: a disorder due to nigral glutathione deficiency, *Neurosci. Lett.*, 1982, **33**, 305–310.
46. P. S. Fitzmaurice, L. Ang, M. Guttman, A. H. Rajput, Y. Furukawa and S. J. Kish, Nigral glutathione deficiency is not specific for idiopathic Parkinson's disease, *Mov. Disord.*, 2003, **18**(9), 969–976.
47. J. Sian, D. T. Dexter, A. J. Lees, S. Daniel, Y. Agid and F. Javoy-Agid, Alterations in glutathione levels in Parkinson's disease and other neurodegenerative disorders affecting basal ganglia, *Ann. Neurol.*, 1994, **36**, 348–355.
48. S. J. Kish, C. Morits and O. Hornykiewicz, Glutathione peroxidase activity in Parkinson's disease brain, *Neurosci. Lett.*, 1985, **58**, 343–346.
49. P. Damier, E. Hirsch and F. Javoy-Agid, Glutathione peroxidase, glial cells and Parkinson ′s disease, *Neuroscience*, 1993, **86**, 321–331.
50. J. Guo, Z. Sun, S. Xiao, D. Liu, G. Jin, E. Wang, J. Zhou and J. Zhou, Proteomic analysis of the cerebrospinal fluid of Parkinson's disease patients, *Cell Res.*, 2009, **19**(12), 1401–1403.
51. V. Srikanth, A. Maczurek, T. Phan, M. Steele, B. Westcott, D. Juskiw and G. Münch, Advanced glycation endproducts and their receptor RAGE in Alzheimer's disease, *Neurobiol. Aging*, 2011, **32**(5), 763–777.
52. K. D. van Dijk, C. E. Teunissen, B. Drukarch, C. R. Jimenez, H. J. Groenewegen, W. H. Berendse and W. D. van de Berg, Diagnostic cerebrospinal fluid biomarkers for Parkinson's disease: a pathogenetically based approach, *Neurobiol. Dis.*, 2010, **39**(3), 229–241.
53. I. Dalle-Donne, R. Rossi, D. Giustarini, A. Milzani and R. Colombo, Protein carbonyl groups as biomarkers of oxidative stress, *Clin. Chim. Acta*, 2003, **329**(1–2), 23–38.
54. M. Bucciantini, E. Giannoni, F. Chiti, F. Baroni, L. Formigli, J. Zurdo, N. Taddei, G. Ramponi, C. M. Dobson and M. Stefani, Inherent toxicity of aggregates implies a common mechanism for protein misfolding diseases, *Nature*, 2002, **416**, 507–511.
55. V. Ruipérez, F. Darios and B. Davletov, Alpha-synuclein, lipids and Parkinson's disease, *Prog. Lipid Res.*, 2010, **49**(4), 420–428.
56. M. Shamoto-Nagai, W. Maruyama, Y. Hashizume, M. Yoshida, T. Osawa, P. Riederer and M. Naoi, In parkinsonian substantia nigra, α-synuclein is modified by acrolein, a lipid-peroxidation product, and accumulates in the dopamine neurons with inhibition of proteasome activity, *J. Neural Transm.*, 2007, **114**(12), 1559–1567.
57. A. Olteanu and G. J. Pielak, Peroxidative aggregation of α-synuclein requires tyrosines, *Protein Sci.*, 2004, **13**, 2852–2856.
58. N. P. Ulrih, C. H. Barry and A. L. Fink, Impact of Tyr to Ala mutations on α-synuclein fibrillation and structural properties, *Biochim. Biophys. Acta*, 2008, **1782**, 581–585.

59. V. Witko-Sarsat, V. Gausson, A. T. Nguyen, M. Touam, T. Drüeke, F. Santangelo and B. Descamps-Latscha, AOPP-induced activation of human neutrophil and monocyte oxidative metabolism: a potential target for *N*-acetylcysteine treatment in dialysis patients, *Kidney Int.*, 2003, **64**(1), 82–91.
60. J. M. García-Moreno, A. Martín de Pablos, M. I. García-Sánchez, C. Méndez-Lucena, F. Damas-Hermoso, M. Rus, J. Chacón and E. Fernández, May serum levels of advanced oxidation protein products serve as a prognostic marker of disease duration in patients with idiopathic Parkinson's disease? *Antioxid. Redox Signaling*, 2013, **18**(11), 1296–1302.
61. H. B. Dunforb, Myeloperoxidase and eosinophil peroxidase: phagocytosis and microbial killing, in *Heme Peroxidases*, ed. H. B. Dunforb, Wiley, New York, 1999, pp. 349–385.
62. Y. W. Yap, M. Whiteman and N. S. Cheung, Chlorinative stress: an under appreciated mediator of neurodegeneration? *Cell. Signalling*, 2007, **19**(2), 219–228.
63. J. E. Harrison and J. Schultz, Studies on the chlorinating activity of myeloperoxidase, *J. Biol. Chem.*, 1976, **251**(5), 1371–1374.
64. B. Descamps-Latscha and V. Witko-Sarsat, Importance of oxidatively modified proteins in chronic renal failure, *Kidney Int.*, 2010, **78**, S108–S113.
65. A. Robaszkiewicz, G. Bartosz and M. Soszyński, *N*-chloroamino acids cause oxidative protein modifications in the erythrocyte membrane, *Mech. Ageing Dev.*, 2008, **129**(10), 572–579.
66. V. Witko-Sarsat, M. Friedlander, C. Capeillère-Blandin, T. Nguyen-Khoa, A. T. Nguyen, J. Zingraff, P. Jungers and B. Descamps-Latscha, Advanced oxidation protein products as a novel marker of oxidative stress in uremia, *Kidney Int.*, 1996, **49**(5), 1304–1313.
67. V. Witko-Sarsat, M. Friedlander, T. Nguyen Khoa, C. Capeillère-Blandin, A. T. Nguyen, S. Canteloup, J. M. Dayer, P. Jungers, T. Drüeke and B. Descamps-Latscha, Advanced oxidation protein products as novel mediators of inflammation and monocyte activation in chronic renal failure, *J. Immunol.*, 1998, **161**(5), 2524–2532.
68. P. L. McGeer, S. Itagaki, B. E. Boyes and E. G. McGeer, Reactive microglia are positive for HLD-DR in the substantia nigra of Parkinson's and Alzheimer's disease brains, *Neurology*, 1988, **38**, 1285–1291.
69. S. Hunot, N. Dugas, B. Faucheux, A. Hartmann, M. Tardieu, P. Debré, Y. Agid, B. Dugas and E. C. Hirsch, FcεRII/CD23 is expressed in Parkinson's disease and induces, *in vitro*, production of nitric oxide and tumor necrosis-alpha in glial cells, *J. Neurosci.*, 1999, **19**, 3440–3447.
70. S. Hunot and E. C. Hirsch, Neuroinflammatory processes in Parkinson's disease, *Ann. Neurol.*, 2003, **53**(suppl. 3), S49–S60.
71. T. Nagatsu and M. Sawada, Cellular and molecular mechanisms of Parkinson's disease: Neurotoxins, Causative Genes, and Inflammatory Cytokines, *Cell. Mol. Neurobiol.*, 2006, **26**, 781–802.

72. V. Witko-Sarsat, T. Nguyen Khoa, P. Jungers, T. Drüeke and B. Descamps-Latscha, Advanced oxidation protein products: oxidative stress markers and mediators of inflammation in uremia, *Adv. Nephrol. Necker Hosp.*, 1998, **28**, 321–341.
73. E. Malle, G. Waeg, R. Schreiber, E. F. Gröne, W. Sattler and H. J. Gröne, Immunohistochemical evidence for the myeloperoxidase/H2O2/halide system in human atherosclerotic lesions: colocalization of myeloperoxidase and hypochlorite-modified proteins, *Eur. J. Biochem.*, 2000, **267**(14), 4495–4503.
74. H. J. Gröne, E. F. Gröne and E. Malle, Immunohistochemical detection of hypochlorite-modified proteins in glomeruli of human membranous glomerulonephritis, *Lab. Invest.*, 2002, **82**(1), 5–14.
75. J. H. Bräsen, T. Häkkinen, E. Malle, U. Beisiegel and S. Ylä-Herttuala, Patterns of oxidized epitopes, but not NF-kappa B expression, change during atherogenesis in WHHL rabbits, *Atherosclerosis*, 2003, **166**(1), 13–21.
76. C. Capeillère-Blandin, V. Gausson, B. Descamps-Latscha and V. Witko-Sarsat, Biochemical and spectrophotometric significance of advanced oxidation protein products, *Biochim. Biophys. Acta*, 2004, **1689**(2), 91–102.
77. G. Münch, B. Kuhla, H. J. Lüth, T. Arendt and S. R. Robinson, Anti-ageing defenses against Alzheimer's disease, *Biochem. Soc. Trans.*, 2003, **31**(pt 6), 1397–1399.
78. C. J. Werner, R. Heyny-von Haussen, G. Mall and S. Wolf, Proteome analysis of human substantia nigra in Parkinson's disease, *Proteome Sci.*, 2008, **14**, 8.
79. P. J. Thornalley, Glutathione-dependent detoxification of alpha-oxoaldehydes by the glyoxalase system: involvement in disease mechanisms and antiproliferative activity of glyoxalase I inhibitors, *Chem.-Biol. Interact.*, 1998, **24**, 137–151.
80. M. Kalousová, J. Skrha and T. Zima, Advanced glycation end-products and advanced oxidation protein products in patients with diabetes mellitus, *Physiol. Res.*, 2002, **51**(6), 597–604.
81. E. Dalfó, M. Portero-Otín, V. Ayala, A. Martínez, R. Pamplona and I. Ferrer, Evidence of oxidative stress in the neocortex in incidental Lewy body disease, *J. Neuropathol. Exp. Neurol.*, 2005, **64**(9), 816–830.
82. E. Guerrero, P. Vasudevaraju, M. L. Hegde, G. B. Britton and K. S. Rao, Recent advances in α-synuclein functions, advanced glycation, and toxicity: implications for Parkinson's disease, *Mol. Neurobiol.*, 2013, **47**(2), 525–536.
83. S. K. Jain, Glutathione and glucose-6-phosphate dehydrogenase deficiency can increase protein glycosylation, *Free Radic. Biol. Med.*, 1998, **24**(1), 197–201.
84. A. Prigione, B. Begni, A. Galbussera, S. Beretta, L. Brighina, R. Garofalo, S. Andreoni, R. Piolti and C. Ferrarese, Oxidative stress in peripheral blood mononuclear cells from patients with Parkinson's disease: negative correlation with levodopa dosage, *Neurobiol. Dis.*, 2006, **23**(1), 36–43.

85. L. Thomson, J. Christie, C. Vadseth, P. N. Lanken, X. Fu, S. L. Hazen and H. Ischiropoulos, Identification of immunoglobulins that recognize 3-nitrotyrosine in patients with acute lung injury after major trauma, *Am. J. Respir. Cell Mol. Biol.*, 2007, **36**(2), 152–157.
86. Y. J. Suzuki, M. Carini and D. A. Butterfield, Protein carbonylation, *Antioxid. Redox Signaling*, 2010, **2**(3), 323–325.
87. J. Everse and P. W. Coates, Neurodegeneration and peroxidases, *Neurobiol. Aging*, 2009, **30**, 1011–1025.
88. L. A. MacMillan-Crow, J. P. Crow, J. D. Kerby, J. S. Beckman and J. A. Thompson, Nitration and inactivation of manganese superoxide dismutase in chronic rejection of human renal allografts, *Proc. Natl. Acad. Sci. U. S. A.*, 1996, **93**(21), 11853–11858.
89. L. A. MacMillan-Crow and J. A. Thompson, Tyrosine modifications and inactivation of active site manganese superoxide dismutase mutant (Y34F) by peroxynitrite, *Arch. Biochem. Biophys.*, 1999, **366**(1), 82–88.
90. K. Aoyama, K. Matsubara, Y. Fujikawa, Y. Nagahiro, K. Shimizu, N. Umegae, N. Hayase, H. Shiono and S. Kobayashi, Nitration of manganese superoxide dismutase in cerebrospinal fluids is a marker for peroxynitrite-mediated oxidative stress in neurodegenerative diseases, *Ann. Neurol.*, 2000, **47**(4), 524–527.
91. E. C. Hirsch and S. Hunot, Neuroinflammation in Parkinson's disease: a target for neuroprotection? *Lancet Neurol.*, 2009, **8**(4), 382–397.
92. B. Kalyanaraman, V. Darley-Usmar, K. J. Davies, P. A. Dennery, H. J. Forman, M. B. Grisham, G. E. Mann, K. Moore, L. J. Roberts 2nd and H. Ischiropoulos, Measuring reactive oxygen and nitrogen species with fluorescent probes: challenges and limitations, *Free Radic. Biol. Med.*, 2012, **52**(1), 1–6.
93. G. A. Qureshi, S. Baig, I. Bednar, P. Södersten, G. Forsberg and A. Siden, Increased cerebrospinal fluid concentration of nitrite in Parkinson's disease, *NeuroReport*, 1995, **6**(12), 1642–1644.
94. A. van der Vliet, J. P. Eiserich, B. Halliwell and C. E. Cross, Formation of reactive nitrogen species during peroxidase-catalyzed oxidation of nitrite. A potential additional mechanism of nitric oxide-dependent toxicity, *J. Biol. Chem.*, 1997, **272**(12), 7617–7625.
95. L. Thomson, J. Christie, C. Vadseth, P. N. Lanken, X. Fu, S. L. Hazen and H. Ischiropoulos, Identification of immunoglobulins that recognize 3-nitrotyrosine in patients with acute lung injury after major trauma, *Am. J. Respir. Cell Mol. Biol.*, 2007, **36**(2), 152–157.
96. M. J. Mihm, B. L. Schanbacher, B. L. Wallace, L. J. Wallace, N. J. Uretsky and J. A. Bauer, Free 3-nitrotyrosine causes striatal neurodegeneration *in vivo*, *J. Neurosci.*, 2001, **21**(11), RC149.
97. E. Fernández, J. M. Garcia-Moreno, A. Martin de Pablos and J. Chacón, May the evaluation of nitrosative stress through selective increase of 3-nitrotyrosine proteins other than nitroalbumin and dominant tyrosine-125/136 nitration of serum α-synuclein serve for diagnosis of sporadic Parkinson's disease? *Antioxid. Redox Signaling*, 2013, **19**(9), 912–918.

98. R. M. Lebovitz, H. Zhang, H. Vogel, J. Cartwright Jr, L. Dionne, N. Lu, S. Huang and M. M. Matzuk, Neurodegeneration, myocardial injury, and perinatal death in mitochondrial superoxide dismutase-deficient mice, *Proc. Natl. Acad. Sci. U. S. A.*, 1996, **93**(18), 9782–9787.
99. H. Neumann, J. L. Hazen, J. Weinstein, R. A. Mehl and J. W. Chin, Genetically encoding protein oxidative damage, *J. Am. Chem. Soc.*, 2008, **130**(12), 4028–4033.
100. B. Blanchard-Fillion, J. M. Souza, T. Friel, G. C. Jiang, K. Vrana, V. Sharov, L. Barrón, C. Schöneich, C. Quijano, B. Alvarez, R. Radi, S. Przedborski, G. S. Fernando, J. Horwitz and H. Ischiropoulos, Nitration and inactivation of tyrosine hydroxylase by peroxynitrite, *J. Biol. Chem.*, 2001, **276**(49), 46017–46023.
101. J. Ara, S. Przedborski, A. B. Naini, V. Jackson-Lewis, R. R. Trifiletti, J. Horwitz and H. Ischiropoulos, Inactivation of tyrosine hydroxylase by nitration following exposure to peroxynitrite and 1-methyl-4-phenyl-1,2,3,6-tetrahydropyridine (MPTP), *Proc. Natl. Acad. Sci. U. S. A.*, 1998, **95**(13), 7659–7663.
102. J. Scholz, K. Toska, A. Luborzewski, A. Maass, V. Schünemann, J. Haavik and A. Moser, Endogenous tetrahydroisoquinolines associated with Parkinson's disease mimic the feedback inhibition of tyrosine hydroxylase by catecholamines, *FEBS J.*, 2008, **275**(9), 2109–2121.
103. M. H. Polymeropoulos, C. Lavedan, E. Leroy, S. E. Ide, A. Dehejia, A. Dutra, B. Pike, H. Root, J. Rubenstein, R. Boyer, E. S. Stenroos, S. Chandrasekharappa, A. Athanassiadou, T. Papapetropoulos, W. G. Johnson, A. M. Lazzarini, R. C. Duvoisin, G. Di Iorio, L. I. Golbe and R. L. Nussbaum, Mutation in the α-synuclein gene identified in families with Parkinson's disease, *Science*, 1997, **276**(5321), 2045–2047.
104. R. Krüger, W. Kuhn, T. Müller, D. Woitalla, M. Graeber, S. Kösel, H. Przuntek, J. T. Epplen, L. Schöls and O. Riess, Ala30Pro mutation in the gene encoding α-synuclein in Parkinson's disease, *Nat. Genet.*, 1998, **18**(2), 106–108.
105. A. Papadimitriou, V. Veletza, G. M. Hadjigeorgiou, A. Patrikiou, M. Hirano and I. Anastasopoulos, Mutated α-synuclein gene in two Greek kindreds with familial PD: incomplete penetrance? *Neurology*, 1999, **52**(3), 651–654.
106. J. Xu, S. Y. Kao, F. J. Lee, W. Song, L. W. Jin and B. A. Yankner, Dopamine-dependent neurotoxicity of α-synuclein : a mechanism for selective neurodegeneration in Parkinson disease, *Nat. Med.*, 2002, **8**(6), 600–606.
107. J. Q. Trojanowski, M. Goedert, T. Iwatsubo and V. M. Lee, Fatal attractions: abnormal protein aggregation and neuron death in Parkinson's disease and Lewy body dementia, *Cell Death Differ.*, 1998, **5**(10), 832–837.
108. R. Jakes, M. G. Spillantini and M. Goedert, Identification of two distinct synucleins from human brain, *FEBS Lett.*, 1994, **345**(1), 27–32.
109. R. A. Ruf, E. A. Lutz, I. G. Zigoneanu and G. J. Pielak, Alpha-Synuclein conformation affects its tyrosine-dependent oxidative aggregation, *Biochemistry*, 2008, **47**(51), 13604–13609.

110. R. Hodara, E. H. Norris, B. I. Giasson, A. J. Mishizen-Eberz, D. R. Lynch, V. M. Lee and H. Ischiropoulos, Functional consequences of α-synuclein tyrosine nitration: diminished binding to lipid vesicles and increased fibril formation, *J. Biol. Chem.*, 2004, **279**(46), 47746–47753.
111. E. Paxinou, Q. Chen, M. Weisse, B. I. Giasson, E. H. Norris, S. M. Rueter, J. Q. Trojanowski, V. M. Lee and H. Ischiropoulos, Induction of α-synuclein aggregation by intracellular nitrative insult, *J. Neurosci.*, 2001, **21**(20), 8053–8061.
112. K. Yanamandra, M. A. Gruden, V. Casaite, R. Meskys, L. Forsgren and L. A. Morozova-Roche, α-synuclein reactive antibodies as diagnostic biomarkers in blood sera of Parkinson's disease patients, *PLoS One*, 2011, **6**(4), e18513.
113. S. Shavali, C. K. Combs and M. Ebadi, Reactive macrophages increase oxidative stress and α-synuclein nitration during death of dopaminergic neuronal cells in co-culture: relevance to Parkinson's disease, *Neurochem. Res.*, 2006, **31**(1), 85–94.
114. Y. Liu, M. Qiang, Y. Wei and R. He, A novel molecular mechanism for nitrated {alpha}-synuclein-induced cell death, *J. Mol. Cell. Biol.*, 2011, **3**(4), 239–249.
115. E. Emmanouilidou, D. Elenis, T. Papasilekas, G. Stranjalis, K. Gerozissis, P. C. Ioannou and K. Vekrellis, Assessment of α-synuclein secretion in mouse and human brain parenchyma, *PLoS One*, 2011, **6**, e22225.
116. R. Barbour, K. Kling, J. P. Anderson, K. Banducci, T. Cole, L. Diep, M. Fox, J. M. Goldstein, F. Soriano, P. Seubert and T. J. Chilcote, Red blood cells are the major source of α-synuclein in blood, *Neurodegener. Dis.*, 2008, **5**(2), 55–59.
117. T. Bartels, J. G. Choi and D. J. Selkoe, α-Synuclein occurs physiologically as a helically folded tetramer that resists aggregation, *Nature*, 2011, **477**, 107–110.
118. N. Gould, D. E. Mor, R. Lightfoot, K. Malkus, B. Giasson and H. Ischiropoulos, Evidence of native α-synuclein conformers in the human brain, *J. Biol. Chem.*, 2014, **289**(11), 7929–7934.
119. U. Dettmer, A. J. Newman, E. S. Luth, T. Bartels and D. Selkoe, *In vivo* cross-linking reveals principally oligomeric forms of α-synuclein and β-synuclein in neurons and non-neural cells, *J. Biol. Chem.*, 2013, **288**, 6371–6385.
120. P. H. Weinreb, W. Zhen, A. W. Poon, K. A. Conway and P. T. J. Lansbury, NACP, a protein implicated in Alzheimer's disease and learning, is natively unfolded, *Biochemistry*, 1996, **35**, 13709–13715.
121. J. M. García-Moreno and J. Chacón, Hipotiroidismo enmascarado por enfermedad de Parkinson, *Rev. Neurol.*, 2002, **35**(8), 741–742.
122. H. Matsui, F. Udaka, M. Oda, A. Tamura, T. Kubori, K. Nishinaka and M. Kameyama, Metaiodobenzylguanidine (MIBG) uptake in Parkinson's disease also decreases at thyroid, *Ann. Nucl. Med.*, 2005, **19**(3), 225–229.
123. H. Tandeter, A. Levy, G. Gutman and P. Shvartzman, Subclinical thyroid disease in patients with Parkinson's disease, *Arch. Gerontol. Geriatr.*, 2001, **33**(3), 295–300.

124. D. N. Tipre and D. S. Goldstein, Cardiac and extracardiac sympathetic denervation in Parkinson's disease with orthostatic hypotension and in pure autonomic failure, *J. Nucl. Med.*, 2005, **46**(11), 1775–1781.
125. E. Fernández, J. M. Garcia-Moreno, A. Martin de Pablos and J. Chacón, May the thyroid gland and thyroperoxidase participate in nitrosylation of serum proteins and sporadic Parkinson's disease? *Antioxid. Redox Signaling*, 2014, **21**(15), 2143–2148.
126. T. Uehara, T. Nakamura, D. Yao, Z. Q. Shi, Z. Gu, Y. Ma, E. Masliah, Y. Nomura and S. A. Lipton, *S*-nitrosylated protein-disulphide isomerase links protein misfolding to neurodegeneration, *Nature*, 2006, **441**(7092), 513–517.
127. A. K. Walker, M. A. Farg, C. R. Bye, C. A. McLean, M. K. Horne and J. D. Atkin, Protein disulphide isomerase protects against protein aggregation and is *S*-nitrosylated in amyotrophic lateral sclerosis, *Brain*, 2010, **133**(pt 1), 105–116.
128. X. Chen, X. Zhang, C. Li, T. Guan, H. Shang, L. Cui, X. M. Li and J. Kong, *S*-nitrosylated protein disulfide isomerase contributes to mutant SOD1 aggregates in amyotrophic lateral sclerosis, *J. Neurochem.*, 2013, **124**(1), 45–58.
129. X. F. Wu, A. F. Wang, L. Chen, E. P. Huang, W. B. Xie, C. Liu, W. Y. Huang, C. X. Chen, P. M. Qiu and H. J. Wang, *S*-Nitrosylating protein disulphide isomerase mediates α-synuclein aggregation caused by methamphetamine exposure in PC12 cells, *Toxicol. Lett.*, 2014, **230**(1), 19–27.
130. P. Kabiraj, J. E. Marin, A. Varela-Ramirez, E. Zubia and M. Narayan, Ellagic acid mitigates SNO-PDI induced aggregation of Parkinsonian biomarkers, *ACS Chem. Neurosci.*, 2014, **5**(12), 1209–1220.
131. K. Chen, T. C. Detwiler and D. W. Essex, Characterization of protein disulphide isomerase released from activated platelets, *Br. J. Haematol.*, 1995, **90**(2), 425–431.
132. P. Root, I. Sliskovic and B. Mutus, Platelet cell-surface protein disulphide-isomerase mediated *S*-nitrosoglutathione consumption, *Biochem. J.*, 2004, **382**(pt 2), 575–580.
133. J. R. Pawloski and J. S. Stamler, Nitric oxide in RBCs, *Transfusion*, 2002, **42**(12), 1603–1609.
134. K. K. Chung, B. Thomas, X. Li, O. Pletnikova, J. C. Troncoso, L. Marsh, V. L. Dawson and T. M. Dawson, *S*-nitrosylation of parkin regulates ubiquitination and compromises parkin's protective function, *Science*, 2004, **304**(5675), 1328–1331.
135. C. Alves da Costa and F. Checler, Apoptosis in Parkinson's disease: is p53 the missing link between genetic and sporadic Parkinsonism? *Cell. Signalling*, 2011, **23**(6), 963–968.
136. C. R. Sunico, T. Nakamura, E. Rockenstein, M. Mante, A. Adame, S. F. Chan, T. F. Newmeyer, E. Masliah, N. Nakanishi and S. A. Lipton, *S*-Nitrosylation of parkin as a novel regulator of p53-mediated neuronal cell death in sporadic Parkinson's disease, *Mol. Neurodegener.*, 2013, **8**, 29.
137. M. J. Farrer, Genetics of Parkinson disease: paradigm shifts and future Prospects, *Nat. Rev. Genet.*, 2006, **7**(4), 306–318.

138. S. G. Rhee, S. W. Kang, W. Jeong, T. S. Chang, K. S. Yang and H. A. Woo, Intracellular messenger function of hydrogen peroxide and its regulation by peroxiredoxins, *Curr. Opin. Cell Biol.*, 2005, **17**(2), 183–189.
139. K. Krapfenbauer, E. Engidawork, N. Cairns, M. Fountoulakis and G. Lubec, Aberrant expression of peroxiredoxin subtypes in neurodegenerative disorders, *Brain Res.*, 2003, **967**(1–2), 152–160.
140. J. Fang, T. Nakamura, D. H. Cho, Z. Gu and S. A. Lipton, *S*-nitrosylation of peroxiredoxin 2 promotes oxidative stress-induced neuronal cell death in Parkinson's disease, *Proc. Natl. Acad. Sci. U. S. A.*, 2007, **104**(47), 18742–18747.
141. A. H. Tsang and K. K. Chung, Oxidative and nitrosative stress in Parkinson's disease, *Biochim. Biophys. Acta*, 2009, **1792**(7), 643–650.
142. M. R. Hara, B. Thomas, M. B. Cascio, B. I. Bae, L. D. Hester, V. L. Dawson, T. M. Dawson, A. Sawa and S. H. Snyder, Neuroprotection by pharmacologic blockade of the GAPDH death cascade, *Proc. Natl. Acad. Sci. U. S. A.*, 2006, **103**(10), 3887–3889.

CHAPTER 14

Dietary Anti-, Pro-Oxidants in the Etiology of Parkinson's Disease

ZEYNEP SENA AGIM* AND JASON R. CANNON*

School of Health Sciences, Purdue University, West Lafayette, IN, 47907, USA
*E-mail: zagim@purdue.edu, cannonjr@purdue.edu

14.1 Introduction

Parkinson's disease (PD) is the second most common neurodegenerative disease worldwide. Only about 10% of PD cases can be attributed to genetic factors such as inherited mutations in the PD-related genes α-synuclein, leucine-rich repeat kinase 2 (LRRK2), DJ-1, parkin, and others. This small proportion of familial PD cases emphasizes the probable contribution of environmental factors to the disease pathology. Several environmental factors are known to increase the risk of PD, including pesticides, heavy metals, and organic solvents.[1] Although many of these compounds have been found to reproduce key features of PD pathology in animal models, significant exposures are typically rare for humans and mostly arise in occupational settings. Thus, attributing specific exposures to a large number of PD cases has been challenging.

In this context, dietary factors have received considerable attention. Association of dietary compounds with potentiation or amelioration of PD-relevant

Issues in Toxicology No. 34
Oxidative Stress and Redox Signalling in Parkinson's Disease
Edited by Rodrigo Franco, Jonathan A. Doorn and Jean-Christophe Rochet

Published by the Royal Society of Chemistry, www.rsc.org

pathology has been examined using cellular and animal models and risk modulation has been evaluated in epidemiological studies.[2] An example of dietary compounds that may elicit dopaminergic dysfunction relevant to PD are heterocyclic amines, compounds that can be present in a variety of foods, notably in charred meat. These compounds and their potential link to PD and essential tremor are discussed in the first part of this chapter.

Natural antioxidants, including polyphenols and vitamins, have often been investigated for their neuroprotective properties in both human and animal studies. Several lines of evidence support a neuroprotective role of antioxidants in PD. First, epidemiological studies have shown that dietary habits can alter the prevalence of PD in different populations. For example, greater adherence to the Mediterranean diet, rich in fruits and vegetables with strong antioxidant properties, showed inverse association with PD risk.[3] Secondly, pretreatment of neuronal and glial cultures with antioxidant compounds inhibited the oxidative damage and resulting cell loss induced by neurotoxicants *in vitro*.[4] Thirdly, *in vivo* studies have revealed that administration of dietary antioxidants is able to attenuate motor impairments in PD animal models.[5]

Although antioxidants are considered suppressors of oxidative stress, they show an inverted U-shaped dose–response curve. In a healthy cell, reactive oxygen species (ROS) are being produced endogenously as byproducts of oxidative phosphorylation in mitochondria. Low levels of ROS help cells adapt to cope with oxidative damage and develop resistance. Antioxidants at lower concentrations have been shown to hinder such beneficial roles of ROS in defense mechanisms.[6] On the other hand, high doses of antioxidants may exert stress-promoting effects by paradoxically acting as pro-oxidants.[7] As detailed in the following sections, many compounds that decrease ROS and enhance antioxidant responses in moderate doses indeed generate more ROS by auto-oxidation, eventually inducing cytotoxicity at higher doses. Therefore, in cell-culture and animal studies as well as clinical trials, the dose of antioxidant administered carries great importance to reveal the protective effect against neurodegeneration.

There are hundreds of antioxidant nutrients in our daily diet, and a large number of these compounds have been evaluated as modulators of neurological disease. In this chapter, we will focus on major polyphenolic compounds and vitamins with antioxidant propensities, along with their mechanisms of action in neuroprotective roles against PD.

14.2 Heterocyclic Amines

Heterocyclic amines (HCAs) were isolated and characterized in cooked meat as byproducts of the cooking process more than 30 years ago. Today, there are more than 20 compounds identified in this structural group. The amount of HCAs varies depending on the type of the meat, cooking temperature, cooking style, doneness, and many other factors. Although they have been classified as mutagens and carcinogens in numerous *in vivo* and *in vitro* studies, their effects in the nervous system, especially in neurodegeneration, are poorly understood.

HCAs can be divided into two structural categories: Beta-carbolines and amino-imidazoazarenes. Beta-carbolines are a group of indole alkaloids, synthesized from L-trytophan, that possess a common tricyclic pyrido[3,4-*b*] indole ring structure. The most basic beta-carboline is 9H-pyrido[3,4-*b*] indole, better known as norharmane. The methylated form of norharmane at carbon 1 is named harmane. Although norharmane and harmane are the most studied beta-carbolines, there are several others such as harmine, harmaline, and harmalol. They can be produced endogenously in the body and also consumed in the diet. The concentration of harmane, like other HCAs, varies according to meat type. Chicken has the highest amount of harmane (mean: 8.48 ng g^{-1} of food), while levels are significantly lower in beefsteak, with an average of 3.80 ng g^{-1} of food.[8] Harmane concentrations directly correlate with meat doneness; as the doneness increases, the concentration increases. Beta-carbolines are also found in alcohol. One study detected the highest levels of harmane in sake and wine among alcohol types (4.1 and 8.5 ng ml^{-1}, respectively), whereas beer is richer in norharmane (2.7 ng ml^{-1}).[9] Conversely, another study showed that beer has more harmane than wine.[10] Thirdly, tobacco has high levels of beta-carbolines varying from 300 ng to almost 10 000 ng per cigarette.[11] Beta-carbolines have also been detected in several other foods and drinks, including but not limited to coffee, soy sauce, and vinegar.[9,12]

A potential association of harmane with essential tremor (ET), the most common movement disorder, has been investigated in several epidemiological studies. Blood harmane levels were increased significantly in ET patients compared to healthy controls.[13] Patients with familial ET have been shown to have higher levels of brain harmane compared to sporadic ET patients and controls, implicating the role of inherited genetic traits in modulation of harmane metabolism.[14] Epidemiological studies also confirmed the association of both harmane and norharmane with PD. In 1995, a small case-control study showed that plasma concentrations of norharmane in PD patients were significantly higher than those in controls.[15] Harmane levels were also higher in cases than controls, but the difference failed to reach significance, which can be attributed to small sample size. A later study with a larger sample size overcame this limitation and found that blood harmane levels were significantly higher in PD patients compared to healthy controls, especially in patients with a family history of both PD and ET.[16] Harmane and norharmane concentrations have also been found to be elevated in cerebral spinal fluid (CSF) of PD patients compared to healthy controls.[17] The mechanism of neurotoxic action of these compounds has not yet been adequately explored.

Amino-imidazoazarenes, a second group of HCAs, are also ingested in diet, mainly in cooked meat and meat products. While levels of HCAs are highly dependent on cooking conditions, 2-amino-1-methyl-6-phenylimidazo[4,5-*b*] pyridine (PhIP) has been identified as the most abundant one, accounting for up to ~75% of the genotoxic material from the crust of cooked meat.[18,19] A recent study in primary midbrain cultures showed that PhIP and its Phase I metabolite, N-OH-PhIP, are selectively toxic to dopaminergic neurons, possibly due to cellular oxidative stress.[20] In the same study, pretreatment of cultures with blueberry extracts that are rich in anthocyanins (an antioxidant discussed later) prevented PhIP-induced dopaminergic neurotoxicity. Both PhIP and N-OH-PhIP cross the blood–brain barrier and form adducts with nucleic and amino acids.[21–23] Given brain entry and implication of key important pathogenic events in PD (oxidative stress and protein adduction), the neurotoxicity of these compounds warrants further investigation. Besides PhIP and its metabolites, there are some early *in vivo* studies that considered how HCAs affect dopamine metabolism. Unilateral infusion of 3-amino-1,4-dimethyl-5H-pyrido[4,3-*b*]indole (Trp-P-1) and 3-amino-1-methyl-5H-pyrido[4,3-*b*]indole (Trp-P-2) in the rat striatum (dose not available) resulted in increased dopamine levels and decreased dopamine metabolism in the striatum.[24] In the same study, systemic administration of Trp-P-1 (200 mg kg^{-1}) caused a significant increase in dopamine levels and decrease in metabolite levels. Trp-P-1 and Trp-P-2 also inhibited type A and B monoamine oxidases, possibly altering monoamine levels.[25,26] It is worth noting that direct infusion of dopaminergic neurotoxicants into the parenchyma may result in high local concentrations and produce non-specific toxicity, whereas systemic administration of the same compound systemically may produce selective dopaminergic neurotoxicity.[27,28] The roles of Trp-P-1 and Trp-P-2 in oxidative stress are still to be investigated, but their involvement in modulation of dopamine levels and occurrence of oxidative stress due to auto-oxidation of excess cytosolic dopamine make them good candidates as pro-oxidants.

14.3 Polyphenolic Compounds

Composed of more than 8000 compounds, polyphenols belong to a structural class of secondary plant metabolites characterized by the presence of multiple phenol units.[29] High levels of polyphenols are found in fruits, vegetables, and herbs. The primary function of polyphenols is to protect plants against ROS generated during photosynthesis.[30] They have come into prominence in human health due to their anti-aging effects.

Polyphenols can be classified according to their structures, in particular the number of phenolic rings and the nature of functional groups that are attached to these rings.[31] They can be structurally divided into two main classes: flavonoids and non-flavonoids. These categories can be further subdivided into several smaller groups described in each section (Figure 14.1).

14.3.1 Flavonoids

Flavonoids are the largest group of polyphenols. The structural backbone of flavonoids consists of two phenolic rings (ring A and B) that are bound to each other by three carbon atoms (C_2, C_3, and C_4) forming an oxygenated heterocycle (ring C) (Figure 14.2). Flavonoids are further subcategorized into isoflavones, flavones, flavanones, flavanols, anthocyanins and flavonols, based on the arrangement of their carbonyl and hydroxyl groups on ring C.

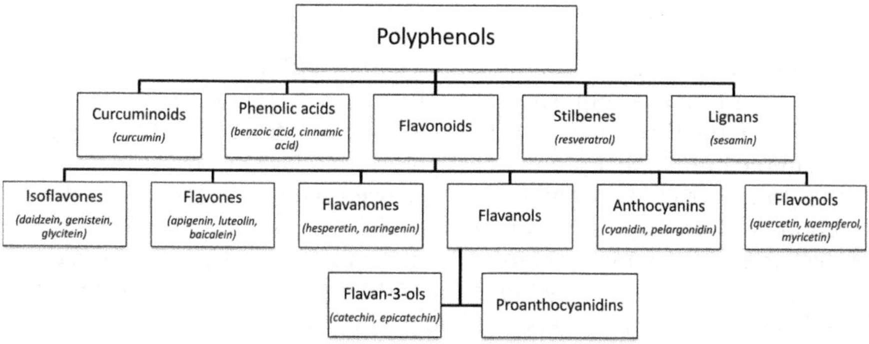

Figure 14.1 Classification of polyphenols. Classes and subclasses of prominent polyphenols are shown.

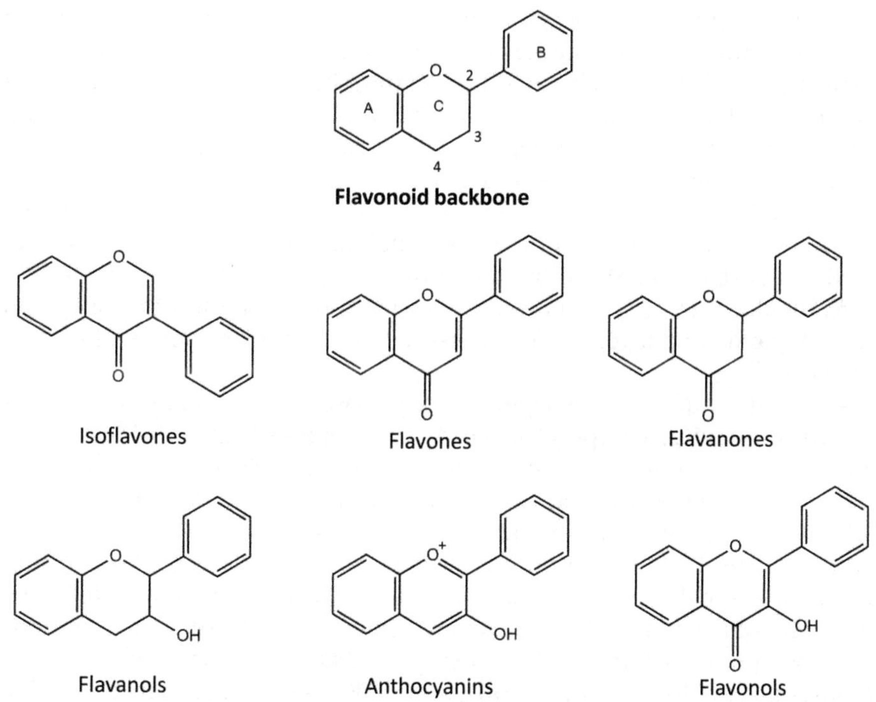

Figure 14.2 Structure of flavonoid backbone and six subclasses of flavonoids (adapted from © Zachariah P. Sellers, ReBecca A. Williams, Jonathan W. Overbay, Jooyoung Cho, Moses Henderson and Tanea T. Reed (2014)). Current Therapeutic Modalities, Enzyme Kinetics, and Redox Proteomics in Traumatic Brain Injury, Traumatic Brain Injury, Dr Farid Sadaka (Ed.), InTech, DOI: 10.5772/57306. Published under CC BY 3.0 license.[378]

14.3.1.1 Isoflavones

Isoflavones have a carbonyl group at C_4 and a double bond between C_2 and $C_{3,}$ and their B ring is connected to C_3 (Figure 14.2). Daidzein, genistein, and glycitein are the three major isoflavones found in legumes, with the highest levels in the human diet present in soybeans and fava beans.[32]

Due to its structural similarity to estrogen, genistein binds to estrogen receptor (ER) and exhibits estrogenic activity. Estrogen has shown neuroprotective effects against dopaminergic neuronal degeneration.[33] Stereotaxic administration of 6-hydroxydopamine (6-OHDA) into the medial forebrain bundle in ovariectomized rats resulted in the loss of dopaminergic neurons from the *substantia nigra*.[5] Removal of ovaries was performed to eliminate the effect of endogenous estrogen. Both motor deficits and loss of dopaminergic cells were reversed by pretreatment with genistein. The effect of genistein was shown to be mediated through ER signalling, since an ER antagonist reversed its protective effects. It is worth noting that the regulation of cell survival mediated by genistein is concentration dependent. Several cancer-chemoprevention studies have shown that administration of higher concentrations of genistein (20–111 μM) inhibited cell growth and induced the activity of pro-apoptotic proteins.[30] In contrast, a lower dose (1 μM) stimulated growth in estrogen-dependent breast cancer cells, MCF-7.[34] However, the estrogenic activity of genistein is not the sole mechanism of action proposed in relation to its neuroprotective effects. Endoplasmic reticulum stress is one of the pathways leading to neuronal apoptosis in PD.[35,36] Genistein used at a dose of 50 nM inhibited neuronal apoptosis induced by endoplasmic reticulum stress.[37] Moreover, genistein's ability to inhibit carbonylation of hemoglobin and myoglobin was associated with the prevention of protein oxidation.[38]

Several lines of evidence suggest that isoflavones also act as anti-inflammatory molecules. Lipopolysaccharide (LPS)-induced nitric oxide production and genes regulated by nitric oxide (IRF-1 and STAT1) were blocked with treatments of genistein, daidzein, and coumestrol in the HAPI rat microglial cell line.[39] It has been shown that pro-inflammatory cytokines, such as interleukin (IL)-1α, IL-2, and tumor necrosis factor alpha (TNFα) released from activated microglia, are neurotoxic to neuronal cell populations. McDowell *et al.* showed that genistein prevented apoptotic death of motor neurons from rat spinal cord induced by such pro-inflammatory factors present in conditioned media from interferon gamma (IFNγ)-treated primary microglial cultures.[40] Genistein also inhibited ROS production and restored mitochondrial membrane potential in the same experimental system. Similarly, daidzein pretreatment suppressed production of pro-inflammatory factors, ROS generation, and p38 MAPK activation in LPS-induced BV2 murine microglial cell lines.[4] Daidzein also attenuated toxicity to the N27 rat dopaminergic neuronal cell line from conditioned media of activated microglial cells.

Isoflavones are neuroprotective in neurotoxicant-induced models of PD both *in vitro* and *in vivo*. Paraquat-induced activation of transcription factor

AP1 and apoptosis in PC12 rat adrenal pheochromocytoma cells was blocked by genistein.[41] Daidzein and genistein isolated from Peuraria roots (*Pueraria thomsonii* Benth.) diminished caspase-3 and caspase-8 activation and subsequent cell loss in 6-OHDA-treated PC12 cells, but did not protect against ROS production, suggesting protection is independent of ROS formation.[42] Intraperitoneal genistein administration in 1-methyl-4-phenyl-1,2,3,6-tetrahydropyridine (MPTP) and 6-OHDA PD models also prevented neuropathological characteristics of the disease, including a decrease in the concentrations of dopamine and its metabolites in the striatum and dopaminergic cell loss in the *substantia nigra pars compacta*.[43,44] Inducible nitric oxide synthase (iNOS) mRNA expression, nitrite content, and lipid peroxidation in the striatum of maneb- and paraquat-treated rats decreased after genistein treatment.[45]

14.3.1.2 Flavones

Flavones are structurally very similar to isoflavones; the key difference between the two is that the B ring is connected to C_2 instead of C_3 (Figure 14.2). Celery, green pepper, parsley and various herbs are a notable source of flavones.[46] The major types of dietary flavones are baicalein, luteolin, apigenin, and acacetin.

Baicalein, isolated from roots of the traditional Chinese herb *Scutellaria baicalensis*, is the flavone studied the most in relation to PD. At low micromolar levels, baicalein binds tightly to α-synuclein in cell-free systems, causing inhibition of fibril formation and dissociation of existing fibrils.[47–51] The products of these disaggregated fibrils are soluble oligomers, which are further stabilized by baicalein *via* covalent interactions.[47] Anti-fibrillization effects of baicalein have been suggested in cell culture as well, using SH-SY5Y human neuroblastoma and HeLa human epithelial adenocarcinoma cells.[52] In another study, baicalein treatment of N2A murine neuroblastoma cells expressing E46K mutant α-synuclein reduced fibrillization of the mutant protein and beta-sheet conformation.[53] This property of baicalein plays a critical role in the rescue of α-synuclein fibril-induced toxicity and the prevention of subsequent damage to lipid membranes.[54]

Several studies in cell culture confirmed a neuroprotective role of baicalein in PD neurotoxicant models. Mitochondrial dysfunction, ROS production, activation of caspase cascades, and phosphorylation of JNK, an apoptotic mediator, were inhibited by baicalein in 6-OHDA-treated SH-SY5Y cells.[55] In rotenone-treated PC12 cells, baicalein was also neuroprotective, preventing ATP deficiency and mitochondrial membrane potential dissipation.[56]

In rat and mouse PD models, administration of baicalein restored neurotoxicant-induced motor deficits and cell loss. Pretreatment with baicalein for one week significantly increased catecholamine levels in the striatum and tyrosine hydroxylase-immunoreactive (TH-IR) neurons in the *substantia nigra*, decreased lipid peroxidation, and improved scores in both the spontaneous motor activity test and the pole test in MPTP-treated mice.[57] A later study further confirmed the neuroprotective role of baicalein in the MPTP-induced

PD mouse model, suggesting that it is involved in modulating dopamine catabolism and turnover.[58] Interestingly, in 6-OHDA-lesioned rats, baicalein treatment had no effect on apomorphine-induced rotation but attenuated muscle tremors.[59] This attenuation may be due to modulatory effects of the flavone on the balance between glutamate and γ-aminobutyric acid (GABA), and inhibition of glutamate-induced intracellular calcium release.[60] Baicalein has been shown to reduce presynaptic glutamate release and increase postsynaptic glutamate receptor-1 (GluR1) expression, resulting in increased striatal glutamate concentrations.[61]

Baicalein acts on multiple signalling pathways and molecules. The proteasome inhibitors MG132 and MG115 caused DNA damage and cell loss accompanied by several markers of apoptosis in PC12 cells.[62] Attenuation of these changes by baicalein suggests that it can act on the ubiquitin–proteasome system. Low dose baicalein (1 mg kg^{-1} body weight for 7 days) also prevented dopaminergic neuronal loss and activation of glial cells by suppressing activities of JNK and ERK.[63] In addition, baicalein has been shown to regulate the Keap1-Nrf2 pathway, the major cytoprotective mechanism against oxidative stress. Under basal conditions, Keap1 is bound to Nrf2, and Nrf2 is polyubiquitinated.[64] In the case of endogenous or exogenous insults, Keap1 is modified at cysteine residues, resulting in the release of Nrf2. Free Nrf2 translocates into the nucleus where it binds to antioxidant response elements (ARE) on DNA, enhancing the expression of target genes such as NAD(P)H quinone oxidoreductase 1 and heme oxygenase 1 (HMOX1). Increased Nrf2/HO-1 protein expression, induced nuclear translocation of Nrf2, and downregulation of Keap1 were neuroprotective mechanisms of baicalein.[65] Baicalein also prevented detrimental consequences derived from endoplasmic reticulum stress inducers, thapsigargin and brefeldin A.[66]

Luteolin, another flavone, protected against damaging effects of PD neurotoxicants *in vivo* and *in vitro*. The LPS-induced decrease in TH-positive neurons and increases in active microglia, TNFα, nitric oxide, and superoxide were inhibited by luteolin pretreatment in primary mesencephalic neuron-glia cultures.[67] Pretreatment also rescued H_2O_2-induced toxicity in primary cortical neurons by a mechanism involving anti-inflammatory activity and decreased ROS production.[68] Luteolin also has anti-apoptotic activity, as observed by the suppression of Bax and p53 and enhancement of Bcl2 expression.[69] It also acts on stress response pathways: PC12 cells exposed to 20 μM luteolin prior to 6-OHDA treatment exhibited less pronounced activation of the unfolded protein response and Nrf2-ARE pathways compared to cells treated with the toxin alone.[70] These findings were replicated in two cell lines, PC12 cells and C6 rat glioma cells, treated with MPP^+, a toxic metabolite of MPTP.[71]

Apigenin, an inhibitor of the MAPK/ERK signalling pathway,[72] exhibited protection against activation of caspases and endoplasmic reticulum stress-associated proteins (such as CHOP), ROS accumulation, and loss of mitochondrial membrane potential in HT22 murine hippocampal neuronal cells.[66] Apigenin was also neuroprotective against induced motor deficits and pathological characteristics in the MPTP mouse model.[73] Although acacetin has not been extensively studied, significant evidence exists regarding its neuroprotective capabilities in PD. Acacetin diminished the release of proinflammatory cytokines and expression of iNOS and COX2 in LPS-induced BV2 microglial cells,[74] and also attenuated degeneration of dopaminergic neurons, depletion of dopamine in *substantia nigra* and striatum, and accumulation of active microglia in the MPTP mouse model.[75]

14.3.1.3 Flavanones

Citrus fruits, including grapefruit, orange, and lemon, are rich in flavanones. Flavanones have a carbonyl group at C_4 with a missing double bond between C_2 and C_3, and a B ring that is connected to C_2 (Figure 14.2). The two major flavanones are naringenin and hesperetin. They are abundantly found in glycosylated forms in nature, referred to as naringin and hesperidin, respectively.

Naringenin's antioxidant capacity has been shown to lower the risk of chronic neurodegenerative diseases.[76] Naringin and naringenin regulate expression and activity of the antioxidant enzymes SOD and catalase in high cholesterol-fed animals.[77,78] Naringenin's neuroprotective capacity has been validated by several studies. In primary mesencephalic cultures from rat embryos, naringenin prevented MPP^+-induced neuronal loss.[79] Rotenone-treated SH-SY5Y cells showed increased phosphorylation of JNK and p38 and alterations in Bcl2 and Bax proteins; naringin treatment protected against these cytotoxic effects, leading to a decrease in cell death.[80] Naringenin also restored Nrf2 protein levels and activation of ARE pathway genes in 6-OHDA-treated SH-SY5Y cells, suggesting a possible role in stress-response pathways.[81] Naringenin also prevented nitrosative stress in LPS/IFN-induced primary glial cells.[82] Additionally, naringenin attenuated iNOS expression, nitric oxide accumulation, and TNFα production and decreased p38 MAPK phosphorylation.

Administration of naringenin and naringin also displayed protection against known PD neurotoxicants *in vivo*. Pretreatment with naringenin for 4 days prior to unilateral stereotaxic injection of 6-OHDA into rats reduced loss of nigral TH positive cells and the depletion of striatal dopamine levels.[83] Interestingly, co-administration of naringenin with iron dextran injections for 4 weeks in rats improved antioxidant enzyme activities and attenuated oxidative damage in the cerebral cortex.[84] A possible role of naringenin in iron chelation was previously suggested as it suppressed the Fenton reaction characteristic of iron-ATP.[85] It is noteworthy to mention that pre- or co-treatment of naringenin with toxicants is not only protective, but also able to reverse ongoing insults. For example, 10 days of naringenin treatment reversed motor deficits and the down-regulation of parkin, DJ-1, and TH in *substantia nigra* and striatum, while reducing caspase and ubiquitin levels induced by 11 days of rotenone exposure in rat models.[86]

Information in the literature on hesperetin and hesperidin is limited compared to naringenin. Hesperidin interfered with the loss of mitochondrial membrane potential, ROS generation, altered Bax and Bcl2 expression, cytochrome c release, and caspase activation induced by rotenone in human SK-N-SH neuroblastoma cells.[87] Furthermore, 28 days of hesperidin post-treatment prevented the decrease in antioxidant enzymes activities (glutathione peroxidase and catalase), increase in ROS levels, and decrease in striatal dopamine and metabolites observed in aged mice that received intracerebroventricular 6-OHDA injection.[88]

Although experimental models of naringenin and hesperetin support their use as a neuroprotectant in PD, the bioavailability of these compounds in the central nervous system is an issue. Youdim *et al.* used *in vitro* models of the blood–brain barrier to test their penetration.[89] Both compounds have a high uptake rate in cells. *In vivo* data on penetration of naringin through the blood–brain barrier is more limited.[90] However, the availability and solubility of naringenin and naringin can be increased by co-administration with P-glycoprotein inhibitors, including a cyclodextrin derivative and cyclosporin A.[90,91]

14.3.1.4 Flavanols

Flavanols can exist as monomers or polymers. Monomeric flavanols have a C_3 hydroxyl group, a structural feature that results in the name: Flavan-3-ols (Figure 14.2). The most abundant types of flavan-3-ols are catechin, epicatechin, gallocatechin, epigallocatechin, and their gallic acid derivatives: catechin gallate (CG), epicatechin gallate (ECG), gallocatechin gallate (GCG), and epigallocatechin gallate (EGCG), respectively. Flavan-3-ols can also form polymers up to 60 units, which are called proanthocyanidins. Cacao and green and black teas contain very high levels of flavanols. Other sources for flavanols are fruits such as apples, grapes, and apricots.[92]

The neuroprotective effects of flavanols are concentration and structure dependent. First, not all flavanols protect similarly against PD-relevant neurotoxicants. For example, in 6-OHDA-treated PC12 cells, while pretreatment with EGCG or ECG was protective at 50–400 μM, epicatechin and (+)-catechin had no effect.[93] In contrast, (+)-catechin and (−)-catechin have been shown to have similar potency in reducing LDH release (indicator of cell death) and ROS production in SH-SY5Y cells treated with MPP^+.[94] Secondly, flavanols at high doses appear to be neurotoxic. At 0.1–10 μM, EGCG was able to block apoptosis and reverse changes in pro-apoptotic and anti-apoptotic gene expression.[95] Microarray data analysis showed increased expression of pro-apoptotic genes and decreased anti-apoptotic genes at a concentration of 50 μM.[96] At the same concentration, EGCG potentiated rotenone-induced production of intracellular superoxide in SH-SY5Y cells.[97] EGCG altered expression of various genes acting on several pathways such as DNA binding, cell adhesion, ion channels, apoptosis, and survival.[98] EGCG in low concentration attenuated cell death in 6-OHDA-treated SH-SY5Y cells mediated through reduced STAT3 levels.[99] Further evidence on modulation of STAT transcription factors by flavanols comes from a study testing the effect of catechin on oxidative stress.[100] Under normal conditions, STAT5 induces expression of miR-182, which inhibits FOXO1, limiting enhancement of pro-apoptotic gene expression. Under oxidative stress, STAT5 is inactivated and disruption of this pathway results in promotion of apoptosis. This study found that catechin treatment restores STAT5 activation, diminishing apoptosis.

The neuroprotective properties of flavanols, especially EGCG, have been extensively studied using *in vitro* PD models in the last decade. Dopaminergic neuron loss has been partially ameliorated by catechin in primary mesencephalic cultures treated with H_2O_2, 4-HNE, rotenone, MPP^+, or 6-OHDA.[79] EGCG also attenuated mitochondrial membrane depolarization and caspase-3 activation in paraquat-treated PC12 cells.[101] Dichlorodiphenyltrichloroethane-induced death of SH-SY5Y cells was prevented by pretreatment with 1, 3, and 10 μM EGCG.[102] Finally, LPS-induced microglial activation and subsequent inflammation-mediated dopaminergic neuronal loss were also inhibited by EGCG *in vitro.*[103]

Flavanols also prevented PD-relevant neuropathology and behavioral characteristics *in vivo*. MPTP-induced decreases in striatal TH and dopamine levels and increases in the levels of antioxidants SOD and catalase were prevented with EGCG pretreatment.[104] Since EGCG did not inhibit monoamine oxidase B (MAO-B) enzyme activity, its neuroprotective effects are not likely due to alterations in MPTP metabolism. EGCG treatment attenuated nitrosative stress, iron accumulation, and α-synuclein aggregation in MPTP-induced PD mouse models.[105,106] The concentration-dependent protection of EGCG was also confirmed *in vivo*. Microglial activation and TH-IR cell loss induced by MPTP in mouse *substantia nigra* were diminished to a greater extent with 5 mg kg^{-1} EGCG than with 1 or 10 mg kg^{-1} doses.[107] More evidence suggested that flavanols, catechin, and EGCG are neuroprotective in rats treated with 6-OHDA or LPS.[108,109] *Drosophila* models have also been used; ECG supplementation at 0.25, 0.5, and 1 μg ml^{-1} delayed lipid peroxidation and apoptosis, and improved climbing activity, in flies expressing human α-synuclein.[110] Flies expressing mutant LRKK2 or lacking parkin expression exhibited dopaminergic and mitochondrial dysfunction, which was suppressed by EGCG in an AMP kinase-dependent manner.[111]

In addition to the free-radical scavenging and transition-metal chelating activities of flavanols,[106,112] these compounds have been shown to exhibit anti-fibril activity. EGCG binds to natively unfolded α-synuclein and prevents its conversion to fibrils.[113] Moreover, EGCG was found to be able to convert large and mature α-synuclein fibrils into smaller and amorphous protein aggregates, which are non-toxic.[114,115] EGCG's effect on α-synuclein fibrils is not limited to disaggregation. EGCG has also been shown to immobilize the C-terminal region of preformed oligomers and reduce the degree of their binding to the plasma membrane.[100] This action prevents permeabilization of vesicle membranes and subsequent cytotoxicity.

Proanthocyanidins, polymers of flavan-3-ols, are neuroprotective in neurodegeneration models, though fewer studies exist for proanthocyanidins than for other flavan-3-ol compounds. Oligonol, which is found primarily in lychees, grapes, and apples, improved locomotive activity in a SAMP8 mouse model exhibiting accelerated aging with overproduction of amyloid precursor protein and oxidative damage.[116] Another proanthocyanidin is theaflavin, abundantly found in tea. Theaflavin pretreatment in an MPTP mouse model attenuated oxidative stress and neuroinflammation, and mitigated nigral TH and dopamine active transporter loss.[117,118] It also improved motor behavior seen in MPTP-treated mice.[119] Like EGCG, theaflavin has been shown to enhance assembly of α-synuclein protein into non-toxic aggregates.[120] Theaflavin is considered to be a more stable molecule with less susceptibility to air oxidation over EGCG.[120] Extracts from grape seeds and plum, which are rich in proanthocyanidins, prevented rotenone-induced dopaminergic toxicity in primary midbrain cultures.[121]

14.3.1.5 Anthocyanins

Anthocyanins have a positively charged oxygen on ring C (Figure 14.2). This positive charge makes absorption of light possible, resulting in the bright colors that occur in many fruits and vegetables with high anthocyanin content. Anthocyanins also have a C_3 carbohydrate; this sugar is generally glucose, but it can also be galactose, rhamnose, or arabinose. The presence of the sugar increases the solubility of these compounds in water. After entering the body, anthocyanins are metabolized and the sugar moiety is removed through hydrolysis. The aglycone forms are called anthocyanidins, which are less stable than the sugar-containing forms. Glycosylation of anthocyanins at a different position (generally C_3, but can be at C_5 or C_7) creates diversity among the structural group, giving rise to more than 300 members. The most commonly found anthocyanidins are cyanidin, pelargonidin, peonidin, delphinidin, malvidin, and petunidin. They are abundantly found in berries, black currant, grapes, red wine, plums and red cabbage.[122,123]

Few studies have been performed on the effects of individual anthocyanins in PD models. One showed that pretreatment with pelargonidin at low doses (0.1–3 μM) inhibited oxidation of dopamine induced by peroxynitrite.[124] Pelargonidin-regulated inhibition prevents generation of DHBT-1 and 5-S-Cys-DA (autoxidation products of dopamine), which are highly neurotoxic. Two mechanisms of action are proposed for pelargonidin and peroxynitrite: nitration of the antioxidant or deactivation of peroxynitrite by electron donation. Pelargonidin also inhibited lipid peroxidation, formation of nitric oxide and free radicals, and neuronal loss in the *substantia nigra* of rats receiving unilateral intrastriatal 6-OHDA injections.[125] On the other hand, a recent study has shown that pretreatment with pomegranate juice (rich in pelargonidin) prior to rotenone administration enhanced oxidative stress and nigrostriatal degeneration in rats, suggesting that it can also act as a pro-oxidant.[126] In primary midbrain cultures, malvidin-3-*O*-glucoside, cyanidin-3-*O*-sophoroside, and delphinidin-3-*O*-glucoside prevented TH-positive neuronal loss induced by rotenone, while cyanidin-3-*O*-glucoside did not, suggesting that the identity of the C3 sugar attached might affect the level of neuroprotection.[121]

Most studies that examined neuroprotection utilized whole fruit extracts containing anthocyanins along with other polyphenols. Consumption of blueberries and strawberries, rich in anthocyanins, reversed age-related behavioral deficits in rats.[127] Another study used three *Vaccinium* species (bilberry, blueberry, and cranberry), which are particularly rich in cyanidin-3-glucoside, to test for the ability to interfere with 6-OHDA oxidation in a cell-free system.[128] Extracts from these fruits inhibited cytochrome c-enhanced oxidation of 6-OHDA, likely due to direct scavenging activity. Pretreatment with bilberry extracts prevented decreases in dopamine and metabolite levels in midbrains of mice induced with psychological and emotional stress (by cutting whiskers).[129] Mulberry extracts prevented 6-OHDA-induced ROS and nitric oxide generation, alteration in Bcl2 and Bax levels, mitochondrial membrane depolarization, and caspase activation in SH-SY5Y cells and 6-OHDA- and MPP^+-induced dopaminergic neuronal loss in primary midbrain cultures.[130] The same study examined the effects of mulberry extract in MPTP mouse models and observed mitigation of both dopaminergic neuronal loss in the midbrain and behavioral deficits such as bradykinesia. However, since mulberry extract contains other polyphenols (rutin and quercetin), it is difficult to interpret whether neuroprotection is due to anthocyanins alone or a mixture of polyphenols. Similar controversy arises from an *in vivo* study performed in a PD *Drosophila* model with commercially available whole grape extract.[131] Although the extract significantly protected mitochondria against oxidative stress and improved locomotor activity in flies, other compounds in the extract such as resveratrol might be responsible for this neuroprotection. One possible improvement that can be made in this type of study design is to evaluate which type(s) of polyphenol(s) play(s) a more central role in neuroprotection. To this end, Strathearn *et al.* compared neuroprotective effects of several fruit and plant extracts with different levels of anthocyanins in rotenone-treated primary midbrain cultures and showed that extracts rich in anthocyanins have greater neuroprotective abilities compared to extracts enriched with other polyphenols, suggesting that the prevention of neurodegeneration in these models might be due to anthocyanin.[121] For more information about single berry fruits on neurodegeneration, refer to Subash *et al.*[132]

Studies suggest that anthocyanins can cross the blood–brain barrier and accumulate in the brain. The RBE4 rat cerebral capillary endothelial cell line is a commonly used *in vitro* blood–brain barrier model. Cyanidin-3-glucoside has been shown to pass across in a time-dependent manner, probably *via* the action of glutamate transporter 1, which has the advantage of carrying a glucose molecule.[133] Anthocyanin levels in kidney, liver, and brain have been measured in rats fed blackberries for 15 days.[134] Low levels of anthocyanins were detected in the brain, further confirming the bioavailability of these antioxidant compounds in the central nervous system.[135–137]

14.3.1.6 Flavonols

The major sources of flavonols in the human diet are capers, onions, leeks, broccoli, curly kale, and spinach. These compounds are also found in some fruits and in tea at lower concentrations.[138] Flavonols have a double bond between C_2 and C_3 and a ketone group at position 4 of ring C. At C_3, they are also hydroxylated and glycosylated (Figure 14.2). The sugar molecule is generally glucose or rhamnose, but can also be glucuronic acid or galactose. Like anthocyanins, many flavonols are naturally found in their glycosylated form. The most abundant flavonols are quercetin, kaempferol, and myricetin.

Several lines of *in vitro* evidence confirmed neuroprotection by quercetin in PD-relevant neurotoxicant models. Quercetin prevents MPP^+ toxicity in PC12 cells.[139] Similar results have been obtained in PC12 cells pretreated for 3 h prior to MPP^+ treatment.[140] 6-OHDA-induced toxicity to PC12 cells and dopaminergic neuron loss in zebrafish, accompanied by overproduction of nitric oxide and overexpression of iNOS in PC12 cells and pro-inflammatory factors in zebrafish, was diminished by pretreatment with quercetin.[141] Similarly, quercetin prevented SH-SY5Y cell loss elicited by 6-OHDA.[142] Quercetin is also effective in preventing glial activation and subsequent toxicity to neuronal cells. LPS induces IL1α and TNFα production in glial cells. When activated microglia were co-cultured with neurons, neuronal loss was observed due to the presence of pro-inflammatory cytokines.[143] Quercetin inhibited both pro-inflammatory factors and neuronal cell death in this study. This compound also protected against MPP^+-induced increases in mRNA and protein levels of pro-inflammatory cytokines and iNOS in N9 murine microglial cells, and the death of PC12 cells cultured with conditioned media from N9 cells activated in the presence of MPP^+.[144] In primary mesencephalic cultures from rat, quercetin interfered with MPP^+-induced cell death.[79]

Rodent PD models have been utilized to test the neuroprotective effect of quercetin *in vivo*. Quercetin administration for 14 days after intracisternal 6-OHDA injection in rats prevented neuronal loss, increased striatal dopamine and antioxidant levels, and decreased carbonyl content in the striatum.[145] An even shorter post-treatment (4 days) increased mitochondrial complex I activity and rescued behavioral deficits, such as apomorphine-induced rotations.[146] In the MPTP mouse model, quercetin also showed neuroprotective action, demonstrated by increased antioxidant enzyme activity, increased dopamine levels, and reduced lipid peroxidation in striatum.[147]

Although many studies have shown the neuroprotective action of quercetin, others have suggested the opposite effect. At 50 μM, quercetin has been shown to inhibit synthesis of heat shock proteins and to prevent their protective mechanism against MPP^+- and paraquat-induced toxicity in primary neuronal cultures.[148] Another study revealed that at low concentrations (0.1 and 0.3 μM) quercetin diminished 5-S-Cys-DA induced toxicity, but at 3 μM it exacerbated toxicity.[124] Ossola *et al.* investigated the balance between neuroprotection and neurotoxicity in a time- and dose-dependent manner.[149] While earlier exposure of quercetin ameliorated 6-OHDA toxicity in SH-SY5Y cells, prolonged treatment increased it. One explanation for this effect is that after scavenging free radicals, quercetin is converted to an ortho-quinone that is highly toxic to neuronal cultures and can enhance neurotoxicity as shown at high doses.[150]

Besides antioxidant and transition metal-chelating activity, quercetin has been shown to bind estrogen receptor alpha and beta with low affinity, suggesting that it has estrogenic activity.[151] Quercetin also has an antagonistic effect on α-synuclein fibrillization.[152] Oxidized quercetin caused a stronger inhibition of fibrillization and disaggregation of fibrils compared to quercetin, probably because of its greater polarity and hydrophilicity.[153]

Quercetin's derivatives also displayed protection against toxicity in PD models. Quercetin-3-*O*-galactoside (hyperoside) was protective against H_2O_2 and *tert*-butyl hydroperoxide toxicity in PC12 cells.[154] Another glycosylated derivative, rutin (quercetin-3-*O*-rutinoside), decreased ROS levels and inhibited molecular characteristics of apoptosis (depleted mitochondrial membrane potential, altered Bax/Bcl-2 ratio, *etc.*) in rotenone-treated SH-SY5Y cells.[155] Rutin also reduced expression of Park2, Park5, and Park7 and activation of caspase-3 and caspase-7, resulting in increased expression of the TH gene in PC12 cells treated with 6-OHDA.[156] In the same study, isoquercetin showed similar protection as rutin. The same group confirmed neuroprotection of isoquercetin by observing reversal of changes in antioxidant enzyme activities and lipid peroxidation induced by 6-OHDA.[157]

Quercetin has been proposed to increase the effectiveness of L-dopa treatment in PD patients. It inhibited COMT and MAO-B activities, which are key enzymes in dopamine metabolism.[158] Quercetin was also shown to pass through cellular barriers in cultured RBE4 cells, an *in vitro* model of the blood–brain barrier.[133] However, all *in vitro* immortalized barrier lines are thought to be considerably 'leakier' than the blood–brain barrier *in vivo*.[159] Thus, the results of these models need to be confirmed *in vivo*.[160] Nonetheless, quercetin and similar molecules carry a high potential to be used in clinical therapeutics in the treatment or early prevention of PD.

As mentioned earlier, other flavonols abundant in plant-derived foods are kaempferol and myricetin. Pretreatment with kaempferol improved motor function and diminished molecular toxicity in the MPTP mouse model of PD.[161] Kaempferol-protected PC12 and primary neuronal cells against H_2O_2 toxicity in a DJ-1-dependent manner.[162] Protection by kaempferol has also been shown to involve autophagy since increased LC3-II protein was observed upon kaempferol treatment of SH-SY5Y cells exposed to rotenone.[163] Moreover, myricetin-attenuated production of ROS, restored mitochondrial membrane potential, and inhibited cell loss in MPP^+-treated MES23.5 dopaminergic cells.[164] Like quercetin, myricetin and its two derivatives (myricetin-3-*O*-rhamnoside and dihydromyricetin) have been shown to act as COMT inhibitors and work to increase the bioavailability of L-dopa.[165] Therefore, myricetin could potentially be used as an effective adjunct to L-dopa therapy, but more research is needed to test toxicity, dose, and metabolism.

14.3.2 Non-Flavonoids

Although non-flavanoids are a very diverse group encompassing a range of constituents and their derivatives, for the sake of simplicity and focusing on the relation with PD, we divide this group into four main structural categories: curcuminoids, phenolic acids, stilbenes and lignans.

14.3.2.1 Curcuminoids

Curcuminoids are linear diarylheptanoids, carrying two aryl groups at two ends of the molecule connected with a seven-carbon chain. The presence of different radical groups provides variety among members of the curcuminoid family. The most abundant form of curcuminoids is curcumin. Curcumin is the main constituent in the spice turmeric (*Curcuma longa*), which is a member of the ginger family (Zingiberaceae) and is widely used in India and China. It is responsible for the yellow color of the spice. Antioxidant, metal chelating, anti-inflammatory, and many other properties of curcumin have been widely studied in the last few decades. In this section, we will summarize the findings on the neuroprotective effects of curcumin and its derivatives in PD studies.

Curcumin has been shown to prevent cytotoxicity elicited by several PD-relevant neurotoxicants *in vitro*. Pretreatment with curcumin attenuated the activity of apoptotic and oxidative stress biomarkers in MPP^+-treated PC12 cells.[166] 6-OHDA in MES23.5 cells disrupted mitochondrial membrane polarization, increased ROS levels, and decreased antioxidant activities, but these negative effects were prevented with curcumin treatment.[167] NF-κB translocation was blocked by curcumin, which also decreased expression of NF-κB, resulting in the mitigation of LPS-induced dopaminergic toxicity in rat midbrain neuroglial cultures.[168] Conditioned medium from LPS-activated microglia caused axonal degeneration when locally applied to axons *via* the MyD88/p38 MAPK and JNK phosphorylation pathways.[169] Curcumin also inhibited degeneration, suggesting that it can serve as a modulator of the p38 MAPK and JNK pathways. Further support for the involvement of curcumin in the JNK pathway came from a study that showed its neuroprotective effects in MPP^+-treated SH-SY5Y cells.[170] Another mechanism by which curcumin may be neuroprotective is through the modulation of autophagy, as it was found to interfere with mTOR signalling, recovering macroautophagy in A53T α-synuclein-overexpressing SH-SY5Y cells.[171]

Several lines of *in vivo* evidence show that curcumin is protective in PD models. Curcumin pretreatment attenuated a loss of dopaminergic neurons in the *substantia nigra* and dopamine depletion in the striatum in rodents that received a unilateral 6-OHDA infusion into the medial forebrain bundle.[83] In the same model, hippocampal injury was prevented *via* a mechanism involving the BDNF/TrkB-dependent pathway.[172] Curcumin also prevented neuroinflammation and dopaminergic toxicity when administered at a high dose (200 mg kg^{-1}) after 6-OHDA infusion.[173] High doses of intraperitoneal curcumin (80 mg kg^{-1}) prevented dopamine depletion and inhibited MAO-B enzyme activity in the MPTP mouse model.[174] Curcumin also inhibited the increase in LRRK2 kinase activity associated with H_2O_2-induced toxicity in a fly model.[175] Pre- and post-treatment with curcumin in an MPTP mouse model diminished neuroinflammation and nitrosative stress along with improving motor behavior.[176] Homocysteine-induced deficits in locomotor activity were restored with prolonged curcumin administration in rats.[177]

In addition to ROS scavenging and stimulation of anti-apoptotic pathways, curcumin has been shown to disrupt α-synuclein fibrillization in PD models. Here, the treatment inhibited aggregation of α-synuclein and increased its solubility in a dose-dependent manner.[178] At least two studies showed that curcumin and its derivative, curcumin-glucoside, bind to oligomeric intermediates, but not monomers, of α-synuclein.[179,180] By binding to oligomers, these compounds alter the hydrophobic surface exposure of aggregates, an action that likely disrupts the structure of fibrils. In SH-SY5Y cells incubated with toxic oligomeric α-synuclein, curcumin treatment destabilized preformed fibrils.[181] Additional findings on anti-fibrillization mechanisms elicited by curcumin are reviewed by Ono *et al.*[182]

Curcumin also has transition metal-chelating properties. Iron and copper accumulate in PD brains, and this accumulation may increase ROS levels and inhibit DNA repair enzymes. Curcumin prevented metal accumulation due to its chelating activity.[183] *In vivo* evidence has shown that curcumin reduced the number of iron-stained cells in 6-OHDA-infused rats, suggesting that it can serve as an iron chelator in neuroprotection against PD models.[184] The role of curcumin in protecting against metal-induced toxicity is reviewed by Jomova *et al.*[185]

The clinical relevance of curcumin as a PD therapeutic is restricted due to its low solubility and limited bioavailability related to the fact that it undergoes rapid degradation at physiological pH. Curcumin derivatives and analogs have been found to have greater potency to attenuate PD phenotypes due to their improved bioavailability. For example, glutamic acid derivatives of curcumin showed improved protection in preventing GSH depletion induced by buthionine sulfoximine (BSO) in N27 cells.[186] The di-glutamoyl ester derivative, which has greater bioavailability compared to curcumin, protected against peroxynitrite-induced mitochondrial dysfunction elicited by MPP^+.[187] Furthermore, the biphenyl zingerone analog was found to have the highest affinity to α-synuclein among curcumin and other analogs.[188] A pyrazole derivative, CNB-001, also exhibited protection against mitochondrial dysfunction and apoptotic cascades in rotenone-treated SH-SY5Y cells.[189] Another strategy to increase the efficacy of curcumin as a therapeutic is through co-administration with other compounds. For example, beta-cyclodextrin, which is an oligosaccharide used to encapsulate molecules with low water solubility, administered with curcumin improved inhibition of α-synuclein assembly and the disaggregation of α-synuclein fibrils.[190] With major advances in nanobiotechnology in recent years, new improvements have been made to increase the bioavailability of curcumin. The use of a nanoparticulate carrier for oral administration of curcumin increased bioavailability and concentration at the target site.[191] Finally, intraventricular injection of liposomal-formulated curcumin improved motor activity and inhibited apoptosis in DJ-1 knock out rats, suggesting future promising advancements in nanotech-based drug discovery and delivery.[192]

14.3.2.2 Phenolic Acids

Phenolic acids are divided into two subcategories based on the length of the carbon chain attached to their phenol ring. Benzoic acid is a simple aromatic carboxylic acid, while cinnamic acid has a longer carbon chain attached to the aromatic ring. Gallic acid is one of the most common derivatives of benzoic acid. Caffeic acid and chlorogenic acid are the most abundant cinnamic acids in cinnamon and coffee.[193,194]

Gallic acid is a trihydroxybenzoic acid and active constituent of grape seed extract.[195] It is also found in gallnuts, sumac, and other plants. High doses (200 mg kg^{-1}) have been shown to improve motor dysfunction in a unilateral 6-OHDA rat model.[196] In the same model, pretreatment decreased lipid peroxidation and increased thiol and glutathione peroxidase content not only in the striatum but also in the hippocampus.[197] Gallic acid also prevented paraquat-induced dopaminergic neurotoxicity, prolonged life span, and improved climbing abilities in *Drosophila*.[198] Gallic acid has been shown to bind and stabilize soluble non-toxic oligomers of α-synuclein transiently, preventing the formation of toxic fibrils.[199] It can also disaggregate preformed fibrils.[200] Another derivative of benzoic acid is sodium benzoate, which is a metabolite in cinnamon and widely used as a non-toxic food preservative.[201] Data on its neuroprotective efficacy are limited. Sodium benzoate was shown to increase DJ-1 levels and diminish LPS-induced inhibition of DJ-1 expression in primary astrocytes and neuronal cultures.[202] Sodium benzoate also enhanced BDNF levels, which might be the mechanism of protection in neurodegeneration.[203]

Among cinnamic acids, caffeic acid is one of the most mass abundant in foods. It is the main phenolic acid in coffee.[193] It can be converted to ferulic acid by COMT activity. Formation of an ester bond between caffeic acid (or ferulic acid) and quinic acid produces chlorogenic acid, which is also found in coffee extract.[204] Caffeic acid is well investigated in *in vitro* and *in vivo* PD models. In the MPTP mouse model, it has been shown that the dose of caffeic acid required to achieve neuroprotection upon administration of the compound after toxic insult is greater than the dose that is sufficient to see the same neuroprotective effect with pretreatment.[205] Caffeic acid (and two other cinnamic acids, ferulic and coumaric acids) prevented impaired motor activity in paraquat- and iron-treated *Drosophila*.[206] When compared to other cinnamic acids, it has been shown to have the greatest potency in rat hepatocytes against glyoxal- and methylglyoxal-induced oxidative stress.[207] Caffeic acid at 10–60 μM inhibited fibrillization of α-synuclein.[208] It also diminished endoplasmic reticulum stress in human primary glioblastoma cell lines[209] and inhibited 5-S-Cys-dopamine-induced toxicity.[124,210]

Caffeic acid phenethyl ester (CAPE), a derivative of caffeic acid, is an ester of caffeic acid and phenethyl alcohol. It is naturally found in honeybees' propolis. It has been shown to block 6-OHDA-induced neurotoxicity in cerebellar granule neurons and rat mesencephalic neurons.[211,212] CAPE also attenuated dopaminergic neurodegeneration, dopamine loss, and an increase in iNOS levels in the MPTP mouse model.[213] Inhibition of the mitochondrial permeability transition, cytochrome c release, and caspase activity further confirmed neuroprotection in CAPE-pretreated 6-OHDA-infused rats.[214]

Chlorogenic acid, the ester of caffeic acid and quinic acid, has been found to exhibit antioxidant and anti-apoptotic activity in PD models. It has been reported to inhibit α-synuclein overexpression-induced toxicity in PC12 cells.[215] Furthermore, the compound has free-radical scavenging activity, as fewer single-stranded DNA breaks resulted from free-radical exposure *in vitro*.[216] For more details on the cognitive and neuroprotective effects of chlorogenic acid, refer to an excellent review written by Heitman *et al.*[217]

Unfortunately, the literature lacks detailed data on the blood–brain barrier permeability and metabolism of cinnamic acids. It is worth noting that coffee is associated with decreased risk of PD in several epidemiological studies, and this inverse relation is proposed to be due to caffeine content.[218] However, further research is needed to evaluate caffeic acid and other cinnamic acids found in coffee to evaluate possible clinical relevance with PD.

14.3.2.3 Stilbenes

In European countries, such as France and Italy, people consume large amounts of food with high saturated fatty acid contents, which should increase their risks of cardiovascular and neurodegenerative diseases. However, in these same countries, people were not found to be at higher risk of these diseases. This notion is referred to as the 'French Paradox'.[219] The French diet can also be categorized as a variety of the Mediterranean diet, characteristically rich in vegetables, fruits, and red wine.[220] One explanation for the 'French Paradox' is found in the higher consumption of red wine, which is rich in polyphenols, especially resveratrol, the most mass-abundant stilbene.

Resveratrol, found in grapes and red wine, has been found to have extensive health benefits.[221] Its antioxidant activity has been shown to occur through free-radical scavenging, modulation of neuronal homeostasis, and glutamate receptor and ion-channel modulation.[222] Resveratrol provides protection at various doses ranging from 10 nM to 100 μM in various *in vitro* PD neurotoxicant models such as cells exposed to MPP^+, dopamine, or rotenone, and in α-synuclein overexpression models.[121,139,223–225] *In vivo*, resveratrol administration prevented MPTP- and unilateral 6-OHDA-induced neurotoxicity by increasing antioxidant enzyme activities and decreasing ROS, lipid peroxidation, and inflammatory reactions.[226–229] Modulation of these biomarkers and molecular processes resulted in an enriched dopaminergic neuron population and enhanced motor activity. Resveratrol-induced anti-apoptotic activity is mediated by several signalling proteins and pathways, including PGC-1α, the FOXO protein family, Akt, SIRT1, SOCS1, and HO-1.[230–233] For more detail, refer to Ebrahimi and Schluesener.[29]

Resveratrol also exhibited anti-inflammatory activity in glial cells. It attenuated activation of microglia induced by LPS, resulting in rescue of PC12 neuronal damage.[143] In addition, resveratrol interfered with NADPH oxidase-mediated generation of ROS in primary neuron-glial cultures, resulting in attenuation of LPS-induced dopaminergic neuronal loss.[234] LPS-mediated generation of inflammatory molecules from primary mouse astrocytes was inhibited by resveratrol.[235] These findings suggest that neuroprotection by resveratrol is mediated *via* inhibition of glial cell activation.

Resveratrol derivatives have also been shown to be protective in PD models. Oxyresveratrol, found in mulberry, has been reported to protect against 6-OHDA-induced toxicity in SH-SY5Y cells.[236] Pinostilbene, a methylated derivative, was protective in the same conditions, with a higher level of neuroprotection than resveratrol in the native form.[237] Tetrahydroxystilbene, a major constituent in the traditional Chinese herb *Polygonum multiflorum* Thunb, attenuated MPP^+-induced apoptosis and oxidative stress in PC12 cells.[238] Tetrahydroxystilbene glucoside diminished oxidative and nitrosative stress *via* activation of the PI3K/Akt signalling pathway *in vitro.*[239–241] Resveratrol packaged in liposomes showed greater potency of protection against neurotoxicity induced by 6-OHDA, probably due to increased bioavailability *in vivo.*[242]

Despite the high content of antioxidants in red wine, epidemiological studies failed to show an inverse association between consumption and PD risk.[2] The other components of alcoholic beverages make assessing neuroprotection of any single compound difficult. For example, ethanol has been shown to reduce inflammation and pro-inflammatory molecules *in vivo*, but it enhances free-radical production,[219,243] suggesting that factors such as the duration and amount of alcohol consumption might be critical in assessing neuroprotection.

14.3.2.4 Lignans

Lignans are relatively large polyphenolic molecules. Sesame and flax seeds, cereals, and soybeans are rich in lignans, such as sesamin, schisandrin, and nordihydroguaiaretic acid.

Sesamin is one of the most common lignans, found primarily in sesame seed. It has been reported to reduce MPP^+ toxicity *in vitro*, preventing increases in ROS, iNOS, and pro-inflammatory proteins.[244,245] Similar protection in MPP^+ cellular models has been observed in N9 microglial cells and PC12 neuronal cells cultured in conditioned medium from activated microglia.[144] Sesamin pretreatment exhibited anti-inflammatory features in 6-OHDA-infused mice[246] and restored TH positive neuronal loss in rotenone PD rat models.[247]

Many members of the lignan family are not found in a typical human diet. However, their anti-apoptotic and anti-inflammatory roles in PD models are worth noting due to the potential therapeutic value. Nordihydroguaiaretic acid (NDGA) is the main metabolite of the creosote bush, *Larrea tridentata*. It has been known as a lipoxygenase inhibitor and has been shown to protect against nitric oxide- and BSO-induced glutathione depletion in mesencephalic cultures.[248,249] In a human α-synuclein expressing fly model, dietary supplementation of NDGA delayed loss of climbing ability.[250] One suggested mechanism for NDGA protection was inhibition of membrane perturbation induced by wild type and mutant α-synuclein.[54]

Schisandrin B, found in the traditional Chinese medicinal herb *Schisandra chinensis* (Turcz.) Baill, suppressed glutathione depletion and enhanced glutathione reductase activity in paraquat-induced PC12 cells.[251] It also displayed anti-inflammatory activity in LPS-induced microglia.[252]

Magnolol, found in the bark of the Houpu magnolia (*Magnolia officinalis*) or in *M. grandiflora*, reversed locomotor deficits and the loss of striatal TH in the 6-OHDA PD model.[253,254] One of the four lignans isolated from *Magnolia fargesii* (Magnoliaceae), epimagnolin B, inhibited nitric oxide production and iNOS and COX expression in LPS-induced microglia.[255]

14.4 Vitamins

14.4.1 Vitamin A

Vitamin A and its derivatives are unsaturated nutritional organic compounds, found at high levels in many foods. Vitamin A can be divided into two groups according to the source: animal and plant derived. Animal-derived vitamin A derivatives are lipid soluble. The major derivative of this group is retinol. Retinol can be metabolically converted to retinaldehyde and retinoic acid. Major sources of retinol are cod liver, milk products, and egg.[256] Plant-derived vitamin A derivatives, also known as provitamin A, are water-soluble and found in fruits and vegetables, including carrots, sweet potato, spinach and kale. Carotenoids, alpha and beta-carotenes, are the most common types of the provitamin A group and are hydrolyzed to retinoids in the small intestine.

Retinoic acid treatment has been shown to increase locomotor activity and striatal dopamine levels, even when administered after 70 days of rotenone injection.[257] Early post-treatment of 9-cis-retinoic acid after unilateral 6-OHDA infusion resulted in higher striatal dopamine levels and dopaminergic neuron counts in the *substantia nigra* compared to rats without protective treatment.[258] Retinoic acid prevented proteasome inhibition-mediated cell death *via* the Akt signalling pathway.[259] The role of retinoic acid receptors in normal brain function is well known. Defects in these receptors are associated with neurodegeneration,[260] suggesting that the protective role of retinoic acid is likely mediated by the activation of retinoic acid receptors. Mice lacking the dopamine receptor D2 lack retinoic acid receptors and exhibit locomotor dysfunction similar to that seen in PD.[261]

Retinoic acid is widely used in neuronal cell cultures to enhance neurite outgrowth and the expression of neuronal markers. Comparison of sensitivity to neurotoxicants in undifferentiated and cells differentiated with retinoic acid provided more evidence for its relation with PD. Retinoic acid-differentiated SH-SY5Y cells exhibited higher tolerance for 6-OHDA and MPP^+ toxicity.[262] Differentiated cells have been found to be more vulnerable to rotenone toxicity, while they were more resistant to proteasome inhibition-induced cell death, implying that retinoic acid signalling might carry importance in repairing damage caused by neurotoxicants.[263]

Lycopene is a provitamin A derivative whose high concentration is responsible for the bright red color of tomatoes. Unlike carotenoids, it does not have vitamin A activity, meaning that it is not converted to retinol in the small intestine. Lycopene exerts antioxidant activity and has shown neuroprotection in PD toxicant models. Lycopene treatment reduced MPP^+ cytotoxicity in SH-SY5Y cells and improved mitochondrial function.[264] In rotenone-treated rats, lycopene suppressed oxidative stress markers and also reversed cognitive decline.[265] One study used unilateral 6-OHDA-infused rats supplemented with a tomato-enriched diet for 14 days (7 days before and after infusion).[266] This diet also attenuated degeneration of nigral DA neurons and decreases in striatal dopamine and metabolite levels. Lyophilized tomato powder treatments also alleviated MPTP-induced striatal dopamine depletion in a mouse model.[267]

Astaxanthin (ASX) is a carotenoid, but like lycopene it is not converted into retinol. Its long conjugated carbon chain provides a striking red color. The double bonds in the carbon chain make ASX a strong antioxidant, though less potent than other carotenoids. The major sources of ASX in the diet are salmon and shrimp.[268] Due to its antioxidant activity, it has been reported to reduce neurodegeneration in PD models. In SH-SY5Y cells, it inhibited increases in apoptotic markers and enhanced cell viability, reducing 6-OHDA toxicity.[268] ASX pretreatment was protective in MPP^+-induced toxicity in cell culture and also in an MPTP mouse model, emphasized by a notable increase in dopaminergic neurons in the *substantia nigra*.[269] Although ASX seems to be a promising antioxidant in animal or *in vitro* models relevant to PD, the molecular mechanism underlying its neuroprotective effect is not known. Two studies showed that its mechanism of action in MPP^+-treated cells *in vitro* is regulated *via* the HO-1/NOX2 and Sp1/NR1 axes.[270,271]

It is also worth noting that carotenoids have been shown to have metal ion chelating ability, mitigating the damage caused by ROS production.[272] Moreover, the most common carotenoid, beta-carotene, inhibited the formation of α-synuclein fibrils and destabilized preformed fibrils *in vitro*, like many other dietary antioxidants explained so far.[273]

So far, in this section, we have summarized findings of neuroprotective activity of vitamin A derivatives in PD culture and animal models. However, interestingly, epidemiological studies have repeatedly failed to show an association of vitamin A intake with PD risk. Serum and plasma levels of vitamin A and derivatives were not significantly different between PD patients and controls.[274–276] Case-control or cohort studies that collected food questionnaires from PD patients and healthy controls showed no association between vitamin A intake and PD.[277–280] Meta-analyses that performed systematic reviews of published data suggested that there is no significant association of beta-carotene with PD.[281] Another meta-analysis tested the statistics between vitamin A and its derivatives and further validated a lack of association.[282]

14.4.2 Vitamin B

B vitamins are water-soluble vitamins that are not stored in the body. Since the excretion of B vitamins from the body is high compared to other vitamins, continuous intake is necessary. Major dietary sources of B vitamins are vegetables (*e.g.* potato), fruits (*e.g.* banana), and grains. This group exhibits a high degree of structural variability, containing more than 10 compounds, with eight being considered of major importance to human health. Commercially available vitamin B complexes contain these eight B vitamins. However, for simplicity and relevance to neurodegeneration, we will focus primarily on four B vitamins in this section: B2 (riboflavin), B6 (pyridoxine), B9 (folate), and B12 (cobalamin).

B vitamins regulate homocysteine metabolism. Homocysteine is produced by the demethylation of methionine by enzymatic reactions involving the conversion of S-adenosylmethionine to S-adenosylhomocysteine. Homocysteine is then used to synthesize cysteine to be used primarily in protein synthesis. The conversion of methionine to homocysteine and from homocysteine to cysteine is rapid, typically resulting in low homocysteine levels. These levels are maintained by two B-vitamin dependent mechanisms. The first is the re-methylation of homocysteine and its subsequent conversion to methionine, regulated by vitamin B9 and B12. Secondly, is the conversion of homocysteine to cysteine by transsulfuration, modulated by vitamin B6. Deficiency in these three vitamins results in elevated blood or organ homocysteine levels.[283] An excess of homocysteine produces cellular toxicity. It has been reported that homocysteine causes oxidative damage *via* autoxidation and increasing intracellular calcium levels.[284,285] When added to dopaminergic cell cultures, homocysteine inhibited mitochondrial complex I and increased oxidative stress.[286] Homocysteine treatment also exacerbated MPTP-induced dopaminergic neuronal loss in the *substantia nigra* of rats and worsened motor behaviors.[286]

Elevated homocysteine levels in plasma and serum have been detected in PD patients, an observation explained by several underlying mechanisms. High levels of homocysteine can be caused by L-dopa treatment[287] *via* the enzyme COMT. The supplementation of L-dopa treatment with folic acid or COMT inhibitors decreased drug-induced toxicity.[288] Polymorphisms in COMT and MTHFR have been shown to be associated with PD, possibly stemming from reduced enzyme activities.[289] For example, mutations in the MTHFR gene were associated with increased DNA damage in PD patients.[290] Vitamin B2 deficiency in rodents resulted in perturbed iron regulation, suggesting a possible role of riboflavin as an iron chelator.[291,292] Importantly, high homocysteine levels were associated with vitamin B-deficiency in PD patients.

Particularly low B12 levels were observed in PD patients, and were associated with the presence of neuropathy, damage to peripheral nerves.[293,294] Homocysteine increases are reversed by supplementation of vitamin B, individually or as a complex.[295] Folic acid, B12 and lipoic acid supplementation as a complex for 12 months decreased homocysteine in PD patients.[296]

Cellular and *in vivo* animal models of vitamin B-enriched treatment partially prevented PD-like neuropathology and behavioral deficits. Pretreatment with a mixture of B vitamins prevented oxidative stress, mitochondrial dysfunction, and α-synuclein accumulation in a cellular model of chronic rotenone exposure.[297] Intake of folate and vitamin B complexes in 6-ODHA rats improved motor features.[298] Moreover, high dose folate in the diet promoted scavenging of peroxynitrite-derived free radicals.[299] Lipoic acid, often obtained from supplements, can act together with B vitamins due to its strong antioxidant activity. It has been reported to protect against MPTP-, rotenone-, and 6-OHDA-induced cytotoxicity in dopaminergic neurons.[300–302] The proposed mechanism of protection is through the Keap1/Nrf2 signalling pathway, resulting in reduction of ROS generation and apoptosis and the reversal of mitochondrial dysfunction.[303]

Epidemiological findings on dietary vitamin B intake are inconsistent, but overall there is no convincing evidence of an association of dietary intake with PD risk. The Honolulu Heart Study, including more than 8000 Japanese-Okinawan males, has shown no significant correlation of vitamin B intake with PD.[280] Two other large cohort studies, examining folate intake in the last 12 months, separately failed to identify a significant relationship between folate intake and PD in females and males.[304] Intakes of B6 and B12 have not been found to be associated with PD.[304] Although some case-control and cohort studies suggest association between low B6 intake and increased risk of PD,[305,306] a systematic review of the literature led to the conclusion that identified associations between dietary vitamin B intake and PD are inconsistent and often result from poor study design.[307] There are two possible explanations for disagreements in the literature. Since B vitamins are water soluble and are not stored in the body, the dietary intake might have a very slight impact on the development of disease pathogenesis. Secondly, most studies showing the protective effect of B vitamins were performed in models of PD neurotoxicants and in the presence of elevated homocysteine. This suggests that neuroprotection by vitamin B only takes place in PD-like pathology-induced models, which may bear limited translational relevance.

14.4.3 Vitamin C

Vitamin C (ascorbate or anion of ascorbic acid) is typically found in fruits and vegetables. It crosses the blood–brain barrier *via* a sodium-dependent vitamin C transporter. In the last few decades, ascorbic acid has been shown to modulate various neurotransmitter systems, including dopaminergic, GABAergic, glutamatergic, and cholinergic pathways.[308]

Vitamin C is the most potent antioxidant among other vitamins in many assays conducted in cell-culture and animal models. The number of studies directly related to PD neurotoxicants is relatively small compared to other compounds. In a model of H_2O_2 exposure, ascorbic acid decreased DNA adduct formation.[309] It also protected primary midbrain cultures against BSO, a glutathione synthesis inhibitor, but the protection was gradually decreased upon addition of nitric oxide, suggesting that timing is crucial for the neuroprotective effect of ascorbic acid.[248,310] In a similar model, glutathione depletion-induced cytotoxicity and ROS depletion was ameliorated by ascorbic acid, a possible upstream player in free-radical scavenging.[311] Ascorbic acid also decreases oxidative stress induced by dopamine and 6-OHDA, including increased H_2O_2 levels and reduced O_2 consumption associated with 6-OHDA auto-oxidation.[312] In SH-SY5Y cells, dopamine enhanced apoptotis *via* PARP and p53 signalling, effects prevented by ascorbic acid.[313] Ascorbic acid also prevented the inhibition of mitochondrial complex-I caused by DHBT-1.[314] In animal models, ascorbic acid-enriched diets delayed climbing ability deficits in a human α-synuclein-expressing *Drosophila* model.[315] DJ-1β mutant flies displayed increased H_2O_2 and oxidized protein levels.[316] While vitamin C supplementation reversed these alterations, it also increased catalase activity. Thus, vitamin C might serve as a preventive agent or early therapeutic in PD instead of a curative drug due its ability to enhance dopaminergic neuron survival in PD models.

Ascorbic acid has been shown to enhance dopaminergic neuron production in precursor cultures. Dopaminergic differentiation in mesencephalic precursors from embryonic day 12 rats was enhanced by ascorbic acid treatment.[317] Additionally, expression levels of a neuronal marker (MAP2) and an astrocyte marker (GFAP) were increased by ascorbic acid in bFGF-expanded mesencephalic precursors, also increasing the portion of dopaminergic neurons in the culture.[318]

In general, no association was found between vitamin C intake and PD risk. Several case-control, cross-sectional, and cohort studies could not detect a significant association between dietary vitamin C intake and PD.[281,319,320] To our knowledge, only two studies observed increased risk of PD in individuals that consumed lower levels of vitamin C.[321,322] Moreover, serum levels of vitamin C in PD patients were not significantly different compared to healthy controls.[275,323] The postmortem analysis of caudate and *substantia nigra* also failed to detect changes in ascorbic acid between PD and control groups.[324] Access to the brain is restricted due to vitamin C's high water solubility. These findings strongly suggest that dietary intake of vitamin C is not directly related with the development of disease.

14.4.4 Vitamin D

Vitamin D is a lipid-soluble secosteroid, subclass steroid with a broken aromatic ring found in fortified milk, liver, and saltwater fish at high levels and in mushrooms at low levels.[325] Vitamin D has two major forms: Vitamin D2 and D3, derived from plant and animal sources, respectively. Human diet typically consists of a higher intake of vitamin D3 than vitamin D2. Upon ingestion or production induced by ultraviolet light exposure from sunlight, vitamin D3 is converted to 1,25-dihydroxyvitamin D3 (which will be referred to as vitamin D3 in the text for simplicity) by 25-hydroxylase in the liver or 1-alpha-hydroxylase in the kidney.[326]

Cell culture and rodent models of PD have been extensively used to test the neuroprotective activity of vitamin D. Co-treatment of SH-SY5Y cells with vitamin D3 and rotenone decreased rotenone-induced cell loss, ROS production, and enhanced autophagy biomarkers compared to rotenone-only cells.[327] While BSO- and MPP^+-induced apoptosis and an increase in ROS levels with diminished glutathione levels in dopaminergic neurons, low doses of vitamin D3 reduced cytotoxicity.[328] This study also suggested that vitamin D3 treatment produces a U-shaped dose–response curve, revealing that higher doses enhanced neurotoxicity. To understand the signalling pathways and proteins underlying vitamin D3 neuroprotection, transcriptomic analysis was performed after exposing mixed neuron-glial cultures to vitamin D3.[329] The expression profiles of 17 genes were altered. These genes were involved in protection against senile plaques, brain morphogenesis, and cell survival.

Unilateral 6-OHDA rodent models were used to test vitamin D3 neuroprotection. Pretreatment with vitamin D3 increased dopamine and metabolite levels in the *substantia nigra*, resulting in improved locomotor activity.[330] To see the same neuroprotection (*i.e.* increase in TH-IR neurons in midbrain), longer treatment was required for post-treatment compared to pre-treatment before stereotaxic infusion of 6-OHDA.[331] Four weeks of vitamin D3 treatment after insult enhanced dopamine release in the lesioned striatum.[332] Similar long-term treatment promoted an increase in striatal dopamine levels and GDNF in the *substantia nigra*.[333] Vitamin D3 has also been shown to diminish neuroinflammation. MPTP-induced microglial activation and pro-inflammatory cytokine expression were prevented by vitamin D3.[334] Zinc infusion to the *substantia nigra* of rats promoted lipid peroxidation and induced apoptosis and the depletion of striatal dopamine.[335] Systemic administration of vitamin D3 blocked zinc-induced oxidative stress, suggesting that it can also serve as a transition metal ion chelator.

Extensive epidemiological studies have been performed to test whether evidence of neuroprotection in animal models can be supported by clinical findings. Vitamin D serum and plasma levels have been associated with the risk of PD in several different populations. In Iranian, Chinese, North American, Japanese, and Finnish populations, PD patients have been found to have lower levels of vitamin D compared to age-matched healthy controls.[336–340] The odds ratio calculated from a meta-analysis revealed that vitamin D deficiency leads to a two-fold increase in PD risk.[341] Vitamin D deficiency is also associated with the severity of symptoms. PD patients with better cognition and mood have been shown to have a higher plasma level of vitamin D, compared with patients with worse conditions.[342] Serum vitamin D was inversely correlated with PD severity assessed by evaluation of balance in patients.[343]

Stooped posture, a phenotypic characteristic of PD, has been found to be significantly more common in PD patients with vitamin D deficiency.[344]

Serum and plasma levels of vitamin D3 are affected by sunlight exposure, as previously explained. Epidemiological studies have shown that time spent outdoors is inversely associated with PD risk in a Chinese population.[345] Danish men working outdoors have been found to have a low risk of PD compared to those working indoors.[346] It is possible that outdoor occupations may be more active. Since physical activity has been shown to decrease the risk of PD,[347] the study set the level of physical activity as a potential cofounder, further strengthening the inverse association between sun exposure and the disease. PD patients exposed to sun during shorter time periods have a higher risk of hip fracture, commonly seen in elderly PD patients.[348,349]

Vitamin D binds to the vitamin D receptor (VDR) to modulate several signalling pathways as reviewed by Kesby *et al.*[350] Polymorphisms in VDR genes have been associated with PD risk. The most validated variants are called BsmI and FoxI. Individuals carrying BsmI and FoxI variants have been shown to have a higher risk of PD in Korean, Han Chinese, and Hungarian populations as well as in meta-analyses.[351–355] The exact effects of these variants are not known, but they have been suggested to alter mRNA stability or protein activity.[356]

14.4.5 Vitamin E

Vitamin E is a lipid-soluble compound that includes tocopherols and tocotrienols. The most common type of vitamin E in the human diet is tocopherol. Good sources of vitamin E are oil extract from corn, soybeans, sunflower, palm, and almond.[357]

In vitro models of PD support neuroprotection by vitamin E. It has been reported to abolish α-synuclein inclusions and ROS levels in α-synuclein-overexpressing GT1-7 hypothalamic neuronal cell lines.[358] In cerebellar granule cells, it inhibited MPP^+- and paraquat-induced cell death.[359,360] Neuroprotection in an MPP^+ model is regulated *via* suppression of cyclin D and E2F-1 expression.[361] In addition, tocotrienol was found to bind to estrogen receptor beta and inhibit cell death through promoting PI3K/Akt signalling.[362] Rodent postnatal midbrain organotypic slices were used to test the long-term effect of vitamin E in rotenone neurotoxicity.[363] In this model, rotenone-induced oxidative damage and decreased TH levels in dopaminergic neurons, along with morphological changes and cytotoxicity, were prevented by α-tocopherol, suggesting that it plays a role as an anti-oxidant. Vitamin E can also prevent metal-induced damage in cell cultures. Manganese exposure in mesencephalic culture generated ROS, resulting in apoptosis.[364] Pretreatment with vitamin E promoted cell survival through NFκB signalling. Moreover, ferric and ferrous iron-induced toxicity to SK-N-SH cells was alleviated by pretreatment with vitamin E, indicated by an inhibition of ROS formation and cell death, along with a partial prevention of α-synuclein aggregation.[365] Finally, palm oil extract containing alpha tocopherol and tocotrienol protected astrocytes from glutamate-induced injury.[366] The data summarized above indicate that vitamin E appears to act as an antioxidant, anti-inflammatory compound and transition metal ion chelator in neuroprotection against PD.

Fly and rodent models support *in vitro* findings suggesting protective effects of vitamin E. DJ-1β-mutant *Drosophila* exhibited low levels of catalase and increased H_2O_2, oxidized protein, and ROS; all of these biochemical changes were reversed by supplementation with α-tocopherol.[316] Co-administration of vitamin E with rotenone for more than a month improved locomotor activity, alleviated catalepsy, and reversed alterations in lipid peroxidation and antioxidant levels in rats.[367] In MPTP-treated mice, enhanced COX2 activity, lipid peroxidation, and nitrite/nitrate levels have been observed. These changes were also inhibited with vitamin E administration.[368] Hippocampal and nigral cell losses induced in iron-injected rats were restored by α-tocopherol.[369] Combinations of pre- and post-treatment of D-α-tocopheryl acid succinate, an ester of α-tocopherol, lessened dopamine cell loss and apomorphine-induced rotations in 6-OHDA-treated rats.[370] Treatment with α-tocopherol before 6-OHDA infusion was found to be sufficient to prevent neurotoxicity.[371]

Dietary vitamin E intake was inversely associated with PD in Japanese and Dutch populations.[320,372] A meta-analysis of case-control, cohort, and cross-sectional studies concluded that moderate intake was correlated with reduced PD risk.[281] The same association was confirmed by another epidemiological study, emphasizing that the protective effect was lost in cases of higher intake.[279] Similar to dietary intake, serum and plasma levels of vitamin E are negatively associated with PD risk.[373] In a Chinese case-control study, serum and urine levels of vitamin E and metal were analyzed.[374] Interestingly, vitamin E was inversely correlated with serum copper but positively associated with urine copper levels, suggesting that vitamin E acts as a modulator of copper metabolism. More studies are needed to determine the mechanism.

The Deprenyl and Tocopherol Antioxidative Therapy of Parkinsonism (DATATOP) study tested vitamin E supplementation against PD in 800 patients.[375] Although the dosage was high (2000 IU alpha-tocopherol daily), the treatment failed to delay or prevent the progression of the disease in patients. Thus, studies to date suggest that vitamin E may be a potential protective factor in disease prevention. However, there are no data to suggest that it is a disease-modifying therapeutic post-diagnosis.

14.5 Summary

One of the primary mechanisms that has been repeatedly implicated in the pathogenesis of PD is oxidative stress. Drugs that have strong antioxidant activity appear promising in modifying the risk of developing PD. Support for neuroprotection in patients diagnosed with the disease is much more limited.

The toxicokinetics of any drug must be evaluated carefully. Dietary antioxidants and vitamins have an advantage because in some cases there are well-established U-shaped dose–response curves, where recommended levels are well known. Most dietary compounds thought to be neuroprotective in PD have been shown to modulate neuroinflammation, iron disposition, and signalling pathways in cell survival, which are mechanisms disrupted in the diseased nervous system. However, there are often major issues in human use for PD:

- Some compounds have high water solubility, resulting in excretion prior to diffusion into target tissue.
- The presence of large aromatic rings limits blood–brain barrier penetrance. BBB penetration is highly correlated with lipophilicity and polarity of the compounds (less polar derivatives have greater penetrance through the barrier, and *vice versa*).[89,376]
- The majority of neuroprotective compounds that are natural dietary constituents likely exhibit U-shaped dose–response curves, and as a result individuals with low blood levels may exhibit increased risk of PD, whereas those supplementing well above suggested daily intake may exhibit toxicity.
- By the time behavioral phenotypes are prominent in patients, more than half of the dopaminergic neurons in affected brain regions are likely lost, though this number may be less than previously thought.[377] Any dietary factor supplemented should at least have significant enough neuroprotective capacity to slow the disease progression, an achievement that has yet to be reached.

References

1. J. R. Cannon and J. T. Greenamyre, The role of environmental exposures in neurodegeneration and neurodegenerative diseases, *Toxicol. Sci.*, 2011, **124**(2), 225–250.
2. Z. S. Agim and J. R. Cannon, Dietary factors in the etiology of Parkinson's disease, *BioMed Res. Int.*, 2015, **2015**, 672838.
3. R. N. Alcalay, Y. Gu, H. Mejia-Santana, L. Cote, K. S. Marder and N. Scarmeas, The association between Mediterranean diet adherence and Parkinson's disease, *Mov. Disord.*, 2012, **27**(6), 771–774.
4. S. J. Chinta, A. Ganesan, P. Reis-Rodrigues, G. J. Lithgow and J. K. Andersen, Anti-inflammatory role of the isoflavone diadzein in lipopolysaccharide-stimulated microglia: implications for Parkinson's disease, *Neurotoxic. Res.*, 2013, **23**(2), 145–153.
5. S. Kyuhou, Preventive effects of genistein on motor dysfunction following 6-hydroxydopamine injection in ovariectomized rats, *Neurosci. Lett.*, 2008, **448**(1), 10–14.
6. M. Ristow, Unraveling the truth about antioxidants: mitohormesis explains ROS-induced health benefits, *Nat. Med.*, 2014, **20**(7), 709–711.

7. J. Bouayed and T. Bohn, Exogenous antioxidants-Double-edged swords in cellular redox state: Health beneficial effects at physiologic doses *versus* deleterious effects at high doses, *Oxid. Med. Cell. Longevity*, 2010, **3**(4), 228–237.
8. E. D. Louis, W. Zheng, W. Jiang, K. T. Bogen and G. A. Keating, Quantification of the neurotoxic beta-carboline harmane in barbecued/grilled meat samples and correlation with level of doneness, *J. Toxicol. Environ. Health, Part A*, 2007, **70**(12), 1014–1019.
9. J. Adachi, Y. Mizoi, T. Naito, K. Yamamoto, S. Fujiwara and I. Ninomiya, Determination of beta-carbolines in foodstuffs by high-performance liquid chromatography and high-performance liquid chromatography-mass spectrometry, *J. Chromatogr.*, 1991, **538**(2), 331–339.
10. T. R. Bosin and K. F. Faull, Harman in alcoholic beverages: pharmacological and toxicological implications, *Alcohol.: Clin. Exp. Res.*, 1988, **12**(5), 679–682.
11. Y. Totsuka, H. Ushiyama, J. Ishihara, R. Sinha, S. Goto and T. Sugimura, *et al.,* Quantification of the co-mutagenic beta-carbolines, norharman and harman, in cigarette smoke condensates and cooked foods, *Cancer Lett.*, 1999, **143**(2), 139–143.
12. T. Herraiz and C. Chaparro, Human monoamine oxidase enzyme inhibition by coffee and beta-carbolines norharman and harman isolated from coffee, *Life Sci.*, 2006, **78**(8), 795–802.
13. E. D. Louis, Environmental epidemiology of essential tremor, *Neuroepidemiology*, 2008, **31**(3), 139–149.
14. E. D. Louis, P. Factor-Litvak, X. Liu, J. P. Vonsattel, M. Galecki and W. Jiang, *et al.,* Elevated brain harmane (1-methyl-9H-pyrido[3,4-*b*]indole) in essential tremor cases *vs.* controls, *Neurotoxicology*, 2013, **38**, 131–135.
15. W. Kuhn, T. Muller, H. Grosse and H. Rommelspacher, Plasma harman and norharman in Parkinson's disease, *J. Neural Transm. Suppl.*, 1995, **46**, 291–295.
16. E. D. Louis, M. Michalec, W. Jiang, P. Factor-Litvak and W. Zheng, Elevated blood harmane (1-methyl-9H-pyrido[3,4-*b*]indole) concentrations in Parkinson's disease, *Neurotoxicology*, 2014, **40**, 52–56.
17. W. Kuhn, T. Muller, H. Grosse and H. Rommelspacher, Elevated levels of harman and norharman in cerebrospinal fluid of parkinsonian patients, *J. Neural Transm. (Vienna)*, 1996, **103**(12), 1435–1440.
18. J. S. Felton, M. G. Knize, N. H. Shen, B. D. Andresen, L. F. Bjeldanes and F. T. Hatch, Identification of the mutagens in cooked beef, *Environ. Health Perspect.*, 1986, **67**, 17–24.
19. J. S. Felton, M. G. Knize, N. H. Shen, P. R. Lewis, B. D. Andresen and J. Happe, *et al.,* The isolation and identification of a new mutagen from fried ground beef: 2-amino-1-methyl-6-phenylimidazo[4,5-*b*]pyridine (PhIP), *Carcinogenesis*, 1986, **7**(7), 1081–1086.
20. A. M. Griggs, Z. S. Agim, V. R. Mishra, M. A. Tambe, A. E. Director-Myska and K. W. Turteltaub, *et al.,* 2-Amino-1-methyl-6-phenylimidazo[4,5-*b*] pyridine (PhIP) is selectively toxic to primary dopaminergic neurons *in vitro*, *Toxicol. Sci.*, 2014, **140**(1), 179–189.

21. S. F. Teunissen, M. L. Vlaming, H. Rosing, J. H. Schellens, A. H. Schinkel and J. H. Beijnen, Development and validation of a liquid chromatography-tandem mass spectrometry assay for the analysis of 2-amino-1-methyl-6-phenylimidazo[4,5-*b*]pyridine (PhIP) and its metabolite 2-hydroxyamino-1-methyl-6-phenylimidazo[4,5-*b*]pyridine (N-OH-PhIP) in plasma, urine, bile, intestinal contents, faeces and eight selected tissues from mice, *J. Chromatogr. B Analyt. Technol. Biomed. Life Sci.*, 2010, **878**(25), 2353–2362.
22. G. Nauwelaers, E. E. Bessette, D. Gu, Y. Tang, J. Rageul and V. Fessard, *et al.*, DNA adduct formation of 4-aminobiphenyl and heterocyclic aromatic amines in human hepatocytes, *Chem. Res. Toxicol.*, 2011, **24**(6), 913–925.
23. L. Peng and R. J. Turesky, Optimizing proteolytic digestion conditions for the analysis of serum albumin adducts of 2-amino-1-methyl-6-phenylimidazo[4,5-*b*]pyridine, a potential human carcinogen formed in cooked meat, *J. Proteomics*, 2014, **103**, 267–278.
24. H. Ichinose, N. Ozaki, D. Nakahara, M. Naoi, K. Wakabayashi and T. Sugimura, *et al.*, Effects of heterocyclic amines in food on dopamine metabolism in nigro-striatal dopaminergic neurons, *Biochem. Pharmacol.*, 1988, **37**(17), 3289–3295.
25. T. Kojima, M. Naoi, K. Wakabayashi, T. Sugimura and T. Nagatsu, 3-amino-1-methyl-5H-pyrido[4,3-*b*]indole (Trp-P-2) and other heterocyclic amines as inhibitors of mitochondrial monoamine oxidases separated from human brain synaptosomes, *Neurochem. Int.*, 1990, **16**(1), 51–57.
26. W. Maruyama, A. Ota, A. Takahashi, T. Nagatsu and M. Naoi, Food-derived heterocyclic amines, 3-amino-1,4-dimethyl-5H-pyrido[4,3-*b*] indole and related amines, as inhibitors of monoamine metabolism, *J. Neural Transm. Suppl.*, 1994, **41**, 327–333.
27. J. R. Cannon, V. Tapias, H. M. Na, A. S. Honick, R. E. Drolet and J. T. Greenamyre, A highly reproducible rotenone model of Parkinson's disease, *Neurobiol. Dis.*, 2009, **34**(2), 279–290.
28. J. C. Rojas, N. Simola, B. A. Kermath, J. R. Kane, T. Schallert and F. Gonzalez-Lima, Striatal neuroprotection with methylene blue, *Neuroscience*, 2009, **163**(3), 877–889.
29. A. Ebrahimi and H. Schluesener, Natural polyphenols against neurodegenerative disorders: potentials and pitfalls, *Ageing Res. Rev.*, 2012, **11**(2), 329–345.
30. S. Ramos, Effects of dietary flavonoids on apoptotic pathways related to cancer chemoprevention, *J. Nutr. Biochem.*, 2007, **18**(7), 427–442.
31. R. Tsao, Chemistry and biochemistry of dietary polyphenols, *Nutrients*, 2010, **2**(12), 1231–1246.
32. P. B. Kaufman, J. A. Duke, H. Brielmann, J. Boik and J. E. Hoyt, A comparative survey of leguminous plants as sources of the isoflavones, genistein and daidzein: implications for human nutrition and health, *J. Altern. Complement. Med.*, 1997, **3**(1), 7–12.
33. H. Sawada and S. Shimohama, Estrogens and Parkinson disease: novel approach for neuroprotection, *Endocrine*, 2003, **21**(1), 77–79.

34. K. Dampier, E. A. Hudson, L. M. Howells, M. M. Manson, R. A. Walker and A. Gescher, Differences between human breast cell lines in susceptibility towards growth inhibition by genistein, *Br. J. Cancer*, 2001, **85**(4), 618–624.
35. T. Omura, M. Kaneko, Y. Okuma, K. Matsubara and Y. Nomura, Endoplasmic reticulum stress and Parkinson's disease: the role of HRD1 in averting apoptosis in neurodegenerative disease, *Oxid. Med. Cell. Longevity*, 2013, **2013**, 239854.
36. Y. Imai, M. Soda, H. Inoue, N. Hattori, Y. Mizuno and R. Takahashi, An unfolded putative transmembrane polypeptide, which can lead to endoplasmic reticulum stress, is a substrate of Parkin, *Cell.*, 2001, **105**(7), 891–902.
37. N. J. Linford and D. M. Dorsa, 17beta-Estradiol and the phytoestrogen genistein attenuate neuronal apoptosis induced by the endoplasmic reticulum calcium-ATPase inhibitor thapsigargin, *Steroids*, 2002, **67**(13–14), 1029–1040.
38. W. Y. Boadi and D. Johnson, Effects of low doses of quercetin and genistein on oxidation and carbonylation in hemoglobin and myoglobin, *J. Diet. Suppl.*, 2014, **11**(3), 272–287.
39. N. Jantaratnotai, P. Utaisincharoen, P. Sanvarinda, A. Thampithak and Y. Sanvarinda, Phytoestrogens mediated anti-inflammatory effect through suppression of IRF-1 and pSTAT1 expressions in lipopolysaccharide-activated microglia, *Int. Immunopharmacol.*, 2013, **17**(2), 483–488.
40. M. L. McDowell, A. Das, J. A. Smith, A. K. Varma, S. K. Ray and N. L. Banik, Neuroprotective effects of genistein in VSC4.1 motoneurons exposed to activated microglial cytokines, *Neurochem. Int.*, 2011, **59**(2), 175–184.
41. X. Li and A. Y. Sun, Paraquat induced activation of transcription factor AP-1 and apoptosis in PC12 cells, *J. Neural Transm.*, 1999, **106**(1), 1–21.
42. C. M. Lin, R. D. Lin, S. T. Chen, Y. P. Lin, W. T. Chiu and J. W. Lin, *et al.,* Neurocytoprotective effects of the bioactive constituents of Pueraria thomsonii in 6-hydroxydopamine (6-OHDA)-treated nerve growth factor (NGF)-differentiated PC12 cells, *Phytochemistry*, 2010, **71**(17–18), 2147–2156.
43. L. X. Liu, W. F. Chen, J. X. Xie and M. S. Wong, Neuroprotective effects of genistein on dopaminergic neurons in the mice model of Parkinson's disease, *Neurosci. Res.*, 2008, **60**(2), 156–161.
44. T. Baluchnejadmojarad, M. Roghani, M. R. Nadoushan and M. Bagheri, Neuroprotective effect of genistein in 6-hydroxydopamine hemi-parkinsonian rat model, *Phytother. Res.*, 2009, **23**(1), 132–135.
45. S. P. Gupta, S. Patel, S. Yadav, A. K. Singh, S. Singh and M. P. Singh, Involvement of nitric oxide in maneb- and paraquat-induced Parkinson's disease phenotype in mouse: is there any link with lipid peroxidation? *Neurochem. Res.*, 2010, **35**(8), 1206–1213.
46. H. Sakakibara, Y. Honda, S. Nakagawa, H. Ashida and K. Kanazawa, Simultaneous determination of all polyphenols in vegetables, fruits, and teas, *J. Agric. Food Chem.*, 2003, **51**(3), 571–581.

47. M. Zhu, S. Rajamani, J. Kaylor, S. Han, F. Zhou and A. L. Fink, The flavonoid baicalein inhibits fibrillation of alpha-synuclein and disaggregates existing fibrils, *J. Biol. Chem.*, 2004, **279**(26), 26846–26857.
48. M. Kostka, T. Hogen, K. M. Danzer, J. Levin, M. Habeck and A. Wirth, *et al.*, Single particle characterization of iron-induced pore-forming alpha-synuclein oligomers, *J. Biol. Chem.*, 2008, **283**(16), 10992–11003.
49. J. H. Lee, I. H. Lee, Y. J. Choe, S. Kang, H. Y. Kim and W. P. Gai, *et al.*, Real-time analysis of amyloid fibril formation of alpha-synuclein using a fibrillation-state-specific fluorescent probe of JC-1, *Biochem. J.*, 2009, **418**(2), 311–323.
50. X. Meng, L. A. Munishkina, A. L. Fink and V. N. Uversky, Molecular mechanisms underlying the flavonoid-induced inhibition of alpha-synuclein fibrillation, *Biochemistry*, 2009, **48**(34), 8206–8224.
51. D. P. Hong, A. L. Fink and V. N. Uversky, Structural characteristics of alpha-synuclein oligomers stabilized by the flavonoid baicalein, *J. Mol. Biol.*, 2008, **383**(1), 214–223.
52. J. H. Lu, M. T. Ardah, S. S. Durairajan, L. F. Liu, L. X. Xie and W. F. Fong, *et al.*, Baicalein inhibits formation of alpha-synuclein oligomers within living cells and prevents Abeta peptide fibrillation and oligomerisation, *ChemBioChem*, 2011, **12**(4), 615–624.
53. M. Jiang, Y. Porat-Shliom, Z. Pei, Y. Cheng, L. Xiang and K. Sommers, *et al.*, Baicalein reduces E46K alpha-synuclein aggregation *in vitro* and protects cells against E46K alpha-synuclein toxicity in cell models of familiar Parkinsonism, *J. Neurochem.*, 2010, **114**(2), 419–429.
54. M. Caruana, J. Neuner, T. Hogen, F. Schmidt, F. Kamp and C. Scerri, *et al.*, Polyphenolic compounds are novel protective agents against lipid membrane damage by alpha-synuclein aggregates *in vitro*, *Biochim. Biophys. Acta*, 2012, **1818**(11), 2502–2510.
55. H. J. Lee, Y. H. Noh, D. Y. Lee, Y. S. Kim, K. Y. Kim and Y. H. Chung, *et al.*, Baicalein attenuates 6-hydroxydopamine-induced neurotoxicity in SH-SY5Y cells, *Eur. J. Cell Biol.*, 2005, **84**(11), 897–905.
56. X. X. Li, G. R. He, X. Mu, B. Xu, S. Tian and X. Yu, *et al.*, Protective effects of baicalein against rotenone-induced neurotoxicity in PC12 cells and isolated rat brain mitochondria, *Eur. J. Pharmacol.*, 2012, **674**(2–3), 227–233.
57. Y. Cheng, G. He, X. Mu, T. Zhang, X. Li and J. Hu, *et al.*, Neuroprotective effect of baicalein against MPTP neurotoxicity: behavioral, biochemical and immunohistochemical profile, *Neurosci. Lett.*, 2008, **441**(1), 16–20.
58. X. Mu, G. R. He, X. Yuan, X. X. Li and G. H. Du, Baicalein protects the brain against neuron impairments induced by MPTP in C57BL/6 mice, *Pharmacol., Biochem. Behav.*, 2011, **98**(2), 286–291.
59. X. Mu, G. He, Y. Cheng, X. Li, B. Xu and G. Du, Baicalein exerts neuroprotective effects in 6-hydroxydopamine-induced experimental parkinsonism *in vivo* and *in vitro*, *Pharmacol., Biochem. Behav.*, 2009, **92**(4), 642–648.

60. X. Yu, G. R. He, L. Sun, X. Lan, L. L. Shi and Z. H. Xuan, *et al.*, Assessment of the treatment effect of baicalein on a model of Parkinsonian tremor and elucidation of the mechanism, *Life Sci.*, 2012, **91**(1–2), 5–13.
61. X. Xue, H. Liu, L. Qi, X. Li, C. Guo and D. Gong, *et al.*, Baicalein ameliorated the upregulation of striatal glutamatergic transmission in the mice model of Parkinson's disease, *Brain Res. Bull.*, 2014, **103**, 54–59.
62. E. B. Jung and C. S. Lee, Baicalein attenuates proteasome inhibition-induced apoptosis by suppressing the activation of the mitochondrial pathway and the caspase-8- and Bid-dependent pathways, *Eur. J. Pharmacol.*, 2014, **730**, 116–124.
63. E. Lee, H. R. Park, S. T. Ji, Y. Lee and J. Lee, Baicalein attenuates astroglial activation in the 1-methyl-4-phenyl-1,2,3,4-tetrahydropyridine-induced Parkinson's disease model by downregulating the activations of nuclear factor-kappaB, ERK, and JNK, *J. Neurosci. Res.*, 2014, **92**(1), 130–139.
64. H. K. Bryan, A. Olayanju, C. E. Goldring and B. K. Park, The Nrf2 cell defence pathway: Keap1-dependent and -independent mechanisms of regulation, *Biochem. Pharmacol.*, 2013, **85**(6), 705–717.
65. Z. Zhang, W. Cui, G. Li, S. Yuan, D. Xu and M. P. Hoi, *et al.*, Baicalein protects against 6-OHDA-induced neurotoxicity through activation of Keap1/Nrf2/HO-1 and involving PKCalpha and PI3K/AKT signaling pathways, *J. Agric. Food Chem.*, 2012, **60**(33), 8171–8182.
66. A. Y. Choi, J. H. Choi, J. Y. Lee, K. S. Yoon, W. Choe and J. Ha, *et al.*, Apigenin protects HT22 murine hippocampal neuronal cells against endoplasmic reticulum stress-induced apoptosis, *Neurochem. Int.*, 2010, **57**(2), 143–152.
67. H. Q. Chen, Z. Y. Jin, X. J. Wang, X. M. Xu, L. Deng and J. W. Zhao, Luteolin protects dopaminergic neurons from inflammation-induced injury through inhibition of microglial activation, *Neurosci. Lett.*, 2008, **448**(2), 175–179.
68. G. Zhao, C. Yao-Yue, G. W. Qin and L. H. Guo, Luteolin from Purple Perilla mitigates ROS insult particularly in primary neurons, *Neurobiol. Aging*, 2012, **33**(1), 176–186.
69. D. J. Guo, F. Li, P. H. Yu and S. W. Chan, Neuroprotective effects of luteolin against apoptosis induced by 6-hydroxydopamine on rat pheochromocytoma PC12 cells, *Pharm. Biol.*, 2013, **51**(2), 190–196.
70. L. W. Hu, J. H. Yen, Y. T. Shen, K. Y. Wu and M. J. Wu, Luteolin modulates 6-hydroxydopamine-induced transcriptional changes of stress response pathways in PC12 cells, *PloS one*, 2014, **9**(5), e97880.
71. C. J. Wruck, M. Claussen, G. Fuhrmann, L. Romer, A. Schulz and T. Pufe, *et al.*, Luteolin protects rat PC12 and C6 cells against MPP+ induced toxicity *via* an ERK dependent Keap1-Nrf2-ARE pathway, *J. Neural Transm. Suppl.*, 2007, **72**, 57–67.
72. N. D. Stull, J. W. Jung and L. Iacovitti, Induction of a dopaminergic phenotype in cultured striatal neurons by bone morphogenetic proteins, *Brain Res. Dev. Brain Res.*, 2001, **130**(1), 91–98.

73. S. P. Patil, P. D. Jain, J. S. Sancheti, P. J. Ghumatkar, R. Tambe and S. Sathaye, Neuroprotective and neurotrophic effects of Apigenin and Luteolin in MPTP induced parkinsonism in mice, *Neuropharmacology*, 2014, **86**, 192–202.
74. S. K. Ha, E. Moon, P. Lee, J. H. Ryu, M. S. Oh and S. Y. Kim, Acacetin attenuates neuroinflammation *via* regulation the response to LPS stimuli *in vitro* and *in vivo*, *Neurochem. Res.*, 2012, **37**(7), 1560–1567.
75. H. G. Kim, M. S. Ju, S. K. Ha, H. Lee, H. Lee and S. Y. Kim, *et al.*, Acacetin protects dopaminergic cells against 1-methyl-4-phenyl-1,2,3,6-tetrahydropyridine-induced neuroinflammation *in vitro* and *in vivo*, *Biol. Pharm. Bull.*, 2012, **35**(8), 1287–1294.
76. H. J. Heo, D. O. Kim, S. C. Shin, M. J. Kim, B. G. Kim and D. H. Shin, Effect of antioxidant flavanone, naringenin, from Citrus junoson neuroprotection, *J. Agric. Food Chem.*, 2004, **52**(6), 1520–1525.
77. S. M. Jeon, S. H. Bok, M. K. Jang, M. K. Lee, K. T. Nam and Y. B. Park, *et al.*, Antioxidative activity of naringin and lovastatin in high cholesterol-fed rabbits, *Life Sci.*, 2001, **69**(24), 2855–2866.
78. M. K. Lee, S. H. Bok, T. S. Jeong, S. S. Moon, S. E. Lee and Y. B. Park, *et al.*, Supplementation of naringenin and its synthetic derivative alters antioxidant enzyme activities of erythrocyte and liver in high cholesterol-fed rats, *Bioorg. Med. Chem.*, 2002, **10**(7), 2239–2244.
79. L. D. Mercer, B. L. Kelly, M. K. Horne and P. M. Beart, Dietary polyphenols protect dopamine neurons from oxidative insults and apoptosis: investigations in primary rat mesencephalic cultures, *Biochem. Pharmacol.*, 2005, **69**(2), 339–345.
80. H. J. Kim, J. Y. Song, H. J. Park, H. K. Park, D. H. Yun and J. H. Chung, Naringin Protects against Rotenone-induced Apoptosis in Human Neuroblastoma SH-SY5Y Cells, *Korean J. Physiol. Pharmacol.*, 2009, **13**(4), 281–285.
81. H. Lou, X. Jing, X. Wei, H. Shi, D. Ren and X. Zhang, Naringenin protects against 6-OHDA-induced neurotoxicity *via* activation of the Nrf2/ARE signaling pathway, *Neuropharmacology*, 2014, **79**, 380–388.
82. K. Vafeiadou, D. Vauzour, H. Y. Lee, A. Rodriguez-Mateos, R. J. Williams and J. P. Spencer, The citrus flavanone naringenin inhibits inflammatory signalling in glial cells and protects against neuroinflammatory injury, *Arch. Biochem. Biophys.*, 2009, **484**(1), 100–109.
83. V. Zbarsky, K. P. Datla, S. Parkar, D. K. Rai, O. I. Aruoma and D. T. Dexter, Neuroprotective properties of the natural phenolic antioxidants curcumin and naringenin but not quercetin and fisetin in a 6-OHDA model of Parkinson's disease, *Free Radical Res.*, 2005, **39**(10), 1119–1125.
84. Y. Chtourou, H. Fetoui and R. Gdoura, Protective effects of naringenin on iron-overload-induced cerebral cortex neurotoxicity correlated with oxidative stress, *Biol. Trace Elem. Res.*, 2014, **158**(3), 376–383.
85. I. F. Cheng and K. Breen, On the ability of four flavonoids, baicilein, luteolin, naringenin, and quercetin, to suppress the Fenton reaction of the iron-ATP complex, *Biometals*, 2000, **13**(1), 77–83.

86. M. Sonia Angeline, A. Sarkar, K. Anand, R. K. Ambasta and P. Kumar, Sesamol and naringenin reverse the effect of rotenone-induced PD rat model, *Neuroscience*, 2013, **254**, 379–394.
87. K. Tamilselvam, N. Braidy, T. Manivasagam, M. M. Essa, N. R. Prasad and S. Karthikeyan, *et al.*, Neuroprotective effects of hesperidin, a plant flavanone, on rotenone-induced oxidative stress and apoptosis in a cellular model for Parkinson's disease, *Oxid. Med. Cell. Longevity*, 2013, **2013**, 102741.
88. M. S. Antunes, A. T. Goes, S. P. Boeira, M. Prigol and C. R. Jesse, Protective effect of hesperidin in a model of Parkinson's disease induced by 6-hydroxydopamine in aged mice, *Nutrition*, 2014, **30**(11–12), 1415–1422.
89. K. A. Youdim, M. S. Dobbie, G. Kuhnle, A. R. Proteggente, N. J. Abbott and C. Rice-Evans, Interaction between flavonoids and the blood-brain barrier: *in vitro* studies, *J. Neurochem.*, 2003, **85**(1), 180–192.
90. T. H. Tsai, Determination of naringin in rat blood, brain, liver, and bile using microdialysis and its interaction with cyclosporin a, a p-glycoprotein modulator, *J. Agric. Food Chem.*, 2002, **50**(23), 6669–6674.
91. M. Shulman, M. Cohen, A. Soto-Gutierrez, H. Yagi, H. Wang and J. Goldwasser, *et al.*, Enhancement of naringenin bioavailability by complexation with hydroxypropyl-beta-cyclodextrin. [corrected], *PLoS One*, 2011, **6**(4), e18033.
92. S. H. Thilakarathna and H. P. Rupasinghe, Flavonoid bioavailability and attempts for bioavailability enhancement, *Nutrients*, 2013, **5**(9), 3367–3387.
93. C. F. Jin, S. R. Shen Sr and B. L. Zhao, Different effects of five catechins on 6-hydroxydopamine-induced apoptosis in PC12 cells, *J. Agric. Food Chem.*, 2001, **49**(12), 6033–6038.
94. H. L. Ruan, Y. Yang, X. N. Zhu, X. L. Wang and R. Z. Chen, Similar potency of catechin and its enantiomers in alleviating 1-methyl-4-phenylpyridinium ion cytotoxicity in SH-SY5Y cells, *J. Pharm. Pharmacol.*, 2011, **63**(9), 1169–1174.
95. Y. Levites, T. Amit, M. B. Youdim and S. Mandel, Involvement of protein kinase C activation and cell survival/cell cycle genes in green tea polyphenol (-)-epigallocatechin 3-gallate neuroprotective action, *J. Biol. Chem.*, 2002, **277**(34), 30574–30580.
96. O. Weinreb, S. Mandel and M. B. Youdim, cDNA gene expression profile homology of antioxidants and their antiapoptotic and proapoptotic activities in human neuroblastoma cells, *FASEB J*, 2003, **17**(8), 935–937.
97. W. G. Chung, C. L. Miranda and C. S. Maier, Epigallocatechin gallate (EGCG) potentiates the cytotoxicity of rotenone in neuroblastoma SH-SY5Y cells, *Brain Res.*, 2007, **1176**, 133–142.
98. M. B. Youdim, E. Grunblatt, Y. Levites, G. Maor and S. Mandel, Early and late molecular events in neurodegeneration and neuroprotection in Parkinson's disease MPTP model as assessed by cDNA microarray; the role of iron, *Neurotoxic. Res.*, 2002, **4**(7–8), 679–689.

99. L. Wang, S. Xu, X. Xu and P. Chan, (-)-Epigallocatechin-3-Gallate protects SH-SY5Y cells against 6-OHDA-induced cell death through STAT3 activation, *J. Alzheimer's Dis.*, 2009, **17**(2), 295–304.
100. N. Lorenzen, S. B. Nielsen, Y. Yoshimura, B. S. Vad, C. B. Andersen and C. Betzer, *et al.*, How epigallocatechin gallate can inhibit alpha-synuclein oligomer toxicity *in vitro*, *J. Biol. Chem.*, 2014, **289**(31), 21299–21310.
101. R. R. Hou, J. Z. Chen, H. Chen, X. G. Kang, M. G. Li and B. R. Wang, Neuroprotective effects of (-)-epigallocatechin-3-gallate (EGCG) on paraquat-induced apoptosis in PC12 cells, *Cell Biol. Int.*, 2008, **32**(1), 22–30.
102. K. K. Tai and D. D. Truong, (-)-Epigallocatechin-3-gallate (EGCG), a green tea polyphenol, reduces dichlorodiphenyl-trichloroethane (DDT)-induced cell death in dopaminergic SHSY-5Y cells, *Neurosci. Lett.*, 2010, **482**(3), 183–187.
103. R. Li, Y. G. Huang, D. Fang and W. D. Le, (-)-Epigallocatechin gallate inhibits lipopolysaccharide-induced microglial activation and protects against inflammation-mediated dopaminergic neuronal injury, *J. Neurosci. Res.*, 2004, **78**(5), 723–731.
104. Y. Levites, O. Weinreb, G. Maor, M. B. Youdim and S. Mandel, Green tea polyphenol (-)-epigallocatechin-3-gallate prevents N-methyl-4-phenyl-1,2,3,6-tetrahydropyridine-induced dopaminergic neurodegeneration, *J. Neurochem.*, 2001, **78**(5), 1073–1082.
105. J. Y. Choi, C. S. Park, D. J. Kim, M. H. Cho, B. K. Jin and J. E. Pie, *et al.*, Prevention of nitric oxide-mediated 1-methyl-4-phenyl-1,2,3,6-tetrahydropyridine-induced Parkinson's disease in mice by tea phenolic epigallocatechin 3-gallate, *Neurotoxicology*, 2002, **23**(3), 367–374.
106. S. Mandel, O. Weinreb, L. Reznichenko, L. Kalfon and T. Amit, Green tea catechins as brain-permeable, non toxic iron chelators to "iron out iron" from the brain, *J. Neural Transm. Suppl.*, 2006, **71**, 249–257.
107. R. Li, N. Peng, F. Du, X. P. Li and W. D. Le, Epigallocatechin gallate protects dopaminergic neurons against 1-methyl-4- phenyl-1,2,3,6-tetrahydropyridine-induced neurotoxicity by inhibiting microglial cell activation, *J. South. Med. Univ.*, 2006, **26**(4), 376–380.
108. M. D. Teixeira, C. M. Souza, A. P. Menezes, M. R. Carmo, A. A. Fonteles and J. P. Gurgel, *et al.*, Catechin attenuates behavioral neurotoxicity induced by 6-OHDA in rats, *Pharmacol., Biochem. Behav.*, 2013, **110**, 1–7.
109. J. S. Al-Amri, M. M. Hagras and I. M. Mohamed, Effect of epigallocatechin-3-gallate on inflammatory mediators release in LPS-induced Parkinson's disease in rats, *Indian J. Exp. Biol.*, 2013, **51**(5), 357–362.
110. Y. H. Siddique, S. Jyoti and F. Naz, Effect of epicatechin gallate dietary supplementation on transgenic Drosophila model of Parkinson's disease, *J. Diet. Suppl.*, 2014, **11**(2), 121–130.
111. C. H. Ng, M. S. Guan, C. Koh, X. Ouyang, F. Yu and E. K. Tan, *et al.*, AMP kinase activation mitigates dopaminergic dysfunction and mitochondrial abnormalities in Drosophila models of Parkinson's disease, *J. Neurosci.*, 2012, **32**(41), 14311–14317.

112. L. Reznichenko, T. Amit, H. Zheng, Y. Avramovich-Tirosh, M. B. Youdim and O. Weinreb, *et al.,* Reduction of iron-regulated amyloid precursor protein and beta-amyloid peptide by (-)-epigallocatechin-3-gallate in cell cultures: implications for iron chelation in Alzheimer's disease, *J. Neurochem.*, 2006, **97**(2), 527–536.
113. D. E. Ehrnhoefer, J. Bieschke, A. Boeddrich, M. Herbst, L. Masino and R. Lurz, *et al.,* EGCG redirects amyloidogenic polypeptides into unstructured, off-pathway oligomers, *Nat. Struct. Mol. Biol.*, 2008, **15**(6), 558–566.
114. J. Bieschke, J. Russ, R. P. Friedrich, D. E. Ehrnhoefer, H. Wobst and K. Neugebauer, *et al.,* EGCG remodels mature alpha-synuclein and amyloid-beta fibrils and reduces cellular toxicity, *Proc. Natl. Acad. Sci. U. S. A.*, 2010, **107**(17), 7710–7715.
115. S. Y. Bae, S. Kim, H. Hwang, H. K. Kim, H. C. Yoon and J. H. Kim, *et al.,* Amyloid formation and disaggregation of alpha-synuclein and its tandem repeat (alpha-TR), *Biochem. Biophys. Res. Commun.*, 2010, **400**(4), 531–536.
116. K. Tomobe, H. Fujii, B. Sun, H. Nishioka and O. I. Aruoma, Modulation of infection-induced inflammation and locomotive deficit and longevity in senescence-accelerated mice-prone (SAMP8) model by the oligomerized polyphenol Oligonol, *Biomed. Pharmacother.*, 2007, **61**(7), 427–434.
117. A. Anandhan, K. Tamilselvam, T. Radhiga, S. Rao, M. M. Essa and T. Manivasagam, Theaflavin, a black tea polyphenol, protects nigral dopaminergic neurons against chronic MPTP/probenecid induced Parkinson's disease, *Brain Res.*, 2012, **1433**, 104–113.
118. A. Anandhan, M. M. Essa and T. Manivasagam, Therapeutic attenuation of neuroinflammation and apoptosis by black tea theaflavin in chronic MPTP/probenecid model of Parkinson's disease, *Neurotoxic. Res.*, 2013, **23**(2), 166–173.
119. A. Anandhan, U. Janakiraman and T. Manivasagam, Theaflavin ameliorates behavioral deficits, biochemical indices and monoamine transporters expression against subacute 1-methyl-4-phenyl-1,2,3,6-tetrahydropyridine (MPTP)-induced mouse model of Parkinson's disease, *Neuroscience*, 2012, **218**, 257–267.
120. G. Grelle, A. Otto, M. Lorenz, R. F. Frank, E. E. Wanker and J. Bieschke, Black tea theaflavins inhibit formation of toxic amyloid-beta and alpha-synuclein fibrils, *Biochemistry*, 2011, **50**(49), 10624–10636.
121. K. E. Strathearn, G. G. Yousef, M. H. Grace, S. L. Roy, M. A. Tambe and M. G. Ferruzzi, *et al.,* Neuroprotective effects of anthocyanin- and proanthocyanidin-rich extracts in cellular models of Parkinsons disease, *Brain Res.*, 2014, **1555**, 60–77.
122. X. Wu, G. R. Beecher, J. M. Holden, D. B. Haytowitz, S. E. Gebhardt and R. L. Prior, Concentrations of anthocyanins in common foods in the United States and estimation of normal consumption, *J. Agric. Food Chem.*, 2006, **54**(11), 4069–4075.

123. J. M. Kong, L. S. Chia, N. K. Goh, T. F. Chia and R. Brouillard, Analysis and biological activities of anthocyanins, *Phytochemistry*, 2003, **64**(5), 923–933.
124. D. Vauzour, G. Ravaioli, K. Vafeiadou, A. Rodriguez-Mateos, C. Angeloni and J. P. Spencer, Peroxynitrite induced formation of the neurotoxins 5-S-cysteinyl-dopamine and DHBT-1: implications for Parkinson's disease and protection by polyphenols, *Arch. Biochem. Biophys.*, 2008, **476**(2), 145–151.
125. M. Roghani, A. Niknam, M. R. Jalali-Nadoushan, Z. Kiasalari, M. Khalili and T. Baluchnejadmojarad, Oral pelargonidin exerts dose-dependent neuroprotection in 6-hydroxydopamine rat model of hemi-parkinsonism, *Brain Res. Bull.*, 2010, **82**(5–6), 279–283.
126. V. Tapias, J. R. Cannon and J. T. Greenamyre, Pomegranate juice exacerbates oxidative stress and nigrostriatal degeneration in Parkinson's disease, *Neurobiol. Aging*, 2014, **35**(5), 1162–1176.
127. J. A. Joseph, B. Shukitt-Hale, N. A. Denisova, D. Bielinski, A. Martin and J. J. McEwen, *et al.,* Reversals of age-related declines in neuronal signal transduction, cognitive, and motor behavioral deficits with blueberry, spinach, or strawberry dietary supplementation, *J. Neurosci.*, 1999, **19**(18), 8114–8121.
128. Y. Yao and A. Vieira, Protective activities of Vaccinium antioxidants with potential relevance to mitochondrial dysfunction and neurotoxicity, *Neurotoxicology*, 2007, **28**(1), 93–100.
129. M. M. Rahman, T. Ichiyanagi, T. Komiyama, S. Sato and T. Konishi, Effects of anthocyanins on psychological stress-induced oxidative stress and neurotransmitter status, *J. Agric. Food Chem.*, 2008, **56**(16), 7545–7550.
130. H. G. Kim, M. S. Ju, J. S. Shim, M. C. Kim, S. H. Lee and Y. Huh, *et al.,* Mulberry fruit protects dopaminergic neurons in toxin-induced Parkinson's disease models, *Br. J. Nutr.*, 2010, **104**(1), 8–16.
131. J. Long, H. Gao, L. Sun, J. Liu and X. Zhao-Wilson, Grape extract protects mitochondria from oxidative damage and improves locomotor dysfunction and extends lifespan in a Drosophila Parkinson's disease model, *Rejuvenation Res.*, 2009, **12**(5), 321–331.
132. S. Subash, M. M. Essa, S. Al-Adawi, M. A. Memon, T. Manivasagam and M. Akbar, Neuroprotective effects of berry fruits on neurodegenerative diseases, *Neural Regener. Res.*, 2014, **9**(16), 1557–1566.
133. A. Faria, D. Pestana, D. Teixeira, J. Azevedo, V. De Freitas and N. Mateus, *et al.,* Flavonoid transport across RBE4 cells: A blood-brain barrier model, *Cell. Mol. Biol. Lett.*, 2010, **15**(2), 234–241.
134. S. Talavera, C. Felgines, O. Texier, C. Besson, A. Gil-Izquierdo and J. L. Lamaison, *et al.,* Anthocyanin metabolism in rats and their distribution to digestive area, kidney, and brain, *J. Agric. Food Chem.*, 2005, **53**(10), 3902–3908.
135. C. Andres-Lacueva, B. Shukitt-Hale, R. L. Galli, O. Jauregui, R. M. Lamuela-Raventos and J. A. Joseph, Anthocyanins in aged blueberry-fed rats are found centrally and may enhance memory, *Nutr. Neurosci.*, 2005, **8**(2), 111–120.

136. P. E. Milbury and W. Kalt, Xenobiotic metabolism and berry flavonoid transport across the blood-brain barrier, *J. Agric. Food Chem.*, 2010, **58**(7), 3950–3956.
137. L. Ho, M. G. Ferruzzi, E. M. Janle, J. Wang, B. Gong and T. Y. Chen, *et al.*, Identification of brain-targeted bioactive dietary quercetin-3-O-glucuronide as a novel intervention for Alzheimer's disease, *FASEB J.*, 2013, **27**(2), 769–781.
138. S. A. Aherne and N. M. O'Brien, Dietary flavonols: chemistry, food content, and metabolism, *Nutrition*, 2002, **18**(1), 75–81.
139. S. Gelinas and M. G. Martinoli, Neuroprotective effect of estradiol and phytoestrogens on MPP+-induced cytotoxicity in neuronal PC12 cells, *J. Neurosci. Res.*, 2002, **70**(1), 90–96.
140. J. Bournival, P. Quessy and M. G. Martinoli, Protective effects of resveratrol and quercetin against MPP+ -induced oxidative stress act by modulating markers of apoptotic death in dopaminergic neurons, *Cell. Mol. Neurobiol.*, 2009, **29**(8), 1169–1180.
141. Z. J. Zhang, L. C. Cheang, M. W. Wang and S. M. Lee, Quercetin exerts a neuroprotective effect through inhibition of the iNOS/NO system and pro-inflammation gene expression in PC12 cells and in zebrafish, *Int. J. Mol. Med.*, 2011, **27**(2), 195–203.
142. T. M. Kaariainen, M. Piltonen, B. Ossola, H. Kekki, S. Lehtonen and T. Nenonen, *et al.*, Lack of robust protective effect of quercetin in two types of 6-hydroxydopamine-induced parkinsonian models in rats and dopaminergic cell cultures, *Brain Res.*, 2008, **1203**, 149–159.
143. G. Bureau, F. Longpre and M. G. Martinoli, Resveratrol and quercetin, two natural polyphenols, reduce apoptotic neuronal cell death induced by neuroinflammation, *J. Neurosci. Res.*, 2008, **86**(2), 403–410.
144. J. Bournival, M. Plouffe, J. Renaud, C. Provencher and M. G. Martinoli, Quercetin and sesamin protect dopaminergic cells from MPP+-induced neuroinflammation in a microglial (N9)-neuronal (PC12) coculture system, *Oxid. Med. Cell. Longevity*, 2012, **2012**, 921941.
145. N. Haleagrahara, C. J. Siew, N. K. Mitra and M. Kumari, Neuroprotective effect of bioflavonoid quercetin in 6-hydroxydopamine-induced oxidative stress biomarkers in the rat striatum, *Neurosci. Lett.*, 2011, **500**(2), 139–143.
146. S. S. Karuppagounder, S. K. Madathil, M. Pandey, R. Haobam, U. Rajamma and K. P. Mohanakumar, Quercetin up-regulates mitochondrial complex-I activity to protect against programmed cell death in rotenone model of Parkinson's disease in rats, *Neuroscience*, 2013, **236**, 136–148.
147. C. Lv, T. Hong, Z. Yang, Y. Zhang, L. Wang and M. Dong, *et al.*, Effect of Quercetin in the 1-Methyl-4-phenyl-1,2,3,6-tetrahydropyridine-Induced Mouse Model of Parkinson's Disease, *Evidence-Based Complementary Altern. Med.*, 2012, **2012**, 928643.
148. V. Donaire, M. Niso, J. M. Moran, L. Garcia, R. A. Gonzalez-Polo and G. Soler, *et al.*, Heat shock proteins protect both MPP(+) and paraquat neurotoxicity, *Brain Res. Bull.*, 2005, **67**(6), 509–514.

149. B. Ossola, T. M. Kaarainen, A. Raasmaja and P. T. Mannisto, Time-dependent protective and harmful effects of quercetin on 6-OHDA-induced toxicity in neuronal SH-SY5Y cells, *Toxicology*, 2008, **250**(1), 1–8.
150. D. Metodiewa, A. K. Jaiswal, N. Cenas, E. Dickancaite and J. Segura-Aguilar, Quercetin may act as a cytotoxic prooxidant after its metabolic activation to semiquinone and quinoidal product, *Free Radical Biol. Med.*, 1999, **26**(1–2), 107–116.
151. G. G. Kuiper, J. G. Lemmen, B. Carlsson, J. C. Corton, S. H. Safe and P. T. van der Saag, *et al.,* Interaction of estrogenic chemicals and phytoestrogens with estrogen receptor beta, *Endocrinology*, 1998, **139**(10), 4252–4263.
152. S. Di Giovanni, S. Eleuteri, K. E. Paleologou, G. Yin, M. Zweckstetter and P. A. Carrupt, *et al.,* Entacapone and tolcapone, two catechol O-methyltransferase inhibitors, block fibril formation of alpha-synuclein and beta-amyloid and protect against amyloid-induced toxicity, *J. Biol. Chem.*, 2010, **285**(20), 14941–14954.
153. M. Zhu, S. Han and A. L. Fink, Oxidized quercetin inhibits alpha-synuclein fibrillization, *Biochim. Biophys. Acta*, 2013, **1830**(4), 2872–2881.
154. Z. Liu, X. Tao, C. Zhang, Y. Lu and D. Wei, Protective effects of hyperoside (quercetin-3-o-galactoside) to PC12 cells against cytotoxicity induced by hydrogen peroxide and tert-butyl hydroperoxide, *Biomed. Pharmacother.*, 2005, **59**(9), 481–490.
155. S. E. Park, K. Sapkota, J. H. Choi, M. K. Kim, Y. H. Kim and K. M. Kim, *et al.,* Rutin from Dendropanax morbifera Leveille protects human dopaminergic cells against rotenone induced cell injury through inhibiting JNK and p38 MAPK signaling, *Neurochem. Res.*, 2014, **39**(4), 707–718.
156. K. B. Magalingam, A. Radhakrishnan, P. Ramdas and N. Haleagrahara, Quercetin Glycosides Induced Neuroprotection by Changes in the Gene Expression in a Cellular Model of Parkinson's Disease, *J. Mol. Neurosci.*, 2014, **55**(3), 609–617.
157. K. B. Magalingam, A. Radhakrishnan and N. Haleagrahara, Protective effects of flavonol isoquercitrin, against 6-hydroxy dopamine (6-OHDA)-induced toxicity in PC12 cells, *BMC Res. Notes*, 2014, **7**, 49.
158. A. Singh, P. S. Naidu and S. K. Kulkarni, Quercetin potentiates L-Dopa reversal of drug-induced catalepsy in rats: possible COMT/MAO inhibition, *Pharmacology*, 2003, **68**(2), 81–88.
159. M. Gumbleton and K. L. Audus, Progress and limitations in the use of *in vitro* cell cultures to serve as a permeability screen for the blood-brain barrier, *J. Pharm. Sci.*, 2001, **90**(11), 1681–1698.
160. A. Ishisaka, S. Ichikawa, H. Sakakibara, M. K. Piskula, T. Nakamura and Y. Kato, *et al.,* Accumulation of orally administered quercetin in brain tissue and its antioxidative effects in rats, *Free Radical Biol. Med.*, 2011, **51**(7), 1329–1336.
161. S. Li and X. P. Pu, Neuroprotective effect of kaempferol against a 1-methyl-4-phenyl-1,2,3,6-tetrahydropyridine-induced mouse model of Parkinson's disease, *Biol. Pharm. Bull.*, 2011, **34**(8), 1291–1296.
162. W. Qu, L. Fan, Y. C. Kim, S. Ishikawa, S. M. Iguchi-Ariga and X. P. Pu, *et al.,* Kaempferol derivatives prevent oxidative stress-induced cell

death in a DJ-1-dependent manner, *J. Pharmacol. Sci.*, 2009, **110**(2), 191–200.

163. G. Filomeni, I. Graziani, D. De Zio, L. Dini, D. Centonze and G. Rotilio, *et al.,* Neuroprotection of kaempferol by autophagy in models of rotenone-mediated acute toxicity: possible implications for Parkinson's disease, *Neurobiol. Aging*, 2012, **33**(4), 767–785.
164. K. Zhang, Z. Ma, J. Wang, A. Xie and J. Xie, Myricetin attenuated MPP(+)-induced cytotoxicity by anti-oxidation and inhibition of MKK4 and JNK activation in MES23.5 cells, *Neuropharmacology*, 2011, **61**(1–2), 329–335.
165. X. Zhu and Y. H. Jia, Inhibition of catechol-o-methyltransferase (COMT) by myricetin, dihydromyricetin, and myricitrin, *Pharmazie*, 2014, **69**(3), 183–186.
166. J. Chen, X. Q. Tang, J. L. Zhi, Y. Cui, H. M. Yu and E. H. Tang, *et al.,* Curcumin protects PC12 cells against 1-methyl-4-phenylpyridinium ion-induced apoptosis by bcl-2-mitochondria-ROS-iNOS pathway, *Apoptosis*, 2006, **11**(6), 943–953.
167. J. Wang, X. X. Du, H. Jiang and J. X. Xie, Curcumin attenuates 6-hydroxydopamine-induced cytotoxicity by anti-oxidation and nuclear factor-kappa B modulation in MES23.5 cells, *Biochem. Pharmacol.*, 2009, **78**(2), 178–183.
168. S. Yang, D. Zhang, Z. Yang, X. Hu, S. Qian and J. Liu, *et al.,* Curcumin protects dopaminergic neuron against LPS induced neurotoxicity in primary rat neuron/glia culture, *Neurochem. Res.*, 2008, **33**(10), 2044–2053.
169. M. A. Tegenge, L. Rajbhandari, S. Shrestha, A. Mithal, S. Hosmane and A. Venkatesan, Curcumin protects axons from degeneration in the setting of local neuroinflammation, *Exp. Neurol.*, 2014, **253**, 102–110.
170. S. Yu, W. Zheng, N. Xin, Z. H. Chi, N. Q. Wang and Y. X. Nie, *et al.,* Curcumin prevents dopaminergic neuronal death through inhibition of the c-Jun N-terminal kinase pathway, *Rejuvenation Res.*, 2010, **13**(1), 55–64.
171. T. F. Jiang, Y. J. Zhang, H. Y. Zhou, H. M. Wang, L. P. Tian and J. Liu, *et al.,* Curcumin ameliorates the neurodegenerative pathology in A53T alpha-synuclein cell model of Parkinson's disease through the down-regulation of mTOR/p70S6K signaling and the recovery of macroautophagy, *J. Neuroimmune Pharmacol.*, 2013, **8**(1), 356–369.
172. J. Yang, S. Song, J. Li and T. Liang, Neuroprotective effect of curcumin on hippocampal injury in 6-OHDA-induced Parkinson's disease rat, *Pathol., Res. Pract.*, 2014, **210**(6), 357–362.
173. W. Tripanichkul and E. O. Jaroensuppaperch, Ameliorating effects of curcumin on 6-OHDA-induced dopaminergic denervation, glial response, and SOD1 reduction in the striatum of hemiparkinsonian mice, *Eur. Rev. Med. Pharmacol. Sci.*, 2013, **17**(10), 1360–1368.
174. A. Rajeswari and M. Sabesan, Inhibition of monoamine oxidase-B by the polyphenolic compound, curcumin and its metabolite

tetrahydrocurcumin, in a model of Parkinson's disease induced by MPTP neurodegeneration in mice, *Inflammopharmacology*, 2008, **16**(2), 96–99.

175. D. Yang, T. Li, Z. Liu, N. Arbez, J. Yan and T. H. Moran, *et al.,* LRRK2 kinase activity mediates toxic interactions between genetic mutation and oxidative stress in a Drosophila model: suppression by curcumin, *Neurobiol. Dis.*, 2012, **47**(3), 385–392.
176. R. P. Ojha, M. Rastogi, B. P. Devi, A. Agrawal and G. P. Dubey, Neuroprotective effect of curcuminoids against inflammation-mediated dopaminergic neurodegeneration in the MPTP model of Parkinson's disease, *J. Neuroimmune Pharmacol.*, 2012, **7**(3), 609–618.
177. Z. Mansouri, M. Sabetkasaei, F. Moradi, F. Masoudnia and A. Ataie, Curcumin has neuroprotection effect on homocysteine rat model of Parkinson, *J. Mol. Neurosci.*, 2012, **47**(2), 234–242.
178. N. Pandey, J. Strider, W. C. Nolan, S. X. Yan and J. E. Galvin, Curcumin inhibits aggregation of alpha-synuclein, *Acta Neuropathol.*, 2008, **115**(4), 479–489.
179. B. S. Gadad, P. K. Subramanya, S. Pullabhatla, I. S. Shantharam and K. S. Rao, Curcumin-glucoside, a novel synthetic derivative of curcumin, inhibits alpha-synuclein oligomer formation: relevance to Parkinson's disease, *Curr. Pharm. Des.*, 2012, **18**(1), 76–84.
180. P. K. Singh, V. Kotia, D. Ghosh, G. M. Mohite, A. Kumar and S. K. Maji, Curcumin modulates alpha-synuclein aggregation and toxicity, *ACS Chem. Neurosci.*, 2013, **4**(3), 393–407.
181. M. S. Wang, S. Boddapati, S. Emadi and M. R. Sierks, Curcumin reduces alpha-synuclein induced cytotoxicity in Parkinson's disease cell model, *BMC Neurosci.*, 2010, **11**, 57.
182. K. Ono, M. Hirohata and M. Yamada, Alpha-synuclein assembly as a therapeutic target of Parkinson's disease and related disorders, *Curr. Pharm. Des.*, 2008, **14**(30), 3247–3266.
183. M. L. Hegde, P. M. Hegde, K. S. Rao and S. Mitra, Oxidative genome damage and its repair in neurodegenerative diseases: function of transition metals as a double-edged sword, *J. Alzheimer's Dis.*, 2011, **24**(suppl. 2), 183–198.
184. X. X. Du, H. M. Xu, H. Jiang, N. Song, J. Wang and J. X. Xie, Curcumin protects nigral dopaminergic neurons by iron-chelation in the 6-hydroxydopamine rat model of Parkinson's disease, *Neurosci. Bull.*, 2012, **28**(3), 253–258.
185. K. Jomova, D. Vondrakova, M. Lawson and M. Valko, Metals, oxidative stress and neurodegenerative disorders, *Mol. Cell. Biochem.*, 2010, **345**(1–2), 91–104.
186. G. Harish, C. Venkateshappa, R. B. Mythri, S. K. Dubey, K. Mishra and N. Singh, *et al.,* Bioconjugates of curcumin display improved protection against glutathione depletion mediated oxidative stress in a dopaminergic neuronal cell line: Implications for Parkinson's disease, *Bioorg. Med. Chem.*, 2010, **18**(7), 2631–2638.

187. R. B. Mythri, G. Harish, S. K. Dubey, K. Misra and M. M. Bharath, Glutamoyl diester of the dietary polyphenol curcumin offers improved protection against peroxynitrite-mediated nitrosative stress and damage of brain mitochondria *in vitro*: implications for Parkinson's disease, *Mol. Cell. Biochem.*, 2011, **347**(1–2), 135–143.
188. A. Marchiani, S. Mammi, G. Siligardi, R. Hussain, I. Tessari and L. Bubacco, *et al.,* Small molecules interacting with alpha-synuclein: anti-aggregating and cytoprotective properties, *Amino Acids*, 2013, **45**(2), 327–338.
189. R. L. Jayaraj, K. Tamilselvam, T. Manivasagam and N. Elangovan, Neuroprotective effect of CNB-001, a novel pyrazole derivative of curcumin on biochemical and apoptotic markers against rotenone-induced SK-N-SH cellular model of Parkinson's disease, *J. Mol. Neurosci.*, 2013, **51**(3), 863–870.
190. S. Gautam, S. Karmakar, A. Bose and P. K. Chowdhury, beta-cyclodextrin and curcumin, a potent cocktail for disaggregating and/or inhibiting amyloids: a case study with alpha-synuclein, *Biochemistry*, 2014, **53**(25), 4081–4083.
191. E. B. Souto, P. Severino, R. Basso and M. H. Santana, Encapsulation of antioxidants in gastrointestinal-resistant nanoparticulate carriers, *Methods Mol. Biol.*, 2013, **1028**, 37–46.
192. S. Chiu, K. J. Terpstra, Y. Bureau, J. Hou, H. Raheb and Z. Cernvosky, *et al.,* Liposomal-formulated curcumin [Lipocurc] targeting HDAC (histone deacetylase) prevents apoptosis and improves motor deficits in Park 7 (DJ-1)-knockout rat model of Parkinson's disease: implications for epigenetics-based nanotechnology-driven drug platform, *J. Complementary Integr. Med.*, 2013, **10**(1), 75–88.
193. M. N. Clifford, Chlorogenic acids and other cinnamates - nature, occurrence and dietary burden, *J. Sci. Food Agric.*, 1999, **79**(3), 362–372.
194. B. Shan, Y. Z. Cai, M. Sun and H. Corke, Antioxidant capacity of 26 spice extracts and characterization of their phenolic constituents, *J. Agric. Food Chem.*, 2005, **53**(20), 7749–7759.
195. M. Kaur, B. Velmurugan, S. Rajamanickam, R. Agarwal and C. Agarwal, Gallic acid, an active constituent of grape seed extract, exhibits anti-proliferative, pro-apoptotic and anti-tumorigenic effects against prostate carcinoma xenograft growth in nude mice, *Pharm. Res.*, 2009, **26**(9), 2133–2140.
196. M. J. Sameri, A. Sarkaki, Y. Farbood and S. M. Mansouri, Motor disorders and impaired electrical power of pallidal EEG improved by gallic acid in animal model of Parkinson's disease, *Pak. J. Biol. Sci.*, 2011, **14**(24), 1109–1116.
197. M. T. Mansouri, Y. Farbood, M. J. Sameri, A. Sarkaki, B. Naghizadeh and M. Rafeirad, Neuroprotective effects of oral gallic acid against oxidative stress induced by 6-hydroxydopamine in rats, *Food Chem.*, 2013, **138**(2–3), 1028–1033.

198. H. F. Ortega-Arellano, M. Jimenez-Del-Rio and C. Velez-Pardo, Dmp53, basket and drICE gene knockdown and polyphenol gallic acid increase life span and locomotor activity in a Drosophila Parkinson's disease model, *Genet. Mol. Biol.*, 2013, **36**(4), 608–615.
199. Y. Liu, J. A. Carver, A. N. Calabrese and T. L. Pukala, Gallic acid interacts with alpha-synuclein to prevent the structural collapse necessary for its aggregation, *Biochim. Biophys. Acta*, 2014, **1844**(9), 1481–1485.
200. M. T. Ardah, K. E. Paleologou, G. Lv, S. B. Abul Khair, A. S. Kazim and S. T. Minhas, *et al.,* Structure activity relationship of phenolic acid inhibitors of alpha-synuclein fibril formation and toxicity, *Front. Aging Neurosci.*, 2014, **6**, 197.
201. B. Nair, Final report on the safety assessment of Benzyl Alcohol, Benzoic Acid, and Sodium Benzoate, *Int. J. Toxicol.*, 2001, **20**(suppl. 3), 23–50.
202. S. Khasnavis and K. Pahan, Sodium benzoate, a metabolite of cinnamon and a food additive, upregulates neuroprotective Parkinson disease protein DJ-1 in astrocytes and neurons,*J. Neuroimmune Pharmacol.*, 2012, **7**(2), 424–435.
203. A. Jana, K. K. Modi, A. Roy, J. A. Anderson, R. B. van Breemen and K. Pahan, Up-regulation of neurotrophic factors by cinnamon and its metabolite sodium benzoate: therapeutic implications for neurodegenerative disorders, *J. Neuroimmune Pharmacol.*, 2013, **8**(3), 739–755.
204. A. Cano-Marquina, J. J. Tarin and A. Cano, The impact of coffee on health, *Maturitas*, 2013, **75**(1), 7–21.
205. S. J. Tsai, C. Y. Chao and M. C. Yin, Preventive and therapeutic effects of caffeic acid against inflammatory injury in striatum of MPTP-treated mice, *Eur. J. Pharmacol.*, 2011, **670**(2–3), 441–447.
206. M. Jimenez-Del-Rio, C. Guzman-Martinez and C. Velez-Pardo, The effects of polyphenols on survival and locomotor activity in Drosophila melanogaster exposed to iron and paraquat, *Neurochem. Res.*, 2010, **35**(2), 227–238.
207. A. A. Maruf, H. Lip, H. Wong and P. J. O'Brien, Protective effects of ferulic acid and related polyphenols against glyoxal- or methylglyoxal-induced cytotoxicity and oxidative stress in isolated rat hepatocytes, *Chem.-Biol. Interact.*, 2015, **234**, 96–104.
208. N. A. Fazili and A. Naeem, Anti-fibrillation potency of caffeic acid against an antidepressant induced fibrillogenesis of human alpha-synuclein: Implications for Parkinson's disease, *Biochimie*, 2015, **108**, 178–185.
209. S. H. Sohn, E. Ko, S. B. Jeon, B. J. Lee, S. H. Kim and M. S. Dong, *et al.,* The genome-wide expression profile of Scrophularia ningpoensis-treated thapsigargin-stimulated U-87MG cells, *Neurotoxicology*, 2009, **30**(3), 368–376.
210. D. Vauzour, G. Corona and J. P. Spencer, Caffeic acid, tyrosol and p-coumaric acid are potent inhibitors of 5-S-cysteinyl-dopamine induced neurotoxicity, *Arch. Biochem. Biophys.*, 2010, **501**(1), 106–111.

211. C. Noelker, M. Bacher, P. Gocke, X. Wei, T. Klockgether and Y. Du, *et al.,* The flavanoide caffeic acid phenethyl ester blocks 6-hydroxydopamine-induced neurotoxicity, *Neurosci. Lett.*, 2005, **383**(1–2), 39–43.
212. Z. Ma, X. Wei, C. Fontanilla, C. Noelker, R. Dodel and H. Hampel, *et al.,* Caffeic acid phenethyl ester blocks free radical generation and 6-hydroxydopamine-induced neurotoxicity, *Life Sci.*, 2006, **79**(13), 1307–1311.
213. C. V. Fontanilla, Z. Ma, X. Wei, J. Klotsche, L. Zhao and P. Wisniowski, *et al.,* Caffeic acid phenethyl ester prevents 1-methyl-4-phenyl-1,2,3,6-tetrahydropyridine-induced neurodegeneration, *Neuroscience*, 2011, **188**, 135–141.
214. R. Barros Silva, N. A. Santos, N. M. Martins, D. A. Ferreira, F. Barbosa Jr and V. C. Oliveira Souza, *et al.,* Caffeic acid phenethyl ester protects against the dopaminergic neuronal loss induced by 6-hydroxydopamine in rats, *Neuroscience*, 2013, **233**, 86–94.
215. M. Teraoka, K. Nakaso, C. Kusumoto, S. Katano, N. Tajima and A. Yamashita, *et al.,* Cytoprotective effect of chlorogenic acid against alpha-synuclein-related toxicity in catecholaminergic PC12 cells, *J. Clin. Biochem. Nutr.*, 2012, **51**(2), 122–127.
216. M. A. Rathod, D. Patel, A. Das, S. R. Tipparaju, S. S. Shinde and R. F. Anderson, Inhibition of radical-induced DNA strand breaks by water-soluble constituents of coffee: phenolics and caffeine metabolites, *Free Radical Res.*, 2013, **47**(6–7), 480–487.
217. E. Heitman and D. K. Ingram, Cognitive and neuroprotective effects of chlorogenic acid, *Nutr. Neurosci.*, 2017, **20**(1), 32–39.
218. I. A. Ludwig, M. N. Clifford, M. E. Lean, H. Ashihara and A. Crozier, Coffee: biochemistry and potential impact on health, *Food Funct.*, 2014, **5**(8), 1695–1717.
219. M. E. Obrenovich, N. G. Nair, A. Beyaz, G. Aliev and V. P. Reddy, The role of polyphenolic antioxidants in health, disease, and aging, *Rejuvenation Res.*, 2010, **13**(6), 631–643.
220. W. C. Willett, F. Sacks, A. Trichopoulou, G. Drescher, A. Ferro-Luzzi and E. Helsing, *et al.,* Mediterranean diet pyramid: a cultural model for healthy eating, *Am. J. Clin. Nutr.*, 1995, **61**(suppl. 6), 1402S–1406S.
221. C. Ramassamy, Emerging role of polyphenolic compounds in the treatment of neurodegenerative diseases: a review of their intracellular targets, *Eur. J. Pharmacol.*, 2006, **545**(1), 51–64.
222. A. P. Raval, H. W. Lin, K. R. Dave, R. A. Defazio, D. Della Morte and E. J. Kim, *et al.,* Resveratrol and ischemic preconditioning in the brain, *Curr. Med. Chem.*, 2008, **15**(15), 1545–1551.
223. D. Alvira, M. Yeste-Velasco, J. Folch, E. Verdaguer, A. M. Canudas and M. Pallas, *et al.,* Comparative analysis of the effects of resveratrol in two apoptotic models: inhibition of complex I and potassium deprivation in cerebellar neurons, *Neuroscience*, 2007, **147**(3), 746–756.
224. Y. Wu, X. Li, J. X. Zhu, W. Xie, W. Le and Z. Fan, *et al.,* Resveratrol-activated AMPK/SIRT1/autophagy in cellular models of Parkinson's disease, *Neurosignals*, 2011, **19**(3), 163–174.

225. M. K. Lee, S. J. Kang, M. Poncz, K. J. Song and K. S. Park, Resveratrol protects SH-SY5Y neuroblastoma cells from apoptosis induced by dopamine, *Exp. Mol. Med.*, 2007, **39**(3), 376–384.
226. K. T. Lu, M. C. Ko, B. Y. Chen, J. C. Huang, C. W. Hsieh and M. C. Lee, *et al.,* Neuroprotective effects of resveratrol on MPTP-induced neuron loss mediated by free radical scavenging, *J. Agric. Food Chem.*, 2008, **56**(16), 6910–6913.
227. F. Jin, Q. Wu, Y. F. Lu, Q. H. Gong and J. S. Shi, Neuroprotective effect of resveratrol on 6-OHDA-induced Parkinson's disease in rats, *Eur. J. Pharmacol.*, 2008, **600**(1–3), 78–82.
228. M. M. Khan, A. Ahmad, T. Ishrat, M. B. Khan, M. N. Hoda and G. Khuwaja, *et al.,* Resveratrol attenuates 6-hydroxydopamine-induced oxidative damage and dopamine depletion in rat model of Parkinson's disease, *Brain Res.*, 2010, **1328**, 139–151.
229. A. Anandhan, K. Tamilselvam, D. Vijayraja, N. Ashokkumar, S. Rajasankar and T. Manivasagam, Resveratrol attenuates oxidative stress and improves behaviour in 1-methyl-4-phenyl-1,2,3,6-tetrahydropyridine (MPTP) challenged mice, *Ann. Neurosci.*, 2010, **17**(3), 113–119.
230. M. Pallas, G. Casadesus, M. A. Smith, A. Coto-Montes, C. Pelegri and J. Vilaplana, *et al.,* Resveratrol and neurodegenerative diseases: activation of SIRT1 as the potential pathway towards neuroprotection, *Curr. Neurovasc. Res.*, 2009, **6**(1), 70–81.
231. D. Albani, L. Polito, S. Batelli, S. De Mauro, C. Fracasso and G. Martelli, *et al.,* The SIRT1 activator resveratrol protects SK-N-BE cells from oxidative stress and against toxicity caused by alpha-synuclein or amyloid-beta (1-42) peptide, *J. Neurochem.*, 2009, **110**(5), 1445–1456.
232. D. D. Lofrumento, G. Nicolardi, A. Cianciulli, F. De Nuccio, V. La Pesa and V. Carofiglio, *et al.,* Neuroprotective effects of resveratrol in an MPTP mouse model of Parkinson's-like disease: possible role of SOCS-1 in reducing pro-inflammatory responses, *Innate Immun.*, 2014, **20**(3), 249–260.
233. T. K. Lin, S. D. Chen, Y. C. Chuang, H. Y. Lin, C. R. Huang and J. H. Chuang, *et al.,* Resveratrol partially prevents rotenone-induced neurotoxicity in dopaminergic SH-SY5Y cells through induction of heme oxygenase-1 dependent autophagy, *Int. J. Mol. Sci.*, 2014, **15**(1), 1625–1646.
234. F. Zhang, J. S. Shi, H. Zhou, B. Wilson, J. S. Hong and H. M. Gao, Resveratrol protects dopamine neurons against lipopolysaccharide-induced neurotoxicity through its anti-inflammatory actions, *Mol. Pharmacol.*, 2010, **78**(3), 466–477.
235. R. D. Wight, C. A. Tull, M. W. Deel, B. L. Stroope, A. G. Eubanks and J. A. Chavis, *et al.,* Resveratrol effects on astrocyte function: relevance to neurodegenerative diseases, *Biochem. Biophys. Res. Commun.*, 2012, **426**(1), 112–115.
236. J. Chao, M. S. Yu, Y. S. Ho, M. Wang and R. C. Chang, Dietary oxyresveratrol prevents parkinsonian mimetic 6-hydroxydopamine neurotoxicity, *Free Radical Biol. Med.*, 2008, **45**(7), 1019–1026.

237. J. Chao, H. Li, K. W. Cheng, M. S. Yu, R. C. Chang and M. Wang, Protective effects of pinostilbene, a resveratrol methylated derivative, against 6-hydroxydopamine-induced neurotoxicity in SH-SY5Y cells, *J. Nutr. Biochem.*, 2010, **21**(6), 482–489.
238. X. Li, Y. Li, J. Chen, J. Sun, X. Li and X. Sun, *et al.,* Tetrahydroxystilbene glucoside attenuates MPP+-induced apoptosis in PC12 cells by inhibiting ROS generation and modulating JNK activation, *Neurosci. Lett.*, 2010, **483**(1), 1–5.
239. F. L. Sun, L. Zhang, R. Y. Zhang and L. Li, Tetrahydroxystilbene glucoside protects human neuroblastoma SH-SY5Y cells against MPP+ -induced cytotoxicity, *Eur. J. Pharmacol.*, 2011, **660**(2–3), 283–290.
240. L. Tao, X. Li, L. Zhang, J. Tian, X. Li and X. Sun, *et al.,* Protective effect of tetrahydroxystilbene glucoside on 6-OHDA-induced apoptosis in PC12 cells through the ROS-NO pathway, *PLoS One*, 2011, **6**(10), e26055.
241. L. Zhang, L. Huang, L. Chen, D. Hao and J. Chen, Neuroprotection by tetrahydroxystilbene glucoside in the MPTP mouse model of Parkinson's disease, *Toxicol. Lett.*, 2013, **222**(2), 155–163.
242. Y. Wang, H. Xu, Q. Fu, R. Ma and J. Xiang, Protective effect of resveratrol derived from Polygonum cuspidatum and its liposomal form on nigral cells in parkinsonian rats, *J. Neurol. Sci.*, 2011, **304**(1–2), 29–34.
243. A. Wandler, J. M. Bruun, M. P. Nielsen and B. Richelsen, Ethanol exerts anti-inflammatory effects in human adipose tissue *in vitro*, *Mol. Cell. Endocrinol.*, 2008, **296**(1–2), 26–31.
244. E. Mazzio, J. Huber, S. Darling, N. Harris and K. F. Soliman, Effect of antioxidants on L-glutamate and *N*-methyl-4-phenylpyridinium ion induced-neurotoxicity in PC12 cells, *Neurotoxicology*, 2001, **22**(2), 283–288.
245. V. Lahaie-Collins, J. Bournival, M. Plouffe, J. Carange and M. G. Martinoli, Sesamin modulates tyrosine hydroxylase, superoxide dismutase, catalase, inducible NO synthase and interleukin-6 expression in dopaminergic cells under MPP+-induced oxidative stress, *Oxid. Med. Cell. Longevity*, 2008, **1**(1), 54–62.
246. S. Ahmad, M. B. Khan, M. N. Hoda, K. Bhatia, R. Haque and I. S. Fazili, *et al.,* Neuroprotective effect of sesame seed oil in 6-hydroxydopamine induced neurotoxicity in mice model: cellular, biochemical and neurochemical evidence, *Neurochem. Res.*, 2012, **37**(3), 516–526.
247. T. Fujikawa, N. Kanada, A. Shimada, M. Ogata, I. Suzuki and I. Hayashi, *et al.,* Effect of sesamin in Acanthopanax senticosus HARMS on behavioral dysfunction in rotenone-induced parkinsonian rats, *Biol. Pharm. Bull.*, 2005, **28**(1), 169–172.
248. C. Mytilineou, E. T. Kokotos Leonardi, B. C. Kramer, T. Jamindar and C. W. Olanow, Glial cells mediate toxicity in glutathione-depleted mesencephalic cultures, *J. Neurochem.*, 1999, **73**(1), 112–119.
249. S. Canals, M. J. Casarejos, S. de Bernardo, E. Rodriguez-Martin and M. A. Mena, Nitric oxide triggers the toxicity due to glutathione depletion in midbrain cultures through 12-lipoxygenase, *J. Biol. Chem.*, 2003, **278**(24), 21542–21549.

250. Y. H. Siddique, G. Ara, S. Jyoti and M. Afzal, The dietary supplementation of nordihydroguaiaretic acid (NDGA) delayed the loss of climbing ability in Drosophila model of Parkinson's disease, *J. Diet. Suppl.*, 2012, **9**(1), 1–8.
251. P. Y. Lam and K. M. Ko, (-)Schisandrin B ameliorates paraquat-induced oxidative stress by suppressing glutathione depletion and enhancing glutathione recovery in differentiated PC12 cells, *BioFactors*, 2011, **37**(1), 51–57.
252. K. W. Zeng, T. Zhang, H. Fu, G. X. Liu and X. M. Wang, Schisandrin B exerts anti-neuroinflammatory activity by inhibiting the Toll-like receptor 4-dependent MyD88/IKK/NF-kappaB signaling pathway in lipopolysaccharide-induced microglia, *Eur. J. Pharmacol.*, 2012, **692**(1–3), 29–37.
253. H. H. Chen, S. C. Lin and M. H. Chan, Protective and restorative effects of magnolol on neurotoxicity in mice with 6-hydroxydopamine-induced hemiparkinsonism, *Neurodegener. Dis.*, 2011, **8**(5), 364–374.
254. Y. J. Lee, Y. M. Lee, C. K. Lee, J. K. Jung, S. B. Han and J. T. Hong, Therapeutic applications of compounds in the Magnolia family, *Pharmacol. Ther.*, 2011, **130**(2), 157–176.
255. J. Y. Kim, H. J. Lim, Y. Lee da, J. S. Kim, H. Kim do and H. J. Lee, *et al.*, *In vitro* anti-inflammatory activity of lignans isolated from Magnolia fargesii, *Bioorg. Med. Chem. Lett.*, 2009, **19**(3), 937–940.
256. S. K. Schagen, V. A. Zampeli, E. Makrantonaki and C. C. Zouboulis, Discovering the link between nutrition and skin aging, *Derm.-Endocrinol.*, 2012, **4**(3), 298–307.
257. G. K. Ulusoy, T. Celik, H. Kayir, M. Gursoy, A. T. Isik and T. I. Uzbay, Effects of pioglitazone and retinoic acid in a rotenone model of Parkinson's disease, *Brain Res. Bull.*, 2011, **85**(6), 380–384.
258. L. H. Yin, H. Shen, O. Diaz-Ruiz, C. M. Backman, E. Bae and S. J. Yu, *et al.*, Early post-treatment with 9-cis retinoic acid reduces neurodegeneration of dopaminergic neurons in a rat model of Parkinson's disease, *BMC Neurosci.*, 2012, **13**, 120.
259. B. Cheng, A. A. Martinez, J. Morado, V. Scofield, J. L. Roberts and S. K. Maffi, Retinoic acid protects against proteasome inhibition associated cell death in SH-SY5Y cells *via* the AKT pathway, *Neurochem. Int.*, 2013, **62**(1), 31–42.
260. W. Krezel, N. Ghyselinck, T. A. Samad, V. Dupe, P. Kastner and E. Borrelli, *et al.*, Impaired locomotion and dopamine signaling in retinoid receptor mutant mice, *Science*, 1998, **279**(5352), 863–867.
261. G. Wolf, Vitamin A functions in the regulation of the dopaminergic system in the brain and pituitary gland, *Nutr. Rev.*, 1998, **56**(12), 354–355.
262. Y. T. Cheung, W. K. Lau, M. S. Yu, C. S. Lai, S. C. Yeung and K. F. So, *et al.*, Effects of all-trans-retinoic acid on human SH-SY5Y neuroblastoma as *in vitro* model in neurotoxicity research, *Neurotoxicology*, 2009, **30**(1), 127–135.

263. D. Jantas, A. Roman, J. Kusmierczyk, E. Lorenc-Koci, J. Konieczny and T. Lenda, *et al.,* The extent of neurodegeneration and neuroprotection in two chemical *in vitro* models related to Parkinson's disease is critically dependent on cell culture conditions, *Neurotoxic. Res.*, 2013, **24**(1), 41–54.
264. F. Yi, X. He and D. Wang, Lycopene protects against MPP(+)-induced cytotoxicity by maintaining mitochondrial function in SH-SY5Y cells, *Neurochem. Res.*, 2013, **38**(8), 1747–1757.
265. H. Kaur, S. Chauhan and R. Sandhir, Protective effect of lycopene on oxidative stress and cognitive decline in rotenone induced model of Parkinson's disease, *Neurochem. Res.*, 2011, **36**(8), 1435–1443.
266. V. di Matteo, M. Pierucci, G. Di Giovanni, L. K. Dragani, S. Murzilli and A. Poggi, *et al.,* Intake of tomato-enriched diet protects from 6-hydroxy-dopamine-induced degeneration of rat nigral dopaminergic neurons, *J. Neural Transm. Suppl.*, 2009, **73**, 333–341.
267. H. Suganuma, T. Hirano, Y. Arimoto and T. Inakuma, Effect of tomato intake on striatal monoamine level in a mouse model of experimental Parkinson's disease, *J. Nutr. Sci. Vitaminol.*, 2002, **48**(3), 251–254.
268. Y. Ikeda, S. Tsuji, A. Satoh, M. Ishikura, T. Shirasawa and T. Shimizu, Protective effects of astaxanthin on 6-hydroxydopamine-induced apoptosis in human neuroblastoma SH-SY5Y cells, *J. Neurochem.*, 2008, **107**(6), 1730–1740.
269. D. H. Lee, C. S. Kim and Y. J. Lee, Astaxanthin protects against MPTP/MPP+-induced mitochondrial dysfunction and ROS production *in vivo* and *in vitro*, *Food Chem. Toxicol.*, 2011, **49**(1), 271–280.
270. Q. Ye, B. Huang, X. Zhang, Y. Zhu and X. Chen, Astaxanthin protects against MPP(+)-induced oxidative stress in PC12 cells *via* the HO-1/NOX2 axis, *BMC Neurosci.*, 2012, **13**, 156.
271. Q. Ye, X. Zhang, B. Huang, Y. Zhu and X. Chen, Astaxanthin suppresses MPP(+)-induced oxidative damage in PC12 cells through a Sp1/NR1 signaling pathway, *Mar. Drugs*, 2013, **11**(4), 1019–1034.
272. K. Jomova and M. Valko, Advances in metal-induced oxidative stress and human disease, *Toxicology*, 2011, **283**(2–3), 65–87.
273. K. Ono and M. Yamada, Vitamin A potently destabilizes preformed alpha-synuclein fibrils *in vitro*: implications for Lewy body diseases, *Neurobiol. Dis.*, 2007, **25**(2), 446–454.
274. F. J. Jimenez-Jimenez, J. A. Molina, P. Fernandez-Calle, A. Vazquez, M. Pondal and T. del Ser, *et al.,* Serum levels of vitamin A in Parkinson's disease, *J. Neurol. Sci.*, 1992, **111**(1), 73–76.
275. D. King, J. R. Playfer and N. B. Roberts, Concentrations of vitamins A, C and E in elderly patients with Parkinson's disease, *Postgrad. Med. J.*, 1992, **68**(802), 634–637.
276. R. A. Abbott, M. Cox, H. Markus and A. Tomkins, Diet, body size and micronutrient status in Parkinson's disease, *Eur. J. Clin. Nutr.*, 1992, **46**(12), 879–884.
277. W. L. Scheider, L. A. Hershey, J. E. Vena, T. Holmlund, J. R. Marshall and J. L. Freudenheim, Dietary antioxidants and other dietary factors in the etiology of Parkinson's disease, *Mov. Disord.*, 1997, **12**(2), 190–196.

278. C. Anderson, H. Checkoway, G. M. Franklin, S. Beresford, T. Smith-Weller and P. D. Swanson, Dietary factors in Parkinson's disease: the role of food groups and specific foods, *Mov. Disord.*, 1999, **14**(1), 21–27.
279. S. M. Zhang, M. A. Hernan, H. Chen, D. Spiegelman, W. C. Willett and A. Ascherio, Intakes of vitamins E and C, carotenoids, vitamin supplements, and PD risk, *Neurology*, 2002, **59**(8), 1161–1169.
280. R. D. Abbott, G. W. Ross, L. R. White, W. T. Sanderson, C. M. Burchfiel and M. Kashon, *et al.*, Environmental, life-style, and physical precursors of clinical Parkinson's disease: recent findings from the Honolulu-Asia Aging Study, *J. Neurol.*, 2003, **250**(suppl. 3), III30–III39.
281. M. Etminan, S. S. Gill and A. Samii, Intake of vitamin E, vitamin C, and carotenoids and the risk of Parkinson's disease: a meta-analysis, *Lancet Neurol.*, 2005, **4**(6), 362–365.
282. A. Takeda, O. P. Nyssen, A. Syed, E. Jansen, B. Bueno-de-Mesquita and V. Gallo, Vitamin A and carotenoids and the risk of Parkinson's disease: a systematic review and meta-analysis, *Neuroepidemiology*, 2014, **42**(1), 25–38.
283. R. Ansari, A. Mahta, E. Mallack and J. J. Luo, Hyperhomocysteinemia and neurologic disorders: a review, *J. Clin. Neurol.*, 2014, **10**(4), 281–288.
284. G. Blundell, B. G. Jones, F. A. Rose and N. Tudball, Homocysteine mediated endothelial cell toxicity and its amelioration, *Atherosclerosis*, 1996, **122**(2), 163–172.
285. P. I. Ho, D. Ortiz, E. Rogers and T. B. Shea, Multiple aspects of homocysteine neurotoxicity: glutamate excitotoxicity, kinase hyperactivation and DNA damage, *J. Neurosci. Res.*, 2002, **70**(5), 694–702.
286. W. Duan, B. Ladenheim, R. G. Cutler, I. I. Kruman, J. L. Cadet and M. P. Mattson, Dietary folate deficiency and elevated homocysteine levels endanger dopaminergic neurons in models of Parkinson's disease, *J. Neurochem.*, 2002, **80**(1), 101–110.
287. H. W. Shin and Y. H. Sohn, Hyperhomocysteinemia in patients with Parkinson's disease and relationship to vitamin B level, *J. Mov. Disord.*, 2009, **2**(1), 33–36.
288. T. Muller, Role of homocysteine in the treatment of Parkinson's disease, *Expert Rev. Neurother.*, 2008, **8**(6), 957–967.
289. M. Bialecka, M. Kurzawski, A. Roszmann, P. Robowski, E. J. Sitek and K. Honczarenko, *et al.*, Association of COMT, MTHFR, and SLC19A1(RFC-1) polymorphisms with homocysteine blood levels and cognitive impairment in Parkinson's disease, *Pharmacogenet. Genomics*, 2012, **22**(10), 716–724.
290. J. Dorszewska, J. Florczak, A. Rozycka, B. Kempisty, J. Jaroszewska-Kolecka and K. Chojnacka, *et al.*, Oxidative DNA damage and level of thiols as related to polymorphisms of MTHFR, MTR, MTHFD1 in Alzheimer's and Parkinson's diseases, *Acta Neurobiol. Exp.*, 2007, **67**(2), 113–129.
291. D. A. Adelekan and D. I. Thurnham, Effects of combined riboflavin and iron deficiency on the hematological status and tissue iron concentrations of the rat, *J. Nutr.*, 1986, **116**(7), 1257–1265.

292. H. J. Powers, L. T. Weaver, S. Austin, A. J. Wright and S. J. Fairweather-Tait, Riboflavin deficiency in the rat: effects on iron utilization and loss, *Br. J. Nutr.*, 1991, **65**(3), 487–496.
293. E. Moore, A. Mander, D. Ames, R. Carne, K. Sanders and D. Watters, Cognitive impairment and vitamin B12: a review, *Int. Psychogeriatr.*, 2012, **24**(4), 541–556.
294. Y. A. Rajabally and J. Martey, Neuropathy in Parkinson disease: prevalence and determinants, *Neurology*, 2011, **77**(22), 1947–1950.
295. J. DiFrancisco-Donoghue, E. M. Lamberg, E. Rabin, A. Elokda, E. Fazzini and W. G. Werner, Effects of exercise and B vitamins on homocysteine and glutathione in Parkinson's disease: a randomized trial, *Neurodegener. Dis.*, 2012, **10**(1–4), 127–134.
296. S. H. Lee, M. J. Kim, B. J. Kim, S. R. Kim, S. Chun and J. S. Ryu, *et al.*, Homocysteine-lowering therapy or antioxidant therapy for bone loss in Parkinson's disease, *Mov. Disord.*, 2010, **25**(3), 332–340.
297. H. Jia, Z. Liu, X. Li, Z. Feng, J. Hao and X. Li, *et al.*, Synergistic anti-Parkinsonism activity of high doses of B vitamins in a chronic cellular model, *Neurobiol. Aging*, 2010, **31**(4), 636–646.
298. H. Haghdoost-Yazdi, N. Fraidouni, A. Faraji, H. Jahanihashemi and M. Sarookhani, High intake of folic acid or complex of B vitamins provides anti-Parkinsonism effect: no role for serum level of homocysteine, *Behav. Brain Res.*, 2012, **233**(2), 375–381.
299. M. F. McCarty, J. Barroso-Aranda and F. Contreras, High-dose folate and dietary purines promote scavenging of peroxynitrite-derived radicals–clinical potential in inflammatory disorders, *Med. Hypotheses*, 2009, **73**(5), 824–834.
300. M. Jalali-Nadoushan and M. Roghani, Alpha-lipoic acid protects against 6-hydroxydopamine-induced neurotoxicity in a rat model of hemi-parkinsonism, *Brain Res.*, 2013, **1505**, 68–74.
301. S. A. Zaitone, D. M. Abo-Elmatty and A. A. Shaalan, Acetyl-L-carnitine and alpha-lipoic acid affect rotenone-induced damage in nigral dopaminergic neurons of rat brain, implication for Parkinson's disease therapy, *Pharmacol., Biochem. Behav.*, 2012, **100**(3), 347–360.
302. S. Karunakaran, L. Diwakar, U. Saeed, V. Agarwal, S. Ramakrishnan and S. Iyengar, *et al.*, Activation of apoptosis signal regulating kinase 1 (ASK1) and translocation of death-associated protein, Daxx, in substantia nigra pars compacta in a mouse model of Parkinson's disease: protection by alpha-lipoic acid, *FASEB J.*, 2007, **21**(9), 2226–2236.
303. Y. Koriyama, Y. Nakayama, S. Matsugo and S. Kato, Protective effect of lipoic acid against oxidative stress is mediated by Keap1/Nrf2-dependent heme oxygenase-1 induction in the RGC-5 cellline, *Brain Res.*, 2013, **1499**, 145–157.
304. H. Chen, S. M. Zhang, M. A. Schwarzschild, M. A. Hernan, G. Logroscino and W. C. Willett, *et al.*, Folate intake and risk of Parkinson's disease, *Am. J. Epidemiol.*, 2004, **160**(4), 368–375.

305. L. M. de Lau, P. J. Koudstaal, J. C. Witteman, A. Hofman and M. M. Breteler, Dietary folate, vitamin B12, and vitamin B6 and the risk of Parkinson disease, *Neurology*, 2006, **67**(2), 315–318.
306. K. Murakami, Y. Miyake, S. Sasaki, K. Tanaka, W. Fukushima and C. Kiyohara, *et al.*, Dietary intake of folate, vitamin B6, vitamin B12 and riboflavin and risk of Parkinson's disease: a case-control study in Japan, *Br. J. Nutr.*, 2010, **104**(5), 757–764.
307. E. Balk, M. Chung, G. Raman, A. Tatsioni, P. Chew and S. Ip, *et al.*, B vitamins and berries and age-related neurodegenerative disorders, *Evid. Rep. Technol. Assess.*, 2006, **134**, 1–161.
308. F. E. Harrison and J. M. May, Vitamin C function in the brain: vital role of the ascorbate transporter SVCT2, *Free Radical Biol. Med.*, 2009, **46**(6), 719–730.
309. G. Levay and W. J. Bodell, Detection of dopamine–DNA adducts: potential role in Parkinson's disease, *Carcinogenesis*, 1993, **14**(6), 1241–1245.
310. S. Canals, M. J. Casarejos, S. de Bernardo, E. Rodriguez-Martin and M. A. Mena, Glutathione depletion switches nitric oxide neurotrophic effects to cell death in midbrain cultures: implications for Parkinson's disease, *J. Neurochem.*, 2001, **79**(6), 1183–1195.
311. J. Ehrhart and G. D. Zeevalk, Cooperative interaction between ascorbate and glutathione during mitochondrial impairment in mesencephalic cultures, *J. Neurochem.*, 2003, **86**(6), 1487–1497.
312. R. Soto-Otero, E. Mendez-Alvarez, A. Hermida-Ameijeiras, A. M. Munoz-Patino and J. L. Labandeira-Garcia, Autoxidation and neurotoxicity of 6-hydroxydopamine in the presence of some antioxidants: potential implication in relation to the pathogenesis of Parkinson's disease, *J. Neurochem.*, 2000, **74**(4), 1605–1612.
313. Y. Jiang, L. Pei, S. Li, M. Wang and F. Liu, Extracellular dopamine induces the oxidative toxicity of SH-SY5Y cells, *Synapse*, 2008, **62**(11), 797–803.
314. H. Li and G. Dryhurst, Irreversible inhibition of mitochondrial complex I by 7-(2-aminoethyl)-3,4-dihydro-5-hydroxy-2H-1,4-benzothiazine-3-carboxylic acid (DHBT-1): a putative nigral endotoxin of relevance to Parkinson's disease, *J. Neurochem.*, 1997, **69**(4), 1530–1541.
315. S. Khan, S. Jyoti, F. Naz, B. Shakya and A. M. Rahul, *et al.*, Effect of L-ascorbic Acid on the climbing ability and protein levels in the brain of Drosophila model of Parkinson's disease, *Int. J. Neurosci.*, 2012, **122**(12), 704–709.
316. S. Casani, R. Gomez-Pastor, E. Matallana and N. Paricio, Antioxidant compound supplementation prevents oxidative damage in a Drosophila model of Parkinson's disease, *Free Radical Biol. Med.*, 2013, **61**, 151–160.
317. J. Yan, L. Studer and R. D. McKay, Ascorbic acid increases the yield of dopaminergic neurons derived from basic fibroblast growth factor expanded mesencephalic precursors, *J. Neurochem.*, 2001, **76**(1), 307–311.

318. D. H. Yu, K. H. Lee, J. Y. Lee, S. Kim, D. M. Shin and J. H. Kim, *et al.*, Changes of gene expression profiles during neuronal differentiation of central nervous system precursors treated with ascorbic acid, *J. Neurosci. Res.*, 2004, **78**(1), 29–37.
319. G. Logroscino, K. Marder, L. Cote, M. X. Tang, S. Shea and R. Mayeux, Dietary lipids and antioxidants in Parkinson's disease: a population-based, case-control study, *Ann. Neurol.*, 1996, **39**(1), 89–94.
320. Y. Miyake, W. Fukushima, K. Tanaka, S. Sasaki, C. Kiyohara and Y. Tsuboi, *et al.*, Dietary intake of antioxidant vitamins and risk of Parkinson's disease: a case-control study in Japan, *Eur. J. Neurol.*, 2011, **18**(1), 106–113.
321. G. P. Paraskevas, E. Kapaki, O. Petropoulou, M. Anagnostouli, V. Vagenas and C. Papageorgiou, Plasma levels of antioxidant vitamins C and E are decreased in vascular parkinsonism, *J. Neurol. Sci.*, 2003, **215**(1–2), 51–55.
322. G. Logroscino, X. Gao, H. Chen, A. Wing and A. Ascherio, Dietary iron intake and risk of Parkinson's disease, *Am. J. Epidemiol.*, 2008, **168**(12), 1381–1388.
323. P. Fernandez-Calle, F. J. Jimenez-Jimenez, J. A. Molina, F. Cabrera-Valdivia, A. Vazquez and D. Garcia Urra, *et al.*, Serum levels of ascorbic acid (vitamin C) in patients with Parkinson's disease, *J. Neurol. Sci.*, 1993, **118**(1), 25–28.
324. W. H. Church and V. L. Ward, Uric acid is reduced in the substantia nigra in Parkinson's disease: effect on dopamine oxidation, *Brain Res. Bull.*, 1994, **33**(4), 419–425.
325. I. Bendik, A. Friedel, F. F. Roos, P. Weber and M. Eggersdorfer, D: Vitamin a critical and essential micronutrient for human health, *Front. Physiol.*, 2014, **5**, 248.
326. G. C. DeLuca, S. M. Kimball, J. Kolasinski, S. V. Ramagopalan and G. C. Ebers, Review: the role of vitamin D in nervous system health and disease, *Neuropathol. Appl. Neurobiol.*, 2013, **39**(5), 458–484.
327. W. Jang, H. J. Kim, H. Li, K. D. Jo, M. K. Lee and S. H. Song, *et al.*, 1,25-Dyhydroxyvitamin D(3) attenuates rotenone-induced neurotoxicity in SH-SY5Y cells through induction of autophagy, *Biochem. Biophys. Res. Commun.*, 2014, **451**(1), 142–147.
328. K. Shinpo, S. Kikuchi, H. Sasaki, F. Moriwaka and K. Tashiro, Effect of 1,25-dihydroxyvitamin D(3) on cultured mesencephalic dopaminergic neurons to the combined toxicity caused by L-buthionine sulfoximine and 1-methyl-4-phenylpyridine, *J. Neurosci. Res.*, 2000, **62**(3), 374–382.
329. M. F. Nissou, J. Brocard, M. El Atifi, A. Guttin, A. Andrieux and F. Berger, *et al.*, The transcriptomic response of mixed neuron-glial cell cultures to 1,25-dihydroxyvitamin d3 includes genes limiting the progression of neurodegenerative diseases, *J. Alzheimer's Dis.*, 2013, **35**(3), 553–564.
330. J. Y. Wang, J. N. Wu, T. L. Cherng, B. J. Hoffer, H. H. Chen and C. V. Borlongan, *et al.*, Vitamin D(3) attenuates 6-hydroxydopamine-induced neurotoxicity in rats, *Brain Res.*, 2001, **904**(1), 67–75.

331. B. Sanchez, J. L. Relova, R. Gallego, I. Ben-Batalla and R. Perez-Fernandez, 1,25-Dihydroxyvitamin D3 administration to 6-hydroxydopamine-lesioned rats increases glial cell line-derived neurotrophic factor and partially restores tyrosine hydroxylase expression in substantia nigra and striatum, *J. Neurosci. Res.*, 2009, **87**(3), 723–732.
332. W. A. Cass, L. E. Peters, A. M. Fletcher and D. M. Yurek, Calcitriol promotes augmented dopamine release in the lesioned striatum of 6-hydroxydopamine treated rats, *Neurochem. Res.*, 2014, **39**(8), 1467–1476.
333. M. P. Smith, A. Fletcher-Turner, D. M. Yurek and W. A. Cass, Calcitriol protection against dopamine loss induced by intracerebroventricular administration of 6-hydroxydopamine, *Neurochem. Res.*, 2006, **31**(4), 533–539.
334. J. S. Kim, S. Y. Ryu, I. Yun, W. J. Kim, K. S. Lee and J. W. Park, *et al.*, 1alpha,25-Dihydroxyvitamin D(3) Protects Dopaminergic Neurons in Rodent Models of Parkinson's Disease through Inhibition of Microglial Activation, *J. Clin. Neurol.*, 2006, **2**(4), 252–257.
335. A. M. Lin, S. F. Fan, D. M. Yang, L. L. Hsu and C. H. Yang, Zinc-induced apoptosis in substantia nigra of rat brain: neuroprotection by vitamin D3, *Free Radical Biol. Med.*, 2003, **34**(11), 1416–1425.
336. M. Moghaddasi, M. Mamarabadi and M. Aghaii, Serum 25-hydroxyvitamin D3 concentration in Iranian patients with Parkinson's disease, *Iran J. Neurol.*, 2013, **12**(2), 56–59.
337. H. Ding, K. Dhima, K. C. Lockhart, J. J. Locascio, A. N. Hoesing and K. Duong, *et al.*, Unrecognized vitamin D3 deficiency is common in Parkinson disease: Harvard Biomarker Study, *Neurology*, 2013, **81**(17), 1531–1537.
338. Y. Liu and B. S. Zhang, Serum 25-hydroxyvitamin D predicts severity in Parkinson's disease patients, *Neurol. Sci.*, 2014, **35**(1), 67–71.
339. Y. Miyake, K. Tanaka, W. Fukushima, S. Sasaki, C. Kiyohara and Y. Tsuboi, *et al.*, Lack of association of dairy food, calcium, and vitamin D intake with the risk of Parkinson's disease: a case-control study in Japan, *Parkinsonism Relat. Disord.*, 2011, **17**(2), 112–116.
340. P. Knekt, A. Kilkkinen, H. Rissanen, J. Marniemi, K. Saaksjarvi and M. Heliovaara, Serum vitamin D and the risk of Parkinson disease, *Arch. Neurol.*, 2010, **67**(7), 808–811.
341. Z. Lv, H. Qi, L. Wang, X. Fan, F. Han and H. Wang, *et al.*, Vitamin D status and Parkinson's disease: a systematic review and meta-analysis, *Neurol. Sci.*, 2014, **35**(11), 1723–1730.
342. A. L. Peterson, C. Murchison, C. Zabetian, J. B. Leverenz, G. S. Watson and T. Montine, *et al.*, Memory, mood, and vitamin D in persons with Parkinson's disease, *J. Parkinson's Dis.*, 2013, **3**(4), 547–555.
343. A. L. Peterson, M. Mancini and F. B. Horak, The relationship between balance control and vitamin D in Parkinson's disease-a pilot study, *Mov. Disord.*, 2013, **28**(8), 1133–1137.
344. Y. Sato, J. Iwamoto and Y. Honda, Vitamin d deficiency-induced vertebral fractures may cause stooped posture in Parkinson disease, *Am. J. Phys. Med. Rehabil.*, 2011, **90**(4), 281–286.

345. D. Zhu, G. Y. Liu, Z. Lv, S. R. Wen, S. Bi and W. Z. Wang, Inverse associations of outdoor activity and vitamin D intake with the risk of Parkinson's disease, *J. Zhejiang Univ., Sci., B*, 2014, **15**(10), 923–927.
346. L. Kenborg, C. F. Lassen, B. Ritz, E. S. Schernhammer, J. Hansen and N. M. Gatto, *et al.*, Outdoor work and risk for Parkinson's disease: a population-based case-control study, *Occup. Environ. Med.*, 2011, **68**(4), 273–278.
347. G. Logroscino, H. D. Sesso, R. S. Paffenbarger Jr and I. M. Lee, Physical activity and risk of Parkinson's disease: a prospective cohort study, *J. Neurol., Neurosurg. Psychiatry*, 2006, **77**(12), 1318–1322.
348. J. Iwamoto, T. Takeda and H. Matsumoto, Sunlight exposure is important for preventing hip fractures in patients with Alzheimer's disease, Parkinson's disease, or stroke, *Acta Neurol. Scand.*, 2012, **125**(4), 279–284.
349. V. Lyell, E. Henderson, M. Devine and C. Gregson, Assessment and management of fracture risk in patients with Parkinson's disease, *Age Ageing*, 2015, **44**(1), 34–41.
350. J. P. Kesby, D. W. Eyles, T. H. Burne and J. J. McGrath, The effects of vitamin D on brain development and adult brain function, *Mol. Cell. Endocrinol.*, 2011, **347**(1–2), 121–127.
351. J. S. Kim, Y. I. Kim, C. Song, I. Yoon, J. W. Park and Y. B. Choi, *et al.*, Association of vitamin D receptor gene polymorphism and Parkinson's disease in Koreans, *J. Korean Med. Sci.*, 2005, **20**(3), 495–498.
352. C. Li, H. Qi, S. Wei, L. Wang, X. Fan and S. Duan, *et al.*, Vitamin D receptor gene polymorphisms and the risk of Parkinson's disease, *Neurol. Sci.*, 2015, **36**(2), 247–255.
353. Y. H. Lee, J. H. Kim and G. G. Song, Vitamin D receptor polymorphisms and susceptibility to Parkinson's disease and Alzheimer's disease: a meta-analysis, *Neurol. Sci.*, 2014, **35**(12), 1947–1953.
354. H. O. Trowbridge, Immunological aspects of chronic inflammation and repair, *J. Endod.*, 1990, **16**(2), 54–61.
355. X. Han, L. Xue, Y. Li, B. Chen and A. Xie, Vitamin D receptor gene polymorphism and its association with Parkinson's disease in Chinese Han population, *Neurosci. Lett.*, 2012, **525**(1), 29–33.
356. P. W. Jurutka, G. K. Whitfield, J. C. Hsieh, P. D. Thompson, C. A. Haussler and M. R. Haussler, Molecular nature of the vitamin D receptor and its role in regulation of gene expression, *Rev. Endocr. Metab. Disord.*, 2001, **2**(2), 203–216.
357. Q. Jiang, Natural forms of vitamin E: metabolism, antioxidant, and anti-inflammatory activities and their role in disease prevention and therapy, *Free Radical Biol. Med.*, 2014, **72**, 76–90.
358. L. J. Hsu, Y. Sagara, A. Arroyo, E. Rockenstein, A. Sisk and M. Mallory, *et al.*, alpha-synuclein promotes mitochondrial deficit and oxidative stress, *Am. J. Pathol.*, 2000, **157**(2), 401–410.
359. R. A. Gonzalez-Polo, A. Rodriguez-Martin, J. M. Moran, M. Niso, G. Soler and J. M. Fuentes, Paraquat-induced apoptotic cell death in cerebellar granule cells, *Brain Res.*, 2004, **1011**(2), 170–176.

360. R. A. Gonzalez-Polo, G. Soler, A. Rodriguezmartin, J. M. Moran and J. M. Fuentes, Protection against MPP+ neurotoxicity in cerebellar granule cells by antioxidants, *Cell Biol. Int.*, 2004, **28**(5), 373–380.
361. D. Alvira, M. Tajes, E. Verdaguer, S. G. de Arriba, C. Allgaier and C. Matute, *et al.,* Inhibition of cyclin-dependent kinases is neuroprotective in 1-methyl-4-phenylpyridinium-induced apoptosis in neurons, *Neuroscience*, 2007, **146**(1), 350–365.
362. K. Nakaso, N. Tajima, Y. Horikoshi, M. Nakasone, T. Hanaki and K. Kamizaki, *et al.,* The estrogen receptor beta-PI3K/Akt pathway mediates the cytoprotective effects of tocotrienol in a cellular Parkinson's disease model, *Biochim. Biophys. Acta*, 2014, **1842**(9), 1303–1312.
363. C. M. Testa, T. B. Sherer and J. T. Greenamyre, Rotenone induces oxidative stress and dopaminergic neuron damage in organotypic substantia nigra cultures, *Brain Res. Mol. Brain Res.*, 2005, **134**(1), 109–118.
364. K. Prabhakaran, D. Ghosh, G. D. Chapman and P. G. Gunasekar, Molecular mechanism of manganese exposure-induced dopaminergic toxicity, *Brain Res. Bull.*, 2008, **76**(4), 361–367.
365. W. Li, H. Jiang, N. Song and J. Xie, Oxidative stress partially contributes to iron-induced alpha-synuclein aggregation in SK-N-SH cells, *Neurotoxic. Res.*, 2011, **19**(3), 435–442.
366. T. R. Selvaraju, H. Khaza'ai, S. Vidyadaran, M. S. Abd Mutalib and R. Vasudevan, The neuroprotective effects of tocotrienol rich fraction and alpha tocopherol against glutamate injury in astrocytes, *Bosnian J. Basic Med. Sci.*, 2014, **14**(4), 195–204.
367. N. Sharma and B. Nehru, Beneficial Effect of Vitamin E in Rotenone Induced Model of PD: Behavioural, Neurochemical and Biochemical Study, *Exp. Neurobiol.*, 2013, **22**(3), 214–223.
368. G. G. Ortiz, F. P. Pacheco-Moises, V. M. Gomez-Rodriguez, E. D. Gonzalez-Renovato, E. D. Torres-Sanchez and A. C. Ramirez-Anguiano, Fish oil, melatonin and vitamin E attenuates midbrain cyclooxygenase-2 activity and oxidative stress after the administration of 1-methyl-4-phenyl-1,2,3,6-tetrahydropyridine, *Metab. Brain Dis.*, 2013, **28**(4), 705–709.
369. M. O. Bostanci, O. Bas and F. Bagirici, Alpha-tocopherol decreases iron-induced hippocampal and nigral neuron loss, *Cell. Mol. Neurobiol.*, 2010, **30**(3), 389–394.
370. M. Roghani and G. Behzadi, Neuroprotective effect of vitamin E on the early model of Parkinson's disease in rat: behavioral and histochemical evidence, *Brain Res.*, 2001, **892**(1), 211–217.
371. C. Heim, W. Kolasiewicz, T. Kurz and K. H. Sontag, Behavioral alterations after unilateral 6-hydroxydopamine lesions of the striatum. Effect of alpha-tocopherol, *Pol. J. Pharmacol.*, 2001, **53**(5), 435–448.
372. M. C. de Rijk, M. M. Breteler, J. H. den Breeijen, L. J. Launer, D. E. Grobbee and F. G. van der Meche, *et al.,* Dietary antioxidants and Parkinson disease. The Rotterdam Study, *Arch. Neurol.*, 1997, **54**(6), 762–765.
373. C. M. Chen, J. L. Liu, Y. R. Wu, Y. C. Chen, H. S. Cheng and M. L. Cheng, *et al.,* Increased oxidative damage in peripheral blood correlates with severity of Parkinson's disease, *Neurobiol. Dis.*, 2009, **33**(3), 429–435.

374. T. Fukushima, X. Tan, Y. Luo and H. Kanda, Serum vitamins and heavy metals in blood and urine, and the correlations among them in Parkinson's disease patients in China, *Neuroepidemiology*, 2011, **36**(4), 240–244.
375. G. Parkinson Study, Effects of tocopherol and deprenyl on the progression of disability in early Parkinson's disease, *N. Engl. J. Med.*, 1993, **328**(3), 176–183.
376. D. Kavvadias, P. Sand, K. A. Youdim, M. Z. Qaiser, C. Rice-Evans and R. Baur, *et al.,* The flavone hispidulin, a benzodiazepine receptor ligand with positive allosteric properties, traverses the blood-brain barrier and exhibits anticonvulsive effects, *Br. J. Pharmacol.*, 2004, **142**(5), 811–820.
377. S. D. Tabbal, L. Tian, M. Karimi, C. A. Brown, S. K. Loftin and J. S. Perlmutter, Low nigrostriatal reserve for motor parkinsonism in nonhuman primates, *Exp. Neurol.*, 2012, **237**(2), 355–362.
378. Z. P. Sellers, R. A. Williams, J. W. Overbay, J. Cho, M. Henderson and T. T. Reed, Current Therapeutic Modalities, Enzyme Kinetics, and Redox Proteomics in Traumatic Brain Injury, in *Traumatic Brain Injury*, ed. F. Sadaka, InTech, 2014, pp. 39–77.

Subject Index